Third Edition

Historical Geology

Books In The Brooks/Cole Earth Science And Astronomy Series:

Third Edition

Historical Geology

EVOLUTION OF EARTH AND LIFE THROUGH TIME

Reed Wicander
James S. Monroe

CENTRAL MICHIGAN UNIVERSITY

Brooks/Cole
Thomson Learning™

Australia • Canada • Denmark • Japan • Mexico • New Zealand • Phillipines •
Puerto Rico • Singapore • Spain • United Kingdom • United States

Sponsoring Editor: *Nina Horne*
Marketing Team: *Tami Cueny, Dena Donnelly*
Marketing Assistant: *Kelly Fielding*
Editorial Assistant: *John-Paul Ramin*
Production Editor: *Keith Faivre*
Manuscript Editor: *Betty Duncan*
Permissions Editor: *Lillian Campobasso*
Interior Design: *Irene Morris*
Cover Design: *Roy R. Neuhaus*

Cover Photo: *Richard Price/FPG International*
Art Editor: *Lisa Torri*
Interior Illustration: *Precision Graphics, Magellan Geographix, Carlyn Iverson, Carto-Graphics, Publication Services, Victor Royer, John and Judy Waller, Rolin Graphics, Darwen and Vally Hennings*
Photo Research: *Amelia Ames Hill Associates*
Print Buyer: *Kris Waller*
Typesetting: *Progressive Information Technologies*
Printing and Binding: *West Publishing Corporation*

For more information, contact:
BROOKS/COLE
511 Forest Lodge Road
Pacific Grove, CA 93950 USA
www.brookscole.com

For permission to use material from this work, contact us by
Web: www.thomsonrights.com
fax: 1-800-730-2215
phone: 1-800-730-2214

Printed in United States of America

10 9 8 7 6 5 4 3 2 1

Library of Congress Cataloging-in-Publication Data

Wicander, Reed [date]
    Historical geology: evolution of earth and life through time
/ Reed Wicander, James S. Monroe. — 3rd ed.
        p.   cm.
    Includes bibliographical references and index.
    ISBN 0 534-57348-7
    1. Historical geology.   I. Monroe, James S. (James Stewart).
II. Title.
QE28.3.W53   2000
551.7—dc21
                                         99-38895

Photo courtesy of Melanie Wicander

REED WICANDER

REED WICANDER is a geology professor at Central Michigan University where he teaches physical geology, historical geology, prehistoric life, and invertebrate paleontology. He has co-authored several geology textbooks with James S. Monroe. His main research interests involve various aspects of Paleozoic palynology, specifically the study of acritarchs, on which he has published many papers. He is a past president of the American Association of Stratigraphic Palynologists and currently a councilor of the International Federation of Palynological Societies.

JAMES S. MONROE

JAMES S. MONROE is professor emeritus of geology at Central Michigan University where he taught physical geology, historical geology, prehistoric life, and stratigraphy and sedimentology since 1975. He has co-authored several textbooks with Reed Wicander and has interests in Cenozoic geology and geologic education.

Brief Contents

Contents

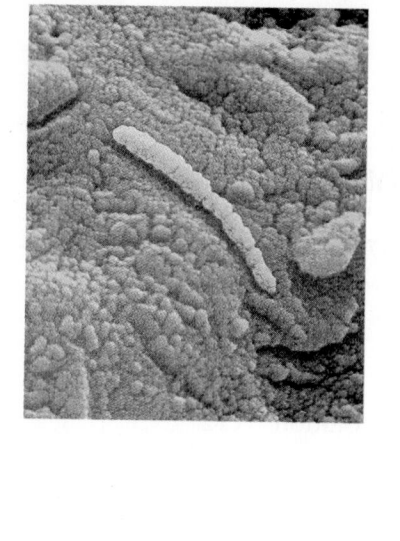

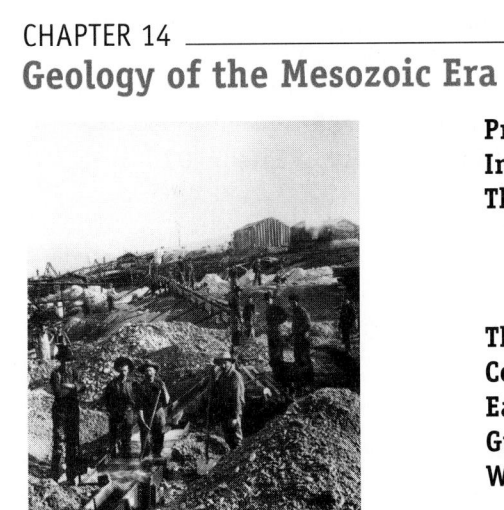

CHAPTER 19
Evolution of the Primates and Humans   510

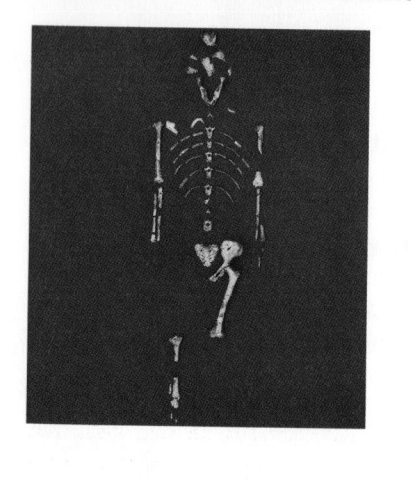

Earth is a dynamic planet that has changed continuously during its 4.6 billion years of existence. The size, shape, and geographic distribution of the continents and ocean basins have changed through time, as have the atmosphere and biota. Historical geologists are concerned with all aspects of Earth and its life history. They seek to determine what events occurred during the past, place those events into an orderly chronological sequence, and provide conceptual frameworks for explaining such events. All of this makes the study of Earth and life history a fascinating endeavor and one that has relevance today.

Historical Geology: Evolution of Earth and Life Through Time, third edition, is designed for a one-semester geology course and is written with students in mind. One of the problems with any science course is that students are overwhelmed by the amount of material that must be learned. Furthermore, most of the material does not seem to be linked by any unifying theme and does not always appear to be relevant to our lives.

One of the goals of this book is to provide students with an understanding of the principles of historical geology and how these principles can be applied to unraveling the history of Earth. It is also our intention to provide students with an overview of the geologic and biologic history of Earth, not as a set of encyclopedic facts to memorize, but rather as a continuum of interrelated events that reflect the underlying geologic and biologic principles and processes that have shaped our planet.

Because of the nature of the subject, all historical geology textbooks share some broad similarities. Most begin with several chapters on concepts and principles, followed by a chronological discussion of Earth history. In this respect we have not departed from convention. We have, however, attempted to place greater emphasis on basic concepts and principles, their historical development, and their importance in deciphering Earth history; in other words, how do we know what we know. For instance, stratigraphic terminology is used sparingly with the idea that understanding the causes and consequences of events is more important than knowing the names of numerous formations. Additionally, we have strived for a balanced coverage of both physical and biological events in our discussion of Earth history.

New Features in the Third Edition

This third edition has undergone considerable rewriting and updating to produce a book that is easier to read with a high level of current information and many new pho-

tographs, prologues, and perspectives. Drawing on the comments and suggestions of reviewers, we have incorporated many new features into this edition. One of the more noticeable changes is that we have added an Epilogue that summarizes the major aspects of the topics covered in this book and places them in perspective in terms of Earth systems. We have also deleted Chapter 2 (Minerals and Rocks) and placed that material in Appendix C as well as adding some of it to Chapter 1. This was done because many adopters said the students were already familiar with that material and they did not need a separate chapter to review it.

New material in this edition of *Historical Geology: Evolution of Earth and Life Through Time* includes a section on Earth systems in Chapter 1, as well as significant changes in what we know about dinosaurs in Chapter 15 and human evolution in Chapter 19. All of the chapters have been updated and many of the sections rewritten to emphasize the dynamic nature of the processes and events shaping Earth history.

Other important changes include a number of new Prologues, such as oil in the Persian Gulf and plate tectonics (Chapter 6), whether life exists on Mars and elsewhere in the universe (Chapter 7), Mary Anning's contributions to paleontology (Chapter 15), and the John Day Fossil Beds, Oregon (Chapter 18). New Perspectives also appear, such as interpreting Earth History (Chapter 1), the geologic history of Uluṟu and Kata Tijuṯa (Chapter 2), fossil frauds and hoaxes (Chapter 3), the rocks of Waterton National Park, Canada, and Glacier National Park, Montana, (Chapter 9), and what did dinosaurs have on the outside? (Chapter 15), as well as several others.

Many photos in the second edition have been replaced, including many of the chapter-opening photographs. In addition, a number of photographs within the chapters have been enlarged to enhance their visual impact. Furthermore, many maps have all been rerendered to enhance their visual appeal and provide additional information.

We feel the rewriting and updating done in the text, as well as the addition of new photographs and enhanced maps, greatly improves the third edition by making it easier to read and comprehend, as well as being a more effective teaching tool. Additionally, improvements have been made in the ancillary package that accompanies the book.

WORLD WIDE WEB ACTIVITIES

One of the new features of this book is a section at the end of each chapter titled World Wide Web Activities. Each section lists several web sites relevant to the chapter. These sites were selected because they provide additional information about the topic covered in the chapter or information about ongoing research or events, such as Paleozoic orogenies, dinosaurs, or evolution. Each site has a short summary, and most sites contain links to other sites students may find of interest.

Rather than publishing the URL (universal resource locator) for each site, students are instructed to log on to **http://www.brookscole.com/geo/** for the current URL of the sites listed as well as additional related web sites. The advantage of logging on to the Brooks/Cole web site is that the URL for each site is kept current and new sites can be added between editions.

We think students will find the World Wide Web Activities section a valuable and useful resource in their geological education.

CD-ROM EXPLORATIONS

The third edition of *Historical Geology: Evolution of Earth and Life through Time* is accompanied by *In-Terra-Active Version 2.0: The Brooks/Cole Geology CD-ROM for Students.* At the end of most chapters, students will find sections under the title CD-ROM Exploration. These engaging, end-of-chapter sections will encourage students to explore the corresponding interactive CD-ROM module(s) to gain greater understanding of the key concepts and processes presented in each chapter.

Thematic Approach

As in the second edition, we develop three major themes in this textbook that are essential to the interpretation and appreciation of historical geology, introduce these themes early, and reinforce them throughout the book. These themes are *time,* the dimension that sets historical geology apart from most of the other sciences; *evolutionary theory,* the explanation for inferred relationships among living and fossil organisms; and *plate tectonics,* a unifying theory for interpreting much of Earth's physical history and, to a large extent, its biological history. In addition, we have emphasized the intimate interrelationship that exists between physical and biological events.

The concept of geologic time and its historical development is presented in Chapter 2. The concept is more fully developed in the following chapter on Rocks, Fossils, and Time. In Chapter 5, we discuss the development of evolutionary theory in the context of the historical period during which it was formulated. Included in this discussion are sections that emphasize the evidence for evolution and the importance of the fossil record in evolutionary studies. Evolutionary theory serves as the unifying theme for subsequent discussions of life history and is constantly reinforced throughout the book.

Particular emphasis is given to the evidence that substantiates plate tectonic theory, why this theory is one of the cornerstones of geology, and why plate tectonic theory serves as a unifying paradigm in explaining many apparently unrelated geologic phenomena. Several examples of how plate tectonics is responsible for seemingly disparate aspects of geology are given in this con-

text; events are viewed as parts of an ever-changing continuum, rather than as discrete events separated from one another in time and space. While we emphasize the geologic evolution of North America, each chapter on geologic history contains an overview of global tectonic events so that the history of North America is discussed in a global context.

Text Organization

This book was written for a one-semester course in historical geology to serve both majors and nonmajors in geology and in the Earth sciences. It may also be used in a one-quarter course by selecting the appropriate material the instructor wishes to cover or emphasize. Many students taking a historical geology course will have had an introductory physical geology course. In this case, Chapter 1 can be used as a review of the principles and concepts of geology and as an introduction to the science of historical geology and the three themes (time, organic evoltuion, and plate tectonics) that the book emphasizes. The text is written at an appropriate level for those students taking historical geology with no prerequisite, but the instructor may have to spend more time expanding some of the concepts and terminology discussed in Chapter 1. Furthermore, Appendix C, "Earth Materials—Minerals and Rocks," has been added to this edition to provide students with a basic understanding of the classification and formation of minerals and rocks. This will be of benefit to both groups of students—a review for students who have taken a physical geology course and an introduction for students who have not had a geology or Earth science course.

Chapter 2 explores the first major theme of the book, the concepts and principles of geologic time. Chapter 3 expands on that theme by integrating geologic time with rocks and fossils. Depositional environments are sometimes covered rather superficially (perhaps little more than a summary table) in some historical geology textbooks. Chapter 4, "The Origin and Interpretation of Sedimentary Rocks," is completely devoted to this topic; it contains sufficient detail to be meaningful, but avoids an overly detailed discussion more appropriate for advanced courses.

The second major theme of the book, evolution, is examined in Chapter 5. Chapter 6 discusses the third major theme, plate tectonics. For those students who have had a physical geology course, much of the chapter will be review. It is, however, an important review because we will often refer to this theory throughout the rest of the book as we discuss how plate tectonics has played a major role in the evolution of our planet.

Chapter 7, "A History of the Universe, Solar System, and Planets," provides a comprehensive, yet easily readable treatment of this topic. The chapter includes discussions of the origin of the universe from the time when physics as we know it began, the theories for the origin of the solar system, the significance of meteorites, and the competing theories for the formation and early history of Earth.

Precambrian time, fully 88% of all geologic time, is sometimes considered in a single chapter. Chapters 8 and 9 are devoted to the geologic and biologic histories of the Archean and Proterozoic eons, respectively.

Chapters 10 through 19 constitute our chronological treatment of the Phanerozoic geologic and biologic history of Earth. These chapters are arranged so that the geologic history of an era is followed by a discussion of the biologic history of that era. We think that this format allows easier integration of life history with geologic history. An Epilogue summarizes the major topics and themes of the book.

Chapter Organization

All chapters have the same organizational format. Each chapter opens with a photograph that relates to the chapter content, a detailed outline, and a Prologue, which is intended to stimulate interest in the chapter by discussing some aspect of the material.

The text is written in a clear informal style, making it easy for students to comprehend. Numerous color diagrams and photographs complement the text, providing a visual representation of the concepts and information presented. Each chapter contains at least one Perspective that presents a brief discussion of an interesting aspect of historical geology or geological research. Each of the chapters on geologic history in the second half of the text contains a final section on mineral resources characteristic of that time period. These sections provide applied economic material of interest to students.

The end-of-chapter materials begin with a concise review of important concepts and ideas in the Summary. The Important Terms, which are printed in boldface type in the chapter text, are listed at the end of each chapter for easy review, and a full glossary of important terms appears at the end of the text. The Review Questions are another important feature of this book; they contain multiple-choice questions with answers as well as short answer and essay questions, and thought-provoking quantitative questions under the Points to Ponder heading. Many new multiple-choice questions as well as short answer and essay questions have been added in each chapter for this edition. Following Points to Ponder, each chapter has an up-to-date list of Additional Readings, most of which are written at a level appropriate for beginning students interested in pursuing a particular topic. Each chapter concludes with World Wide Web Activities that list World Wide Web sites students can visit to get additional information about the topics covered in that chapter, and a CD-ROM Exploration.

Special Features

FIGURES

Many of the figures are original artwork especially designed for this book. Our color paleogeographic maps are designed to illustrate clearly and accurately the geography during the various geologic periods. Many of the illustrations depicting geologic processes or events are block diagrams rather than cross sections so that students can more easily visualize the salient features of these processes and events. Full color scenes showing associations of plants and animals are based on the most current interpretations. Great care was taken to ensure that the art and captions provide an attractive, informative, and accurate illustration program for the third edition.

PROLOGUES

The introductory prologues are designed to provide an introduction to the chapter material by focusing on a particularly interesting aspect of the chapter. Prologue topics include possible life on Mars (Chapter 7), the Mazon Creek fauna and flora (Chapter 11), the Burgess Shale (Chapter 12), and the Eve hypothesis concerning the origin of modern humans (Chapter 19).

SUMMARY TABLES

A number of summary tables appear throughout the book. Of particular assistance to students are the end-of-chapter summary tables. These tables are designed to give an overall perspective of the geologic and biologic events that occurred during a particular time interval and to show how these events are interrelated. The emphasis in these tables is on the geologic evolution of North America. Global tectonic events and sea-level changes are also incorporated into these tables to provide global insights.

PERSPECTIVES

The chapter Perspectives focus on aspects of current geologic research, interesting features related to the material in the chapter, or national parks or monuments. They were chosen to provide students with an overview of the many fascinating aspects of geology. Examples include the abundance of fossils (Perspective 3.1), convergent evolution (Perspective 5.2), the supercontinent cycle (Perspective 6.2), Dinosaur National Monument (Perspective 14.2), and how large was the largest dinosaur? (Perspective 15.1). The Perspectives can be assigned as part of the chapter reading, used as the basis for lecture or discussion topics, or even as the starting point for student papers.

REVIEW QUESTIONS

Most historical geology texts have a set of review questions at the end of each chapter. This book, however, includes not only the usual essay questions but also thought-provoking and quantitative questions in the section titled Points to Ponder. In addition, there is a set of multiple-choice questions. The answers to the multiple-choice questions are at the end of the book so that students can check their answers and increase their confidence before taking an examination.

WORLD WIDE WEB ACTIVITIES

Each chapter contains a section on World Wide Web Activities. Several web sites relevant to the chapter material are listed with instructions on how to reach each web site. These sites were selected to provide additional information about the topics covered in the chapter or to provide the most current information. Each site has a short summary about it and, typically, a few questions for the students to answer.

CD-ROM EXPLORATION

Most chapters conclude with CD-ROM Exploration, a section that encourages students to use the enclosed interactive *In-Terra-Active Version 2.0: The Brooks/Cole Geology CD-ROM for Students* to gain greater understanding of the key concepts and processes presented in each chapter.

APPENDIXES

With few exceptions, units of length, mass, and volume are given in metric units in the text. Appendix A is an *English-Metric Conversion Chart*. The terminology of biological classification is complex, and though we have minimized it in the text, some is necessary. A *Classification of Organisms* appears in Appendix B. Appendix C is *Earth Materials — Minerals and Rocks*, and is an overview on mineral and rock identification and classification.

GLOSSARY

It is important for students to comprehend geologic terminology if they are to understand geologic principles and concepts. Accordingly, an alphabetical and comprehensive glossary of all important terms covered in each chapter appears at the back of the book.

Ancillary Materials

To accompany *Historical Geology: Evolution of Earth and Life Through Time*, 3e, we're pleased to offer a full suite of text and multimedia products.

FOR INSTRUCTORS

Instructor's Edition The full student edition of the text, plus a special front matter insert for instructors containing a visual presentation of the pedagogical features and style of the book and its ancillaries. (ISBN 0-534-37381-X)

Instructor's Manual A comprehensive manual meant to help instructors in preparing for lectures. Includes teaching ideas, lecture outlines, teaching tips, and an expanded list of references.

Thomson Learning Testing Tools A fully integrated suite of computerized testing tools enables instructors to deliver tests via print, LAN, or the Internet, use existing test questions, create questions, and provide feedback to objective test questions. Testing and tutorial results can be integrated into the classroom management tool, which offers scoring, grade book, and reporting capabilities. Hundreds of questions are included. (Dual-platform CD-ROM ISBN: 0-534-37220-1)

Color Transparency Acetates A set of 150 full-color transparency acetates showing important line drawings, maps, charts, and graphs from the text. (ISBN 0-534-37216-3)

Slide Set Full-color slides featuring key photographs in the text. (ISBN 0-534-37218-X)

CNN Today Video for Meteorology, Volume 2 A 45-minute video showing dramatic, in-depth coverage of recent meteorological events from CNN's collection. (ISBN 0-534-37207-4)

GeoLink 3.0 A lecture/presentation tool for instructors. This CD-ROM allows professors to customize lectures, assemble multimedia presentations, and export lecture notes for use on the World Wide Web. An extremely easy-to-use interface enables instructors to incorporate still graphics, digital video, animation, and audio clips into their lectures and presentations. Includes images from Wicander/Monroe and other Brooks/Cole geology titles. (ISBN 0-534-52190-8)

FOR STUDENTS

Study Guide A supplement intended to facilitate and enhance student understanding of the historical geology course. Contains an overview of each chapter in the text, objectives, key terms, study questions, critical thinking questions, and activities. (ISBN 0-534-37218-X)

In-terra-Active CD-ROM A multimedia tool that provides students with an opportunity to explore geological topics through hands-on exercises. The 47 exercises are interactive and are enhanced by 35 minutes of video clips, images, and a searchable glossary. Every copy of the book includes this CD-ROM.

Current Perspectives in Geology, 2000 Edition Edited by McKinney, McHugh, and Meadows. An anthology of more than 40 geology-related articles from respected journals and popular periodicals. Each article starts with a synopsis and is followed by a list of discussion questions. (ISBN 0-534-37213-9)

Geology Workbook for the Web By Bruce Blackerby. Over twenty chapters of Internet exercises on various geology topics, including fossils and evolution, geologic time, the Precambrian, and Paleozoic, Mesozoic, and Cenozoic Earth. Three-hole punched, perforated pages for easy detaching. (ISBN 0-314-21072-5)

Earth Online: An Internet Guide for Earth Science By Michael Ritter. (ISBN 0-534-51707-2)

Acknowledgments

As the authors, we are of course responsible for the organization, style, and accuracy of the text, and any mistakes, omissions, or errors are our responsibility. The finished product is the culmination of many years of work during which we received numerous comments and advice from many geologists who reviewed parts of the text. We wish to express our sincere appreciation to the reviewers who reviewed the first and second editions and made many helpful and useful comments that led to the improvements seen in this third edition.

Thomas W. Broadhead
University of Tennessee at Knoxville

Mark J. Camp
University of Toledo

James F. Coble
Tidewater Community College

William C. Cornell
University of Texas at El Paso

Rex E. Crick
University of Texas at Arlington

Richard Fluegeman, Jr.
Ball State University

Annabelle Foos
University of Akron

Susan Goldstein
University of Georgia

Bryan Gregor
Wright State University

Thor A. Hansen
Western Washington University

Paul D. Howell
University of Kentucky

Jonathan D. Karr
North Carolina State University

R. L. Langenheim, Jr.
University of Illinois at Urbana-Champaign

Michael McKinney
University of Tennessee at Knoxville

Arthur Mirsky
Indiana University, Purdue University at Indianapolis

William C. Parker
Florida State University

Anne Raymond
Texas A & M University

G. J. Retallack
University of Oregon

Mark Rich
University of Georgia

Barbara L. Ruff
University of Georgia

W. Bruce Saunders
Bryn Mawr College

Ronald D. Stieglitz
University of Wisconsin, Green Bay

Carol M. Tang
Arizona State University

Michael J. Tevesz
Cleveland State University

Art Troell
San Antonio College

We also wish to thank Kathy Benison, Richard V. Dietrich (Professor Emeritus), Eric L. Johnson, David J. Matty, Jane M. Matty, and Wayne E. Moore (Professor Emeritus) of the Geology Department of Central Michigan University and Bruce M. C. Pape of the Geography Department for providing us with photographs and answering our questions concerning various topics. We also thank Pam Iacco of the Geology Department, whose general efficiency was invaluable during the preparation of this book. Theresa Barber and Anna Dutton, also of the Geology Department, helped with some of the clerical tasks associated with this book. We are also grateful for the generosity of the various agencies and individuals from many countries who provided photographs.

Special thanks must go to Stacy Purviance, editor at Wadsworth Publishing Company who initiated this third edition, and to Nina Horne, acquisitions editor at Brooks/Cole Publishing Company, who saw this edition through to completion. She was assisted by her editorial assistant, John-Paul Ramin. We are equally indebted to our production editor, Keith Faivre, whose attention to detail and consistency is greatly appreciated. We would also like to thank Betty Duncan for her copyediting skills. We appreciate her help in improving our manuscript. We thank Lillian Campobasso, who coordinated the permissions portion of the text production, and Jane Farnol, who did the index. Because historical geology is largely a visual science, we extend special thanks to Carlyn Iverson, who rendered the reflective art, and to the artists at Precision Graphics, who were responsible for much of the rest of the art program. We especially thank the artists at Magellan Geographix, who rerendered many of the maps. They all did an excellent job, and we enjoyed working with them.

Our families were very patient and encouraging when most of our spare time and energy were devoted to this book. We thank them for their support and understanding.

1

The Dynamic Earth

Apollo 17 view of Earth. Almost the entire coastline of Africa is clearly shown in this view, with Madagascar visible off its eastern coast. The Arabian Peninsula can be seen at the northeastern edge of Africa, while the Asian mainland is on the horizon toward the northeast. The present location of continents and ocean basins is the result of plate movement. The interaction of plates through time has affected the physical and biological history of Earth.

Prologue

If we could film the history of Earth from its beginning 4.6 billion years ago to the present and speed up that film so it could be shown as a feature movie, we would see a planet undergoing remarkable change. Its geography would change as continents moved about its surface, and as a result of these movements, ocean basins would open and close. Mountain ranges would form along continental margins or where continents collided with each other. Oceanic and atmospheric circulation patterns would shift in response to the movement of continents. Massive ice sheets would form, grow, and then melt away. At other times, our film would show extensive swamps or vast interior deserts.

If we focused our imaginary camera closer to Earth's surface, we would capture a breathtaking panorama of different organisms. We would witness the first living cells evolving from a primordial organic soup sometime between 4.6 and 3.6 billion years ago. About 2 billion years later, cells with a nucleus evolved. Next (approximately 700 million years ago), the first multicelled soft-bodied organisms evolved in the oceans, then animals with skeletons, and then animals with backbones.

As we scanned the various continents, we would see an essentially bare landscape until about 450 million years ago. At that time, the landscape would seem to come to life as plants and animals moved out of the water and onto the land. From then on, the landscape would be dominated by insects, amphibians, reptiles, birds, and mammals. In our film, humans and their history would occupy only the last second or so of the movie. However, the impact of human activity on the global ecosystem far exceeds the effects caused by all other lifeforms that came before.

Three interrelated themes run through the preceding discussion. The first is that the crust of Earth is composed of a series of moving plates whose interactions have affected its physical and biological history. The second is that Earth's biota has evolved. The third is that the physical and biological changes that occurred took place over long periods of time.

The first of these themes, *plate tectonics*, provides geologists with a unifying model that explains Earth's internal workings and accounts for a variety of apparently unrelated geologic features and events. The second theme, the theory of *organic evolution*, explains how life has changed through time, based on the idea that all living organisms are the evolutionary descendants of lifeforms that existed in the past. Finally, the third theme, the concept of *geologic time*, allows geologists to show how small, almost imperceptible changes over vast lengths of time have resulted in significant changes.

These three interrelated themes, plate tectonics, organic evolution, and *geologic time*, are central to our understanding of the workings of our planet and appreciation of its history. As you read this book, keep in mind that the different topics you are studying are parts or components of a dynamic, complex integrated system, and not isolated pieces of unrelated information. For example, mountain ranges, such as the Appalachians or Andes are the result of complex interactions involving Earth's interior and surface. The formation of the ranges not only affects the surrounding area but also influences climatic changes, as well as the distribution of plants and animals.

Introduction to Earth Systems

In this book we focus on Earth as a complex, dynamic planet that has changed continually since its origin some 4.6 billion years ago. These changes are the result of internal and external processes that interact and affect each other, leading to the present-day features we observe. In fact, Earth is unique among the planets of our solar system in that it supports life and has oceans of water, a hospitable atmosphere, and a variety of climates. It is ideally suited for life as we know it because of a combination of factors, including its distance from the Sun and the evolution of its interior, crust, oceans, and atmosphere. Over time, changes in Earth's atmosphere, oceans, and, to some extent, its crust have been influenced by life processes. In turn, these physical changes have affected the evolution of life.

If we view Earth as a whole, we can see innumerable interactions occurring between its various components. Furthermore, these components do not act in isolation but are interconnected—such that when one part of the system changes, it affects the other parts of the system.

One way to help understand the complexity of Earth is to think of it as a system. A **system** is defined as a combination of related parts that interact in an organized fashion (Figure 1.1). Earth, when considered as a system, consists of a collection of various subsystems or related parts interacting with each other in complex ways. Information, materials, and energy entering the system from the outside are *inputs*, while information, materials, and energy that leave the system are *outputs*.

An automobile is a good example of a system. Its various subsystems include the engine, transmission, steering, and brakes. These subsystems are interconnected in such a way that a change in any one subsystem affects other subsystems. The main input into the automobile system is gasoline, and its outputs are movement, heat, and pollutants.

Thus, we can view Earth in the same way we view an automobile—that is, as a system of interconnected components that interact and affect each other in many ways. Its complex interactions result in a dynamically changing body that exchanges matter and energy and recycles them into different forms (Table 1.1). These systems can interact with themselves, and many of the interactions are two-way. Also, chains of interactions frequently occur. For example, solar heating warms the land; unequal heating of land and water drive the wind, which in turn drives the ocean currents. When examined in this manner, the continuous evolution of Earth and its life makes geology an exciting and ever-changing science in which new discoveries are continually being made.

What Is Geology?

Geology, from the Greek *geo* and *logos,* is defined as the study of Earth. It is generally divided into two broad areas—physical geology and historical geology. *Physical geology* is the study of Earth materials, such as minerals and rocks, as well as the processes that operate beneath and

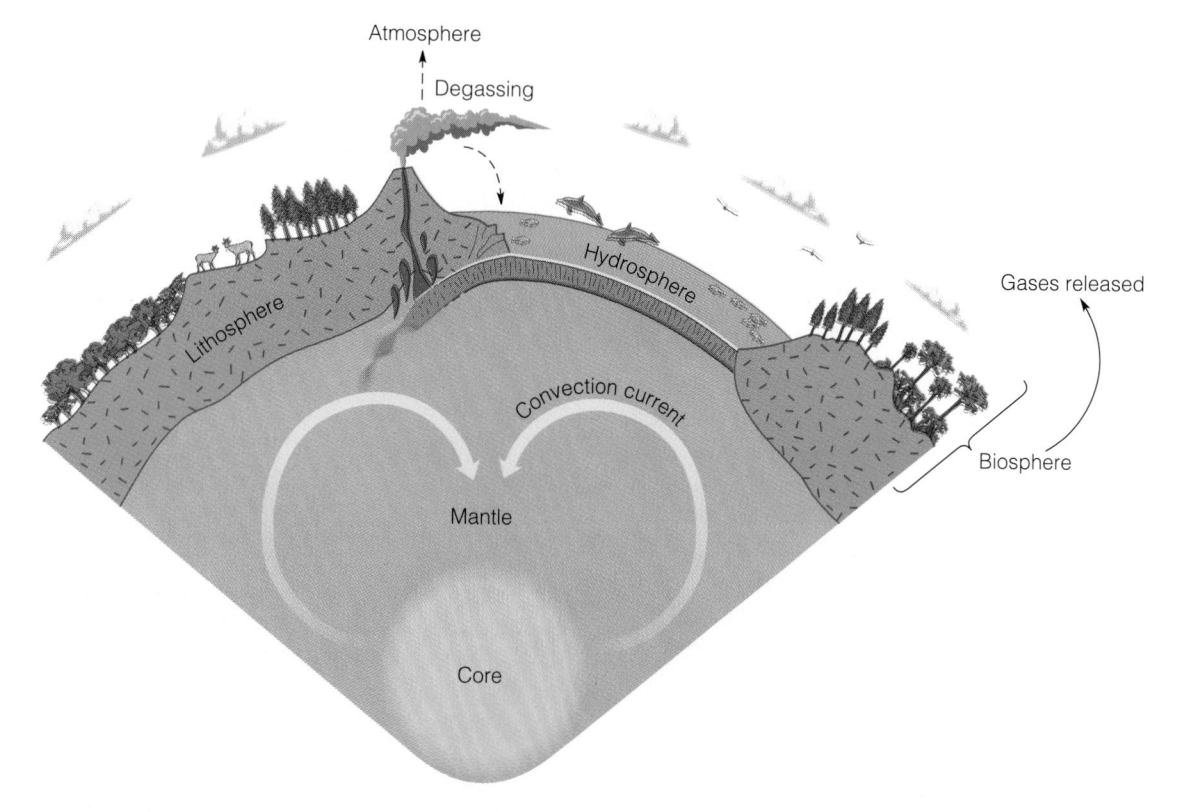

FIGURE 1.1 The atmosphere, biosphere, hydrosphere, lithosphere, mantle, and core can all be thought of as subsystems of Earth. The interactions of these subsystems are what make Earth a dynamic planet, which has evolved and changed since its origin 4.6 billion years ago.

TABLE 1.1

Interactions Between Major Earth Systems

	EARTH–UNIVERSE SYSTEM	ATMOSPHERE	HYDROSPHERE	BIOSPHERE	SOLID EARTH
EARTH–UNIVERSE SYSTEM	Gravitational interaction Recycling of stellar materials	Solid energy input Creation of ozone layer	Heating by solar energy Tides	Solar energy for photosynthesis Daily/seasonal rhythms	Heating by solar energy Meteor impacts
ATMOSPHERE	Escape of heat radiation to space	Interaction of air masses	Winds drive surface currents Evaporation	Gases for respiration Transport of seeds, spores	Wind erosion Transport of water vapor for precipitation
HYDROSPHERE	Tidal friction affects Moon's orbit	Input of water vapor and stored solar heat	Mixing of oceans Deep ocean circulation Hydrologic cycle	Water for cell fluids Medium for aquatic organisms Transport of organisms	Precipitation Water and glacial erosion Solution of minerals
BIOSPHERE	Biogenic gases affect escape of heat radiation to space	Gases from respiration	Removal of dissolved materials by organisms	Predator–prey interactions Food cycles	Modification of weathering and erosion processes Formation of soil
SOLID EARTH	Gravity of Earth affects other bodies	Input of stored solar heat Mountains divert air movements	Source of sediment and dissolved materials	Source of mineral nutrients Modification of habitats by crustal movements	Plate tectonics Crustal movements

Source: Adapted by permission from Stephen Dutch, James S. Monroe, and Joseph Moran, Earth Science (Minneapolis/St. Paul: West Publishing Co.).

upon its surface. *Historical geology* examines the origin and evolution of Earth, its continents, atmosphere, oceans, and life.

Historical geology is, however, more than just a recitation of past events. It is the study of a dynamic planet that has changed continuously during the past 4.6 billion years—although not always at the same rate. Geologists seek not only to place events in an orderly chronologic arrangement but also, more importantly, to explain how and why past events happened (see Perspective 1.1). It is one thing to observe in the fossil record that dinosaurs went extinct but quite another to ask how they became extinct. Geologists have long noted evidence in rocks of numerous mountain-building episodes. Only in the last few decades, however, has a theory emerged to provide geologists with a comprehensive model explaining how mountain building occurs and how individual episodes of mountain building may be related to each other.

In addition to aiding in the interpretation of Earth's history, the basic principles of historical geology have practical applications. William Smith, an English surveyor and engineer, recognized that by studying the sequences of rocks and the fossils they contained, he could predict the kinds and thicknesses of rocks that would have to be excavated in the construction of canals. The same principles Smith used in the late eighteenth and early nineteenth centuries are still used today in mineral and oil exploration.

Historical Geology and the Formulation of Theories

The term **theory** has various meanings. In colloquial usage, it means a speculative or conjectural view of something—hence the widespread belief that scientific theories are little more than unsubstantiated wild guesses. In scientific usage, however, a theory is a coherent explanation for one or several related natural phenomena that is supported by a large body of objective evidence. From a theory are derived predictive statements that can be tested by observation and/or experiment so that their validity can be assessed. The law of universal gravitation is an example of a theory describing the attraction between masses (an apple and Earth in the popularized account of Sir Isaac Newton and his discovery).

Theories are formulated through the process known as the **scientific method.** This method is an orderly, logical approach that involves gathering and analyzing the facts or data about the problem under consideration. Tentative explanations, or **hypotheses,** are then formulated to explain the observed phenomena. Next, the hypotheses are tested to see if what they predicted actually occurs in a given situation. Finally, if one of the hypotheses is found, after repeated tests, to explain the phenomena, then that hypothesis is proposed as a theory. One should remember, however, that in science, even a theory is still subject

Interpreting Earth History

*H*istorical geology is the study of the origin and evolution of Earth. Geologists are interested not only in placing events in a chronological sequence but, more importantly, in explaining how and why past events took place. Recently, historical geology has taken on even greater importance because scientists in many disciplines are looking to the past to help explain current events (such as short- and long-term climatic changes) and using this information to try and predict future trends.

We look at Earth as a system consisting of a collection of various subsystems or related parts interacting with each other in complex ways. By using this systems approach, we can see that the evolution of Earth, far from being a series of isolated events, is a continuum in which the different components both affect and are affected by each other. An example of this is the early history of Earth in which the evolution of the atmosphere, hydrosphere, lithosphere, and biosphere are intimately related (see Chapters 8 and 9). Today, scientists are examining the effect humans are having on short-term climate changes and the environment, as well as what a decrease in global biodiversity means for both humans and the planet in general.

Geologists seek to know not only *what* happened in the past but *why* something happened and what the implications are for Earth today and in the future. Thus, it is important to understand present-day processes and to understand how the various Earth systems are connected. It is also important to have an accurate means of measuring geologic time so as to understand and appreciate the duration of past events and how these events might affect Earth and its inhabitants today.

An important component of historical geology is understanding how we know what we know. How do we know dinosaurs became extinct 65 million years ago, or that glacial conditions prevailed over what is now the Sahara Desert during the Carboniferous Period? How can we be so sure that the early atmosphere was devoid of oxygen and evolved over millions of years to one that today has oxygen? These are all questions that historical geology addresses, and does so by seeking answers in rocks and fossils. As more information becomes available from new observations or scientific techniques, geologists become more confident in their interpretations of past events.

What makes geology (and science in general) so exciting is that there are still so many unanswered questions. For example, there is still heated debate on what caused the Permian mass extinctions, and many new hypotheses have been put forth and are being tested (see Chapter 12). Another exciting area of research is the determination of past environments. By analyzing the fluid inclusions trapped in halites from Permian-aged red beds, geologists can determine how acidic the lakes and ground waters were during that time. This has important implications for how lake and ground water systems have evolved through time, as well as the origin of terrestrial red beds and even Martian red beds (see Chapter 7). New studies indicate that changes in the chemistry of the oceans may have significantly affected the carbon cycle and have important implications in terms of present-day reef ecology and evolution.

What is important to remember is that rocks and fossils provide the clues to Earth's evolution. By applying the principles of historical geology, we can interpret Earth's history. It is also equally important to remember that historical geology is not a static science but one that, like the dynamic Earth which it seeks to understand, is constantly evolving as new information becomes available.

to further testing and refinement as new data become available.

The fact that a scientific theory can be tested and is subject to such testing separates science from other forms of human inquiry. Because scientific theories can be tested, they have the potential of being supported or even proved wrong. Accordingly, science must proceed without any appeal to beliefs or supernatural explanations, not because such beliefs or explanations are necessarily untrue, but because we have no way to investigate them. For this reason, science makes no claim about the existence or nonexistence of a supernatural or spiritual realm.

Each scientific discipline has certain theories that are of particular importance for that discipline. In geology the formulation of plate tectonic theory has changed the way geologists view Earth. Geologists now view Earth history in terms of interrelated events that are part of a global pattern of change.

Earth Materials

The only evidence available for interpreting Earth history is preserved in rocks. Consequently, the origin, distribution, and interrelationships among rocks are fundamental to an understanding of Earth history. We will review the three major rock groups in this chapter and will consider sedimentary rocks in more detail in Chapters 3 and 4 because they are particularly important in historical geology.

A **rock** is an aggregate of minerals. **Minerals** are naturally occurring, inorganic, crystalline solids that have definite physical and chemical properties (see Appendix C). Minerals are composed of elements such as oxygen, silicon, or aluminum, and elements are made up of atoms, the smallest particles of matter that still retain the characteristics of an element. The common minerals are composed mostly of the eight chemical elements making up the bulk of Earth's crust (Table 1.2). More than 3,500 minerals have been identified and described, but only about a dozen make up the bulk of the rocks in the crust (Table 1.3).

ROCKS AND THE ROCK CYCLE

Geologists recognize three major groups of rocks—*igneous, sedimentary,* and *metamorphic*—each of which is characterized by its mode of formation (see Appendix C). Each group contains a variety of individual rock types that differ from one another on the basis of composition or texture (the size, shape, and arrangement of mineral grains).

The **rock cycle** is a way of viewing the interrelationships between Earth's internal and external processes (Figure 1.2). It relates the three rock groups to each other;

TABLE 1.2
Common Elements in Earth's Crust

ELEMENT	SYMBOL	PERCENTAGE OF CRUST (BY WEIGHT)	PERCENTAGE OF CRUST (BY ATOMS)
Oxygen	O	46.6%	62.6%
Silicon	Si	27.7	21.2
Aluminum	Al	8.1	6.5
Iron	Fe	5.0	1.9
Calcium	Ca	3.6	1.9
Sodium	Na	2.8	2.6
Potassium	K	2.6	1.4
Magnesium	Mg	2.1	1.8
All others		1.5	0.1

TABLE 1.3
Common Rock-Forming Minerals
RELATIVE ABUNDANCES OF MINERALS IN EARTH'S CRUST

Plagioclase	39%
Quartz	12
Orthoclase	12
Pyroxenes	11
Micas	5
Amphiboles	5
Clay minerals	5
Olivine	3
Others	8

to surficial processes such as weathering, transportation, and deposition; and to internal processes such as magma generation and metamorphism. Plate movement is the mechanism responsible for recycling rock materials and therefore drives the rock cycle.

Igneous rocks are classified on the basis of their texture and mineral composition and result from the crystallization of molten material called *magma,* or by the accumulation and consolidation of volcanic ejecta such as ash. As a magma cools, minerals crystallize and the resulting rock is characterized by interlocking mineral grains. Magma that cools slowly beneath the surface produces *intrusive igneous rocks* with a phaneritic texture (Figure 1.3a). A phaneritic texture is one in which the individual minerals can be seen without magnification. *Extrusive igneous rocks,* such as those forming from lava, have an aphanitic texture that results from magma cooling rapidly at Earth's surface. Aphanitic textures must be magnified for the individual grains to be seen (Figure 1.3b). When a magma cools so rapidly that crystallization cannot occur, the resulting rock is a natural glass.

Rocks exposed at Earth's surface are broken into particles and dissolved by various weathering processes. The particles and dissolved material may be transported by wind, water, or ice and eventually deposited as *sediment.* This sediment may then be compacted or cemented into a sedimentary rock by a process termed *lithification.*

Two major categories of **sedimentary rocks** are recognized. *Detrital sedimentary rocks* are composed of solid particles that were once part of other rocks and are classified according to particle size (see Table C5). For example, conglomerate is composed of rounded particles larger than 2 millimeters (mm) (Figure 1.3c), while sand is made up of particles 1/16 to 2 mm in diameter.

Chemical sedimentary rocks are derived from dissolved material carried in solution (Figure 1.3d) and are classified on the basis of their chemical composition (see Table C5). Some chemical rocks are the result of precipitation of minerals by organisms or are accumulations of organic material such as shells and plants and form a subcategory called *biochemical sedimentary rocks.*

Sedimentary rocks are very useful for interpreting Earth history. These rocks form at or near Earth's surface, and

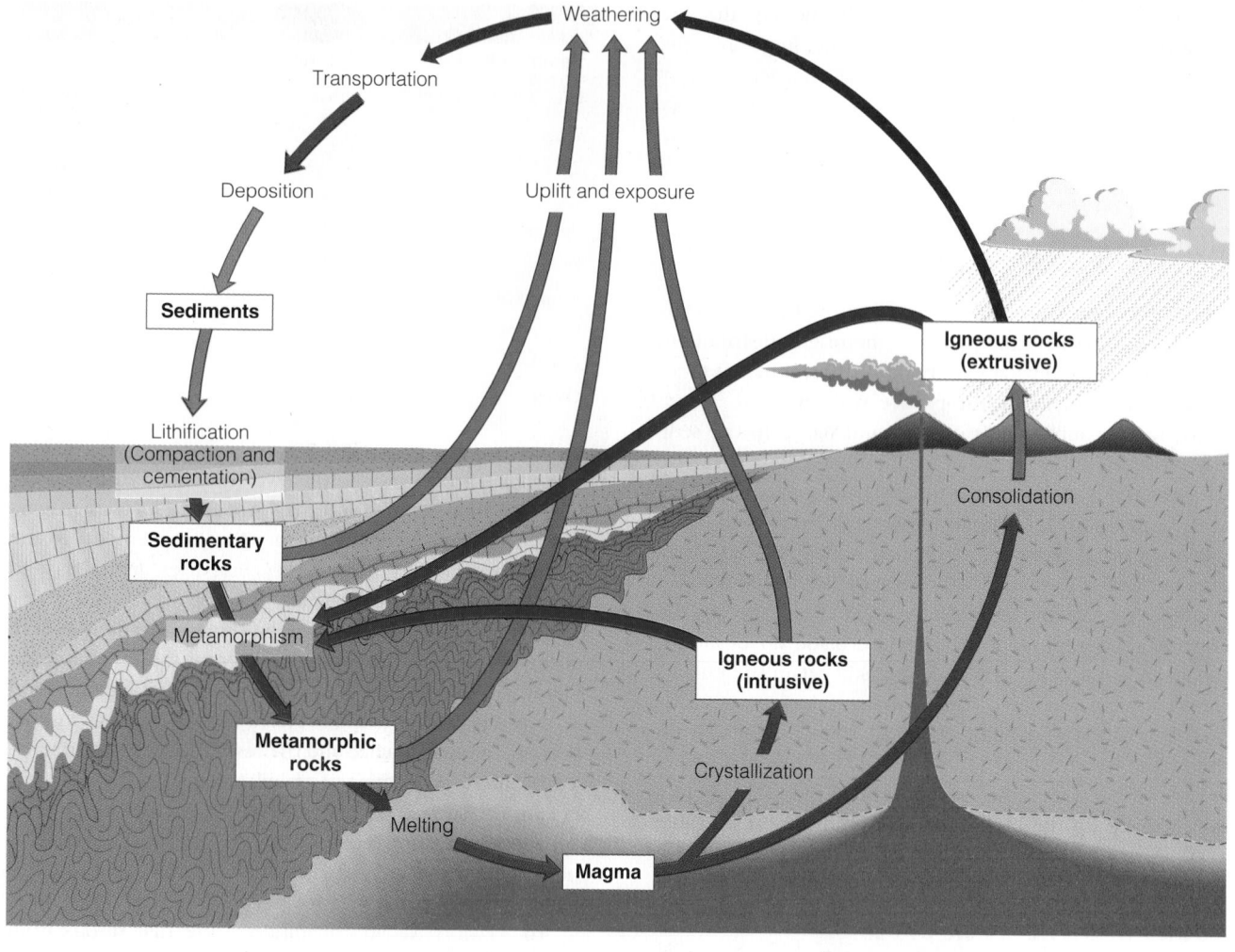

FIGURE 1.2 The rock cycle showing the interrelationships between Earth's internal and external processes and how each of the three major rock groups is related to the others.

as layers of sediment accumulate, they record information about the environment of deposition, the type of transporting agent, and perhaps even something about the source from which the particles were derived (Figure 1.4). If the sedimentary layers contain fossils, geologists can use these fossils to interpret past environmental conditions and identify the different types of organisms that were living at that time; they may also be able to determine the age of the rocks and correlate them with rocks of the same age in different places (Figure 1.5).

Some rock layers contain structures formed at the time of deposition (sedimentary structures). For example, cross-bedding allows geologists to reconstruct the environment at the time the cross-beds formed (Figure 1.6). Furthermore, some rocks by their very nature are good indicators of past environmental conditions. Tillites, for example, indicate glacial conditions (Figure 1.7), while coals signify swampy environments. In Chapters 3 and 4, we will discuss in more detail how sedimentary rocks and sedimentary structures can be used to interpret Earth history.

Metamorphic rocks result from the alteration of other rocks, usually beneath the surface, by heat, pressure, and the chemical activity of fluids (see Table C6). For example, marble, a rock preferred by many sculptors and builders, is a metamorphic rock produced when the agents of metamorphism are applied to the sedimentary rocks limestone or dolostone. Metamorphic rocks are either *foliated* (Figure 1.3e) or *nonfoliated* (Figure 1.3f). Foliation, the parallel alignment of minerals due to pressure, gives the rock a layered or banded appearance. If the metamorphic rock is foliated, the type of foliation determines its name, while nonfoliated metamorphic rocks are named on the basis of their composition.

The sequence of processes shown in the rock cycle (Figure 1.2) illustrates the interrelationship of the three rock groups and typifies an idealized cycle. However, modifications of this idealized cycle are common. For example, an igneous rock may never be exposed at the surface. After forming, it may be subjected to heat and pressure and converted directly to metamorphic rock without ever going through the intermediate sedimentary phase. Sedimentary rocks may never be buried deeply enough to be converted to metamorphic rock or melted to form magma, but instead may undergo several cycles of weathering, erosion, and deposition.

(a) Granite

(b) Basalt

(c) Conglomerate

(d) Limestone

(e) Gneiss

(f) Quartzite

FIGURE 1.3 Hand specimens of common igneous (a, b), sedimentary (c, d), and metamorphic (e, f) rocks. (a) Granite, an intrusive igneous rock. (b) Basalt, an extrusive igneous rock. (c) Conglomerate, a sedimentary rock formed by the consolidation of rock fragments. (d) Limestone, a sedimentary rock formed by the extraction of mineral matter from seawater by organisms or by the inorganic precipitation of the mineral calcite from seawater. (e) Gneiss, a foliated metamorphic rock. (f) Quartzite, a nonfoliated metamorphic rock.

The concept of the rock cycle was first outlined in the late eighteenth century by James Hutton as part of his overall theory of Earth. Hutton thought that Earth's internal heat provided the mechanism for magma generation, uplift, and mountain building. Today, plate tectonic theory offers an explanation for how and where magma is generated and how and why mountain building occurs. Following a brief discussion of Earth's interior and plate tectonics, we will return to the relationship between the rock cycle and plate tectonics.

Earth as a Dynamic Planet

Earth is a dynamic planet that has continuously changed during its 4.6-billion-year existence. The size, shape, and geographic distribution of continents and ocean basins have changed through time, the composition of the atmosphere has evolved, and lifeforms existing today differ from those that lived during the past. We can easily visualize how mountains and hills are worn down by erosion and how landscapes are changed by the forces of wind, water, and ice. Volcanic eruptions and earthquakes reveal an ac-

FIGURE 1.4 Sedimentary rocks preserve evidence of the processes responsible for deposition of the sediment. This rock specimen was most likely deposited in a stream environment as indicated by its color and the small ripple marks showing that the current flowed from left to right.

FIGURE 1.5 This fossiliferous limestone was deposited in a marine environment as indicated by the numerous fossil brachiopods.

FIGURE 1.6   Cross-bedding in this sandstone in Zion National Park, Utah, indicates that it was deposited as sand dunes.

tive interior, and folded and fractured rocks indicate the tremendous power of Earth's internal forces.

Earth consists of three concentric layers: the core, the mantle, and the crust (Figure 1.8). This orderly division results from density differences between the layers as a function of variations in composition, temperature, and pressure.

The **core** has a calculated density of 10 to 13 grams per cubic centimeter (g/cm^3) and occupies about 16% of Earth's total volume. Seismic (earthquake) data indicate that the core consists of a small, solid, inner core and a larger, apparently liquid, outer core. Both are thought to consist largely of iron and a small amount of nickel.

The **mantle** surrounds the core and contains about 83% of Earth's volume. It is less dense than the core (3.3–5.7 g/cm^3) and is thought to be composed largely of *peridotite*, a dark, dense igneous rock containing abundant iron and

magnesium (see Table C4). The mantle can be divided into three distinct zones based on physical characteristics. The lower mantle is solid and forms most of the volume of Earth's interior. The **asthenosphere** surrounds the mantle. It has the same composition as the lower mantle but behaves plastically and slowly flows. Partial melting within the asthenosphere generates magma, some of which rises to the surface because it is less dense than the rock from which

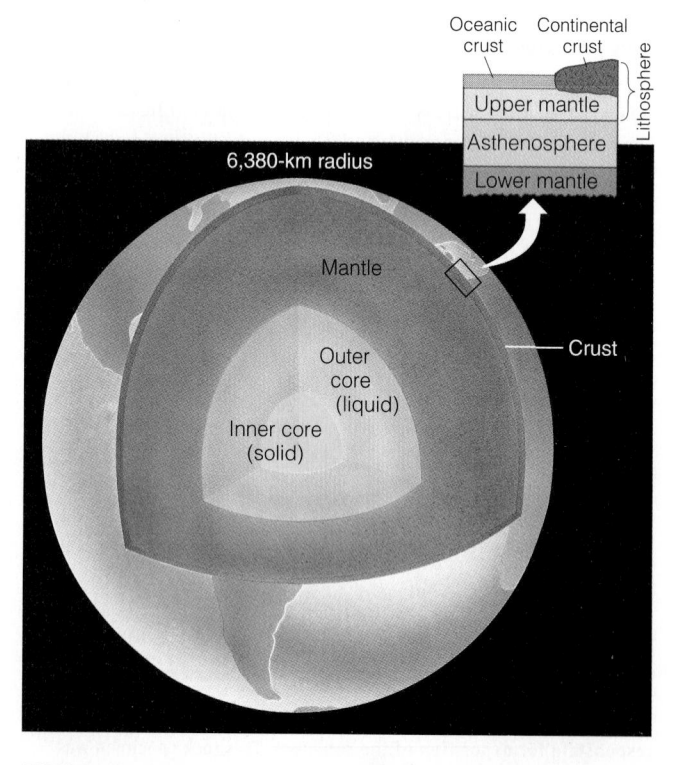

FIGURE 1.7   Glacial till is an unsorted mixture of many different sediment sizes deposited at the front of a melting glacier. The rock tillite (which is lithified till) provides evidence that glaciation occurred where the tillite is found.

FIGURE 1.8   A cross section of Earth illustrating the core, mantle, and crust. The enlarged portion shows the relationship between the lithosphere, composed of the continental crust, oceanic crust, and upper mantle, and the underlying asthenosphere and lower mantle.

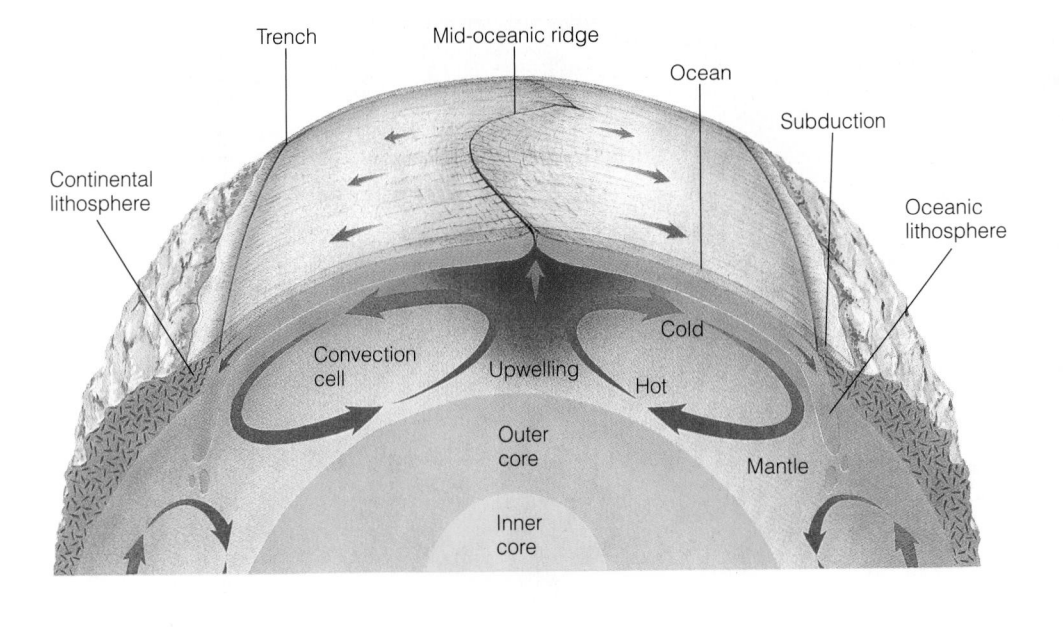

Trench   Mid-oceanic ridge

Ocean

Subduction

Continental
lithosphere

Oceanic
lithosphere

Convection
cell   Upwelling

Cold

Hot

Outer
core

Mantle

Inner
core

FIGURE 1.9 Earth's plates are thought to move as a result of underlying mantle convection cells in which warm material from deep within Earth rises toward the surface, cools, and then, upon losing heat, descends back into the interior. The movement of these convection cells is thought to be the mechanism responsible for the movement of Earth's plates, as shown in this diagrammatic cross section.

it was derived. The upper mantle surrounds the asthenosphere. The solid upper mantle and the overlying crust constitute the **lithosphere,** which is broken into numerous individual pieces called **plates** that move over the asthenosphere as a result of underlying *convection cells* (Figure 1.9). Interactions of these plates are responsible for such phenomena as earthquakes, volcanic eruptions, and the formation of mountain ranges and ocean basins.

The **crust,** Earth's outermost layer, consists of two types. *Continental crust* is thick (20–90 km), has an average density of 2.7 g/cm^3, and contains considerable silicon and aluminum. *Oceanic crust* is thin (5–10 km), denser than continental crust (3.0 g/cm^3), and is composed of the dark igneous rock *basalt* (see Table C4).

Since the widespread acceptance of plate tectonic theory more than 25 years ago, geologists have viewed Earth from a global perspective in which all of its systems are interconnected. Thus, the distribution of mountain chains, major fault systems, volcanoes and earthquakes, the origin of new ocean basins, the movement of continents, and several other geologic processes and features are perceived to be interrelated.

Plate Tectonics

The acceptance of **plate tectonic theory** is recognized as a major milestone in the geologic sciences. It is comparable to the revolution caused by Darwin's theory of evolution in biology. Plate tectonics has provided a framework for interpreting the composition, structure, and internal processes of Earth on a global scale. It has led to the realization that the continents and ocean basins are part of a lithosphere–atmosphere–hydrosphere (water portion of the planet) system that evolved together with Earth's interior (Table 1.4).

According to plate tectonic theory, the lithosphere is divided into plates that move over the asthenosphere (Figure 1.10). Zones of volcanic activity, earthquake activity, or both, mark most plate boundaries. Along these boundaries, plates diverge, converge, or slide sideways past each other.

At **divergent plate boundaries,** plates move apart as magma rises to the surface from the asthenosphere (Figure 1.11). The magma solidifies to form rock, which attaches to the moving plate. The margins of divergent plate boundaries are marked by mid-oceanic ridges in oceanic crust, such as the Mid-Atlantic Ridge, and are recognized by linear rift valleys, where newly forming divergent boundaries occur beneath continental crust.

TABLE 1.4
Plate Tectonics and Earth Systems

SOLID EARTH	Plate tectonics is driven by convection in the mantle and in turn drives mountain-building and associated igneous and metamorphic activity.
ATMOSPHERE	Arrangement of continents affects solar heating and cooling, and thus winds and weather systems. Rapid plate spreading and hot-spot activity may release volcanic carbon dioxide and affect global climate.
HYDROSPHERE	Continental arrangement affects ocean currents. Rate of spreading affects volume of mid-oceanic ridges and hence sea level. Placement of continents may contribute to onset of ice ages.
BIOSPHERE	Movement of continents creates corridors or barriers to migration, the creation of ecological niches, and transport of habitats into more or less favorable climates.
EXTRATERRESTRIAL	Arrangement of continents affects free circulation of ocean tides and influences tidal slowing of Earth's rotation.

Source: Adapted by permission from Stephen Dutch, James S. Monroe, and Joseph Moran, Earth Science (Minneapolis/St. Paul: West Publishing Co.).

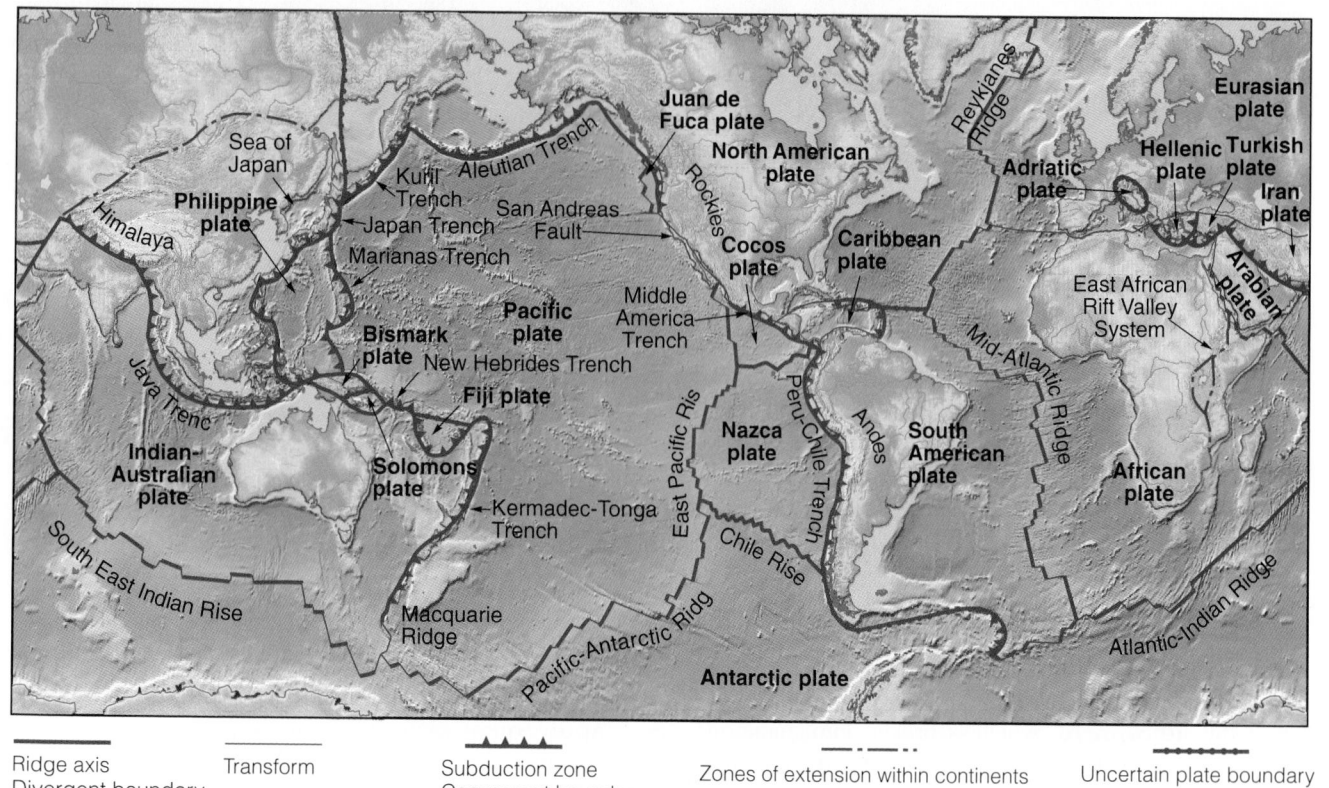

FIGURE 1.10   Earth's lithosphere is divided into rigid plates of various sizes that move over the asthenosphere.

Plates move toward one another along **convergent plate boundaries,** where one plate sinks beneath another plate, along what is known as a **subduction zone** (Figure 1.11). As the plate descends into Earth, it becomes hotter until it melts, thus generating a magma. As this magma rises, it may erupt at the surface, forming a chain of volcanoes. The Andes Mountains on the west coast of South America are a good example of a volcanic mountain range formed as a result of subduction along a convergent plate boundary (Figure 1.10).

Transform plate boundaries are sites where plates slide sideways past each other (Figure 1.11). The San Andreas fault in California is a transform plate boundary separating the North American from the Pacific plate

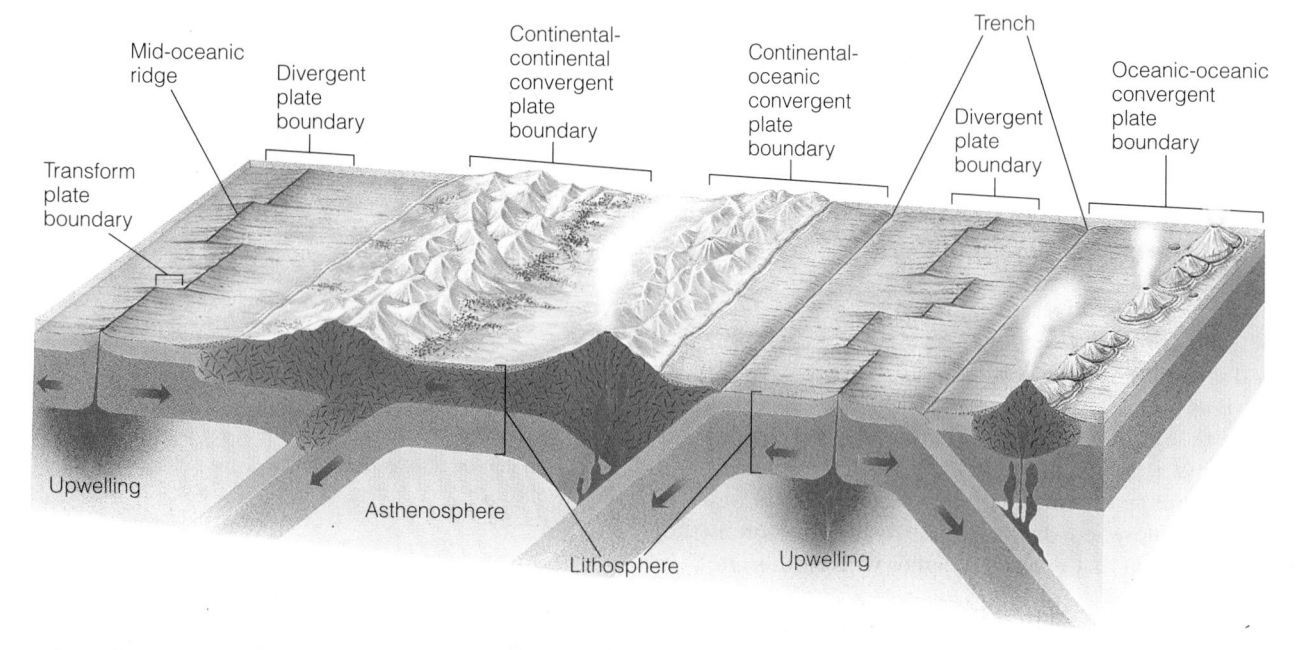

FIGURE 1.11   An idealized cross section illustrating the relationship between the lithosphere and the underlying asthenosphere and the three principal types of plate boundaries: divergent, convergent, and transform.

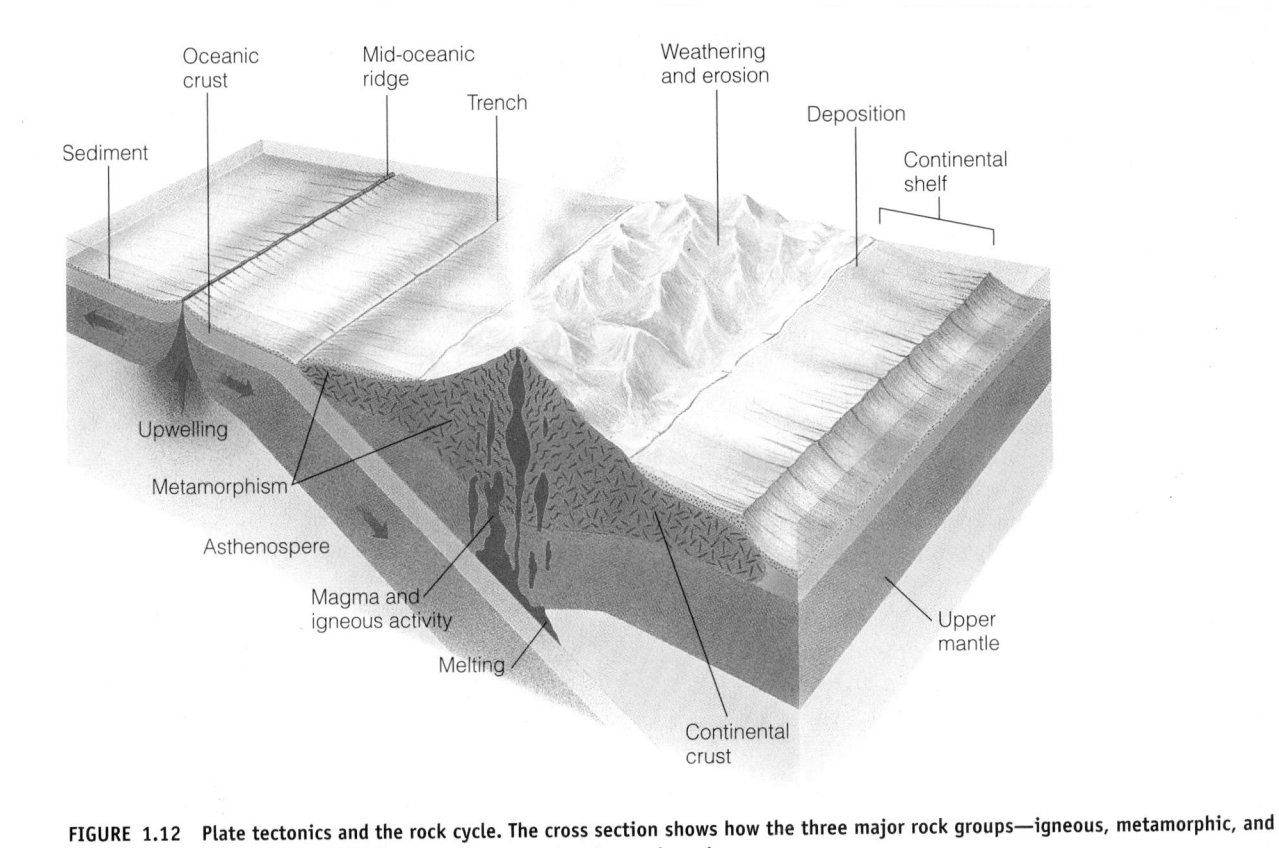

Sediment · Oceanic crust · Mid-oceanic ridge · Trench · Weathering and erosion · Deposition · Continental shelf · Upwelling · Metamorphism · Asthenospere · Magma and igneous activity · Melting · Continental crust · Upper mantle

FIGURE 1.12 Plate tectonics and the rock cycle. The cross section shows how the three major rock groups—igneous, metamorphic, and sedimentary—are recycled through both the continental and oceanic regions.

(Figure 1.10). The earthquake activity along the San Andreas faults results from the Pacific plate moving northward relative to the North American plate.

Although geologists can describe and explain what is happening at plate boundaries, there is still debate about a possible driving mechanism that causes plate movement. Most geologists, however, accept a mechanism involving some type of convection system in which mantle rock is heated and flows like a very viscous liquid (Figure 1.9).

A revolutionary concept when it was proposed in the 1960s, plate tectonic theory has had significant and far-reaching consequences in all fields of geology because it provides the basis for relating many seemingly unrelated phenomena. Besides being responsible for the major features of Earth's crust, plate movements also affect the formation and distribution of Earth's natural resources, as well as influencing the distribution and evolution of the world's biota.

The impact of plate tectonic theory has been particularly notable in the interpretation of Earth history. For example, the Appalachian Mountains in eastern North America and the mountain ranges of Greenland, Scotland, Norway, and Sweden are not the result of unrelated mountain-building episodes but, rather, are part of a larger mountain-building event that involved the closing of an ancient "Atlantic Ocean" and the formation of the supercontinent Pangaea about 245 million years ago (see Chapter 11).

THE ROCK CYCLE AND PLATE TECTONICS

The earlier discussion of the rock cycle was presented without reference to a mechanism responsible for recycling rock materials. Plate tectonics provides such a recycling mechanism.

Interactions among plates determine, to a certain extent, which of the three rock groups will form (Figure 1.12). For example, weathering and erosion produce sediments that are transported by agents such as running water from the continents to the oceans where they are deposited. These sediments, some of which may be lithified and become sedimentary rock, become part of a moving plate along with the underlying oceanic crust. When plates converge, heat and pressure generated along the plate boundary may lead to igneous activity and metamorphism within the descending oceanic plate, thus producing various igneous and metamorphic rocks.

Some of the sediment and sedimentary rock is subducted and melts, while other sediment and sedimentary rocks along the boundary of the nonsubducted plate are metamorphosed by the heat and pressure generated along the convergering plate boundary. Later, the mountain range or chain of volcanic islands formed along the convergent plate boundary will once again be weathered and eroded, and the new sediments will be transported to the ocean to begin yet another cycle.

Organic Evolution

Plate tectonic theory provides us with a model for understanding the internal workings of Earth and their effect on Earth's surface. The theory of **organic evolution** provides the conceptual framework for understanding the history of life. Together, the theories of plate tectonics and organic evolution have changed the way we view our planet, and we should not be surprised at the intimate association between them. The relationship between plate tectonic processes and the evolution of life is incredibly complex, but paleontological data provide indisputable evidence of the influence of plate movement on the distribution of organisms.

The publication in 1859 of Darwin's *On the Origin of Species by Means of Natural Selection* revolutionized biology and marked the beginning of modern evolutionary biology. With its publication, most naturalists recognized that evolution provided a unifying theory that explained an otherwise encyclopedic collection of biologic facts.

The central thesis of organic evolution is that all present-day organisms are related and that they have descended with modifications from organisms that lived during the past. When Darwin proposed his theory of organic evolution, he cited a wealth of supporting evidence including the way organisms are classified, embryology, comparative anatomy, the geographic distribution of organisms, and, to a limited extent, the fossil record. Furthermore, Darwin proposed that *natural selection,* which results in the survival to reproductive age of those organisms best adapted to their environment, is the mechanism that accounts for evolution.

Despite overwhelming evidence to support evolution, Darwin could not explain how variation within a population arose, nor how it was maintained. We now know that mutations occur within genes (the basic units of inheritance) causing changes within an organism and that these changes are passed on to the next generation during reproduction. Thus, mutations provide one source of variation within populations, and natural selection acts on them in determining which variants are favorable for survival. Another important source of variation occurs during sexual reproduction when the offspring receive one-half of their genes from each parent. This allows new mutations to be introduced into the population in many different combinations because potentially the parent with the mutation may mate with a variety of partners.

Perhaps the most compelling evidence in favor of evolution can be found in the fossil record. Just as the rock record allows geologists to interpret physical events and conditions in the geologic past, fossils provide a record of past life and evidence that organisms have evolved.

Fossils, which are the remains or traces of once-living organisms, not only provide evidence that evolution has occurred but also demonstrate that Earth has a history extending beyond that recorded by humans. The succession of fossils in the rock record provides geologists with a means for dating rocks and allowed a relative geologic time scale to be constructed in the 1800s.

Geologic Time and Uniformitarianism

An appreciation of the immensity of geologic time is central to understanding the evolution of Earth and its biota.

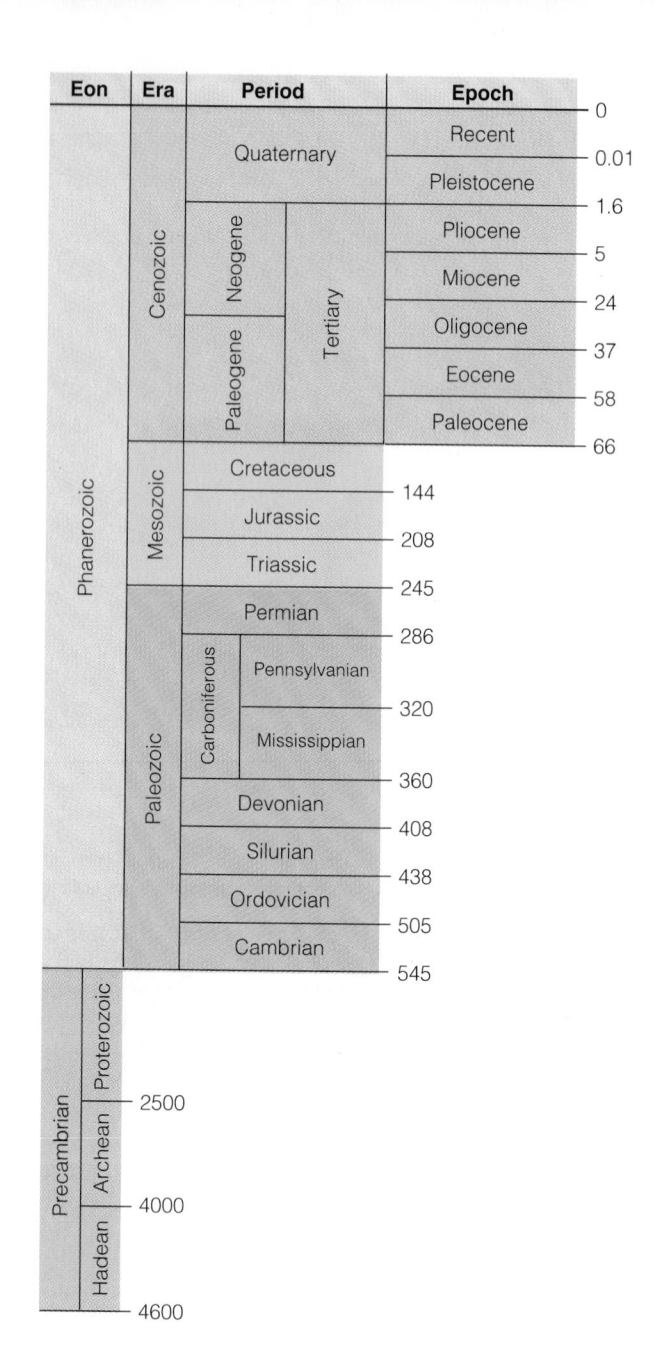

FIGURE 1.13 The geologic time scale. Numbers to the right of the columns are ages in millions of years before the present.

Indeed, time is one of the main aspects that sets geology apart from the other sciences. Most people have difficulty comprehending geologic time because they tend to think in terms of the human perspective—seconds, hours, days, and years. Ancient history is what occurred hundreds or even thousands of years ago. When geologists talk of ancient geologic history, however, they are referring to events that happened hundreds of millions or even billions of years ago. To a geologist, recent geologic events are those that occurred within the last million years or so.

It is also important to remember that Earth goes through cycles of a much longer duration than the human perspective of time. Although they may have disastrous effects on the human species, global warming and cooling are part of a larger cycle that has resulted in numerous glacial advances and retreats during the past 1.6 million years. In fact, geologists can make important contributions to the debate on global warming because of their geologic perspective.

The **geologic time scale** resulted from the work of many nineteenth-century geologists who pieced together information from numerous rock exposures and constructed a sequential chronology based on changes in Earth's biota through time. Subsequently, with the discovery of radioactivity in 1895 and the development of various radiometric dating techniques, geologists have been able to assign age dates in years to the subdivisions of the geologic time scale (Figure 1.13).

One of the cornerstones of geology, the **principle of uniformitarianism,** is based on the premise that present-day processes have operated throughout geologic time. Therefore, to understand and interpret the rock record, we must first understand present-day processes and their results.

Uniformitarianism is a powerful principle that allows us to use present-day processes as the basis for interpreting the past and for predicting potential future events. We should keep in mind that uniformitarianism does not exclude such sudden or catastrophic events as volcanic eruptions, earthquakes, landslides, or flooding. These are processes that shape our modern world, and some geologists in fact view the history of Earth as a series of such short-term or punctuated events. Such a view is certainly in keeping with the modern principle of uniformitarianism.

Furthermore, uniformitarianism does not require that the rates and intensities of geologic processes be constant through time. We know that volcanic activity was more intense in North America 5 to 10 million years ago than it is today and that glaciation has been more prevalent during the last several million years than in the previous 300 million years.

What uniformitarianism means is that even though the rates and intensities of geologic processes have varied during the past, the physical and chemical laws of nature have remained the same. Although Earth is in a dynamic state of change and has been ever since it was formed, the processes that shaped it during the past are the same ones in operation today.

Summary

1. Earth can be viewed as a system of interconnected components that interact and affect each other in many ways. These complex interactions result in a dynamically changing body that exchanges matter and energy and recycles them into different forms.
2. Geology, the study of Earth, is divided into two broad areas: Physical geology is the study of Earth materials as well as the processes that operate within and upon Earth's surface; historical geology examines the origin and evolution of Earth's continents, atmosphere, oceans, and life.
3. The scientific method is an orderly, logical approach that involves gathering and analyzing facts about a particular phenomenon, formulating hypotheses to explain the phenomenon, testing the hypotheses, and finally proposing a theory. A theory is a testable explanation for some natural phenomenon that has a large body of supporting evidence.
4. The rock and fossil record provide the only evidence for interpreting Earth history.
5. Igneous, sedimentary, and metamorphic rocks comprise the three major groups of rocks. Igneous rocks result from the crystallization of magma or consolidation of volcanic ejecta. Sedimentary rocks are formed by the consolidation of rock fragments, precipitation of mineral matter from solution, or compaction of plant or animal remains. Metamorphic rocks are produced from other rocks generally beneath Earth's surface by heat, pressure, and chemically active fluids.
6. The rock cycle illustrates the interactions between internal and external Earth processes and shows how the three rock groups are interrelated.
7. Earth is differentiated into layers. The outermost layer is the crust, which is divided into continental and oceanic crust. Below the crust is the upper mantle. The crust and upper mantle overlie the asthenosphere, a zone that slowly flows. The asthenosphere is underlain by the solid lower mantle. Earth's core consists of an outer liquid portion and an inner solid portion.
8. The lithosphere is broken into a series of plates that diverge, converge, and slide sideways past one another.
9. Plate tectonic theory provides a unifying explanation for many geologic features and events. The interaction between plates is responsible for volcanic eruptions, earthquakes, the formation of mountain ranges and ocean basins, and the recycling of rock material.

10. The central thesis of the theory of organic evolution is that all living organisms evolved (descended with modification) from organisms that existed in the past. Mutations and sexual reproduction provide the source of variation within a population that natural selection acts upon.

11. Time sets geology apart from the other sciences, except astronomy. The geologic time scale is the calendar geologists use to date past events.

12. The principle of uniformitarianism is basic to the interpretation of Earth history. This principle holds that the laws of nature have been constant through time and that the same processes operating today have operated in the past, although at different rates.

Important Terms

asthenosphere

convergent plate
  boundary

core

crust

divergent plate boundary

fossil

geologic time scale

geology

hypothesis

igneous rock

lithosphere

mantle

metamorphic rock

mineral

organic evolution

plate

plate tectonic theory

principle of
  uniformitarianism

rock

rock cycle

scientific method

sedimentary rock

subduction zone

system

theory

transform plate boundary

Review Questions

1. The study of Earth materials is:
   a. _____ paleontology.
   b. _____ stratigraphy.
   c. _____ physical geology.
   d. _____ historical geology.
   e. _____ environmental geology.

2. Which of the following statements about a scientific theory is *not* true?
   a. _____ It is an explanation for some natural phenomenon.
   b. _____ It is testable.
   c. _____ It has a large body of supporting evidence.
   d. _____ It is a conjecture or a guess.
   e. _____ Predictive statements can be derived from it.

3. Which of the following statements about a mineral is *not* true?
   a. _____ It is organic.
   b. _____ It has definite physical and chemical properties.
   c. _____ It is naturally occurring.
   d. _____ It is a crystalline solid.
   e. _____ None of these

4. The rock cycle implies that:
   a. _____ only sedimentary rocks can be derived from igneous rocks.
   b. _____ igneous rocks only form beneath Earth's surface.
   c. _____ only sedimentary rocks can be weathered.
   d. _____ any rock group can be derived from any other rock group.
   e. _____ metamorphic rocks form from the cooling of magma.

5. A combination of related parts interacting in an organized fashion is:
   a. _____ sustainable development.
   b. _____ a system.
   c. _____ uniformitarianism.
   d. _____ a theory.
   e. _____ none of these

6. A plate is composed of the:
   a. _____ core and lower mantle.
   b. _____ lower mantle and asthenosphere.
   c. _____ asthenosphere and upper mantle.
   d. _____ upper mantle and crust.
   e. _____ continental and oceanic crust.

7. The Andes Mountains of South America formed at what type of plate boundary?
   a. _____ Divergent
   b. _____ Transform
   c. _____ Convergent
   d. _____ Hot spot
   e. _____ Answers (a) and (d)

8. Which rocks result from the alteration of other rocks, usually beneath Earth's surface, by heat, pressure, and the chemical activity of fluids?
   a. _____ Igneous
   b. _____ Sedimentary
   c. _____ Metamorphic
   d. _____ All of these
   e. _____ None of these

9. The premise that present-day processes have operated throughout geologic time is known as the principle of:
   a. _____ plate tectonics.
   b. _____ seafloor spreading.
   c. _____ continental drift.
   d. _____ organic evolution.
   e. _____ uniformitarianism.

10. Which layer has the same composition as the mantle but behaves plastically?
    a. _____ Continental crust
    b. _____ Oceanic crust
    c. _____ Outer core
    d. _____ Inner core
    e. _____ Asthenosphere

11. A plate boundary along which plates slide horizontally past each other is a:
    a. _____ divergent boundary.
    b. _____ convergent boundary.
    c. _____ transform boundary.
    d. _____ subduction boundary.
    e. _____ none of these

12. That all living organisms are the descendants of different lifeforms that existed in the past is the central claim of:
    a. _____ organic evolution.
    b. _____ the principle of fossil succession.
    c. _____ the principle of uniformitarianism.
    d. _____ plate tectonics.
    e. _____ none of these

13. Name and define the three major rock groups.

14. Describe the scientific method and explain how it may lead to a scientific theory.

15. Name the major layers of Earth and describe their composition.

16. Why are the concepts of plate tectonics, organic evolution, and geologic time so important to historical geology?

17. Briefly describe plate tectonic theory and explain why it is a unifying theory in geology.

18. Name and describe the three types of plate boundaries.

19. Describe the rock cycle and explain how it is related to plate tectonics.

20. Briefly discuss the theory of organic evolution.

21. Reproduce the geologic time scale and give the dates for the boundaries of each era.

22. What is the principle of uniformitarianism? Does it allow for catastrophic events?

Points to Ponder

1. Propose a pre–plate tectonic hypothesis explaining the formation and distribution of mountain ranges.

2. Why is an accurate geologic time scale particularly important for geologists in their search for mineral and energy resources and for determining Earth history?

3. Discuss how plate tectonics influences organic evolution.

Additional Readings

Ausich, W. I., and N. G. Lane. 1999. *Life of the past*. 4th ed. Englewood Cliffs, N.J.: Prentice Hall.

Berry, W. B. N. 1987. *Growth of a prehistoric time scale*. 2d ed. Palo Alto, Calif.: Blackwell Scientific Publications.

Dietrich, R. V., and B. J. Skinner. 1990. *Gems, granites, and gravels: Knowing and using rocks and minerals*. New York: Cambridge University Press.

Ernst, W. G. 1990. *The dynamic planet*. Irvington, N.Y.: Columbia University Press.

Monroe, J. S., and R. Wicander. 1998. *Physical geology: Exploring the Earth*. 3d ed. Belmont, Calif.: Wadsworth.

Officer, C., and J. Page. 1993. *Tales of the Earth*. New York: Oxford University Press.

World Wide Web Activities

▶ **EARTH SCIENCE RESOURCES ON THE INTERNET**

This site, maintained by the geology department at the University of North Carolina at Chapel Hill, is an excellent starting place to learn how to find geoscience items on the World Wide Web. As you will see, a tremendous amount of information, graphics, and maps is available, and this site will get you started finding many of them and practicing your Net skills.

1. Read the *Starting up* and *How to find things* sections.

2. Visit several of the geology sites mentioned to see the variety of information, videos, and images available.

3. Visit the home page of several of the different geologic societies and organizations to see what the societies do and what information is available from them.

4. Check the home page of several of the different schools and universities to find information on course offerings, faculty, and requirements to earn a degree. Compare the home pages and information available from different universities around the world.

▶ **ONLINE RESOURCES FOR EARTH SCIENTISTS (ORES)**

This large and comprehensive Earth science resource list is operated by Bill Thoen and Ted Smith and provides links to many other resource lists. It is continually updated and is certainly a site worth visiting.

1. Check out the various headings under *Contents* to see what is available in the different geoscience and related areas and explore the various links to other resource lists.

2. When you find a particular link interesting and would like to return to it in the future, make a "Bookmark" of the page (that is, save its URL). This way when you want to go back to the site, all you will have to do is find its name in the "Bookmark Menu" and click it, and your browser will automatically connect you to that site.

▶ **WEST'S GEOLOGY DIRECTORY**

This comprehensive directory of geologic internet links is an excellent place to find links on virtually any geologic topic. The Contents-Index lists numerous topics. Just click on a topic and a list of sites, which you can then click on, will appear. The links are regularly updated and sites are checked, reviewed, and rated where possible. In addition, there is also a link to field trip guides and bibliographies.

CD-ROM Exploration

Explore the following *In-Terra-Active 2.0* CD-ROM module(s) and increase your understanding of key concepts and processes presented in this chapter.

Chapter Concept: **Scientific Method**

Scientists formulate theories through a process known as the scientific method. Organizing observations into a hypothesis is an important first step of this process. This module introduces you to the evidence that scientists use to support plate tectonic theory and enables you to apply your skills to a hypothetical tectonic plate configuration.

Section: **Interior**
Module: **Plate Tectonics: Evidence**
Question: *If you were an astronaut landing on Venus, what evidence would you look for to confirm or deny the existence of plate tectonics there?*

Chapter Concept: **Rock Cycle**

The rock cycle illustrates the relation between the different rock types — igneous, sedimentary, and metamorphic — and Earth's internal and external processes. This module will enable you to use the rock cycle to determine the sequencing of Earth processes and the formation of the different rock types.

Section: **Materials**
Module: **Rock Cycle**

During a visit to Thailand, you find a specimen of limestone and another of marble in the same area.

Questions: *Which rock is older? How can you tell?*

Geologic Time: Concepts and Principles

Horizontal sedimentary rocks exposed at Inspiration Point in Bryce Canyon National Park, Utah.

Prologue

What is time? We seem obsessed with it and organize our lives around it with the help of clocks, calendars, and appointment books. Yet most of us feel we don't have enough of it—we are always running "behind" or "out of time." According to biologists and psychologists, children less than two years old and animals exist in a "timeless present." Some scientists think that our early ancestors may also have lived in a state of timelessness with little or no perception of a past or future. According to Buddhist, Taoist, and Mayan beliefs, time is circular, and like a circle, all things are destined to return to where they once were. Thus, in these belief systems, there is no beginning or end, but rather a cyclicity to everything.

In some respects, time is defined by the methods used to measure it. Many prehistoric monuments are oriented to detect the summer solstice, and sundials were used to divide the day into measurable units. As civilization advanced, mechanical devices were invented to measure time, the earliest being the water clock, first used by the ancient Egyptians and further developed by the Greeks and Romans. The pendulum clock was invented in the seventeenth century and provided the most accurate time keeping for the next two and a half centuries.

Today, the quartz watch is the most popular timepiece. Powered by a battery, a quartz crystal vibrates approximately 100,000 times per second. An integrated circuit counts these vibrations and converts them into a digital or dial reading on your watch face. An inexpensive quartz watch today is more accurate than the best mechanical watch, and precision manufactured quartz clocks are accurate to within one second per ten years.

Precise time keeping is important in our technological world. Ships and aircraft plot their locations by satellite, relying on an extremely accurate time signal. Deep-space and planetary probes, such as the *Voyagers* and the *Mars Pathfinder* and robotic rover *Sojourner*, re-

quire radio commands timed to billionths of a second, and physicists exploring the motion inside the nucleus of an atom deal in trillionths of a second as easily as we talk about minutes.

To achieve such accuracy, scientists use atomic clocks. First developed in the 1940s, these clocks rely on an atom's oscillating electrons, a rhythm so regular that they are accurate to within a few thousandths of a second per day. An atomic clock accurate to within one second per 3 million years was recently installed at the National Institute of Standards and Technology.

While physicists deal with incredibly short intervals of time, astronomers and geologists are concerned with geologic time measured in millions or billions of years. When astronomers look at a distant galaxy, they are seeing what it looked like billions of years ago. When geologists investigate rocks in the walls of the Grand Canyon, they are deciphering events that occurred over an interval of 2 billion years. Geologists can measure decay rates of such radioactive elements as uranium, thorium, and rubidium to determine how long ago an igneous rock formed. Furthermore, geologists know that Earth's rotational velocity has been slowing a few thousandths of a second per century as a result of the frictional effects of tides, ocean currents, and varying thicknesses of polar ice. Five hundred million years ago a day was only 20 hours long, and at the current rate of slowing, 200 million years from now a day will be 25 hours long.

Time is a fascinating topic that has been the subject of numerous essays and books. And though we can comprehend concepts like milliseconds and understand how a quartz watch works, deep time, or geologic time, is still very difficult for most people to comprehend. As we will see in this chapter, geology could not have advanced as a science until the foundation of geologic time was firmly established and accepted.

Introduction

Time is what sets geology apart from most of the other sciences, and an appreciation of the immensity of geologic time is fundamental to understanding both the physical and biologic history of our planet. Most people have difficulty comprehending geologic time because they tend to view time from the perspective of their own existence. Ancient history is what occurred hundreds or even thousands of years ago, but when geologists talk in terms of ancient geologic history, they are referring to events that happened millions or even billions of years ago! (See Perspective 2.1.)

Geologists use two different frames of reference when speaking of geologic time. **Relative dating** involves placing geologic events in a sequential order as determined from their position in the rock record. Relative dating will not tell us how long ago a particular event occurred, only that one event preceded another. The various principles used to determine relative dating were discovered hundreds of years ago and since then have been used to construct the *relative geologic time scale* (Figure 2.1). These principles are still widely used today.

Absolute dating results in specific dates for rock units or events expressed in years before the present. Radio-

FIGURE 2.1 The geologic time scale. Some of the major biological and geological events are indicated along the right-hand side.

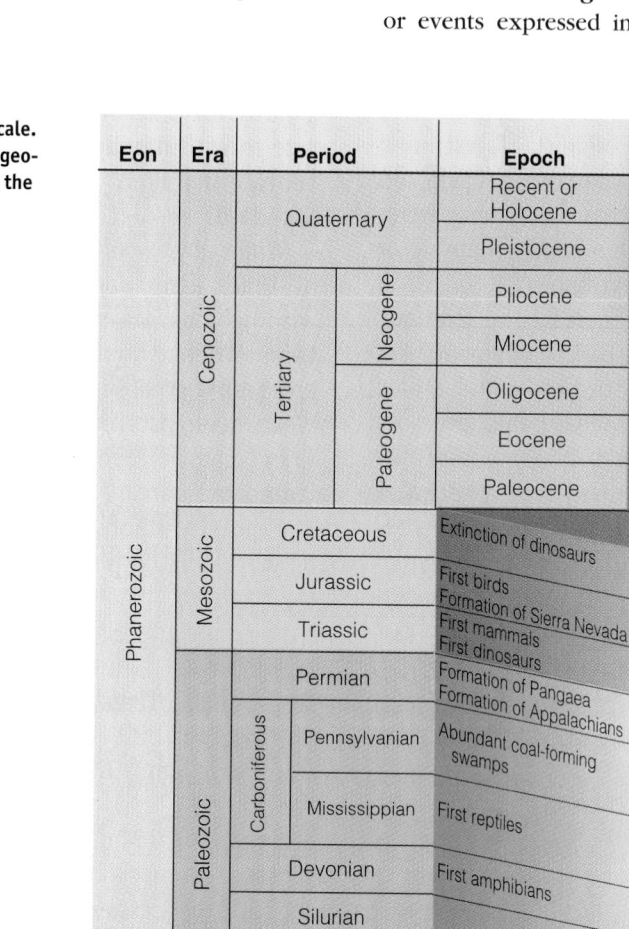

Eon	Era	Period		Epoch	Major Geologic and Biologic Events	Millions of Years Ago
Phanerozoic	Cenozoic	Quaternary		Recent or Holocene	Ice Age ends	0
				Pleistocene	Ice Age begins	0.01
		Tertiary	Neogene	Pliocene	Earliest humans	1.6
				Miocene		5.3
			Paleogene	Oligocene		23.7
				Eocene	Formation of Himalyas Formation of Alps	36.6
				Paleocene		57.8
	Mesozoic	Cretaceous			Extinction of dinosaurs	66
		Jurassic			First birds Formation of Sierra Nevada	144
		Triassic			First mammals First dinosaurs	208
	Paleozoic	Permian			Formation of Pangaea Formation of Appalachians	245
		Carboniferous	Pennsylvanian		Abundant coal-forming swamps	286
			Mississippian		First reptiles	320
		Devonian			First amphibians	360
		Silurian				408
		Ordovician			First land plants	438
		Cambrian			First fish	505
Precambrian	Proterozoic				Earliest shelled animals	545
	Archean					2500
					Earliest fossil record of life	4000
	Hadean					4600

metric dating is the most common method of obtaining absolute-age dates. Such dates are calculated from the natural rates of decay of various radioactive elements present in trace amounts in some rocks. It was not until the discovery of radioactivity near the end of the nineteenth century that absolute ages could be accurately applied to the relative geologic time scale. Today the geologic time scale is really a dual scale: a relative scale based on rock and fossil sequences with radiometric dates expressed as years before present added to it (Figure 2.1).

Early Concepts of Geologic Time and the Age of Earth

The concept of geologic time and its measurement have changed through human history. Early Christian theologians were largely responsible for formulating the idea that time is linear rather than circular. When St. Augustine of Hippo (A.D. 354–430) stated that the Crucifixion was a unique event from which all other events could be measured, he helped establish the idea of the B.C. and A.D. time scale. This prompted many religious scholars and clerics to try and establish the date of creation by analyzing historical records and the genealogies found in Scripture.

A famous and influential Christian scholar, James Ussher (1581–1665), archbishop of Armagh, in Ireland, is popularly credited as being the first to calculate the age of Earth based on recorded history and the genealogies described in Genesis. In 1650 Ussher announced that Earth had been created on October 22, 4004 B.C. This date was later reproduced in many editions of the Bible and incorporated into the dogma of the Christian Church. For nearly a century thereafter, it was considered heresy to assume that Earth and all its features were more than about 6000 years old. Thus, the idea of a very young Earth provided the basis for most chronologies of Earth history prior to the eighteenth century.

During the eighteenth and nineteenth centuries, several attempts were made to determine the age of Earth on the basis of scientific evidence rather than revelation. The French zoologist Georges Louis de Buffon (1707–1788) assumed Earth gradually cooled to its present condition from a molten beginning. To simulate this history, he melted iron balls of various diameters and allowed them to cool to the surrounding temperature. By extrapolating their cooling rate to a ball the size of Earth, he determined that Earth was at least 75,000 years old. While this age was still older than that derived from Scripture, it was still vastly younger than we now know the planet to be.

Other scholars were equally ingenious in attempting to calculate Earth's age. For example, if deposition rates could be determined for various sediments, geologists reasoned that they could calculate how long it would take to deposit any rock layer. They could then extrapolate how old Earth was from the total thickness of sedimentary rock in its crust. Rates of deposition vary, however, even for the same type of rock. Furthermore, it is impossible to estimate how much rock has been removed by erosion or how much a rock sequence has been reduced by compaction. As a result of these uncertainties, estimates ranged from less than 1 million years to more than 1 billion years.

Another attempt at determining the age of Earth involved calculating the age of the oceans. If the ocean basins were filled very soon after the origin of the planet, then they would be only slightly younger than Earth itself. The best-known calculations for the oceans' age were made by the Irish geologist John Joly in 1899. He reasoned that Earth's ocean waters were originally fresh and their present salinity was the result of dissolved salt being carried into the ocean basins by rivers. By measuring the present amount of salt in the world's rivers, and knowing the volume of ocean water and its salinity, Joly calculated that it would have taken at least 90 million years for the oceans to reach their present salinity level. This was still much younger than the now-accepted age of 4.6 billion years for Earth, mainly because Joly had no way to calculate how much salt had been recycled or the amount of salt stored in continental salt deposits or in seafloor clay deposits.

Besides trying to determine Earth's age, the naturalists of the eighteenth and nineteenth centuries were also formulating some of the fundamental geologic principles that are used in deciphering Earth history. From the evidence preserved in the geologic record, it was clear to them that Earth is very old and that geologic processes have operated over long periods of time.

Fundamental Geologic Principles

Before the development of radiometric-dating techniques, geologists had no reliable means of absolute-age dating and therefore depended solely on relative-dating methods. These methods allow events to be placed in sequential order only and do not tell us how long ago an event took place. Though the principles of relative dating may now seem self-evident, their discovery was an important scientific achievement because they provided geologists with a means to interpret geologic history and develop a relative geologic time scale.

STENO'S PRINCIPLES

The seventeenth century was an important time in the development of geology as a science because of the widely circulated writings of the Danish anatomist Nicolas Steno (1638–1686). Steno observed that when streams flood, they spread out across their floodplains and deposit layers of sediment that bury organisms dwelling on the floodplain. Subsequent floods produce

perspective 2.1

The Geologic History of Uluru and Kata Tijuta

Rising majestically above the surrounding flat desert of central Australia is Uluru (Ayers Rock) and its neighbor to the northwest, Kata Tijuta (The Olgas) (Figure 1). Uluru and Kata Tijuta are the Aboriginal names for what most people know as Ayers Rock and The Olgas. Uluru was named Ayers Rock in 1873 by the explorer W. C. Gosse, after Sir Henry Ayers, the then Premier of South Australia. Kata Tijuta was named Mount Olga in 1872 by Ernest Giles, in honor of the then King and Queen of Spain.

Archeological evidence indicates there has been human activity in this region of Australia for at least the past 22,000 years, and probably even earlier. Both Uluru and Kata Tijuta are an important part of Aboriginal culture, and some sites are so sacred that only Aboriginal

people can visit them. When the Australian government gave title to Uluru National Park back to the Aboriginal people in 1985, part of the agreement was to replace the European names of features and localities with Aboriginal (Anangu) ones.

The geologic history of Uluru and Kata Tijuta begins between about 900 million and 600 million years ago (Late Proterozoic) when much of central Australia was beneath sea level and part of the Amadeus Basin. This period of marine deposition was followed by the Petermann Ranges Orogeny that took place during the Cambrian, resulting in a large mountain range. As this mountain range eroded, huge alluvial fans formed at its base. The sediments comprising these al-

luvial fans were eventually lithified into sedimentary rocks that would later become Uluru and Kata Tijuta. The sedimentary rocks of Uluru and Kata Tijuta are coarse-grained and poorly sorted arkoses (Uluru Arkose) and arkosic conglomerates (Mount Currie Conglomerate), indicative of alluvial fan deposition.

Sometime about 500 million years ago, the region was again covered by warm, shallow seas, and the alluvial fan deposits were buried by sandstones, mudstones, and limestones. Another orogeny, the Alice Springs Orogeny, uplifted the area about 400 million years ago, resulting in much folding and faulting. Although the beds around Uluru were tilted almost vertically, the deformation in the Kata Tijuta region was not as severe. The area that is

FIGURE 1 Location of Uluru and Kata Tijuta in central Australia.

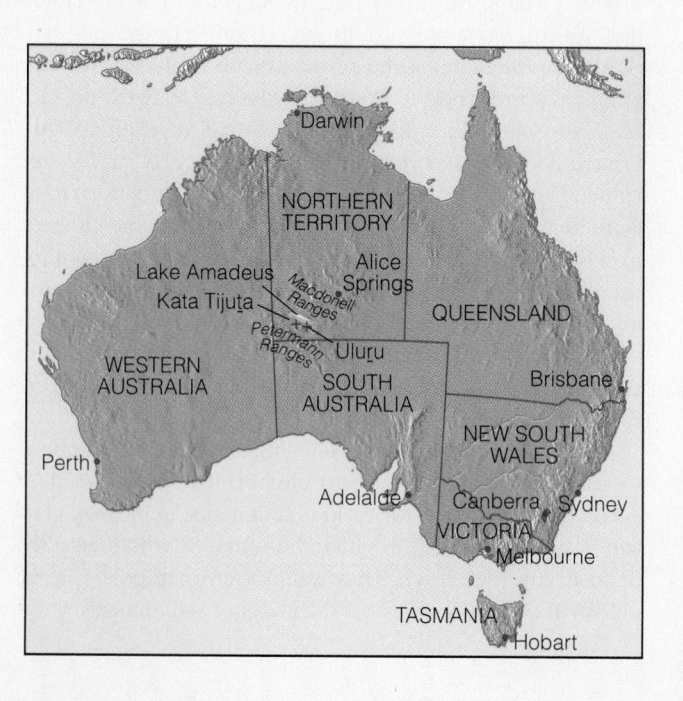

(a)

(b)

FIGURE 2 (a) Uluṟu. (b) Kata Tjuṯa.

far different than it is today. Instead of the arid conditions found presently, a rainy and more temperate climate prevailed, with year-round, flowing streams and marshes a common feature of the landscape. Within the marshes plant material accumulated, forming peat layers that later became converted to thin layers of low-grade coal.

By about 500,000 years ago, the climate became more arid, and has continued to remain arid to the present day. Many of the sand dunes seen in Uluṟu Park today began forming about 30,000 years ago.

The spectacular and varied rock shapes of Uluṟu and Kata Tjuṯa are the result of millions of years of weathering and erosion by water and, to a minor extent, wind acting on the various fractures and joints caused by uplifting of the area. Differences in the composition of the arkoses and conglomerates that make up Uluṟu and Kata Tjuṯa has also played a role in the sculpturing of these colorful and magnificent structures.

now Uluṟu National Park was raised above sea level during the Alice Springs Orogeny and has remained above sea level ever since. Erosion of the area began to shape the uplifted rocks, giving form to what would eventually become Uluṟu and Kata Tjuṯa (Figure 2).

About 70 million years ago, a wide valley was eroded between Uluṟu and Kata Tjuṯa. The climate during this time was

new layers of sediments that are deposited or superposed over previous deposits. When lithified, these layers of sediment become sedimentary rock. Thus, in an undisturbed succession of sedimentary rock layers, the oldest layer is at the bottom, and the youngest layer is at the top. This **principle of superposition** is the basis for determining the relative-age determination of strata and their contained fossils (Figure 2.2).

Steno also observed that, because sedimentary particles settle from water under the influence of gravity, sediment is deposited in essentially horizontal layers, illustrating the **principle of original horizontality** (Figure 2.2). Therefore, a sequence of sedimentary rock layers that is steeply inclined from the horizontal must have been tilted after deposition and lithification.

Steno's third principle, the **principle of lateral continuity,** states that sediment extends laterally in all directions until it thins and pinches out or terminates against the edge of the depositional basin (Figure 2.2).

Establishment of Geology as a Science—The Triumph of Uniformitarianism over Neptunism and Catastrophism

Steno's principles were significant contributions to early geologic thought, but the prevailing concepts of Earth history continued to be those that could be easily reconciled with a literal interpretation of Scripture. Two of these ideas, *neptunism* and *catastrophism,* were particularly appealing and accepted by many naturalists. In the [...] analysis, however, another concept, *uniformitari-* [...] became the underlying philosophy of geology because it provided a better explanation for observed geologic phenomena than either neptunism or catastrophism.

NEPTUNISM AND CATASTROPHISM

The concept of **neptunism** was proposed in 1787 by a German professor of mineralogy, Abraham Gottlob Werner (1749–1817). Although Werner was an excellent mineralogist, he is best remembered for his incorrect interpretation of Earth history. He thought that all rocks, including granite and basalt, were precipitated in an orderly sequence from a primeval, worldwide ocean (Table 2.1).

Werner's subdivision of Earth's crust by supposed relative age attracted a large following in the late 1700s and became almost universally accepted as the standard geologic column. Two factors account for this. First, Werner's charismatic personality, enthusiasm for geology, and captivating lectures popularized the concept. Second, neptunism included a worldwide ocean that easily conformed with the biblical deluge.

Despite Werner's personality and arguments, his neptunian theory failed to explain what happened to the tremendous amount of water that once covered Earth. An even greater problem was Werner's insistence that all ig-

TABLE 2.1	
Werner's Subdivision of the Rocks of Earth's Crust	
Alluvial rocks	Unconsolidated sediments
Secondary rocks	Various fossiliferous, detrital, and chemical rocks such as sandstones, limestones, and coal as well as basalts
Transition rocks	Chemical and detrital rocks including fossiliferous rocks
Primitive rocks	Oldest rocks of Earth (igneous and metamorphic)

neous rocks were precipitated from seawater. It was Werner's failure to recognize the igneous origin of basalt that finally led to the downfall of neptunism.

From the late eighteenth century to the mid-nineteenth century, the concept of **catastrophism,** proposed by the French zoologist Baron Georges Cuvier (1769–1832), dominated European geologic thinking. Cuvier explained the physical and biologic history of Earth as resulting from a series of sudden widespread catastrophes. Each catastrophe accounted for significant and rapid changes in Earth, exterminating existing life in the affected area. Following a catastrophe, new organisms were either created or migrated in from elsewhere.

According to Cuvier, six major catastrophes had occurred in the past. These conveniently corresponded to the six days of biblical creation. Furthermore, the last catastrophe was taken to be the biblical deluge, so catastrophism had wide appeal, especially among theologians.

Eventually, both neptunism and catastrophism were abandoned as untenable hypotheses because their basic assumptions could not be supported by field evidence. The simplistic sequence of rocks predicted by neptunism (Table 2.1) was contradicted by field observations from many different areas. Moreover, basalt was shown to be of igneous origin, and subsequent discoveries of volcanic

rocks interbedded with secondary and primitive deposits proved that volcanic activity had occurred throughout Earth history. As more field observations from widely separated areas were made, naturalists realized that far more than six catastrophes were needed to account for Earth history. With the demise of neptunism and catastrophism, the principle of uniformitarianism, advocated by James Hutton and Charles Lyell, became the guiding philosophy of geology.

UNIFORMITARIANISM

James Hutton (1726–1797) is considered by many to be the father of historical geology. His detailed studies and observations of rock exposures and present-day geologic processes served as the basis for two important geologic principles. The **principle of cross-cutting relationships** holds that an igneous intrusion or a fault must be younger than the rocks it intrudes or displaces (Figure 2.3).

Hutton observed the processes of wave action, erosion by running water, and sediment transport and concluded that given enough time these processes could account for the geologic features in his native Scotland. He thought that "the past history of our globe must be explained by what can be seen to be happening now." This assumption that present-day processes have operated throughout geo-

(a)

(b)

FIGURE 2.3 The principle of cross-cutt~~ing~~ ~~relations~~ **(a) A dark-colored dike has been intruded into ol~~der rocks~~** ~~along~~ **the north shore of Lake Superior, Onta~~rio~~** ~~... A fault~~ **displacing tilted beds along Templin H~~ighway~~**

logic time was the basis for the **principle of uniformitarianism.**

Although Hutton developed a comprehensive theory of uniformitarian geology, it was Charles Lyell (1797–1875) who became the principal advocate and interpreter of uniformitarianism. The term itself was coined by William Whewell in 1832.

Hutton viewed Earth history as cyclical: The continents are worn down by erosional processes, the eroded sediment is deposited in the sea, and uplift of the seafloor creates new continents, thus completing a cycle. He thought the mechanism for uplift was thermal expansion from Earth's hot interior. Hutton's field observations, and experiments performed by his contemporaries involving the melting of basalt samples, convinced him that igneous rocks were the result of cooling magmas. This interpretation of the origin of igneous rocks, called *plutonism,* eventually displaced the neptunian view that igneous rocks precipitated from seawater.

Hutton also recognized the importance of unconformities in his cyclical view of Earth history. At Siccar Point, Scotland, he observed steeply inclined rocks that had been eroded and covered by flat-lying younger rocks (Figure 2.4). It was clear to him that severe upheavals had tilted the lower rocks and formed mountains. These were then worn away and covered by younger, flat-lying rocks. The erosional surface meant there was a gap in the rock record, and the rocks above and below this surface provided evidence that both mountain building and erosion had occurred. Although Hutton did not use the word "unconformity," he was the first to understand and explain the significance of such gaps in the rock record.

Hutton was also instrumental in establishing the concept that geologic processes had vast amounts of time in which to operate. Because Hutton relied on known processes to account for Earth history, he concluded that Earth must be very old. However, he estimated neither how old Earth was nor how long it took to complete a cycle of erosion, deposi-

tion, and uplift. He merely allowed that "we find no vestige of a beginning, and no prospect of an end," which was in keeping with a cyclical view of Earth history.

Unfortunately, Hutton was not a particularly good writer, so his ideas were not widely disseminated or accepted. In fact, neptunism and catastrophism continued to be the dominant geologic concepts well into the 1800s. In 1830, however, Charles Lyell published a landmark book, *Principles of Geology,* in which he championed Hutton's concept of uniformitarianism.

Lyell clearly recognized that small, imperceptible changes brought about by present-day processes could, over long periods of time, have tremendous cumulative effects. Not only did Lyell effectively reintroduce and establish the concept of unlimited geologic time, but he also discredited catastrophism as a viable explanation of geologic phenomena and firmly established uniformitarianism as the guiding philosophy of geology. The recognition of virtually limitless time was also instrumental in the acceptance of Darwin's theory of evolution (see Chapter 5).

Perhaps because uniformitarianism is such a general concept, scientists have interpreted it in different ways. Lyell's concept of uniformitarianism embodied the idea of a steady-state Earth in which present-day processes have operated at the same rate in the past as they do today. For example, the frequency of earthquakes and volcanic eruptions for any given period of time in the past has been the same as it is today. If the climate in one part of the world became warmer, another area would have to become cooler so that overall the climate remains the same. By such reasoning, Lyell claimed that conditions for Earth as a whole had been constant and unchanging through time.

MODERN VIEW OF UNIFORMITARIANISM

Geologists today assume that the principles, or laws, of nature are constant but the rates and intensities of change have varied through time. For example, volcanic activity was more intense in North America during the Miocene Epoch than it is today, while glaciation has been more prevalent during the last 1.6 million years than in the previous 200 million years. Because the rates and intensities of geologic processes have varied through time, some geologists prefer to use the term *actualism* rather than uniformitarianism to remove the idea of "uniformity" from the concept. Most geologists, though, still use the term *uniformitarianism* because it indicates that, even though the rates and intensities of change have varied in the past, the laws of nature have remained the same.

Uniformitarianism is a powerful concept that allows us through analogy and inductive reasoning to use present-day processes as the basis for interpreting the past and for predicting potential future events. It does not eliminate the occurrence of occasional, sudden, short-term events such as volcanic eruptions, earthquakes, floods, or even meteorite impacts as forces that shape our modern world. In fact, some geologists view Earth history as a series of such short-term, or punctuated, events; and this view is

FIGURE 2.4   Angular unconformity at Siccar Point, Scotland. James
Hutton first realized the significance of unconformities at this site in

certainly in keeping with the modern principle of uniformitarianism. Earth is in a state of dynamic change and has been since it formed. While the rates of change may have varied in the past, the natural laws governing the processes have not.

Lord Kelvin and a Crisis in Geology

Lord Kelvin (1824–1907), an English physicist, claimed in a paper written in 1866 to have destroyed the uniformitarian foundation on which Huttonian-Lyellian geology was based. Kelvin did not accept Lyell's strict uniformitarianism, in which chemical reactions in Earth's interior were supposed to continually produce heat, allowing for a steady-state Earth. Kelvin rejected this idea as perpetual motion—impossible according to the known laws of physics—and accepted instead, the assumption that Earth was originally molten.

Kelvin knew from deep mines in Europe that Earth's temperature increases with depth, and he reasoned that Earth is losing heat from its interior. By knowing the size of Earth, the melting temperature of its rocks, and the rate of heat loss, Kelvin extrapolated back to the time when Earth was entirely molten. From these calculations, he concluded that Earth could be no older than 400 million years or younger than 20 million years. This wide discrepancy in age reflected uncertainties in average temperature increases with depth and the various melting points of Earth's constituent materials.

After finally establishing that Earth was very old and showing how present-day processes can be extended back over long periods of time to explain geologic features, geologists were in a quandary. Either they had to accept Kelvin's dates and squeeze events into a shorter time frame, or they had to abandon the concept of seemingly limitless time that was the underpinning of Huttonian-Lyellian geology and one of the foundations of Darwinian evolution. Some geologists objected to such a young age for Earth, but their objections seem to have been based more on faith than on hard facts. Kelvin's quantitative measurements and arguments seemed flawless and unassailable to many scientists.

Kelvin's reasoning and calculations were sound, but his basic premises were false, thereby invalidating his conclusions. Kelvin was unaware that Earth has an internal heat source, radioactivity, that has allowed it to maintain a fairly constant temperature through time.* Kelvin's 40-year campaign for a young Earth ended with the discovery of radioactivity near the end of the nineteenth century. His "unassailable calculations" were no longer valid and his proof for a geologically young Earth collapsed. Kelvin's theory, like neptunism, catastrophism, and a worldwide flood, became an interesting footnote in the history of geology.

While the discovery of radioactivity destroyed Kelvin's arguments, it provided geologists with a clock that could measure Earth's age and validate what geologists had been saying—that Earth was indeed very old! Less than ten years after the discovery that radium generated heat, radiometric calculations were providing ages of billions of years for some of Earth's oldest rocks.

Absolute-Dating Methods

Although most of the 92 naturally occurring elements are stable, some are radioactive and spontaneously decay to other more stable isotopes of elements, releasing energy in the process. The discovery, in 1903 by Pierre and Marie Curie, that radioactive decay produces heat meant that geologists finally had a mechanism for explaining Earth's internal heat that did not rely on residual cooling from a molten origin. Furthermore, geologists now had a powerful tool to date geologic events accurately and verify the long time periods postulated by Hutton and Lyell.

ATOMS AND ISOTOPES

All matter is made up of chemical elements, each of which is composed of extremely small particles called *atoms*. The nucleus of an atom is composed of *protons* (positively charged particles) and *neutrons* (neutral particles) with *electrons* (negatively charged particles) encircling it. The number of protons defines an element's *atomic number* and helps determine its properties and characteristics.

The combined number of protons and neutrons in an atom is its *atomic mass number*. However, not all atoms of the same element have the same number of neutrons in their nuclei. These variable forms of the same element are called *isotopes*. Most isotopes are stable, but some are unstable. It is the decay rate of unstable isotopes that geologists measure to determine absolute age of rocks.

RADIOACTIVE DECAY AND HALF-LIVES

Radioactive decay is the process whereby an unstable atomic nucleus is spontaneously transformed into an atomic nucleus of a different element. Three types of radioactive decay are recognized, all of which result in a change of atomic structure (Figure 2.5). In **alpha decay,** two protons and two neutrons are emitted from the nucleus, resulting in a loss of two atomic numbers and four atomic mass numbers. In **beta decay,** a fast-moving electron is emitted from a neutron in the nucleus, changing that neutron to a proton and consequently increasing the atomic number by one, with ⸺⸺⸺⸺ mass number change. **Electron cap**⸺⸺⸺ proton captures an electron f⸺⸺⸺ thereby converts to a neutron ⸺⸺⸺ atomic number but not changi⸺⸺⸺

Some elements undergo on⸺⸺⸺ ing from an unstable form to ⸺⸺⸺

---

*Actually, Earth's temperature has decreased through time but at a considerably slower rate than would lend any credence to Kelvin's calculations (see p. 197, Chapter 8).

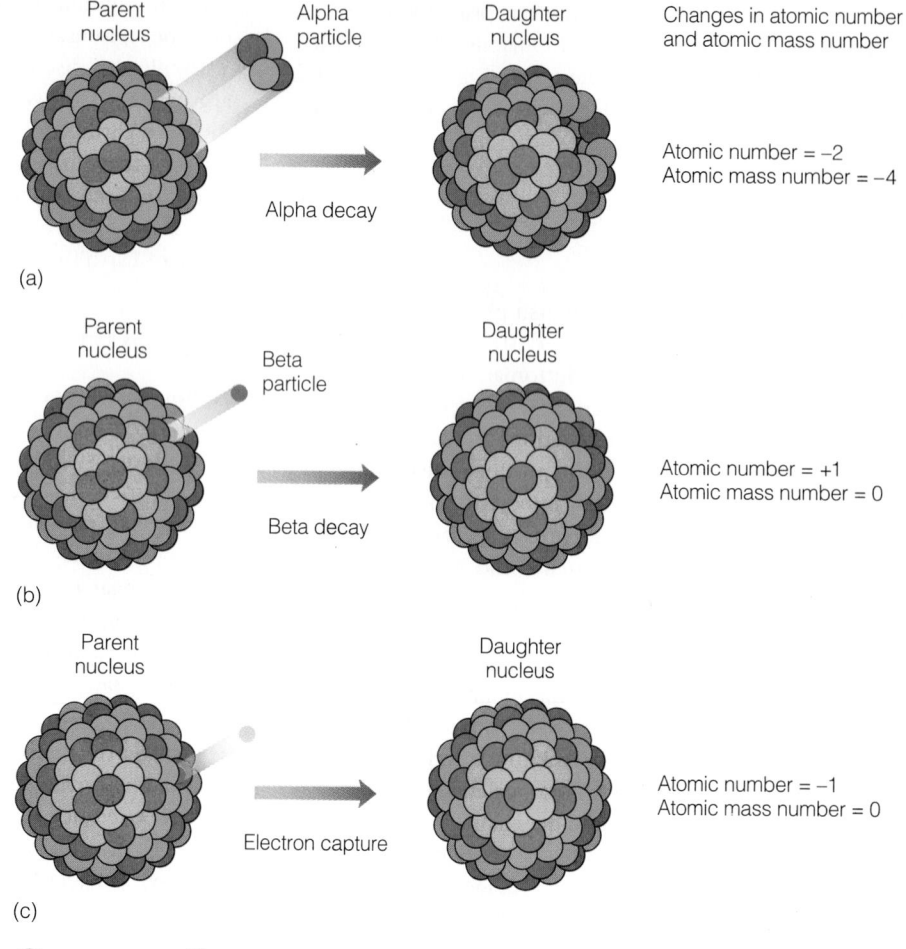

FIGURE 2.5 Three types of radioactive decay. (a) Alpha decay, in which an unstable parent nucleus emits two protons and two neutrons. **(b)** Beta decay, in which an electron is emitted from the nucleus. **(c)** Electron capture, in which a proton captures an electron and is thereby converted to a neutron.

Parent nucleus
Alpha particle
Daughter nucleus
Changes in atomic number and atomic mass number

Alpha decay

Atomic number = –2
Atomic mass number = –4

(a)

Parent nucleus
Beta particle
Daughter nucleus

Beta decay

Atomic number = +1
Atomic mass number = 0

(b)

Parent nucleus
Daughter nucleus

Electron capture

Atomic number = –1
Atomic mass number = 0

(c)

Protron   Neutron   Electron

rubidium 87 decays to strontium 87 by a single beta emission, and potassium 40 decays to argon 40 by a single electron capture. Other radioactive elements undergo several decay steps. Uranium 235 decays to lead 207 by seven alpha steps and six beta steps, while uranium 238 decays to lead 206 by eight alpha and six beta steps (Figure 2.6).

When discussing decay rates, it is convenient to refer to them in terms of half-lives. The **half-life** of a radioactive element is the time it takes for one-half of the atoms of the original unstable **parent element** to decay to atoms of a new, more stable **daughter element.** The half-life of a given radioactive element is constant and can be precisely measured. Half-lives of various radioactive elements range from less than a billionth of a second to 49 billion years.

Radioactive decay occurs at a geometric rate rather than a linear rate. Therefore, a graph of the decay rate produces a curve rather than a straight line (Figure 2.7). For example, an element with 1,000,000 parent atoms will have 500,000 parent atoms and 500,000 daughter atoms after one half-life. After two half-lives, it will have 250,000 parent atoms (one-half of the previous parent atoms, which is equivalent to one-fourth of the original atoms) and 750,000 daughter atoms. After three

half-lives, it will have 125,000 parent atoms (one-half of the previous parent atoms, or one-eighth of the original parent atoms) and 875,000 daughter atoms, and so on, until the number of parent atoms remaining is so few that they cannot be accurately measured by present-day instruments.

By measuring the parent–daughter ratio and knowing the half-life of the parent (which has been determined in the laboratory), geologists can calculate the age of a sample containing the radioactive element. The parent–daughter ratio is usually determined by a *mass spectrometer,* an instrument that measures the proportions of elements of different masses.

Sources of Uncertainty   The most accurate radiometric dates are obtained from igneous rocks. As a magma cools and begins to crystallize, radioactive parent atoms are separated from previously formed daughter atoms. Because they are the right size, some radioactive parent atoms are incorporated into the crystal structure of certain minerals. The stable daughter atoms, though, are a different size than the radioactive parent atoms and consequently cannot fit into the crystal structure of the same

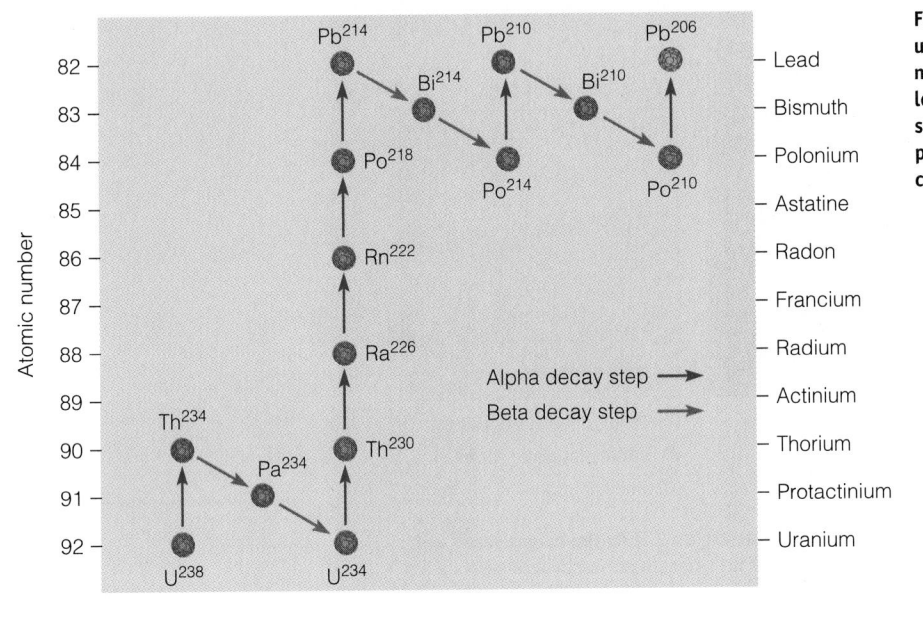

FIGURE 2.6 Radioactive decay series for uranium 238 to lead 206. Radioactive uranium 238 decays to its stable end product, lead 206, by eight alpha and six beta decay steps. A number of different isotopes are produced as intermediate steps in the decay series.

mineral as the parent atoms. Therefore, a mineral crystallizing in a cooling magma will contain radioactive parent atoms but no stable daughter atoms (Figure 2.8). Thus, the time that is being measured is the time of crystallization of the mineral containing the radioactive atoms, and not the time of formation of the radioactive atoms.

Except in unusual circumstances, sedimentary rocks cannot be radiometrically dated because one would be measuring the age of a particular mineral rather than the time that it was deposited as a sedimentary particle. One of the few instances in which radiometric dates can be obtained on sedimentary rocks is when the mineral glauconite is present. Glauconite is a greenish mineral

containing radioactive potassium 40, which decays to argon 40 (Table 2.2). It forms in certain marine environments as a result of chemical reactions with clay minerals during the conversion of sediments to sedimentary rock. Thus, glauconite forms when the sedimentary rock forms, and a radiometric date indicates the time of the sedimentary rock's origin. Being a gas, however, the daughter product argon can easily escape from a mineral. Therefore, any date obtained from glauconite, or any other mineral containing the potassium 40 and argon 40 pair, must be considered a minimum age.

To obtain accurate radiometric dates, geologists must be sure that they are dealing with a *closed system,* meaning

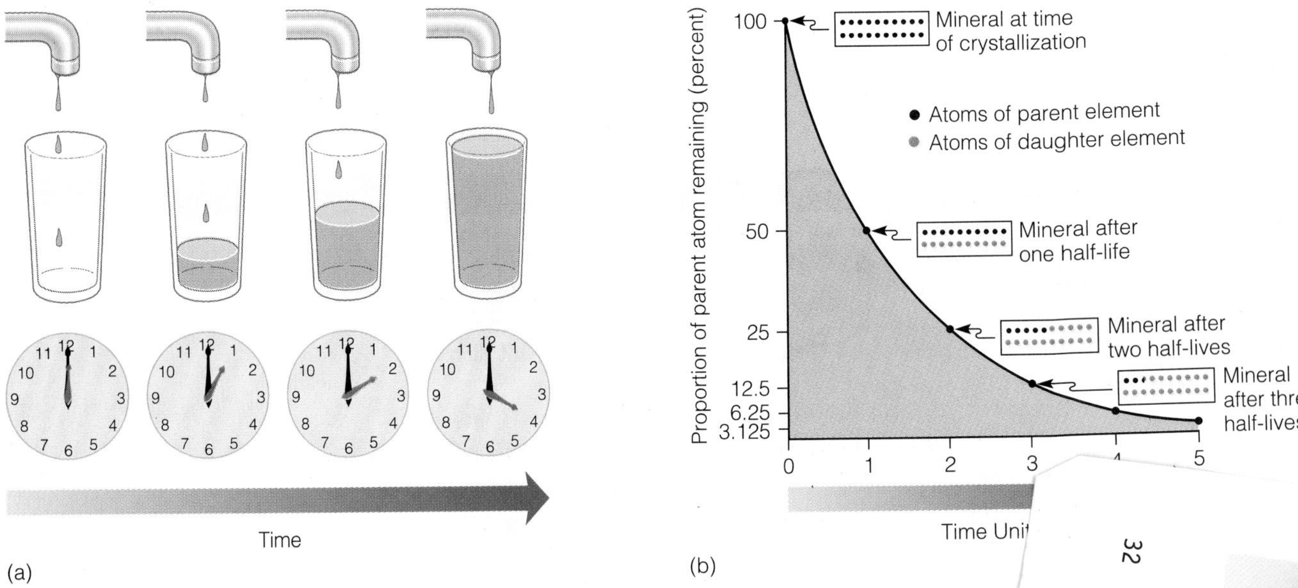

(a)

Time

(b)

Proportion of parent atom remaining (percent)

Mineral at time of crystallization

● Atoms of parent element
● Atoms of daughter element

Mineral after one half-life

Mineral after two half-lives

Mineral after three half-lives

Time Unit

FIGURE 2.7 (a) Uniform, linear depletion is characteristic of many familiar processes. (b) Geometric radioacti[...] time unit represents one half-life, and each half-life is the time it takes for one-half of the parent element to [...]

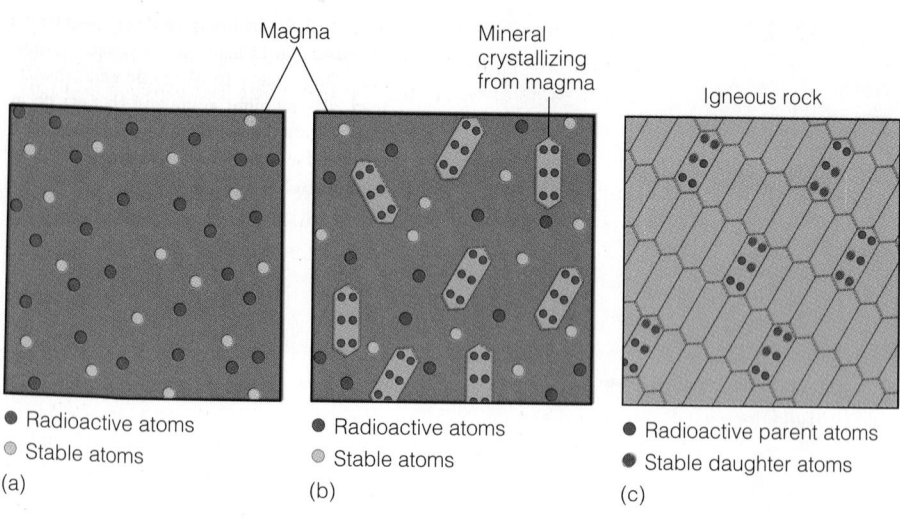

Magma | Mineral crystallizing from magma | Igneous rock

- Radioactive atoms
- Stable atoms

(a)

- Radioactive atoms
- Stable atoms

(b)

- Radioactive parent atoms
- Stable daughter atoms

(c)

FIGURE 2.8 (a) A magma contains both radioactive and stable atoms. (b) As the magma cools and begins to crystallize, some radioactive atoms are incorporated into certain minerals because they are the right size and can fit into the crystal structure. Therefore, at the time of crystallization, the mineral will contain 100% radioactive parent atoms and 0% stable daughter atoms. (c) After one half-life, 50% of the radioactive parent atoms will have decayed to stable daughter atoms.

that neither parent nor daughter atoms have been added or removed from the system since crystallization and that the ratio between them results only from radioactive decay. Otherwise, an inaccurate date will result. If daughter atoms have leaked out of the mineral being analyzed, the calculated age will be too young; if parent atoms have been removed, the calculated age will be too great.

Leakage may take place if the rock is heated or subjected to intense pressure as can sometimes occur during metamorphism. If this happens, some of the parent or daughter atoms may be driven from the mineral being analyzed, resulting in an inaccurate age determination. If the daughter product was completely removed, then one would be measuring the time since metamorphism (a useful measurement itself) and not the time since crystallization of the mineral (Figure 2.9). Because heat affects the parent–daughter ratio,

metamorphic rocks are difficult to date accurately. Remember that while the parent–daughter ratio may be affected by heat, the decay rate of the parent element remains constant, regardless of any physical or chemical changes.

It is sometimes possible to cross-check the radiometric date obtained by measuring the parent–daughter ratio of two different radioactive elements in the same mineral. For example, naturally occurring uranium consists of both uranium 235 and uranium 238 isotopes. Through various decay steps, uranium 235 decays to lead 207, whereas uranium 238 decays to lead 206 (Figure 2.6). If the minerals containing both uranium isotopes have remained closed systems, the ages obtained from each parent–daughter ratio should agree closely. If they do, they are said to be *concordant*, reflecting the time of crystallization of the magma. If

TABLE 2.2

Five of the Principal Long-Lived Radioactive Isotope Pairs Used in Radiometric Dating

ISOTOPES		HALF-LIFE OF PARENT (YEARS)	EFFECTIVE DATING RANGE (YEARS)	MINERALS AND ROCKS THAT CAN BE DATED
Parent	*Daughter*			
Uranium 238	Lead 206	4.5 billion	10 million to 4.6 billion	Zircon Uraninite
Uranium 235	Lead 207	704 million		
Thorium 232	Lead 208	14 billion		
Rubidium 87	Strontium 87	48.8 billion	10 million to 4.6 billion	Muscovite Biotite Potassium feldspar Whole metamorphic or igneous rock
Potassium 40	Argon 40	1.3 billion	100,000 to 4.6 billion	Glauconite Hornblende Muscovite Whole volcanic rock Biotite

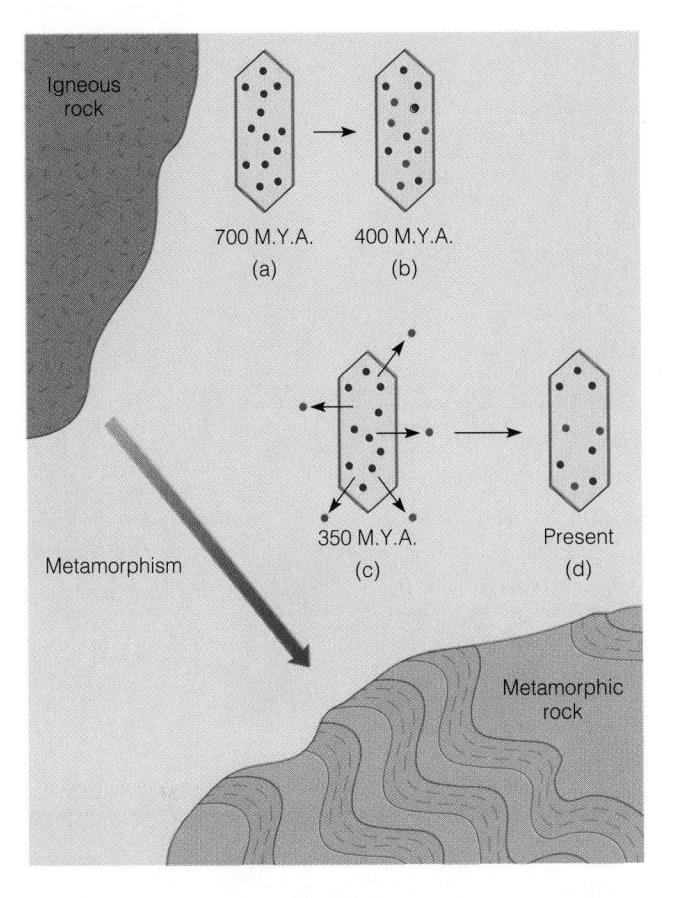

FIGURE 2.9 The effect of metamorphism in driving out daughter atoms from a mineral that crystallized 700 million years ago (M.Y.A). The mineral is shown immediately after crystallization (a), then at 400 million years (b), when some of the parent atoms had decayed to daughter atoms. Metamorphism at 350 M.Y.A. (c) drives the daughter atoms out of the mineral into the surrounding rock. (d) Assuming the rock has remained a closed chemical system throughout its history, dating the mineral today yields the time of metamorphism, while dating the whole rock provides the time of its crystallization, 700 M.Y.A.

the ages do not closely agree, then they are said to be *discordant*, and other samples must be taken and ratios measured to see which, if either, date is correct.

Recent advances and the development of new techniques and instruments for measuring various isotope ratios have enabled geologists to analyze not only increasingly smaller samples, but with a greater precision than ever before. Presently, the measurement error for many radiometric dates is typically less than 0.5% of the age, and in some cases is even better than 0.1%. Thus, for a rock 540 million years old (near the beginning of the Cambrian Period), the possible error could range from nearly 2.7 million years, to as low as less than 540,000 years.

LONG-LIVED RADIOACTIVE ISOTOPE PAIRS

Table 2.2 shows the five common, long-lived parent–daughter isotope pairs used in radiometric dating. Long-lived pairs have half-lives of millions or billions of years. All of these

were present when Earth formed and are still present in measurable quantities. Other shorter-lived radioactive isotope pairs have decayed to the point that only small quantities near the limit of detection remain.

The most commonly used isotope pairs are the uranium–lead and thorium–lead series, which are used principally to date ancient igneous intrusives, lunar samples, and some meteorites. The rubidium–strontium pair is also used for very old samples and has been effective in dating the oldest rocks on Earth as well as meteorites. The potassium–argon method is typically used for dating fine-grained volcanic rocks from which individual crystals cannot be separated; hence, the whole rock is analyzed. Because argon is a gas, great care must be taken to ensure that the sample has not been subjected to heat, which would allow argon to escape; such a sample would yield an age that is too young. Other long-lived radioactive isotope pairs exist, but they are rather rare and used only in special situations.

FISSION TRACK DATING

The emission of atomic particles resulting from the spontaneous decay of uranium within a mineral damages its crystal structure. The damage appears as microscopic linear tracks that are visible only after etching the mineral with hydrofluoric acid. The age of the sample is determined on the basis of the number of fission tracks present and the amount of uranium the sample contains: the older the sample, the greater the number of tracks (Figure 2.10).

Fission track dating is of particular interest to archaeologists and geologists because the technique can be used to date samples ranging from only a few hundred to hundreds of millions of years in age. It is most useful for dating samples between about 40,000 and 1.5 million years ago, a period for which other dating techniques are not always particularly suitable. One of the problems in fission track dating occurs when the rocks have later been subjected to high temperatures. If this happens, the damaged crystal structures are repaired by annealing, and consequently the tracks disappear. In such instances, the calculated age will be younger than the actual age.

RADIOCARBON- AND TREE-RING-DATING METHODS

Carbon is an important element in nature and is one of the basic elements found in all forms of life. It has three isotopes; two of these, carbon 12 and 13, are stable, whereas carbon 14 is radioactive. Carbon 14 has a half-life of 5730 years plus or minus 30 years. The **carbon 14 dating technique** is based on the ratio of car̶̶̶̶̶̶̶on 12 and is generally used to date once-livi̶̶̶

The short half-life of carbon̶̶̶ nique practical only for speci̶̶̶ 70,000 years. Consequently, th̶̶̶ is especially useful in archaeo̶̶̶ unravel the events of the Pleist̶̶̶

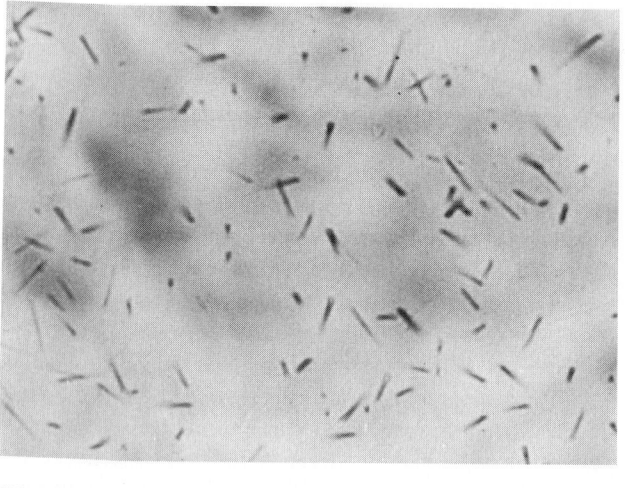

FIGURE 2.10 Each fission track (about 16 μm in length) in this apatite crystal is the result of the radioactive decay of a uranium atom. The crystal, which has been etched with hydrofluoric acid to make the fission tracks visible, comes from one of the dikes at Shiprock, New Mexico, and has a calculated age of 27 million years.

Carbon 14 is constantly formed in the upper atmosphere when cosmic rays, which are high-energy particles (mostly protons), strike the atoms of upper-atmospheric gases, splitting their nuclei into protons and neutrons. When a neutron strikes the nucleus of a nitrogen atom (atomic number 7, atomic mass number 14), it may be absorbed into the nucleus and a proton emitted. Thus, the atomic number of the atom decreases by one, while the atomic mass number stays the same. Because the atomic number has changed, a new element, carbon 14 (atomic number 6, atomic mass number 14), is formed. The newly formed carbon 14 is rapidly assimilated into the carbon cycle and, along with carbon 12 and 13, is absorbed in a nearly constant ratio in all living organisms (Figure 2.11). When an organism dies, however, carbon 14 is not replenished, and the ratio of carbon 14 to carbon 12 decreases as the carbon 14 decays back to nitrogen by a single beta-decay step (Figure 2.5).

Currently, the ratio of carbon 14 to carbon 12 is remarkably constant in both the atmosphere and living organisms. There is good evidence, though, that the production of carbon 14, and thus the ratio of carbon 14 to carbon 12, has varied somewhat during the past several thousand years. This was determined by comparing ages established by carbon 14 dating of wood samples against those established by counting annual tree rings in the same samples. As a result, carbon 14 ages have been corrected to reflect such variations in the past.

Tree-ring dating is another useful method for dating geologically recent events. The age of a tree can be determined by counting the growth rings in the lower part of the stem. Each ring represents one year's growth, and the pattern of wide and narrow rings can be compared among trees to establish the exact year in which the rings were

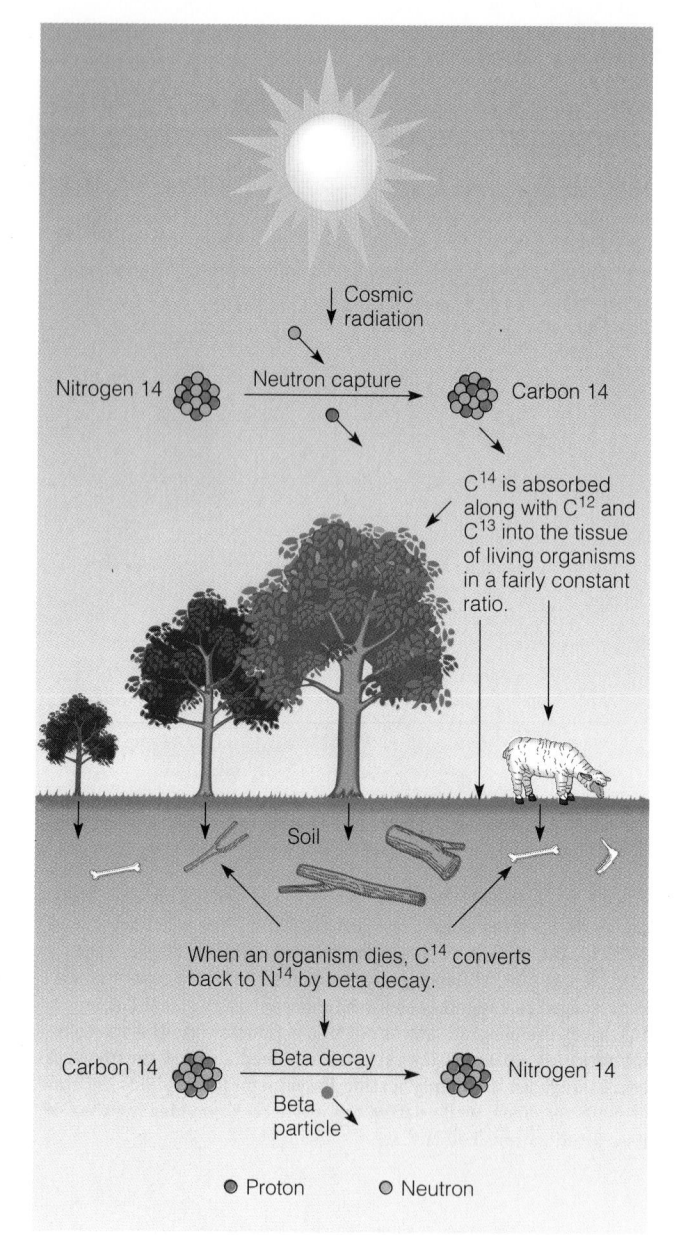

FIGURE 2.11 The carbon cycle showing the formation, dispersal, and decay of carbon 14.

formed. The procedure of matching ring patterns from numerous trees and wood fragments in a given area is referred to as *cross-dating*. By correlating distinctive tree-ring sequences from living to nearby dead trees, a time scale can be constructed that extends back to about 14,000 years ago (Figure 2.12). By matching ring patterns to the composite ring scale, wood samples whose ages are not known can be accurately dated.

The applicability of tree-ring dating is somewhat limited because it can only be used where continuous tree records are found. It is therefore most useful in arid regions, particularly the southwestern United States.

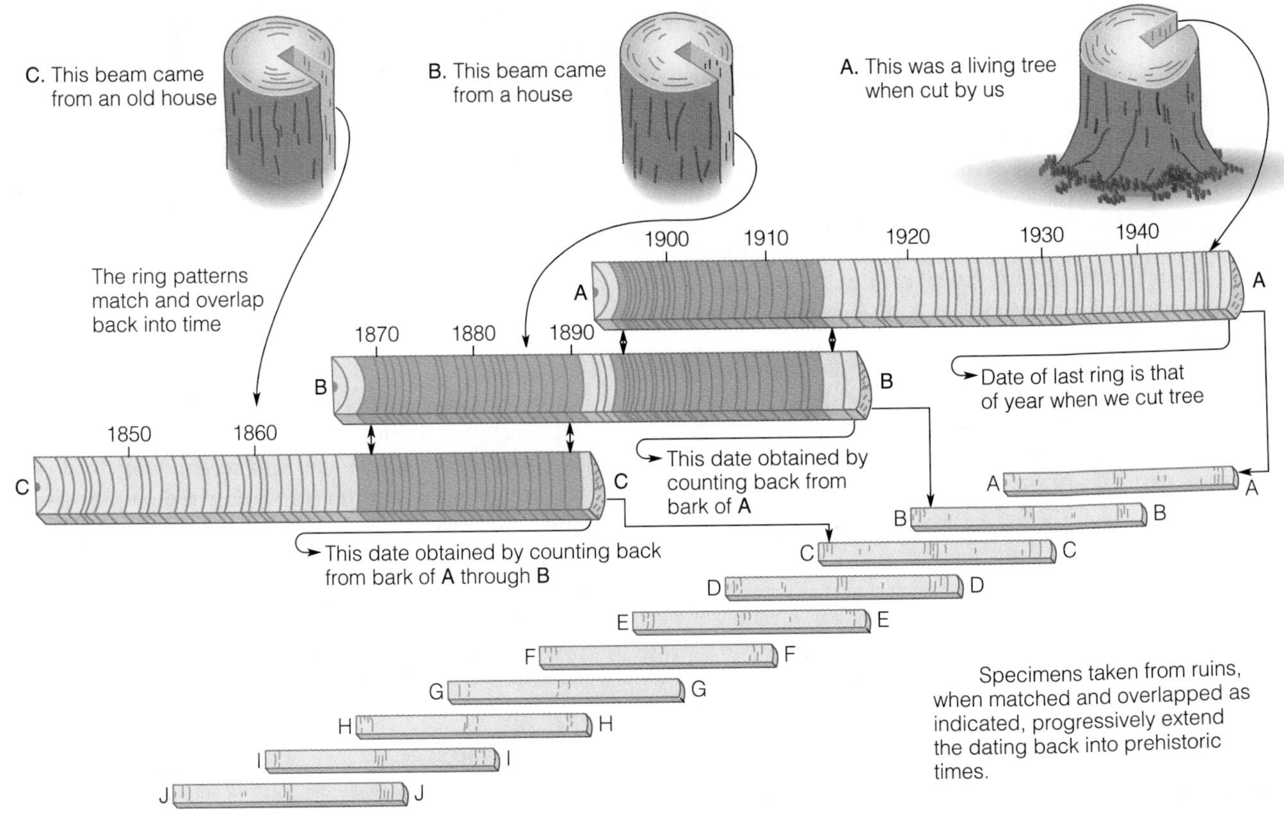

FIGURE 2.12 In the cross-dating method, tree-ring patterns from different woods are matched against each other to establish a ring-width chronology backward in time.

Summary

1. Early Christian theologians were responsible for formulating the idea that time is linear and that Earth was very young. Archbishop Ussher calculated an age of about 6000 years based on his interpretation of Scripture.

2. During the eighteenth and nineteenth centuries, attempts were made to determine Earth's age based on scientific evidence rather than revelation. While some attempts were quite ingenious, they yielded a variety of ages that now are known to be much too young.

3. Considering the religious, political, and social climate of the seventeenth, eighteenth, and early nineteenth cen-

turies, it is easy to see why concepts such as neptunism, catastrophism, and a very young Earth were eagerly embraced. As geologic data accumulated, it became apparent that these concepts were not supported by evidence, and that Earth must be older than 6000 years.

4. James Hutton thought that present-day processes operating over long periods of time could explain all the geologic features of Earth. He also viewed Earth history as cyclical and thought Earth to be very old. Hutton's observations were instrumental in establishing the basis for the principle of

5. Uniformitarianism, as articulated by Charles Lyell, soon displaced neptunism and catastrophism as the guiding doctrine of geology. The principle of uniformitarianism holds that the laws of nature have been constant through time and that the same processes operating today also operated in the past, although not necessarily at the same rates.

6. Besides uniformitarianism, the principles of superposition, original horizontality, lateral continuity, and cross-cutting relationships are basic for determining relative geologic ages and for interpreting Earth history.

7. Radioactivity was discovered during the late nineteenth century, and soon thereafter radiometric-dating techniques allowed geologists to determine absolute ages for geologic events.

8. Absolute-age dates for rock samples are usually obtained by determining how many half-lives of the radioactive parent element have elapsed since the sample originally crystallized. A half-life is the time it takes for one-half of the radioactive parent element to decay to a daughter element.

9. The most accurate radiometric dates are obtained from long-lived radioactive isotope pairs in igneous rocks. The five common long-lived radioactive isotope pairs are uranium 238–lead 206, uranium 235–lead 207, thorium 232–lead 208, rubidium 87–strontium 87, and potassium 40–argon 40. The most reliable dates are those obtained by using at least two different radioactive-decay series in the same rock.

10. Carbon 14 dating can be used only for organic matter such as wood, bones, and shells and is effective back to about 70,000 years ago. Unlike the long-lived isotopic pairs, the carbon 14 dating technique determines age by the ratio of radioactive carbon 14 to stable carbon 12.

Important Terms

absolute dating

alpha decay

beta decay

carbon-14-dating technique

catastrophism

daughter element

electron capture decay

fission track dating

half-life

neptunism

parent element

principle of cross-cutting relationships

principle of lateral continuity

principle of original horizontality

principle of superposition

principle of uniformitarianism

radioactive decay

relative dating

tree-ring dating

Review Questions

1. Placing geologic events in sequential order as determined by their position in the geologic record is called:
   a. _____ absolute dating.
   b. _____ uniformitarianism.
   c. _____ relative dating.
   d. _____ correlation.
   e. _____ historical dating.

2. If a rock is heated during metamorphism and the daughter atoms migrate out of a mineral that is subsequently radiometrically dated, an inaccurate date will be obtained. This date will be _____ the actual date.
   a. _____ younger than
   b. _____ older than
   c. _____ the same as
   d. _____ It cannot be determined.
   e. _____ None of these

3. What fundamental geologic principle states that the oldest layer is on the bottom of a succession of sedimentary rocks and the youngest is on top?
   a. _____ Lateral continuity
   b. _____ Fossil succession
   c. _____ Original horizontality
   d. _____ Superposition
   e. _____ Cross-cutting relationships

4. In which type of radioactive decay is a fast-moving electron emitted from a neutron in the nucleus, resulting in an atomic number increase of one and no change in the atomic mass number?
   a. _____ Radiocarbon
   b. _____ Alpha
   c. _____ Fission track
   d. _____ Beta
   e. _____ Electron capture

5. Who is generally considered to be the father of historical geology?
   a. _____ Cuvier
   b. _____ Hutton
   c. _____ Lyell
   d. _____ Whewell
   e. _____ Werner

6. The atomic number of an element is determined by the number of _____ in its nucleus.
   a. _____ protons.
   b. _____ neutrons.
   c. _____ electrons.
   d. _____ protons and neutrons.
   e. _____ protons and electrons.

7. If a radioactive element has a half-life of 4 million years, the amount of parent material remaining after 12 million years of decay will be what fraction of the original amount?
   a. _____ 1/32
   b. _____ 1/16
   c. _____ 1/8
   d. _____ 1/4
   e. _____ 1/2

8. What is being measured in radiometric dating is:
   a. _____ the time when a radioactive isotope formed.
   b. _____ the time of crystallization of a mineral containing an isotope.
   c. _____ the amount of the parent isotope only.
   d. _____ when the dated mineral became part of a sedimentary rock.
   e. _____ when the stable daughter isotope was formed.

9. How many half-lives are required to yield a mineral with 625 atoms of uranium 238 and 19,375 atoms of lead 206?
   a. _____ 4

   b. _____ 5
   c. _____ 6
   d. _____ 8
   e. _____ 10

10. Which of the following is *not* a long-lived radioactive isotope pair?
    a. _____ Uranium–lead
    b. _____ Rubidium–strontium
    c. _____ Thorium–lead
    d. _____ Potassium–argon
    e. _____ Carbon 14–carbon 12

11. The author of *Principles of Geology* and the principal advocate and interpreter of uniformitarianism was:
    a. _____ Hutton.
    b. _____ Steno.
    c. _____ Kelvin.
    d. _____ Lyell.
    e. _____ Werner.

12. In carbon 14 dating, which ratio is being measured?
    a. _____ C^{14}/N^{14}
    b. _____ C^{12}/C^{13}
    c. _____ C^{12}/C^{14}
    d. _____ C^{12}/N^{14}
    e. _____ Parent–daughter ratio

13. Discuss Lord Kelvin's method of determining the age of Earth and explain what led to its rejection.

14. Why is it difficult to date sedimentary and metamorphic rocks radiometrically?

15. Why is the principle of uniformitarianism important to geologists?

16. What are the fundamental principles used in relative dating? Why are they so important in historical geology?

17. Compare the modern view of uniformitarianism to Lyell's view.

18. If you wanted to calculate the absolute age of an intrusive body, what information would you need?

19. What are some of the uncertainties that must be considered in radiometric dating?

20. How does the carbon 14 dating technique differ from uranium–lead dating methods?

Points to Ponder

1. Discuss some of the reasons why the production of carbon 14 has varied during the past several thousand years, thus affecting the ages determined by carbon 14 dating.

2. Can the same principles geologists use to interpret the geologic history of Earth be used by astronauts to determine the geologic history of other planets? Explain.

3. A zircon mineral in an igneous rock contains both uranium 235 and uranium 238. In measuring the parent–daughter ratio for both uranium isotopes in the zircon, it was discovered that uranium 238 had undergone two half-lives in decaying to lead 206, while uranium 235 had undergone ten half-lives during its decay to lead 207. Using Table 2.2, determine the age of the igneous rock based on each long-lived radioactive isotope pair. Are they the same age? If not, why? What explanations can you give for the difference in geologic age between the two samples?

4. Calculate the age of a rock containing the following atoms of radioactive parent element A and stable daughter B. *Radioactive parent element A:* 1,125,000 atoms; *stable daughter element B:* 34,875,000 atoms. The half-life of element A is 6.25 million years. What is the age of the rock containing these parent and daughter elements?

Additional Readings

Albritton, C. C., Jr. 1984. Geologic time. *Journal of Geological Education* 32, no. 1: 29–37.

Berry, W. B. N. 1987. *Growth of a prehistoric time scale.* 2d ed. Palo Alto, Calif.: Blackwell Scientific Publications.

Boslough, J. 1990. The enigma of time. *National Geographic* 177, no. 3: 109–132.

Geyh, M. A., and H. Schleicher. 1990. *Absolute age determination.* New York: Springer-Verlag.

Gould, S. J. 1987. *Time's arrow, time's cycle.* Cambridge, Mass.: Harvard University Press.

Harland, W. B., R. L. Armstrong, A. V. Cox, L. E. Craig, A. G. Smith, and D. G. Smith. 1990. *A geologic time scale 1989.* New York: Cambridge University Press.

Itano, W. M., and N. F. Ramsey. 1993. Accurate measurement of time. *Scientific American* 269, no. 1: 56–57.

Ramsey, N. F. 1988. Precise measurement of time. *American Scientist* 76, no. 1: 42–49.

Renne, P. R., D. B. Karner, and K. A. Ludwig. 1998. Absolute ages aren't exactly. *Science* 282, no. 5395: 1840–1841.

Shea, J. H. 1982. Twelve fallacies of uniformitarianism. *Geology* 10, no. 9: 455–460.

Svitil, K. A. 1998. Probing the past. *Discover* 19, no. 11: 96–101.

For these web site addresses, along with current updates and exercises, log on to

http://www.brookscole.com/geo/

World Wide Web Activities

▶ **UNIVERSITY OF CALIFORNIA MUSEUM OF PALEONTOLOGY**

Visit this site for an excellent description of any aspect of geologic time, paleontology, and evolution. Click on the *On-line Exhibits* to go to the *Paleontology without Walls* home page, which is an introduction to the UCMP Virtual Exhibits. Click on the *Geologic Time* site. It will take you to the *Geology and Geologic Time* home page, which discusses geologic time. You can then click on one of the eras to learn more about what occurred during that time interval. Click on the Cenozoic site to learn more about how geologic time scales are constructed.

▶ **RADIOCARBON WEB-INFO**

Maintained by the radiocarbon labs of Waikato and Oxford Universities, this site contains information about the basis

of carbon 14 dating, applications, measurement methods, and other carbon 14 websites.

1. Click on the *Basis of the Method* site. What is the carbon 14 method? Who developed it? How does it work?

2. Click on the *Measurements methods* site. What are the three principal methods of measuring residual carbon 14 activity?

3. Click on the *Applications* site. What are the various ways carbon 14 can be used to date objects and events? Click on some of the archaeology sites to see how carbon 14 dating is being used in current projects such as the radiocarbon dating of the Dead Sea Scrolls.

4. Click on the *Corrections to C14 dates* site. What type of corrections need to be made to carbon 14 dates?

▶ GEOLOGIC TIME SCALE

This site, maintained by the Univeristy of Alberta, provides a short discussion on relative and absolute geologic time, followed by an illustration of the geologic time scale.

▶ FOSSILS, ROCKS, AND TIME

This online USGS publication, written by Lucy E. Edwards and John Pojeta Jr., covers the topics of geologic time, fossils, and their importance in the evolution of the geologic time scale. The publication is intended for general audiences and presents the topic of geologic time in a very understandable manner.

CD-ROM Exploration

Explore the following *In-Terra-Active 2.0* CD-ROM module(s) and increase your understanding of key concepts and processes presented in this chapter.

Chapter Concept: **Relative Dating**

Ordering the rock record from oldest to youngest according to principles such as superposition and cross-cutting relations provided the basis for the geologic time scale. In this module you will explore the different processes that produce the rock record and apply your knowledge to interpreting the relative geologic history of two landscape cross-sections.

Section: **Materials**
Module: **Geologic Time: Relative**

How old is old? One of geology's greatest legacies is that it answers this question, showing us just how deep-reaching Earth's history really is. In relative time, the immediate goal is not dating an event or object in years, but understanding its place in a geologic sequence. We get our best information from outcrops and other rock records. Study the photo.

Questions: *Which part of the rock in this picture is oldest? Which is youngest?*

Chapter Concept: **Absolute Dating**

Absolute dating of the geologic time scale depends on our understanding of radioactivity, the decay of various radioactive elements, and their potential for measuring geologic time. This module will extend your understanding of the different types of radioactive decay and enable you to apply this to isotopic data from two different igneous rocks as you determine their ages.

Section: **Materials**
Module: **Geologic Time: Absolute**

In 1903, the discovery of radioactivity provided a compelling explanation of where some of Earth's heat originates. It also gave geologists an extremely useful tool for dating materials, especially minerals. However, this tool is not foolproof. When you enter the field of radiometric dating, you learn that some specimens yield more reliable dates than others.

Question: *What preconditions must a mineral meet for its absolute age dating to be accurate?*

3

Rocks, Fossils, and Time

Fossilized Sequoia stump in Florrisant Fossil Beds National Monument, Colorado. Volcanic mudflows 3 to 6 m thick buried the lower parts of a number of trees at this site.

Prologue

Most people today are aware that sand deposited on a beach may become sandstone and that shells and bones in rocks are the remains of once-living animals. These concepts are well entrenched in modern thinking. During most of human existence, however, people had little or no idea of how rocks formed or what fossils were.

Humans have used rocks and fossils since prehistoric times for tools, weapons, and ornaments. Fossils, for example, were a trade item in the ancient world, and even during the Middle Ages some fossils were believed to have curative powers (Figure 3.1). The Greek philosopher Xenophanes, who lived during the sixth century B.C., was apparently the first to recognize that fossil seashells and bones represent the remains of once-living organisms; the Chinese came to the same conclusion by the first century B.C. Through most of historic time in the West, though, fossils were variously believed to be inorganic objects formed inside rocks by spontaneous generation or growing seeds; images produced in rocks by some kind of molding force; or even objects placed in the rocks by the Creator to test our faith or by Satan to sow seeds of doubt.

One of the reasons fossils were misunderstood was that no one knew how rocks formed. Some of the ancient Greeks and Egyptians had some idea of rock-forming processes, and Leonardo da Vinci (1452–1519) realized that sand transported and deposited by rivers could become sandstone. Nevertheless, it was not until Nicolas Steno (see Chapter 2) elucidated his principles in 1669 that a framework for interpreting rocks existed. In short, before Steno's work, little was known about Earth history.

By the late 1700s and early 1800s, scientists had gained considerable insight into Earth history. Old ideas are difficult to displace, however, and the idea of a young Earth shaped by catastrophes (perhaps a single catastrophe) persisted. One of the most extreme views of Earth history was proposed in 1857 by Philip Henry Gosse, a British naturalist. According to Gosse, Earth was created a few thousands of years ago with the appearance of a lengthy history. In other words, rocks that appeared to have been deposited by rivers or that appeared to have formed from lava flows were created with that appearance.

The events denoted by their appearance had not really happened. Fossils were not the remains of organisms that once lived on Earth but were simply another created aspect of rocks.

Gosse's proposal was immediately rejected by most people, who reasoned that if he were correct, the historical record preserved in rocks was the product of a deceptive Creator. In addition to the philosophical implications of Gosse's proposal, it is important to understand why it cannot be considered scientific. No conceivable observation or test could demonstrate whether it is right or wrong; the geologic record would be exactly the same if it had been created as Gosse claimed, or if it resulted from billions of years of Earth evolution. Furthermore, Gosse believed Earth to be no more than a few thousand years old, but had to admit that no evidence in the geologic record allowed him to deduce the time of creation. As a result, critics were quick to point out that Gosse's Earth with the appearance of great age could have been created at any time during the past, including only moments ago along with all records and memories. In brief, Gosse's idea is completely untestable and thus not scientific.

Even before Gosse proposed his idea, scientists had concluded that Earth had a long history and that that history could be read from the only available record—the geologic record.

FIGURE 3.1   Fossil teeth of rays, a type of fish, were called toadstones during the Middle Ages. The stones were believed to come from the heads of living toads (*left*) and were considered to be a powerful cure for several ailments (*right*).

Introduction

Earth is a dynamic planet. Volcanic eruptions, earthquakes, floods, and shoreline erosion are manifestations of some of the internal and surface processes that continuously modify it. The fact that Earth has changed, or evolved, throughout its existence is obvious from the evidence preserved in rocks, or what is known as the **geologic record.** Although sedimentary rocks have a special place in historical geology, all rock types are important when deciphering the geologic record. Igneous rocks record volcanic activity related to rifting and plate convergence, and studies of metamorphic rocks reveal something of the dynamic processes within the crust.

Igneous and metamorphic rocks are by far the most abundant, comprising about 95% of all rocks in the crust. Nevertheless, sediments or sedimentary rocks are most commonly encountered at or near the surface where they cover most of the seafloor and about two-thirds of the continents. Furthermore, sediments and sedimentary rocks preserve evidence of the surface processes responsible for their origin, so geologists can make inferences about the presence and distribution of rivers, shorelines, glaciers, lakes, and deserts. And many sedimentary rocks contain the remains or traces of organisms, or what are called *fossils,* thus providing details about ancient climates and most of our record of prehistoric life.

Our main objective in this chapter is to explore the principles and concepts geologists use to make sense of the geologic record. Just as the written historical record must be analyzed and interpreted, the geologic record requires analyses if we are to make inferences about geologic history—that is, if we are to read Earth history from rocks.

Stratigraphy

The branch of geology known as **stratigraphy** is concerned with the composition, origin, age relationships, and geographic extent of layered or stratified rocks. Any of the three families of rocks can show stratification; a succession of lava flows or ash falls will be layered, many metamorphic rocks also show this feature, and nearly all sedimentary rocks are stratified (Figure 3.2).

Our main interest here is with stratification in sedimentary rocks, but some of the principles of stratigraphy apply to any sequence of layered rocks. Where sedimentary rocks are observed in surface exposures (Figure 3.2), the vertical relationships among the individual layers or strata are easily determined. Lateral relationships, equally important in stratigraphic studies, usually must be determined from a number of separate exposures.

VERTICAL RELATIONSHIPS

In many vertical successions of sedimentary rocks, individual layers are separated from one another by surfaces known as *bedding planes.* The rocks above and below a bedding plane differ in composition, texture, or both

(a)

(b)

(c)

FIGURE 3.2 Stratification. (a) These lava flows in Oregon are stratified. (b) Stratification in metamorphic rocks, the Siamo Slate of Michigan. (c) Stratified sedimentary rocks in the Valley of the Gods, Utah.

(Figure 3.3). Such abrupt changes in rock properties indicate rapid changes in sedimentation or perhaps a period of nondeposition or erosion followed by renewed deposition. In marked contrast, some sedimentary rock layers grade upward from one rock type into another, thus repre-

(a)

(b)

FIGURE 3.3 Vertical relationships among sedimentary rocks. (a) Several layers of sandstone and shale separated by planar surfaces known as bedding planes. (b) At this outcrop light-colored mudrocks are overlain by conglomerate. The surface between the two rock types is irregular, indicating that the mudrocks were eroded prior to deposition of the overlying gravel.

senting gradually changing conditions of sedimentation. Regardless of the nature of the vertical relationships among rock layers, the correct vertical sequence of strata at the time of deposition must be known.

Superposition

Nicolas Steno (see Chapter 2) realized that correct relative ages of horizontal (undeformed) strata could be determined by their position in a sequence of rocks (see Figure 2.2). If strata have been deformed, however, relative age determinations are more difficult. Geologists commonly solve relative age problems in deformed sedimentary rocks by looking for sedimentary structures, structures that formed in sediments at the time of deposition or shortly thereafter. In Figure 3.4 the strata contain wave-formed ripple marks, the crests of which point in the direction of the youngest bed, or the original top of the rock sequence. Several other sedimentary structures and fossil organisms in their original living positions

may also be used to determine which bed is youngest in a sequence (Figure 3.4).

In addition to deformation, the association of sedimentary and igneous rocks may cause problems in relative dating. Buried lava flows and sills both look much the same in a sequence of strata (Figure 3.5). However, a buried lava flow is older than the rocks above it, while a sill is younger than the bed immediately above it. Similar problems arise where sedimentary rocks are adjacent to large, irregularly shaped igneous bodies such as batholiths (Figure 3.6).

To resolve relative age problems such as these, geologists look closely at the sedimentary rocks in contact with the igneous rocks to see if they show signs of baking by heat, that is, contact metamorphism. Any sedimentary rock showing such effects is older than the igneous rock with which it is in contact. In Figure 3.5, for example, a sill produces a zone of baking in the rocks at its upper and lower margins, whereas a lava flow bakes only those rocks below.

Another way to determine relative ages is by use of the **principle of inclusions.** According to this principle, inclusions, or fragments in a rock unit, are older than the rock unit itself. For instance, the batholith shown in Figure 3.6a contains sandstone inclusions, and the sandstone unit shows the effects of baking. Accordingly, we conclude that the sandstone is older than the batholith. In Figure 3.6b, on the other hand, the sandstone contains granite rock fragments, indicating that the batholith was a source rock for the sandstone and must be older than the sandstone.

Unconformities

Our discussion thus far has been concerned with **conformable** sequences of strata, sequences in which no depositional breaks of any consequence occur. A sharp bedding plane separating strata may represent a depositional break of minutes, hours, years, or even tens of years but is inconsequential when considered in the context of geologic time. Surfaces of discontinuity representing significant amounts of geologic time are **unconformities,** and any interval of geologic time not represented by strata in a particular area is a *hiatus* (Figure 3.7). Thus, an unconformity is a surface of nondeposition or erosion that separates younger strata from older rocks. As such, it represents a break in our record of geologic time.

Three major types of unconformities are recognized (Figure 3.8). A **disconformity** is a surface of erosion or nondeposition between younger and older beds that are parallel with one another. An **angular unconformity** is an erosional surface on tilted or folded strata over which younger strata have been deposited. Both younger and older strata may dip, but if their dip angles are different (generally the older strata dip more steeply), an angular unconformity is present. A **nonconformity** is a type of unconformity in which an erosion surface cut into metamorphic or intrusive rocks is covered by sedimentary rocks. Unconformities of regional extent may change later-

FIGURE 3.4 Determining relative ages of deformed sedimentary rock layers. (a) These rock layers contain a variety of sedimentary structures, all of which are discussed more fully in Chapter 4. Note the wave-formed ripple marks with their sharp crests pointing up toward the next younger layer. (b) We can use the wave-formed ripple marks, and other sedimentary structures, to determine the relative ages of the beds even when they have been deformed.

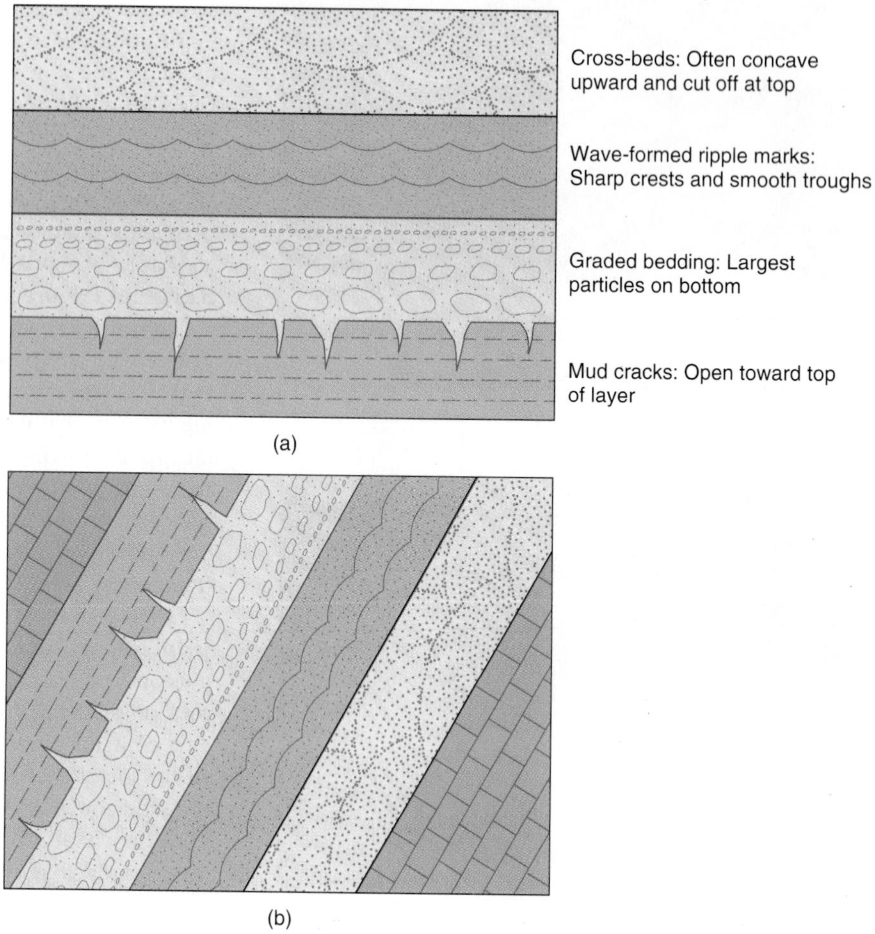

Undisturbed layers

Cross-beds: Often concave upward and cut off at top

Wave-formed ripple marks: Sharp crests and smooth troughs

Graded bedding: Largest particles on bottom

Mud cracks: Open toward top of layer

(a)

(b)

FIGURE 3.5 Relative ages of lava flows, sills, and associated sedimentary rocks may be difficult to determine. (a) A buried lava flow (4) baked the underlying bed, and bed 5 contains inclusions of the lava flow. The lava flow is younger than bed 3 and older than beds 5 and 6. (b) The rock units above and below the sill (3) have been baked, indicating that the sill is younger than beds 2 and 4, but its age relative to bed 5 cannot be determined.

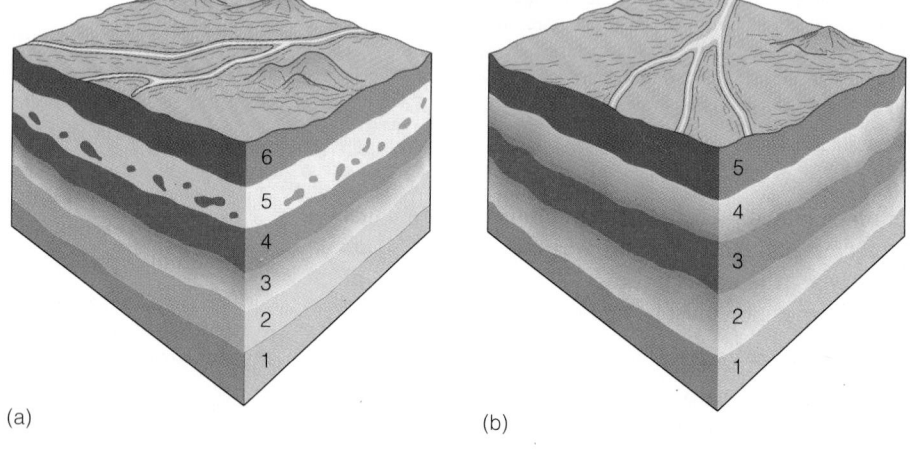

(a)

(b)

FIGURE 3.6 (a) The batholith is younger than the sandstone because the sandstone has been baked at its contact with the granite and the granite contains sandstone inclusions. (b) Granite inclusions in the sandstone indicate that the batholith was the source of the sandstone and therefore is older.

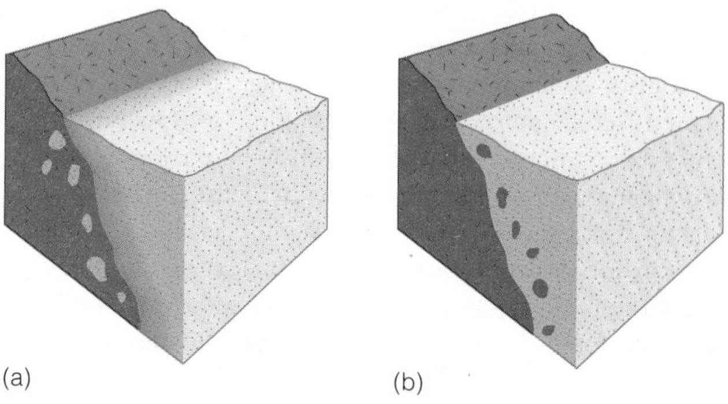

(a)

(b)

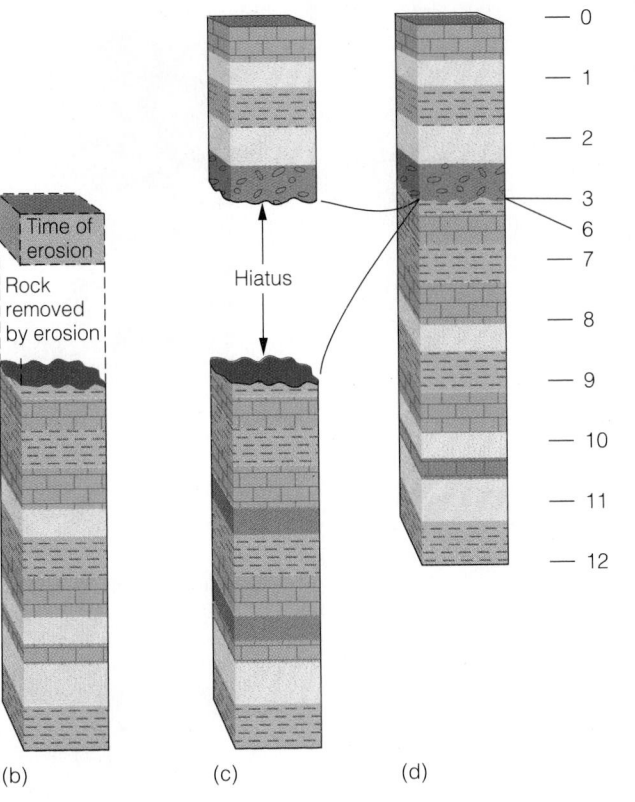

M.Y.A

(a) (b) (c) (d)

Time of erosion

Rock removed by erosion

Hiatus

FIGURE 3.7 Simplified diagram showing the development of an unconformity and a hiatus. (a) Deposition began 12 million years ago (M.Y.A.) and continued more or less uninterrupted until 4 M.Y.A. (b) A 1-million-year episode of erosion occurred, and during that time strata representing 2 million years of geologic time were eroded. (c) A hiatus of 3 million years exists between the older strata and the strata that formed during a renewed episode of deposition that began 3 M.Y.A. (d) The actual stratigraphic record. The unconformity is the surface separating the strata and represents a major break in our record of geologic time.

ally from one type to another and do not encompass equivalent amounts of time everywhere (Figure 3.9).

LATERAL RELATIONSHIPS—FACIES

In 1669 Nicolas Steno formulated the principle of lateral continuity (see Chapter 2). He recognized that sediment layers extend outward in all directions until they terminate. They may terminate abruptly where they abut the edge of a depositional basin, where eroded, or where truncated by faults (Figure 3.10). Sedimentary rocks may also terminate laterally when a rock unit becomes progressively thinner until it pinches out. *Intertonguing* occurs when a unit splits laterally into thinner units, each of which pinches out. And finally, a unit may change in composition or texture, or both, by lateral gradation until it is no longer recognizable (Figure 3.10e).

Intertonguing and lateral gradation result from different depositional processes operating in adjacent environments. For example, sand may be deposited in the high-energy nearshore part of the continental shelf, while mud and carbonate sediments accumulate simultaneously in the laterally adjacent low-energy, offshore area (Figure 3.11). Deposition in each environment produces a body of sediment, each of which is characterized by a distinctive set of physical, chemical, and biological attributes. Such distinctive bodies of sediment or sedimentary rock are **sedimentary facies.**

Any aspect of a sedimentary rock unit that makes it recognizably different from laterally adjacent rocks of the same age, or approximately the same age, can be used to establish a sedimentary facies. In Figure 3.11 we recognize

three sedimentary facies, a sandstone facies, a shale facies, and a limestone facies. Armanz Gressly in 1838 was the first to use the term *facies.* Gressly carefully traced sedimentary rock units in the Jura Mountains of Switzerland and noticed lateral changes, from shale to limestone, for example. He concluded that these lateral changes resulted from deposition in different depositional environments that existed simultaneously.

TRANSGRESSIONS AND REGRESSIONS

Many rock units in the interiors of continents show clear evidence of having been deposited in marine depositional environments. These deposits consist of vertical sequences of facies that resulted from deposition during times when sea level rose or fell with respect to continents. When sea level rises with respect to a continent, the shoreline moves inland, giving rise to a **marine transgression** (Figure 3.12). As the shoreline moves inland, the depositional environments parallel to the shoreline do likewise. Remember that each laterally adjacent environment in Figure 3.12 is the depositional site of a different sedimentary facies. As a result of a marine transgression, the facies that formed in the offshore environments are superposed over the facies that was deposited in the nearshore environment.

Another important aspect of marine transgressions is that individual rock units become younger in a landward direction. The sandstone, shale, and limestone facies in Figure 3.12 were deposited continuously as the shoreline moved landward, but none of the facies was deposited

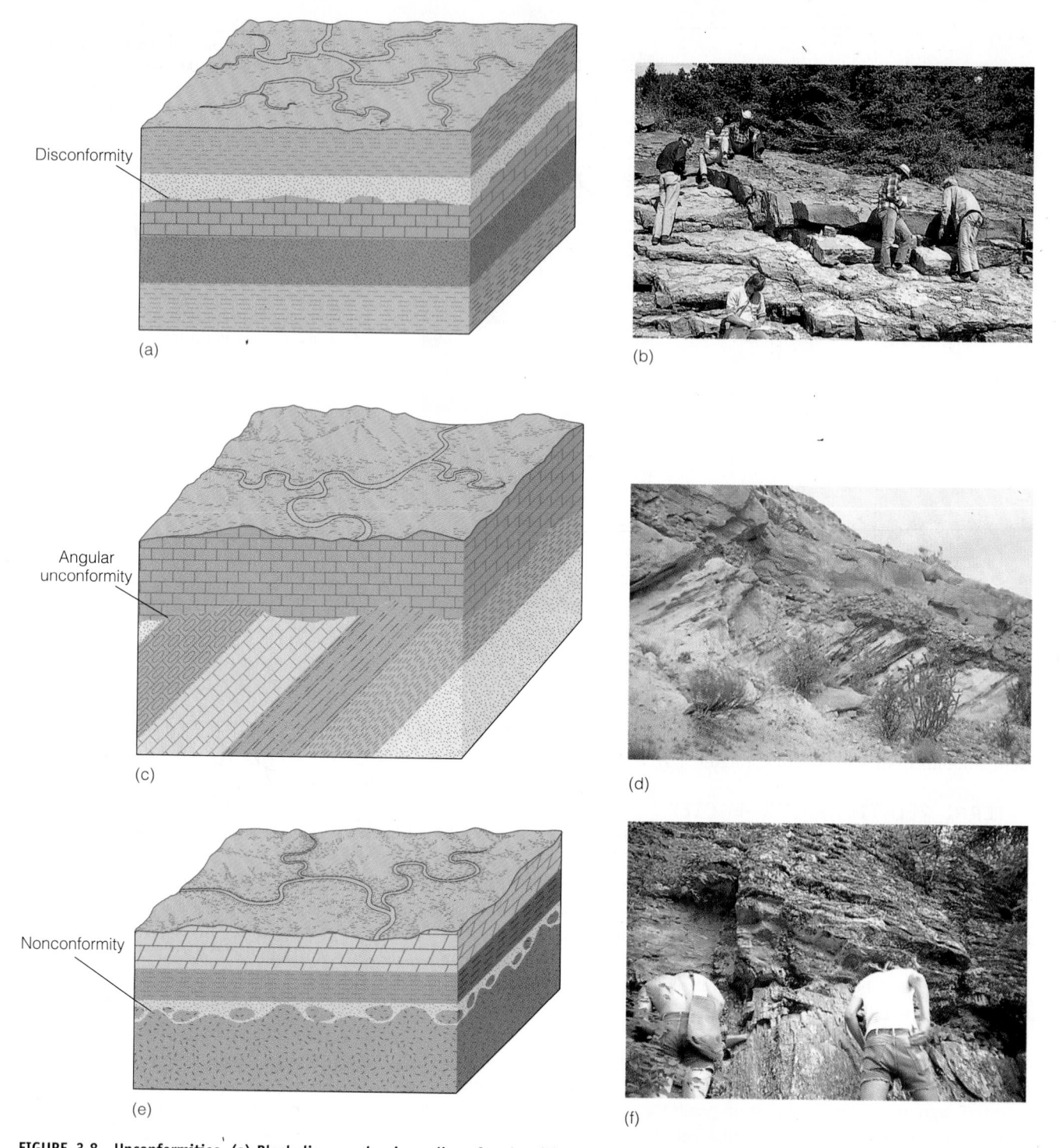

FIGURE 3.8 Unconformities. (a) Block diagram showing a disconformity. (b) Disconformity between Mississippian and Jurassic strata in Montana. The geologist at the upper left is sitting on Jurassic strata, and his right foot is resting upon Mississippian rocks. (c) Angular unconformity. (d) Angular unconformity in New Mexico (e) Nonconformity. (f) Nonconformity between metamorphic rocks and the overlying Cambrian age sandstone in South Dakota.

simultaneously over the entire area it now covers. In other words, the facies are **time transgressive**, meaning that their ages vary from place to place.

The opposite of a marine transgression is a **marine regression.** If sea level falls with respect to a continent, the shoreline and environments that parallel the shoreline move seaward (Figure 3.12). The vertical sequence pro-

duced by a marine regression has facies of the nearshore environment superposed over facies of offshore environments. In addition, individual rock units become younger in a seaward direction.

The significance of lateral and vertical facies relationships was first recognized by Johannes Walther (1860–1937). When Walther traced rock units laterally, he rea-

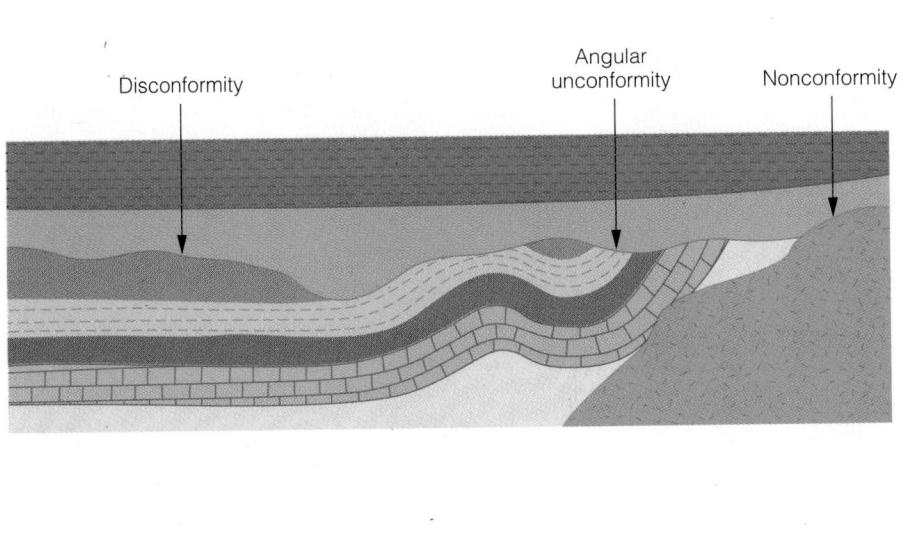

FIGURE 3.9 An unconformity of regional extent may change from one type to another as shown in this cross section.

Disconformity

Angular unconformity

Nonconformity

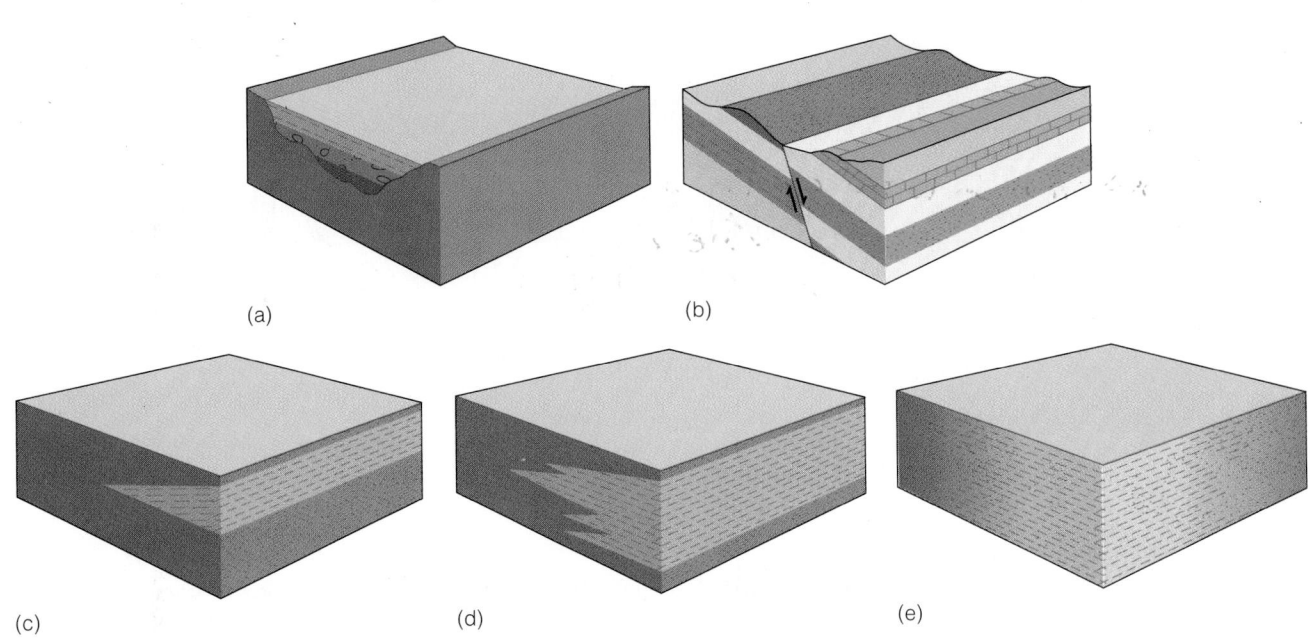

(a)

(b)

(c)

(d)

(e)

FIGURE 3.10 (a) Lateral termination of rock units at the edge of a depositional basin. (b) Faulting and erosion causing lateral termination. Notice that the beds above the sandstone do not appear on the left side of the fault. (c) Lateral termination by pinchout. (d) Intertonguing. (e) Lateral gradation.

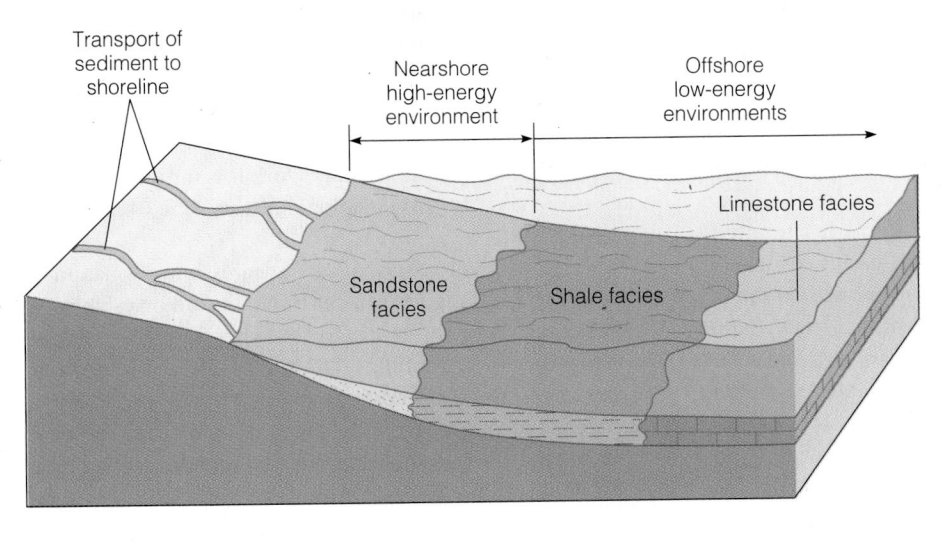

FIGURE 3.11 Simultaneous deposition in different environments adjacent to one another. Each distinct depositional unit is recognized as a sedimentary facies.

Transport of sediment to shoreline

Nearshore high-energy environment

Offshore low-energy environments

Limestone facies

Sandstone facies

Shale facies

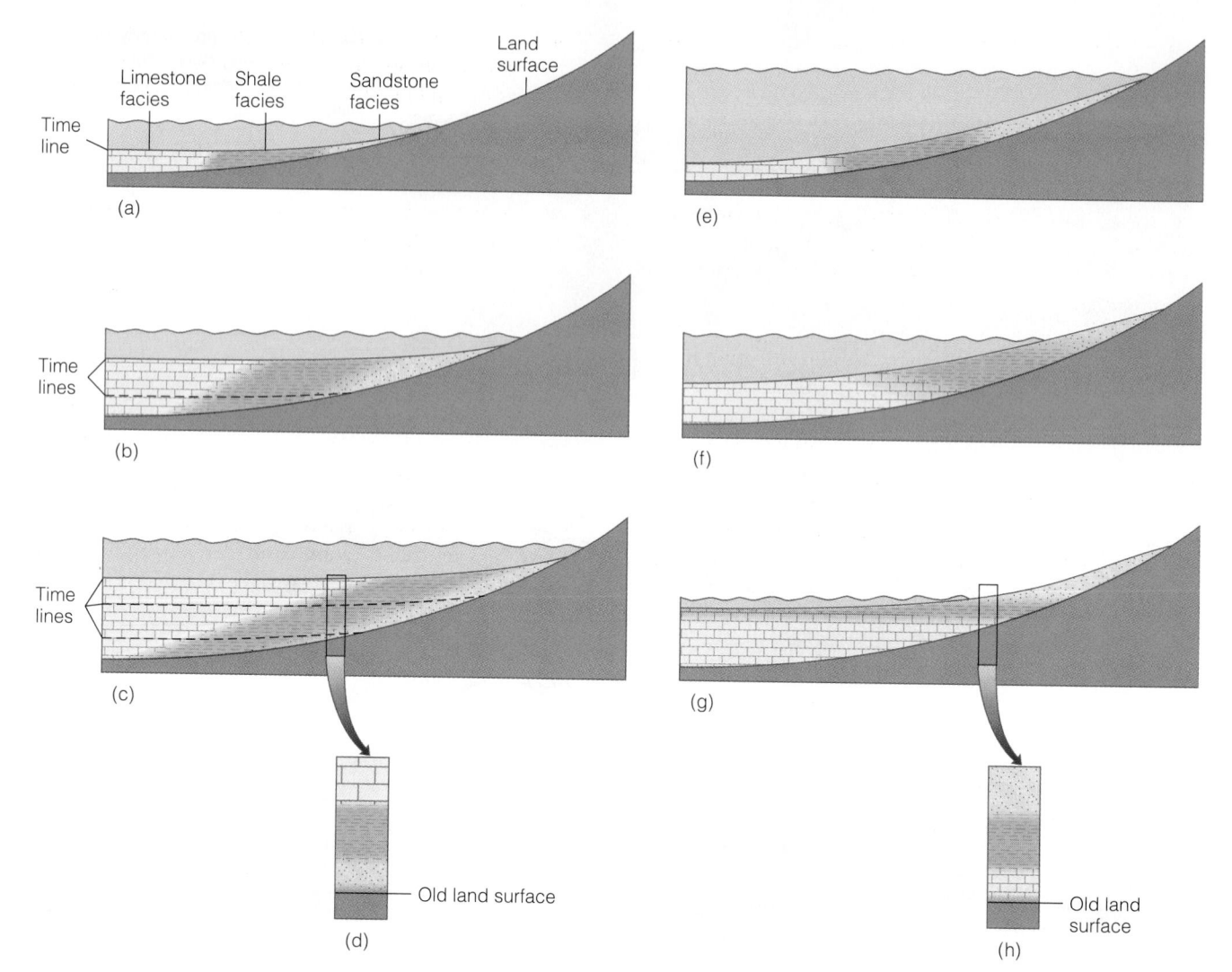

FIGURE 3.12 (a), (b), and (c) Three stages of a marine transgression. (d) Diagrammatic representation of the vertical sequence of facies resulting from the transgression. (e), (f), and (g) Three stages of a marine regression. (h) Diagrammatic representation of the vertical sequence of facies resulting from the regression.

soned, as Gressly had, that each sedimentary facies he encountered had been deposited in laterally adjacent but different environments. Furthermore, he realized that the same facies that he found laterally were also superposed upon one another in a vertical sequence. Walther's observations have since been formulated into **Walther's law,** which holds that the same facies following one another in a conformable vertical sequence will also replace one another laterally.

The value of Walther's law is well illustrated by its application to vertical facies relationships resulting from marine transgression or regression. In practice, one commonly cannot trace individual rock units far enough laterally to demonstrate facies changes. It is much easier to observe vertical facies relationships and then to work out the lateral relationships from them using Walther's law. Note, however, that Walther's law applies only to vertical successions of facies that contain no unconformities. Should an unconformity be present, the facies immediately above and below the surface of unconformity are unrelated, and Walther's law is not applicable.

EXTENT, RATES, AND CAUSES OF TRANSGRESSIONS AND REGRESSIONS

The seas have clearly occupied large parts of all continents at several times during the past. In North America, six major episodes of transgression followed by regression have taken place since the beginning of the Paleozoic Era. During these transgressions and regressions, widespread sequences of strata bounded by unconformities were deposited. These transgressive–regressive sequences will serve as the stratigraphic framework for our discussions of Paleozoic and Mesozoic geologic history (see Chapters 10, 11, and 14).

The maximum extent of a transgression or maximum withdrawal of the sea during a regression can be established with some certainty, but rates and causes of these events are problematic. Shoreline movements probably occur at rates of centimeters per year, but these rates are deceptive because they are determined by dividing distance by time. For example, if a shoreline moved inland 1000 km in 20 million years, the rate of transgression averages 5 cm per year. However, major transgressions and regressions are not simply events during which the shoreline moves steadily inland or seaward. As a matter of fact, both are characterized by numerous reversals in the overall transgressive or regressive trend.

Most geologists agree that uplift or subsidence of the continents, rates of seafloor spreading, and the amount of seawater contained in glaciers are sufficient to cause a transgression or regression. If a continent is uplifted, it rises with respect to sea level, the shoreline moves seaward, and a regression ensues. The opposite type of movement, subsidence, results in a transgression (Figure 3.13).

Seafloor spreading can cause transgressions and regressions by changing the volumes of the ocean basins. During relatively rapid episodes of extensive seafloor spreading, such as during the Cretaceous, the mid-oceanic ridges expand and displace seawater onto the continents. At other times, when seafloor spreading is comparatively slow, the ridges subside, the volume of the ocean basins increases, and the seas retreat from the continents (Figure 3.14).

Worldwide changes in sea level related to additions or removal of water from the oceans may also cause marine transgressions and regressions. During a widespread glacial episode, large quantities of seawater are on land as glacial ice, and, consequently, sea level falls. When glaciers melt, the water returns to the seas, and sea level rises (Figure 3.13d). A number of Pennsylvanian transgressions and regressions can be attributed to sea-level changes caused by glaciation (see Chapter 11).

Fossils

In Chapter 2 and preceding sections of this chapter, several principles, including superposition and cross-cutting relationships, were discussed. These principles are essential for interpreting geologic history but can be applied only on a local scale. For instance, consider two vertical sequences of strata in different areas (Figure 3.15). The relative ages in each area are easily determined, but with the data given, the age of the rocks in one area relative to

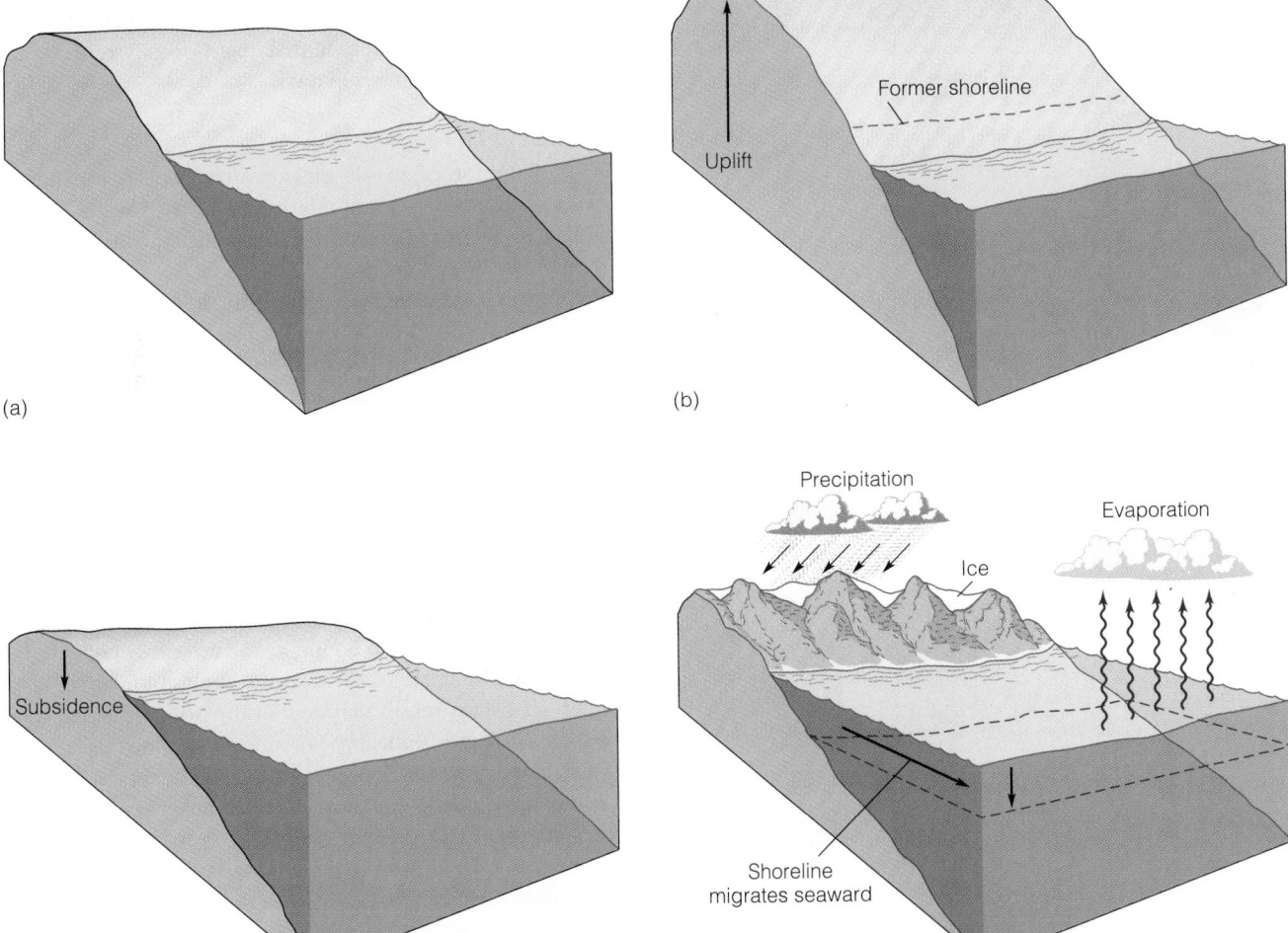

(a)

(b)

(c)

(d)

FIGURE 3.13 Causes of transgressions and regressions. (a) Reference diagram showing position of shoreline. (b) Uplift of a continent causes the shoreline to migrate seaward, resulting in a marine regression. (c) A marine transgression takes place when a continent subsides. (d) When a large volume of seawater is stored on land as glacial ice, sea level is lowered, and a marine regression occurs. When glaciers melt, however, the water returns to the sea, sea level rises, and a transgression occurs.

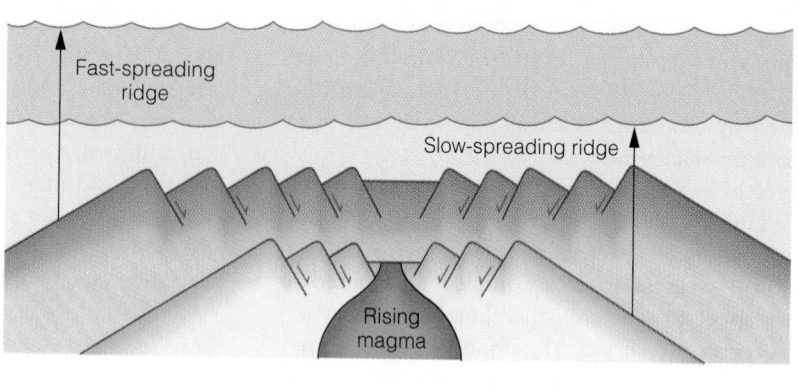

FIGURE 3.14 The cross-sectional area of a mid-oceanic ridge depends on its spreading rate. The ridge forms due to expansion by heating. A fast-spreading ridge has a larger cross section and consequently displaces more seawater, resulting in a rise in sea level and marine transgressions onto the continents.

those in the other area cannot be determined. The solution to this problem involves the use of fossils.

Fossils are the remains or traces of prehistoric organisms that have been preserved in rocks (see Perspective 3.1). Besides their use in determining relative ages of strata, fossils are important in determining environments of

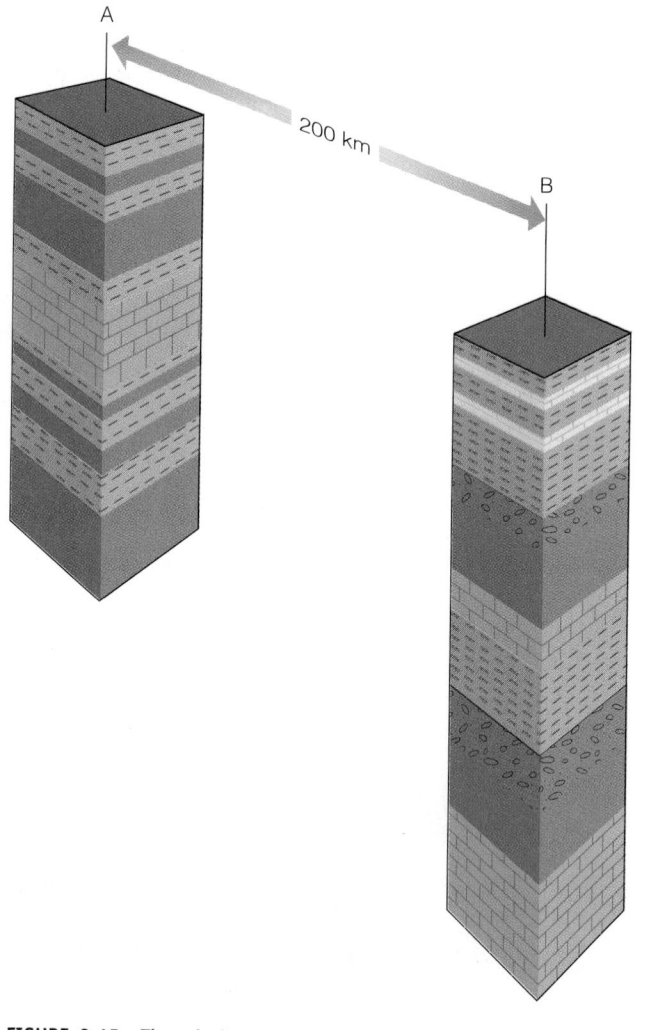

FIGURE 3.15 The relative ages of strata in either section A or B can be determined by superposition; but with the data given, the ages of rocks in section A cannot be determined relative to those in section B. The rocks in section A may be older than any in B, the same age as some in B, or younger than the rocks in B.

deposition (see Chapter 4), and they provide some of the evidence for evolution (see Chapter 5).

During most of historic time, most people did not recognize that fossils were the remains of organisms (see the Prologue). Such perceptive observers as Leonardo da Vinci in 1508, Robert Hooke in 1665, and Nicolas Steno in 1667 recognized the true nature of fossils, but their views were largely ignored. By the late eighteenth century though, the evidence had become overwhelming that fossils had indeed once been living organisms. By the nineteenth century, it was also apparent that many fossils represented types of organisms now extinct.

FOSSILIZATION

Our definition of fossils includes the phrase "remains or traces." Remains, commonly called **body fossils** (Figure 3.16a and b), are usually the hard skeletal elements such as bones, teeth, and shells. In exceptional cases, preservation by freezing and mummification provides us with partial or nearly complete body fossils with flesh, hair, and internal organs. **Trace fossils** include tracks, trails, burrows, borings, nests, or any other indication of the activities of organisms (Figure 3.16c and d). Fossilized feces, called **coprolites,** provide evidence of activity of organisms (Figure 3.16e) and may provide important information on the diet and size of the animal that produced them.

Fossils in general are quite common (see Perspective 3.2). The remains of various microorganisms are by far the most abundant and in many ways the most useful. Shells of marine animals are also common and are easily collected in many areas (Figure 3.16b). Despite their abundance, however, fossils represent very few of the organisms that lived at any one time, since any potential fossil must escape the ravages of destructive processes such as waves or running water, scavengers, exposure to the atmosphere, and bacterial action. Hard skeletal elements are more resistant to destructive processes than soft parts, but even skeletal elements must be buried in some protective medium such as mud or sand if they are to be preserved; and even if buried, they may be dissolved by groundwater or destroyed by alteration of the host rock.

Considering all the ways in which potential fossils can be destroyed, it comes as no surprise that the fossil record is biased toward those organisms with hard parts and those

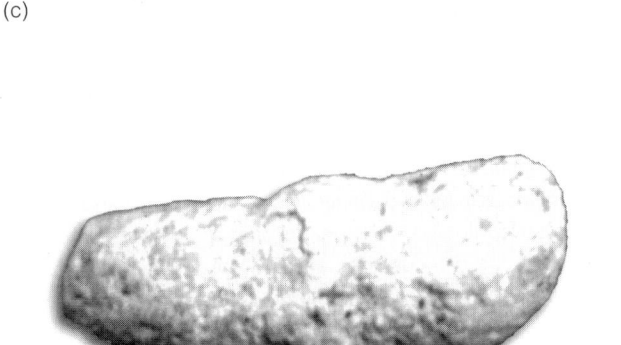

(a)

(b)

(c)

(d)

FIGURE 3.16 Body fossils are the actual remains of organisms such as bones (a) and shells (b). Trace fossils include crawling traces of invertebrate animals (c) and tracks made by birds (d). The slab in (d) is actually a layer of rock that was deposited upon the layer containing the tracks. Accordingly, it is a cast of the tracks. (e) Fossilized feces (coprolite) of a carnivorous mammal. Specimen measures 5.5 cm long.

(e)

organisms that lived in areas of active sedimentation. Accordingly, most fossils are preserved in sediments or sedimentary rocks. Volcanic ash falls may also contain fossils, but fossils are rare to nonexistent in other types of igneous rocks and in metamorphic rocks. Despite its biases, the fossil record is our only record of prehistoric life. Indeed, it is only through fossils that we have any knowledge of such extinct lifeforms as trilobites and dinosaurs.

TYPES OF FOSSIL PRESERVATION

Fossils may be preserved as *unaltered remains,* in which case they retain their original structure and composition. The pollen and spores of plants have a tough outer cuticle that is chemically resistant to change (Figure 3.17a). Partial or complete insects may be trapped in the resin secreted by

some plants, especially by coniferous trees. When this resin is buried, it hardens to become amber and preserves in exquisite detail the remains contained therein (Figure 3.17b).

Unusual types of preservation of unaltered remains include preservation in tar, mummification, and freezing. The La Brea Tar Pits of southern California contain numerous tar-impregnated bones of Pleistocene animals that otherwise retain their original composition and structure (Figure 3.18). Mummification involves air drying and shriveling of soft parts such as muscles and tendons before burial (Figure 3.19a).

Perhaps the most interesting example of preservation of unaltered remains is by freezing. Only a few types of animals have been found frozen, including mammoths, woolly rhinoceroses, musk oxen, and a few others, and all are

Fossil Frauds and Hoaxes

*E*veryone now accepts that fossils are truly the remains of organisms that existed during the past. Occasionally, though, a fossil, or what is thought to be a fossil, is reported that is either an outright forgery or something placed in rocks where it would not naturally occur. For instance, in 1866 a modern human skull was reported found in Pliocene rocks in California. Dubbed the *Calaveras skull,* it was claimed to be a fossil that upset standard geologic interpretations by occurring in rocks too old. However, it was determined that a shopkeeper had placed a skull from a recent Indian burial in a mineshaft to fool a local doctor. In short, it was a practical joke.

In 1869, a 3-meter-long, 1360-kilogram stone man was unearthed near Cardiff, New York, and was claimed to be a petrified man (Figure 1). It was soon referred to as the *Cardiff Giant,* and Mr. Newell who discovered the "fossil" quickly developed a thriving business by charging an admission

fee to view the object. The "petrified man," it turned out, was carved from gypsum in Iowa, shipped to New York, and then buried on Mr. Newell's farm where it was later "discovered" when Mr. Newell directed workers to dig a well.

The hoax began in Iowa when George Hull, Mr. Newell's brother-in-law, had a disagreement with a local clergyman over the biblical reference to giants. The clergyman claimed the reference was literal, whereas Hull held that it was not. In any event, Hull had the giant stone figure carved at some great expense. Although one scientist thought the object was genuine, most dismissed it as preposterous. In fact, Hull eventually made a public confession of how the object originated.

Albert Koch perpetrated several fossil hoaxes during the 1840s and 1850s. He collected real fossils but took great liberties in his reconstructions. In one instance he placed mastodon (fossil elephant) tusks

on top of the skull, thereby rendering a bizarre beast. Koch also collected fossil remains of whales and used bones from more than one individual to construct a so-called sea serpent (Figure 2). He displayed his creatures in both the United States and Europe and made considerable money from them. Scientists were well aware that Koch's creatures were hoaxes, so when he did report some legitimate fossil evidence no one took him seriously.

Dr. Johann Bartholomew Adam Beringer (1667–1740), a member of the faculty of the University of Würtzburg, Germany, during the early 1700s, was the victim of a cruel hoax that ruined his reputation. Three brothers working for Beringer made a series of remarkable discoveries of stones showing in raised-relief plants, animals, the Sun and its rays, letters, words, and various other objects (Figure 3). Beringer was convinced that the stones were genuine and published a

FIGURE 1 The Cardiff Giant, sometimes referred to as the American Goliath, is the figure of a man carved from gypsum. It is now in the Farmer's Museum in Cooperstown, New York.

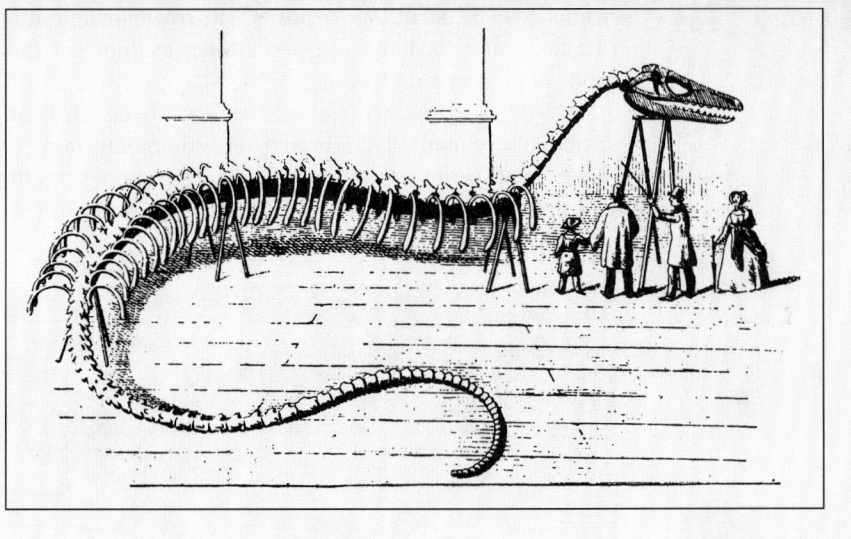

FIGURE 2 This "sea serpent" on display in New York was constructed by Albert Koch from the bones of several fossil whales.

book with numerous illustrations. Unfortunately, the hoax was revealed, and even Beringer had to admit that the stones were forgeries. Two of Beringer's colleagues at the university had carved the stones and given them to his workmen. Although Beringer kept his post at the university, he quietly began buying up his books and destroying them. Those responsible for the hoax left Würtzburg in disgrace.

Charles Dawson, an amateur paleontologist, discovered parts of a human brain case and lower jaw and some artifacts in 1912 at Piltdown, Kent County, England. Dawson sent the remains to the British Museum where Sir Arthur Smith-Woodward examined them. Even though the brain case and jaw seemed to fit, there were anomalies: The brain case was much like

that of a present-day human's, whereas the jaw appeared to be quite primitive. In any case, *Piltdown man*, as it was called, was widely accepted as evidence that humans had existed in England before the Ice Age. In fact, some claimed it was a so-called missing link between apes and humans. Some critics remained unconvinced, but additional remains, discovered by Dawson in 1915, seemed to be compelling evidence for the genuine nature of the fossils.

In 1953, a small hole drilled in the jaw showed that the dark color of the fossil was only on the surface; apparently the remains had been chemically stained to make them look ancient. Furthermore, a fluorine analysis revealed that the brain case and jaw were of different ages. Piltdown man was a hoax. The brain case was

from a fairly recent human, perhaps a Roman-aged burial, and the jaw was from a present-day orangutan. The teeth had been filed to simulate a human wear pattern, and the jaw joints had been broken off, otherwise it would have been obvious that the jaw did not belong with the skull.

Charles Dawson and Sir Arthur Smith-Woodward have most often been cited as perpetrators of the hoax, but others including Sir Arthur Conan Doyle and the French paleontologists Teilhard de Chardin have also been mentioned. However, recent evidence indicates that Martin A. C. Hinton, a volunteer at the British Museum, had a dispute with Smith-Woodward and as a result concocted the hoax. In any event, it was the most successful hoax in the history of paleontology.

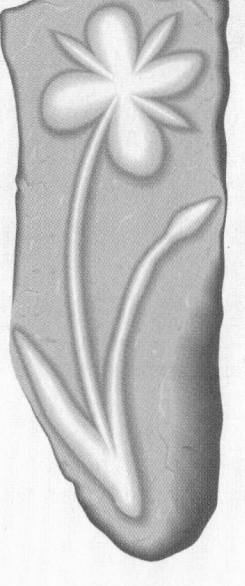

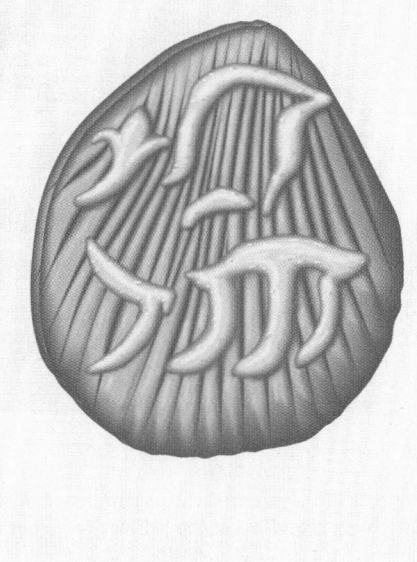

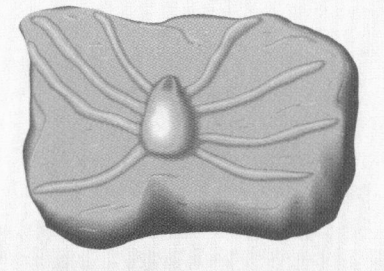

FIGURE 3 Illustrations of some of the figured stones given to Dr. Johann Bartholomew Adam Beringer by his fossil collectors.

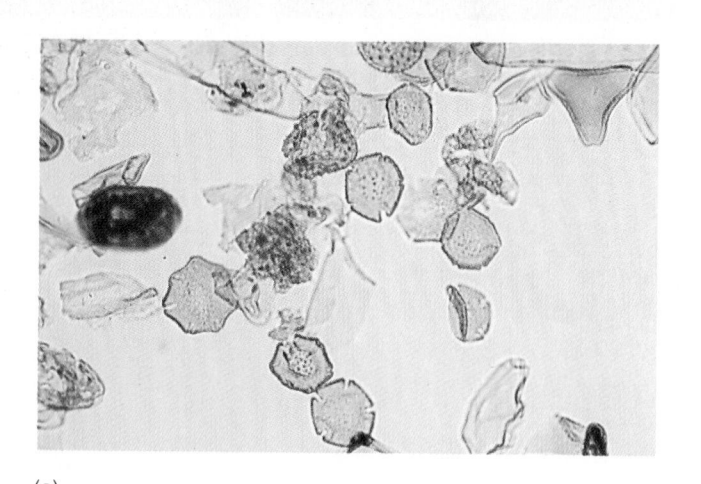

(a)

(b)

FIGURE 3.17 (a) Late Eocene pollen from the Eurla Basin, Australia. (b) Insects in amber.

from Late Pleistocene–age deposits. The frozen mammoths of Siberia are no doubt the best-known example of this mode of preservation (Figure 3.19b).

Altered remains are fossils that have been changed structurally, chemically, or both. Quite commonly, mineral matter is added to the pores and cavities of bones, teeth, and shells after burial. This process, called *permineraliza-*

FIGURE 3.18 Excavation of mammoth bones from tar pit 9 at Rancho La Brea in Los Angeles, California, in 1913.

(a)

(b)

FIGURE 3.19 (a) Mummified remains of the duck-billed dinosaur *Anatosaurus*. (b) Frozen baby mammoth found in Siberia in 1977. The baby was six or seven months old, 1.15 m long, 1.0 m tall, and when alive had a hairy coat. Most of the hair has fallen out except around the feet. The carcass is about 40,000 years old.

perspective 3.2

The Abundance of Fossils

*O*nly a tiny proportion of all organisms that ever lived are fossilized, but fossils are nevertheless quite common, much more so than most people realize. The reason for this abundance? So many billions of organisms existed over so many millions of years that even if only 1 in 10,000 was fossilized, the total number of fossils is truly astonishing.

By far the most common types of easily collected fossils are those of marine *invertebrates,* animals lacking a segmented vertebral column, such as clams, oysters, sea lilies, brachiopods, and corals. Dozens or hundreds of these fossils can be collected in many areas by anyone interested in doing so. In fact, they are so common that, when weathering out of their host rocks, they litter the surface of some localities.

Animals possessing a segmented vertebral column are known as *vertebrates* and include various fish, amphibians, reptiles, birds, and mammals. Their fossils are not as common as those of invertebrates, but even these are better represented by fossils than is generally believed. Thousands of fossilized fish skeletons are found on single surfaces within 50-million-year-old lake deposits in Wyoming, for example; and dinosaur bone fragments and complete or partial teeth are easily collected in some places. It is true that entire skeletons of dinosaurs or other land-dwelling animals are rare, but some remarkable concentrations of their fossils have been discovered.

Excavations beginning in 1934 at what is known as the Howe Quarry in Wyoming revealed an extraordinary concentration of Jurassic-age dinosaur fossils (Figure 1). The deposit, which measures only about 18×16 m, yielded 4000 bones from at least 20 large dinosaurs. But the most interesting aspect of this discovery was the 12 legs found preserved in an upright posi-

FIGURE 1 Howe Quarry in Wyoming, where more than 4000 dinosaur bones have been recovered.

tion penetrating the muddy sedimentary deposits, indicating that the dinosaurs became mired in mud and perished. The parts of the animals above the surface decomposed and formed a heap of bones, whereas the legs were preserved in the position in which the dinosaurs were trapped.

In 1981 in northwestern Montana, a Cretaceous-age bone bed was discovered that contains an estimated 10,000 duck-billed dinosaurs. Apparently, a vast herd was overcome by ash and gases from a nearby volcano and subsequently buried in volcanic ash. Individual dinosaurs varied from juveniles measuring 3.6 m long to adults 7.6 m long, representing the standing age of the herd at the time of death.

Ten million years ago in what is now northeastern Nebraska, a vast grassland was inhabited by short-legged aquatic rhinoceroses, camels, three-toed horses, saber-toothed deer, land turtles, and

many other animals. Thousands of animals perished when a vast cloud of volcanic ash covered the area. Many of the animals died during the initial ash fall, but their remains lay at the surface and partly decomposed before they were buried. In contrast, hundreds of rhinoceroses were killed later when the ash was probably redistributed by wind, and they were buried rapidly as indicated by the large number of complete skeletons (Figure 2).

Scientists known as *paleontologists,* who study the history of life as revealed by fossils, are not often afforded opportunities to find and recover so many well-preserved fossils as in these examples. Ancient calamities provide paleontologists with a unique glimpse of what life was like millions of years ago.

FIGURE 2 Paleontologists excavating rhinoceros (foreground) and horse (background) skeletons from volcanic ash near Orchard, Nebraska.

tion, increases the preservation potential of fossils by increasing their durability.

The shells of many marine invertebrates are composed of an unstable form of calcium carbonate known as aragonite. When buried, these shells commonly recrystallize in the more stable form of calcium carbonate called calcite. A few plants and animals have shells or skeletal elements composed of opal, a variety of silicon dioxide (SiO_2). These also commonly recrystallize. Although recrystallization does not change the chemical composition or outward appearance, the microscopic crystal structure of the shell is altered.

We know that clams, snails, and many other invertebrates have skeletons of calcium carbonate, yet in some rocks we find these shells composed of silicon dioxide (SiO_2) or pyrite (FeS_2). These shells have been altered by *replacement,* during which the original skeletal material is replaced by a compound of a different composition. Insect skeletons replaced by silicon dioxide are a particularly interesting example of this type of preservation (Figure 3.20a).

Wood may be preserved by the complete replacement of woody tissue by silicon dioxide in the form of chert or opal and, more rarely, by calcite (Figure 3.20b). Wood preserved in this manner is often called *petrified,* a term that means to become stone. Particularly when silicon dioxide is involved, the replacement may be so exact that even details of cell structure are finely preserved. In many cases the form of the woody tissue actually remains unaltered, and the pore spaces have simply been filled in by silicon dioxide.

The altered remains of plants may be preserved by *carbonization.* During carbonization, the volatile elements (hydrogen, oxygen, and nitrogen) of organic material, such as a leaf, vaporize, leaving a thin carbon film (Figure 3.21a). Fish, insects, and soft-bodied animals such as worms may also be carbonized (Figure 3.21b).

In addition to unaltered and altered remains, fossils may be preserved as **molds** and **casts** (Figure 3.22). Molds and casts of shelled marine organisms such as clams and brachiopods are quite common. Molds are formed when the remains of a buried organism are dissolved, thereby leaving a cavity with the external shape of the organism (an external mold). If the mold is later filled with sediment or mineral matter, an external cast is formed. Internal molds are cavities showing the inner features of shells, and internal casts are formed when such cavities are filled with sediment or mineral matter.

FOSSILS AND UNIFORMITARIANISM

Uniformitarianism in its simplest form can be expressed as "the present is the key to the past." This is an oversimplification, but observations of the present-day processes leading to the burial of organic remains do give us a better understanding of how fossils were preserved.

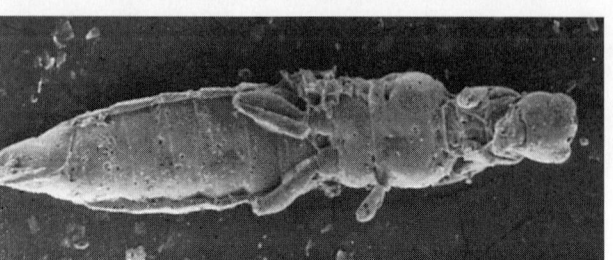

(a)

(b)

FIGURE 3.20 (a) Insect replaced by silicon dioxide in the Miocene Button Bed of the Barstow Formation, southern California. (b) Three-million-year-old petrified redwood tree in California.

Fossilization begins with the death of an organism and its burial in sediment or volcanic ash. Only rarely is the cause of death of a fossil organism apparent, but the conditions of burial commonly can be inferred by comparison with present-day processes. For example, Charles Darwin visited Uruguay during the 1830s and learned that millions of horses and cattle had died during a prolonged drought. When the drought broke, the rivers flooded and buried thousands of these animals in floodplain deposits. Similarly, many of the fossil mammals in the White River Badlands of South Dakota were buried in flood-plain deposits.

Numerous additional examples exist of the burial of organic remains. Present-day small animals are trapped and

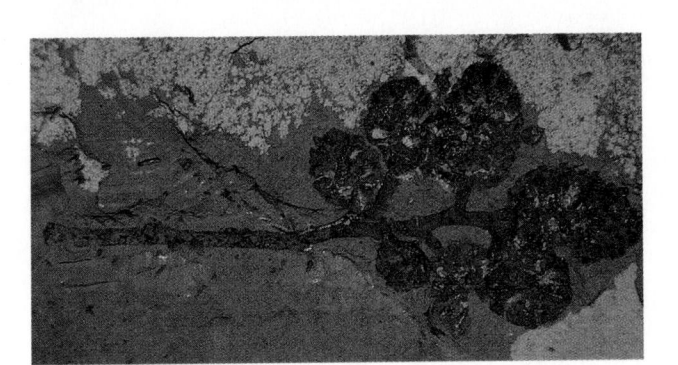

(a)

(b)

FIGURE 3.21 Altered remains preserved by carbonization. (a) Leaf from the Tertiary John Day Formation of Oregon. (b) Insect from the Oligocene Renova Formation, Montana.

die in the sticky residue of southern California oil seeps, a single storm on the continental shelf may cause the burial of shallow marine invertebrates, and lake sediments contain leaves, insects, and occasionally fish. Following the 1980 eruption of Mount St. Helens in Washington, mudflows transported and buried logs and stumps and partially buried trees in growth position. The same or very similar processes occurred numerous times during the Eocene Epoch in what is now Yellowstone National Park, Wyoming. Events such as the burial of Pompeii by volcanic ash in A.D. 79 remind us that humans too are potential fossils.

It is important to note from these examples that the burial of organisms by natural processes is a common occurrence. The conditions of burial range from slow sedimentation in lakes to rapid burial by ash falls and mudflows, but all are consistent with the concept of uniformitarianism. We have every reason to think that organisms in the fossil record were buried under similar circumstances.

Fossils and Time

The utility of fossils in relative dating and geologic mapping was clearly demonstrated in England and France in the early nineteenth century. William Smith (1769–1839), an English civil engineer involved in surveying and building canals in southern England, independently recognized the principle of superposition, reasoning that fossils lowest in a sequence of strata are oldest and those higher in

FIGURE 3.22 Molds and casts. (a) Burial of a shell in sediment. (b) The shell dissolves leaving a cavity, or mold. (c) The mold is filled by sediment, thus forming a cast. (d) This fossil turtle shell shows how both a mold and a cast form. Note that some of the original turtle shell remains as the outermost part of the fossil. The hollow interior of the shell served as a mold that was filled with sediment, thus forming a cast of the shell's interior. Later, much of the shell was lost, revealing the cast showing details of the inner part of the shell.

(a)

(b)

(c)

(d)

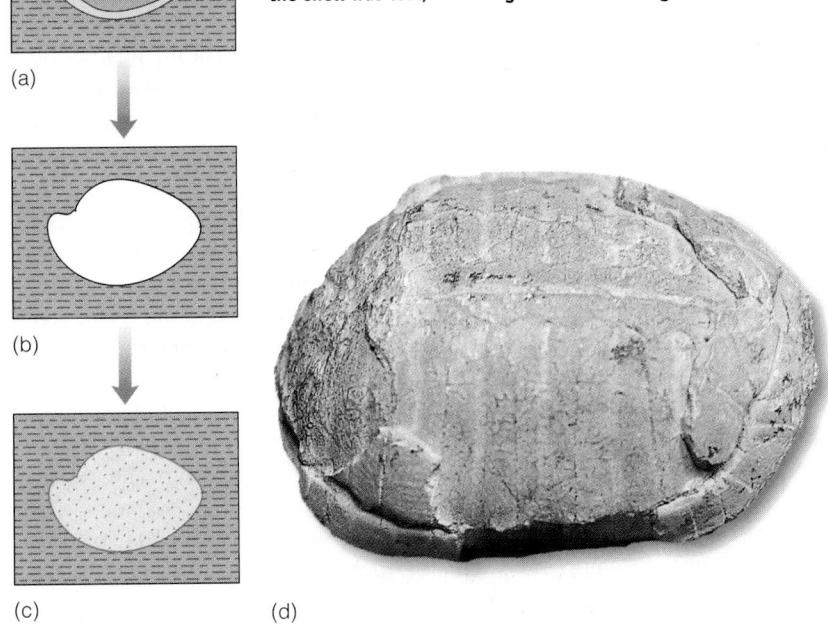

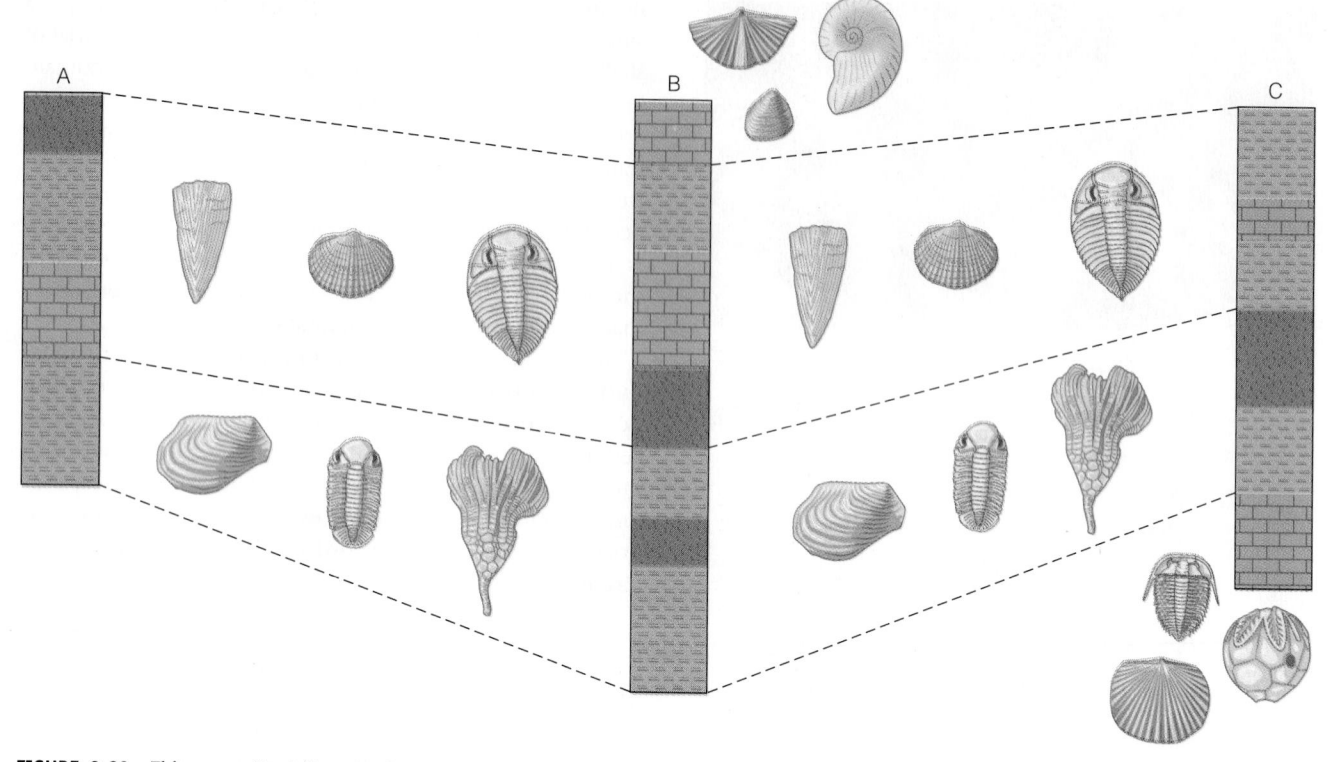

FIGURE 3.23 This generalized diagram shows how geologists use the principle of fossil succession to identify strata of the same age in different areas. The rocks in the three sections encompassed by the dashed lines contain similar fossils and are thus the same age. Note that the youngest rocks in this region are in section B, whereas the oldest rocks are in section C.

the sequence are younger. He made observations in many natural exposures, mines, and quarries and discovered that the relative sequence of fossils, and particularly groups or assemblages of fossils, is consistent from area to area. In short, he discovered a method whereby the relative ages of strata in widely separated areas regardless of their composition could be determined by their fossil content (Figure 3.23).

By recognizing the relationship between strata and fossils, Smith was able to predict the rock units that would be encountered in digging a canal. This knowledge also helped him determine the best route for a canal and the best areas for bridge foundations. His observations proved to be of economic significance, and his services were in great demand.

The discoveries of Smith, and those of other geologists such as Alexandre Brongniart in France, served as the basis for what is now called the **principle of fossil succession.** According to this principle, fossil assemblages succeed one another through time in a regular and determinable order. This is an important principle because, though superposition can be used to determine relative ages in a single locality, the relative ages of rocks in distant areas cannot be deduced by using this principle. Nor can rock type be used for relative ages because the same type of rock—sandstone, for instance—has formed repeatedly through time. Fossils have also formed continuously, but because organisms differ through time fossil assemblages

are unique. In other words, a group of fossil organisms has a distinctive aspect compared with those younger or older. Accordingly, if rocks containing similar fossils are matched up, we can assume that they are of the same relative age.

The Relative Geologic Time Scale

Recall from Chapter 2 that many eighteenth-century geologists, such as the Neptunists, believed that the relative ages of rocks could be determined by their composition. This same thinking prevailed into the nineteenth century, but eventually it became apparent that composition is not a sufficient criterion for determining relative ages, and geologists in Great Britain and continental Europe began using stratigraphic position (superposition) and fossil content (fossil succession) for their subdivisions. Their efforts, largely between 1830 and 1842, resulted in the recognition of rock bodies called *systems* and the construction of a geologic column that is the basis for the relative geologic time scale (Figure 3.24). A short discussion of how some of the systems were recognized and defined will be helpful in understanding how the geologic column and relative geologic time scale became established.

In the 1830s, Adam Sedgwick described and named the Cambrian System for rocks exposed in northern Wales, while Sir Roderick Impey Murchison, working in southern

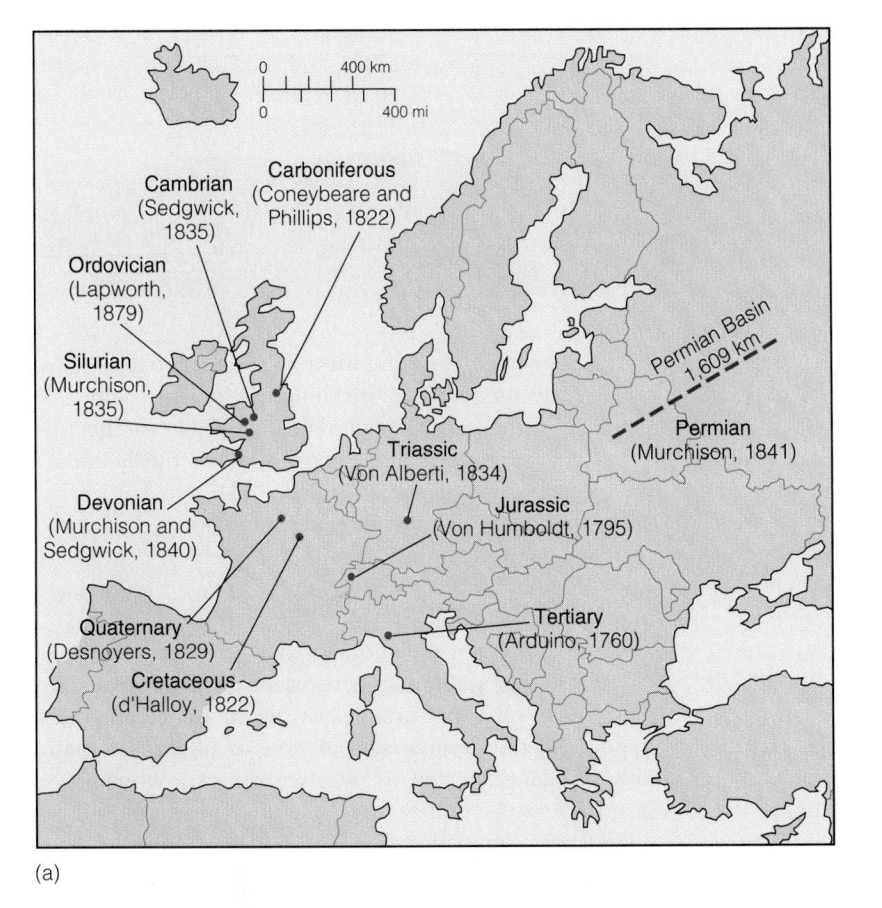

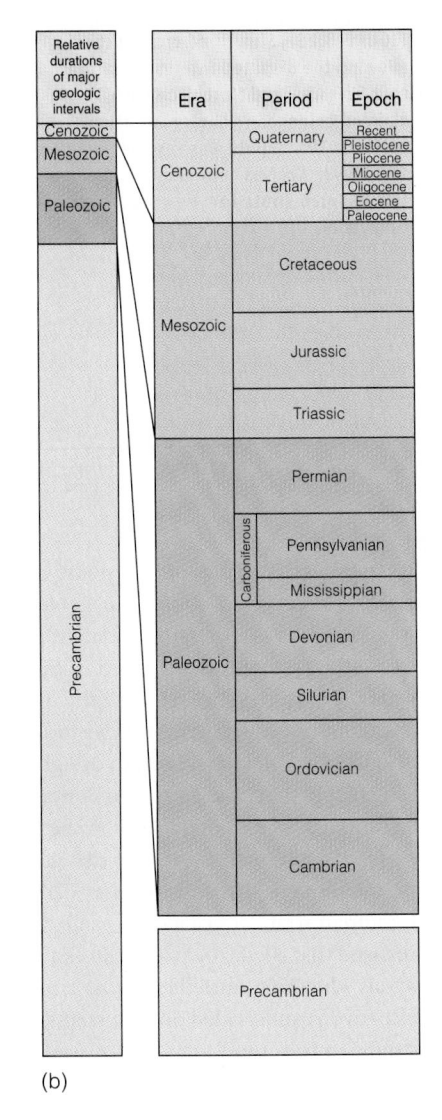

FIGURE 3.24 Development of the geologic column and relative geologic time scale. (a) Geologists in Great Britain and continental Europe defined the geologic systems at the locations shown. (b) A geologic column was constructed by arranging the systems in their correct relative sequence, so the geologic column is in effect a relative geologic time scale.

Wales, named the Silurian System. Unfortunately, Sedgwick's strata contain few fossils, and he paid little attention to the few present. His Cambrian System based on rock type could not be recognized beyond the area where it was described. Murchison, on the other hand, carefully described fossils characteristic of his Silurian System, so strata of Silurian age could be recognized elsewhere.

In 1835 Sedgwick and Murchison jointly published the results of their studies. Stratigraphic position clearly demonstrated that Silurian strata were younger than Cambrian strata, but it was also apparent that Sedgwick's and Murchison's systems partially overlapped (Figure 3.25). Each man claimed that the strata in the overlapping interval belonged to his system. This boundary dispute resulted in a feud that ended their long-standing friendship and was not resolved until 1879, when Charles Lapworth suggested that the disputed strata be assigned to a new system, the Ordovician. In the final analysis, three systems were named,

their stratigraphic positions were known, and distinctive fossils of each system were recognized.

Before Sedgwick's and Murchison's feud, they named the Devonian System for rocks exposed near Devonshire, England (Figure 3.24). Rocks defined as Devonian contained fossils that differed from those in underlying Silurian strata and were overlain by rocks of the Carboniferous System,* which had been described earlier. In 1840–1841, Murchison visited western Russia, where he identified strata as Silurian, Devonian, and Carboniferous by their fossils. Overlying the Carboniferous strata were fossil-bearing rocks that he assigned to a Permian System (Figure 3.24).

We need not discuss the specifics of where and when the other systems were defined, except to note that superposition and fossil content were the criteria used in their

---

*In North America two systems, the Mississippian and the Pennsylvanian, are recognized and correspond to the Lower and Upper Carboniferous, respectively.

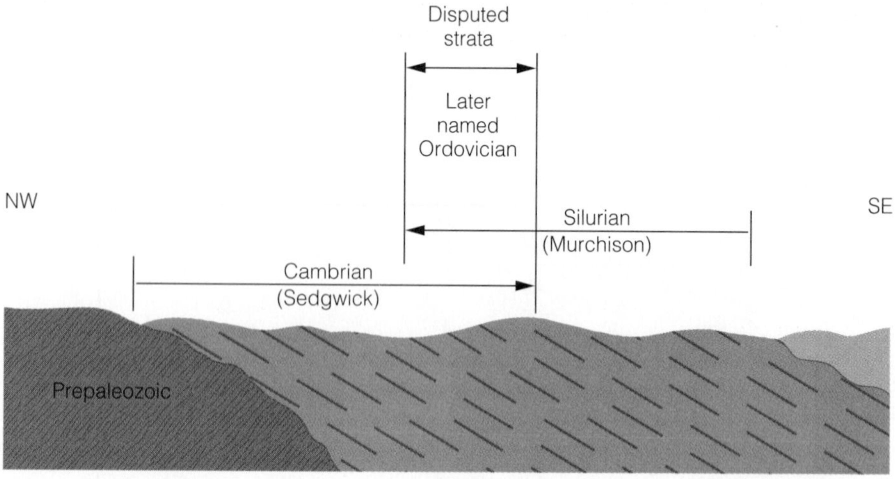

FIGURE 3.25 Simplified cross section showing the disputed interval of strata that Sedgwick and Murchison each claimed belonged to their respective systems. The dispute was resolved in 1879 when Charles Lapworth assigned the disputed strata to a new system, the Ordovician.

recognition. The important point is that geologists were piecing together a composite geologic column. This geologic column is in effect a relative geologic time scale because the systems are arranged in their correct chronologic order. Thus, we can refer to the Devonian System when speaking of the stratigraphic position of rocks, and we can also use Devonian as a term to designate a particular interval of geologic time, the Devonian Period. We will have more to say about systems and periods in the following section.

One additional aspect of the geologic column should be mentioned. Notice in Figure 3.24 that all rocks beneath Cambrian strata are called Precambrian. Long ago geologists realized that these rocks contain few fossils (none were positively identified until the 1950s) and that they could not be effectively subdivided into systems. Terminology for Precambrian rocks and time is discussed in Chapters 8 and 9.

Stratigraphic Terminology

The recognition of systems and a relative geologic time scale brought some order to stratigraphy. Problems re-

mained, however, because many rock units are time transgressive. Consequently, a rock unit may belong to one system in a particular area but also be included in another system elsewhere, or it may simply straddle the boundary between systems (Figure 3.26). To deal with both rocks and time, modern stratigraphic terminology includes two fundamentally different kinds of units: those defined by their content and those expressing or related to geologic time (Table 3.1).

Units defined by their content include lithostratigraphic and biostratigraphic units. **Lithostratigraphic* units** are defined by physical attributes of the rocks, such as rock type, with no consideration of time of origin. The basic lithostratigraphic unit is the **formation,** which is a mappable rock unit with distinctive upper and lower boundaries. Formations may be lumped together into larger units called *groups* or *supergroups* or divided into smaller units called *members* and *beds* (Figure 3.27 and Table 3.1).

Fossil content is the only criterion used to define **biostratigraphic units,** which are bodies of strata containing recognizably distinct fossils. Biostratigraphic unit boundaries do not necessarily correspond to lithostratigraphic boundaries. The fundamental biostratigraphic unit is the **biozone.** Several types of biozones are recognized, three of which are discussed in the section on correlation.

The category of units expressing or related to geologic time includes time-stratigraphic units (also known as chronostratigraphic units) and time units (Table 3.1). **Time-stratigraphic units** are units of rock that were deposited during a specific interval of time. The **system** is the fundamental time-stratigraphic unit. It is based on rocks in a particular area, the *stratotype,* and is recognized beyond the stratotype area primarily on the basis of fossil content.

Time units are simply units designating specific intervals of geologic time. For example, the Cambrian Period is defined as the time during which strata of the Cam-

FIGURE 3.26 The Chattanooga Shale, shown here in Tennessee, straddles the boundary between the Devonian and Mississippian Systems.

_____
**Lith-* and *litho-* are prefixes meaning "stone" or "stonelike."

TABLE 3.1

Classification of Stratigraphic Units

UNITS DEFINED BY CONTENT		UNITS EXPRESSING OR RELATED TO GEOLOGIC TIME	
Lithostratigraphic Units	*Biostratigraphic Units*	*Time-Stratigraphic Units*	*Time Units*
Supergroup	Biozones	Eonothem............................... Eon	
Group		Erathem .... Era	
Formation		System.................................... Period	
Member		Series...................................... Epoch	
Bed		Stage..................................... Age	

brian System were deposited. The basic time unit is the **period,** but smaller units including epoch and age are also recognized. The time units period, epoch, and age correspond to the time-stratigraphic units system, series, and stage, respectively (Table 3.1). Time units of higher rank than period also exist. Eras include several periods, whereas eons include two or more eras. The time-stratigraphic terms corresponding to era and eon are rarely used (Table 3.1).

Time-stratigraphic units and time units and their relationships are particularly confusing to students. Time-stratigraphic units are defined by their position in a sequence, so terms such as lower, medial, and upper may be applied. Thus, we may refer to Upper Devonian strata, meaning that these strata occur in the upper part of the Devonian System. In contrast, time units are simply subdivisions of geologic time and may be designated as early, middle, and late; the Late Devonian Period was the time

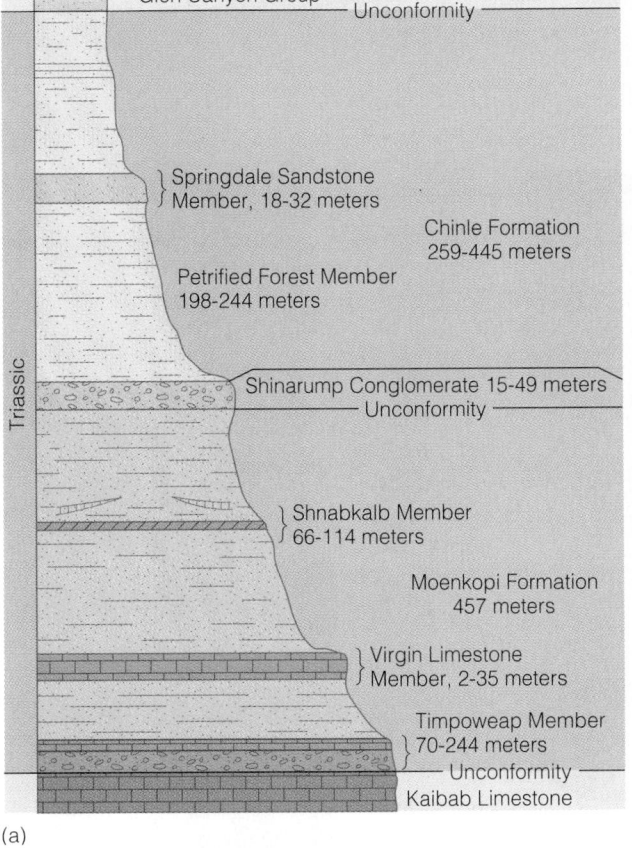

(a)

(b)

FIGURE 3.27 Lithostratigraphic units. (a) Illustration of the formations and members of the Glen Canyon Group in the Zion National Park, Utah. (b) View of three formations in the Grand Canyon, Arizona. The Tapeats Sandstone (lower cliff) is overlain by the Bright Angel Shale (forming the slope), and Muav Limestone (upper cliff). Each formation is easily recognized by its composition and position in the sequence. These Cambrian-aged formations were deposited during a widespread marine transgression.

during which rocks of the Upper Devonian System was deposited.

Correlation

If we could not demonstrate the equivalency of rock units in different areas, stratigraphic studies would be limited in scope. **Correlation** is the process of demonstrating equivalency. Correlation of lithostratigraphic units is the recognition of units of similar lithology (composition and texture) and stratigraphic position over broad geographic areas. If exposures are adequate, units may simply be traced laterally, even if occasional gaps exist (Figure 3.28). Other criteria used to correlate are lithologic simi-

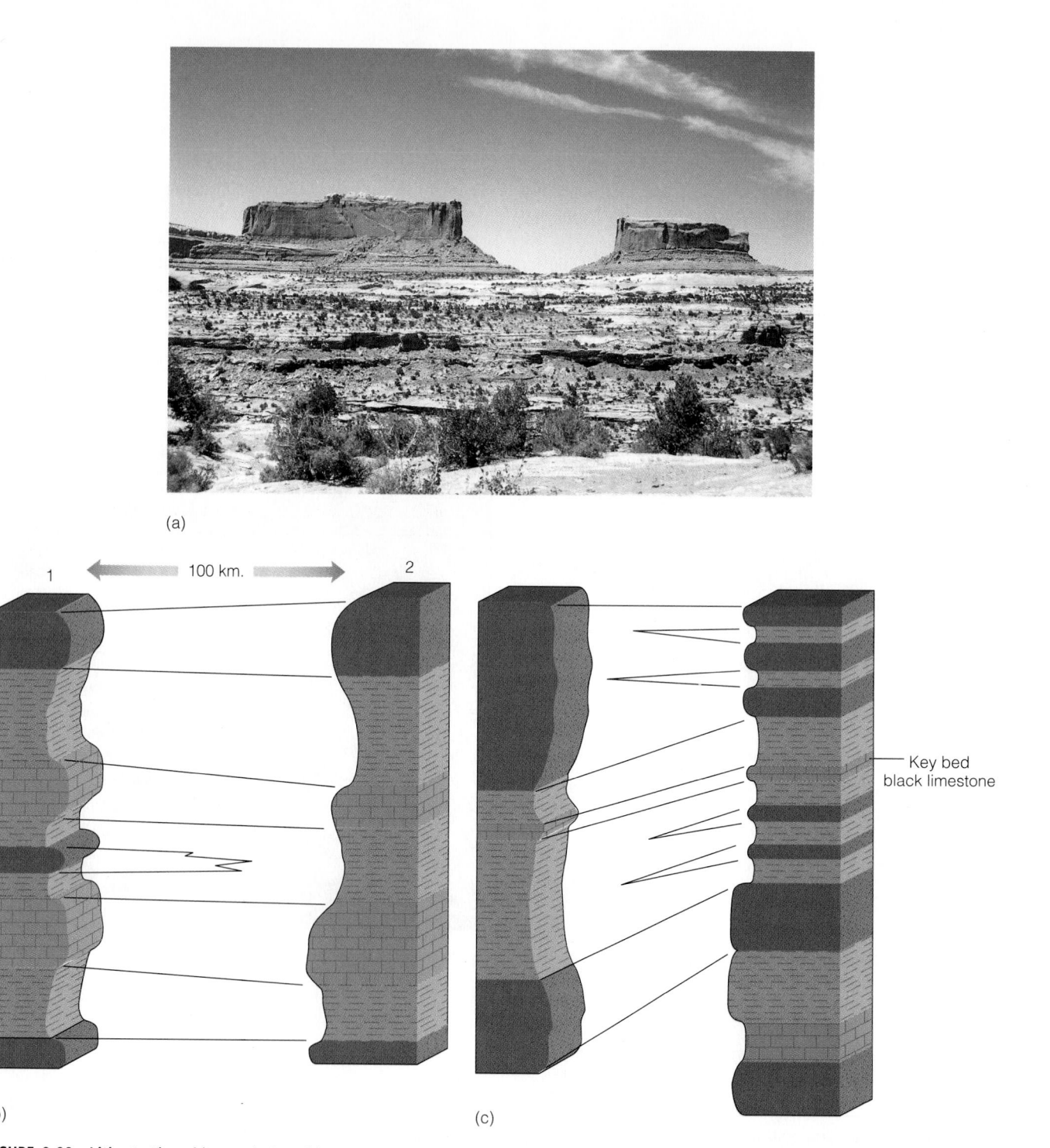

FIGURE 3.28 Lithostratigraphic correlation. (a) In areas of adequate exposures, rock units can be traced laterally even if occasional gaps exist. These rocks on the skyline in Utah were originally part of a continuous layer. (b) Correlation by similarities in rock type and position in a sequence. The sandstone in column 1 is assumed to intertongue, or grade laterally, into the shale at column 2. (c) Correlation using a key bed, a distinctive black limestone in this case.

larity, position in a sequence, and *key beds,* which are beds sufficiently distinctive to be identified in different areas. Geologists correlate rocks below the surface by using rock cores and well cuttings from drilling operations and geophysical information derived by lowering instruments down drill holes.

Lithostratigraphic correlation demonstrates the geographic extent of rock units, but it does not imply time equivalence. To demonstrate that strata in different areas are the same age involves time-stratigraphic correlation. Because most rock units are time transgressive, we usually cannot use rock type in this kind of correlation. In most cases, biozones are used, although several other methods are useful as well.

One type of biozone, the **range zone,** is defined by the total geologic range of a particular fossil group (a species or a group of related species, called a *genus,* for example) (Figure 3.29). Fossils that are easily identified, are geographically widespread, and have rather short geologic ranges are particularly useful. The brachiopod genus *Lingula* is easily identified and widespread, but its geologic range of Ordovician to Recent makes it of little use in biostratigraphy. In contrast, the trilobite *Paradoxides* and the brachiopod *Atrypa* meet all of these criteria and are therefore good **guide fossils** (Figure 3.30). **Concurrent range zones** are established by plotting the overlapping ranges of fossils that have different geologic ranges. The first and last occurrences of fossils are used to establish zone boundaries (Figure 3.31).

Correlation of range zones or concurrent range zones generally yields correlation lines that are considered time

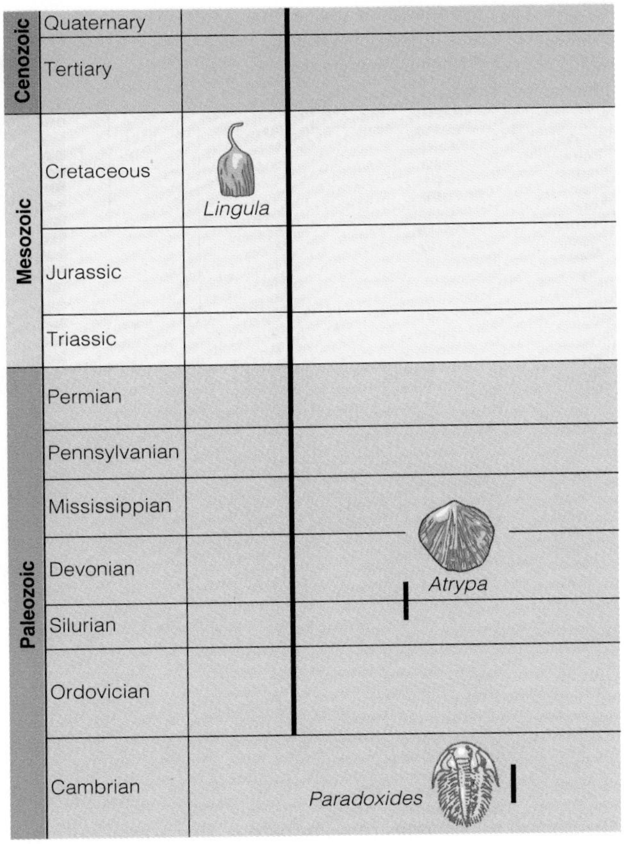

FIGURE 3.30 Comparison of the times of existence (geologic ranges) of three marine invertebrate animals. Geologic ranges are shown by heavy vertical lines. *Lingula* is of little use in biostratigraphy because it has a long geologic range. *Atrypa* and *Paradoxides* are good guide fossils because they are widespread, easily identified, and have short geologic ranges.

equivalent. In other words, the strata encompassed by the correlation lines so established are thought to be the same age. Geologists are aware, however, that zones are not of precisely the same age everywhere, because no fossil organism appeared and disappeared simultaneously over its entire geographic range. Even so, correlation of range zones can yield an error no greater than the range of the fossil used, and the use of concurrent range zones yields even more precise correlations.

Time equivalence can also be based on some physical events of short duration. Ash falls (Figure 3.32), for example, may cover large areas, and they occur instantaneously when considered in the context of geologic time. Furthermore, ash falls are not restricted to a particular depositional environment, so they may be continuous from continental into marine depositional environments.

Quantifying the Relative Geologic Time Scale

Several of the correlation techniques discussed here demonstrate time equivalence between sections, but the ages determined are relative only. During the 1840s,

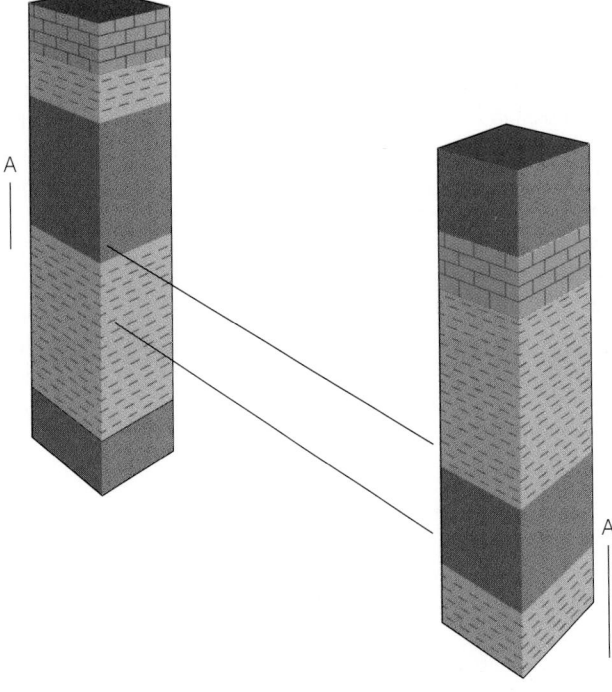

FIGURE 3.29 Correlation between two sections based on the range zone of fossil A.

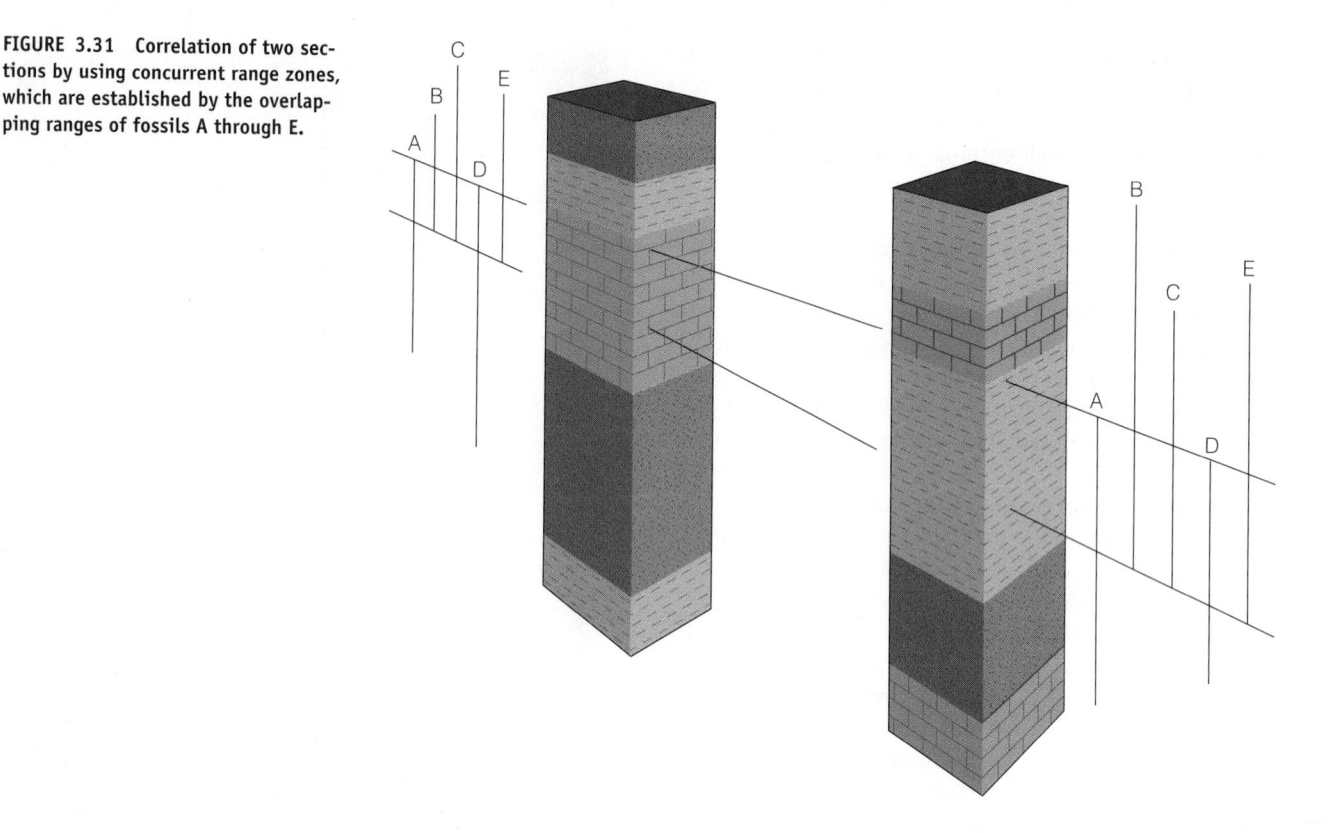

FIGURE 3.31 Correlation of two sections by using concurrent range zones, which are established by the overlapping ranges of fossils A through E.

Murchison used fossils to demonstrate that strata in Russia, Germany, and England belong to the Permian System and are therefore of the same age. Furthermore, the relative age of Permian rocks was known with respect to Carboniferous and Triassic strata, but no one knew when the Permian began or how long it lasted.

A number of minerals in sedimentary rocks can be dated radiometrically, but the ages tell only the age of the source rock that supplied the minerals to the deposit. All this tells us is that the sedimentary rock and its fossils are younger than the age determined. Glauconite, a mineral that forms in some sediments shortly after deposition, can be dated by the potassium–argon method (see Chapter 2). Unfortunately, glauconite easily loses argon, so absolute ages determined by this method must be considered minimum ages. In most cases, absolute ages of sedimentary rocks must be determined indirectly by dating associated igneous and metamorphic rocks.

According to the principle of cross-cutting relationships, an intrusive igneous body is younger than the rock it intrudes. Therefore, an accurately dated intrusive igneous body provides a minimum age for the sedimentary rock it intrudes and a maximum age for the sedimentary rocks un-

conformably overlying it (Figure 3.33). Dating of regionally metamorphosed rocks give a maximum age for any overlying sedimentary rocks (Figure 3.33).

One of the best ways to get good radiometric dates in a sedimentary sequence is by using interbedded volcanic rocks. An ash fall or lava flow provides an excellent marker bed that is a time-equivalent surface, providing a minimum age for the sedimentary rocks below and a maximum age for the rocks above. Ash falls are particularly useful because they may fall over both marine and nonmarine sedimentary environments and can provide a connection between these different environments. Multiple ash falls, lava flows, or a combination of both in a stratigraphic sequence are particularly useful in determining absolute ages of sedimentary rocks and their contained fossils.

Thousands of absolute ages are now known for sedimentary rocks of known relative ages, and these absolute dates have been added to the relative geologic time scale. In this way, geologists have been able to determine the absolute ages of the various systems and to determine their durations.

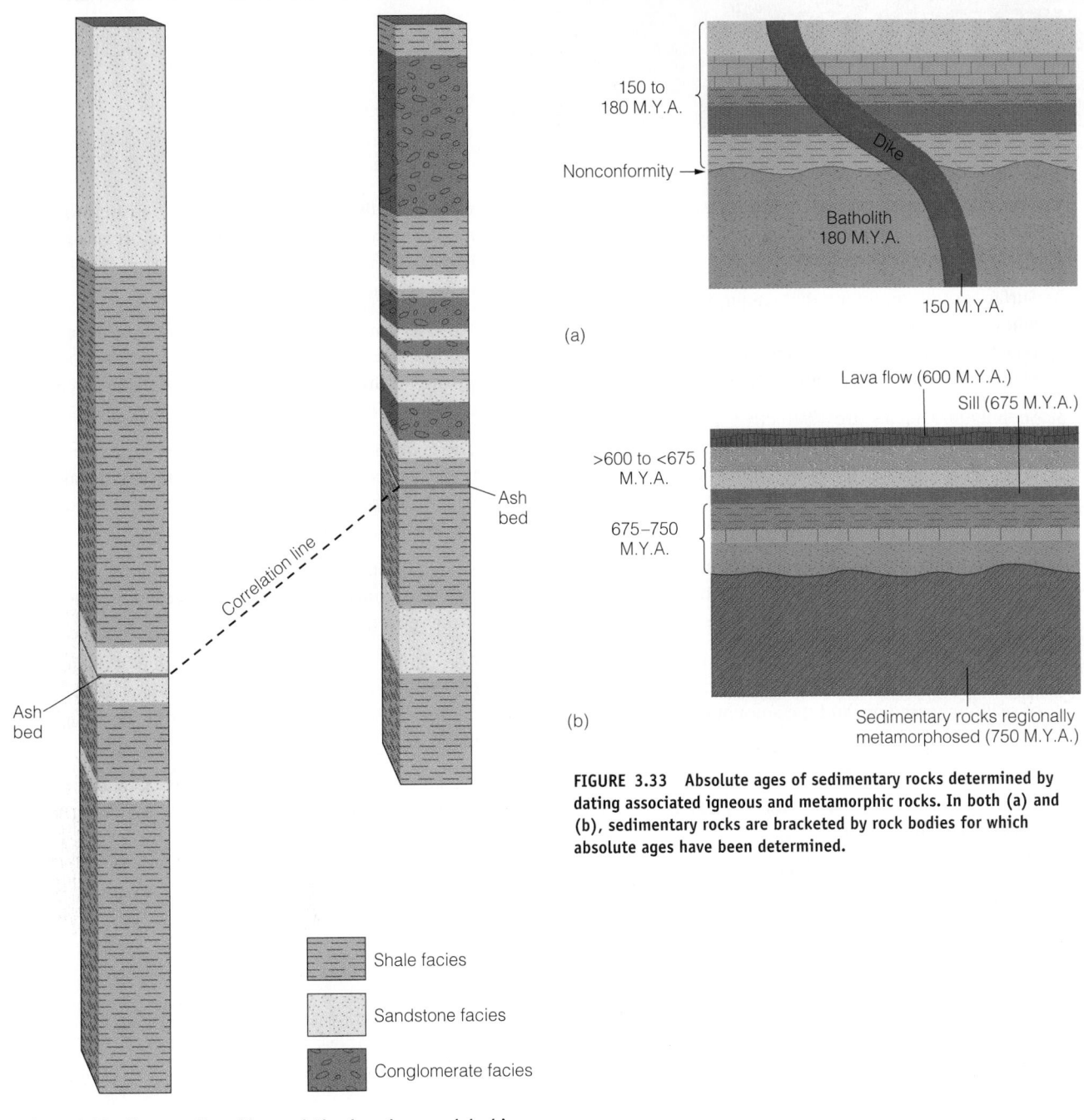

Correlation line

Ash bed

Ash bed

150 to 180 M.Y.A.

Nonconformity

Dike

Batholith 180 M.Y.A.

150 M.Y.A.

(a)

Lava flow (600 M.Y.A.)

Sill (675 M.Y.A.)

>600 to <675 M.Y.A.

675–750 M.Y.A.

(b)

Sedimentary rocks regionally metamorphosed (750 M.Y.A.)

FIGURE 3.33 Absolute ages of sedimentary rocks determined by dating associated igneous and metamorphic rocks. In both (a) and (b), sedimentary rocks are bracketed by rock bodies for which absolute ages have been determined.

Shale facies

Sandstone facies

Conglomerate facies

FIGURE 3.32 Time-stratigraphic correlation by using an ash bed in Miocene deposits in Montana.

Summary

1. The relative ages of rock units in any sequence of strata can be determined by superposition even if the strata are complexly deformed.

2. Surfaces of discontinuity that encompassed significant amounts of geologic time are common in the geologic record. Such surfaces are unconformities and result from times of nondeposition, erosion, or both.

3. Sedimentary facies are distinctive sedimentary rock units resulting from deposition in adjacent but different environments. The same facies that replace one another laterally may be superposed in a vertical sequence as a result of marine transgression or regression.

4. In a conformable sequence of facies, Walther's law can be applied to determine what the lateral facies relationships were at the time of deposition.

5. The causes of marine transgressions and regressions include uplift and subsidence of the continents, rates of seafloor spreading, and the amount of seawater contained in glaciers.

6. Most fossils are preserved in sedimentary rocks as unaltered remains, altered remains, molds and casts, or traces of organic activity.

7. Although fossils of some organisms are quite common, very few of the organisms that lived at any one time were actually preserved as fossils. Furthermore, the fossil record is biased toward those organisms with hard skeletal elements and those organisms that lived in areas of active sedimentation.

8. Fossil assemblages succeed one another through time in a predictable sequence. William Smith's work with fossils in Great Britain is the basis for what is now called the principle of fossil succession.

9. The geologic column is a composite sequence of rock bodies called systems arranged in chronological order. Superposition and fossil succession were used to determine the relative ages of the systems.

10. Stratigraphic terminology includes two fundamentally different kinds of units: those based on content and those related to geologic time.

11. Correlation is the stratigraphic practice of demonstrating equivalency of units in different areas. Demonstrating time equivalence is most commonly done by correlating strata with similar fossils.

12. Most absolute ages of sedimentary rocks and their contained fossils are obtained indirectly by dating associated igneous or metamorphic rocks.

Important Terms

angular unconformity
biostratigraphic unit
biozone
body fossil
cast
concurrent range zone
conformable
coprolite
correlation

disconformity
formation
fossil
geologic record
guide fossil
lithostratigraphic unit
marine regression
marine transgression
mold

nonconformity
period
principle of fossil
   succession
principle of inclusions
range zone
sedimentary facies
stratigraphy
system

time-stratigraphic unit
time transgressive
time unit
trace fossil
unconformity
Walther's law

Review Questions

1. According to Walther's law,
   a. _____ the relative ages of rocks can be determined by radioactive decay.
   b. _____ fossils of marine invertebrates are more common than those of land-dwelling vertebrates.
   c. _____ the facies in a conformable vertical sequence of sedimentary rocks will replace one another laterally.
   d. _____ an unconformity is present between Paleozoic rocks and Mesozoic rocks.
   e. _____ the formation is the basic time-stratigraphic unit.

2. Which one of the following is a trace fossil?
   a. _____ Roman coin
   b. _____ worm burrow
   c. _____ cast of a clam shell
   d. _____ dinosaur tooth
   e. _____ elephant tusk

3. If sedimentary rocks are cut by a dike 225 million years old, they must be _____ million years old:
   a. _____ about 400
   b. _____ more than 225
   c. _____ between 200 and 350
   d. _____ younger than 150
   e. _____ no more than 65

4. The process whereby rocks of the same relative age are matched up from area to area is known as:
   a. _____ correlation.
   b. _____ disconformity.
   c. _____ stratigraphic inference.
   d. _____ concurrent dating.
   e. _____ Steno's principle.

5. The superposition of offshore facies over nearshore facies can be accounted for by a(an):
   a. _____ marine transgression.
   b. _____ angular unconformity.
   c. _____ range zone.
   d. _____ lithostratigraphic inclusion.
   e. _____ biostratigraphic upheaval.

6. A rock unit known as a formation:
   a. _____ is defined by its geologic age.
   b. _____ contains fossils useful for correlation.
   c. _____ is composed of alternating layers of volcanic ash and granite.
   d. _____ must be bounded above and below by unconformities.
   e. _____ is the basic lithostratigraphic unit.

7. Superposition and fossil succession were used to establish:
   a. _____ absolute ages for sedimentary rocks lacking fossils.
   b. _____ the relative geologic time scale.
   c. _____ the degree to which rocks have been altered by metamorphism.
   d. _____ relative ages of buried lava flows and sills.
   e. _____ the duration of the Precambrian.

8. Which one of the following might cause a marine regression?
   a. _____ subsidence of a continent
   b. _____ rapid seafloor spreading
   c. _____ rising sea level
   d. _____ widespread glaciation
   e. _____ extensive volcanism

9. The addition of mineral matter to the pores and cavities in a bone or shell is known as:
   a. _____ cementation.
   b. _____ carbonization.
   c. _____ recrystallization.

d. _____ silicification.

e. _____ permineralization.

10. Which one of the following statements is correct?

    a. _____ Most fossils are found in ash falls and lava flows.

    b. _____ Biostratigraphic zone boundaries do not necessarily correspond to lithostratigraphic boundaries.

    c. _____ Rocks deposited during a marine transgression are the same age over their entire geographic extent.

    d. _____ A period consists of two or more eras.

    e. _____ Good guide fossils are those of organisms that existed during at least several geologic periods.

11. Which one of the following is a time unit?

    a. _____ Cambrian System

    b. _____ Lower Cretaceous

    c. _____ Eocene Series

    d. _____ Upper Silurian

    e. _____ Pennsylvanian Period

12. A vertical sequence of sedimentary rocks containing no depositional breaks of any consequence is said to be:

    a. _____ deformed.

    b. _____ time transgressive.

    c. _____ unaltered.

    d. _____ conformable.

    e. _____ regressive.

13. Explain how geologists can determine whether volcanic rock interlayered with sedimentary rocks is a buried lava flow or a sill.

14. The principles of superposition and fossil succession were used to establish the geologic column. Define both principles and explain how they were used to determine the relative ages of the systems.

15. Explain how the principle of uniformitarianism is used to understand fossilization and the origin of sedimentary rocks.

16. In a vertical sequence of strata, the following are observed: lava flow (absolute age 255 million years) followed upward by sandstone, shale, and an ash fall (absolute age 240 million years). What is the approximate absolute age of the sedimentary rocks? Explain how you found your answer.

17. What are the basic time-stratigraphic units and their corresponding time units? Why did geologists find it necessary to establish these two kinds of units?

18. Describe two ways in which the relative ages of sedimentary rocks in different areas can be determined.

19. What sequence of events can lead to the formation of an angular unconformity?

20. What features of fossils are necessary to make them good guide fossils? Of what use are guide fossils?

21. Describe a concurrent range zone and explain how it can be used in correlation.

22. A conformable vertical sequence of facies consists of sandstone followed upward by shale and, finally, limestone. All facies contain fossil clams, oysters, and corals. Diagram the lateral facies relationships and show which facies was deposited nearest the shoreline.

23. Compare the processes of permineralization, replacement, and recrystallization of fossils.

24. Discuss the probable causes of marine transgressions and regressions.

Points to Ponder

1. Discuss several processes operating at present that might lead to the preservation of fossils. How can information about these present-day processes be used to help interpret how fossils were preserved during the past?

2. If you were to visit another planet with fossil-bearing sedimentary rocks, how would you go about establishing a geologic column and relative geologic time scale?

3. While visiting one of the national parks, you observe the following vertical sequence of rock layers: marine-fossil-bearing sandstone overlain by shale and limestone, all of which are inclined 40 degrees from horizontal; lying over these rocks is a horizontal layer of shale baked by an overlying basalt and sandstone. Decipher the sequence of events that took place to yield this rock succession.

4. How is it possible for geologists to determine that dinosaurs became extinct 66 million years ago in view of the fact that the sedimentary rocks containing dinosaur fossils can rarely be radiometrically dated?

Additional Readings

Boggs, S., Jr. 1995. *Principles of sedimentology and stratigraphy.* 2d ed. Columbus, Ohio: Merrill.

Donovan, S. K., ed. 1991. *The processes of fossilization.* New York: Columbia University Press.

Fortey, R. 1991. *Fossils: The key to the past.* Cambridge, Mass.: Harvard University Press.

Friedman, G. M., J. E. Sanders, and D. C. Kopaska-Merkel. 1992. *Principles of sedimentary deposits.* New York: Macmillan.

Grimaldi, D. A. 1996. Captured in amber. *Scientific American* 274, no.4: 84–91.

Mayr, H. 1992. *A guide to fossils.* Princeton, N.J.: Princeton University Press.

Pinna, G. 1990. *Illustrated encyclopedia of fossils.* New York: Facts on File.

Prothero, D. R. 1990. *Interpreting the stratigraphic record.* New York: Freeman.

Prothero, D. R., and F. Schwab. 1996. *Sedimentary geology.* New York: Freeman.

World Wide Web Activities

▶ MOHAWK VALLEY FOSSILS

This site is maintained by the Union College Geology Department, Schenectady, New York. It has information, maps, and diagrams depicting the rocks exposed in the Mohawk River Valley. Click on the box labeled *lithology*.

1. What is meant by "time corrected" in the lithologic cross-section of the rocks in the Mohawk River Valley?

2. What is the geologic age of these rocks?

3. Where was the source area for the detrital sediments?

4. What kind of evidence indicates that the shale was deposited where little or no oxygen was present?

5. Do the rocks here show facies changes? Explain.

6. How can the bentonite beds be used to demonstrate that rocks in different areas are the same age?

▶ THE FOSSIL COMPANY—ABOUT FOSSILS

Click on any of the fossil icons to see images and learn about fossil mammals, starfish, ammonites, plants, reptiles, and many others.

▶ ASHFALL FOSSIL BEDS STATE HISTORICAL PARK

The Nebraska Game and Parks Commission maintains this site, which tells the fascinating story of the mass death of a variety of mammals and other animals 10 million years ago in Nebraska.

1. What are some of the common fossil mammals at this park?

2. How were these animals killed and buried?

3. What kinds of evidence indicates that predators were present at this site following the deaths of the fossil mammals, even though no fossil predators have been found so far?

▶ FOSSILS—WINDOWS TO THE PAST

At this site, fossils are defined, the conditions for their preservation are discussed, and the significance of fossils is explained. See the images and discussions of permineralization, casts and molds, freezing, drying and desiccation, and preservation in amber.

▶ STRATIGRAPHY OF THE PALEOZOIC

This is one of many sites maintained by the University of California Museum of Paleontology. It also has sites on Mesozoic and Cenozoic stratigraphy. Click on *Pennsylvanian*.

1. When was the Pennsylvanian?

2. What term is used in Europe for the Mississippian and Pennsylvania? Why is it so named?

Click on the *Cambrian,* and then click on *stratigraphy* and read about how stratigraphic boundaries are established.

▶ FOSSILS, ROCKS, AND TIME

The United States Geological Survey maintains this site that deals with deciphering Earth history. Go to the *Table of Contents* and see the section on *Rocks and Layers* where images and discussions of basic principles are found. Also see the section on *Putting Events in Order,* which also has examples and images. See the sections on *The Relative Time Scale* and *Numeric Time Scale* and discover why a distinction is made between the two.

Explore the following *In-Terra-Active 2.0* CD-ROM module(s) and increase your understanding of key concepts and processes presented in this chapter.

Chapter Concept: **Stratigraphic Relationships**

The relative geologic time scale is the product of early geologists piecing together the rock record on the basis of stratigraphic relationships between different strata and compiling this record worldwide. In this module, you will explore the different processes that produce the rock record and apply your knowledge to interpreting the relative geologic history of two landscape cross-sections.

Section: **Materials**

Module: **Geologic Time: Relative**

How old is old? One of geology's greatest benefits is that it answers this question by showing us just how deep-reaching Earth's history really is. In relative time, the immediate goal is not dating an event or object in years, but understanding its place in a geologic sequence. We get our best information from outcrops and other rock records. Study the photo.

Questions: *Which part of the rock in this picture is oldest? Which is youngest?*

4

Origin and Interpretation of Sedimentary Rocks

Cambrian-aged sedimentary rocks of the Munising Formation exposed along the shore of Lake Superior at Pictured Rocks National Lakeshore, Michigan. Deposition took place when a shallow sea was present in this area.

About 50 million years ago, parts of present-day Wyoming, Utah, and Colorado were covered by two large lakes. Sand, mud, and dissolved minerals were carried into these lakes where they were deposited and subsequently lithified to form what we now call the Green River Formation. These rocks contain the fossilized remains of fish, insects, and plants and are also a source of large quantities of oil, combustible gases, and other resources.

Thousands of fossil fish are found on single surfaces within the Green River Formation, indicating that mass mortality must have taken place repeatedly (Figure 4.1a). What caused these mass deaths is not known with certainty, but some geologists think that blooms of blue-green algae produced toxic substances that killed the fish. Others propose that rapidly changing water temperature or excessive salinity at times of increased evaporation was responsible. Whatever the cause, fish died by the thousands, on numerous occasions, and settled to the lake bottom where decomposition was inhibited because the water contained little or no oxygen.

One area of the formation in Wyoming where fossil plants are particularly abundant has been designated as Fossil Butte National Monument. Fossils of palms and ferns indicate a warm, humid climate where ancestors of horses, rhinoceroses, and elephants lived. Although plant fossils are especially common at Fossil Butte, the remains of numerous animals are also present, including exquisitely preserved remains of various fish (Figure 4.1b).

The Green River Formation is also well known for its huge deposits of oil shale and for evaporite rocks of economic importance. Oil shale (Figure 4.2) consists of small clay particles, carbonate minerals, and an organic substance known as *kerogen*. When the appropriate processes are used, liquid oil and combustible gases can be extracted from the kerogen of oil shale. To be considered a true oil shale, though, the rock must yield a minimum of 10 gallons of oil per ton or rock.

The use of oil shale as a source of fuel is not new. During the Middle Ages, people in Europe used oil shale as solid fuel for domestic purposes, and during the 1850s, small oil shale industries existed in the eastern United States but were discontinued when drilling and pumping of oil began in 1859. Oil shale is found on all continents, but the Green River Formation contains the most extensive deposits. According to one estimate, 80

(a)

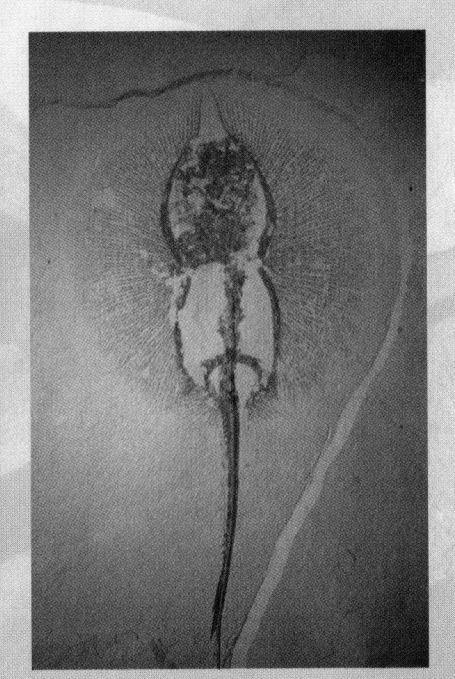

(b)

FIGURE 4.1 Fossil fish from the Green River Formation. (a) This single surface is covered with numerous fish fossils, indicating a mass mortality. (b) Well-preserved stingray. This is a remarkable fossil because rays have a skeleton of cartilage, so their remains are rarely found in the fossil record.

FIGURE 4.2 Layers of oil shale of the Green River Formation are exposed along these hillsides. These rocks were deposited in lakes about 50 million years ago.

billion barrels of oil could be recovered from the Green River Formation, with existing technology. Currently, no oil is produced from oil shale in the formation because it would be more expensive than oil recovered by conventional drilling and pumping.

To produce oil from oil shale, the rock is heated to nearly 500° C in the absence of oxygen, and hydrocarbons are driven off as gases and recovered by condensation. During this process, 25 to 75% of the organic matter of oil shale can be converted to oil and combustible gases. For each ton of rock processed, anywhere from 10 to 140 gallons of oil can be recovered.

Another rock of interest in the Green River Formation is the evaporite composed of the sodium carbonate mineral *trona*, which is mined as a source of sodium compounds. The trona and associated evaporites such as rock salt are interpreted as deposits that accumulated in lakes during times of intense evaporation.

Introduction

In Chapter 3 we noted that sedimentary rocks have a special place in historical geology because (1) they preserve evidence of the surface processes responsible for the deposition of sediment and (2) many contain fossils. Weathering of any kind of rock yields solid particles known as detritus and dissolved minerals, thus producing the raw materials for sedimentary rocks (Figure 4.3). Detrital sediment includes gravel, sand, silt, and clay, which simply are size designations, whereas chemical sediment forms from minerals extracted from solution. Any sediment can be acted upon by the processes of lithification (compaction and cementation) and be converted to either detrital sedimentary rock or chemical sedimentary rock. A subcategory of chemical sedimentary rocks called *biochemical* is also recognized and draws attention to the fact that organisms are important in their origin. Many varieties of limestone, for instance, are biochemical.

In this chapter our main objective is to investigate the processes leading to the accumulation of sedimentary deposits. Sedimentary rocks in the geologic record acquired their various characteristics partly as a consequence of the physical, chemical, and biological processes that operated in the original depositional environment. Accordingly, our task is to determine what the original environment was by examining and interpreting the properties of sedimentary rocks.

Features of Sedimentary Rocks

The first step in any investigation of sedimentary rocks is observation and data gathering. Accordingly, geologists visit sedimentary rock exposures, carefully examine and describe them, measure the thickness, trace units laterally, and collect samples for further analysis in the lab. While in the field, geologists commonly make some preliminary interpretations. For instance, the color of sedimentary rocks is a useful environmental parameter: red rocks may indicate deposition in a continental environment, whereas greenish rocks are more typical of marine environments. Exceptions are numerous, so color must be interpreted

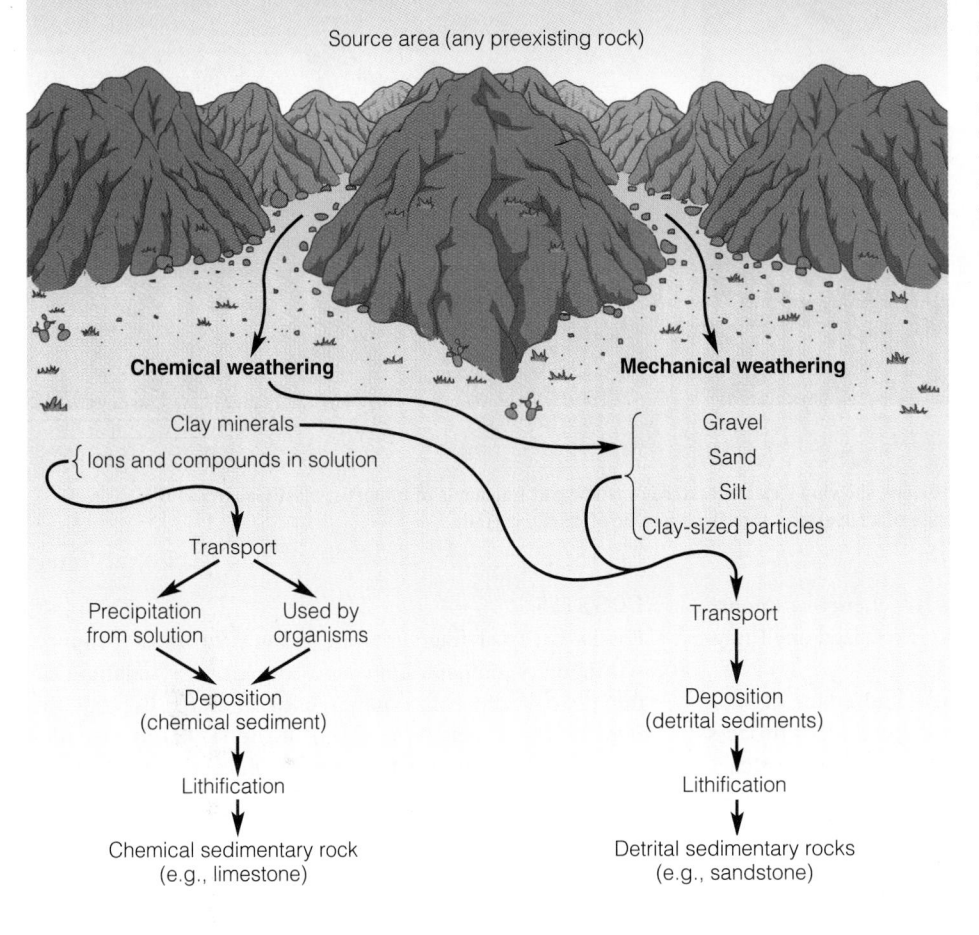

Source area (any preexisting rock)

Chemical weathering   Mechanical weathering

Clay minerals
{ Ions and compounds in solution

Gravel
Sand
Silt
Clay-sized particles

Transport

Precipitation   Used by
from solution   organisms

Deposition
(chemical sediment)

Lithification

Chemical sedimentary rock
(e.g., limestone)

Transport

Deposition
(detrital sediments)

Lithification

Detrital sedimentary rocks
(e.g., sandstone)

FIGURE 4.3 The derivation of sediment from preexisting rocks. Whether yielded by chemical or mechanical weathering, solid particles and materials in solution are transported and deposited as sediment, which if lithified becomes sedimentary rock. This illustration simply shows part of the rock cycle in more detail (see Figure 1.2).

with caution. Geologists may also recognize types of sedimentary particles or fossils that indicate deposition in a particular environment.

After field studies have been completed, the data can be more fully analyzed in the lab. Analyses may include microscopic and chemical examination of rock samples, identification of fossils, statistical determination of ancient current directions, and construction of diagrammatic representations of vertical and lateral facies relationships. And finally, when all available data have been analyzed, an environmental interpretation is made.

LITHOLOGY

The term **lithology** refers to the physical characteristics of rocks such as composition and texture. More than 100 different minerals have been identified in detrital sedimentary rocks, but only quartz, feldspars, and clays are very common. Unfortunately, detrital rock composition reveals little about transport or depositional history. Composition is, however, very important in determining the source area or areas that yielded the sedimentary particles.

Environmental inferences can be made from the composition of chemical sedimentary rocks. Limestone, for example, is composed of the mineral calcite ($CaCO_3$), a mineral that most commonly indicates a warm, shallow marine envi-

ronment; some limestones, however, form in lakes. Evaporites such as gypsum and rock salt indicate depositional environments in which evaporation rates were high—conditions that prevail in some lake and marine environments such as the Great Salt Lake, Utah, and the Persian Gulf area.

Sedimentary rocks may have a **clastic texture,** meaning they are composed of clasts (the broken particles of preexisting rocks) (Figure 4.4a). All detrital sedimentary rocks and some chemical rocks are composed of clasts. Conglomerate and sandstone, for example, contain gravel- and sand-sized clasts, respectively, and clastic limestone may be composed of broken shell fragments. Many chemical sedimentary rocks are composed of an interlocking mosaic of mineral crystals and are said to have a **crystalline texture** (Figure 4.4b).

Other important textural features include grain size and sorting. **Grain size** is simply a measure of a sedimentary particle's size; gravel (>2 mm), sand (1/16–2 mm), silt (1/256–1/16 mm), and clay (<1/256 mm). High-energy processes such as running water and waves are needed to transport large particles; thus, detrital gravel and sand tend to be deposited in stream channels or on beaches. Silt and clay, on the other hand, can be transported by weak currents and accumulate under low-energy conditions such as in lakes and lagoons. Clastic limestones are typical of

(a)

(b)

FIGURE 4.4 (a) Photomicrograph of a sandstone showing a clastic texture consisting of fragments of minerals, mostly quartz in this case. (b) Photomicrograph of the crystalline texture of a limestone showing a mosaic of calcite crystals.

beaches and offshore bars where wave energy is intense and may be composed of sand- and gravel-sized shell fragments.

During transport, the sharp corners and edges of sedimentary particles are abraded and smoothed. This process is particularly effective on sand- and gravel-sized particles, which become **rounded** as they collide with one another (Figure 4.5). Particles possessing sharp corners and edges are called angular.

Sorting refers to variation in grain sizes. Sedimentary rocks can be characterized as well sorted if the range of grain sizes is not great and as poorly sorted if the range is great (Figure 4.5). Sorting results from processes that selectively transport and deposit particles by size. Wind-blown dunes are composed of well-sorted sand, because wind cannot effectively transport gravel, and it blows silt and clay beyond the areas of sand accumulation. Glaciers, however, are unselective, because their transport power allows them to move many sizes of sediment, and their deposits are typically poorly sorted.

FOSSILS

Fossils or fossil fragments are among the most common sedimentary grains in limestone. Much of the sediment on the deep seafloor is composed of microfossils, and the framework of reefs is made up of the skeletons of corals and other organisms. Diatomite, a rock composed of the shells of microscopic plants called *diatoms,* is mined and used for abrasives and as a filtering agent in gas purification (Figure 4.6).

Two factors must be considered when using fossils in environmental analyses. The first is whether the fossil organisms lived where they were buried or whether they were transported there. The second is what kind of habitat the organisms originally occupied. Studies of a fossil's structure and its living descendants, if there are any, are helpful in this endeavor. For example, clams with thick shells typically live in shallow, turbulent water. In contrast, organisms dwelling in low-energy environments commonly have thin, fragile shells. Also, organisms that carry on photosynthesis are restricted to the zone of sunlight penetra-

(a)

(b)

FIGURE 4.5 Rounding and sorting of sedimentary particles. (a) A deposit of well-rounded gravel. (b) Angular gravel particles with sharp corners and edges. Both (a) and (b) have a wide range of grain sizes and are thus poorly sorted.

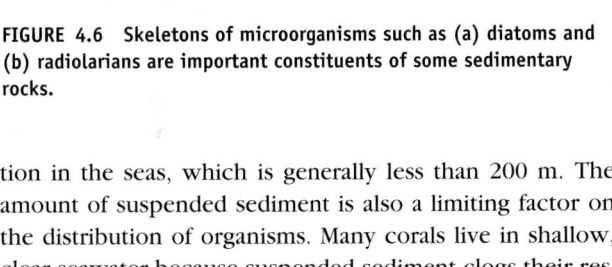

(a)

(b)

FIGURE 4.6 Skeletons of microorganisms such as (a) diatoms and (b) radiolarians are important constituents of some sedimentary rocks.

tion in the seas, which is generally less than 200 m. The amount of suspended sediment is also a limiting factor on the distribution of organisms. Many corals live in shallow, clear seawater because suspended sediment clogs their respiratory and food-gathering organs, and some have photosynthesizing algae living in their tissues.

Microfossils are particularly useful for environmental studies because the remains of numerous individual organisms can be recovered from small rock samples. In oil-drilling operations, for instance, small rock chips called *well cuttings* are brought to the surface. These samples rarely contain entire macrofossils but may contain numerous microfossils that aid in age and environmental interpretations. Trace fossils (see Figure 3.16) are not transported from the place where they formed, and certain traces are known to be characteristic of particular environments.

SEDIMENTARY STRUCTURES AND PALEOCURRENTS

Sedimentary rocks commonly contain structures that formed in sediment at the time of deposition or shortly

thereafter. These **sedimentary structures** are the manifestations of physical and biological processes that operated in the depositional environments and thus provide information on what the depositional processes were. The processes whereby many sedimentary structures originate are well known because they can be observed forming in present-day depositional environments, and many can be formed experimentally. Several other structures of no importance in determining depositional environment are also known (see Perspective 4.1).

Most sedimentary rocks have a layered appearance known as **bedding** or **stratification.** Individual layers less than 1 cm thick are *laminae* and *beds* if they are thicker. Some beds show a decrease in grain size from bottom to top, a feature known as **graded bedding** (Figure 4.7). Graded bedding is common in turbidity current deposits. *Turbidity currents* are underwater flows of sediment-water mixtures that have greater densities than sediment-free water. They move downslope to the bottom of the sea or lake, where their velocity diminishes and the heaviest particles settle out first, followed by progressively smaller particles (Figure 4.7). Turbidity currents also commonly produce scour marks in the muddy sediments over which they flow. When the turbidity current loses velocity, the scour marks fill with sand, producing *flute casts* in the sand layer at the base of the deposit.

Cross-bedding forms when layers are deposited at an angle to the surface upon which they accumulate. Cross-beds result from transport and deposition by wind or water currents and dip in the direction of flow (Figure 4.8). Since their orientation depends on the direction of flow, cross-beds are good indicators of ancient current directions, or **paleocurrents.**

Bedding planes may be marked by sedimentary structures such as **ripple marks** and **desiccation cracks** (Figures 4.9 and 4.10). Current ripples form in response to water or wind currents that move in one direction, and their asymmetric profiles allow geologists to determine paleocurrent directions (Figure 4.9b and c). Wave-formed ripples result from the to-and-fro motion of waves and tend to be symmetric in profile (Figure 4.9d and e). Crests of wave-formed ripples are oriented more or less parallel to shorelines. When clay-rich sediments dry, they shrink and crack into polygonal forms bounded by desiccation cracks (Figure 4.10). These desiccation cracks indicate alternating periods of wetting and drying such as occur along lake margins or on river floodplains.

Biogenic sedimentary structures are produced by organisms and include tracks, trails, and burrows; such organically produced marks are also called *trace fossils* (see Chapter 3). Extensive burrowing by organisms, a process known as **bioturbation,** may alter the physical and chemical properties of sediments and modify or destroy other sedimentary structures (Figure 4.10).

It is important to realize that no single sedimentary structure is unique to a particular depositional environment. Current ripples, for example, are common in stream channels

perspective 4.1

Secondary Sedimentary Structures and Pseudofossils

*R*ipple marks, cross-bedding, and desiccation cracks are all termed *primary sedimentary structures* because they formed when sediments were deposited. All are very useful for determining depositional environments. Sedimentary rocks also possess a variety of features known as *secondary sedimentary structures* that formed in sediments long after deposition. In addition, sedimentary rocks contain a variety of features commonly mistaken for fossils; thus, they are known as *pseudofossils*. Although secondary sedimentary structures tell nothing of how a rock layer was deposited, they are interesting in their own right, and some are quite attractive. Many of these structures are known, but only one variety, known as a concretion, is discussed here.

A mass of material that can easily be separated from the enclosing rock is a *concretion* (Figure 1). The most common type is concretions in sandstone, which form when cementation is more thorough in localized areas of a sand deposit. Their size ranges from 1 cm to 9 m in diameter, and although commonly spherical, their shapes can be quite irregular. Because concretions are more thoroughly cemented and thus harder than the surrounding rock, they accumulate at the surface during weathering (Figure 1a).

Imatra stones, or *marlekor,* are particularly interesting concretions (Figure 1b). They are typically disc shaped, composed of calcium carbonate and silt, and are found in clay layers of glacial lake deposits. Most are simple forms only 2 to

3 cm across, but some unusual shapes develop when simple forms grow together or where outgrows occur. The middle imatra stone in Figure 1b, for instance, has such a regular geometric shape that it and similar ones are mistaken for fossils or human artifacts.

Perhaps the most attractive concretions are *septaria* or *septarian nodules.* These spheroidal concretions have a series of cracks that widen toward the concretion's center and are crossed by smaller cracks more or less parallel with the margin (Figure 2a). The cracks apparently formed when the original concretion shrank as a result of dehydration; later, they were filled or nearly filled with min-

FIGURE 1 Concretions. (a) Spherical concretions from the Pumpkin Patch in the Ocotillo Wells State Vehicular Recreation Area in the Anza Borrego Desert, California. (b) These disc-shaped concretions called *imatra stones,* or *marlekor,* are from glacial lake deposits in Connecticut.

(a)

(b)

(a)

(b)

FIGURE 2   (a) Cross section showing the internal structure of a septarian nodule. (b) Surface view of septarian nodule resembling a turtle shell.

eral crystals, mostly calcite. Septaria commonly have a surface pattern resembling a turtle shell (Figure 2b). Indeed, some amateur fossil collectors have mistakenly identified them as fossil turtles.

Pseudofossils are so numerous that we will only mention several types in passing and discuss a few in any detail. Any natural object resembling a fossil but not a fossil is a *pseudofossil*. Thus, pseudofossils might be inorganic marks, mineral growths, deformation features, primary sedimentary structures, or erosion features in many types of rocks. Oval concretions, for example, are sometimes identified as fossil eggs, and oddly eroded rocks might resemble shells, teeth, or bones.

Even though these objects superficially resemble eggs, bones, and so on, close examination usually shows, among other things, that their internal structures do not match their external shapes—that is, they are merely eroded portions of common rocks. Professional paleontologists are familiar with pseudofossils and rarely fooled by them, but casual fossil collectors commonly misidentify them.

Dendrites resemble fossil ferns (Figure 3a). Although they look like thin films of carbon that resemble plants, most if not all of them consist of manganese oxide

(a)

(b)

FIGURE 3   (a) Dendrites. These are mineral growths found on planes between rock layers or along fractures. (b) Cone-in-cone. This specimen measures about 2 cm high.

minerals rather than carbon. They are found commonly on the planes separating adjacent layers of rock or along fractures. Cone-in-cone structures consist of a succession of nested cones with ribbed or grooved sides (Figure 3b). These structures look remarkably like fossil horn corals, but they formed by a concretionary growth process mostly in mudrocks or muddy limestones.

Some oddly shaped concretions in Cretaceous rocks in Oklahoma, and elsewhere, have been mistakenly identified as human shoe prints and claimed by some people to be evidence for the coexistence of humans and dinosaurs (Figure 4). These "prints" stand in relief compared to the surrounding rock and result from nothing more than differential weathering of the concretion-bearing host rock. Their slight resemblance to human shoe prints is purely coincidental.

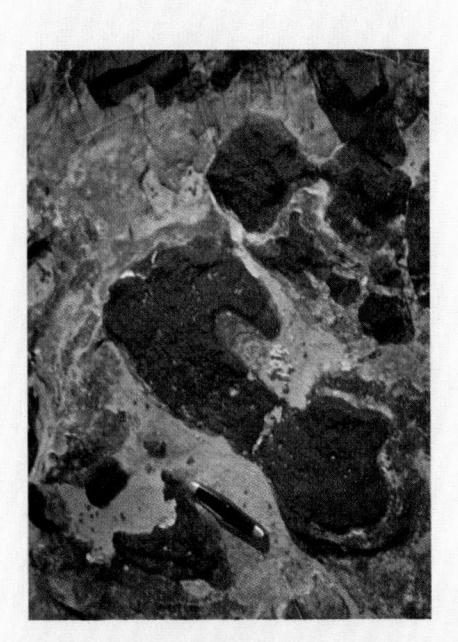

FIGURE 4   Concretions eroded from Cretaceous rocks in Oklahoma. These were mistakenly identified as human shoe prints. The specimens measure about 20 to 25 cm long.

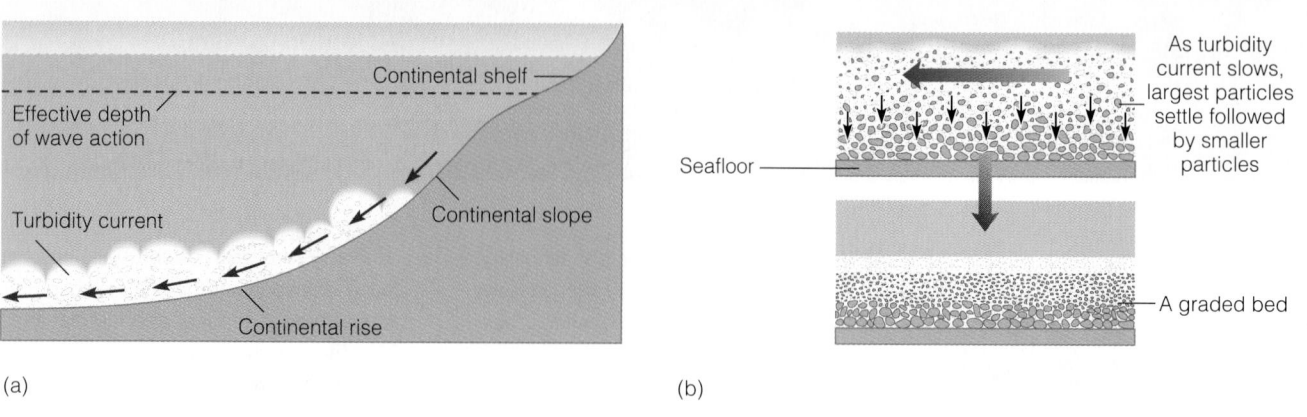

FIGURE 4.7 Deposition of a graded bed by a turbidity current. (a) A turbidity current generated on the continental slope moves downslope to the continental rise and seafloor. (b) Origin of a graded bed.

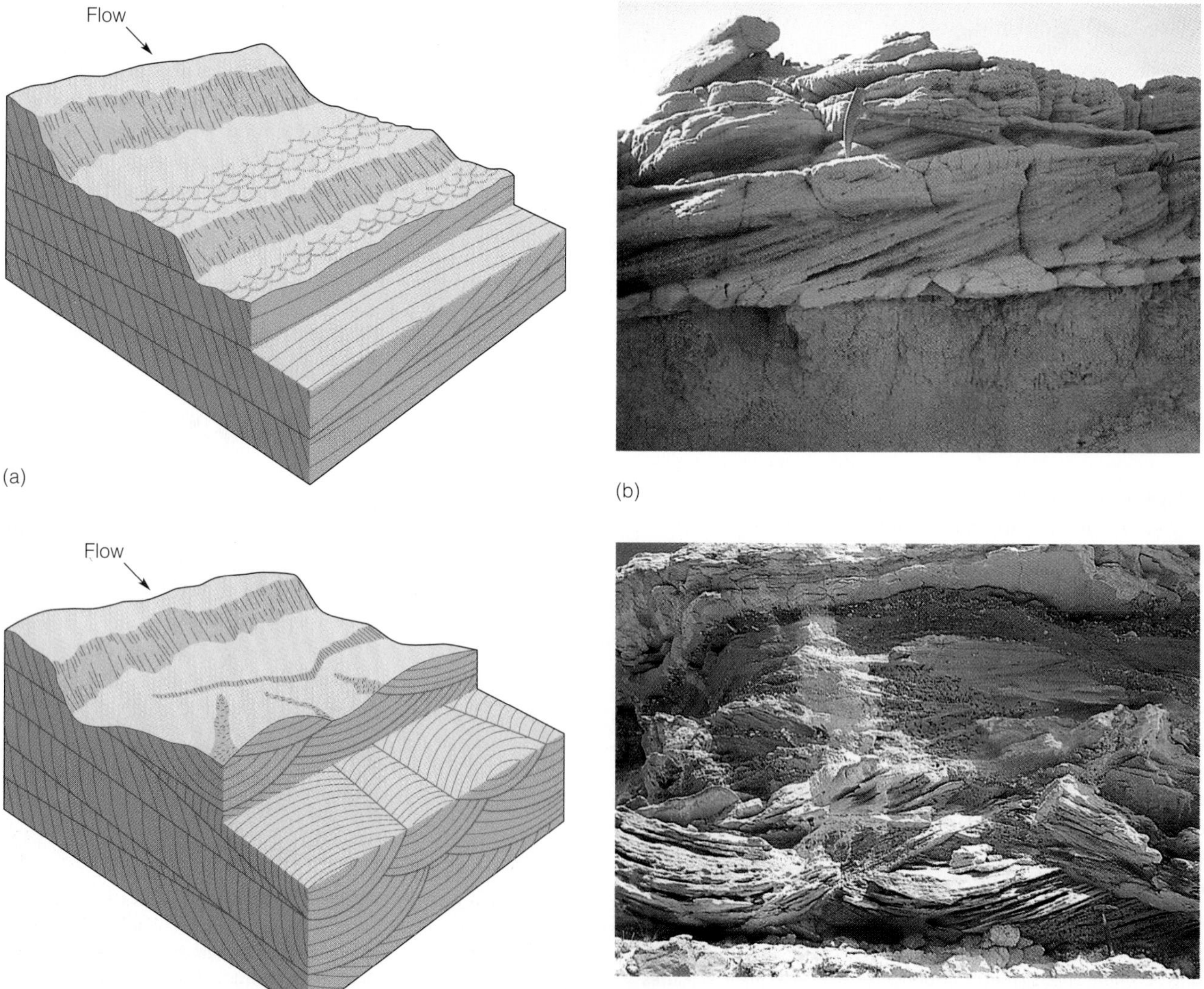

FIGURE 4.8 Two types of cross-bedding. (a) Tabular cross-bedding formed by migration of sand waves. (b) Tabular cross-bedding in the Upper Cretaceous Two Medicine Formation, Montana. (c) Trough cross-bedding formed by migrating dunes. (d) Trough cross-bedding in the Pliocene Six-Mile Creek Formation, Montana. This view shows the cross-beds illustrated on the right side of diagram (c).

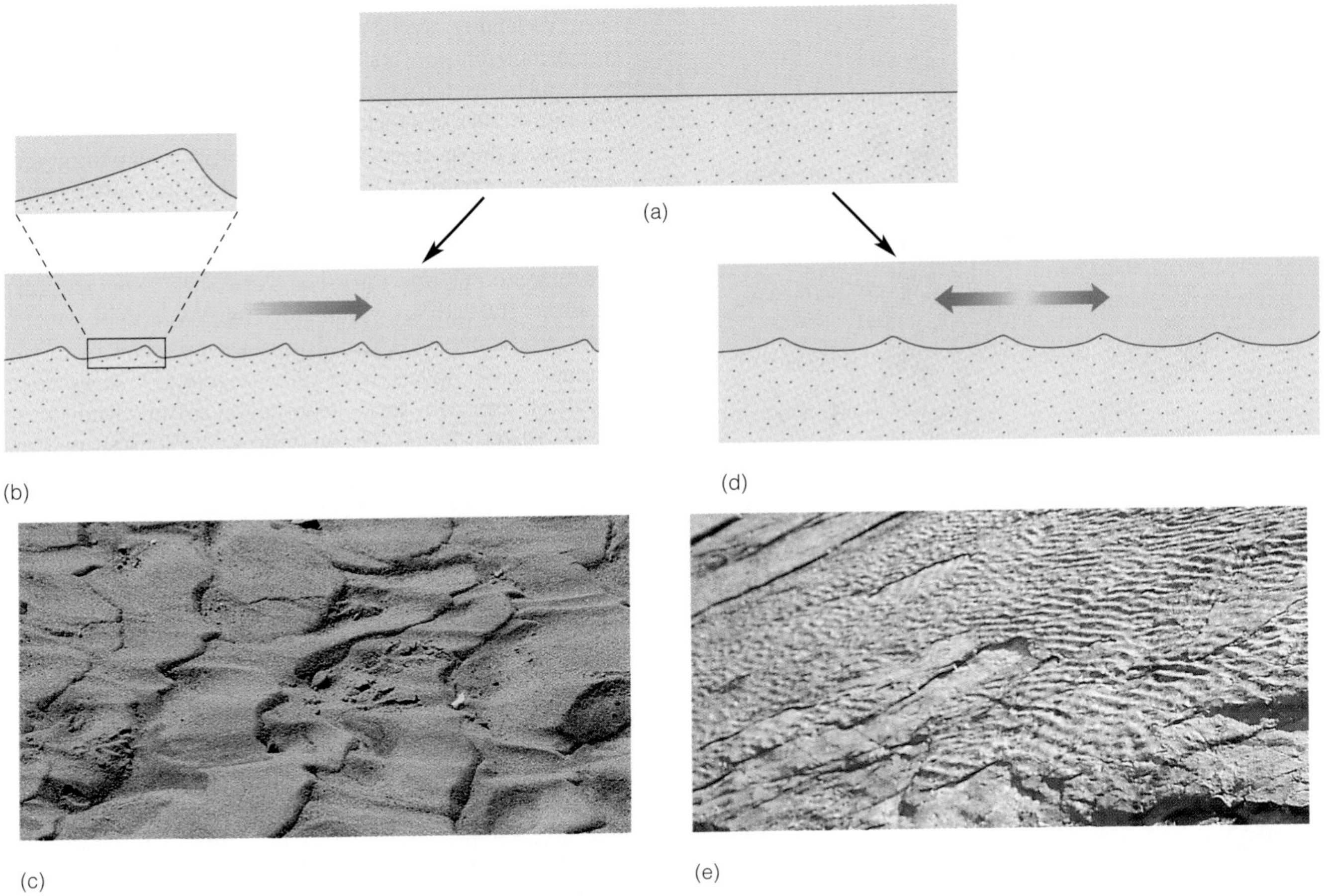

FIGURE 4.9 Ripple marks. (a) Undisturbed sand layer. (b) Current ripple marks form in response to flow in one direction, as in stream channels. The enlargement of one ripple shows its internal structure. Note that individual layers within the ripple are inclined, showing an example of cross-bedding. (c) Current ripples in a small stream channel. (d) The to-and-fro currents of waves in shallow water deform the surface of the sand layer into wave-formed ripple marks. (e) Wave-formed ripple marks on ancient rocks.

but may also be found in tidal channels, on the seafloor, or near lake margins. Associations of sedimentary structures are particularly useful in environmental analyses, especially when considered with other aspects of a rock unit such as composition, texture, and fossil content (Figure 4.11).

GEOMETRY

The three-dimensional shape, or the geometry, of a rock body may be helpful in environmental analyses, but must be interpreted with caution. The same geometry may be produced in different sedimentary environments. More-

FIGURE 4.10 (a) Desiccation cracks. (b) Bioturbation in siltstone caused by burrowing organisms. The vertical dark areas in the rock are burrows.

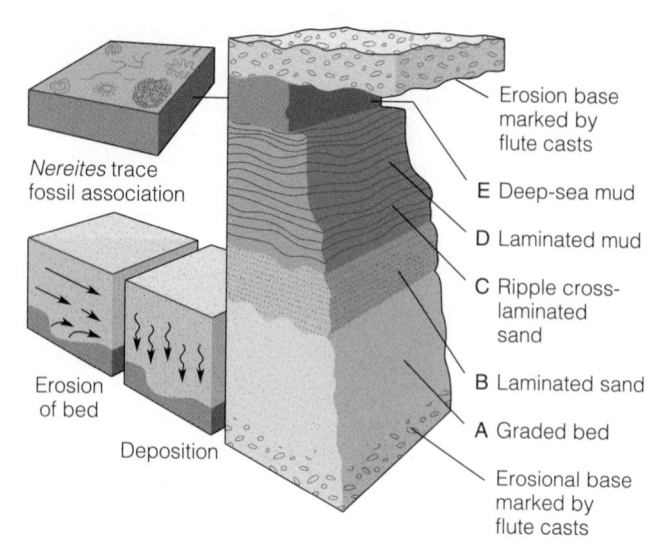

Nereites trace fossil association

Erosion of bed

Deposition

Erosion base marked by flute casts

E Deep-sea mud

D Laminated mud

C Ripple cross-laminated sand

B Laminated sand

A Graded bed

Erosional base marked by flute casts

FIGURE 4.11 Associations of sedimentary structures are useful for environmental interpretations, especially when considered along with other rock properties such as textures and fossils. This example shows an idealized vertical sequence deposited by a turbidity current. The flute casts are produced by erosion, but as a turbidity current slows, units A through D are deposited in succession. The deep-sea mud (unit E) forms as particles settle from seawater. It commonly contains the *Nereites* trace fossil association consisting of horizontal grazing traces.

over, rock-body geometry can be modified by sediment compaction, erosion, and deformation. Nevertheless, geometry can be useful when considered in conjunction with other properties of a rock body.

Some of the most extensive sedimentary rock units in the stratigraphic record are those deposited during marine transgressions and regressions (Figure 3.12). Such rock units may cover tens or hundreds of thousands of square kilometers, but they are not very thick compared with their other dimensions of length and width. Rock units with these dimensions are said to have a *blanket*, or *sheet*, geometry.

Some sand bodies have an *elongate*, or *shoestring*, *geometry*, especially those deposited in stream channels or barrier islands. Delta deposits tend to be lens shaped in cross profile or long profile but lobate when observed from above. Buried reefs are irregular, but many are elongated.

Depositional Environments

Any area in which sediment accumulates is a **depositional environment.** Geologists generally recognize three major depositional settings: continental, transitional, and marine, each of which contains several specific environments (Figure 4.12). As a starting point in environmental interpretations, geologists rely upon the concept of uniformitarianism. We assume, for example, that current ripple marks and cross-beds formed in the past just as they do in present-day depositional environments. Thus, the study of

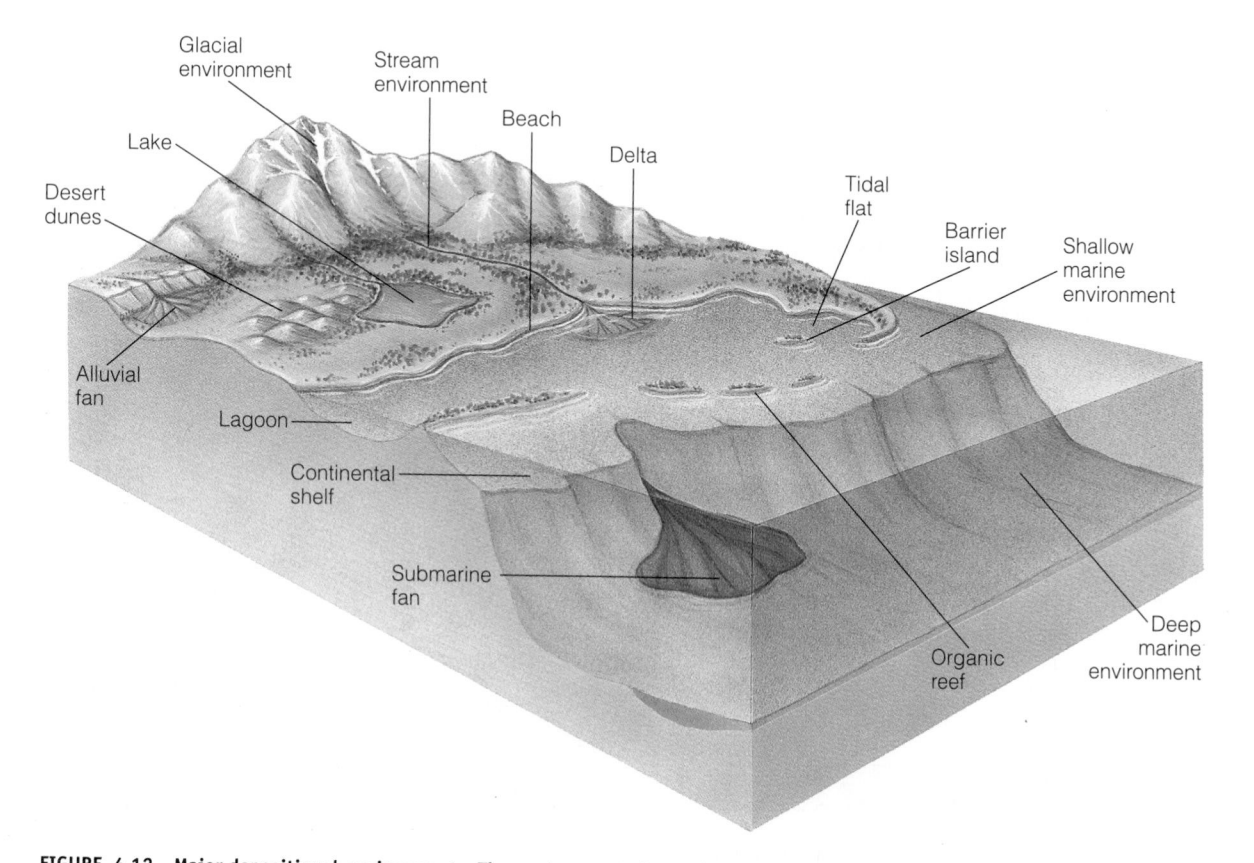

FIGURE 4.12 Major depositional environments. The environments located along the marine shoreline are transitional from continental to marine. The shallow marine environment corresponds to the continental shelf and can be the site of either detrital or carbonate deposition.

present-day depositional processes and environments is fundamental to our understanding of ancient processes and environments.

CONTINENTAL ENVIRONMENTS

Depositional environments on continents include stream (fluvial) systems, deserts, and glacial environments (Table 4.1). **Fluvial** refers to stream activity and to the deposits that accumulate in stream systems. The two types of stream systems recognized, braided and meandering, do most of their geologic work of erosion, sediment transport, and deposition when they flood (Figure 4.13). Consequently, their deposits do not record continuous day-to-day activities but, rather, the periodic events of sedimentation during floods. The deposits of braided and meandering streams differ considerably, however, as shown in the idealized facies models in Figure 4.13b and d. Deposits of braided streams consist mostly of horizontally bedded gravel and cross-bedded sand, while mud is conspicuous by its near-absence. In contrast, meandering stream deposits are dominated by mud

TABLE 4.1		
Depositional Environments Discussed in the Text		
Continental environments	Fluvial	Braided stream
		Meandering stream
	Desert	Sand dunes
		Alluvial fans
		Playa lakes
	Glacial	Ice deposition (moraines)
		Fluvial deposition (outwash)
		Glacial lakes
Transitional environments	Delta	Stream dominated
		Wave dominated
		Tide dominated
	Beach	
	Barrier island	Beach
		Sand dune
		Lagoon
Marine environments	Continental shelf	Detrital deposition
		Carbonate deposition
	Carbonate platform	
	Continental slope and rise	Submarine fans (turbidities)
	Deep-ocean basin	Oozes
		Pelagic clay
	Evaporite environments*	

* Evaporites may be deposited in a variety of environments including playa lakes, saline lakes, lagoons, and tidal flats in arid regions, and in marine environments.

with subordinate sand bodies with elongate, or shoestring, geometries (Figure 4.13c and d).

One of the most distinctive aspects of meandering stream sand bodies is the **point bar** sequence. When meandering streams flood, they erode the steep, outer banks of meanders and simultaneously deposit sediment on the inner, gently sloping banks as point bars. Successive episodes of erosion and deposition result in an eroded surface overlain by a sequence of point bars, each of which consists of a cross-bedded sand body. In addition to deposition of point bars, flooding meandering rivers transport silt and clay into the floodplain, where they settle out to form layers of mud.

Windblown desert dunes are recognized largely on the basis of textures, sedimentary structures, and facies relationships. Dune sands are typically well sorted and cross-bedded, on a scale measured in meters or tens of meters (Figure 4.14a). Large-scale cross-bedding is also found in dunelike sands of the continental shelf, but these sand deposits contain marine fossils and are associated with other marine rocks. Desert dunes, in contrast, may contain the remains or traces of land animals.

The association of windblown dunes with other deposits typical of deserts is especially helpful. For example, **alluvial fans** composed of sand and gravel commonly accumulate on the margins of desert basins, and **playa lakes,** temporary lakes in which mud and evaporites accumulate, may form on the desert floor (Figure 4.14b).

Glaciers effectively erode, transport, and deposit. Many of the surficial deposits in the northern tier of states and in parts of Canada were deposited by Pleistocene glaciers or in associated subenvironments (Figure 4.15). All sediments deposited in glacial environments are collectively called **drift,** but we must distinguish between two types of drift. **Till** is unsorted, unstratified drift deposited directly by glacial ice, mostly as ridgelike deposits called *moraines* (Figure 4.15). A second type of drift, called **outwash,** is deposited by fluvial processes, mostly in braided streams, that issue from melting glaciers (Figure 4.15). These streams are heavily loaded with sediment, much of which is deposited as sheets of sand and gravel. Since the terminus of a glacier may advance or retreat, it is not uncommon for till and outwash to be interbedded.

Deposits of glacial lakes typically accumulate finely laminated muds consisting of alternating light and dark laminae; each light–dark couplet is a **varve** (Figure 4.15d). Each varve represents an annual deposit: the light layers form in the spring and summer and consist of silt and clay; dark layers form in the winter when fine-grained clay and organic matter settle from suspension as the lake freezes over. Another distinctive feature of varved deposits is *dropstones,* which are pieces of gravel dropped from floating ice (Figure 4.15d).

TRANSITIONAL ENVIRONMENTS

Depositional environments in which deposits are acted upon by both marine processes and processes typical of

(a)

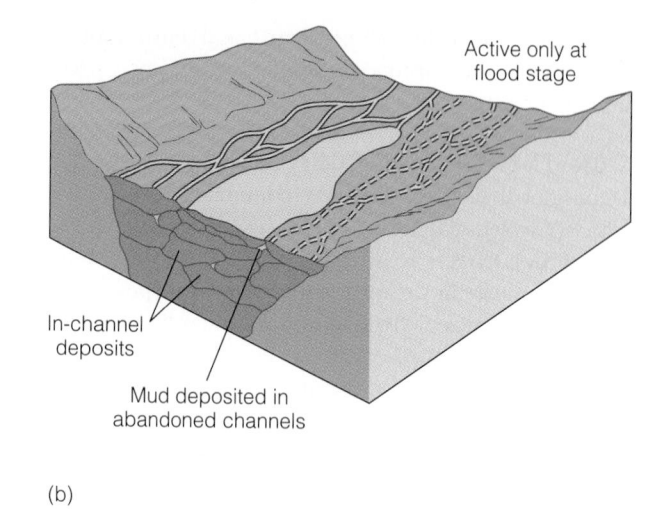

Active only at
flood stage

In-channel
deposits

Mud deposited in
abandoned channels

(b)

(c)

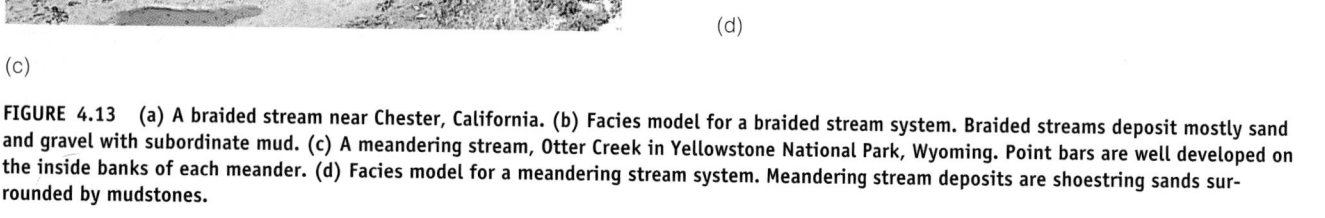

New
meander
belt

Channel deposits

Floodplain deposits

(d)

FIGURE 4.13 (a) A braided stream near Chester, California. (b) Facies model for a braided stream system. Braided streams deposit mostly sand and gravel with subordinate mud. (c) A meandering stream, Otter Creek in Yellowstone National Park, Wyoming. Point bars are well developed on the inside banks of each meander. (d) Facies model for a meandering stream system. Meandering stream deposits are shoestring sands surrounded by mudstones.

continents are transitional (Figure 4.12; Table 4.1). For instance, fluvial processes account for the deposition of a delta along a seashore, but such marine processes as waves and tides modify the deposit.

Deltas form by a rather simple process; where a stream enters a standing body of water, a lake or the sea, its velocity diminishes and deposition ensures. Deltas in lakes are common, but marine deltas are much more common in the geologic record, much larger, and far more important economically. As a result of delta deposition along a lake or seashore, the shoreline builds out, or **progrades,** into the lake or sea.

Prograding deltas deposit a characteristic vertical sequence in which bottomset beds are overlain successively by foreset beds and topset beds (Figure 4.16). These vertical sequences result from the lateral migration of environments, illustrating Walther's law (see Chapter 3). The flow velocity of a stream entering the sea diminishes rapidly, but the finest sediments are carried some distance beyond the

stream's mouth, where they settle from suspension and form bottomset beds. Nearer the stream's mouth, sand and silt are deposited as gently inclined foreset beds. The topset beds are traversed by a network of distributary channels in which fluvial deposits of silt and sand accumulate. Between the distributary channels are low marshy areas where fine-grained sediment and organic matter are deposited.

It should be clear from the preceding discussion that deltas are composites of marine and continental deposits. The facies of the bottomset and foreset beds are marine and commonly contain marine fossils. However, the foreset beds may also contain pieces of waterlogged wood and other remains of land plants. The distributary channel deposits of the topset beds show clear evidence of deposition by fluvial processes, and any fossils they contain will be of land plants and animals.

Most marine deltas are much more complex than the simple three-part division described above implies. De-

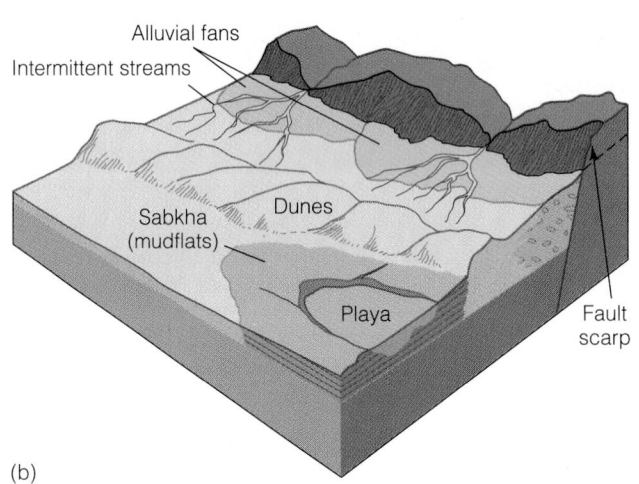

FIGURE 4.14 (a) Large-scale cross-bedding in dune sands of the Navajo Sandstone in Zion National Park, Utah. (b) Desert basin showing the association of alluvial fans, windblown dunes, and playa lakes.

(b)

pending on the relative importance of fluvial, wave, and tidal processes, three types of deltas are recognized (Figure 4.17). A *stream-dominated delta,* such as the Mississippi delta, consists of long fingerlike sand bodies, each deposited in a distributary channel that progrades far seaward. The Nile delta of Egypt is *wave dominated.* It also possesses distributary channels, but the entire seaward margin of the delta progrades, giving it a roughly triangular shape when viewed from above. *Tide-dominated deltas,* such as the Ganges–Brahmaputra of Bangladesh, are continually modified into tidal sand bodies that parallel the direction of tidal flow.

Shorelines where marine processes effectively redistribute sediments as fast as they are supplied are characterized by linear beaches and barrier islands paralleling the shoreline (Figure 4.12). Beaches are simply narrow sand bodies, constructed and modified by waves and nearshore currents, grading seaward into interbedded silt and sand. On their landward sides, beaches are commonly bordered by windblown sand dunes.

Elongate sand bodies separated from the mainland by a lagoon are **barrier islands** (Figure 4.18). These are high-energy environments where windblown dune and beach deposits accumulate; and during storms, waves may overtop barrier islands and deposit washover lobes in the lagoon (Figure 4.18). Sand dunes consist of well-sorted, cross-bedded sand. Quartz is the dominant min-

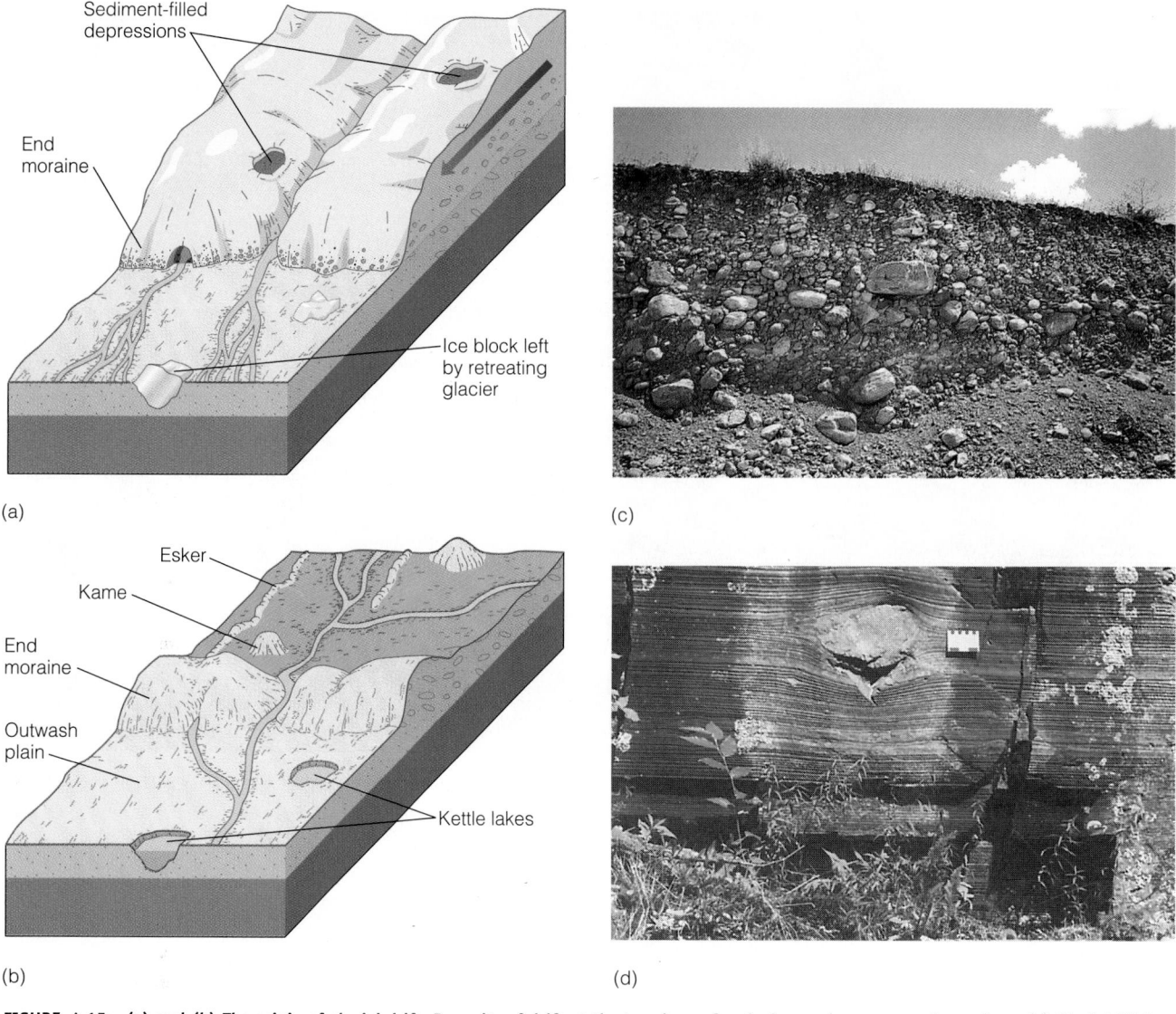

(a)

(b)

(c)

(d)

FIGURE 4.15 (a) and (b) The origin of glacial drift. Deposits of drift at the terminus of a glacier are known as end moraines. (c) Glacial till in an end moraine. (d) Ice-rafted dropstone in Late Proterozoic varved sediments of the Gowganda Formation, Canada.

eral in dune sands, but sand-sized fragments of marine shells are commonly present, too. In fact, the presence of marine fossils and facies relationships are the criteria used to distinguish these dunes from desert dunes. Beach sands are typically coarser grained than dune sands, contain fragmented marine fossils, and grade seaward into sediments of the shoreface environment (Figure 4.18).

MARINE ENVIRONMENTS

Marine environments include such depositional settings as the continental shelf, slope, and rise and the deep-sea floor (Figure 4.12; Table 4.1). Much of the detritus eroded from continents is deposited in marine environments, but other types of sediment may accumulate as well. For example, limestone is common in warm, shallow seas, and evaporites may be deposited in arid regions.

Detrital Marine Environments The continental shelf is a gently seaward-sloping surface bounded at its outer edge by a shelf–slope break. Beyond this break lie the continental slope and continental rise and finally the deep-ocean basins (Figure 4.19). Continental shelves are subdivided into a shallow, high-energy inner shelf and a deeper, low-energy outer shelf (Figure 4.19). Waves and tides more or less continuously stir up and sort inner-shelf sediments until mostly sand is left. This sand is commonly shaped into large, cross-bedded dunes and sand waves. The outer shelf is a low-energy area largely unaffected by tides and waves, except during large storms. Mud is the most common sediment on the outer shelf, but layers of silt and sand are present too, especially near the boundary between the inner and outer shelf.

Much of the sediment derived from the continents crosses the continental shelf in submarine canyons and

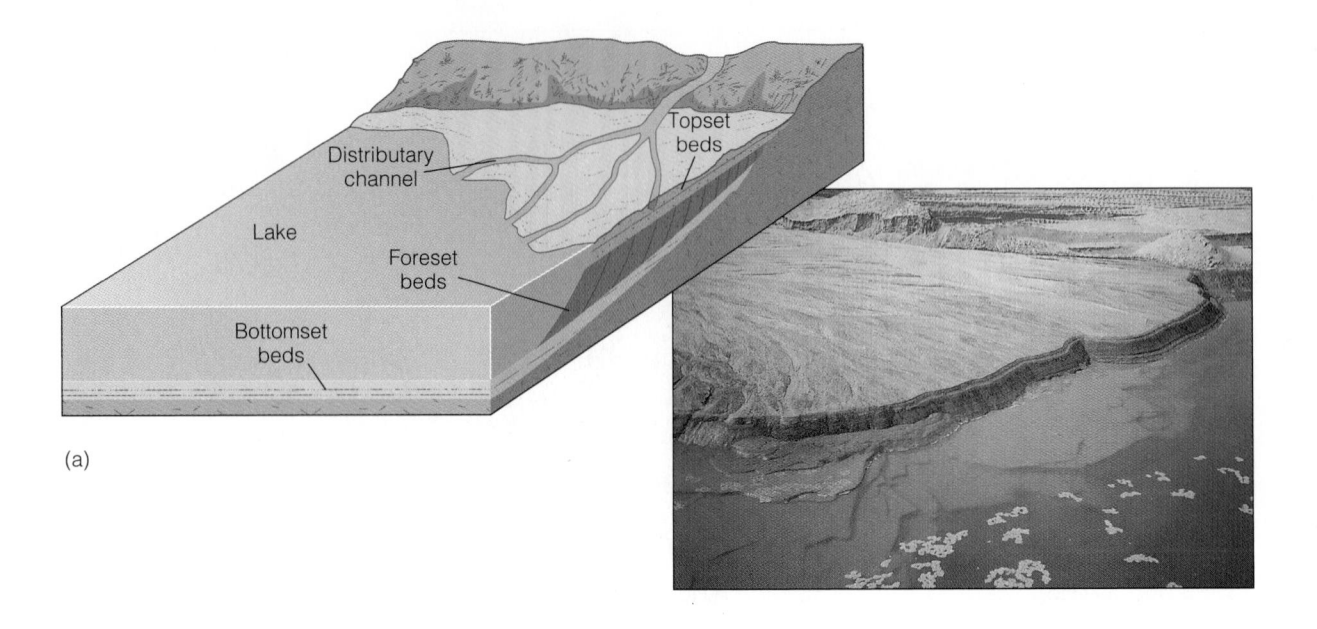

(a)

(b)

FIGURE 4.16 (a) Internal structure of the simplest type of prograding delta. (b) A small delta in which bottomset, foreset, and topset beds are visible.

eventually comes to rest on the continental slope and rise as a series of overlapping **submarine fans** (Figure 4.19). Once sediment has passed the shelf edge, it is transported and deposited by gravity processes such as slumps, submarine debris flows, and particularly by turbidity currents (Figure 4.7).

Beyond the continental slope and rise system, the seafloor is covered mostly by fine-grained deposits. Sources of fine-grained sediments include windblown dust and ash from oceanic islands or continental volcanoes and meteorite dust from space; the organically productive surface waters of the oceans contribute the shells of floating microorganisms. The seafloor is covered mostly by three types of sediment—pelagic clay and two types of oozes composed of the shells of single-celled plants and animals (Figure 4.20).

Carbonate Depositional Environments The only common carbonate rocks, those containing minerals with the carbonate (CO_3) ion, are limestone and dolostone, but most dolostone was originally limestone. When limestone is converted to dolostone during a process known as *dolomitization*, the mineral calcite ($CaCO_3$) is converted to dolomite ($CaMg(CO_3)_2$). Accordingly, our discussion of carbonate depositional environments is mostly a consideration of the origin of limestone.

In some respects limestones are similar to detrital sedimentary rocks. For example, many limestones are composed of particles, or grains, and microcrystalline carbonate mud called *micrite*. These grains and micrite are the carbonate equivalents of detrital gravel, sand, and mud. Limestones composed of shell fragments or spherical grains known as *oolites* (Figure 4.21) are deposited where currents or waves are intense. In contrast, micrite and small grains such as fecal pellets accumulate in low-energy environments such as lagoons.

Despite their textural similarities, limestones and detrital rocks also have some fundamental differences. Composition is one obvious difference; but more important, carbonate grains and micrite originate within the basin where they are deposited. These carbonate constituents may be locally transported and shaped into dunes and ripples, but they do not indicate a remote source area.

Limestone forms in some lakes, but by far most of it has been deposited in warm, shallow seas on carbonate shelves. These shelves may be attached to a continent or lie across the tops of offshore banks (Figure 4.22). In either case, deposition takes place in shallow water where little or no detrital mud is present. Carbonate barriers (Figure 4.22) form in high-energy areas and may be reefs or banks of skeletal particles or oolites. **Reefs** are wave-resistant structures composed of a framework of corals, mollusks, sponges, and many other organisms. Reef rock tends to be structureless, whereas banks of skeletal particles and oolites are characterized by horizontal beds, cross-beds, and ripple marks. Much of the carbonate sediment of lagoons is fossiliferous micrite and sand-sized fecal-pellet accumulations. Mats of algae, desiccation cracks, and, in arid regions, evaporite minerals characterize the tidal flats of carbonate depositional environments.

Limestone and dolostone rock units covering tens of thousands of square kilometers are common in the geologic record. Many of these carbonate units were deposited during major transgressions or regressions and therefore have overall sheetlike geometries. However, individual depositional units within carbonates may be shoestring shaped (oolite and skeletal sand barriers).

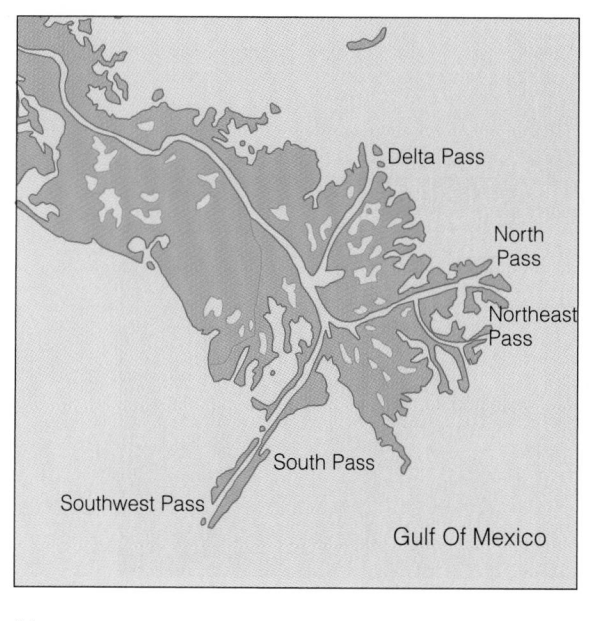

(a)

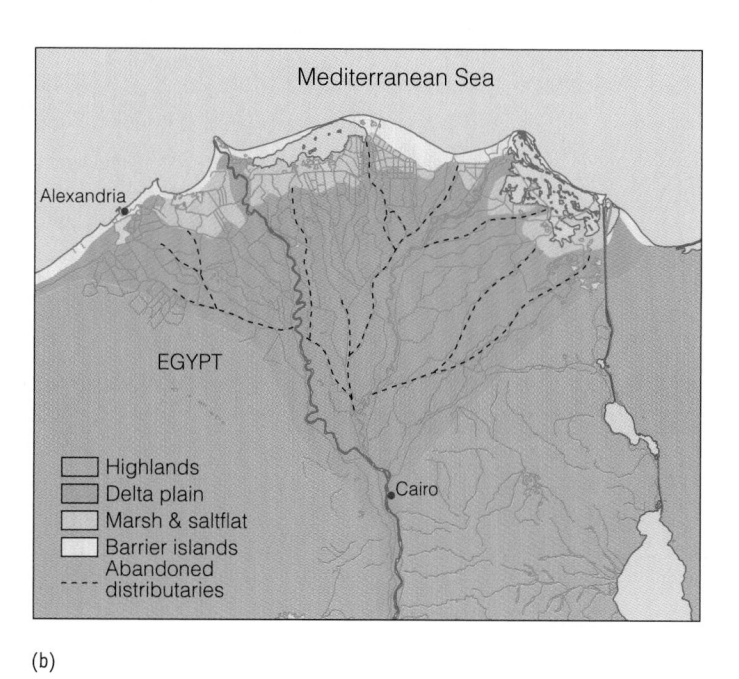

(b)

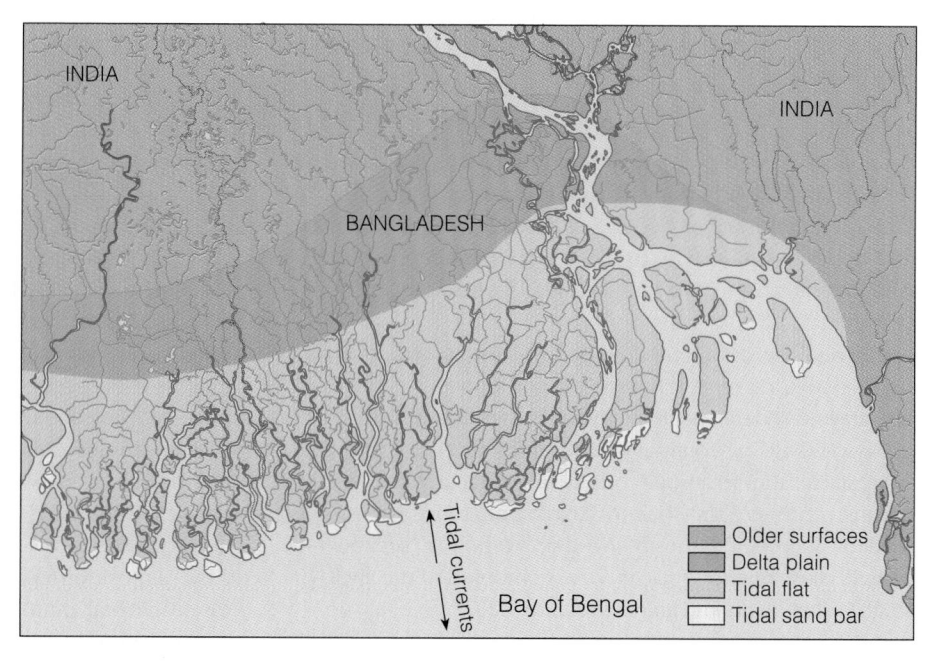

(c)

FIGURE 4.17 (a) The Mississippi River delta of the U. S. Gulf Coast is stream dominated, and (b) the Nile delta of Egypt is wave dominated. (c) The Ganges–Brahmaputra delta of Bangladesh is tide dominated.

Evaporite Environments Evaporites, mostly rock gypsum* and rock salt, are not particularly common when compared with the abundance of sandstone, shale, and limestone. Nevertheless, some evaporite deposits are of impressive dimensions, and some are locally important as resources. Evaporite rocks form in saline lakes such as the Dead Sea, in playa lakes, and in some marine environments.

Marine evaporites are the most common type in the geologic record, but little agreement exists on the specific environment in which they formed. Geologists agree, though, on some of the conditions necessary for evaporite deposition. Obviously, evaporation rates would have to be high so that seawater would become increasingly saline until it reached a point at which minerals would begin to precipitate. Such conditions are best met in nearshore environments such as lagoons or embayments in arid regions. In an embayment, for example, the inflow of normal seawater is restricted, salinity increases because of evaporation, and evaporite minerals form.

---

*Gypsum ($CaSO_4 \cdot 2H_2O$) is the common sulfate mineral precipitated from seawater; but when deeply buried, gypsum loses its water and is converted to anhydrite ($CaSO_4$).

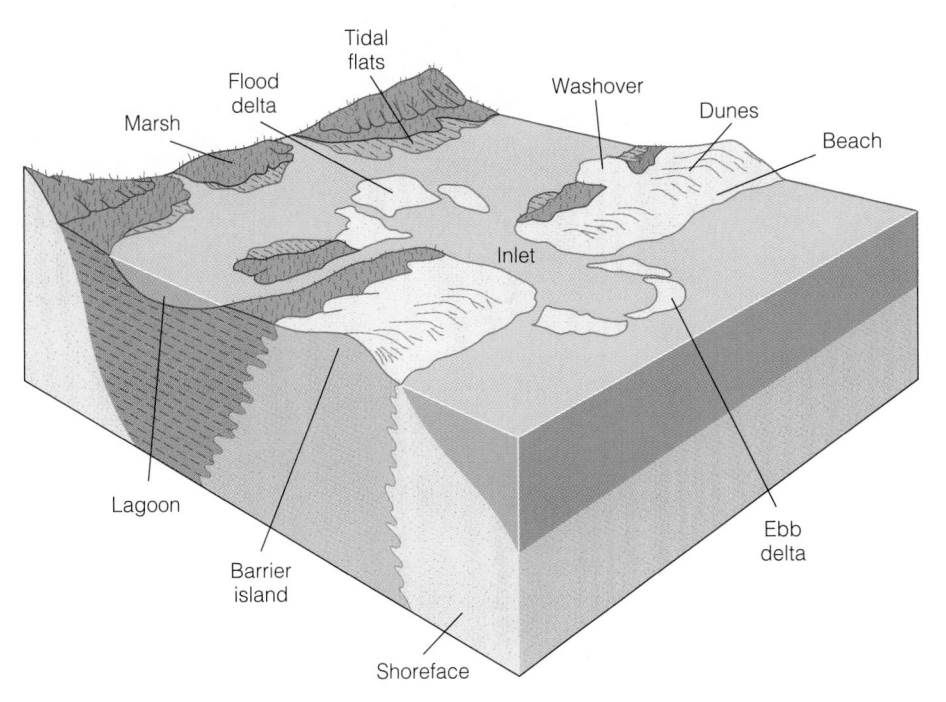

FIGURE 4.18 A barrier island and associated environments.

Marsh
Flood delta
Tidal flats
Washover
Dunes
Beach
Inlet
Lagoon
Barrier island
Shoreface
Ebb delta

Some marine evaporites appear to have been deposited in long, narrow seas that formed in response to continental rifting. When a continent is rifted, the crust is stretched and thinned and eventually subsides below sea level. Chapter 6 addresses this subject further. If this newly formed sea is in an arid region, evaporite deposition occurs. The Louann Salt (see Chapters 14 and 16) of the Gulf Coast of the United States and several other evaporite deposits are thought to have formed in such a plate tectonic setting.

Environmental Interpretations and Historical Geology

The ultimate goal in historical geology is to determine (1) what events took place during the past and (2) the order in which they occurred (see Perspective 4.2 on page 94). Such determinations are based, in part, on analyses of sedimentary rocks and inferences that can be

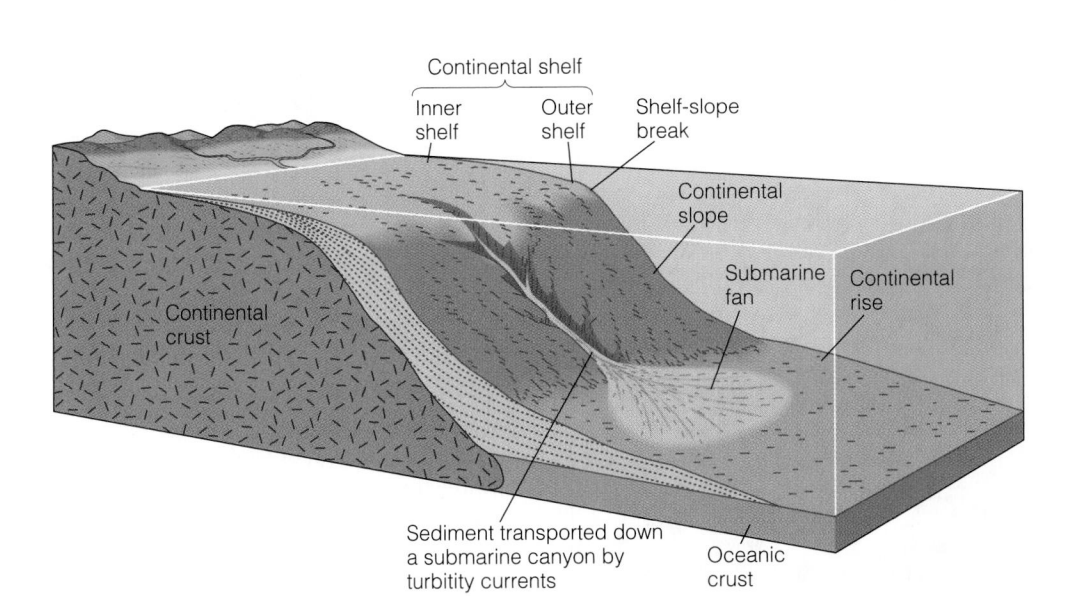

Continental shelf
Inner shelf
Outer shelf
Shelf-slope break
Continental slope
Submarine fan
Continental rise
Continental crust
Sediment transported down a submarine canyon by turbidity currents
Oceanic crust

FIGURE 4.19 Depositional environments on the continental shelf, slope, and rise. Submarine canyons are the main avenues of sediment transport across the shelf. Turbidity currents carry sediment beyond the shelf into the deeper waters of the slope and rise where they accumulate as submarine fans.

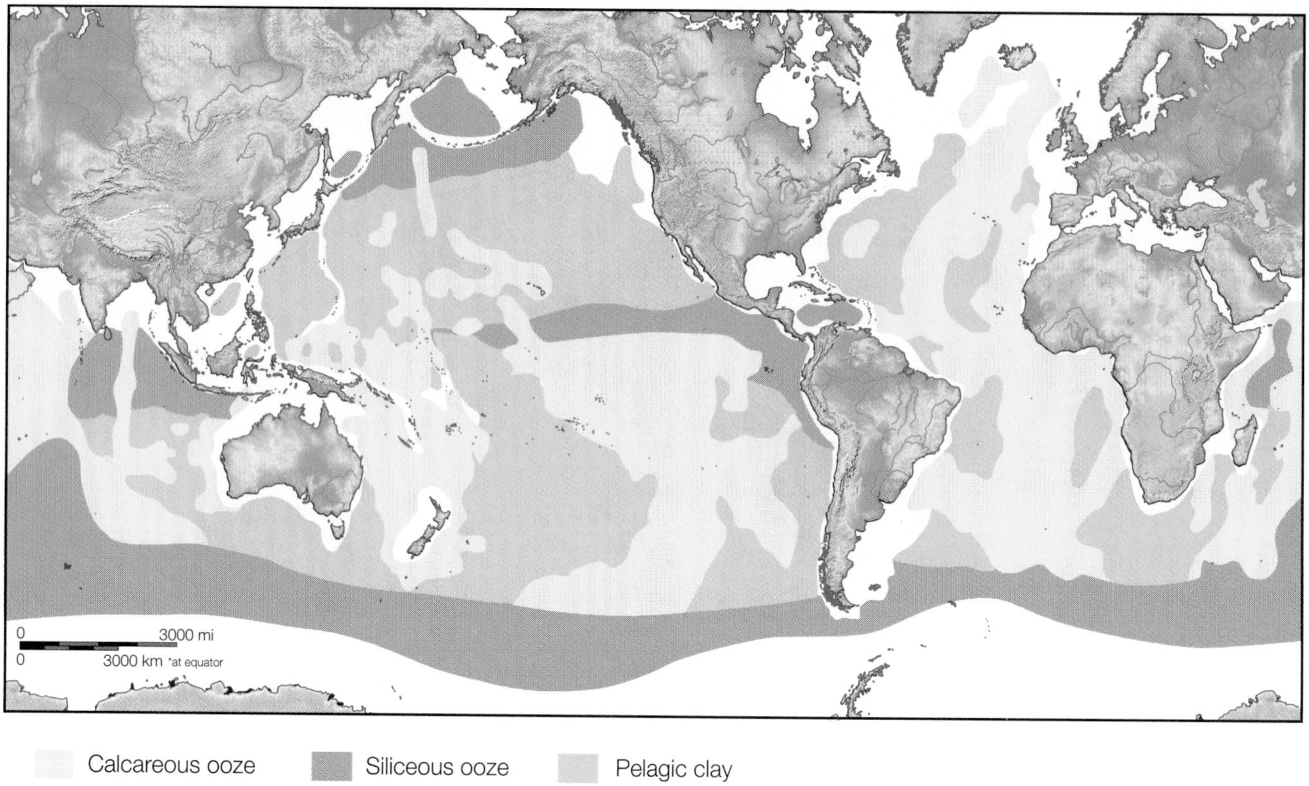

Calcareous ooze   Siliceous ooze   Pelagic clay

FIGURE 4.20 **The distribution of sediments on the deep-sea floor.**

drawn from such analyses. In Chapter 3 we discussed the use of sedimentary rocks and their contained fossils in establishing a meaningful stratigraphic succession, and this chapter has been devoted to environmental analyses. In short, we now have the information necessary to determine what specific events occurred and when. For example, Lower Silurian strata in New Jersey and Pennsylvania possess characteristics of grain size, rock types, and sedimentary structures that indicate deposition in a braided stream environment (Figure 4.23). Vertical facies relationships, rock types, fossils, and sedimentary structures of Ordovician carbonate rocks in Arkansas are

FIGURE 4.21 **Small, spherical or ovate calcium carbonate grains, such as those in this oolitic limestone, are common constituents of some limestones. Most oolites measure 0.5 to 1.0 mm in diameter.**

interpreted as transgressive carbonate shelf sediments (Figure 4.24). Depositional environments will figure importantly in the chapters on geologic history. These interpretations are based on the kinds of data and analyses covered in this chapter. Furthermore, environmental interpretations are not of academic interest only; they allow geologists to more effectively predict the locations and geometries of rock units that contain important mineral resources or oil and gas.

Paleogeography and Paleoclimates

Earth's surface patterns have not remained unchanged through time. During the Late Paleozoic, for example, a single supercontinent, Pangaea, existed, but it began breaking up during the Triassic and has continued to do so ever since. Thus, the present geographic distribution of continents evolved through time. **Paleogeography** involves a study of Earth's or a region's surface patterns for a particular time in the past.

The paleogeography of several Rocky Mountain states has been reconstructed in some detail in Figure 4.25. Such reconstructions must be made for a particular time, so geologists must determine depositional environments, work out the physical continuity of rock units, and determine time-stratigraphic relationships. Several paleogeographic maps have been constructed for this region, two of which are shown in Figure 4.25. According to the interpretation of

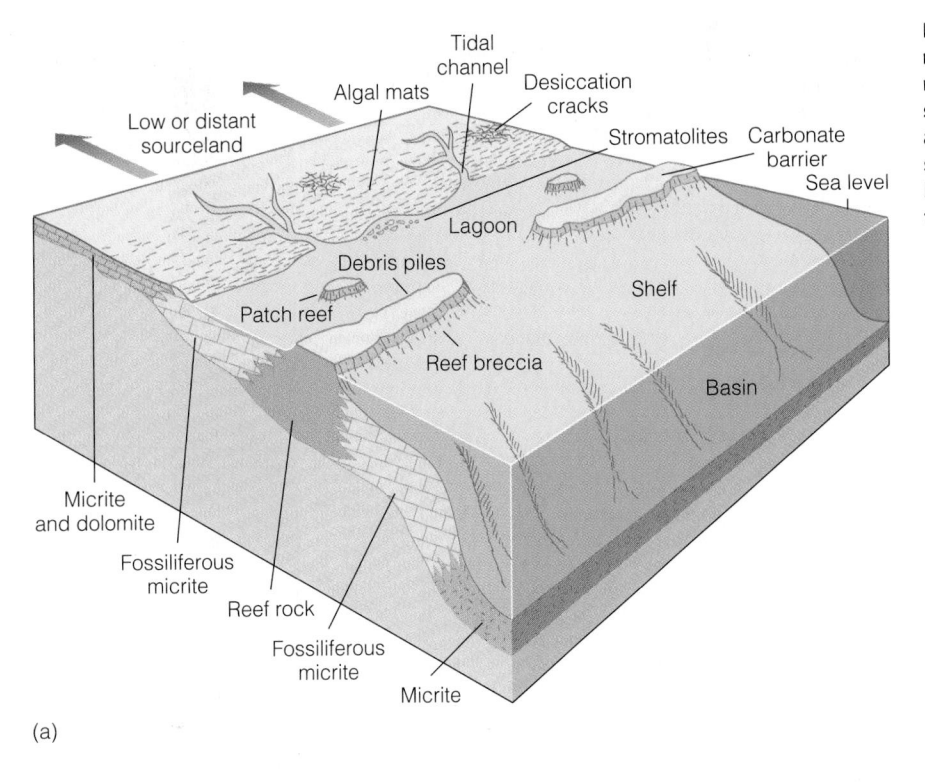

Low or distant sourceland

Tidal channel

Algal mats

Desiccation cracks

Stromatolites

Carbonate barrier

Sea level

Lagoon

Debris piles

Shelf

Patch reef

Reef breccia

Basin

Micrite and dolomite

Fossiliferous micrite

Reef rock

Fossiliferous micrite

Micrite

(a)

FIGURE 4.22 Carbonate depositional environments. (a) A shelf attached to a continent. Carbonate deposition is occurring on such attached shelves in southern Florida and the Persian Gulf. (b) Generalized cross section of the Great Bahama Bank. The carbonate shelf lies on top of a bank that rises from the seafloor.

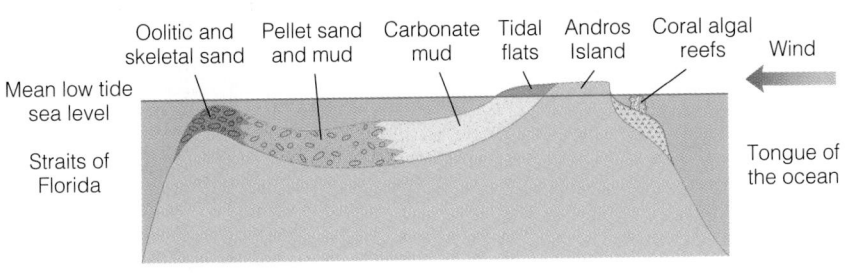

Oolitic and skeletal sand

Pellet sand and mud

Carbonate mud

Tidal flats

Andros Island

Coral algal reefs

Wind

Mean low tide sea level

Straits of Florida

Tongue of the ocean

(b)

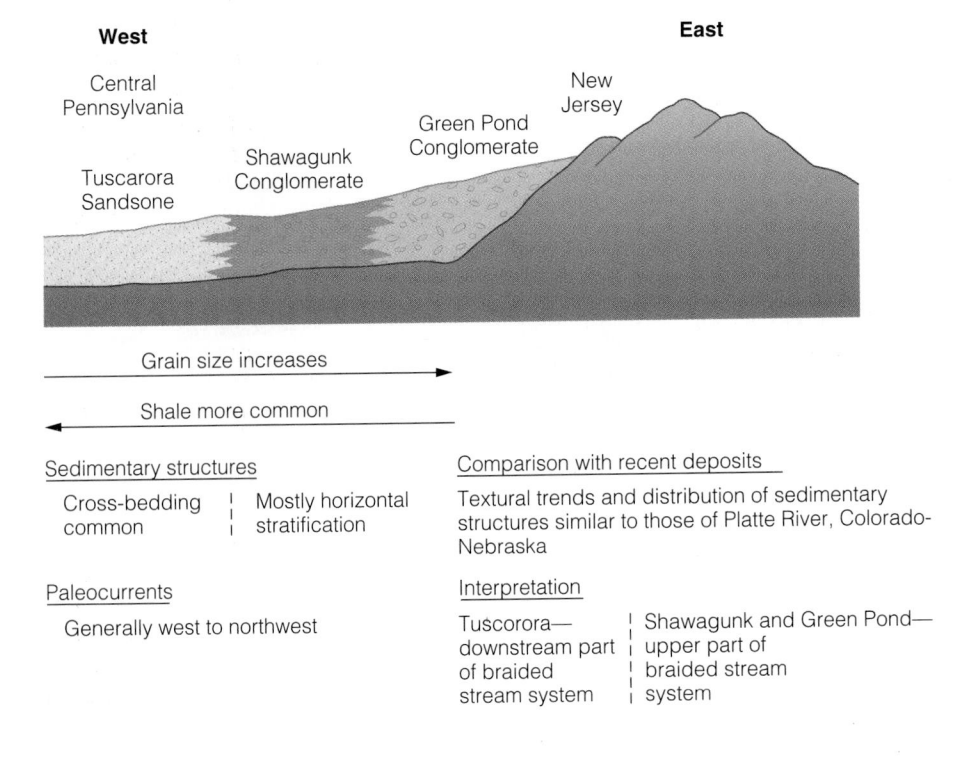

West

Central Pennsylvania

Tuscarora Sandsone

Shawagunk Conglomerate

Green Pond Conglomerate

New Jersey

East

Grain size increases

Shale more common

Sedimentary structures

Cross-bedding common | Mostly horizontal stratification

Paleocurrents

Generally west to northwest

Comparison with recent deposits

Textural trends and distribution of sedimentary structures similar to those of Platte River, Colorado-Nebraska

Interpretation

Tuscorora—downstream part of braided stream system | Shawagunk and Green Pond—upper part of braided stream system

FIGURE 4.23 Simplified cross section showing the probable lateral relationships and environmental interpretation for Lower Silurian strata in the eastern United States.

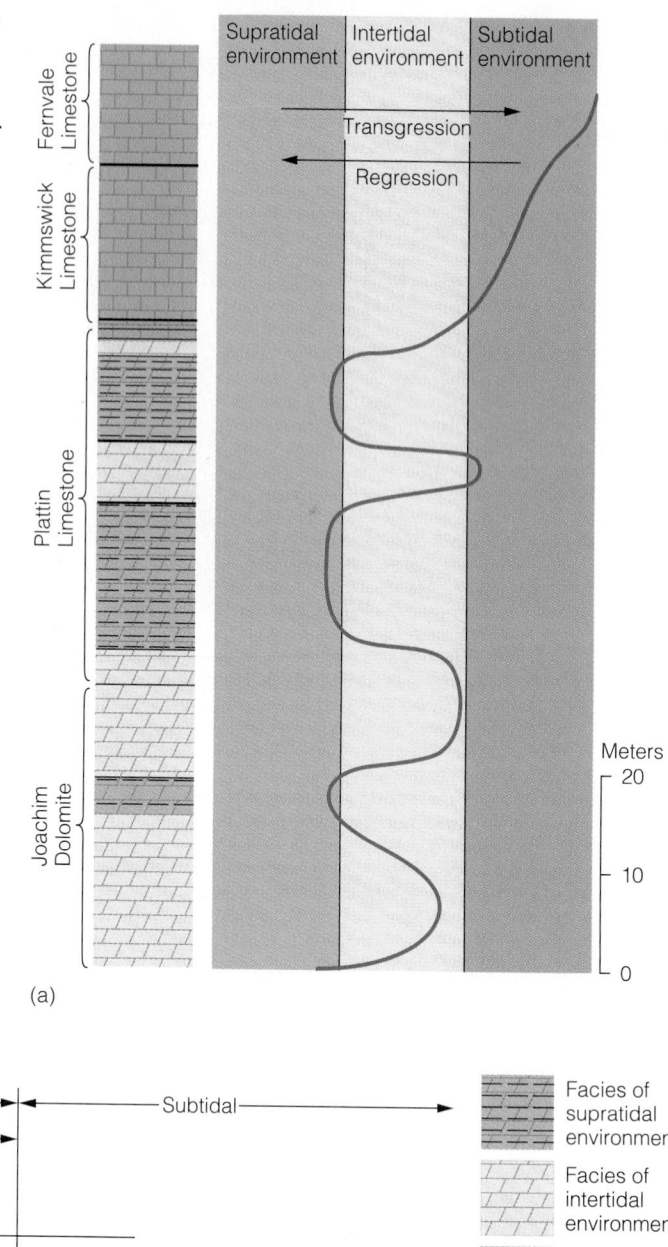

FIGURE 4.24 Middle and Upper Ordovician strata in northern Arkansas. (a) Vertical stratigraphic relationships and inferred environments of deposition. Notice that the trend was transgressive even though several minor transgressions and regressions occurred. (b) Simplified cross section showing lateral facies relationships.

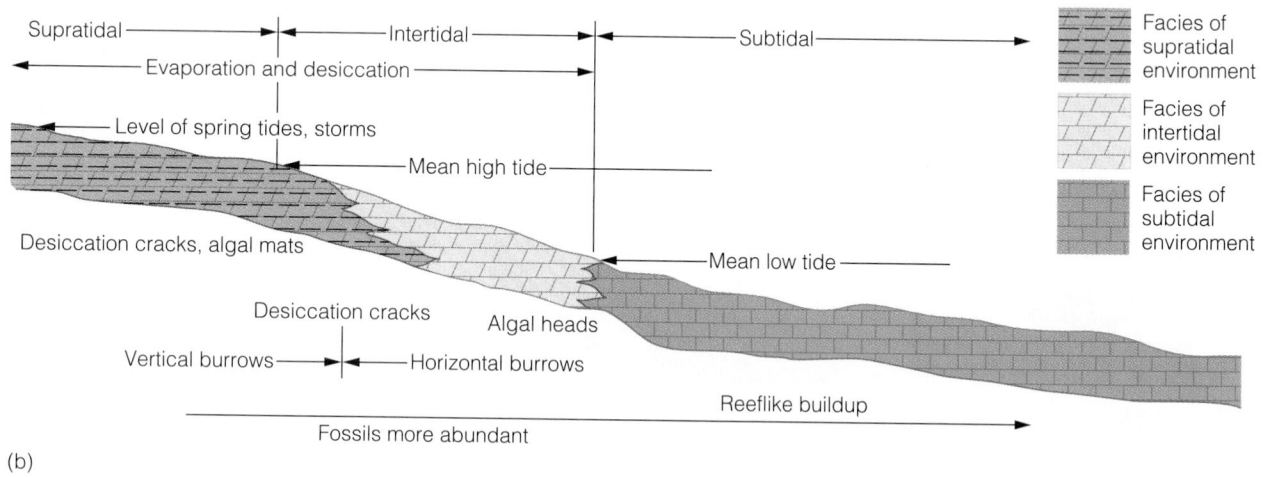

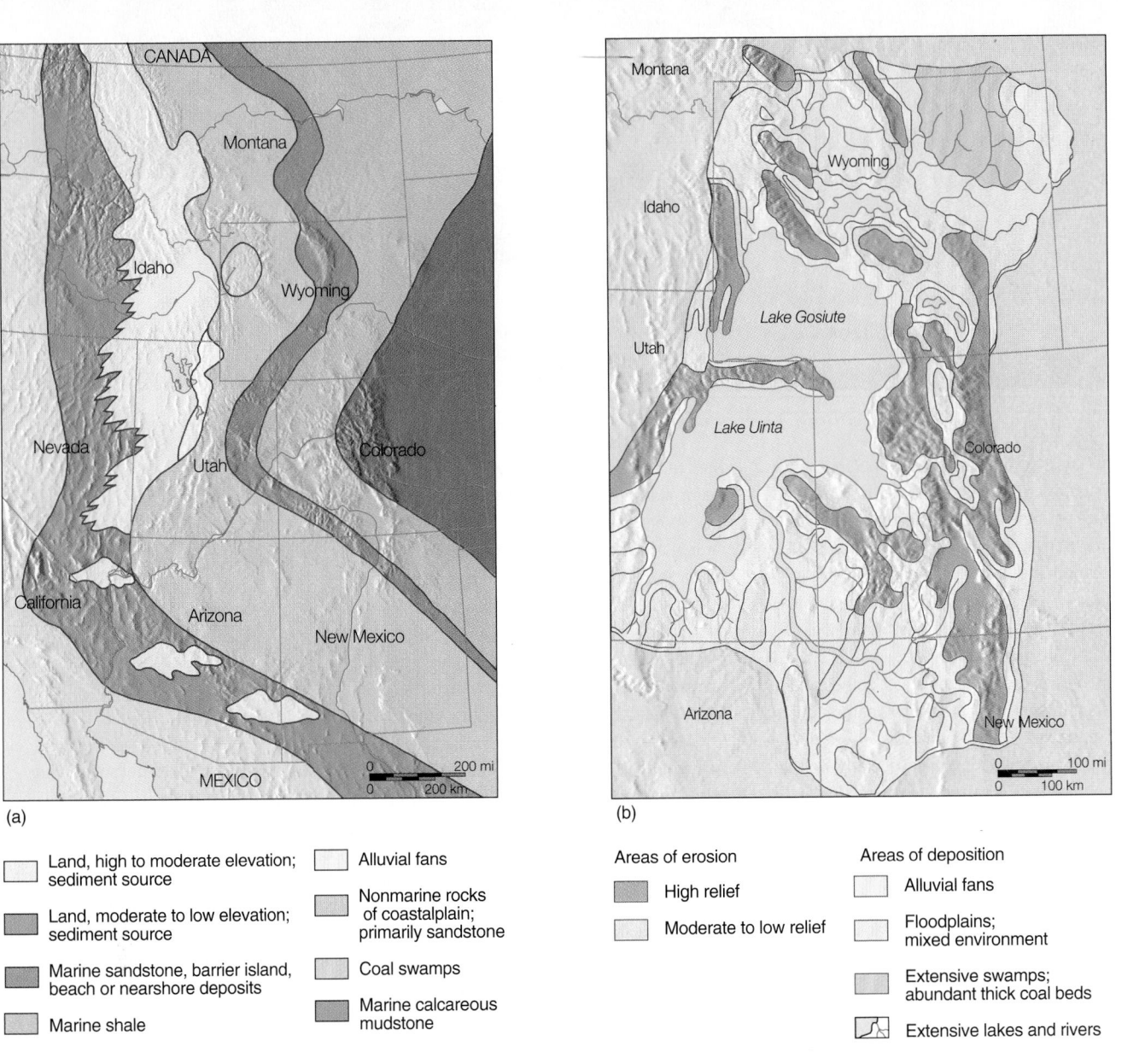

(a)

Land, high to moderate elevation; sediment source

Land, moderate to low elevation; sediment source

Marine sandstone, barrier island, beach or nearshore deposits

Marine shale

Alluvial fans

Nonmarine rocks of coastalplain; primarily sandstone

Coal swamps

Marine calcareous mudstone

Areas of erosion

High relief

Moderate to low relief

Areas of deposition

Alluvial fans

Floodplains; mixed environment

Extensive swamps; abundant thick coal beds

Extensive lakes and rivers

FIGURE 4.25 Maps showing (a) Late Cretaceous and (b) Eocene paleogeography of several Rocky Mountain states.

Late Cretaceous geography in Figure 4.25a, a broad coastal plain sloped gently eastward from a mountainous region to the sea. Furthermore, the rocks tell us that lush vegetation, dinosaurs, flying reptiles, lizards, turtles, and early mammals lived on this coastal plain. By Eocene time (Figure 4.25b), much of this area was the site of a large lake with adjacent streams, floodplains, alluvial fans, and mountainous regions.

Sedimentary rocks and fossils also tell something of ancient climates, or **paleoclimates.** Pennsylvanian and Per-

mian strata of the southwestern United States indicate desert conditions, whereas extensive coal beds in the east indicate warm, moist swamps (see Chapter 11). The recognition of glacial deposits including tillite (lithified glacial till) and varved deposits (Figure 4.15) in Precambrian, Ordovician, Pennsylvanian, and Pleistocene strata has allowed geologists to outline several areas of glacial climates. Deposits of evaporites indicate hot, arid climates.

perspective 4.2

Stratigraphy and Depositional Environments of the Rocks in Badlands National Park, South Dakota

Rocks in Badlands National Park, South Dakota, are some of the most colorful and interesting exposed in North America. *Badlands* develop in dry areas with sparse vegetation and nearly impermeable yet easily eroded rocks. Rain falling on such unprotected rocks rapidly runs off and intricately dissects the surface. As a result, numerous, closely spaced small gullies and deep ravines—separated by sharp angular slopes, ridges, and pinnacles—develop. Although badlands are particularly well developed here, scattered areas of similar topography are found from Arizona to southern Canada—in, for instance, Dinosaur Provincial Park, Alberta; Theodore Roosevelt National Park, North Dakota; and Bryce Canyon National Park, Utah.

Four rock units are exposed in Badlands National Park (Figure 1). The oldest is the Pierre Shale of Late Cretaceous age, which was deposited in what is known as the Cretaceous Interior Seaway during a major marine transgression (see Chapter 14). As its name implies, it consists largely of rocks composed of mud. The formation is particularly well known for its fossils—especially ammonites, which are extinct relatives of squid and octopus. By the end of the Cretaceous, a marine regression was underway during which the Cretaceous Interior Seaway withdrew from North America. As this sea regressed, a variety of rocks were deposited in transitional environments such as deltas and in continental environments.

Geologic activity of another kind was also beginning during the Late Cretaceous to the west of what is now Badlands National Park. Deformation was beginning that resulted in the uplift of the Rocky Mountains, including the Black Hills of South Dakota. The Pierre Shale was now exposed on land where deep yellow soils formed on it. It too was deformed, but only slightly as it was tilted several degrees from horizontal. Millions of years passed before deposition resumed when sediments of the White River Group consisting of three formations began accumulating (Figure 1). All these formations were deposited in continental environments, but the precise conditions of deposition differed. Furthermore, they lie with angular unconformity upon the

FIGURE 1   Rocks of Badlands National Park, South Dakota. Weathering and erosion of the Brule Formation (on the skyline) yield typical badlands topography, whereas smooth rounded slopes develop on the underlying Chadron Formation. The gently dipping yellowish rocks in the gully belong to the Pierre Shale.

Pierre Shale, and it is these rock units upon which the scenic badlands topography developed, particularly in the Brule and Sharps Formations (Figure 1).

During the Late Eocene and continuing into the Oligocene Epoch, deposition of the Chadron Formation took place. Sand and mud that eroded from the Black Hills was carried eastward and deposited in a complex of meandering streams and their adjacent floodplains and small lakes. Ancient stream channels are now seen as lens-shaped white sandstones, whereas floodplain deposits are mostly mud and lake deposits are thin layers of limestone. Apparently, the climate was quite humid. Crocodiles, turtles, and a variety of mammals such as now-extinct oreodonts, rhinoceros-sized titanotheres, ancestors of present-day horses and camels, and many others lived in the area.

Following deposition of the Chadron Formation, but still during the Oligocene Epoch, deposition took place that gave rise to the Brule Formation and finally the Sharps Formation. These formations also contain a variety of land-dwelling animal fossils, especially mammals. But the climate had changed; it was much drier and cooler when the Brule was deposited. Following deposition of the Brule Formation, the climate changed again, and the overlying Sharps Formation was deposited in stream channels and on floodplains. Volcanic ash falls are present in the Sharps Formation, and white layers of volcanic ash are also visible in several other parts of the White River Group.

Since deposition of the Sharps Formation, no additional sedimentation has occurred in the park area. In fact, since the Oligocene, the rocks have been deeply eroded, and, most notably, badlands topography has developed. The scenic badlands is reason enough to visit the park, but it has more to offer; it has the most complete succession of mammal fossils known anywhere in the world. The Park Service has left a number of fossils exposed but protected for viewing by park visitors.

Summary

1. To determine what environment sedimentary rocks were deposited in, geologists must observe and interpret the physical and biological features of those rocks.

2. Sedimentary structures and fossils are the most useful sedimentary rock features for environmental analyses, but composition, texture, and rock-body geometry are helpful as well.

3. Three major depositional settings are recognized: continental, transitional, and marine. Each setting includes several specific depositional environments.

4. Fluvial systems may be braided, where deposits are mostly gravel and sand with subordinate mud, or meandering where deposits are mostly mud with elongate sand bodies.

5. The association of alluvial fans, playa lake deposits, and windblown dunes is typical of desert depositional environments. Deposits in areas affected by glaciation are mostly till and outwash.

6. Marine deltas form where a stream supplies sediments to a shoreline. Deltas prograde seaward and produce vertical sequences of bottomset beds overlain by foreset beds and topset beds.

7. A barrier island complex includes lagoon, beach, and dune subenvironments, each characterized by a unique

association of lithologies, fossils, and sedimentary structures.

8. Waves and tides are important on continental shelves, but beyond the shelf–slope break, gravity processes—especially turbidity currents—predominate. Detrital shelf deposits consist of innershelf sands and outer-shelf muds. Deposits on the continental slope–rise systems are mostly submarine fans.

9. Sediments on the deep-sea floor are mostly fine-grained clay and ooze because no mechanism exists to transport large amounts of coarse detritus from continents into deep seas.

10. Most carbonates are deposited in warm, shallow seas. Many limestone textural features are indicators of depositional conditions. Carbonate shelves may be attached to a landmass or lie on top of banks rising from the seafloor.

11. The best conditions for the origin of evaporites occur along marine shorelines in arid regions. Marine embayments and lagoons are areas in which many present-day evaporites are deposited.

12. Studies of sedimentary rocks provide information needed to decipher a region's paleogeography and tell us something about ancient climates.

Important Terms

alluvial fan
barrier island
bedding (stratification)
biogenic sedimentary
 structure
bioturbation
clastic texture
cross-bedding

crystalline texture
delta
depositional environment
desiccation crack
drift
fluvial
graded bedding
grain size

lithology
outwash
paleoclimate
paleocurrent
paleogeography
playa lake
point bar
progradation

reef
ripple mark
rounding
sedimentary structure
sorting
submarine fan
till
varve

Review Questions

1. Which one of the following is a biochemical sedimentary rock?
   a. _____ Limestone composed of shell fragments
   b. _____ Sandstone with a large amount of volcanic ash
   c. _____ Rock gypsum
   d. _____ Any sedimentary rock composed mostly of detrital sediment
   e. _____ A rock with at least 50% gravel-sized clasts

2. When sand and gravel are transported, their sharp corners and edges are worn smooth by a processes known as:
   a. _____ sorting.
   b. _____ selective transport.
   c. _____ rounding.
   d. _____ lithification.
   e. _____ compaction.

3. Graded bedding is characterized by:
   a. _____ layers deposited at an angle.
   b. _____ cracks that extend downward in muddy sediments.
   c. _____ an assortment of vertical burrows.

   d. _____ bedding planes with wavelike structures.
   e. _____ an upward decrease in grain size in a layer.

4. Which one of the following is a transitional depositional environment?
   a. _____ Continental slope
   b. _____ Barrier island
   c. _____ Fluvial
   d. _____ Glacial
   e. _____ Carbonate platform

5. The deposits of braided streams consist mostly of:
   a. _____ sand and gravel.
   b. _____ limestone and rock salt.
   c. _____ point bars and mud.
   d. _____ bottomset, foreset, and topset beds.
   e. _____ overlapping submarine fans.

6. The only common carbonate rocks are:
   a. _____ sandstone and conglomerate.
   b. _____ rock gypsum and rock salt.
   c. _____ shale and coal.

d. _____ limestone and dolostone.

e. _____ pelagic clay and ooze.

7. Which one of the following sedimentary structures is a good paleocurrent indicator?

a. _____ Lamination

b. _____ Current ripple marks

c. _____ Graded bedding

d. _____ Trace fossils

e. _____ Desiccation cracks

8. A detrital sedimentary rock in which all particles are about the same size is said to be:

a. _____ well sorted.

b. _____ graded.

c. _____ moderately rounded.

d. _____ cemented.

e. _____ an evaporite.

9. Pelagic clay and oozes are:

a. _____ deposited in high-energy environments such as stream channels.

b. _____ the types of sediments covering most of the deep-sea floor.

c. _____ annual deposits that accumulate in glacial lakes.

d. _____ deposited mostly by turbidity currents.

e. _____ found most commonly on continental shelves.

10. Point bar deposits are characteristic of:

a. _____ meandering streams.

b. _____ alluvial fans.

c. _____ deltas.

d. _____ barrier islands.

e. _____ moraines.

11. One component of many limestones is micrite, which is:

a. _____ detrital sediment.

b. _____ a calcium sulfide mineral.

c. _____ microcrystalline carbonate mud.

d. _____ fragments of plants.

e. _____ outwash.

12. A fundamental principle used to interpret ancient depositional environments is:

a. _____ cross-cutting relationships.

b. _____ uniformitarianism.

c. _____ chemical continuity.

d. _____ paleogeography.

e. _____ sedimentary textures.

13. Some ancient deposits clearly indicate deposition of sand in dunes. How can you determine whether the dunes were deposited in a desert or on a barrier island?

14. Describe the vertical sequence of facies that develops when a delta progrades into a lake or the sea.

15. What types of detrital deposits are typical of the continental shelf, slope, and rise?

16. Draw a generalized cross section of a carbonate shelf and show the types of sediments you would expect to find on various parts of the shelf.

17. What are the common evaporite deposits, and under what conditions do they form?

18. Explain how fossils, and especially microfossils, can be used to help determine ancient depositional environments.

19. Describe or illustrate three sedimentary structures that can be used to determine paleocurrent directions. How does each of these structures form?

20. How does graded bedding originate, and in what depositional environment is it particularly common?

21. What types of detrital sediments are common on continental shelves, and what processes are important in determining the properties of these deposits?

22. What types of sediments cover most of the deep-sea floor? What are the sources of these sediments?

23. In what ways are limestones and detrital sedimentary rocks similar and dissimilar?

24. Define grain size and sorting and explain how each is used in environmental analyses. Why is sedimentary rock composition of little use in this endeavor?

25. How do the deposits of braided and meandering streams differ?

Points to Ponder

1. A rock unit consists mostly of crudely bedded mudstone with subordinate sandstones. The sandstones are cross-bedded, contain fossil land mammals and flowering plants, and each has a shoestring-shaped geometry. What is the most likely environment of deposition? Give evidence to support your answer.

2. Explain why rock type is of little use in determining depositional environments, whereas sedimentary structures and fossils are quite helpful in this endeavor.

3. What sequence of events can account for a section of sedimentary rocks consisting of oolitic limestone overlain in succession by micrite (carbonate mud) with abundant fecal pellets and body fossils and, finally, algal laminated dolostone?

Additional Readings

Boggs, S., Jr. 1992. *Petrology of sedimentary rocks.* New York: Macmillan.

————. 1995. *Principles of sedimentology and stratigraphy.* 2d ed. Columbus, Ohio: Merrill.

Collinson, J. D., and D. B. Thompson. 1989. *Sedimentary structures.* 2d. ed. London: Allen & Unwin.

Davis, R. A., Jr. 1992. *Depositional systems: An introduction to sedimentology and stratigraphy.* 2d ed. Englewood Cliffs, N. J.: Prentice Hall.

Friedman, G. M., J. E. Sanders, and D. C. Kopaska-Merkel. 1992. *Principles of sedimentary deposits.* New York: Macmillan.

Galloway, W. E. 1996. *Terrigenous clastic depositional systems: Applications to fossil fuel and groundwater resources.* New York: Springer.

Julien, P. Y. 1995. *Erosion and sedimentation.* New York: Cambridge University Press.

McLane, M. 1995. *Sedimentology.* New York: Oxford University Press.

Prothero, D. R., and F. Schwab. 1996. *Sedimentary geology.* New York: Freeman.

Ricci Lucchi, F. 1995. *Sedimentographica: Photographic atlas of sedimentary structures.* New York: Columbia University Press.

Scoffin, T. P. 1987. *An introduction to carbonate sediments and rocks.* New York: Chapman & Hall.

Tucker, M. E. 1996. *Sedimentary rocks in the field.* New York: Wiley.

World Wide Web Activities

For these web site addresses, along with current updates and exercises, log on to

http://www.brookscole.com/geo/

▶ **SEDIMENTARY ROCKS**

This site has images and information on 11 common sedimentary rocks. It is part of the Soil Science 233 Rock and Minerals Reference web site. Click on any of the *rock names* and compare the information and images to the information in this chapter.

▶ **GEOLOGICAL SURVEY OF NEWFOUNDLAND AND LABRADOR**

Information on a variety of topics is contained at this site. Click on the label *Education,* and then click *Geology and Geologic Features.* At this point click on *Sedimentary Rocks.*

1. What is matrix in conglomerate?

2. See the image of limestone and explain what stromatolites are.

Return to *Geology and Geologic Features* and click on *Sedimentary Structures.*

1. Go to the image of cross-beds and explain how one can tell that these beds are overturned, that is, they are upside down. Why is it important in historical geology to know if beds are right-side up or overturned?

2. How do load casts form? Explain how these can be used to determine if beds are overturned?

3. Go to the image of ripple marks. Were these ripple marks formed by a current moving in one direction as in a stream channel or by waves? How do you know?

▶ **GEOLOGY 110 WEB PAGES OF SEDIMENTARY STRUCTURES**

This site contains links to two slide collections of sedimentary structures that are used in Geology 110 at Duke University. Click on either the *Sedimentary Structures, Part 1* or *Sedimentary Structures, Part 2* headings to view slides of a variety of sedimentary structures. Use this collection to review the different sedimentary structures discussed in this chapter.

▶ **SEDIMENTARY ROCKS**

This site, part of the Georgia Science On-Line Project, was created by Dr. Pamela J. W. Gore for Geology 101 at DeKalb College, Clarkston, Georgia. It has good images of all common sedimentary rocks, along with descriptions of their textures and composition. See the images of various sedimentary rocks.

1. What are the three basic components of terrigenous (also called detrital and clastic) sedimentary rocks?

2. What types of siliceous rocks are listed, and what are they composed of?

3. Give the names of two varieties of organic sedimentary rocks.

▶ THE ROCK CYCLE: SEDIMENTARY ROCKS

This site has a brief discussion on the origin of sedimentary rocks, but has important information on the differences between carbonate and siliciclastic sediments. Siliciclastic refers to sedimentary rocks composed of silicate minerals and clasts, that is, that have a clastic texture. For our purposes, we can consider siliciclastic sedimentary rocks to be synonymous with detrital sedimentary rocks.

Many carbonate rocks, just as siliciclastic (detrital) rocks, are composed of grains, matrix, and cement, but they also differ in several important ways. Read the discussion of differences between the two and become aware of what kinds of environmental interpretations can be made from aspects of both rock types.

CD-ROM Exploration

Explore the following *In-Terra-Active 2.0* CD-ROM module(s) and increase your understanding of key concepts and processes presented in this chapter.

Chapter Concept: **Environmental Interpretations from the Rock Record**

The lithology of sedimentary rocks not only reflects the lithology of the sediment source but also the processes involved in weathering of the source area, transportation, and eventual deposition. In this module you will explore the different processes that produce the sedimentary rock record and apply your knowledge to sedimentary materials and processes in different parts of the sedimentary rock system.

Section: **Materials**
Module: **Sedimentary Rocks**

The compositions, textures, and structures preserved in sedimentary rocks provide geologists with very clear clues as to the type of setting in which the sedimentary material was originally deposited. The purpose of the module is to explore the links between starting material, transportation and deposition process, and final sedimentary rock.

Question: *What can we conclude about the sedimentary environment in which this sandstone accumulated in northwest Wyoming?*

Evolution

An armadillo (top) and a restoration of an extinct armored mammal known as a glyptodont. Evidence such as the geographic distribution of armadillos and glyptodonts and the fact that they share several unusual characteristics was some of the evidence that convinced Charles Darwin that organisms had evolved.

Prologue

On December 27, 1811, Charles Robert Darwin sailed from Devonport, England, aboard the H.M.S. *Beagle* as an unpaid naturalist. Nearly five years later the *Beagle* returned to England, and Darwin never ventured far from home again. Nevertheless, his 64,360-kilometer voyage (Figure 5.1) was his most important experience, an experience that changed his view of nature and ultimately revolutionized all of science.

When Darwin departed from Devonport, he was a little-known recent graduate in theology from Christ's College. In fact, his father, Dr. Robert Darwin, had sent him to Christ's College as a last resort because Charles showed little aptitude for academics, except science, and he had already withdrawn from medical studies at Edinburgh. He completed his studies in theology at Christ's College but was rather indifferent to religion. He did, however, fully accept the biblical account of creation as historical fact, including the concept of *fixity of species*, meaning that existing species had been created in their present form. That is, no appreciable change took place in organisms. Darwin's belief in biblical creation initially endeared him to the *Beagle*'s captain, Robert Fitzroy, but his views changed during the voyage, and his relationship with Captain Fitzroy became strained.

When the voyage began, Darwin was nominally a clergyman with interests in botany, zoology, and geology. He suffered from prolonged bouts of seasickness but still managed to keep detailed notes on his observations and to collect, catalog, and dissect specimens; he eventually became a professional naturalist. The *Beagle* made several lengthy stops in South America where Darwin explored rain forests, experienced an earthquake, and collected fossils. Some of the fossil mammals he collected differed from existing sloths, armadillos, and llamas but were clearly similar to them and implied that living species descended with change from ancient species. In short, he began to question the concept of fixity of species.

Darwin was particularly fascinated by the plants and animals of the Cape Verde Islands and the Galapagos Islands, which are comparable distances west of Africa and South America, respectively (Figure 5.1). These islands had their own unique plants and animals, yet these plants and animals most closely resembled those of the nearby continent. The Galapagos, for instance, are populated by 13 species of

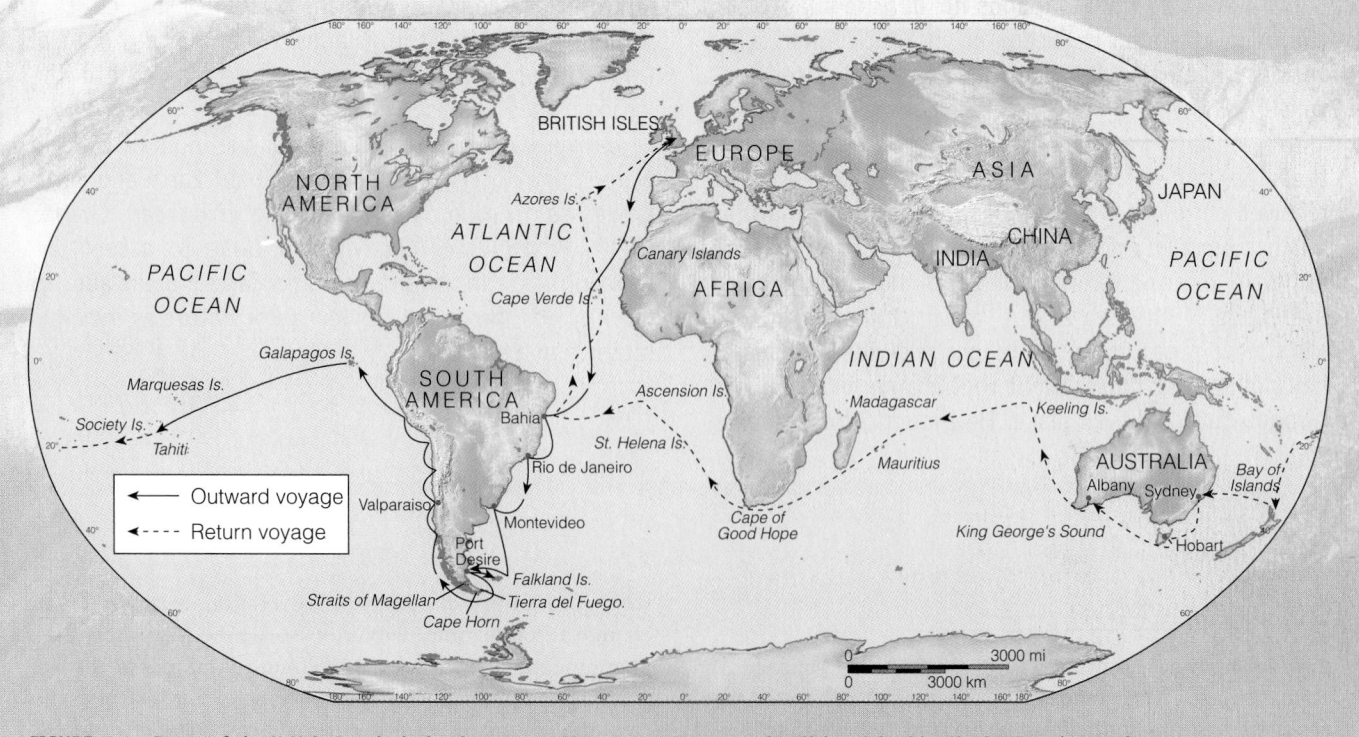

FIGURE 5.1 Route of the H.M.S. *Beagle* during its 1831–1836 voyage. Important localities visited by Charles Darwin are shown.

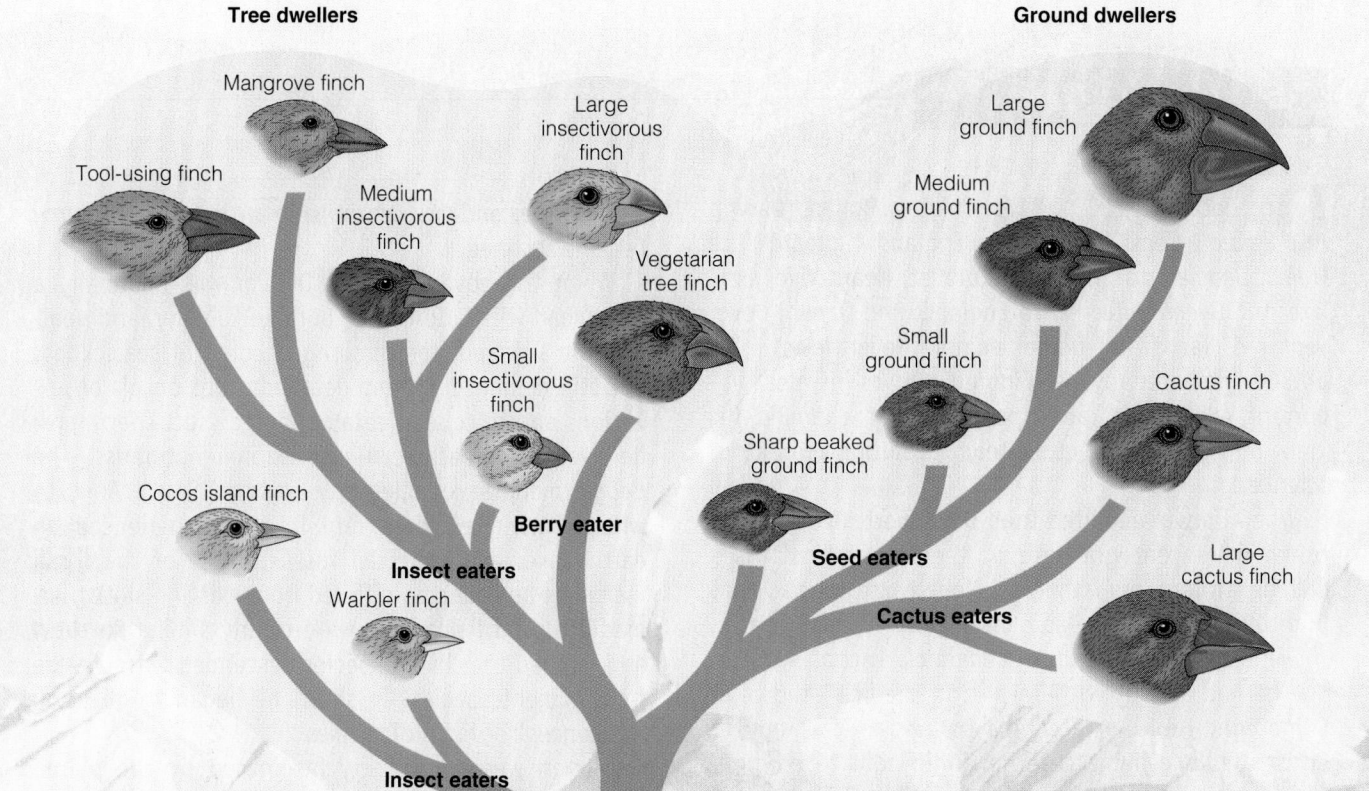

Tree dwellers

Mangrove finch

Tool-using finch

Medium insectivorous finch

Large insectivorous finch

Vegetarian tree finch

Small insectivorous finch

Berry eater

Cocos island finch

Insect eaters

Warbler finch

Insect eaters

Ground dwellers

Large ground finch

Medium ground finch

Small ground finch

Cactus finch

Sharp beaked ground finch

Seed eaters

Cactus eaters

Large cactus finch

FIGURE 5.2 Darwin's finches arranged to show evolutionary relationships. The six species on the right are ground dwellers, and the others are adapted for life in trees. Notice that the shapes of the beaks of finches in both groups vary depending on diet.

finches (Figure 5.2), but only 1 species exists in South America. The Galapagos finches are adapted to the different habitats occupied in South America by various species such as parrots, flycatchers, and toucans.

To account for his observations of organisms on these oceanic islands, Darwin reasoned that each had received colonists from the nearby continent and that "such colonists would be liable to modification—the principle of inheritance still betraying their original birthplace."* This statement reveals a change in Darwin's view of nature; he no longer accepted the idea of fixity of species but thought that change in organisms through time did take place. He thought, for example,

that the finches on the Galapagos Islands were the modified descendants of an ancestral species that somehow reached the islands from South America. Once there, this species had differentiated into the various types he observed.

Darwin's views on the nature of the biological world were changing, as were his ideas about Earth and Earth processes. During the voyage he had read Charles Lyell's *Principles of Geology* and came to accept the concept of uniformitarianism and Earth's great age. In other words, he began to view all of nature as dynamic rather than static.

*C. Darwin, *The Origin of Species* (New York: New American Library, 1958), p. 377.

Introduction

Evolution comes from the Latin term meaning "unrolled." Roman books were written on parchment and rolled on wooden rods, so they were unrolled, or evolved, as they

were read. In present-day usage, evolution usually refers to change through time. Evolution so defined is a pervasive phenomenon; Earth has evolved and continues to do so, and languages and social systems evolve. For instance, the meanings of many English words have changed through

time; *brave* once referred to "cowardice," and *nice* has had a variety of meanings different from what is now accepted. Change through time is demonstrable. Historic records document changes in social systems and languages, and the fossil record reveals a succession of different organisms that have existed through time. Moreover, some of these fossils show clear evidence of relationships among similar but not identical organisms.

The biological concept of evolution, most appropriately known as *organic evolution,* is concerned with inheritable changes. Such evolutionary changes can be observed on a small scale in present-day organisms. For example, the proportion of hereditary determinants in populations of moths in Great Britain changed so that their color changed from light to dark in several generations. Large-scale evolution is not directly observable but can be inferred from fossil evidence and from studies of comparative anatomy, embryology, and biochemistry of living organisms.

As we discussed in Chapter 1, theories are scientific explanations of natural phenomena that have a large body of supporting evidence. The **theory of evolution** is no exception. This theory's central claim is that all existing organisms are the evolutionary descendants of lifeforms that lived during the past. As a scientific theory, it meets the usual criteria for theories; it is a naturalistic explanation, and it can be tested so that its validity can be assessed.

The Evolution of an Idea

Many people think the idea of evolution began with Charles Darwin, and indeed he is justly credited with elucidating the idea and providing evidence. However, some naturalists seriously considered the biological concept of evolution even before Darwin was born. As early as 600 B.C., the ancient Greeks speculated on possible interrelationships among organisms, and evolution in one form or another was discussed by several theologians and philosophers during the Middle Ages. Overall, though, people in neither the ancient world nor the medieval gave much thought to evolution. In fact, the prevailing belief well into the eighteenth century was that all important knowledge was contained in the works of Aristotle and the Bible. One did not have to observe nature to understand it because all the answers were in these two sources, particularly in the first two chapters of Genesis. Literally interpreted, Genesis was taken as the final word on the origin and diversity of life and much of Earth history. According to this view, all species of plants and animals were perfectly fashioned by God during the days of creation and have remained fixed and immutable ever since. To question divine creation in any way was regarded as heresy.

In eighteenth-century Europe, the social and intellectual climate changed, and the absolute authority of the Church in all matters declined. Ironically, the very naturalists who were determined to find physical evidence supporting

Genesis as a factual, historical account were responsible for finding more and more evidence that could not be explained by a literal reading of Scripture. This was particularly true of the evidence for a worldwide catastrophic flood. For example, the sedimentary deposits that traditionally had been attributed to the biblical deluge showed clear indications of deposition in environments similar to those of the present. Many naturalists reasoned that this evidence truly reflected the conditions that prevailed when the strata were deposited, for to infer otherwise implied a deception on the part of the Creator, a thesis they could not accept.

It should not be concluded from this short discussion that eighteenth-century naturalists abandoned the Judeo-Christian tradition. As a matter of fact, most did not. Rather, they came to accept Genesis as a symbolic account containing important spiritual messages but not a factual account of creation, a view shared today by most theologians and biblical scholars.

These changing attitudes can be attributed in large part to the Enlightenment, an eighteenth-century philosophical movement that relied on rationalism. In science this meant that one could discover natural laws to explain observations of natural systems; all did not depend on divine revelation. The sciences, particularly biology and chemistry, began to acquire a modern look.

In this changed intellectual atmosphere, the concepts of uniformitarianism and Earth's great age were gradually accepted. The French zoologist Georges Cuvier demonstrated that many types of plants and animals had become extinct, and that fossils differed through time. In view of the fossil evidence, as well as observations and studies of living organisms, many naturalists became convinced that species were not immutable. Change from one species to another—that is, evolution—became an acceptable idea. What was lacking, though, was an overall theoretical framework to explain evolution.

LAMARCK AND THE GIRAFFE'S NECK

Erasmus Darwin (1731–1802), a physician, poet, and naturalist and Charles Darwin's grandfather, was the first to propose a process by which gradual evolution could occur. But it was Jean Baptiste de Lamarck (1744–1829), a French botanist-geologist, who was first taken seriously by his colleagues. Pointing out that organisms are adapted to their environments, he proposed that evolution is an adaptive process whereby organisms acquire traits or characteristics during their lifetimes and pass these acquired traits on to their descendants.

According to Lamarck's theory of **inheritance of acquired characteristics** ancestral short-necked giraffes, for example, stretched their necks to browse on leaves of trees (Figure 5.3). The environmentally imposed characteristic of neck-stretching was thus expressed in future generations as a longer neck. In other words, the capacity for giraffes to produce offspring with longer necks was acquired through the habit of neck-stretching.

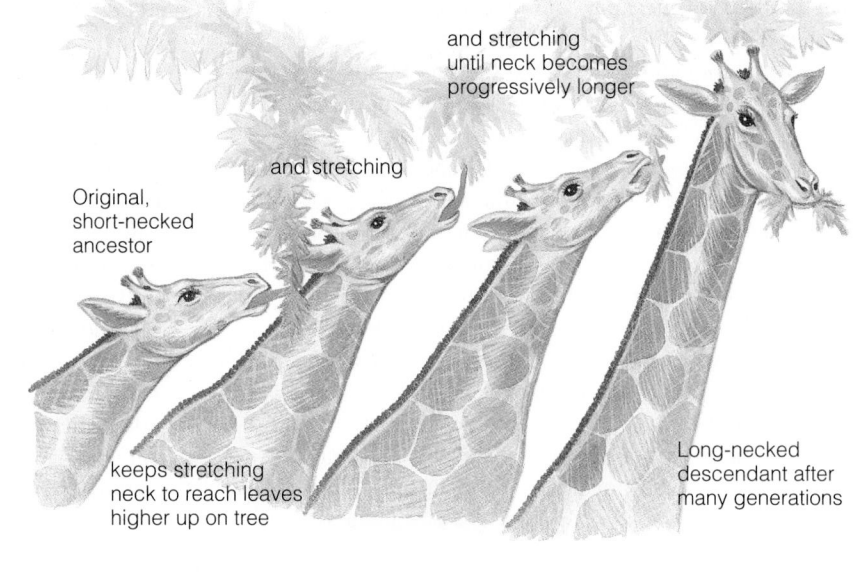

FIGURE 5.3 According to Lamarck's theory of inheritance of acquired characteristics, ancestral short-necked giraffes stretched their necks to reach leaves higher up on trees, and their descendants were born with longer necks.

Original, short-necked ancestor

and stretching

and stretching until neck becomes progressively longer

keeps stretching neck to reach leaves higher up on tree

Long-necked descendant after many generations

Considering the data available in the early 1800s, Lamarck formulated a reasonable theory. Little was known about how traits are passed from parent to offspring, so Lamarck's inheritance mechanism seemed logical and was accepted by many naturalists as a viable mechanism for evolution. In fact, Lamarck's concept of inheritance was not completely refuted until more modern concepts of inheritance were developed. Based on the work of Gregor Mendel in the 1860s, and developments in the early 1900s, we now know that all organisms possess hereditary determinants called *genes* that cannot be modified by any effort of the organism during its lifetime. Numerous attempts have been made to validate Lamarckian inheritance, including one during this century in the former Soviet Union (see Perspective 5.1), but all have failed.

CHARLES DARWIN'S CONTRIBUTION

In 1859 Charles Robert Darwin published *The Origin of Species* in which his ideas on evolution were outlined (Figure 5.4). Most naturalists almost immediately recognized that evolution provides a unifying theory that explains an otherwise encyclopedic collection of biologic facts. Even most, but not all, churches and theologians eventually accepted the idea of evolution as a naturalistic explanation that does not conflict with the purpose and intent of Scripture.

While 1859 may mark the beginning of modern evolutionary biology, Darwin had actually formulated his ideas more than 20 years earlier, but aware of the furor his ideas would generate, had been reluctant to publish. As a matter of fact, when the furor did erupt, Darwin, a semi-invalid most of his life, never defended his theories in public; he left their defense to others.

As discussed in the Prologue, Darwin concluded during the voyage of the *Beagle* that species are not fixed. In fact, by the time Darwin returned home in 1836, he had what he thought was good evidence that species had indeed changed through time, but he had no idea what might bring about change. By 1838, however, his observations of the selection practiced by animal and plant breeders, and a chance reading of Thomas Malthus's essay on population, gave him the elements necessary to formulate his theories.

FIGURE 5.4 Charles Darwin in 1850.

The Tragic Lysenko Affair

*A*n unquestioning obedience to the dictates of some pseudoscience, or to a discredited scientific theory, can have devastating results. One of the most tragic examples of adherence to a disproved scientific theory involved Trofim Denisovich Lysenko (1898–1976) who became president of the Soviet Academy of Agricultural Sciences in 1938. Lysenko endorsed the theory of inheritance of acquired characteristics according to which plants and animals could be changed in desirable ways simply by exposing them to a new environment. For example, according to Lysenko, seeds exposed to dry conditions would acquire a resistance to drought, and this trait would be inherited by future generations. Lysenko accepted inheritance of acquired characteristics because of its apparent compatibility with Marxist–Leninist philosophy. As president of the academy, and with the endorsement of the Central Committee of the Soviet Union, Lysenko did not allow Soviet scientists to conduct any other research concerning inheritance mechanisms, and those who publicly disagreed with him lost their jobs or were even sent to labor camps.

Unfortunately for the Soviet people, inheritance of acquired characteristics had been discredited as a scientific theory more than 50 years before. The results of Lysenko's belief in the political correctness of this theory were widespread crop failures, starvation, and misery for millions of people. In fact, Lysenko's dismantling of genetic research was so complete that the former Soviet Union is still striving to catch up to the rest of the world in this field.

An interesting sidenote to this tragedy is that the only reputable Western biologist to support Lysenko was England's J. B. S. Haldane. Haldane, whose numerous contributions to genetics and evolutionary theory won him international recognition and respect, discovered Marxism during the 1930s and joined the Communist party in 1942. Unfortunately, his belief in the political correctness of the Marxist–Leninist doctrine appears to have blinded Haldane to the scientific absurdity of Lysenko's claims.

The lesson to be drawn from this example is that scientific research must be based on scientific realities. Science proceeds on the basis of the scientific method, not philosophical or political beliefs. Lysenko's ideas on inheritance were not mandated as the correct way to proceed in agricultural research because of their scientific merit but because they were deemed compatible with a belief system.

Animal and plant breeders selected organisms for desirable traits, bred organisms with those traits, and thereby induced a great amount of change. This practice of **artificial selection** yielded a fantastic variety of domestic pigeons, dogs, and plants (Figure 5.5). If this much change could be produced artificially, was a process that selected among variant types perhaps also acting on natural populations?

Darwin came to appreciate fully the power of selection after reading Malthus's book on population. Malthus argued that far more animals are born than reach maturity, yet the numbers of adult animals of a species remain rather constant. He reasoned that competition for resources resulted in a high infant mortality rate, thus limiting the size of the population. Darwin proposed that a natural process was selecting only a few animals for survival, a process he called **natural selection.**

NATURAL SELECTION

For 20 years Darwin kept his ideas on evolution and natural selection largely to himself. But in 1858 he received a letter from Alfred Russell Wallace, a young naturalist working in southern Asia. Wallace had also read Malthus's essay on population and had come to precisely the same conclusion as Darwin. Friends convinced Darwin that his paper and Wallace's should be read at the same session of the Linnaean Society of London. Later some of Darwin's corre-

FIGURE 5.5 Artificial selection has yielded more than 300 domesticated breeds of pigeons, three of which are shown here. The common rock dove (upper left) is thought to be the parent stock of the domestic breeds.

spondence was published, establishing his scientific priority, and indeed Wallace even insisted that Darwin be given credit in recognition of his earlier discovery and more thorough documentation of the theory.

The Darwin–Wallace theory of natural selection can be summarized as follows:

1. All populations contain heritable variations—size, speed, agility, coloration, digestive enzymes, and so forth.
2. Some variations are more favorable than others; that is, some variant types have an edge in the competition for resources and avoidance of predators.
3. Not all young survive to reproductive maturity.
4. Those with favorable variations are more likely to survive and pass on their favorable variations.

Evolution by natural selection is then largely a matter of reproductive success, for only those that reproduce can pass on favorable variations. Of course, favorable variations do not guarantee survival for an individual, but within a population, those with favorable variations are more likely to survive and reproduce.

Natural selection is the process proposed by Darwin and Wallace to account for evolution. But is their statement substantially different from Lamarck's evolution by the inheritance of acquired characteristics? In other words,

how would Darwin and Wallace explain the giraffe's long neck?

Suppose in an ancestral population of giraffes some individuals had longer necks than most of the others (Figure 5.6). These long-necked animals could obviously browse higher on trees. The important point here is that some simply have longer necks and therefore enjoy a selective advantage in that they can acquire resources more effectively. Consequently, they are more likely to survive, reproduce, and pass on their favorable variations. Even in the second generation, a few more giraffes might have longer necks, which would give a competitive edge. And so it would go, generation by generation, thereby increasing the proportion of giraffes with longer necks.

A common misconception about natural selection is that among animals only the biggest, strongest, and fastest are likely to survive. It is true that size and strength are important when male bighorn sheep compete for mates, but females also pass on their genes. Speed is an advantage to some predators and their prey, but not all predators are fast. Consider badgers, for example, and that some small cats use stealth and pouncing rather than speed to capture prey. In fact, in some animal populations, natural selection might favor the smallest, or most easily concealed, or those that more readily adapt to a new food source or have the ability to detoxify some natural or human-made substance.

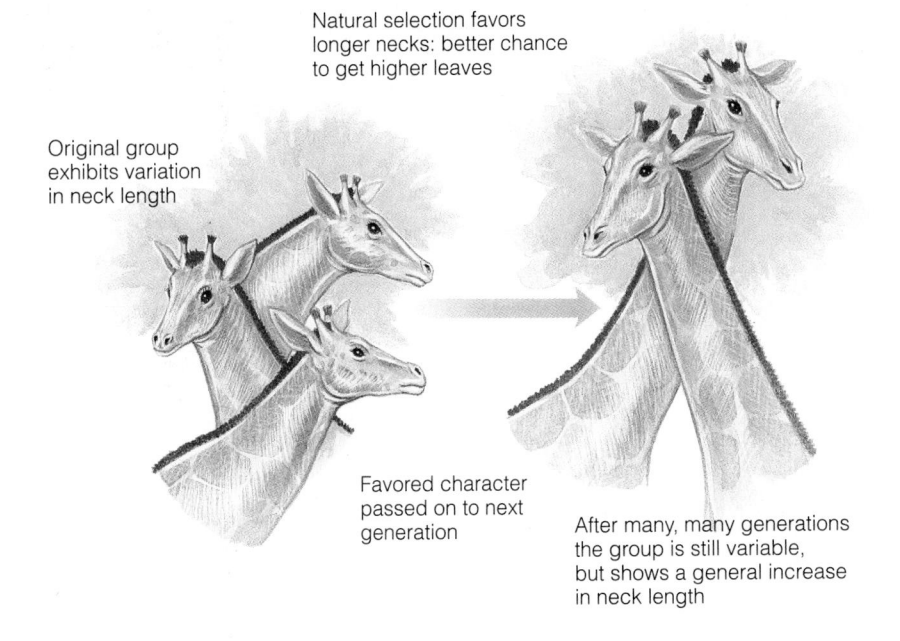

Natural selection favors
longer necks: better chance
to get higher leaves

Original group
exhibits variation
in neck length

Favored character
passed on to next
generation

After many, many generations
the group is still variable,
but shows a general increase
in neck length

FIGURE 5.6 According to the Darwin–Wallace theory of natural selection, the giraffe's long neck evolved as animals with favorable variations were selected for survival.

In short, favorable variations need not be obvious features of animals or any other organism.

Darwin and Wallace proposed two separate theories, descent with modification (evolution) and natural selection. The first, *descent with modification,* means that organisms existing at present are the modified descendants of organisms that lived during the past. That is, in a line of descent, change has occurred. *Natural selection,* on the other hand, is the theoretical mechanism to account for descent with modification. Each theory is independently testable and must be evaluated separately. Darwin and Wallace were attempting to explain how organisms changed through time; they were not, however, trying to explain how life originated.

Mendel and the Birth of Genetics

Critics of natural selection were quick to point out that Darwin and Wallace could not account for the origin of variation or explain how it was maintained in populations. They reasoned that should a variant trait arise, it would simply blend with other traits in the population and be lost. At the time, this was a valid criticism because so little was known about the heritability of traits, and neither Darwin nor Wallace could effectively answer it. Actually, information that could have given them the answer existed, but it remained in obscurity until 1900.

A monastery may seem to be an unlikely setting for the discovery of the rules of inheritance, and an Austrian monk may seem an unlikely candidate for the "father of genetics." Nevertheless, during the 1860s Gregor Mendel, who later became abbot of the monastery, was doing research that answered some of the inheritance problems that plagued Darwin and Wallace. Unfortunately, Mendel's work was published in an obscure journal and went largely unnoticed. Ironically, Mendel even sent Darwin a copy of his manuscript, but it apparently lay unread on his bookshelf.

Mendel performed a series of controlled experiments with true-breeding strains of garden peas (strains that when self-fertilized always display the same trait, such as a particular flower color). In one experiment, he transferred pollen from white-flowered plants to red-flowered plants, which produced a second generation of all red-flowered plants. But when left to self-fertilize, these plants yielded a third generation with a ratio of red-flowered plants to white-flowered plants of slightly over 3 to 1 (Figure 5.7).

Mendel concluded from his experiments that traits such as flower color are controlled by a pair of factors, or what we now call **genes.** He also concluded that genes controlling the same trait occur in alternate forms, or **alleles;** that one allele may be dominant over another; and that offspring receive one allele of each pair from each parent. When an organism produces sex cells—pollen and ovules in plants, and sperm and eggs in animals—only one allele for a trait is present in each sex cell (Figure 5.7). For example, if R represents the allele for red flower color, and r represents white, the offspring may inherit the combinations of alleles symbolized as RR, Rr, or rr. And since R is dominant over r, only those offspring with the rr combination will have white flowers.

The most important aspects of Mendel's work can be summarized as follows: The factors (genes) controlling traits do not blend during inheritance but are transmitted as discrete units; even though particular traits may not be expressed in each generation, they are not lost. Therefore, some variation in populations is accounted for by alternate expression of genes (alleles), and variation can be maintained.

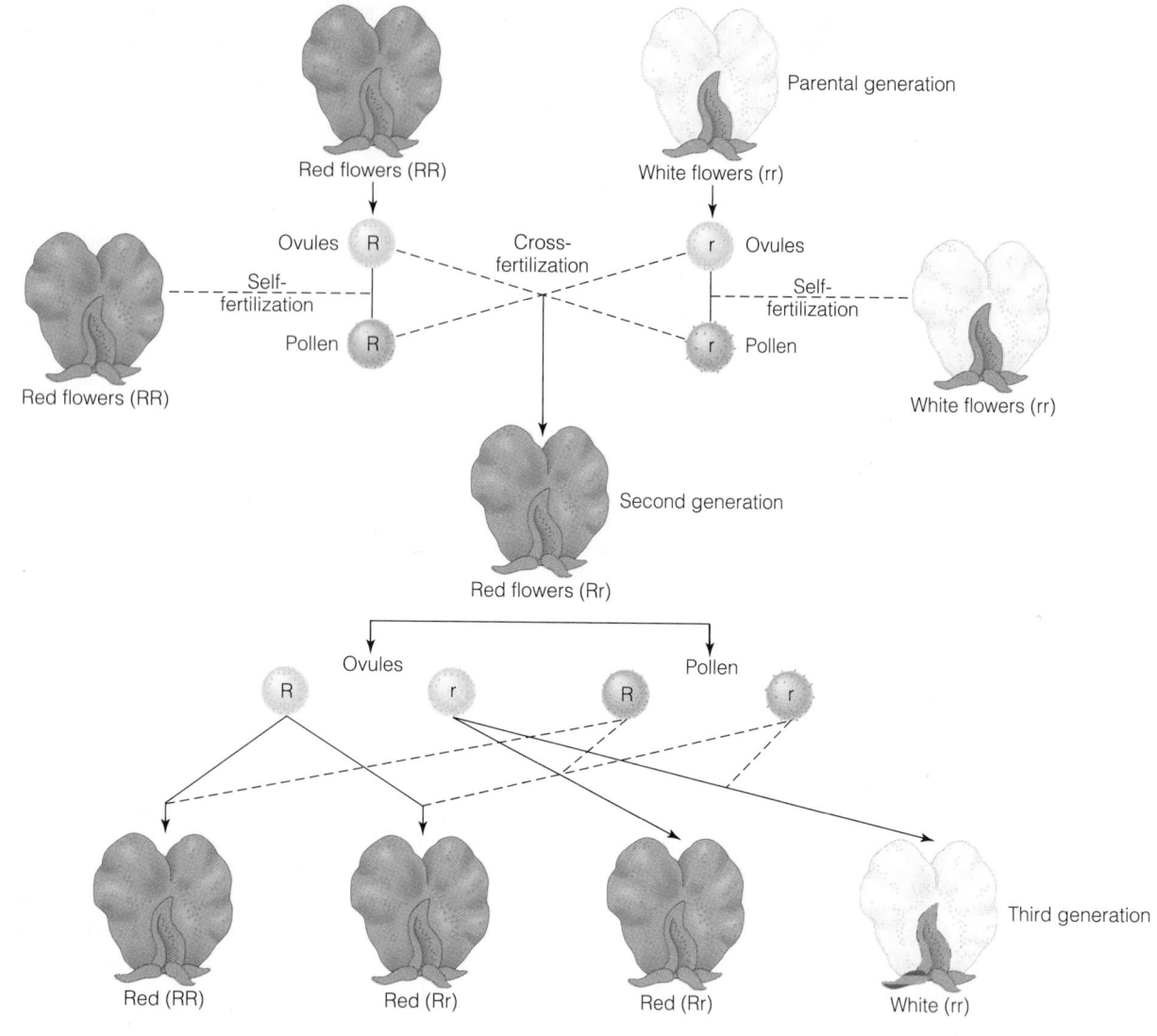

FIGURE 5.7 In his experiments with flower color in garden peas, Mendel used true-breeding strains. Such plants, shown here as the parental generation, when self-fertilized always yield offspring with the same trait as the parent. However, if the parental generation is cross-fertilized, all plants in the second generation will receive the combination of alleles indicated by Rr; these plants will have red flowers because R is dominant over r. The second generation of plants produces ovules and pollen with the alleles shown and, when left to self-fertilize, produces a third generation with a ratio of three plants with red flowers to one plant with white flowers.

Even though Mendelian genetics explains much about heredity, we now know the situation is much more complex. For example, our discussion has relied upon a single gene controlling a trait, but most traits are in fact controlled by many genes. Nevertheless, Mendel discovered the answers Darwin and Wallace needed, though they went unnoticed until three independent researchers rediscovered them in 1900.

GENES AND CHROMOSOMES

The cells of all organisms contain threadlike structures known as **chromosomes,** complex, double-stranded, helical molecules of **deoxyribonucleic acid (DNA)** (Figure 5.8). Specific segments, or regions, of the DNA molecule are the basic hereditary units, the genes. The number of chromosomes is specific for a species but varies among

species. For example, the fruit fly *Drosophila* has 8 chromosomes, humans have 46, and domestic horses have 64. However, chromosomes occur in pairs, pairs that carry genes controlling the same characteristics. Remember that the genes on chromosome pairs may occur in different forms called alleles.

In sexually reproducing organisms, the production of eggs and sperm results when parent cells undergo a type of cell division known as **meiosis.** The meiotic process yields cells containing only one chromosome of each pair (Figure 5.9a). Accordingly, eggs and sperm have only one-half the chromosome number of the parent cell; human eggs and sperm have 23 chromosomes, one of each pair.

During reproduction, a sperm fertilizes an egg, producing a fertilized egg with the full chromosome number for that species—46 in humans. As Mendel deduced from his

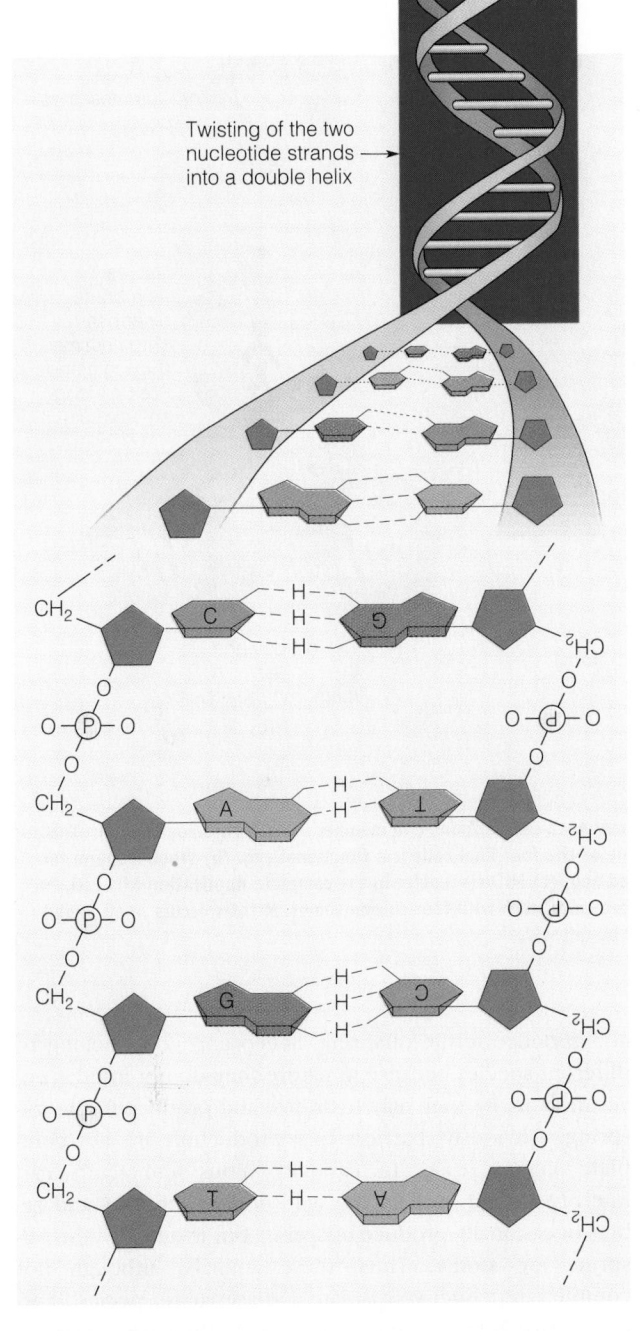

Twisting of the two nucleotide strands into a double helix →

FIGURE 5.8 Chromosomes are double-stranded, helical molecules of deoxyribonucleic acid (DNA) shown here diagrammatically. Specific segments of chromosomes are genes. The two strands of the molecule are joined by hydrogen bonds (H).

garden pea experiments, one-half of the genes of a fertilized egg come from each parent. The fertilized egg, however, develops and grows by a cell division process called **mitosis** that does not reduce the chromosome number (Figure 5.9c).

According to the **chromosomal theory of inheritance,** chromosomes carrying genes are passed from one generation to the next. However, changes known as **mutations** may take place in genes, and any such change occur-

The Modern Synthesis

As noted previously, the Darwin–Wallace concept of natural selection was challenged because it could not explain how variation arose nor how it was maintained in populations. The developing science of genetics partly answered the inheritance problems, but early in this century, geneticists believed that mutation rather than natural selection was the mechanism whereby evolution occurred.

During the 1930s and 1940s, the ideas developed by geneticists, paleontologists, population biologists, and others were brought together to form a **modern synthesis** or neo-Darwinian view of evolution. The chromosome theory of inheritance was incorporated into evolutionary thinking; mutation was seen as one source of variation in populations; Lamarckian concepts of inheritance were completely rejected; and the importance of natural selection was reaffirmed. The modern synthesis also emphasized that evolution is a gradual process, a point that has been challenged by some recent investigators, as will be discussed later.

SOURCES OF VARIATION

The raw materials for evolution by natural selection are variations within populations. Most variations can be accounted for by sexual reproduction and the reshuffling of alleles from generation to generation. Considering that each of thousands of genes may have several alleles and that offspring receive one-half of their genes from each parent, the potential for variation is enormous.

New variation may be introduced into a population by a change in genes, a mutation. To understand mutations, we must explore the function of chromosomes. One function of chromosomes is to direct the synthesis of molecules called proteins. During protein synthesis, information in the chromosome structure directs the formation of a protein by selecting the appropriate amino acids in a cell and arranging these amino acids into a sequence. The proteins produced determine the characteristics of an organism. Any change in the information directing protein synthesis is a mutation.

Two additional aspects of mutations are important. First, only mutations in sex cells are inheritable. And second, mutations are random with respect to fitness. That is, no evidence indicates a predetermination of a mutation's effect; it may be harmful (many are), neutral, or beneficial. However, the attributes of harmful versus beneficial can be considered only with respect to the environment.

If a species is well adapted to its environment, most mutations would not be particularly useful and perhaps would be harmful. But if the environment changes, what was once a harmful mutation may become beneficial. For instance, some plants have developed a tolerance for con-

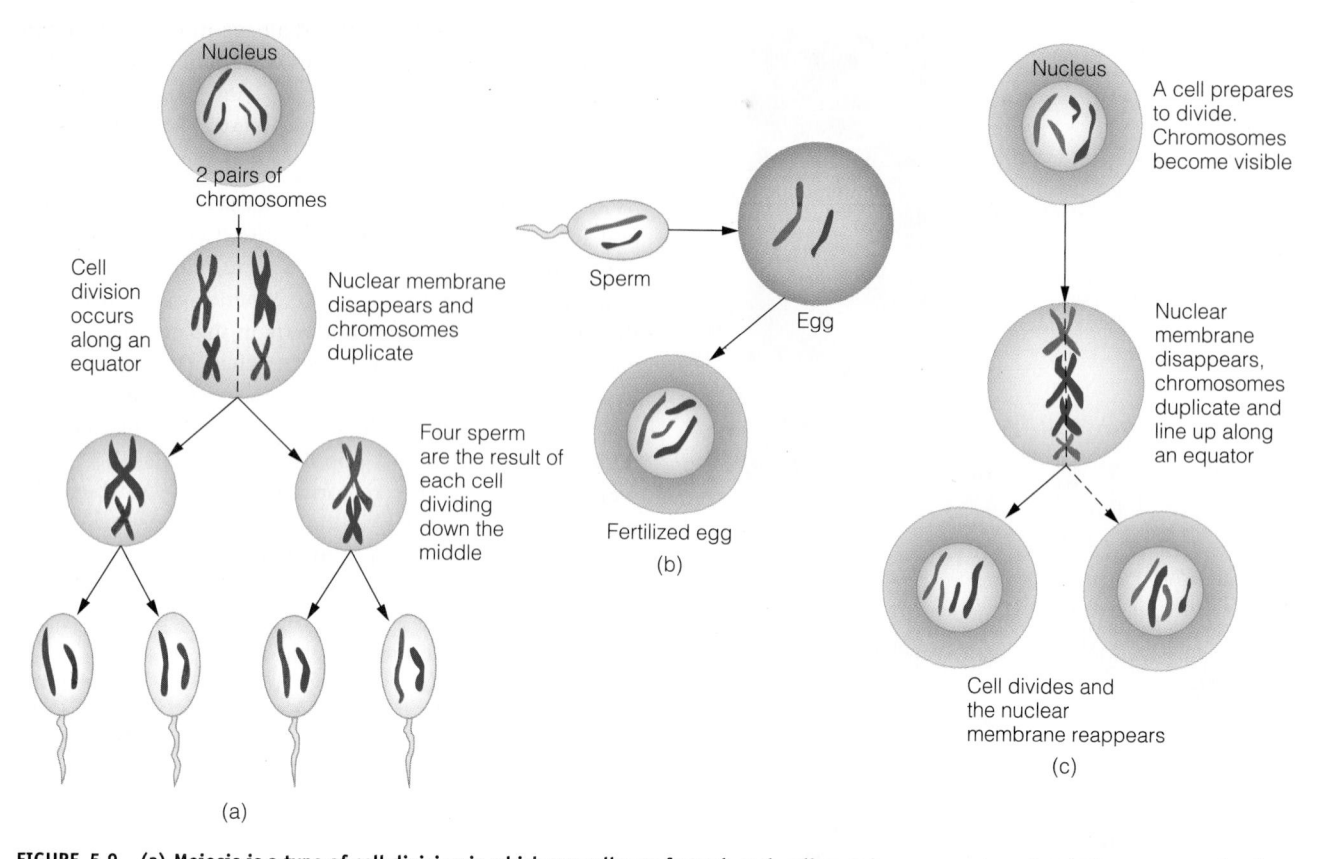

FIGURE 5.9 (a) Meiosis is a type of cell division in which sex cells are formed; each cell contains one member of each chromosome pair. The formation of sperm cells is shown; eggs form in a similar manner, but only one of the four final cells is a functional egg. (b) When a sperm fertilizes an egg, the full complement of chromosomes is present in the fertilized egg. (c) Mitosis results in the complete duplication of a cell. In this simplified example, a cell with four chromosomes (two pairs) produces two cells, each with four chromosomes. Mitosis occurs in all body cells except sex cells. Once an egg is fertilized, the developing embryo grows by mitosis.

taminated soils around mines. Plants of the same species from the normal environment do poorly or die in these contaminated soils, while contaminant-resistant plants do poorly in the normal environment. The mutations for contaminant resistance probably occurred repeatedly in the population, but their adaptive significance was not realized until contaminated soils were present.

Variations in the genetic makeup of a population can also change by chance alone. This phenomenon known as *genetic drift* is most important in small populations where the loss of a few individuals might eliminate or drastically reduce the frequency of genes for a particular trait. Whether the gene in question is adaptive or more advantageous than other genes is irrelevant. Genetic drift can take place when, for instance, a few individuals that become separated from a larger population start a new population, as when a few birds or plants reach an oceanic island. In addition, some kind of natural disaster such as disease might eliminate all but a few members of a population, be it large or small, thus accounting for a marked change in gene frequencies.

SPECIATION AND THE RATE OF EVOLUTION

In classifying organisms, the term **species** is used for populations of similar individuals that in nature can interbreed

and produce fertile offspring. Sheep and goats belong to different species because in nature they do not interbreed, yet in captivity they may hybridize and produce fertile offspring. Obviously, barriers to reproduction are not complete in this case. In captivity, lions and tigers can interbreed, and even in the wild domestic horses and zebras occasionally produce offspring, but in all cases the offspring are sterile. Thus, these animals, although not complete reproductively isolated, are separate species. If reproductive barriers are complete, as in the case of species as different as horses and deer, for example, no interbreeding is possible under any circumstances.

The process of speciation involves a change in the genetic makeup of a population—that is, a change in the frequency of alleles in its available genes, or simply its **gene pool.** Marked change may also take place in form and structure (**morphology**), but morphologic change is not necessary in the origin of a new species. In fact, a descendant species might look very much like its ancestral species.

According to the concept of **allopatric speciation,** species arise when a small part of a population is geographically isolated from its parent population by some kind of barrier. A rising mountain range, a stream, or an invasion of part of a continent by the sea may effectively iso-

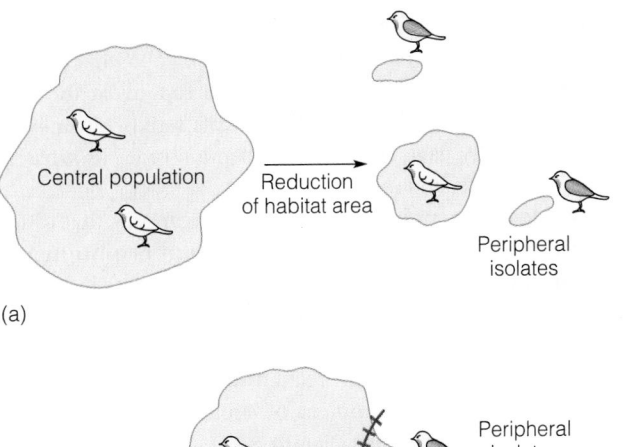

(a)

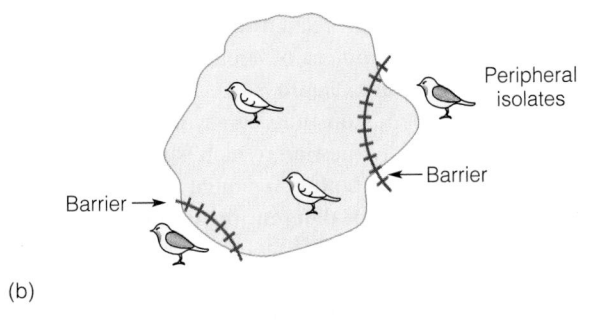

(b)

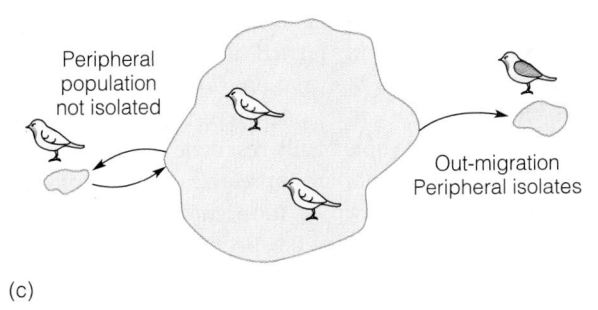

(c)

FIGURE 5.10 Allopatric speciation. (a) Reduction of the area occupied by a species may leave small, isolated populations, *peripheral isolates,* at the periphery of the once more extensive range. In this example, members of both peripheral isolates have evolved into new species. (b) Barriers have formed across parts of a central population's range, thereby isolating small populations. (c) Out-migration and the origin of a peripheral isolate.

late parts of a once-interbreeding population (Figure 5.10). Isolation may also occur when a few individuals of a species are accidentally introduced to a remote area.

Once a small population becomes isolated, it no longer shares in the parent population's gene pool, and it is sub-

jected to different selection pressures. Under these conditions, the isolated population evolves independently and eventually may give rise to a reproductively isolated species. Numerous examples of allopatric speciation have been well documented, especially in extremely remote areas such as oceanic islands; the finches of the Galapagos Islands are a good example (Figure 5.2).

Most investigators agree that allopatric speciation is a common phenomenon, but they disagree about how rapidly new species evolve. Darwin proposed that the origin of a new species is generally a gradual process. That is, the gradual accumulation of minor changes brings about a transition from one species to another, a process commonly called **phyletic gradualism** (Figure 5.11a). According to this concept, an ancestral species changes gradually and continuously and grades imperceptibly into a descendant species.

Another view, known as **punctuated equilibrium,** holds that a species changes little or not at all during most of its history and then evolves rapidly to give rise to a new species. In other words, long periods of equilibrium are occasionally punctuated by short periods of rapid evolution. Accordingly, a new species arises rapidly, perhaps in a few thousand years, but once it has evolved, it remains much the same for the rest of its existence (Figure 5.11b).

Proponents of punctuated equilibrium claim that the fossil record supports their hypothesis. They argue that few examples of gradual transitions from one species to another are found in the fossil record. According to this view, transitional forms connecting ancestral and descendant species are unlikely to be preserved in the fossil record because species arise rapidly in small, geographically isolated populations. And, as a matter of fact, many species do appear abruptly in the fossil record with no evidence of direct ancestors.

Critics, however, point out that neither Darwin nor those who formulated the modern synthesis insisted that all evolutionary change was gradual and continuous, a view shared by many present-day biologists and paleontologists. Indeed, Darwin allowed for times during which evolutionary change in small populations could be quite rapid. Furthermore, deposition of sediments in most environments is not continuous; thus, the lack of gradual transitions in many cases is simply an artifact of the fossil record.

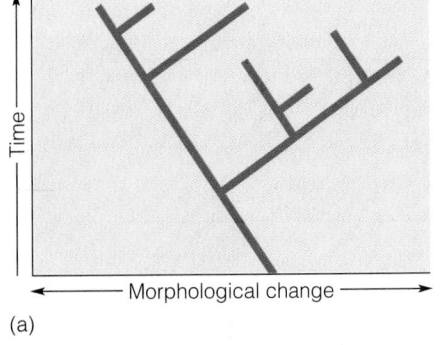

(a)

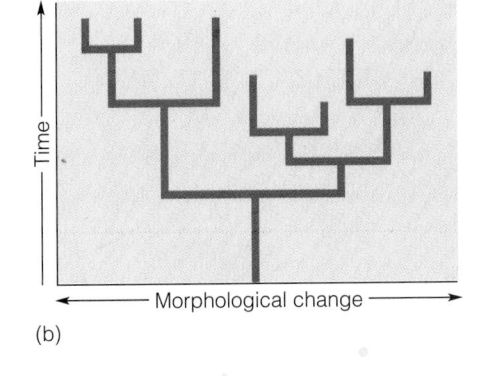

(b)

FIGURE 5.11 Comparison of two models for differentiation of organisms from a common ancestor. (a) In the phyletic gradualism model, slow, continuous change takes place as one species evolves into another. (b) According to the punctuated equilibrium model, change occurs rapidly, and new species evolve rapidly. However, little or no change occurs in a species during most of its existence.

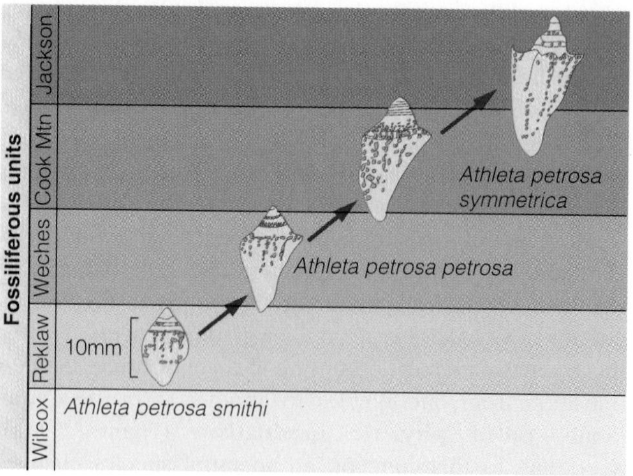

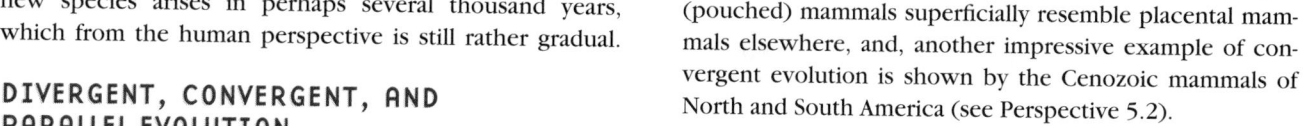

FIGURE 5.12 Gradual evolution of the small snail *Athleta* showing changes in size, shape, and the development of spines on the shell.

And finally, despite the incomplete nature of the fossil record, a number of examples of gradual transitions from ancestral to descendant species are known (Figure 5.12). Today, most scientists consider punctuated equilibrium as one possible way for evolution to take place. Furthermore, one must remember that, even according to this concept, a new species arises in perhaps several thousand years, which from the human perspective is still rather gradual.

DIVERGENT, CONVERGENT, AND PARALLEL EVOLUTION

When an interbreeding population gives rise to diverse descendant types of organisms, the process is referred to as **divergent evolution** (Figure 5.13). Divergence into numerous related types involves an *adaptive radiation*, which occurs when species of related ancestry exploit different aspects of the environment. At some time in the past, a population of finches, apparently from South America, reached the Galapagos Islands. This ancestral population diversified into the adaptive types of species of finches that Darwin observed (Figure 5.2).

The fossil record provides many examples of divergence and adaptive radiation. The diversification and adap-

tive radiation of placental mammals from a common ancestor is an excellent example (Figure 5.14). Following the extinction of dinosaurs and some related reptiles at the end of the Mesozoic Era, placental mammals rapidly diversified and eventually gave rise to such adaptive types as whales, bats, elephants, and monkeys.

While divergent evolution leads to organisms that differ markedly from their ancestors, **convergent evolution** and **parallel evolution** are processes whereby similar adaptations arise in different groups (Figure 5.13). Convergent evolution can be defined as the development of similiar characteristics in distantly related organisms, and parallel evolution as the development of similar characteristics in closely related organisms (Figure 5.13). The distinction between the two depends on the degree of relatedness between the organisms in question, which is not always easy to determine. However, both convergent and parallel evolution occur as a result of different organisms adapting to similar ways of life.

The shells of clams and brachiopods, although not identical, are similar, as they represent an adaptive response to a similar lifestyle (Figure 5.15). Because clams and brachiopods are very distantly related, their similar shells are an example of convergent evolution. Convergence also accounts for the fact that many Australian marsupial (pouched) mammals superficially resemble placental mammals elsewhere, and, another impressive example of convergent evolution is shown by the Cenozoic mammals of North and South America (see Perspective 5.2).

Parallel evolution is also a common phenomenon, but, as previously noted, it is not always easy to distinguish from convergent evolution. For example, jerboas and kangaroo rats are closely related, but each independently developed similar structures (Figure 5.16).

EVOLUTIONARY TRENDS

The evolutionary history, or *phylogeny,* of a group can be worked out in some detail if sufficient fossil material is known. Furthermore, constructing phylogenies reveals the *evolutionary trends* that characterize a group of organisms. The phylogeny of ammonites, an extinct group related to the living squid and octopus, is well known; one

FIGURE 5.13 In this example, divergent evolution results as species A diverges and gives rise to two descendant species, B and C. Parallel evolution also involves divergence from a common ancestor, A, but descendant species, B and C, then develop in a similar way as they adapt to similar lifestyles. Convergent evolution is much like parallel evolution except that species B and C are distantly related but resemble one another in some aspects because of similar adaptations.

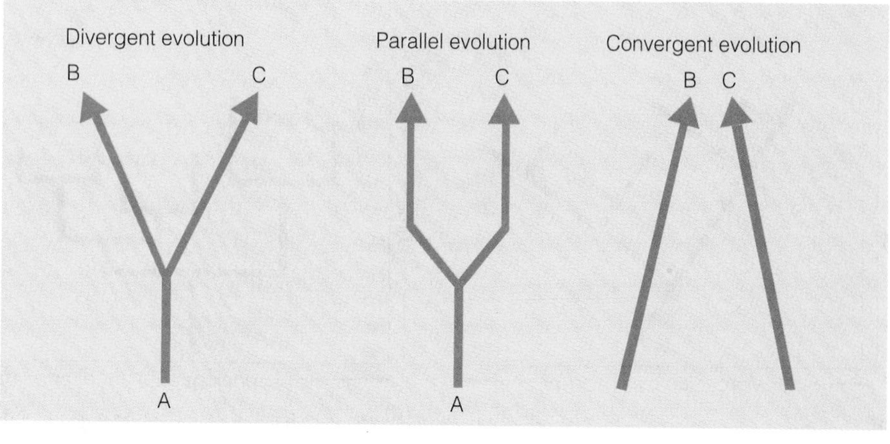

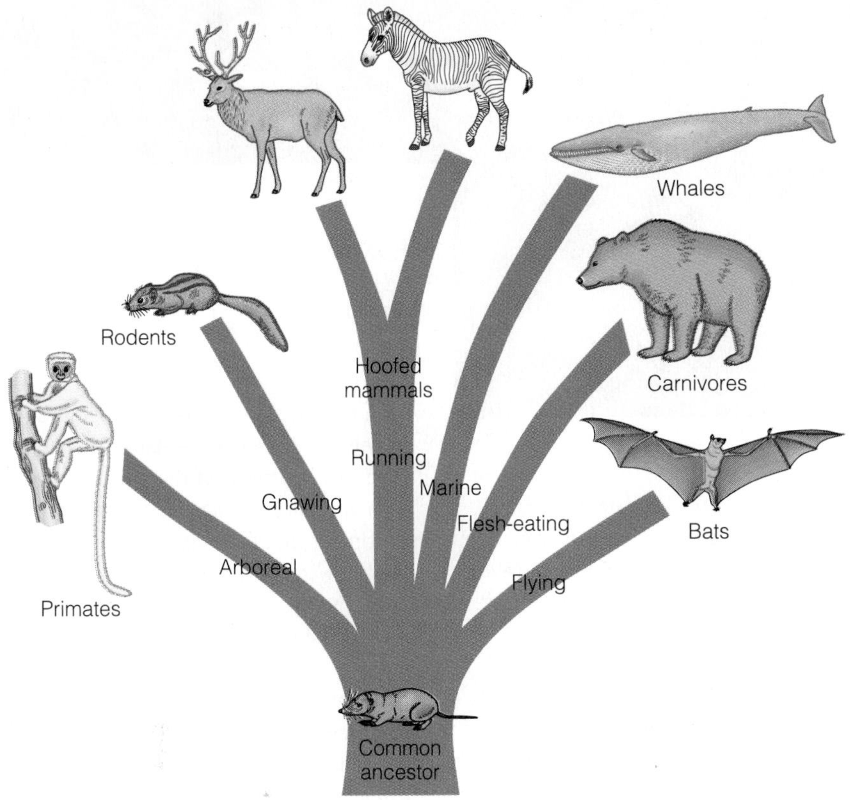

FIGURE 5.14 Divergent evolution of the placental mammals. An ancestor, probably a small, shrewlike animal, gave rise to diverse descendant groups, each of which adapted to different lifestyles.

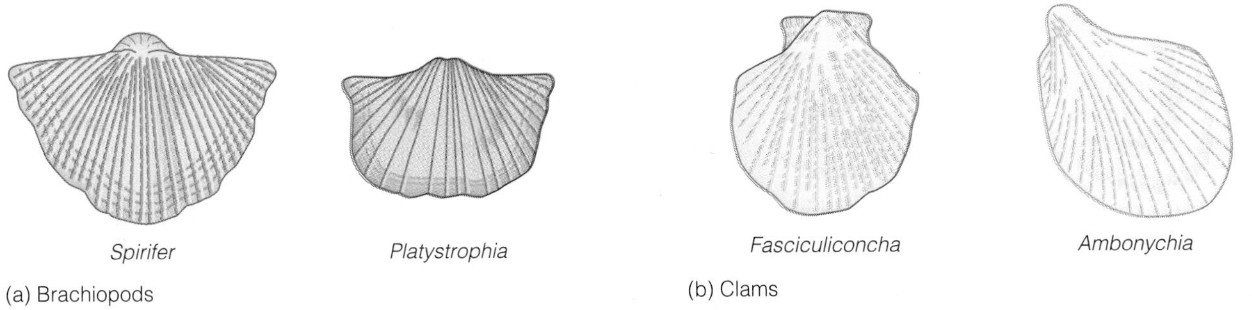

Whales

Carnivores

Rodents

Hoofed mammals

Running

Gnawing

Marine

Flesh-eating

Bats

Arboreal

Flying

Primates

Common ancestor

Spirifer

Platystrophia

Fasciculiconcha

Ambonychia

(a) Brachiopods

(b) Clams

FIGURE 5.15 (a) Brachiopods and (b) clams are distantly related but have adapted to a very similar habitat. Convergence is shown by the similarities in their shells; in both groups the shell consists of two parts hinged together. However, the details of the shells of each group differ in a number of aspects.

Kangaroo rat

Jerboa

FIGURE 5.16 Parallel evolution. The kangaroo rat and jerboa are closely related but have independently developed similar features.

perspective 5.2

Convergent Evolution

*T*he development of similar features in distantly related organisms resulting from comparable lifestyles is termed *convergent evolution*. It is such a common phenomenon that paleontologists must be careful or they might infer close evolutionary relationships where none exist. In Figure 5.15, for instance, clam and brachiopod shells closely resemble one another, but we know from fossil and living species that many differences are also present both in the shells and the animals' internal anatomy. Thus, even though they look alike, their common ancestor is extremely remote, and their similarities can be accounted for by convergent evolution. Likewise, dolphins (marine mammals) and ichthyosaurs (marine reptiles), despite outward appearances, are not closely related but developed along similar evolutionary pathways (see Figure 5.22). Some

present-day lizards have no legs, and all snakes are legless—yet each evolved from distantly related ancestors. And the eyes of vertebrate animals and those of squid and octopus are similar.

Most familiar examples of convergent evolution are found in vertebrate animals, but convergence can and does take place in all organisms. For instance, the biting mouthparts of some bugs and mosquitos are similar but not identical. And many desert plants of Africa and the American Southwest have developed in a similar fashion. A good example is the cactus *Echinocereus* of America, which resembles *Euphorbia* of Africa (Figure 1); both plants evolved independently from leafy plants that are only distantly related. Chaparral, a many-branched woody plant of the American West, has counterparts in Chile, Africa, and along the Mediterranean coast.

A remarkable example of convergence in animals is seen in the marsupial (pouched) mammals of the Australian region and placental mammals on the other continents. Both marsupials and placentals are warm blooded, have hair, and nurse their young, so they share a common ancestor, but their evolutionary paths diverged during the Cretaceous Period. Nevertheless, striking similarities have developed so that we now recognize mouselike, doglike, catlike, and squirrel-like marsupials. The only marsupials for which there are no similar counterparts elsewhere are kangaroos.

During most of the Cenozoic Era, South America was isolated from all other continents, much like Australia is today. As a result of this isolation, its mammalian fauna of marsupials and placentals evolved independently. A number of mammals on each continent adapted to similar environments and developed many features in common. If we could somehow

(a)

(b)

FIGURE 1   Each of these plants evolved from distantly related ancestors, so their similarities resulted from convergent evolution. (a) *Echinocereus* of the American Southwest and (b) *Euphorbia* from southwestern Africa.

visit South America about 30 million years ago, we would see animals resembling camels, horses, rhinoceroses, and others of North America (Figure 2). A particularly interesting example of convergence among North and South American mammals is the development of saberlike canine teeth in cats and catlike marsupials.

All living carnivorous mammals have well-developed canine teeth, the daggerlike teeth used to make the kill and rip and tear flesh from the prey. Sabertoothed cats of the Pleistocene with canines 15 cm long are the best-known carnivores with enlarged canines, but they were not the only ones with this specialization. Similar canines developed independently four times during the Cenozoic Era: once in an extinct group of mammals known as creodonts, once in a South American marsupial, and twice in the cats (Figure 2). Parallel evolution best explains the development of enlarged canines in two closely related groups of cats, whereas convergent evolution accounts for the same features in cats, creodonts, and marsupials—all of which are distantly related.

Besides enlarged canine teeth, other remarkable similarities are seen in these animals. Skull modifications were necessary so that the mouth could be opened widely enough for effective use of the long canines. Thus, it is not surprising that the skulls of all saber-toothed carnivorous mammals are similarly constructed because there are only a few ways to modify the basic mammalian skull to accommodate enlarged canines. Considering that saberlike canine teeth existed in one kind of mammal or another during most of the Cenozoic Era, it is surprising that no animal with this specialization exists at the present.

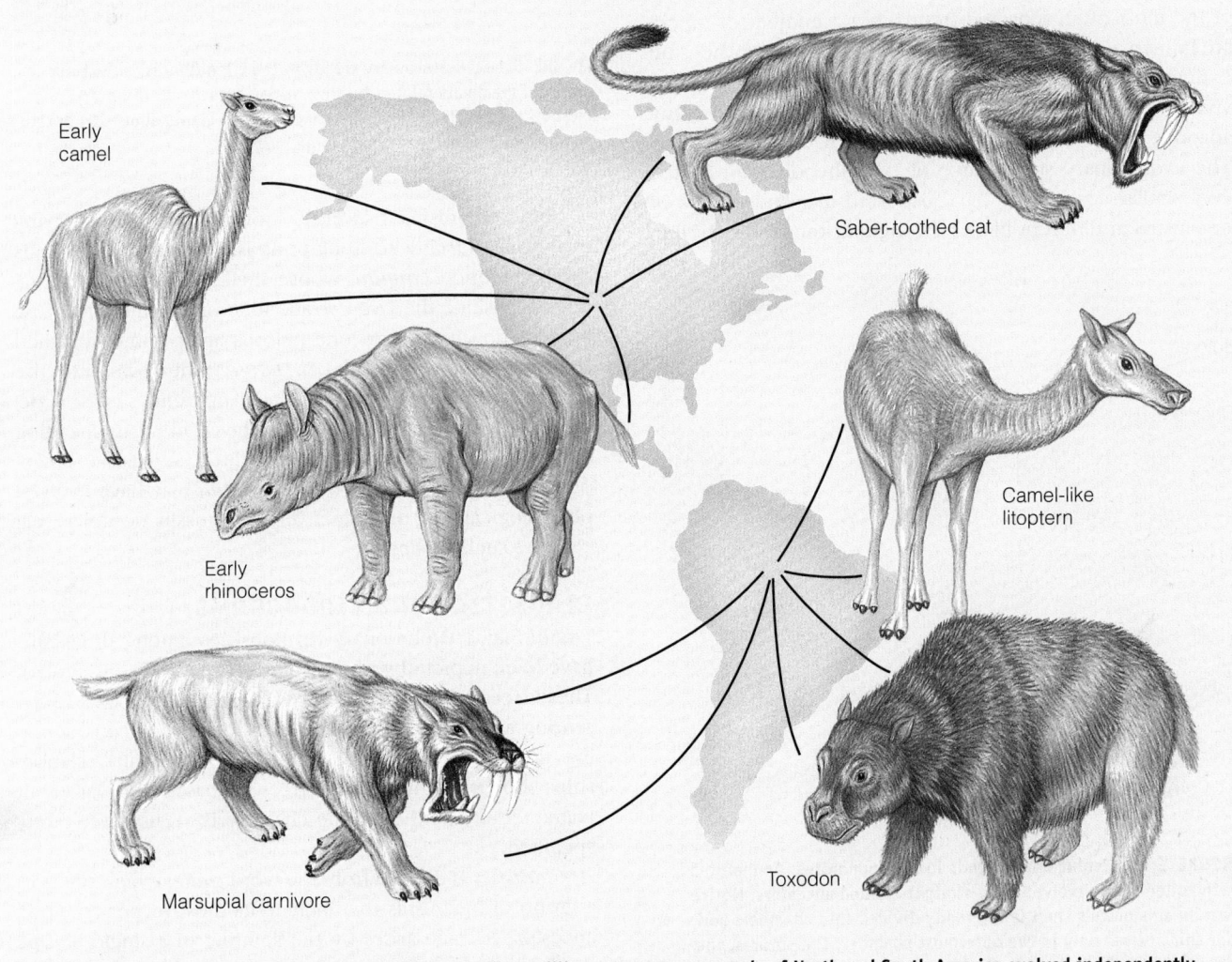

Early camel

Saber-toothed cat

Early rhinoceros

Camel-like litoptern

Marsupial carnivore

Toxodon

FIGURE 2 Before the origin of the Isthmus of Panama a few million years ago, mammals of North and South America evolved independently. Similar adaptations of animals on each continent resulted from convergent evolution.

evolutionary trend in this group was an increasingly complex shell structure (Figure 5.17).

Titanotheres were mammals that lived in North America and eastern Asia from Early Eocene to Early Oligocene time. Numerous fossils record their phylogeny and several evolutionary trends (Figure 5.18). One trend was simply an increase in size, but titanotheres also developed large horns, and the shape of the skull changed.

Increase in size is one of the most common evolutionary trends. However, trends are complex in that they do not always proceed at the same rate, they may be reversed, or several may occur in the same group. Horses evolved from small ancestors that lived during the Eocene. One trend was an increase in size, but it was not a steady, uniform increase, and, in fact, some horses actually show a reversal in this trend. Horse evolution is also characterized by a reduction in the number of toes, lengthening of the legs and feet, and changes in the teeth and skull. All of these trends are well documented by fossils, but they did not take place at the same rate (see Chapter 18).

One can view evolutionary trends as a series of adaptations to changing environments or adaptations that occur in response to exploitation of new habitats. The same trends occur repeatedly in the fossil record. Increase in size is one of the most often repeated trends, but a number of others are known as well. Ammonites, for example, nearly became extinct at the end of the Paleozoic Era, but those that survived into the Mesozoic diversified and gave rise to many adaptive types very similar to those that had existed earlier. The evolutionary significance of this phenomenon is that very similar adaptations have occurred in many groups of organisms in different places throughout the history of life.

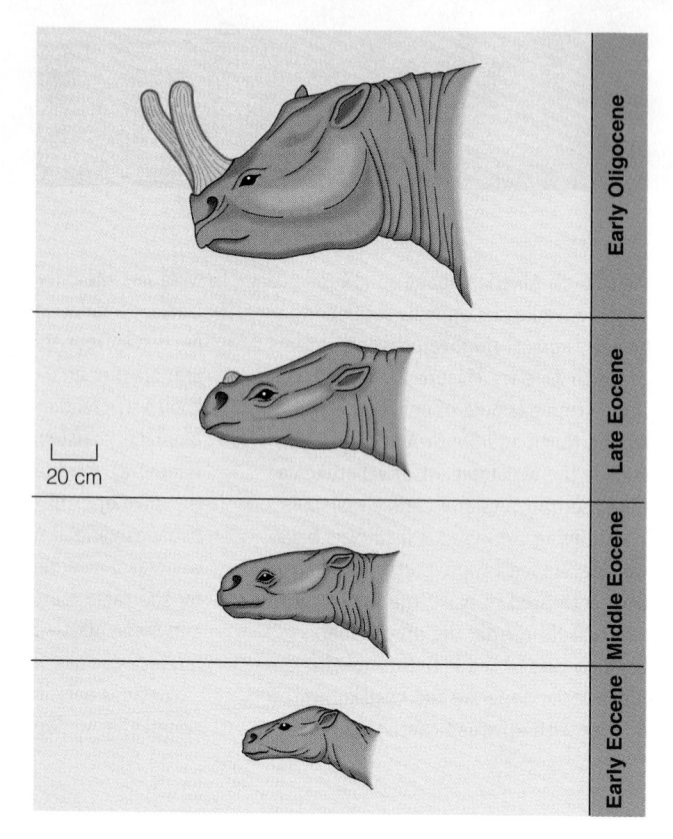

FIGURE 5.18 Evolutionary trends in the titanotheres, an extinct group of mammals related to present-day horses and rhinoceroses. Titanotheres evolved from small ancestors to giants about 2.4 m tall at the shoulder and developed large horns.

Some organisms, however, show little evidence of any evolutionary trends for long periods. One good example is the brachiopod *Lingula,* whose shell has not changed significantly since the Ordovician. Such organisms that show little or no change for long periods are commonly called **living fossils** (Figure 5.19). *Lingula* burrows into the muddy sediments of shallow seas. Apparently, once it developed this adaptive strategy, it stayed in an unchanging environment and has changed very little, at least in the appearance of its shell. Of course, *Lingula* may have evolved physiologically or reproductively, but fossils do not reveal changes such as these.

CLADISTICS AND CLADOGRAMS

Traditionally, evolutionary relationships among organisms have been depicted with *phylogenetic trees* (Figure 5.20a). These trees are constructed with the horizontal axis representing anatomical differences and the vertical axis denoting time. The patterns of ancestor–descendant relationships shown in phylogenetic trees are based on a variety of characteristics, although the characteristics used are rarely specified.

Cladistics is derived from the word *clade,* which refers to a group of organisms that include its most recent common ancestor. A clade arises by the splitting of a single lineage into two discrete lineages, and a *cladogram* is a diagrammatic representation of these relationships. Cladograms are

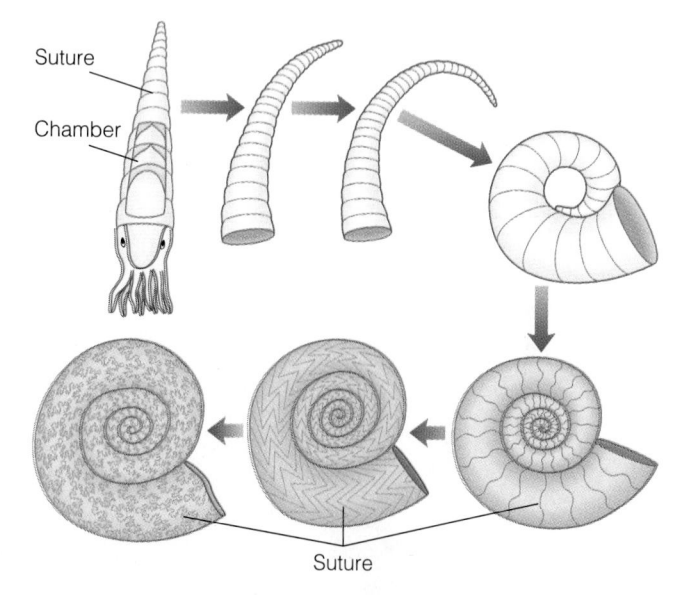

FIGURE 5.17 Evolutionary trends in the ammonites. Ammonites with coiled shells evolved from straight-shelled ancestors. Notice that an ammonite's shell is internally divided into chambers and that the animal lives only in the outermost chamber. The sutures, the lines formed where chamber walls meet the wall of the outer shell, became increasingly complex as ammonites evolved.

FIGURE 5.19 Two examples of so-called living fossils. (a) *Neopilina* is a deep-sea-dwelling mollusk that closely resembles Paleozoic mollusks. (b) *Latimeria* is a member of a group of fishes that were thought to have become extinct at the end of the Mesozoic Era. A specimen was caught in 1938 off the east coast of Africa, and since then several more have been caught.

(a) (b)

constructed on the basis of derived as opposed to primitive characteristics. Bone and paired limbs are found in all land-dwelling vertebrates (amphibians, reptiles, birds, and mammals) and are thus considered primitive because all members of this group have them. Accordingly, they are of little use in establishing relationships among vertebrate animals. Hair and mammary glands, however, are derived characteristics, sometimes called *evolutionary novelties,* because they are present in only one of the subcaldes, the mammals (Figure 5.20b). If one considers only mammals, hair and mammary glands are primitive characteristics, but live birth is derived because it serves to distinguish egg-laying mammals from those that give birth to live young.

One can construct a cladogram for any association of organisms. For bats, dogs, and birds, we can construct three

cladograms and evaluate which most likely depicts their evolutionary relationships. Figure 5.21 shows three possible cladograms, each a different interpretation of the relationships among these animals. Bats and birds fly, so we might think that they are more closely related to one another than either is to dogs (Figure 5.21a). But if we concentrate on evolutionary novelties such as hair and birth of live young, rather than laying eggs, we conclude that bats and dogs are closely related, more so than either is to birds. Thus, the cladogram in Figure 5.21c is the best interpretation of the relationships among these animals.

Cladograms usually do not have an absolute time scale associated with them, but they still show the relative times of appearance of the groups possessing derived characteristics. In short, they elucidate evolutionary relationships.

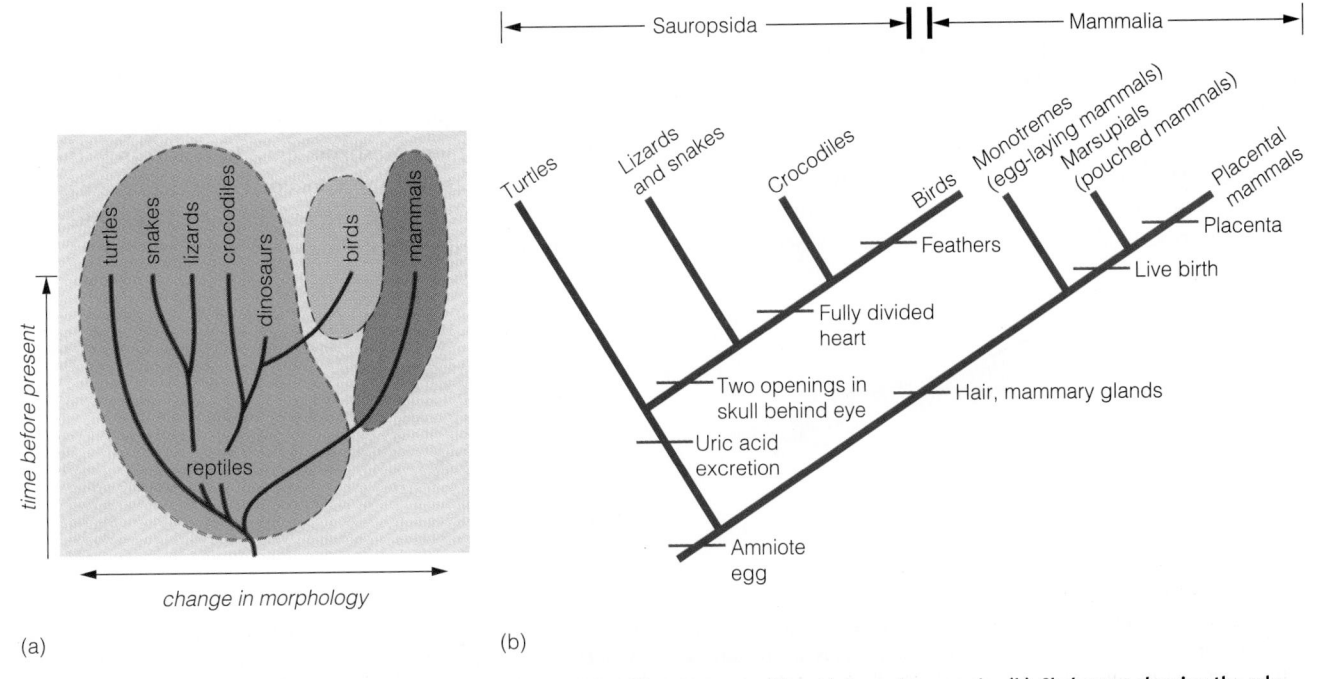

(a) (b)

FIGURE 5.20 (a) A phylogenetic tree showing the inferred relationships among reptiles, birds, and mammals. (b) Cladogram showing the relationships among living reptiles, birds, and mammals. Some of the characteristics used to differentiate subclades are indicated.

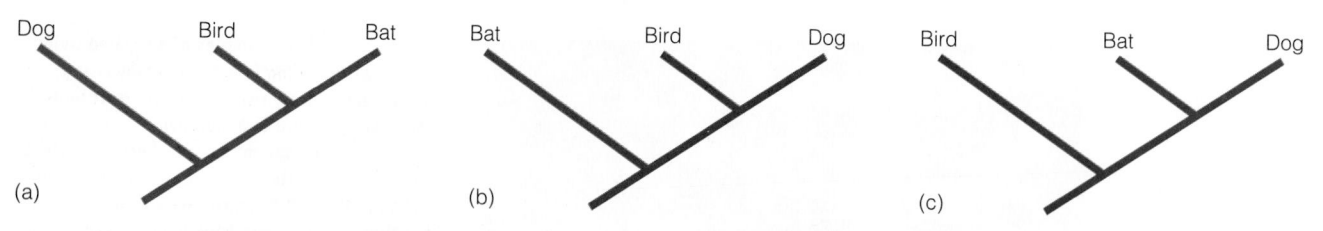

(a) Dog Bird Bat

(b) Bat Bird Dog

(c) Bird Bat Dog

FIGURE 5.21 Cladograms showing three hypotheses for the relationships among birds, bats, and dogs. Derived characteristics such as hair and giving birth to live young indicate that dogs and bats are most closely related, as shown in (c).

For instance, in Figure 5.20b, all mammals are more closely related among themselves than to any of the sauropsids, and crocodiles are more closely related to birds than either is to turtles, snakes, or lizards.

Cladistics and cladograms work well for living organisms and are now widely used for fossil plants and animals. With fossils, though, care must be taken in recognizing primitive versus derived characteristics, especially in groups with poor fossil records. Even in the better-known evolutionary lineages, the earliest diversification is usually the most poorly represented by fossils. Furthermore, the traits evaluated in cladistic analyses are assumed to have arisen solely by descent from a common ancestor, so paleontologists must be especially careful of characteristics that resulted from convergent evolution.

EXTINCTIONS

James Audubon (1785–1851) once estimated that a single flock of passenger pigeons numbered more than 1 billion individuals. In 1857 a committee of the Ohio Senate said passenger pigeons needed no protection. In 1914 the last passenger pigeon died in the Cincinnati Zoo. Steller's sea cow was discovered in 1742 and 27 years later had been hunted to extinction. Rhinoceroses in the wild are now on the verge of extinction.

The preceding examples of extinction were brought about by humans, and, indeed, humans have a degree of control over their environment unprecedented in the history of life. They can bring about changes that might not have taken place otherwise or that would have occurred at very different rates. Judging from the fossil record, extinction is the rule in the history of life. In fact, it seems that

extinction is the ultimate fate of all species. According to one estimate, more than 99% of all species that ever existed are now extinct.

The term *extinct* seems so final, and, in a very real sense, it is. But we must distinguish between two types of extinction. In one case a species evolves into a new species that differs enough from its ancestral group that the parent species can be considered extinct. Perhaps this is best called *pseudoextinction* because, strictly speaking, no extinction occurred. The second case of extinction is terminal. That is, a species dies out without giving rise to anything else. Many examples of terminal extinction are known from the fossil record, flying reptiles known as pterosaurs being a well-known example.

Extinction by one method or another is a continual occurrence in the history of life. But so is the origin of new species that usually quickly exploit the opportunities created by the disappearance of other species. In some cases of extinction, an ecologically equivalent organism may not appear for some time. Ichthyosaurs were Mesozoic marine reptiles that lived like and superficially resembled present-day porpoises and dolphins (Figure 5.22). Yet, whereas ichthyosaurs are unknown after the Mesozoic, porpoises and dolphins did not occupy their niche until some 30 million years later.

At times extinction rates have been greatly accelerated in events called *mass extinctions*. The two greatest crises in the history of life were the mass extinctions at the end of the Paleozoic and Mesozoic eras. In both cases, the diversity of life was sharply reduced. These extinctions, and some of lesser magnitude, will be discussed in greater detail in the following chapters.

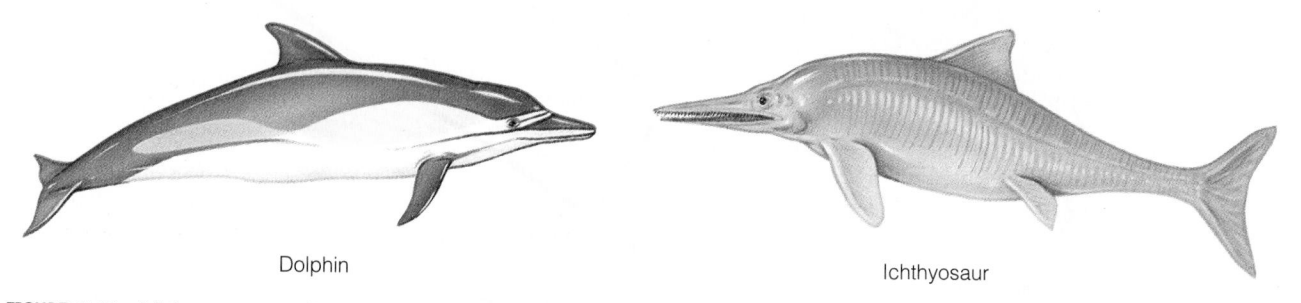

Dolphin

Ichthyosaur

FIGURE 5.22 Ichthyosaurs were fish-eating, marine reptiles that became extinct at the end of the Mesozoic Era. Following their extinction, no air-breathing animal occupied their niche until dolphins and porpoises appeared about 30 million years later. Notice that ichthyosaurs and dolphins resemble one another in several features. Both groups adapted to a similar lifestyle and provide an excellent example of convergent evolution.

Evidence for Evolution

When Darwin proposed his theory, he cited such supporting evidence as classification, embryology, comparative anatomy, the geographic distribution of organisms, and, to a limited extent, the fossil record. Darwin had little knowledge of the mechanism of inheritance or biochemistry, however, and molecular biology was completely unknown during his lifetime. Studies in these areas coupled with a better understanding of the fossil record have added to the body of evidence. Today, most scientists accept that the theory of evolution is as well supported by evidence as any major theory.

CLASSIFICATION—A NESTED PATTERN OF SIMILARITIES

The Swedish naturalist Carolus Linnaeus (1707-1778) proposed a formal classification of organisms. According to the Linnaean system, an organism is given a two-part name; the coyote, for example, is referred to by its genus (plural, *genera*) name and its species name, *Canis latrans.*

Table 5.1 shows the basic arrangement of the Linnaean classification although the scheme now used contains more categories. The arrangement is hierarchical. That is, as one proceeds up the list, the categories become more inclusive. The coyote *(Canis latrans)* and the wolf *(Canis lupus)* are members of different species, but they share many characteristics, so both belong to the same genus. The red fox *(Vulpes fulva)* also shares some, but fewer, characteristics with coyotes and wolves, so it, along with other related doglike animals, make up the family Canidae. In turn, all canids share some characteristics with cats, bears, weasels, and raccoons, all of which belong to the order Carnivora. Likewise, all carnivores, rodents, bats, primates, and elephants have hair and mammary glands and constitute the class Mammalia (Figure 5.23). And so it goes, up to kingdom, the most inclusive category in the classification scheme.

It should be clear from the preceding discussion that shared characteristics are the bases for inclusion in a specific category. Linnaeus clearly recognized similarities among organisms, but he simply intended to classify species that he believed were specially created and immutable. He viewed his classification as a reflection of the Great Chain of Being, a sequence from simple to complex, with humans at the apex.

Following the publication of *The Origin of Species,* biologists soon realized that shared characteristics among organisms constitute a strong argument for evolution. In our example, coyotes and wolves share many characteristics because they evolved from a common ancestor in the not too distant past. They also share some characteristics with bears and cats because all members of the order Carnivora had a common ancestor in the more remote past.

BIOLOGICAL EVIDENCE FOR EVOLUTION

According to evolutionary theory, all lifeforms are related, and all existing organisms descended with modification from ancestors that lived during the past. If this statement is correct, then evidence of fundamental similarities among all lifeforms should exist, and closely related organisms should be more similar to one another than to more distantly related organisms. As a matter of fact, all living things, from bacteria to whales, are composed of the same chemical elements—mostly carbon, nitrogen, hydrogen, and oxygen. Furthermore, the chromosomes of all organisms are composed of DNA, except bacteria, which have RNA; and in all cells, proteins are synthesized in essentially the same way.

Thousands of biochemical tests of numerous organisms have provided compelling evidence for evolutionary relationships. Blood proteins, for example, are similar among all mammals, but they also indicate that among the primates humans are most closely related to the great apes, followed, in order, by the Old World monkeys, the New World monkeys, and the lower primates such as lemurs. Biochemical tests indicate that all birds are related among

---

TABLE 5.1

The Classification Scheme Now in Use Showing the Hierarchical Arrangement of the Categories

The boxes include those animals to which the coyote, *Canis latrans,* is most closely related at various levels in the classification scheme (see Figure 5.23).

	Coyote	Wolf	Red fox	Bobcat	Horse	Snapping turtle	Amphioxus	Starfish	White clover
Kingdom	Animalia	Animalia	Animalia	Animalia	Animalia	Animalia	Animalia	Animalia	Plantae
Phylum	Chordata	Chordata	Chordata	Chordata	Chordata	Chordata	Chordata	Echinodermata	Pterophyta
Subphylum	Vertebrata	Vertebrata	Vertebrata	Vertebrata	Vertebrata	Vertebrata	Cephalochordata	Eleutherozoa	
Class	Mammalia	Mammalia	Mammalia	Mammalia	Mammalia	Reptilia	Leptocardii	Asteroidea	Angiospermae
Order	Carnivora	Carnivora	Carnivora	Carnivora	Perissodactyla	Chelonia		Forcipulata	Rosales
Family	Canidae	Canidae	Canidae	Felidae	Equidae	Chelydridae		Asteriidae	Leguminosae
Genus	*Canis*	*Canis*	*Vulpes*	*Lynx*	*Equus*	*Chelydra*	*Branchiostoma*	*Asterias*	*Trifolium*
Species	*latrans*	*lupus*	*vulpes*	*rufus*	*caballus*	*serpentinia*	*virginiae*	*forbesi*	*repens*

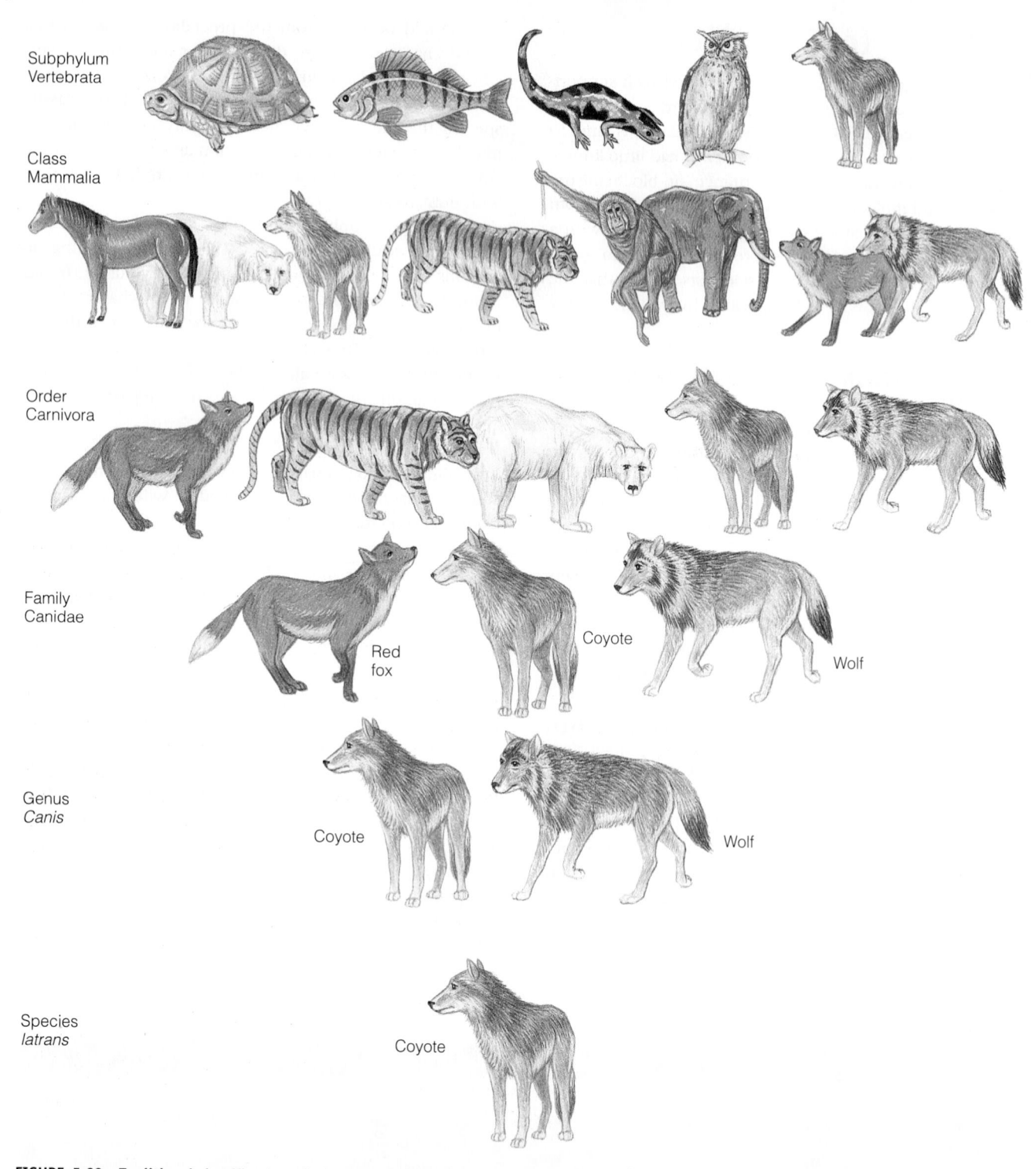

FIGURE 5.23 Traditional classification of organisms based on shared characteristics. All members of the subphylum Vertebrata—fishes, amphibians, reptiles, birds, and mammals—have a segmented vertebral column. Among the vertebrates, only those warm-blooded animals having hair or fur and mammary glands are mammals. Eighteen orders of mammals are recognized, including the order Carnivora shown here. All members of this order have teeth specialized for a diet of meat. The family Canidae includes only the doglike carnivores, and the genus *Canis* includes closely related species. The coyote, *Canis latrans,* stands alone as a species.

themselves and are more closely related to turtles and crocodiles than to snakes and lizards, a finding corroborated by the fossil record.

Studies of developing embryos reveal that some organisms are similar to one another until very late in embryonic development, while others are less similar. Fish, chimpanzee, and human embryos are very similar in the earliest stages, but differences soon become apparent in fish embryos. Chimpanzees and humans, in contrast, retain remarkable embryonic similarities until very late in the

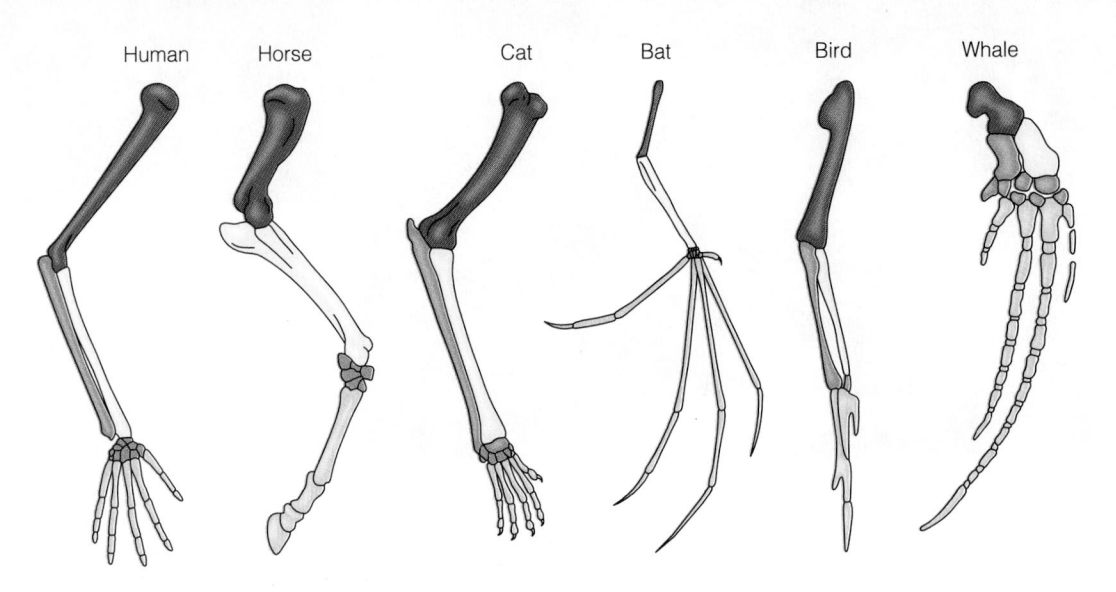

FIGURE 5.24 Homologous organs such as the forelimbs of vertebrates may serve different functions but are composed of the same elements and undergo similar embryological development.

developmental process. The inference is that chimpanzees and humans are more closely related to one another than they are to fish.

Evidence for evolution is also provided by comparing anatomical structures of organisms (see Perspective 5.3). The forelimbs of humans, whales, dogs, and birds (Figure 5.24) are superficially dissimilar, yet all are composed of the same bones and have basically the same arrangement of muscles, nerves, and blood vessels; all are similarly arranged with respect to other structures, and all have a similar pattern of embryonic development. Such structures are said to be **homologous.** Descent with modification from a common ancestor—that is, a common genetic heritage—is the best explanation for the existence of homologous organs.

Not all similarities among organisms are evidence of evolutionary relationships; in fact, some similarities are rather superficial. Insects are not closely related to bats and birds, but all three of these animals have wings. Such structures that serve the same function are **analogous organs** (Figure 5.25). The wings of bats and birds, but not of insects, are homologous, too, and provide evidence for a common ancestry. Insect wings, on the other hand, represent convergence—a similar solution to an adaptive problem, flight—but they differ from bat and bird wings in both structure and embryological development.

Another type of evidence for evolution is provided by observations of small-scale evolution in living organisms (Figure 5.26). We have already mentioned one example, the adaptations of plants to contaminated soils. New insecticides and pesticides must be developed continually as insects and rodents develop resistance to existing ones. Development of antibiotic-resistant strains of bacteria constitutes a continuing problem in medicine.

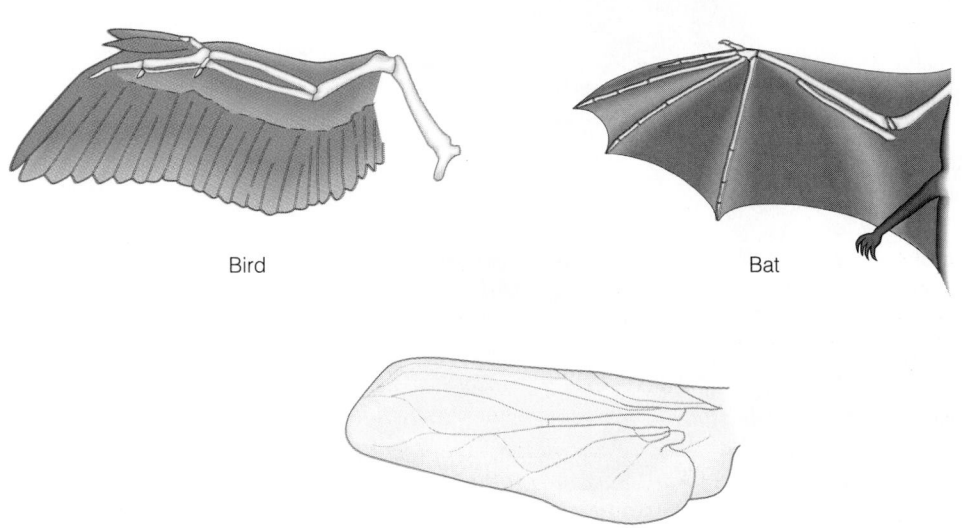

FIGURE 5.25 The fly's wings serve the same function as wings of birds and bats but have a different structure and different embryological development. Organs that serve the same function are analogous, but the bird and bat wings are also homologous.

Evidence for Evolution 121

perspective 5.3

The Evolutionary Significance of Vestigial Structures

*P*robably every living species has *vestigial structures*—nonfunctional or partly functional remnants of organs that were functional in their ancestors. Vestigial structures are leftovers in the evolutionary sense, structures that are probably being lost. For example, one trend in human evolution has been shortening of the jaw, which leaves too little room for the third molars, the so-called wisdom teeth. Some people never develop wisdom teeth, but in others, these teeth might be present but not erupt through the gums, or they may come in crooked and useless. Anyone who has suffered the pain of impacted wisdom teeth can attest to the fact that the human jaw has too little room for these teeth.

Many mammals have dewclaws, which are small, functionless, vestiges of fingers and toes. Dewclaws are particularly obvious in dogs (Figure 1). The ancestors of dogs were rather flat-footed animals with five functional digits on each foot, but as dogs evolved, they became toe-walkers, and only four digits contact the ground. Consequently, the first digit of each foot, the thumbs and big toes, lost their function and were reduced.

Vestigial toes are not restricted to dogs. Among the even-toed hoofed mammals, deer and pigs walk on two toes but retain obvious vestiges of two additional toes. Horses have only one functional toe in each foot, the third, which corresponds to the middle finger and middle toe of the human hand and foot. Horses, however, retain remnants of the second and fourth toes as so-called splint-bones, one on each side of the main toe (Figure 2).

These splint-bones are not usually visible because they lie beneath the skin, but a few horses are born with extra toes.

Whales adapted to an aquatic lifestyle and in the process lost the rear limbs possessed by their land-dwelling ancestors. Yet whales still have a remnant of the pelvic girdle, and a few whales have been caught that have rear limbs. And fossil whales with fully developed tiny vestigial rear limbs are known. Furthermore, one major group of whales have no teeth as adults but as embryos develop a full set of teeth similar to those of their land-

dwelling ancestors. These teeth do not erupt through the gums; they are resorbed before birth and thus serve no function at all.

The kiwi, a small flightless bird living only in New Zealand, has tiny remnants of the typical bones of a bird wing. Boa constrictors lack limbs yet retain vestiges of a pelvis and tiny bones that might be vestiges of rear legs. A fossil discovered in 1978 was identified as a meter-long lizard with tiny rear limbs, but recent studies of its anatomy reveal that it was a snake with 2.5-cm-long vestigial rear limbs.

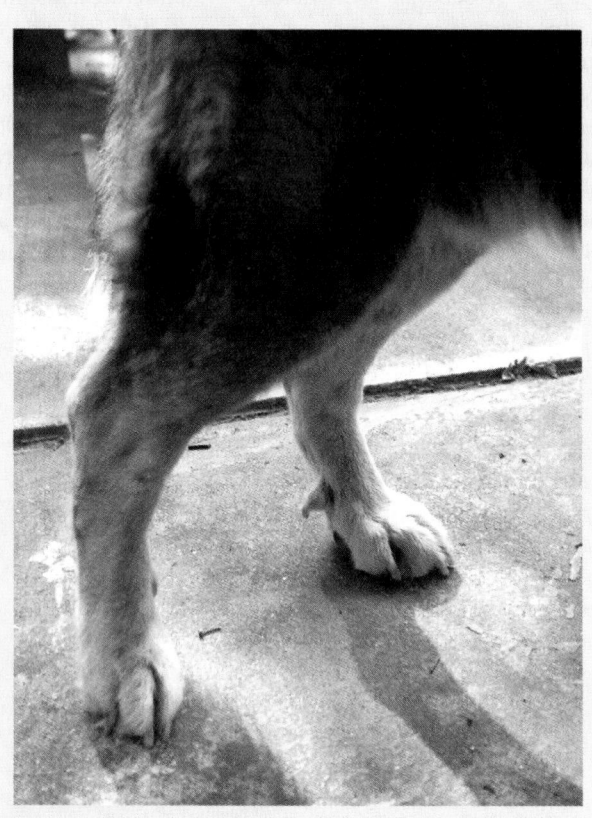

FIGURE 1 **Left hindfoot of a dog showing the dewclaw, which is a vestige of the big toe.**

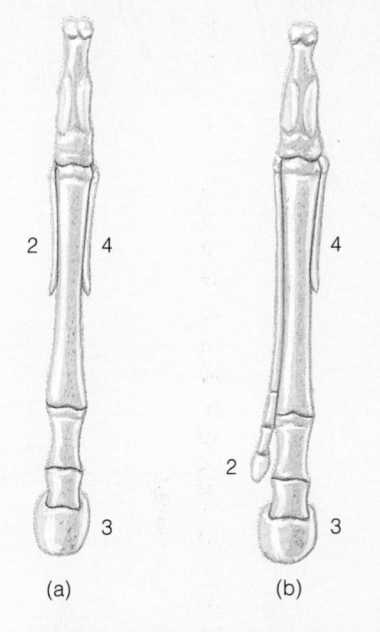

FIGURE 2 (a) Normal condition of a horse's back foot; the only functional toe is the third, but side splints representing toes 2 and 4 are still present. (b) Horse with an extra toe.

It seems doubtful that anyone would seriously argue that dogs were specially created with dewclaws or that extra toes in horses really serve some function. Their presence makes sense, however, if they are the remnants of once-functional organs. Charles Darwin had this to say: "They [vestigial organs] have been partially retained by the power of inheritance, and relate to a former state of things."*

---

*The Origin of Species (New York: New American Library, 1958), pp. 419–420.

FIGURE 5.26 Small-scale evolution of the peppered moths of Great Britain. In pre-industrial England most of these moths were gray and blended well with the lichens on trees. Black varieties were known, but rare, since they stood out against the light background and were easily spotted by birds. With industrialization and pollution, the trees became soot-covered and dark, and the frequency of dark-colored moths increased while that of light-colored moths decreased. However, in rural areas unaffected by pollution, the light–dark frequency did not change.

The important points regarding small-scale evolution are that variation existed, and that some variant types had a better chance to survive and reproduce. The source of variation is irrelevant; it may have already existed, or it may have been introduced by mutations. Regardless of its source, the survival and reproduction of some variant types caused the genetic makeup of the population, its gene pool, to change through time.

FOSSILS AND EVOLUTION

Fossil marine animals far from the sea, even high in mountains, led many early naturalists to conclude that they were deposited during a worldwide flood. As early as 1508, Leonardo da Vinci realized that the fossil distribution was not what one would expect in a rising flood. Nevertheless, the flood explanation persisted, and John Woodward (1665–1728) proposed that organic remains had separated out of the floodwater according to their density. In other words, a prediction of Woodward's hypothesis was that fossils should be arranged in a vertical sequence, with the densest at the bottom followed upward by those of decreasing density. This hypothesis was quickly rejected because field observations clearly did not support the prediction; fossils of various densities are found throughout the fossil record.

The fossil record does show a sequence, but not one based on density, size, shape, or habitat. Rather, it shows a sequence of appearances of different lifeforms through time (Figure 5.27). The fact is that older and older fossiliferous strata contain organisms increasingly different from those living today. One-celled organisms appeared before multicelled organisms, plants before animals, and invertebrates before vertebrates. Among the vertebrates, fish appear first, followed in order by amphibians, reptiles,

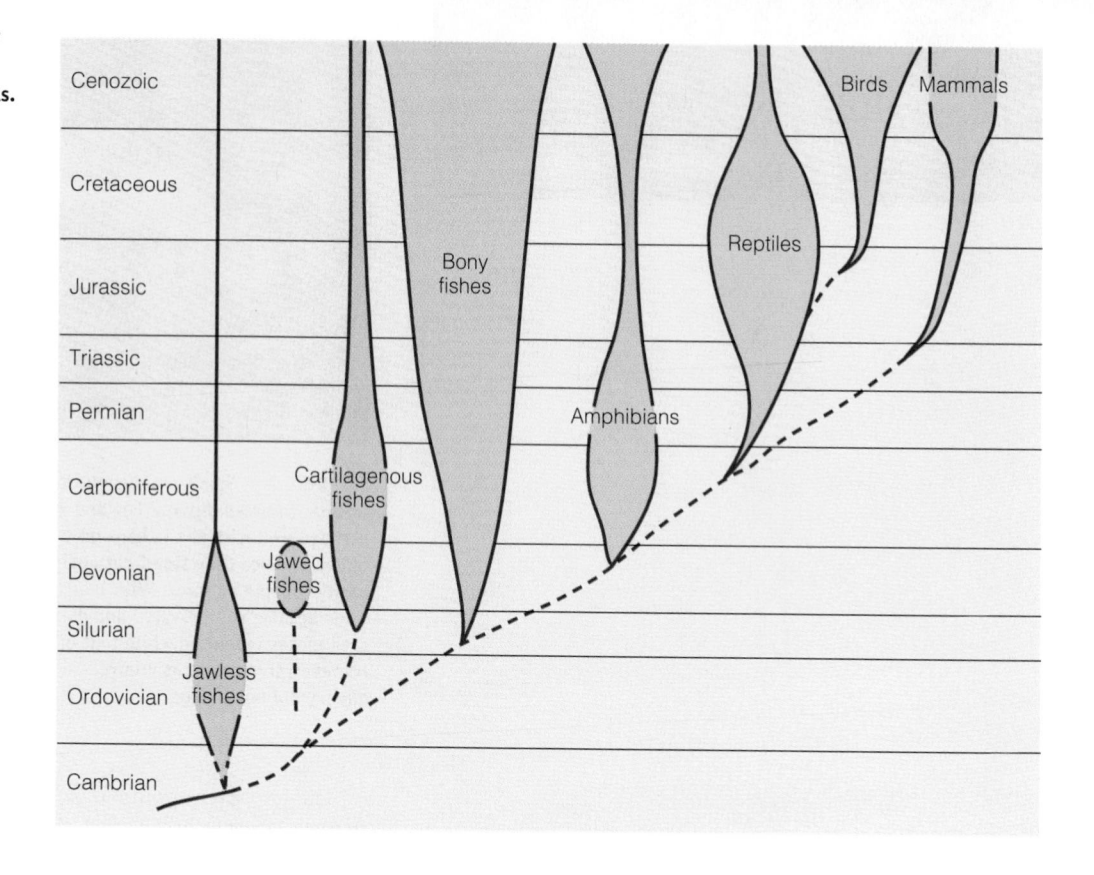

FIGURE 5.27 The times of appearance of the major groups of vertebrate animals.

mammals, and birds (Figure 5.27). Once established, many primitive groups have persisted to the present, but their order of first appearance is consistent with evolutionary theory.

If the fossil record provided no more than the sequence shown in Figure 5.27, however, it could be interpreted as a succession of independently created groups. But fossils also provide evidence for the evolutionary origins of some of these groups and evidence for evolution within groups.

In Perspective 3.2, we discussed the fact that fossils are much more abundant than most people realize. Even though fossilization is a rare event, so many organisms have lived during so many millions of years that if only a tiny fraction was preserved, the total number is staggering. So, one might ask, "Where are the fossils showing the diversification of horses, rhinoceroses, and tapirs from a common ancestor, the origin of birds from reptiles, and the evolution of whales from presumed land-dwelling ancestors?" And even though it is true that the origin and earliest diversification of an evolutionary lineage is usually poorly represented by fossils, in these cases and a number of others, fossils are known that clearly show relationships between ancestors and descendants.

The evolution of horses and their living relatives, the rhinoceroses and tapirs, is well documented by fossils (Figure 5.28) and studies of living animals. This may seem to be an odd assortment of animals, but they all share several characteristics that imply they are related by evolutionary descent. If so, one would predict that as each family is

traced back in the fossil record, differentiating one from the other would become increasingly difficult. In fact, the earliest members of each family are differentiated from one another by minor differences in size and characteristics of their teeth.

Fossils from the Jurassic Solnhofen Limestone of Germany show the first appearance of features we associate with birds. Several fossil specimens showing feather impressions have been discovered, but in almost every other known physical feature, these fossils are more similar to small carnivorous dinosaurs. As a matter of fact, two specimens were initially misidentified as dinosaurs.

This early birdlike creature, *Archaeopteryx* (ancient feather), illustrates the concept of **mosaic evolution.** All organisms are mosaics of characteristics, some of which are retained from the ancestral condition, whereas others are more recently evolved. *Archaeopteryx,* for example, retains dinosaurlike teeth, tail, hind limb structure, and brain size but also possesses feathers and a wishbone, characteristics typical of birds.

Until only several years ago, the origin of whales was rather poorly understood. Fossil whales were common enough, but those linking fully adapted marine animals to their presumed land-dwelling ancestors were mostly lacking. It turns out that this important transition took place in a part of the world where the fossil record was not well known. Fortunately for paleontologists this situation has changed as several fossils have been found that show good evidence of how and when whales evolved. We shall have

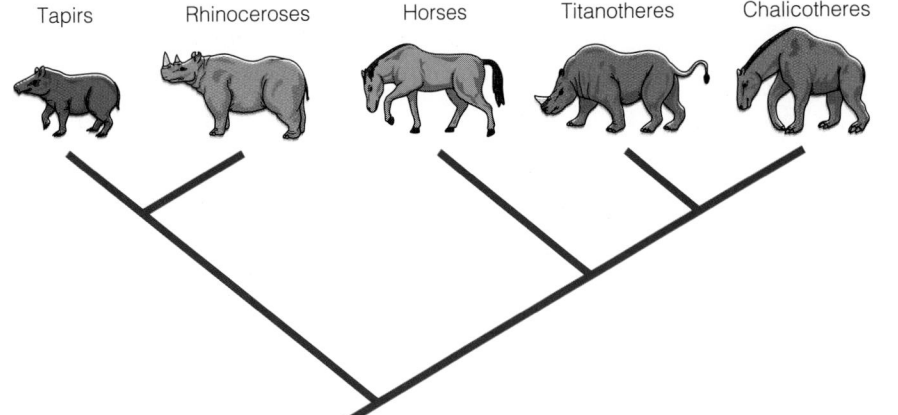

Tapirs   Rhinoceroses   Horses   Titanotheres   Chalicotheres

FIGURE 5.28 Cladogram showing the relationships among odd-toed hoofed mammals (order Perissodactyla), which includes present-day horses, rhinoceroses, and tapirs. Titanotheres and chalicotheres are extinct. These relationships are well documented by fossils. The evolution of these animals is discussed more fully in Chapter 18.

more to say about the evolution of birds, horses, and whales in Chapters 15 and 18.

Many people have the idea that the fossil record provides the main evidence for evolution. Fossils are unquestionably important in evolutionary studies, but one must not overlook the fact that fossils are only one of several kinds of evidence that support the concept of evolution. Biochemistry, comparative anatomy, genetics, molecular biology, and small-scale evolution of living organisms are equally important.

Summary

1. The first formal theory of evolution to be taken seriously was proposed by Jean Baptiste de Lamarck. Inheritance of acquired characteristics was his mechanism for evolution.

2. In 1859 Charles Darwin and Alfred Russell Wallace simultaneously published their views on evolution. They proposed natural selection as the mechanism for evolutionary change.

3. Darwin's observations of variation in natural populations and artificial selection, as well as his reading of Thomas Malthus's essay on population, helped him formulate the idea that natural processes select favorable variants for survival.

4. Gregor Mendel's breeding experiments with garden peas provided some of the answers regarding how variation is maintained and passed on. Mendel's work is the basis for modern genetics.

5. The hereditary determinates in all organisms are genes, which are specific segments of chromosomes. Only the genes on chromosomes in sex cells—sperm and eggs in animals, and pollen ovules in plants—are inheritable.

6. Most variation in populations results from sexual reproduction during which half the genetic makeup of an organism comes from each parent and from genetic changes known as mutations.

7. Evolution by natural selection is a two-step process. First, variation must be present and maintained in interbreeding populations, and second, favorable variants must be selected for survival.

8. An important way in which new species evolve is by allopatric speciation. When a group is isolated from its parent population, gene flow is restricted or eliminated, and the isolated group is subjected to different selection pressures.

9. Divergent evolution involves an adaptive radiation during which an ancestral stock gives rise to diverse species. The development of similar adaptive types in different groups of organisms results from parallel and convergent evolution.

10. The evidence for evolution includes the way organisms are classified, comparative anatomy and embryology, biochemical similarities, small-scale evolution, and the fossil record.

Important Terms

allele
allopatric speciation
analogous organ
artificial selection
chromosomal theory of
    inheritance
chromosome
convergent evolution

divergent evolution
DNA (deoxyribonucleic
    acid)
gene
gene pool
homologous organ
inheritance of acquired
    characteristics

living fossil
meiosis
mitosis
modern synthesis
morphology
mosaic evolution
mutation

natural selection
parallel evolution
phyletic gradualism
punctuated equilibrium
species
theory of evolution

Review Questions

1. Biologists and paleontologists think that most new
   species arise by:
   a. _____ mosaic evolution.
   b. _____ allopatric speciation.
   c. _____ differential reproduction.
   d. _____ homologous differentiation.
   e. _____ inheritance of vestigial structures.

2. Organisms that have changed little during a long time of
   existence are known as:
   a. _____ conservative species.
   b. _____ adaptive generalists.
   c. _____ living fossils.
   d. _____ alleles.
   e. _____ mosaic animals.

3. According to the concept of punctuated equilibrium:
   a. _____ species change gradually and continually over
       millions of years.
   b. _____ most species that ever existed are now extinct.
   c. _____ the most common evolutionary trend is size in-
       crease.
   d. _____ new species arise rapidly, in perhaps only a few
       thousand years.
   e. _____ organs with same basic structure and develop-
       ment are analogous.

4. Clams and brachiopods are very distantly related but possess similarities because of:
   a. _____ convergent evolution.
   b. _____ adaptive divergence.
   c. _____ artificial selection.
   d. _____ parallel adaptation.
   e. _____ inheritance of the same alleles.

5. The theory of natural selection was proposed by Charles
   Darwin and _____ .
   a. _____ William Smith
   b. _____ Erasmus Darwin
   c. _____ Alfred Wallace

d. _____ Charles Lyell
e. _____ Jean Baptiste de Lemarck

6. The type of cell division by which sex cells such as eggs
   and sperm are produced is known as:
   a. _____ meiosis.
   b. _____ divergence.
   c. _____ morphology.
   d. _____ speciation.
   e. _____ mosaic evolution.

7. In the classification scheme now used, an organism is referred by its _____ names:
   a. _____ family and phylum
   b. _____ genus and species
   c. _____ order and class
   d. _____ subphylum and phylum
   e. _____ kingdom and family

8. Chromosomes are complex molecules of:
   a. _____ deoxyribonucleic acid.
   b. _____ potassium, aluminum, and silicon.
   c. _____ carbon dioxide and helium.
   d. _____ calcium carbonate.
   e. _____ genes and homologous organs.

9. Divergent evolution involves the:
   a. _____ interbreeding by unrelated species.
   b. _____ development of similar features in distantly re-
       lated groups of plants.
   c. _____ inheritance of acquired characteristics.
   d. _____ diversification of a species into two or more de-
       scendant species.
   e. _____ replacement of one species by another follow-
       ing an extinction.

10. Organs such as bird and bat wings that have a similar de-
    velopment pattern and similar relationship to other or-
    gans are referred to as:
    a. _____ artificial.

b. _____ mosaic.

c. _____ allopatric.

d. _____ alleles.

e. _____ homologous.

11. An adaptive radiation takes place when:

a. _____ species of related ancestry exploit different aspects of the environment.

b. _____ distantly related organisms interbreed and yield two or more new species.

c. _____ brachiopods and clams develop similar characteristics.

d. _____ placental and marsupial mammals develop into increasingly similar types of animals.

e. _____ a mass extinction is about to occur.

12. According to studies of living animals and fossils, horses are most closely related to which one of the following?

a. _____ Pigs

b. _____ Birds

c. _____ Rhinoceroses

d. _____ Camels

e. _____ Reptiles

13. Explain how the fossil record provides evidence for evolution. Give an example.

14. Compare evolution by inheritance of acquired characteristics with evolution by natural selection.

15. What is a mutation, and how is it possible for a mutation to be harmful under some circumstances but beneficial under others?

16. How does sexual reproduction maintain variation in an interbreeding population?

17. Compare the concepts of punctuated equilibrium and phyletic gradualism. Is there any evidence for either? If so, what?

18. Discuss the concept of allopatric speciation and give an example of where it has occurred.

19. How did the experiments by Gregor Mendel help solve some of the criticisms of natural selection that plagued Charles Darwin and Alfred Wallace?

20. Give examples of parallel and convergent evolution. Why do they occur?

21. What criteria are used to distinguish homologous organs from analogous organs?

22. Explain how the classification of organisms according to Linnaeus's scheme provides evidence for evolution.

23. What is meant by the term *mosaic evolution?*

24. Draw three cladograms showing possible relationships among sharks, whales, and bears. Which cladogram most likely depicts evolutionary relationships among these animals, and what criteria did you use to make your decision?

Points to Ponder

1. A cladogram showing relationships among carnivorous mammals indicates that cats, hyenas, and mongooses are more closely related to one another than they are to other carnivores. Yet, hyenas resemble dogs and mongooses resemble weasels. What kinds of evidence from existing animals and from fossils would lend credence to the idea that cats, hyenas, and mongooses constitute a closely related group?

2. Fossils are known that show the origin of some groups of animals such as the evolution of mammals from reptiles, the evolution of ammonites, and various others, but in general transitions from one group to another are poorly represented in the fossil record. Why do you think this is so?

3. Descent with modification (evolution) by natural selection works on variations in populations. Give some examples of favorable, neutral, and harmful variations that might occur in a population of mammals. Would only favorable variations be selected for survival? Explain. Also, what role might chance play in the evolution of a population?

Additional Readings

Allen, K., and D. Briggs, eds. 1993. *Evolution and the fossil record.* Washington, D.C.: Smithsonian Institution Press.

Bowler, P. J. 1990. *Charles Darwin: The man and his influence.* Cambridge, Mass.: Blackwell.

Briggs, D., and P. R. Crowther. 1990. *Paleobiology: A synthesis.* London: Blackwell Scientific.

Eldridge, N. 1991. *Fossils: The evolution and extinction of species.* New York: Abrams.

Fitch, W. M., and F. J. Ayala, eds. 1995. *Tempo and mode in evolution.* Washington, D.C.: National Academy Press.

Grant, P. R. 1991. Natural selection and Darwin's finches. *Scientific American* 265, no. 4: 82–87.

McKinney, M. L. 1993. *Evolution of life: Processes, patterns, and prospects.* Englewood Cliffs, N.J.: Prentice Hall.

McNamara, K. T., ed. 1990. *Evolutionary trends.* Tucson: University of Arizona Press.

Nielsen, C. 1995. *Animal evolution: Interrelationships of the living phyla.* New York: Oxford University Press.

Ridley, M. 1996. *Evolution.* 2d ed. London: Blackwell Scientific.

Ross, R. M., and W. D. Allmon. 1990. *Causes of evolution.* Chicago: University of Chicago Press.

World Wide Web Activities

For these web site addresses, along with current updates and exercises, log on to

http://www.brookscole.com/geo/

▶ **UNIVERSITY OF CALIFORNIA MUSEUM OF PALEONTOLOGY**

This huge site has excellent discussions and images on a variety of topics, including On-Line Exhibits with subheadings of Time Periods, Phylogeny, Evolution, and Geology.

1. Go to the Education and Public Outreach section and click on *Learning from Fossils* in which you will find *A Collection of Classroom Activities.* This site is particularly appropriate for students contemplating a career in teaching biology, earth science, or geology.

2. Return to the main page and click *Evolution.* Read about Charles Darwin and the theory of evolution. Click on *Systematics* and see what phylogenetic systematics is about. Then go to *cladistics* to see how paleontologists made testable hypotheses about relationships among organisms. You can also explore the classification and evolution of numerous organisms from bacteria to dinosaurs to plants.

3. The main page also has a designation *Paleontology* which has links to many other sites with discussions and images of fossils.

4. On the main page click *On-Line Exhibits* to go to the Paleontology without Walls home page, which is an introduction of the UCMP Virtual Exhibits. Click on the *Phylogeny* icon. This will take you to the Phylogeny of Life home page. Click on the *Metazoa, All Animals* icon. This will take you to the Introduction to the Metazoa home page. From here you can click one of four sites to learn more about the life history and ecology of any animal group, its morphology, systematics, and its fossil record. For example, find out about trilobites and brachiopods.

▶ **UNIVERSITY OF MIAMI DEPARTMENT OF BIOLOGY**

On the University of Miami Department of Biology page click *courses* and then scroll down to *Biology 160—Evolution and Biodiversity.* Among the topics for this course with extensive notes are:

Historical Perspective, Darwin and Natural Selection

The Species Concept, Allopatric and Sympatric Speciation (how do these models of speciation differ?)

Adaptive Radiation

Biosystematics I—Analogy, Homology, and Convergent Evolution

Biosystematics II—Phenetics and Cladistics (in this section scroll down and click "The Cladistics School of Classification" and see the interactive cladogram that includes Humans)

▶ **BOWDEN GRANDVIEW SCHOOL**

The Bowden Grandview School in Alberta, Canada, is a technologically oriented high school, but its site on natural selection, evolution, and related topics provides a good review even for students familiar with these concepts. Click on *Our Technology Classrooms,* then click *High School Subject Areas,* then click *Biology Links.* Among the biology links are natural selection, evolution, formation of Pangaea, and many more. Under natural selection click *Natural Selection—The Process* and see what the genetic basis for natural selection is. Also click on *Evolution—The*

Evidence and click on *The Fossil Record* and read the section on Radioisotopic Dating of Fossils. How are radioisotopic ages assigned to fossils?

▶ AMERICAN MUSEUM OF NATURAL HISTORY: UNDERSTANDING CLADISTICS

The American Museum of Natural History in New York City maintains a large site at which you can explore a variety of topics, including cladistics, evolution, and the fossil record. At this site see what cladograms are, how they are constructed, and why they are used. Click on *Expedition Guide* to see what features are used to make a cladogram for vertebrate animals.

CHAPTER 6

Plate Tectonics: A Unifying Theory

This is the view southward across the Thingvellir graben on the crest of the Mid-Atlantic ridge, where it passes through part of Iceland. The photo was taken from the American plate, looking across the graben toward the Atlantic plate. Lake Thingvallavatn, in the background, occupies part of the graben.

It is certainly not surprising that oil and politics are closely linked. The Iran-Iraq War of 1980–1989 and the Gulf War of 1990–1991 were both fought over oil. Most people, however, are not aware of why there is so much oil in this region of the world.

Although large concentrations of petroleum occur in many areas of the world, more than 50% of all proven reserves are in the Persian Gulf region. However, this region did not become a significant petroleum-producing area until the economic recovery following World War II. Following the war, Western Europe and Japan in particular became dependent on Persian Gulf oil and still rely heavily on this region for most of their supply. The United States is also dependent on imports from the Persian Gulf, but receives significant quantities of petroleum from other sources such as Mexico and Venezuela.

Why is there so much oil in the Persian Gulf region? The answer lies in the paleogeography and plate movements of this region during the Mesozoic and Cenozoic eras (Figure 6.1). During the Mesozoic Era, and particularly the Cretaceous Period when most of the petroleum formed, the Persian Gulf area was a broad, stable marine shelf extending eastward from Africa. This passive continental margin lay near the equator where countless microorganisms lived in the surface waters. The remains of these organisms accumulated with the bottom sediments and were buried, beginning the complex process of petroleum generation and the formation of source beds.

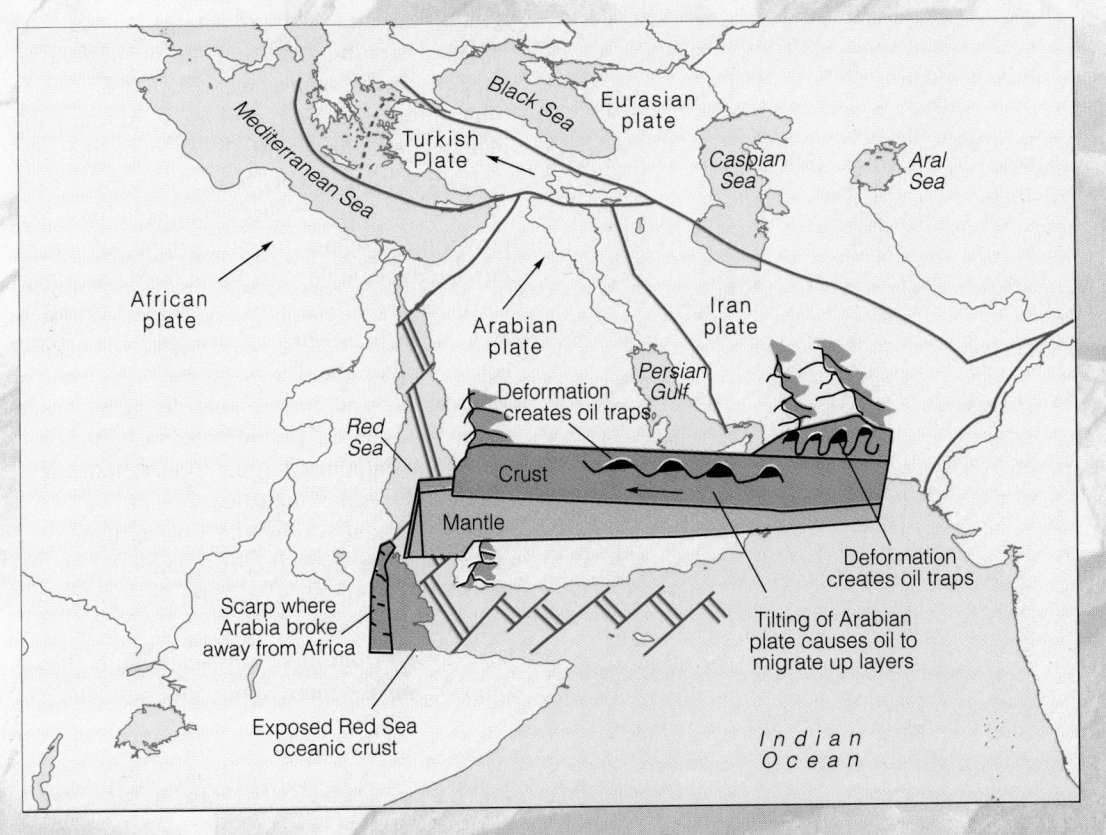

FIGURE 6.1 Petroleum geologists divide oil-producing regions into two categories: the Middle East and everything else. In this oblique view, an imaginary trench has been cut across Arabia and the Persian Gulf. The collision between Arabia and Iran has tilted the Arabian plate and crumpled rocks on the edges of both plates. The tilting of Arabia allows oil to migrate upslope to accumulate in traps created by folding. Along almost the entire length of the plate boundary, conditions are ideal for maximum accumulation of oil. Elsewhere, the convergence of Arabia and the Eurasian plate is squeezing Turkey westward. Also, the opposing coastlines of the Red Sea fit almost perfectly, except at the southern end. There, a portion of Red Sea ocean floor is exposed, one of the few places oceanic crust is exposed on dry land. The true edge of the African crust is marked by a steep scarp, which matches the southwestern corner of the Arabian plate almost perfectly.

As a consequence of rifting in the Red Sea and Gulf of Aden during the Cenozoic Era, the Arabian plate is moving northeast away from Africa and subducting beneath Iran. As the sediments of the passive continental margin were initially subducted, during the early stages of collision between Arabia and Iran, the heating broke down the organic molecules and led to the formation of petroleum. The tilting of the Arabian block to the northeast allowed the newly formed petroleum to migrate upward into the interior of the Arabian plate. The continued subduction and collision with Iran folded the rocks, creating traps for petroleum to accumulate. Thus, the vast area south of the collision zone (known as the Zagros suture) is oil producing.

Introduction

The decade of the 1960s was a time of geologic as well as social and cultural revolution. The ramifications of the newly proposed plate tectonic theory radically changed the way in which geologists viewed our planet. Earth could now be treated as a system in which seemingly unrelated geologic phenomena were related and interconnected (Table 6.1). No longer could Earth be regarded as an unchanging planet on which continents and ocean basins remained fixed through time but, instead, was recognized as a dynamically changing planet.

Plate tectonics has been the dominant process affecting the evolution of Earth. The interactions between moving plates determines the locations of continents, ocean basins, and mountain systems, which in turn affects atmospheric and oceanic circulation patterns that ultimately determine global climates. Plate movements have also profoundly influenced the geographic distribution, evolution, and extinction of plants and animals.

Although plate tectonic processes are extremely slow by human standards, they nevertheless have profound effects on our lives. Geologists now realize that most earthquakes and volcanic eruptions occur at or near plate boundaries and are not merely random occurrences. Furthermore, the formation and distribution of many geologic resources, such as metal ores, are related to plate tectonic processes, and geologists are now incorporating plate tectonic theory into their prospecting efforts.

Plate tectonic theory has led to a greater understanding of how Earth has evolved and continues to do so. This powerful, unifying theory accounts for many apparently unrelated geologic events, allowing geologists to view Earth history in terms of interrelated events that are part of a global panorama of dynamic change through time. For example, the Paleozoic history of the Appalachian Mountains is no longer considered an isolated regional event but, rather, part of a global interaction between plates that culminated in the formation of a large supercontinent at the end of the Paleozoic Era.

We will first review the various hypotheses that preceded plate tectonic theory, examining the evidence that led some people to accept the idea of continental movement and others to reject it. Because plate tectonic theory has evolved from numerous scientific inquiries and observations, only the more important ones will be covered.

Early Ideas About Continental Drift

The earliest maps showing the east coast of South America and the west coast of Africa probably provided people with the first evidence that continents may have once been joined together, then broken apart and moved to their present positions. It was not until 1858, however, that the Frenchman Antonio Snider-Pellegrini suggested in his book,

TABLE 6.1

Plate Tectonics and Earth Systems

SOLID EARTH	Plate tectonics is driven by convection in the mantle and in turn drives mountain-building processes and associated igneous and metamorphic activity.
ATMOSPHERE	Arrangement of continents affects solar heating and cooling and thus winds and weather systems. Rapid plate spreading and hot-spot activity may release volcanic carbon dioxide and affect global climate.
HYDROSPHERE	Continental arrangement affects ocean currents. Rate of spreading affects volume of mid-ocean ridges and hence sea level. Placement of continents may contribute to onset of ice ages.
BIOSPHERE	Movement of continents creates corridors or barriers to migration and thus creates ecological niches. Habitats may be transported into more or less favorable climates.
EXTRATERRESTRIAL	Arrangement of continents affects free circulation of ocean tides and influences tidal slowing of Earth's rotation.

Creation and Its Mysteries Revealed, that all continents were linked together during the Pennsylvanian Period and later split apart. He based his conclusions on the similarity between plant fossils in the Pennsylvanian-aged coal beds of Europe and North America and attributed the separation of the continents to the biblical deluge. Some years after Snider-Pellegrini, another Frenchman, Elisée Reclus, discussed the idea of continental movement in his book, *The Earth,* published in 1872. Reclus thought continental drift, not the Great Flood, was responsible for mountain building, volcanism, and earthquake activity.

During the late nineteenth century, the Austrian geologist Edward Suess noted the similarity between the Late Paleozoic plant fossils of India, Australia, South Africa, and South America as well as evidence of glaciation in the rock sequences of these southern continents. The plant fossils comprise a unique flora that occurs in the coal layers just above the glacial deposits of these southern continents. This flora is very different from the contemporaneous coal swamp flora of the northern continents and is collectively known as the ***Glossopteris* flora** after its most conspicuous genus (Figure 6.2).

In his book, *The Face of the Earth,* published in 1885, Suess proposed the name *Gondwanaland* (or **Gondwana**) for a supercontinent composed of the aforementioned southern continents. Gondwana is a province in India where abundant fossils of the *Glossopteris* flora are found in coal seams. Suess thought these continents were connected by land bridges over which plants and animals migrated. Thus, in his view, the similar fossils of the continents were due to the appearance and disappearance of the connecting land bridges.

In 1910 the American geologist Frank Taylor published a pamphlet presenting his own theory of continental drift. He explained the formation of mountain ranges as a result of the lateral movement of continents. He also envisioned the present-day continents as parts of larger polar continents that eventually broke apart and migrated toward the equator after Earth's rotation supposedly slowed due to gigantic tidal forces. According to Taylor, these tidal forces were generated when Earth captured the Moon about 100 million years ago.

Although we now know that Taylor's mechanism is incorrect, one of his most significant contributions was his suggestion that the Mid-Atlantic Ridge, discovered by the 1872–1876 British HMS *Challenger* expeditions, might mark the site along which an ancient continent broke apart to form the present-day Atlantic Ocean.

ALFRED WEGENER AND THE CONTINENTAL DRIFT HYPOTHESIS

Alfred Wegener, a German meteorologist (Figure 6.3), is generally credited with developing the hypothesis of **continental drift.** In his monumental book, *The Origin of Continents and Oceans* (first published in 1915), Wegener proposed that all landmasses were originally united into a single supercontinent that he named **Pangaea,** from the Greek meaning "all land." Wegener portrayed his grand concept of continental movement in a series of maps showing the breakup of Pangaea and the movement of the various continents to their present-day locations.

(a)

(b)

FIGURE 6.2 Representative members of the *Glossopteris* flora. Fossils of these plants are found on all five of the Gondwana continents. *Glossopteris* leaves from (a) the Upper Permian Dunedoo Formation and (b) the Upper Permian Illawarra Coal Measures, Australia.

FIGURE 6.3 Alfred Wegener, a German meteorologist, proposed the continental drift hypothesis in 1912 based on a tremendous amount of geologic, paleontologic, and climatologic evidence. He is shown here waiting out the Arctic winter in an expedition hut.

Wegener amassed a tremendous amount of geologic, paleontologic, and climatologic evidence in support of continental drift, but the initial reaction of scientists to his then-heretic ideas can best be described as mixed. Wegener noted that marine and nonmarine rock sequences of the same age are found on widely separated continents (Figure 6.4); mountain ranges and glacial deposits match up when continents are united into a single landmass (Figure 6.5); the shorelines of continents fit together, forming a large supercontinent; and many of the same extinct plant and animal groups are found today on widely separated continents, indicating the continents must have been in proximity at one time (Figure 6.6). Wegener argued that this vast amount of evidence from a variety of sources surely indicated that the continents must have been close together at one time in the past.

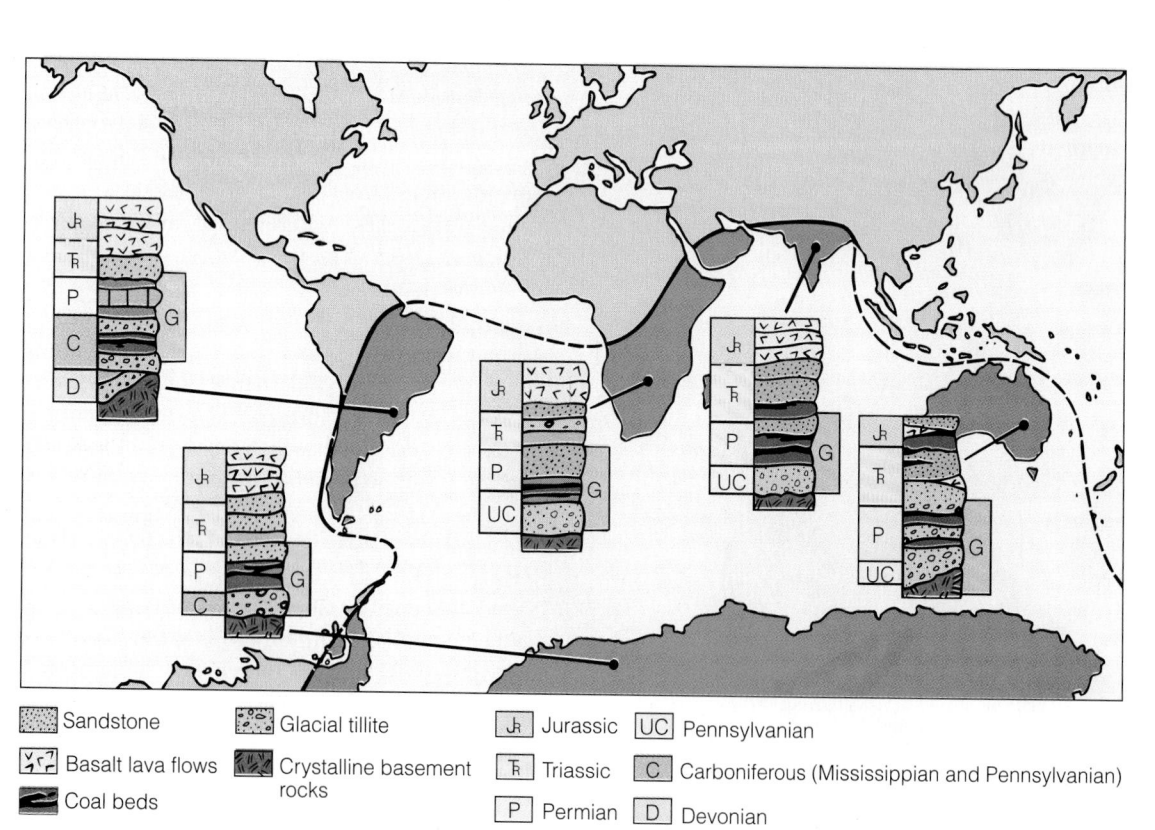

Sandstone	Glacial tillite	Jr Jurassic UC Pennsylvanian
Basalt lava flows	Crystalline basement rocks	Tr Triassic C Carboniferous (Mississippian and Pennsylvanian)
Coal beds		P Permian D Devonian

FIGURE 6.4 Marine, nonmarine, and glacial rock sequences of Pennsylvanian to Jurassic age are nearly the same for all Gondwana continents. Such close similarity strongly suggests that they were joined together at one time. The range indicated by G is that of the *Glossopteris* flora.

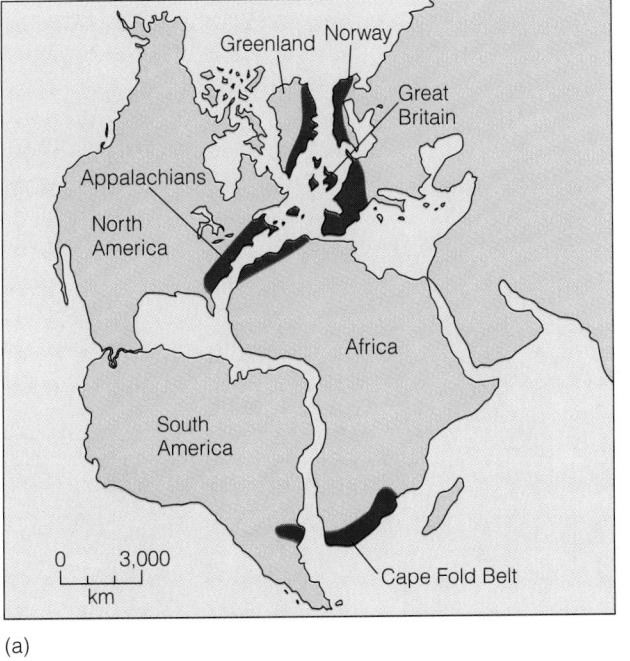

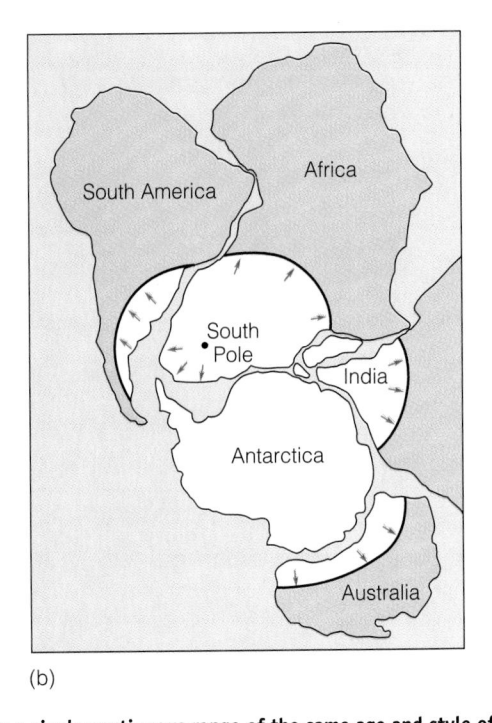

(a)   (b)

FIGURE 6.5 (a) When continents are brought together, their mountain ranges form a single continuous range of the same age and style of deformation throughout. (b) Furthermore, if the continents are brought together so that South Africa is located at the South Pole, then the glacial movement indicated by the striations makes sense. In this situation, the glacier, located in a polar climate, moved radially outward from a thick central area toward its periphery. Such evidence as this indicates that the continents were at one time joined together and were subsequently separated.

ADDITIONAL SUPPORT FOR CONTINENTAL DRIFT

With the publication of *The Origin of Continents and Oceans* and its subsequent four editions, some scientists began to take Wegener's unorthodox views seriously. Alexander du Toit, a South African geologist, was one of his more ardent supporters. He further developed Wegener's arguments and introduced more geologic evidence in support of continental drift. In 1937 du Toit published *Our Wandering Continents,* in which he contrasted the glacial deposits of Gondwana with coal deposits of the same age found in the continents of the Northern Hemisphere. To resolve this apparent climatologic paradox, du Toit placed the southern continents of Gondwana at or near the South Pole and arranged the northern continents together such that the coal deposits were located at the equator. He named this northern landmass **Laurasia.**

Du Toit also provided additional paleontologic support for continental drift by noting that fossils of the Permian freshwater reptile *Mesosaurus* occurred in rocks of the same age in both Brazil and South Africa (Figure 6.6). Because the physiology of freshwater and marine animals is completely different, it is hard to imagine how a freshwater reptile could have swum across the Atlantic Ocean and then found a freshwater environment nearly identical to its former habitat. Furthermore, if *Mesosaurus* could have swum across the ocean, its fossil remains should occur in other localities besides Brazil and South Africa. It is more logical to assume that *Mesosaurus* lived in lakes in

what are now adjacent areas of South America and Africa but were then united into a single continent.

Despite what seemed to be overwhelming evidence presented by Wegener and later by du Toit and others, most geologists simply refused to entertain the idea that continents might have moved in the past. The geologists were not necessarily being obstinate about accepting new ideas; rather, they found the proposed mechanisms for continental drift inadequate and unconvincing.

Paleomagnetism and Polar Wandering

Interest in continental drift revived during the 1950s as a result of new evidence from paleomagnetic studies. **Paleomagnetism** is the remanent magnetism in ancient rocks recording the direction of Earth's magnetic poles at the time of the rock's formation. Earth can be thought of as a giant dipole magnet in which the magnetic poles essentially coincide with the geographic poles (Figure 6.7). Such an arrangement means that the strength of the magnetic field is not constant but varies, being weakest at the equator and strongest at the poles. Earth's magnetic field is thought to result from the different rotation speeds of the outer core and mantle.

When a magma cools, the magnetic iron-bearing minerals align themselves with Earth's magnetic field, recording both its direction and strength. The temperature at which

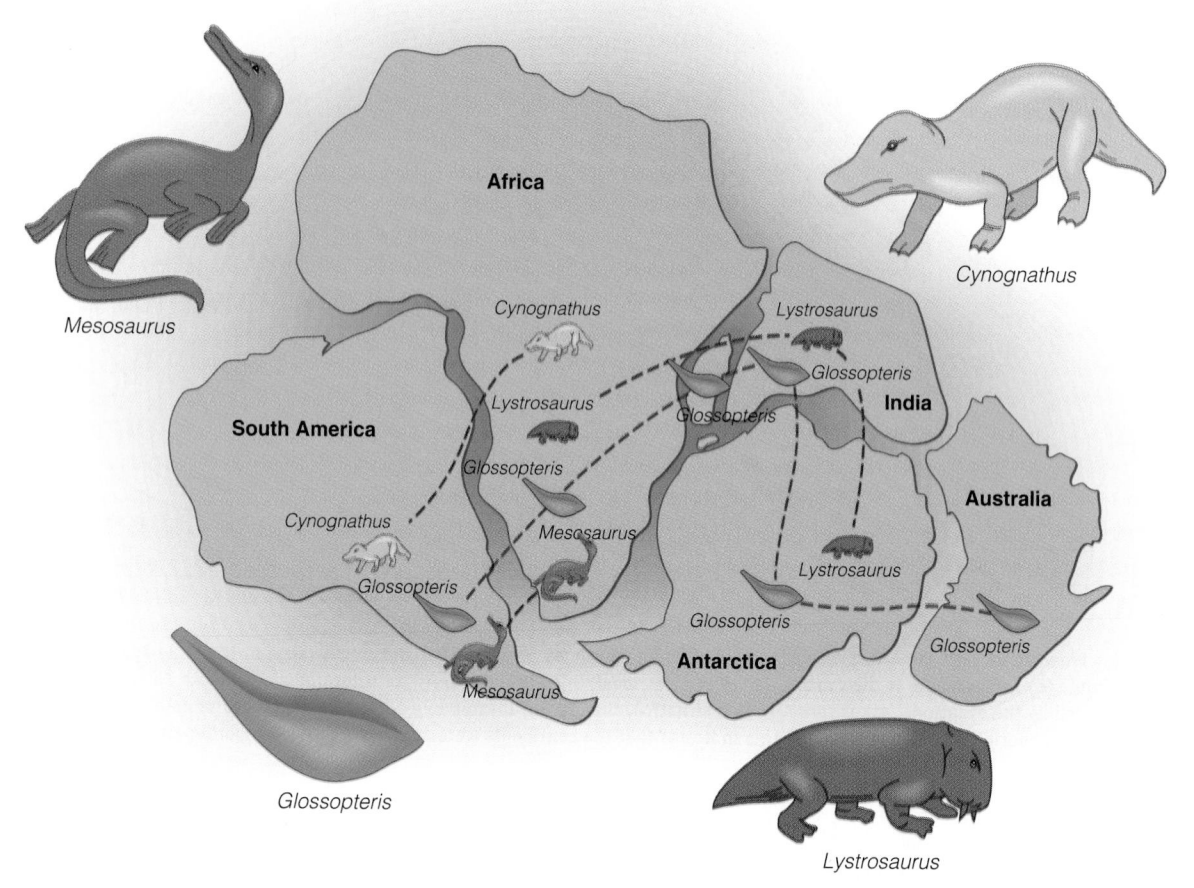

FIGURE 6.6 Some of the animals and plants whose fossils are found today on the widely separated continents of South America, Africa, India, Australia, and Antarctica. These continents were joined together during the Late Paleozoic to form Gondwana, the southern landmass of Pangaea. *Glossopteris* and similar plants are found in Pennsylvanian- and Permian-aged deposits on all five continents. *Mesosaurus* is a freshwater reptile whose fossils are found in Permian-aged rocks in Brazil and South Africa. *Cynognathus* and *Lystrosaurus* are land reptiles who lived during the Early Triassic Period. Fossils of *Cynognathus* are found in South America and Africa, and fossils of *Lystrosaurus* have been recovered from Africa, India, and Antarctica.

iron-bearing minerals gain their magnetization is called the **Curie point.** As long as the rock is not subsequently heated above the Curie point, it will preserve that remanent magnetism. Thus, an ancient lava flow provides a record of the orientation and strength of Earth's magnetic field at the time the lava flow cooled.

As paleomagnetic research progressed in the 1950s, some unexpected results emerged. When geologists measured the paleomagnetism of recent rocks, they found it was generally consistent with Earth's current magnetic field. The paleomagnetism of ancient rocks, though, showed different orientations. For example, paleomagnetic studies of Silurian lava flows in North America indicated that the north magnetic pole was located in the western Pacific Ocean at that time, while the paleomagnetic evidence from Permian lava flows pointed to yet another location in Asia. When plotted on a map, the paleomagnetic readings of numerous lava flows from all ages in North America trace the apparent movement of the magnetic pole through time (Figure 6.8). This paleomagnetic evidence from a single continent could be interpreted in

three ways: The continent remained fixed, and the north magnetic pole moved; the north magnetic pole stood still, and the continent moved; or both the continent and the north magnetic pole moved.

Upon analysis, magnetic minerals from European Silurian and Permian lava flows pointed to a different magnetic pole location than those of the same age from North America (Figure 6.8.) Furthermore, analysis of lava flows from all continents indicated each continent had its own series of magnetic poles. Does this mean there were different north magnetic poles for each continent? That would be highly unlikely and difficult to reconcile with the theory accounting for Earth's magnetic field.

The best explanation for such data is that the magnetic poles have remained at their present locations at the geographic north and south poles and the continents have moved. When the continental margins are fitted together so that the paleomagnetic data point to only one magnetic pole, we find, just as Wegener did, that the rock sequences and glacial deposits match up, and the fossil evidence is consistent with the reconstructed paleogeography.

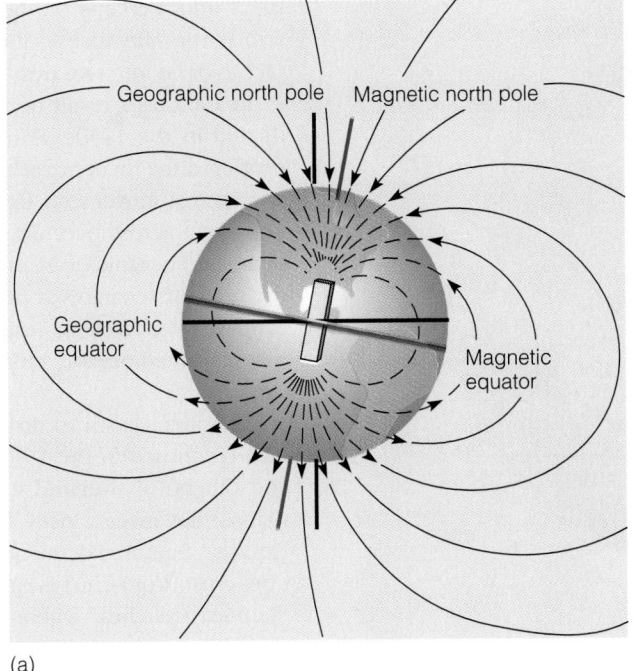

(a)

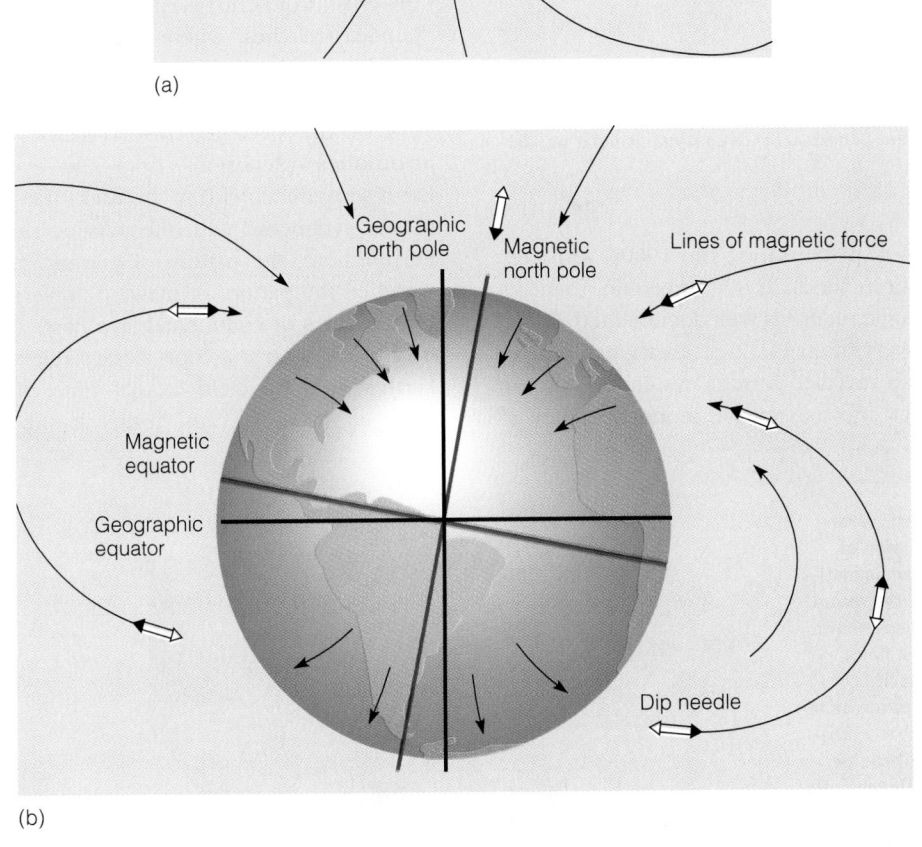

(b)

FIGURE 6.7 (a) Earth's magnetic field has lines of force just like those of a bar magnet. (b) The strength of the magnetic field changes uniformly from the magnetic equator to the magnetic poles. This change in strength causes a dip needle to parallel Earth's surface only at the magnetic equator, whereas its inclination with respect to the surface increases to 90 degrees at the magnetic poles.

Magnetic Reversals and Seafloor Spreading

Geologists refer to Earth's present magnetic field as being normal, that is, with the north and south magnetic poles located approximately at the north and south geographic poles. At various times in the geologic past, Earth's magnetic field has completely reversed. The existence of such **magnetic reversals** was discovered by dating and determining the orientation of the remanent magnetism in lava flows on land (Figure 6.9). Once their existence was well established, magnetic reversals were also discovered in ocean basalts as part of the extensive mapping of the

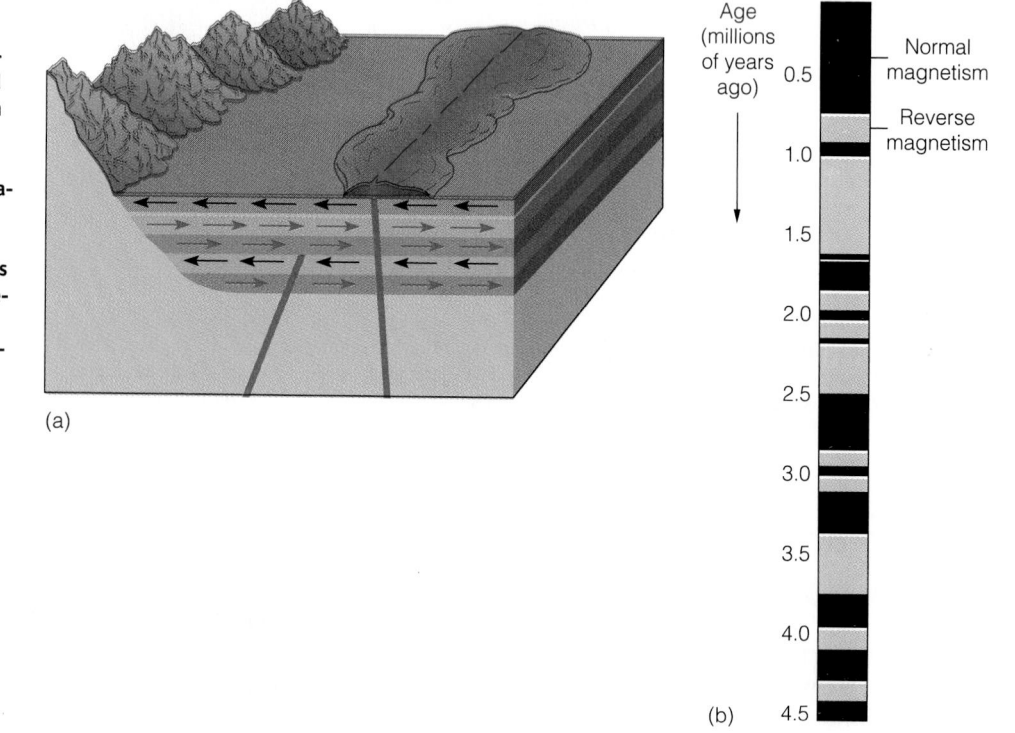

FIGURE 6.8 The apparent paths of polar wandering for North America and Europe. The apparent location of the north magnetic pole is shown for different periods on each continent's polar wandering path.

ocean basins that took place during the 1960s. Although the cause of magnetic reversals is still uncertain, their occurrence in the geologic record is well documented.

Besides the discovery of magnetic reversals, mapping of the ocean basins also revealed a ridge system 65,000 km long, constituting the most extensive mountain range in the world. Perhaps the best-known part of the ridge system is the Mid-Atlantic Ridge, which divides the Atlantic Ocean basin into two nearly equal parts (Figure 6.10).

In 1962, as a result of the oceanographic research conducted in the 1950s, Harry Hess of Princeton University proposed the theory of **seafloor spreading** to account for continental movement. Hess suggested that continents do not move across oceanic crust, but rather that the continents and oceanic crust move together. He suggested that the seafloor separates at oceanic ridges where new crust is formed by upwelling magma. As the magma cools, the newly formed oceanic crust moves laterally away from the ridge.

As a mechanism to drive this system, Hess revived the idea (proposed in the 1930s and 1940s by Arthur Holmes and others) of **thermal convection cells** in the mantle; that is, hot magma rises from the mantle, intrudes along rift zone fractures defining oceanic ridges, and thus forms new crust. Cold crust is subducted back into the mantle at deep-sea trenches, where it is heated and recycled, thus completing a thermal convection cell (see Figure 1.9).

How could Hess's theory be confirmed? Magnetic surveys of the oceanic crust revealed striped **magnetic anomalies** (deviations from the average strength of Earth's magnetic field) in the rocks that were both parallel to and symmetric with the oceanic ridges (Figure 6.11). Furthermore, the pattern of oceanic magnetic anomalies matched the pattern of magnetic reversals already known from studies of continental lava flows (Figure 6.9). When magma wells up and cools along a ridge summit, it records Earth's magnetic field at that time as either normal or reversed. As new crust forms at the summit, the pre-

FIGURE 6.9 (a) Magnetic reversals recorded in a succession of lava flows are shown diagrammatically by red arrows, and the record of normal polarity events is shown by black arrows. The lava flows containing a record of such magnetic–polarity events can be radiometrically dated so that a magnetic time scale as in (b) can be constructed. (b) Magnetic reversals for the last 4.5 million years as determined from lava flows on land. Black bands represent normal magnetism and blue bands represent reverse magnetism.

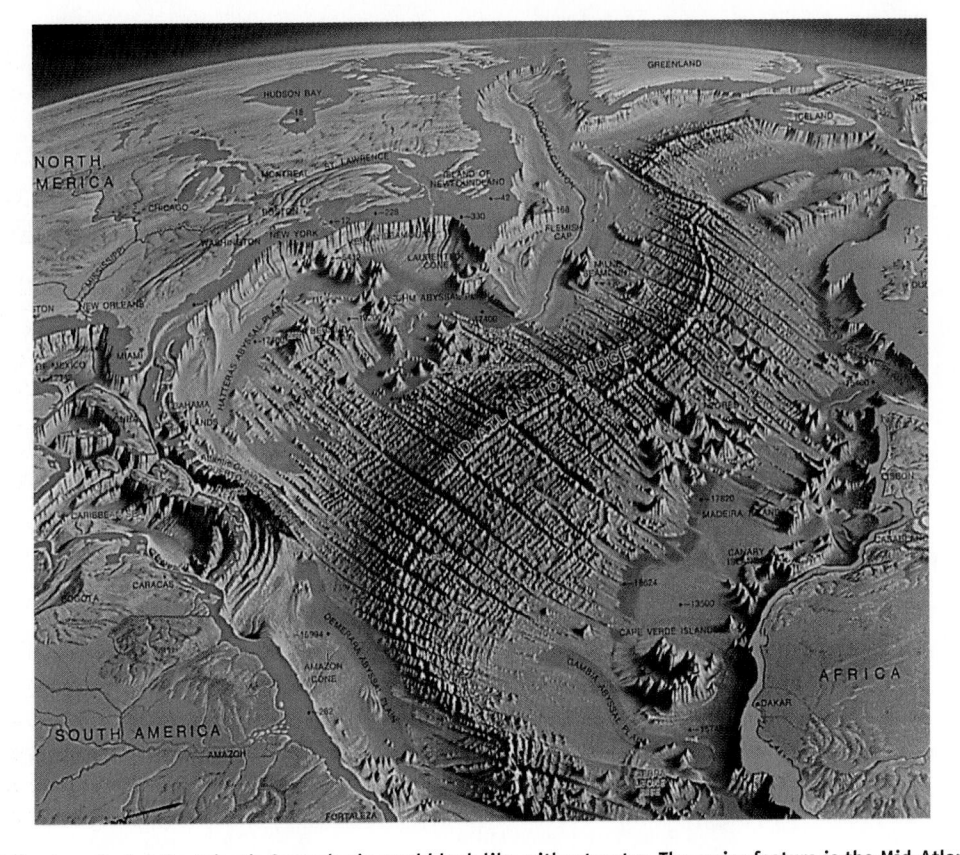

FIGURE 6.10 Artistic view of what the Atlantic Ocean basin would look like without water. The major feature is the Mid-Atlantic Ridge.

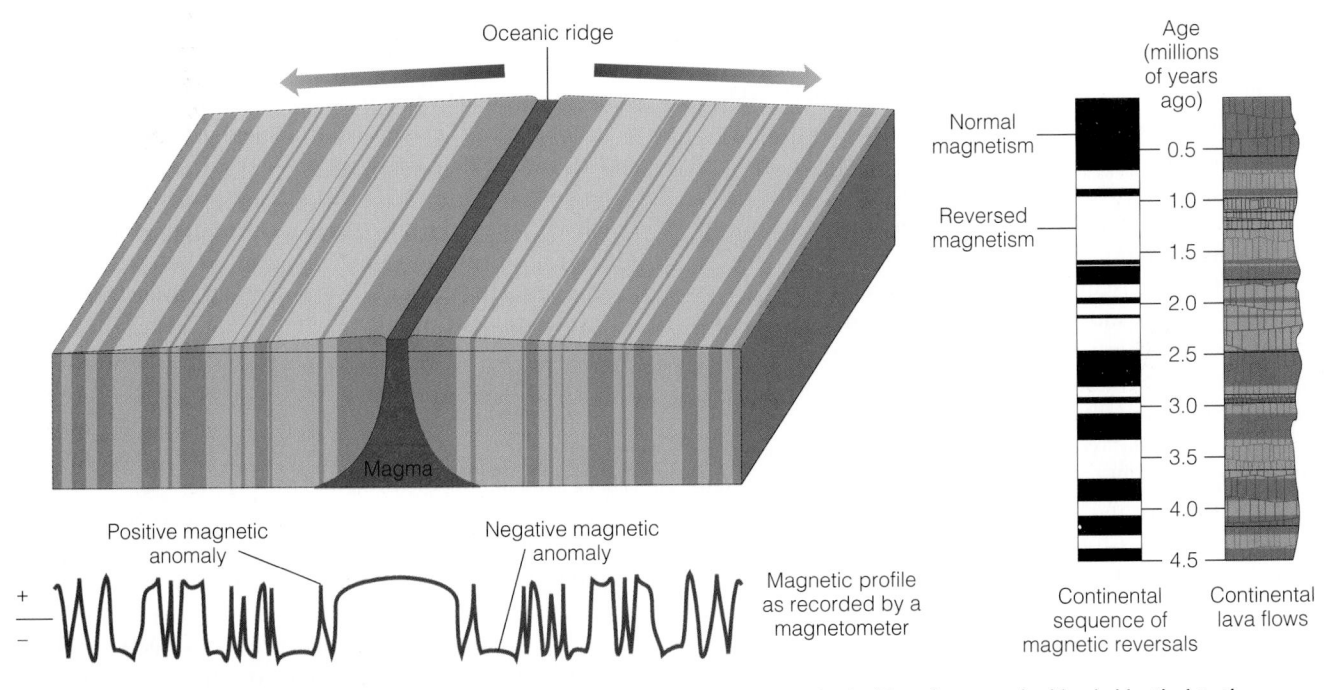

FIGURE 6.11 The sequence of magnetic anomalies preserved within the oceanic crust on both sides of an oceanic ridge is identical to the sequence of magnetic reversals already known from continental lava flows. Magnetic anomalies are formed when basaltic magma intrudes into oceanic ridges; when the magma cools below the Curie point, it records Earth's magnetic polarity at the time. Seafloor spreading splits the previously formed crust in half so that it moves laterally away from the oceanic ridge. Repeated intrusions record a symmetric series of magnetic anomalies that reflect periods of normal and reversed polarity. The magnetic anomalies are recorded by a magnetometer, which measures the strength of the magnetic field.

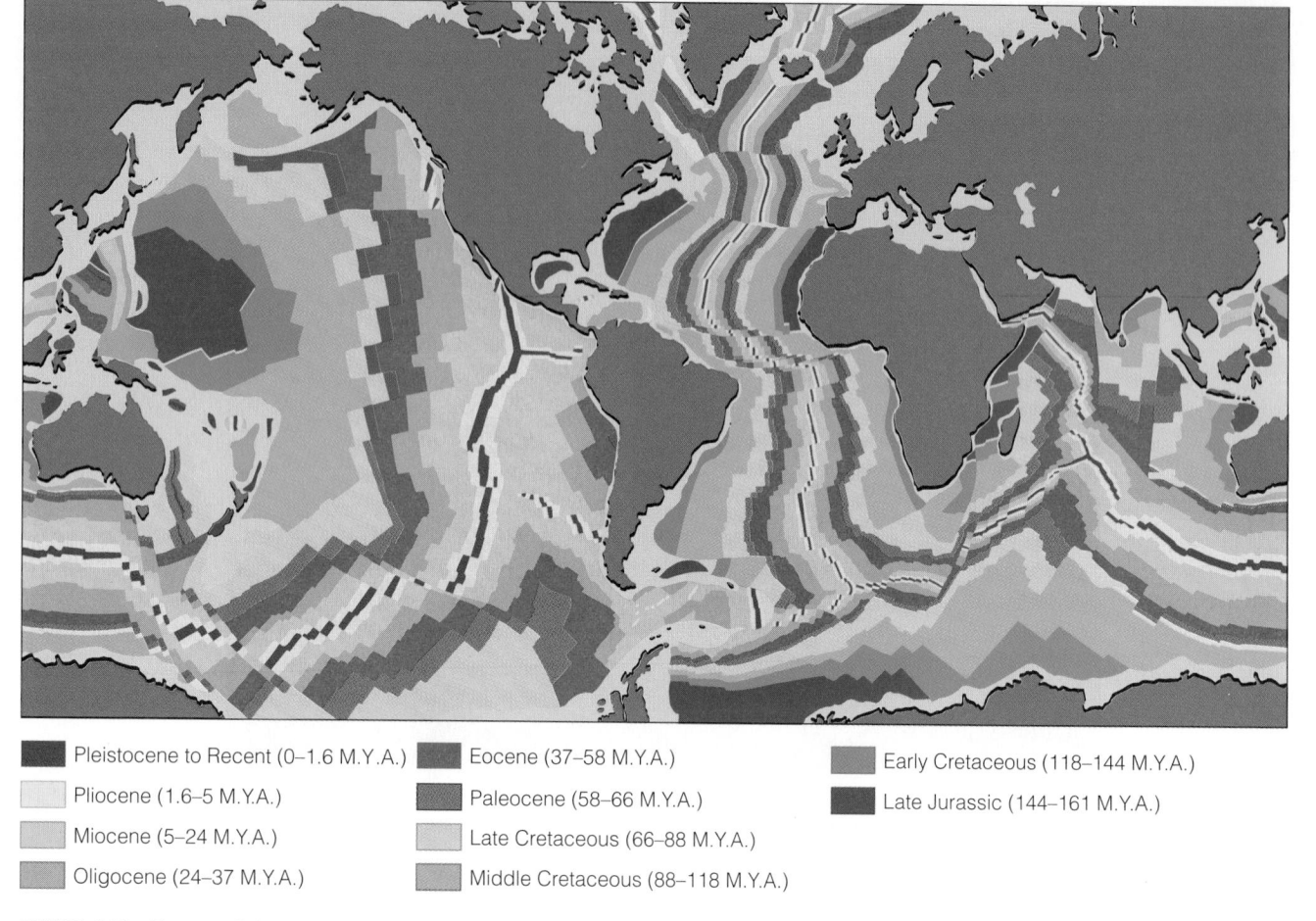

■ Pleistocene to Recent (0–1.6 M.Y.A.)	■ Eocene (37–58 M.Y.A.)	■ Early Cretaceous (118–144 M.Y.A.)
■ Pliocene (1.6–5 M.Y.A.)	■ Paleocene (58–66 M.Y.A.)	■ Late Jurassic (144–161 M.Y.A.)
■ Miocene (5–24 M.Y.A.)	■ Late Cretaceous (66–88 M.Y.A.)	
■ Oligocene (24–37 M.Y.A.)	■ Middle Cretaceous (88–118 M.Y.A.)	

FIGURE 6.12 The age of the world's ocean basins established from magnetic anomalies demonstrates that the youngest oceanic crust is adjacent to the spreading ridges and that its age increases away from the ridge axis.

viously formed crust moves laterally away from the ridge. These magnetic stripes, representing times of normal or reversed polarity, are parallel to and symmetric around oceanic ridges (where upwelling magma forms new oceanic crust), conclusively confirming Hess's theory of seafloor spreading.

One of the consequences of the seafloor-spreading theory is its confirmation that ocean basins are geologically young features whose openings and closings are partially responsible for continental movement (Figure 6.12). Radiometric dating reveals that the oldest oceanic crust is less than 180 million years old, whereas the oldest continental crust is 3.96 billion years old. Although geologists do not universally accept the idea of thermal convection cells as a driving mechanism for plate movement, most accept that plates are created at oceanic ridges and destroyed at deepsea trenches, regardless of the driving mechanism involved.

Plate Tectonics and Plate Boundaries

Plate tectonic theory is based on a simple model of Earth. The rigid lithosphere, consisting of both oceanic and continental crust, as well as the underlying upper mantle, consists of numerous variable-sized pieces called **plates** (Figure 6.13). The plates vary in thickness; those composed of upper mantle and continental crust are as much as 250 km thick, whereas those of upper mantle and oceanic crust are up to 100 km thick.

The lithosphere overlies the hotter and weaker semiplastic asthenosphere. It is thought that movement resulting from some type of heat-transfer system within the asthenosphere causes the overlying plates to move. As plates move over the asthenosphere, they separate, mostly at oceanic ridges; in other areas such as at oceanic trenches, they collide and are subducted back into the mantle.

Most geologists accept plate tectonic theory, in part, because the evidence for it is overwhelming and because it is a unifying theory that accounts for a variety of apparently unrelated geologic features and events. Consequently, geologists now view many geologic processes, such as mountain building, seismicity, and volcanism, from the perspective of plate tectonics. Furthermore, because all inner planets have had a similar origin and early history, geologists are interested in determining whether plate tectonics is unique to Earth or if it operates in the same way on other planets (see Perspective 6.1).

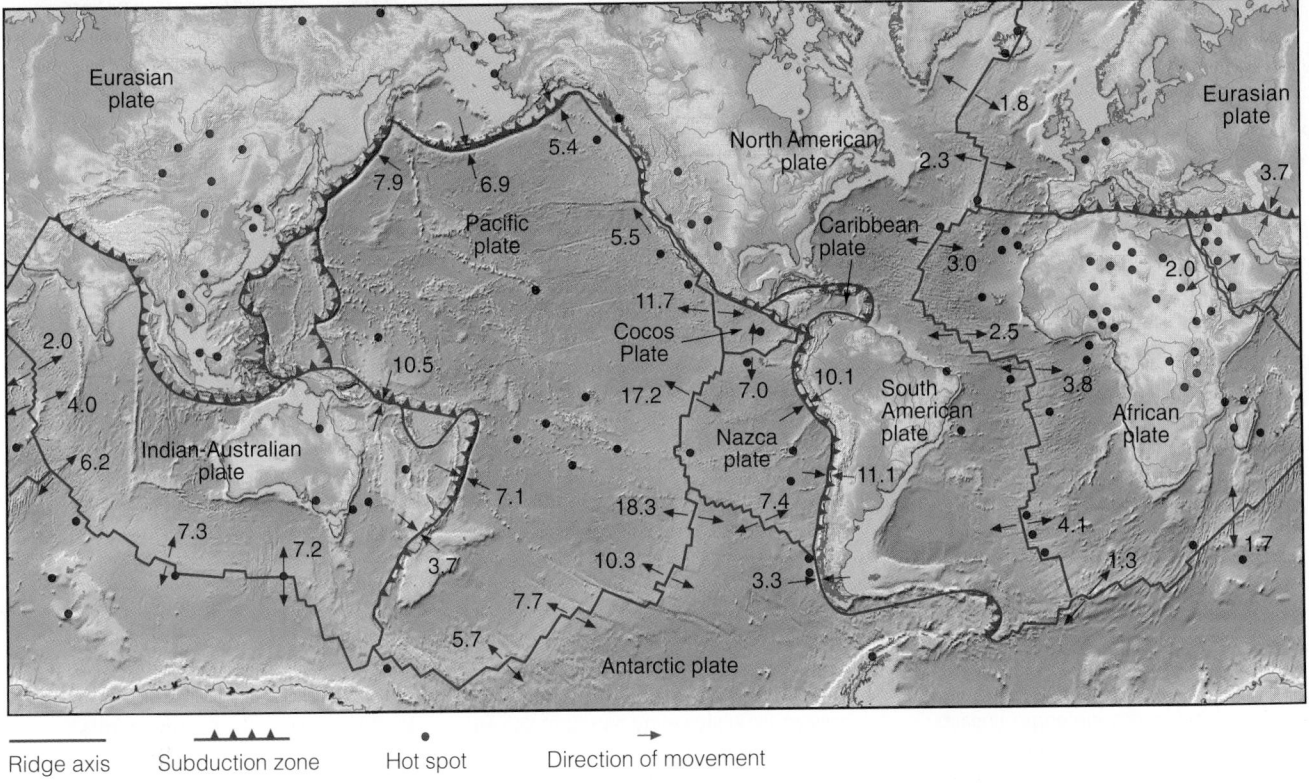

<table>
<tr><td>———</td><td>▲▲▲▲</td><td>•</td><td>→</td></tr>
<tr><td>Ridge axis</td><td>Subduction zone</td><td>Hot spot</td><td>Direction of movement</td></tr>
</table>

FIGURE 6.13 A map of the world showing the plates, their boundaries, relative motion and rates of movement in centimeters per year, and hot spots.

Because it appears that plate tectonics has operated since at least the Proterozoic (see Chapter 9), it is important that we understand how plates move and interact with each other and how ancient plate boundaries are recognized. After all, the movement of plates has had a profound effect on the geologic and biologic history of this planet.

Geologists recognize three major types of plate boundaries: *divergent, convergent,* and *transform* (Table 6.2). It is along these boundaries that new plates are formed, consumed, or slide laterally past one another. To understand the implications of plate interactions as they have affected

Earth history, geologists must study present plate boundaries.

DIVERGENT PLATE BOUNDARIES

Divergent plate boundaries, or **spreading ridges,** occur where plates are separating and new oceanic lithosphere is forming. Divergent boundaries are places where the crust is extended, thinned, and fractured as magma—derived from the partial melting of the mantle—rises to the surface. The magma is almost entirely basaltic and intrudes into vertical fractures to form dikes and lava flows. As successive injections of magma cool

TABLE 6.2			
Types of Plate Boundaries			
TYPE	EXAMPLE	LANDFORMS	VOLCANISM
Divergent			
Oceanic	Mid-Atlantic Ridge	Mid-oceanic ridge with axial rift valley	Basalt
Continental	East African Rift Valley	Rift valley	Basalt and rhyolite, no andesite
Convergent			
Oceanic–oceanic	Aleutian Islands	Volcanic island arc, offshore oceanic trench	Andesite
Oceanic–continental	Andes	Offshore oceanic trench, volcanic mountain chain, mountain belt	Andesite
Continental–continental	Himalayas	Mountain belt	Minor
Transform	San Andreas fault	Fault valley	Minor

perspective 6.1

Tectonics of the Terrestrial Planets

The four inner, or terrestrial, planets—Mercury, Venus, Earth, and Mars—all had a similar early history involving accretion, differentiation into a metallic core and silicate mantle and crust, and formation of an early atmosphere by outgassing. Their early history was also marked by widespread volcanism and meteorite impacts, both of which helped modify their surfaces. The volcanic and tectonic activity and resultant surface features (other than meteorite craters) of these planets are clearly related to the way they transport heat from their interiors to their surfaces.

Earth appears to be unique in that its outer portion is broken into a series of plates. The formation and destruction of these plates at spreading ridges and subduction zones transfer the majority of Earth's internally produced heat. In addition, movement of the plates—together with lifeforms, the formation of sedimentary rocks, and water—is responsible for the cycling of carbon dioxide between the atmosphere and lithosphere and thus the maintenance of a habitable climate on Earth.

Heat is transferred between the interior and surface of both Mercury and Mars, mainly by lithospheric conduction. This method is sufficient for these planets because both are significantly smaller than Earth or Venus. Because Mercury and Mars have a single globally continuous plate, they have exhibited fewer types of volcanic and tectonic activity than has Earth. The initial interior warming of Mercury and Mars produced tensional features such as normal faults and widespread volcanism, while their subsequent cooling produced folds and faults resulting from compressional forces, as well as volcanic activity.

Mercury's surface is heavily cratered and shows little in the way of primary volcanic structures. It does, however, have a global system of lobate scarps. These have been interpreted as evidence that Mercury shrank soon after its crust hardened, resulting in crustal cracking.

Mars has numerous features that indicates an extensive early period of volcanism. These include Olympus Mons, the solar system's largest volcano, lava flows, and uplifted regions thought to have resulted from mantle convection. In addition to volcanic features, Mars also displays abundant evidence of tensional tectonics, including numerous faults and large fault-produced valley structures.

Although Mars was tectonically active during the past, no evidence indicates that plate tectonics comparable to that on Earth has ever occurred there.

Venus underwent essentially the same early history as the other terrestrial planets, including a period of volcanism, but it is more Earth-like in its tectonics than either Mercury or Mars. Initial radar mapping in 1990 by the *Magellan* spacecraft revealed a surface of extensive lava flows, volcanic domes, folded mountain ranges, and an extensive and intricate network of faults, all of which attest to an internally active planet (Figure 1).

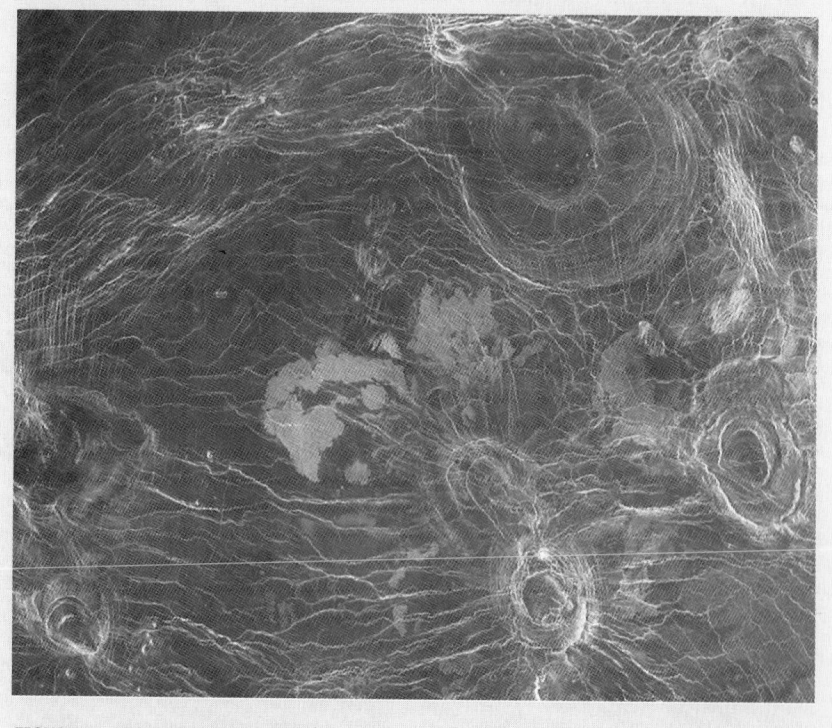

FIGURE 1   This radar image of Venus made by the *Magellan* spacecraft reveals circular and oval-shaped volcanic features. A complex network of cracks and fractures extends outward from the volcanic features. Geologists think these features were created by blobs of magma rising from the interior of Venus with dikes filling some of the cracks.

FIGURE 6.14 Pillow lavas forming along the Mid-Atlantic Ridge. Their distinctive bulbous shape is the result of underwater eruption.

and solidify, they form new oceanic crust and record the intensity and orientation of Earth's magnetic field (Figure 6.11). Divergent boundaries most commonly occur along the crests of oceanic ridges, for example, the Mid-Atlantic Ridge. Oceanic ridges are thus characterized by rugged topography with high relief resulting from displacement of rocks along large fractures, shallow-focus earthquakes, high heat flow, and basaltic flows or pillow lavas (Figure 6.14).

Divergent boundaries are also present under continents during the early stages of continental breakup (Figure 6.15). When magma wells up beneath a continent, the crust is initially elevated, stretched, and thinned, producing fractures and rift valleys (Figure 6.15a). During this stage, magma typically intrudes into the faults and frac-

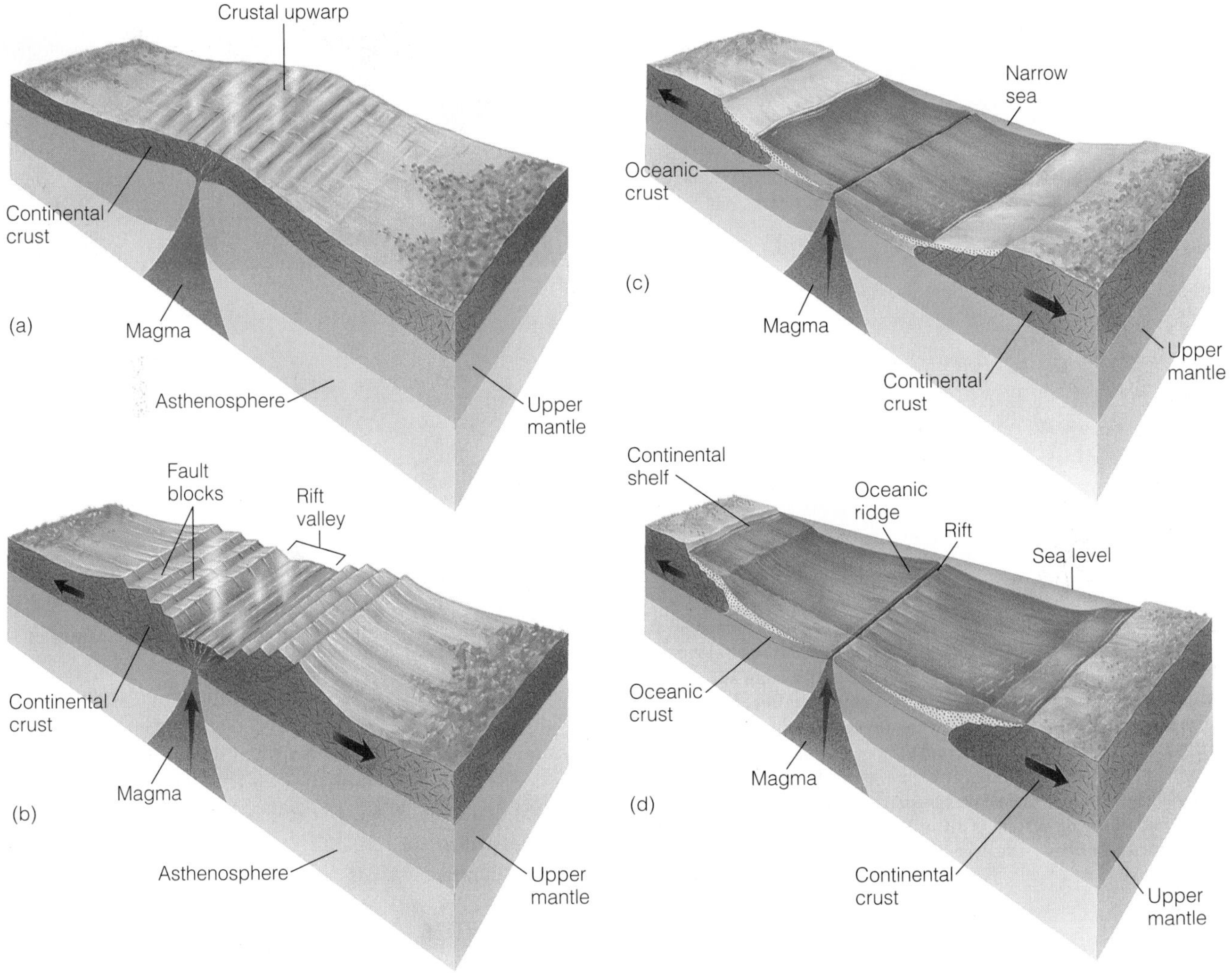

FIGURE 6.15 History of a divergent plate boundary. (a) Heat from rising magma beneath a continent causes it to bulge, producing numerous cracks and fractures. (b) As the crust is stretched and thinned, rift valleys develop, and lava flows onto the valley floors. (c) Continued spreading further separates the continent until a narrow seaway develops. (d) As spreading continues, an oceanic ridge system forms, and an ocean basin develops and grows.

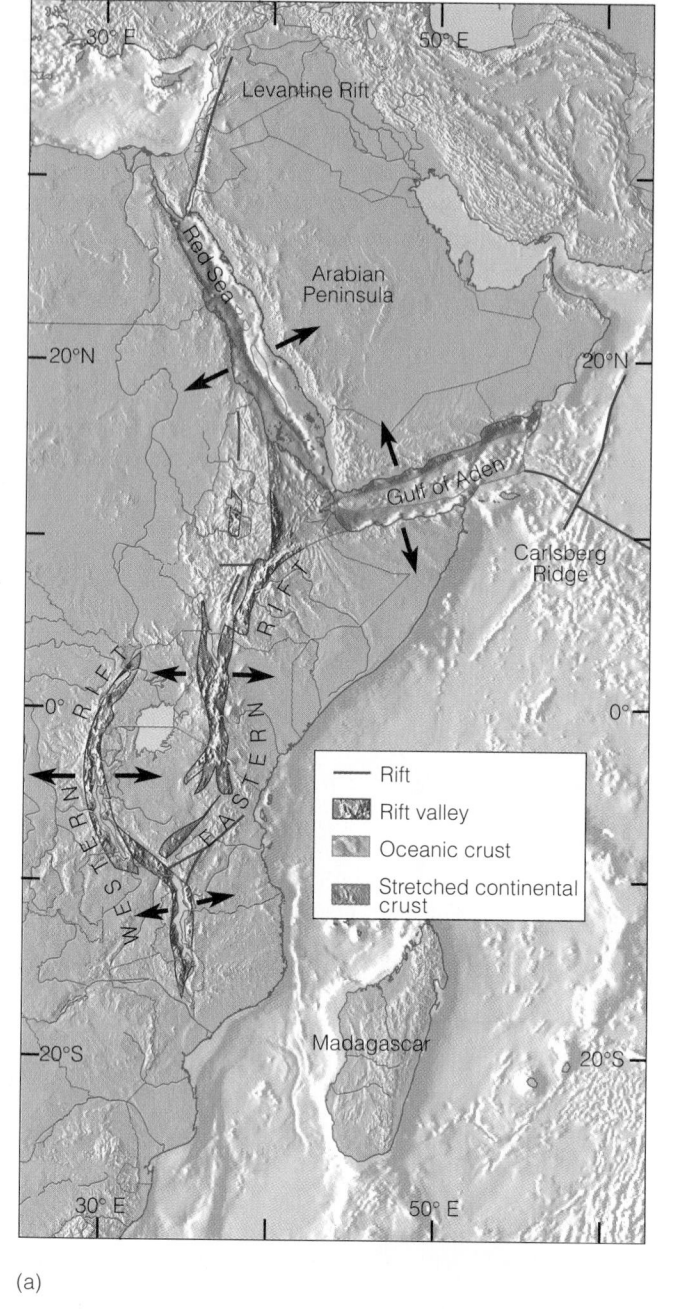

(a)

(b)

FIGURE 6.16 (a) The East African rift valley is being formed by the separation of eastern Africa from the rest of the continent along a divergent plate boundary. The Red Sea represents a more advanced stage of rifting, in which two continental blocks are separated by a narrow sea. (b) View looking down the Great Rift Valley of Africa. Little Magadi, seen in the background, is one of numerous soda lakes forming in the valley. Because of high evaporation rates and lack of any drainage outlets, these lakes are very saline. The Great Rift Valley is part of the system of rift valleys resulting from stretching of the crust as plates move away from each other in eastern Africa.

tures, forming sills, dikes, and lava flows; the latter often cover the rift valley floor (Fig. 6.15b). The East African rift valleys are an excellent example of this stage of continental breakup (Fig. 6.16).

If spreading proceeds, some rift valleys will continue to lengthen and deepen until they form a narrow linear sea separating two continental blocks (Figure 6.15c). The Red Sea separating the Arabian Peninsula from Africa (Figure 6.16) and the Gulf of California, which separates Baja California from mainland Mexico, are good examples of this more advanced stage of rifting.

As a newly created narrow sea continues enlarging, it may eventually become an expansive ocean basin such as the Atlantic, which separates North and South America from Europe and Africa by thousands of kilometers (Figure 6.15d). The Mid-Atlantic Ridge is the boundary between these diverging plates; the American plates are moving westward, and the Eurasian and African plates are moving eastward.

AN EXAMPLE OF ANCIENT RIFTING

What features in the rock record can geologists use to recognize ancient rifting? Associated with regions of continental rifting are faults, dikes, sills, lava flows, and thick sedimentary sequences within rift valleys. The Triassic

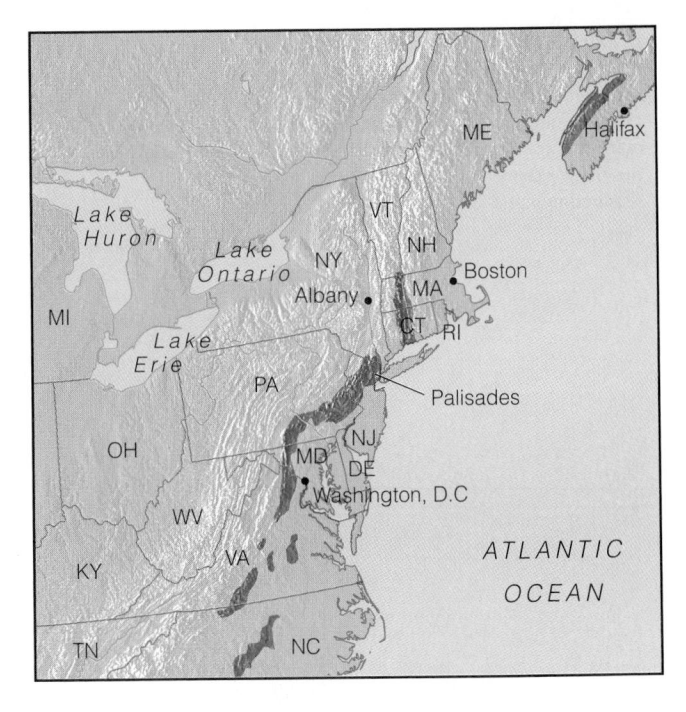

(a)

(b)

FIGURE 6.17 (a) Areas where Triassic fault-block basin deposits crop out in eastern North America. (b) Palisades of the Hudson River. This sill was one of many that were intruded into the fault-block basin sediments during the Late Triassic rifting that marked the separation of North America from Africa.

fault basins of the eastern United States are a good example of ancient continental rifting (Figure 6.17). These fault basins mark the zone of rifting that occurred when North America split apart from Africa. They contain thousands of meters of continental sediment and are riddled with dikes and sills (see Chapter 14).

CONVERGENT PLATE BOUNDARIES

Whereas new crust forms at divergent plate boundaries, older crust must be destroyed and recycled in order for the entire surface area of Earth to remain the same. Otherwise, we would have an expanding Earth. Such plate destruction occurs at **convergent plate boundaries** where two plates collide and the leading edge of one plate descends beneath the margin of the other by a process known as *subduction*. As the subducting plate moves down into the asthenosphere, it is heated and eventually incorporated into the mantle.

Convergent boundaries are characterized by deformation, volcanism, mountain building, metamorphism, sismicity, and important mineral deposits. Three types of convergent plate boundaries are recognized: oceanic-oceanic, oceanic-continental, and continental-continental.

Oceanic-Oceanic Plate Boundaries When two oceanic plates converge, one is subducted beneath the other along an **oceanic–oceanic plate boundary** (Figure 6.18). The subducting plate bends downward to form the outer wall of an oceanic trench. A *subduction complex,* composed of wedge-shaped slices of highly folded and faulted marine sediments and oceanic lithosphere scraped off the descending plate, forms along the inner wall. As the subducting plate descends into the mantle, it is heated and partially melted, generating magma commonly of andesitic composition. This magma is less dense than the surround-

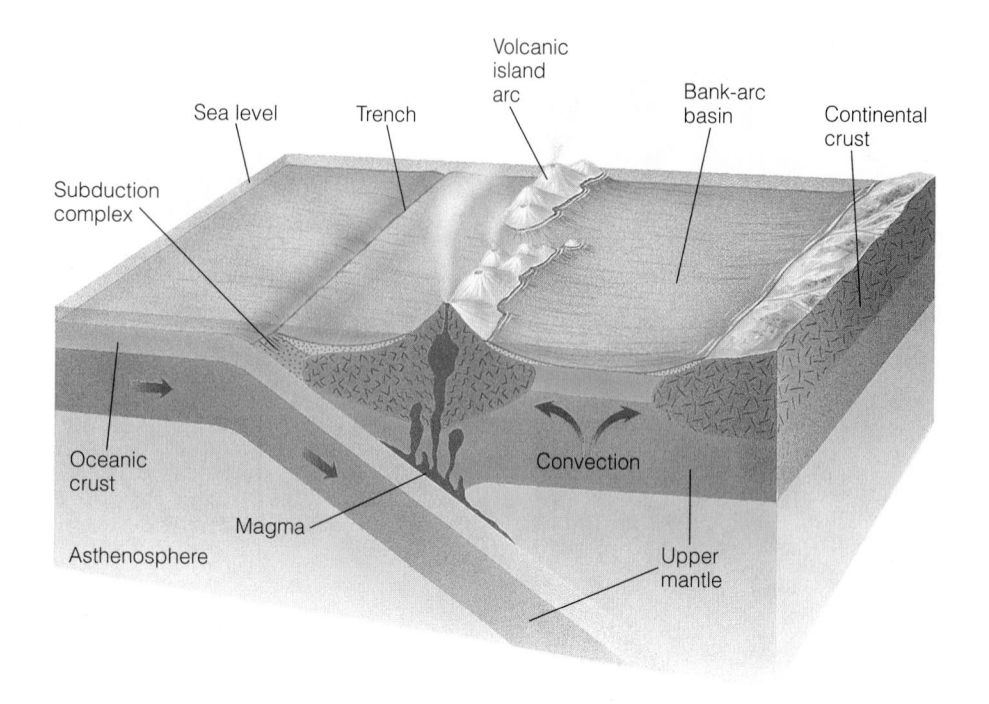

FIGURE 6.18 Oceanic–oceanic plate boundary. An oceanic trench forms where one oceanic plate is subducted beneath another. On the nonsubducted plate, a volcanic island arc forms from the rising magma generated from the subducting plate.

ing mantle rocks and rises to the surface of the nonsubducted plate, forming a curved chain of volcanic islands called a **volcanic island arc** (any plane intersecting a sphere makes an arc). This arc is nearly parallel to the oceanic trench and is separated from it by a distance of up to several hundred kilometers—the distance depends on the angle of dip of the subducting plate (Figure 6.18).

In those areas where the rate of subduction is faster than the forward movement of the overriding plate, the lithosphere on the landward side of the volcanic island arc may be subjected to tensional stress and stretched and thinned, resulting in the formation of a *back-arc basin.* This back-arc basin may grow by spreading if magma breaks through the thin crust and forms new oceanic crust (Figure 6.18). A good example of a back-arc basin associated with an oceanic-oceanic plate boundary is the Sea of Japan between the Asian continent and the islands of Japan.

Most present-day active volcanic island arcs are in the Pacific Ocean basin and include the Aleutian Islands, the Kermadec-Tonga arc, and the Japanese and Philippine Islands. The Scotia and Antillean (Caribbean) island arcs are present in the Atlantic Ocean basin.

Oceanic–Continental Plate Boundaries

When an oceanic and continental plate converge, the denser oceanic plate is subducted under the continental plate along an **oceanic–continental plate boundary** (Figure 6.19). Just as at oceanic-oceanic plate boundaries, the descending oceanic plate forms the outer wall of an oceanic trench.

As the cold, wet, and slightly denser oceanic plate descends into the hot asthenosphere, partial melting occurs and magma is generated. This magma rises beneath the

overriding continental plate and can extrude at the surface, producing a chain of andesitic volcanoes (also called a volcanic arc), or intrude into the continental margin as plutons, especially batholiths.

An excellent example of an oceanic–continental plate boundary is the Pacific coast of South America where the oceanic Nazca plate is currently being subducted under South America. The Peru-Chile Trench marks the site of subduction, and the Andes Mountains are the resulting volcanic mountain chain on the nonsubducting plate. This particular example demonstrates the effect plate tectonics has on our lives. For instance, earthquakes are commonly associated with subduction zones, and the western side of South America is the site of frequent and devastating earthquakes.

Continental–Continental Plate Boundaries

Two continents approaching each other will initially be separated by an ocean floor that is being subducted under one continent. The edge of that continent will display the features characteristic of oceanic-continental convergence. As the ocean floor continues to be subducted, the two continents will come closer together until they eventually collide. Because continental lithosphere, which consists of continental crust and upper mantle, is less dense than oceanic lithosphere (oceanic crust and upper mantle), it cannot sink into the asthenosphere. Although one continent may partly slide under the other, it cannot be pulled or pushed down into a subduction zone (Figure 6.20).

When two continents collide, they are welded together along a zone marking the former site of subduction. At this **continental–continental plate boundary,** an interior mountain belt is formed consisting of deformed sedimentary

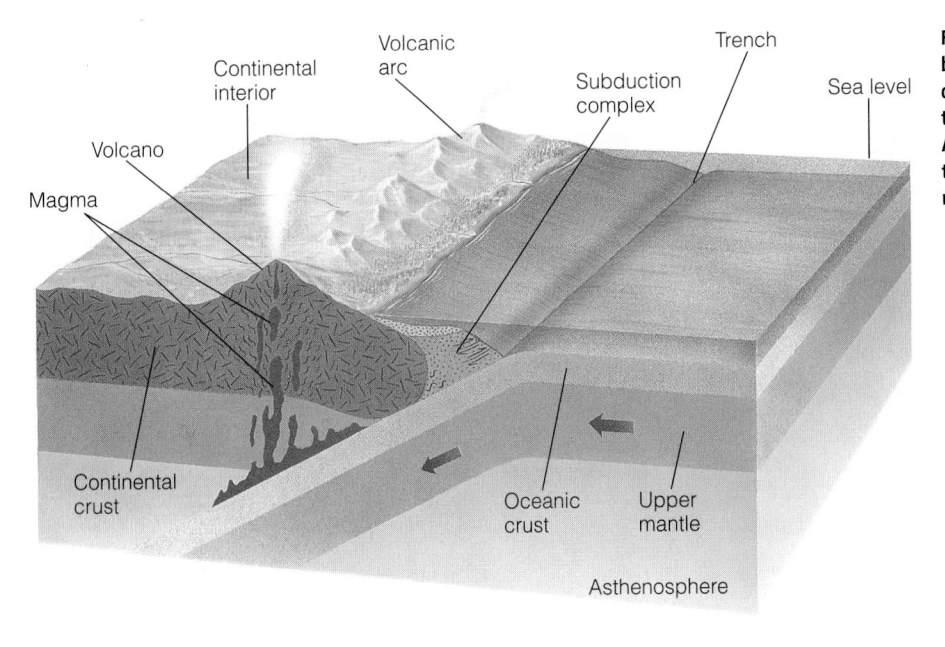

FIGURE 6.19 Oceanic–continental plate boundary. When an oceanic plate is subducted beneath a continental plate, partial melting occurs, generating a magma. An andesitic volcanic mountain range is then formed on the continental plate as a result of rising magma.

rocks, igneous intrusions, metamorphic rocks, and fragments of oceanic crust. In addition, the entire region is subjected to numerous earthquakes. The Himalayas in central Asia, the world's youngest and highest mountain system, resulted from the collision between India and Asia, which began about 40 to 50 million years ago and is still continuing.

ANCIENT CONVERGENT PLATE BOUNDARIES

How can former subduction zones be recognized in the geologic record? One clue is provided by igneous rocks. The magma erupted at the surface, forming island arc volcanoes and continental volcanoes, is of andesitic composition. Another clue can be found in the zone of intensely deformed rocks that occurs between the deep-sea trench where subduction is taking place and the area of igneous activity. Here, sediments and submarine rocks are folded,

faulted, and metamorphosed into a chaotic mixture of rocks termed a *mélange.*

During subduction, pieces of oceanic lithosphere are sometimes incorporated into the mélange and accreted onto the edge of the continent. Such slices of oceanic crust and upper mantle are called **ophiolites** (Figure 6.21). They consist of a layer of deep-sea sediments that include graywackes (poorly sorted sandstones containing abundant feldspars and rock fragments, usually in a clay-rich matrix), black shales, and cherts. These deep-sea sediments are underlain by pillow lavas, a sheeted dike complex, massive gabbro, and layered gabbro, all of which form the oceanic crust. Beneath the gabbro is peridotite, which probably represents the upper mantle. Ophiolites are key features in recognizing plate convergence along a subduction zone.

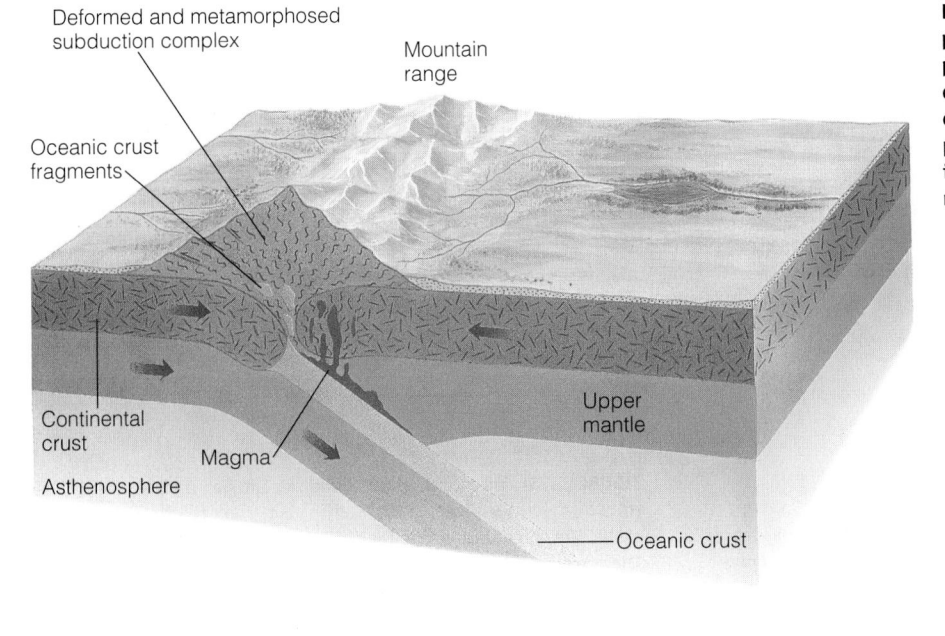

FIGURE 6.20 Continental–continental plate boundary. When two continental plates converge, neither is subducted because of their great thickness and low and equal densities. As the two continental plates collide, a mountain range is formed in the interior of a new and larger continent.

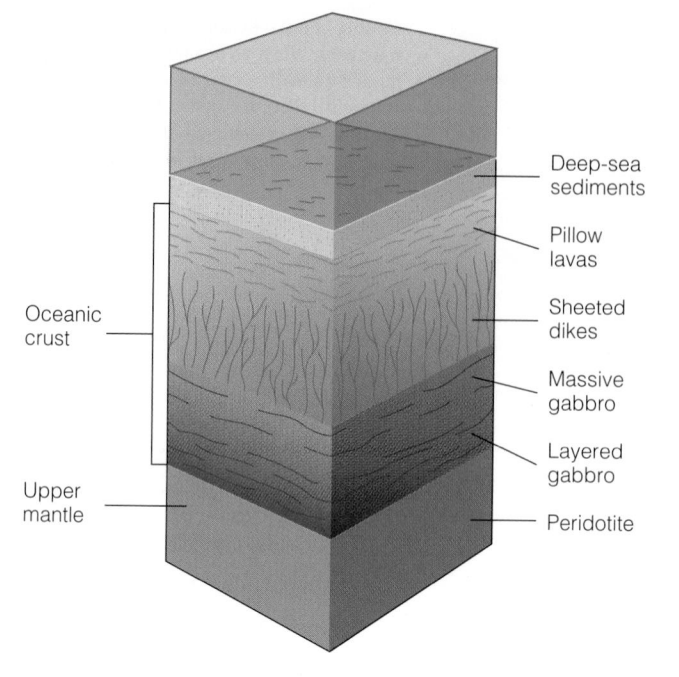

FIGURE 6.21 Ophiolites are sequences of rock on land consisting of deep-sea sediments, oceanic crust, and upper mantle.

Elongate belts of folded and faulted marine sedimentary rocks, andesites, and ophiolites are found in the Appalachians, Alps, Himalayas, and Andes Mountains. The combination of such features is good evidence that these mountain ranges resulted from deformation along convergent plate boundaries.

TRANSFORM PLATE BOUNDARIES

The third type of plate boundary is a **transform plate boundary.** These occur along fractures in the seafloor, known as transform faults, where plates slide laterally past each other, roughly parallel to the direction of plate move-

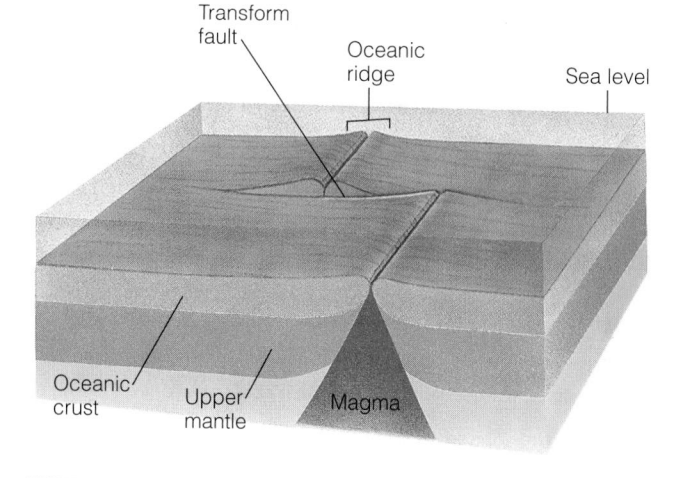

FIGURE 6.22 Horizontal movement between plates occurs along a transform fault. The majority of transform faults connect two oceanic ridge segments. Note that relative motion between the plates only occurs between the two ridges.

ment. Although lithosphere is neither created nor destroyed along a transform boundary, the movement between plates results in a zone of intensely shattered rock and numerous shallow-focus earthquakes.

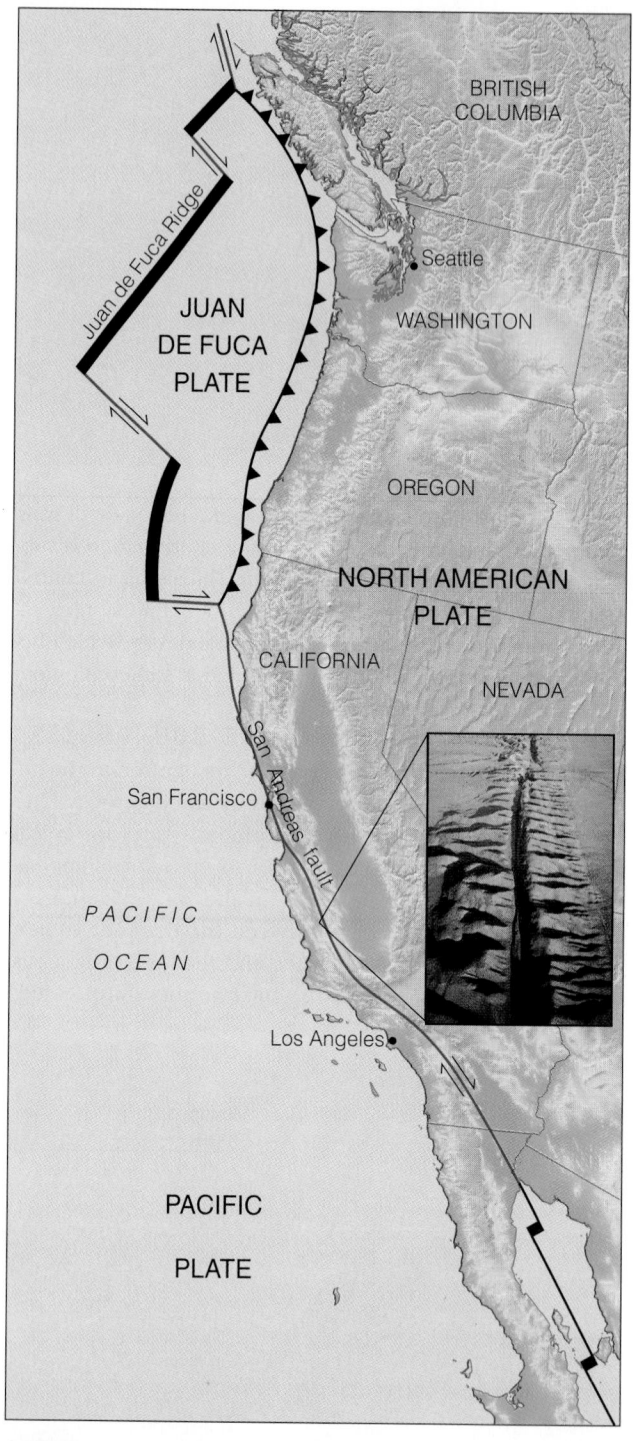

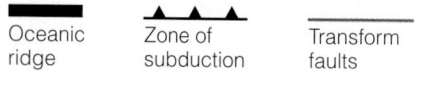

FIGURE 6.23 Transform plate boundary. The San Andreas fault is a transform fault separating the Pacific plate from the North American plate. Movement along this fault has caused numerous earthquakes. The photograph shows a segment of the San Andreas fault as it cuts through the Carrizo Plain, California.

Transform faults are particular types of faults that "transform" or change one type of motion between plates into another type of motion. Most commonly, transform faults connect two oceanic ridge segments (Figure 6.22), but they can also connect ridges to trenches, and trenches to trenches. While the majority of transform faults are in oceanic crust and are marked by distinct fracture zones, they can also extend into continents.

One of the best-known transform faults is the San Andreas fault in California. It separates the Pacific plate from the North American plate and connects spreading ridges in the Gulf of California with the Juan de Fuca and Pacific plates off the coast of northern California (Figure 6.23). Many of the earthquakes affecting California are the result of movement along this fault.

Unfortunately, transform faults generally do not leave any characteristic or diagnostic features except for the obvious displacement of the rocks that they are associated with. This displacement is commonly large, on the order of tens to hundreds of kilometers. Such large displacements in ancient rocks can sometimes be related to transform fault systems.

Hot Spots and Mantle Plumes

Before leaving the topic of plate boundaries, we should briefly mention an intraplate feature found beneath both oceanic and continental plates. **Hot spots** are locations where stationary columns of magma, originating deep within the mantle (mantle plumes), slowly rise to the surface and form volcanoes or flood basalts and manifest themselves as volcanoes (Figure 6.24). Because the mantle plumes apparently remain stationary (although some evi-

dence suggests that they might not) while the plates move over them, the resulting hot spots leave a trail of extinct, progressively older volcanoes called *aseismic ridges.* Some examples of aseismic ridges and hot spots are the Emperor Seamount–Hawaiian Island chain and Yellowstone National Park in Wyoming.

Plate Movement and Motion

How fast and in what direction are Earth's various plates moving, and do they all move at the same rate? Rates of plate movement can be calculated in several ways. The least accurate method is to determine the age of the sediments immediately above any portion of the oceanic crust and divide that age by the distance from the spreading ridge. Such calculations give an average rate of movement.

A more accurate method of determining both the average rate of movement and the relative motion is by dating the magnetic anomalies in the crust of the seafloor (Figure 6.12). The distance from an oceanic ridge axis to any magnetic anomaly indicates the width of new seafloor that formed during that time interval. Thus, for a given interval of time, the wider the strip of seafloor, the faster the plate has moved. In this way, not only can the present average rate of movement and relative motion be determined (Figure 6.25), but the average rate of movement during the past can also be calculated by dividing the distance between anomalies by the amount of time elapsed between anomalies.

Geologists can not only calculate the average rate of plate movement from magnetic anomalies, but they can also determine plate positions at various times in the past using magnetic anomalies. Because magnetic anomalies are

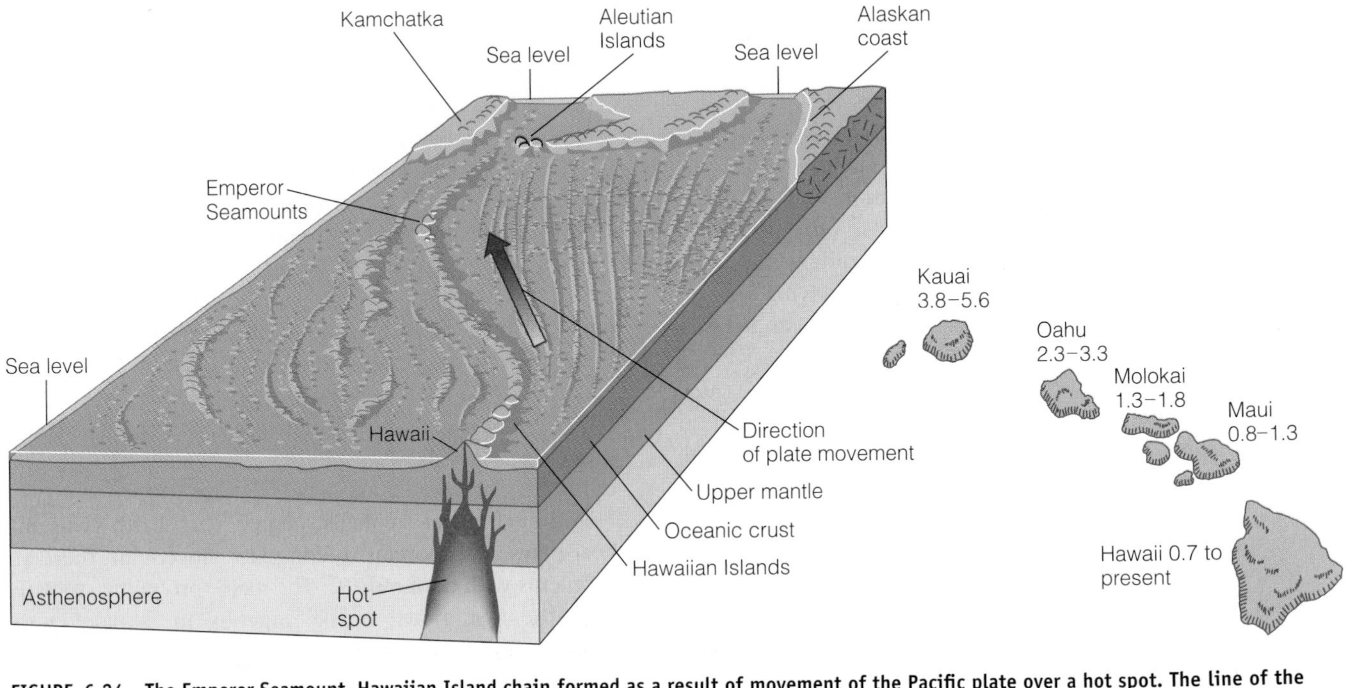

FIGURE 6.24 The Emperor Seamount–Hawaiian Island chain formed as a result of movement of the Pacific plate over a hot spot. The line of the volcanic islands traces the direction of plate movement. The numbers indicate the age of the islands in millions of years.

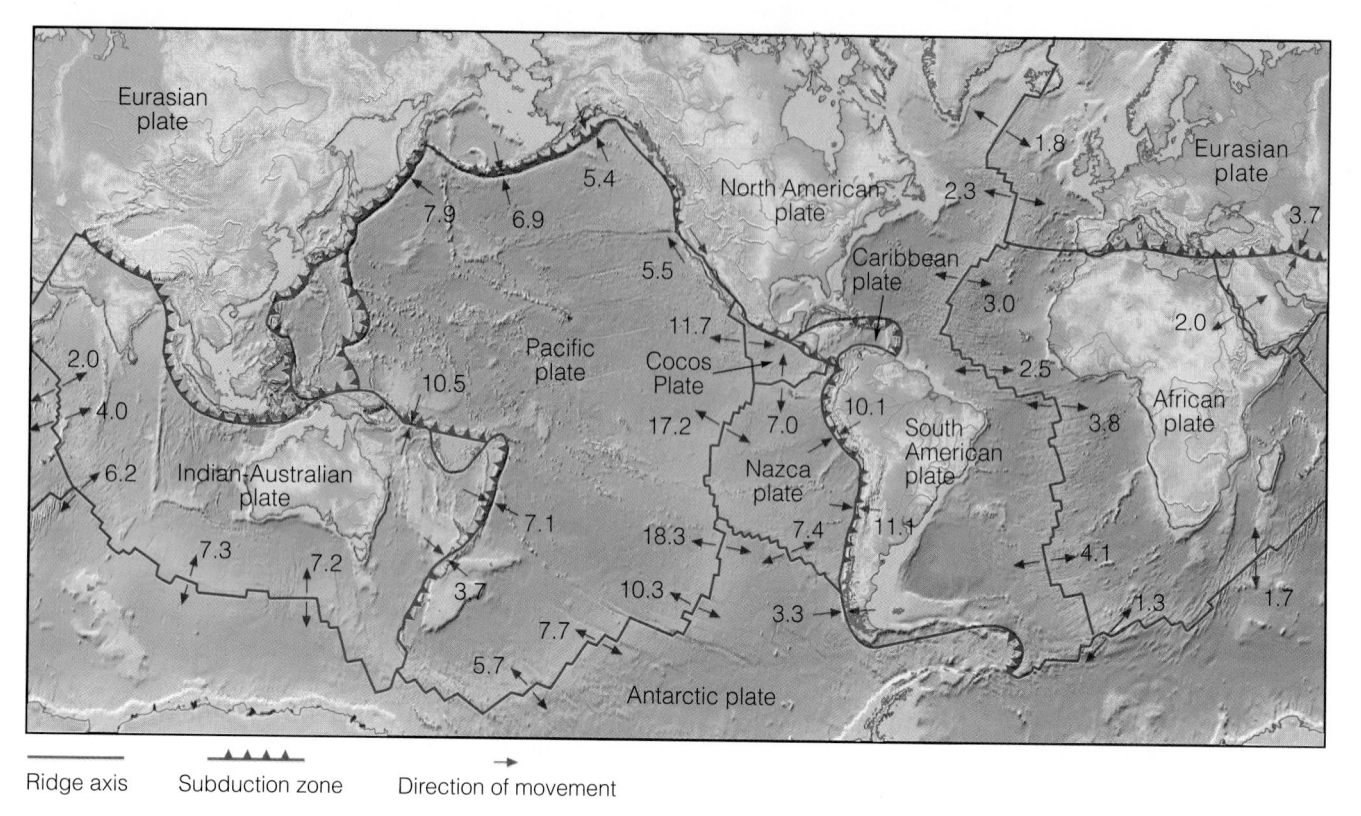

Ridge axis   Subduction zone   Direction of movement

FIGURE 6.25 This map shows the average rate of movement in centimeters per year and relative motion of Earth's plates.

parallel and symmetric with respect to spreading ridges, all one has to do to determine the position of continents when particular anomalies formed is to move the anomalies back to the spreading ridge, which will also move the continents with them (Figure 6.26). Unfortunately, subduction destroys oceanic crust and the magnetic record it carries. Thus, we have an excellent record of plate movements since the breakup of Pangaea, but not as good an understanding of plate movement before that time.

Satellite-laser ranging techniques can also be used to determine the average rate of movement as well as the relative motion between any two plates. Laser beams from a station on one plate are bounced off a satellite (in geosynchronous orbit) and returned to a station on a different plate. As the plates move away from each other, the laser beam takes more time to go from the sending station to the stationary satellite and back to the receiving station. This difference in elapsed time is used to calculate the rate of movement and relative motion between plates.

Plate motions derived from magnetic reversals and satellite-laser ranging techniques give only the relative motion of one plate with respect to another. Hot spots allow geologists to determine absolute motion because they provide an apparently fixed reference from which the rate and direction of plate movement can be measured. The Emperor Seamount—Hawaiian Island chain (Fig. 6.24) formed as a result of movement over a hot spot. Thus, the line of the volcanic islands traces the direction of plate movement, and dating the volcanoes enables geologists to determine the rate of movement.

The Driving Mechanism of Plate Tectonics

A major obstacle to the acceptance of continental drift was the lack of a driving mechanism to explain continental movement. When it was shown that continents and ocean floors moved together, not separately, and that new crust formed at spreading ridges by rising magma, most geologists accepted some type of convective heat system as the basic process responsible for plate motion. The question still remains, however: What exactly drives the plates?

Two models involving thermal convection cells have been proposed to explain plate movement (Figure 6.27). In one model, thermal convection cells are restricted to the asthenosphere; in the second model, the entire mantle is involved. In both models, spreading ridges mark the ascending limbs of adjacent convection cells, and trenches occur where the convection cells descend back into Earth's interior. The locations of spreading ridges and trenches are therefore determined by the convection cells themselves, and the lithosphere is considered to be the top of the thermal convection cell.

Although most geologists agree that Earth's internal heat plays an important role in plate movement, there are problems with both models. The major problem associated with the first model is the difficulty in explaining the source of heat for the convection cells and why they are restricted to the asthenosphere. In the second model, the source of heat comes from the outer core, but it is still not

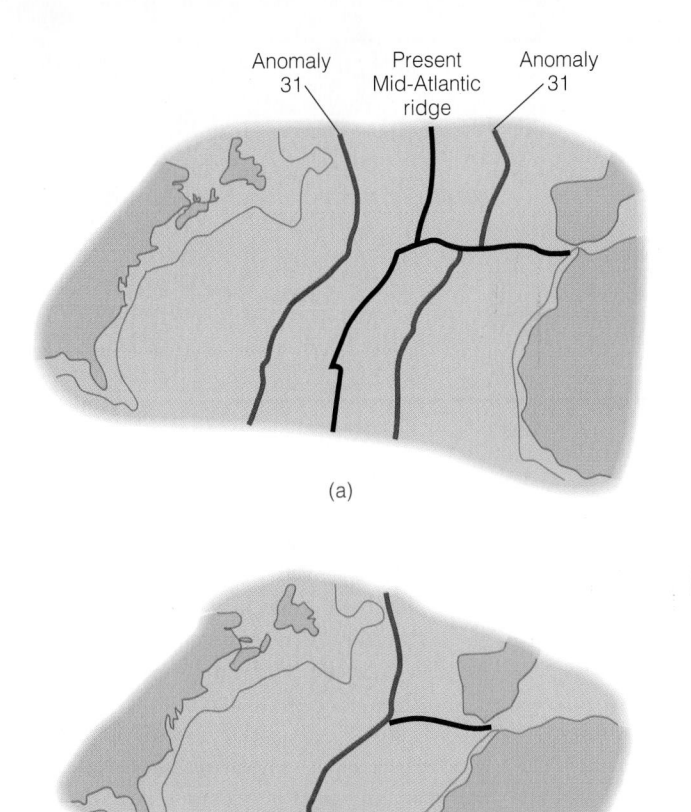

(a)

(b)

FIGURE 6.26 Reconstructing plate positions using magnetic anomalies. (a) The present North Atlantic, showing the present ridge and magnetic anomaly 31, which formed 67 million years ago. (b) The Atlantic 67 million years ago. Anomaly 31 marks the plate boundary 67 million years ago. By moving the anomalies back together, along with the plates they are on, we reconstruct the former positions of the continents.

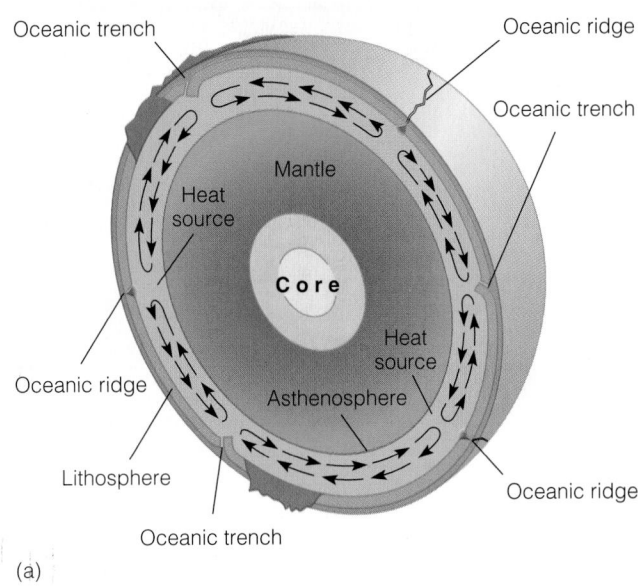

(a)

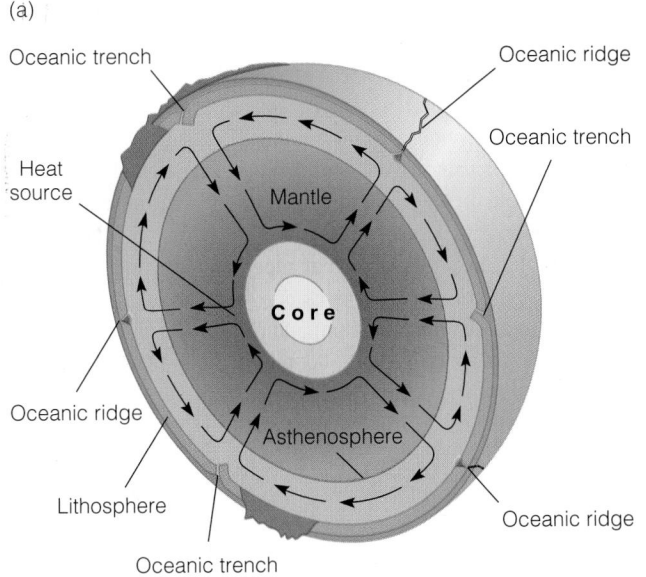

(b)

FIGURE 6.27 Two models involving thermal convection cells have been proposed to explain plate movement. (a) In one model, thermal convection cells are restricted to the asthenosphere. (b) In the other model, thermal convection cells involve the entire mantle.

known how heat is transferred from the outer core to the mantle. Nor is it clear how convection can involve both the lower mantle and the asthenosphere.

Some geologists think that in addition to thermal convection, plate movement also occurs because of a mechanism involving "slab-pull" or "ridge-push," both of which are gravity driven but still depend on thermal differences within Earth (Figure 6.28). In slab-pull, the subducting cold slab of lithosphere, being denser than the surrounding warmer asthenosphere, pulls the rest of the plate along with it as it descends into the asthenosphere. Operating in conjunction with slab-pull is ridge-push, which involves rising magma along oceanic ridges. The rising magma pushes the oceanic ridges higher than the surrounding oceanic crust, and gravity then pushes the oceanic lithosphere toward the trenches.

Currently, geologists are fairly certain that some type of convection system is involved in plate movement, but the extent to which other mechanisms such as slab-pull and ridge-push are involved is still unresolved. Consequently, a comprehensive theory of plate movement has not yet been developed, and much still remains to be learned about Earth's interior.

Plate Tectonics and Mountain Building

An **orogeny** is an episode of intense rock deformation or mountain building. Orogenies are the consequence of compressive forces related to plate movement. As one plate is subducted under another, sedimentary and volcanic rocks are folded and faulted along the plate margin while the more deeply buried rocks are subjected to regional metamorphism. Magma generated within the mantle either rises to the surface to erupt as andesitic volcanoes

The Super-continent Cycle

At the end of the Paleozoic Era, all continents were amalgamated into the supercontinent Pangaea. Pangaea began fragmenting during the Triassic Period and continues to do so, thus accounting for the present distribution of continents and oceans. It now appears that another supercontinent existed at the end of the Proterozoic Eon, and there is some evidence for even earlier supercontinents. Recently, it has been proposed that supercontinents consisting of all or most of Earth's landmasses form, break up, and reform in a cycle spanning about 500 million years.

The supercontinent cycle hypothesis is an expansion on the ideas of the Canadian geologist J. Tuzo Wilson. During the early 1970s, Wilson proposed a cycle (now known as the Wilson cycle) that includes continental fragmentation, the opening and closing of an ocean basin, and reassembly of the continent. According to the *supercontinent cycle hypothesis,* heat accumulates beneath a supercontinent because rocks of continents are poor conductors of heat. As a result of the heat accumulation, the supercontinent domes upward and fractures. Basaltic magma rising from below fills the fractures. As a basalt-filled fracture widens, it begins subsiding and forms a long narrow ocean such as the present-day Red Sea. Continued rifting eventually forms an expansive ocean basin such as the Atlantic.

According to proponents of the supercontinent cycle, one of the most convincing arguments for their hypothesis is the "surprising regularity" of mountain building caused by compression during continental collisions. These mountain-building episodes occur about every 400 to 500 million years and are followed by an episode of rifting about 100 million years later. In other words, a supercontinent fragments and its individual plates disperse following a rifting episode, an interior ocean forms, and then the dispersed fragments reassemble to form another supercontinent (Figure 1).

Besides accounting for the dispersal and reassembly of supercontinents, the supercontinent cycle hypothesis also explains two distinct types of orogens (linear parts of Earth's crust that were deformed during mountain building): *interior orogens* resulting from compression induced by continental collisions and *peripheral orogens* that form at continental margins in response to subduction and igneous activity. Following the fragmentation of a supercontinent, an interior ocean forms and widens as plates separate. After about 200 million years, plate separation ceases as the oceanic crust becomes cooler and denser and begins to subduct at the margins of the interior ocean basin, thus transforming inward-facing passive continental margins into active continental margins. Rising magma resulting from subduction forms plutons and volcanoes along these continental margins, and an interior orogeny develops when plates converge, causing compression-induced deformation, crustal thickening, and metamorphism. The present-day Appalachian Mountains of the eastern United States and Canada originally formed as an interior orogen (see Chapters 10 and 11).

Peripheral orogenies are caused by convergence of oceanic plates with the margins of a continent, igneous activity resulting from subduction, and collisions of island arcs with the continent. Subduction-related volcanism in peripheral orogens is rather continuous, but collisions of island arcs with the supercontinent are episodic. Peripheral orogenies can develop at any time during the supercontinent cycle. A good example of a presently active peripheral orogeny is the Andes of western South America where the Nazca plate is being subducted beneath the South American plate.

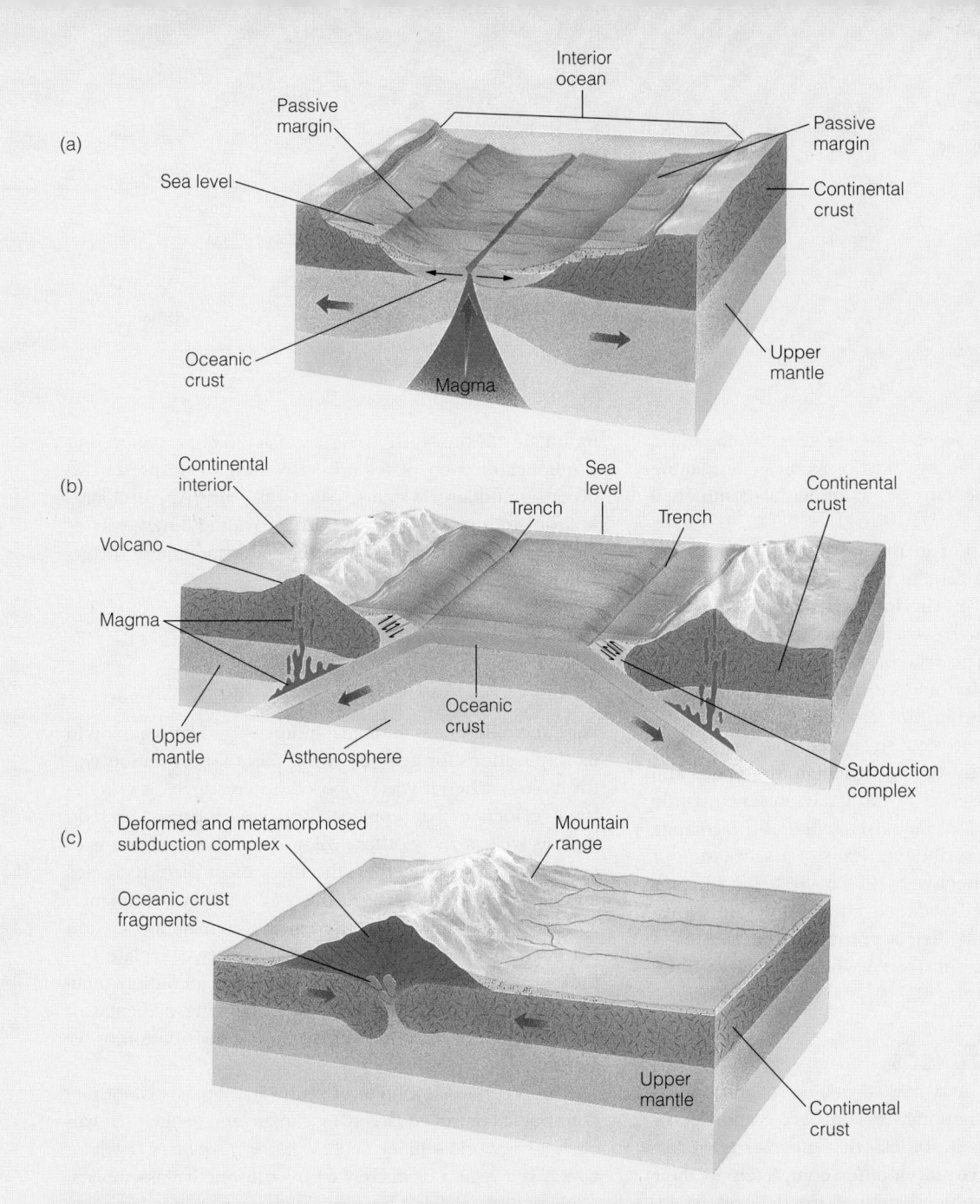

(a)

Interior ocean

Passive margin

Sea level

Passive margin

Continental crust

Oceanic crust

Magma

Upper mantle

(b)

Continental interior

Sea level

Continental crust

Volcano

Trench

Trench

Magma

Upper mantle

Asthenosphere

Oceanic crust

Subduction complex

(c)

Deformed and metamorphosed subduction complex

Mountain range

Oceanic crust fragments

Upper mantle

Continental crust

FIGURE 1   The supercontinent cycle. (a) Breakup of a supercontinent and the formation of an interior ocean basin. (b) Subduction along the margins of the interior ocean basin begins approximately 200 million years later, resulting in volcanic activity and deformation along an active oceanic–continental plate boundary. (c) Continental collisions and the formation of a new supercontinent result when all of the oceanic crust of the interior ocean basin is subducted.

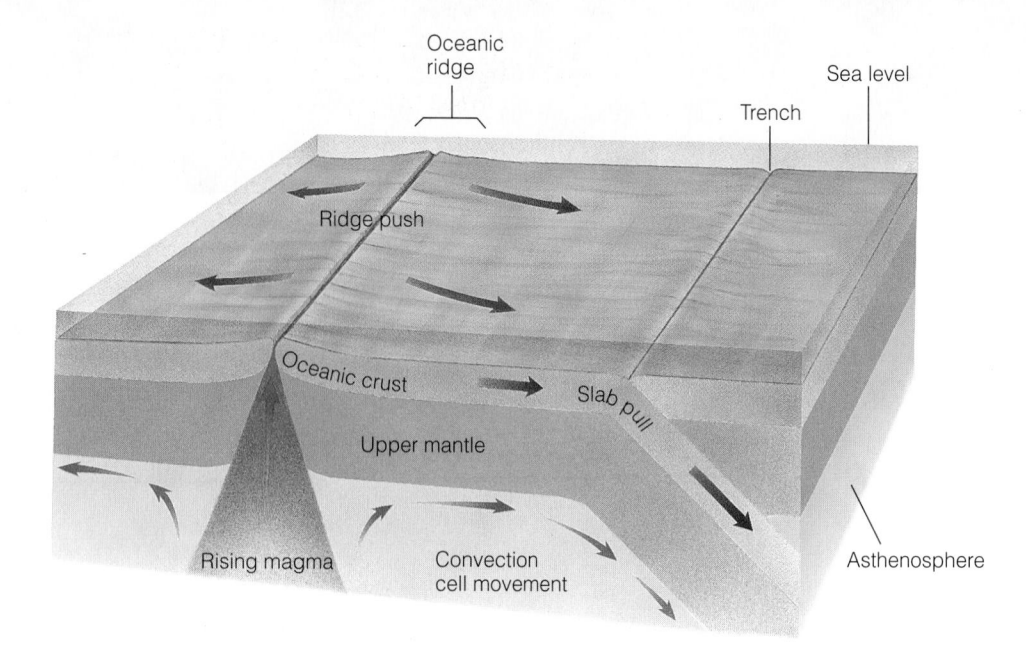

FIGURE 6.28 Plate movement is also thought to occur because of gravity-driven "slab-pull," or "ridge-push" mechanisms. In slab-pull, the edge of the subducting plate descends into the interior, and the rest of the plate is pulled downward. In ridge-push, rising magma pushes the oceanic ridges higher than the rest of the oceanic crust. Gravity thus pushes the oceanic lithosphere away from the ridges and toward the trenches.

or cools and crystallizes beneath the surface, forming intrusive igneous bodies. Typically, most orogenies occur along either oceanic-continental or continental-continental plate boundaries.

As we discussed earlier in this chapter, ophiolites are evidence of ancient convergent plate boundaries. The slivers of ophiolites found in the interiors of such mountain ranges as the Alps, Himalayas, and Urals mark the sites of former ocean basins. The relationship between mountain building and the opening and closing of ocean basins is called the **Wilson cycle** in honor of the Canadian geologist J. Tuzo Wilson, who first suggested that an ancient ocean had closed to form the Appalachian Mountains and then reopened and widened to form the present Atlantic Ocean. According to some geologists, much of the geology of continents can be described in terms of a succession of Wilson cycles (see Perspective 6.2)! We will see in Chapter 14, however, that new evidence concerning the movement of microplates and their accretion at the margin of continents must also be considered when dealing with the tectonic history of a continent.

MICROPLATE TECTONICS

During the late 1970s and 1980s, geologists discovered that portions of many mountain systems are composed of small accreted lithospheric blocks that are clearly of foreign origin. These **microplates** differ completely in their fossil content, stratigraphy, structural trends, and paleomagnetic properties from the rocks of the surrounding mountain system. In fact, these microplates are so different from adjacent rocks that most geologists think that they formed elsewhere and were carried great distances as parts of other plates until they collided with other microplates or continents.

To date, most microplates have been identified in mountains of the North American Pacific coast region, but a number of such plates are suspected to be present in other mountain systems, such as the Appalachians. Microplate tectonics is, however, providing a new way of viewing Earth and gaining a better understanding of the geologic history of the continents.

Plate Tectonics and the Distribution of Life

Plate tectonic theory is as revolutionary and far-reaching in its implications for geology as the theory of evolution was for biology when it was proposed. Interestingly, it was the fossil evidence that convinced Wegener, Suess, and du Toit, as well as many other geologists, of the correctness of continental drift. Together, the theories of plate tectonics and evolution have changed the way we view our planet, and we should not be surprised at the intimate association between them. While the relationship between plate tectonic processes and the evolution of life is incredibly complex, paleontologic data provide convincing evidence of the influence of plate movement on the distribution of organisms.

The present distribution of plants and animals is not random but is controlled largely by climate and geographic barriers. The world's biota occupy *biotic provinces,* each of which is a region characterized by a distinctive assemblage of plants and animals. Organisms within a province have similar ecologic requirements, and the boundaries separating provinces are therefore natural ecologic breaks. Climatic or geographic barriers are the most common province boundaries, and these are largely controlled by plate movements.

Because adjacent provinces usually have less than 20% of their species in common, global diversity is a direct reflection of the number of provinces; the more provinces there are in the world, the greater the global diversity. When continents break up, for example, the opportunity for new provinces to form increases, with a resultant increase in diversity. Just the opposite occurs when continents come together. Plate tectonics thus plays an important role in the distribution of organisms and their evolutionary history.

Complex interaction of wind and ocean currents have a strong influence on the world's climates. These currents are influenced by the number, distribution, topography, and orientation of continents. For example, the southern Andes Mountains act as an effective barrier to moist, easterly blowing Pacific winds, resulting in a desert east of the southern Andes that is virtually uninhabitable. Temperature is one of the major limiting factors for organisms, and province boundaries often reflect temperature barriers. Because atmospheric and oceanic temperatures decrease from the equator to the poles, most species exhibit a strong climatic zonation. This biotic zonation parallels the world's latitudinal atmospheric and oceanic circulation patterns. Changes in climate thus have a profound effect on the distribution and evolution of organisms.

The distribution of continents and ocean basins not only influences wind and ocean currents but also affects provinciality by creating physical barriers to or pathways for the migration of organisms. Intraplate volcanoes, island arcs, mid-oceanic ridges, mountain ranges, and subduction zones all result from the interaction of plates, and their orientation and distribution strongly influence the number of provinces and hence total global diversity. Thus, provinciality and diversity will be highest when there are numerous small continents spread across many zones of latitude.

When a geographic barrier separates a once-uniform fauna, species may undergo divergence. If conditions on opposite sides of the barrier are sufficiently different, then species must adapt to the new conditions, migrate, or become extinct. Adaptation to the new environment by various species may involve enough change that new species eventually evolve. The marine invertebrates found on opposite sides of the Isthmus of Panama provide an excellent example of divergence caused by the formation of a geographic barrier. Prior to the rise of this land connection between North and South America, a homogeneous population of bottom-dwelling invertebrates inhabited the shallow seas of the area. After the rise of the Isthmus of Panama by subduction of the Pacific plate about 5 million years ago, the original population was divided. In response to the changing environment, new species evolved on opposite sides of the isthmus (Figure 6.29).

The formation of the Isthmus of Panama also had an impact on the evolution of the North and South American mammalian faunas (see Chapter 18). During most of the Cenozoic Era, South America was an island continent, and its mammalian fauna evolved in isolation from the rest of the world's faunas. When North and South America were connected by the Isthmus of Panama, most of the indige-

(a)

Pacific species   Caribbean species

(b)

FIGURE 6.29   (a) The Isthmus of Panama forms a barrier that divides a once-uniform fauna. (b) Divergence of molluscan species after the formation of the Isthmus of Panama. Each pair belongs to the same genus but is a different species.

nous South American mammals were replaced by migrants from North America. Surprisingly, only a few South American mammal groups migrated northward.

Plate Tectonics and the Distribution of Natural Resources

Besides being responsible for the major features of Earth's crust and influencing the distribution and evolution of the world's biota, plate movements also affect the formation and distribution of natural resources. Consequently, geolo-

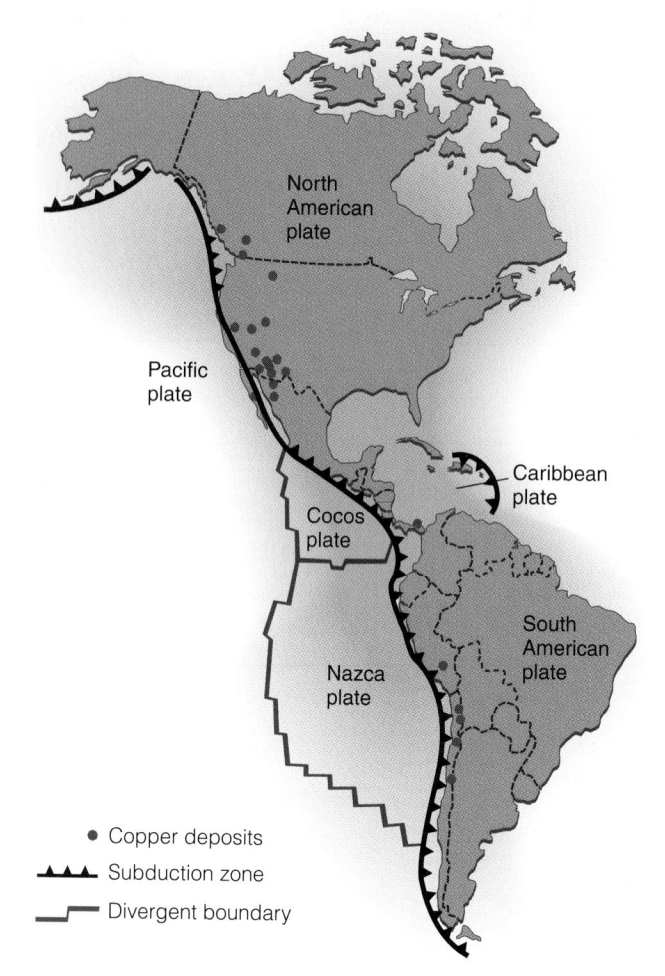

● Copper deposits
▲▲▲ Subduction zone
┌─┐└─ Divergent boundary

FIGURE 6.30 Important copper deposits are located along the west coasts of North and South America.

gists are using plate tectonic theory in their search for new mineral deposits and in explaining the occurrence of known deposits.

Many metallic mineral deposits such as copper, gold, lead, silver, tin, and zinc are related to igneous and associated hydrothermal activity, so it is not surprising that a close relationship exists between plate boundaries and the occurrence of these valuable deposits.

The magma generated by partial melting of a subducting plate rises toward the surface, and as it cools, it precipitates and concentrates various metallic sulfide ores. Many of the world's major metallic ore deposits, such as the porphyry copper deposits of western North and South America (Figure 6.30) are associated with convergent plate boundaries.

Divergent plate boundaries also yield valuable resources. Hydrothermal vents are the sites of considerable metallic mineral precipitation. The island of Cyprus in the Mediterranean is rich in copper and has been supplying all or part of the world's needs for the last 3000 years. The concentration of copper on Cyprus formed as a result of precipitation adjacent to hydrothermal vents.

Studies indicate that minerals of such metals as copper, gold, iron, lead, silver, and zinc are currently forming as sulfides in the Red Sea. The Red Sea is opening as a result of plate divergence and represents the earliest stage in the growth of an ocean basin.

It is becoming increasingly clear that if we are to keep up with the continuing demands of a global industrialized society, the application of plate tectonic theory to the origin and distribution of mineral resources is essential. We will discuss natural resources and their formation as part of the geologic history of Earth in later chapters in this book.

Summary

1. The concept of continental movement is not new. The earliest maps showing the similarity between the east coast of South America and the west coast of Africa provided people with the first evidence that continents may once have been united and subsequently separated from each other.

2. Alfred Wegener is generally credited with developing the hypothesis of continental drift. He provided abundant geologic and paleontologic evidence to show that the continents were once united into one supercontinent he named Pangaea. Unfortunately, Wegener could not explain how the continents moved, and most geologists ignored his ideas.

3. The hypothesis of continental drift was revived during the 1950s when paleomagnetic studies of rocks indicated the presence of multiple magnetic north poles instead of just one as there is today. This paradox was resolved by moving the continents into different positions, making the paleomagnetic data consistent with a single magetic north pole.

4. Magnetic surveys of the oceanic crust revealed magnetic anomalies in the rocks, indicating that Earth's magnetic field had reversed itself in the past. Because the anomalies are parallel and form symmetric belts adjacent to the oceanic ridges, new oceanic crust must have formed as the seafloor was spreading.

5. Radiometric dating reveals the oldest oceanic crust is less than 180 million years old, while the oldest continental crust is 3.96 billion years old. Clearly, the ocean basins are recent geologic features.

6. Plate tectonic theory became widely accepted by the 1970s because of the overwhelming evidence supporting it and because it provides geologists with a powerful theory for explaining such phenomena as volcanism, seismicity, mountain building, global climatic changes, the distribution of the world's biota, and the distribution of some mineral resources.

7. Three types of plate boundaries are recognized: divergent boundaries, where plates move away from each other; convergent boundaries, where two plates collide; and transform boundaries, where two plates slide past each other.

8. Ancient plate boundaries can be recognized by their associated rock assemblages and geologic structures. For divergent boundaries, these may include rift valleys with thick sedimentary sequences and numerous dikes and sills. For convergent boundaries, ophiolites and andesitic rocks are two characteristic features. Transform faults generally do not leave any characteristic or diagnostic features in the rock record.

9. Although a comprehensive theory of plate movement has yet to be developed, geologists think that some type of convective heat system is involved.

10. The average rate of movement and relative motion of plates can be calculated several ways. The results of these different methods all agree and indicate that the plates move at different average velocities.

11. Absolute motion of plates can be determined by the movement of plates over mantle plumes. A mantle plume is an apparently stationary column of magma that rises to the surface where it becomes a hot spot and forms a volcano.

12. Geologists now realize that plates can grow when microplates collide with the margins of continents.

13. A close relationship exists between the formation of some mineral deposits and plate boundaries. Furthermore, the formation and distribution of some natural resources are related to plate movements.

Important Terms

continental–continental plate boundary

continental drift

convergent plate boundary

Curie point

divergent plate boundary

Glossopteris flora

Gondwana

hot spot

Laurasia

magnetic anomaly

magnetic reversal

microplate

oceanic–continental plate boundary

oceanic–oceanic plate boundary

ophiolite

orogeny

paleomagnetism

Pangaea

plate

plate tectonic theory

seafloor spreading

spreading ridge

thermal convection cell

transform fault

transform plate boundary

volcanic island arc

Wilson cycle

Review Questions

1. The man credited with developing the continental drift hypothesis is:
   a. _____ du Toit.
   b. _____ Hess.
   c. _____ Vine.
   d. _____ Wegener.
   e. _____ Wilson.

2. The name of the supercontinent that formed at the end of the Paleozoic Era is:
   a. _____ Laurasia.
   b. _____ Gondwana.
   c. _____ Panthalassa.
   d. _____ Atlantis.
   e. _____ Pangaea.

3. Iron-bearing minerals in a magma gain their magnetism and align themselves with the magnetic field when they cool through the:
   a. _____ negative magnetic anomaly.
   b. _____ Curie point.
   c. _____ positive magnetic reversal.
   d. _____ thermal convection point.
   e. _____ hot-spot point.

4. Along what type of plate boundary does subduction occur?
   a. _____ Divergent
   b. _____ Transform
   c. _____ Convergent
   d. _____ Answers (a) and (b)
   e. _____ Answers (b) and (c)

5. The San Andreas fault is an example of a(an) _____ boundary.
   a. _____ divergent
   b. _____ convergent
   c. _____ transform
   d. _____ oceanic-continental
   e. _____ continental-continental

6. Magnetic surveys of the ocean basins indicate that:
   a. _____ the oceanic crust is youngest adjacent to mid-oceanic ridges.
   b. _____ the oceanic crust is oldest adjacent to mid-oceanic ridges.
   c. _____ the oceanic crust is youngest adjacent to the continents.
   d. _____ the oceanic crust is the same age everywhere.
   e. _____ Answers (b) and (c)

7. The East African rift valleys are good examples of a(an) _____ plate boundary.
   a. _____ continental-continental
   b. _____ oceanic-oceanic
   c. _____ oceanic-continental
   d. _____ divergent
   e. _____ transform

8. Which of the following allows geologists to determine absolute plate motion?
   a. _____ Hot spots
   b. _____ The age of the sediment directly above any portion of the oceanic crust
   c. _____ Magnetic reversals in the oceanic crust
   d. _____ Satellite-laser ranging techniques
   e. _____ All of these.

9. The driving mechanism of plate movement is thought to be:

a. _____ isostasy.
b. _____ Earth's rotation.
c. _____ magnetism.
d. _____ thermal convection cells.
e. _____ none of these.

10. The formation and distribution of copper deposits are associated with _____ boundaries.
    a. _____ divergent
    b. _____ convergent
    c. _____ transform
    d. _____ Answers (a) and (b)
    e. _____ Answers (b) and (c)

11. The most common biotic province boundaries are:
    a. _____ geographic barriers.
    b. _____ biologic barriers.
    c. _____ climatic barriers.
    d. _____ answers (a) and (b).
    e. _____ answers (a) and (c).

12. A distinctive assemblage of deep-sea sediments, oceanic crustal rocks, and upper mantle constitutes a:
    a. _____ microplate.
    b. _____ back-arc basin.
    c. _____ subduction complex.
    d. _____ fore-arc basin.
    e. _____ None of these

13. Explain how rates of movement of plates can be determined.

14. Explain how plate tectonics affects the evolution of life.

15. Explain why global diversity increases with an increase in biotic provinces. How does plate movement affect the number of biotic provinces?

16. Discuss how a geologist could use plate tectonic theory to help locate mineral deposits.

17. What evidence convinced Wegener that the continents were once joined together and subsequently broke apart?

18. Briefly discuss microplate tectonics and its contributions to plate tectonic theory.

19. How would you recognize ancient convergent plate boundaries?

20. What is the significance of polar wandering in relation to continental drift?

21. Why is plate tectonics such a powerful unifying theory?

Points to Ponder

1. If movement along the San Andreas fault, which separates the Pacific plate from the North American plate, averages 5.5 cm per year, how long will it take before Los Angeles is opposite San Francisco?

2. The average rate of movement away from the Mid-Atlantic Ridge between North America and Africa is 3.0 cm per year (Figure 6.13), and the average sedimentation rate in the open ocean is 0.275 cm per 1000 years. Calculate what the age of the oceanic crust is adjacent to Norfolk, Virginia, and what the thickness of the sediment is overlying the oceanic crust at that location. Does your calculation agree with the age of the oceanic crust at that location as shown in Figure 6.12?

3. What features would an astronaut look for on the Moon or another planet to find out if plate tectonics is currently active or if it was active during the past?

Additional Readings

Bolina, R. 1996. How to move a continent. *Earth* 5, no. 2:14.

Bonatti, E. 1987. The rifting of continents. *Scientific American* 256, no. 3: 96-103.

Brimhall, G. 1991. The genesis of ores. *Scientific American* 264, no. 5: 84-91.

Condie, K. 1989. *Plate tectonics and crustal evolution.* 3d ed. New York: Pergamon Press.

Cromie, W. J. 1989. The roots of midplate volcanism. *Mosaic* 20, no. 4: 19-25.

Dalziel, I. W. D. 1995. Earth before Pangaea. *Scientific American* 272, no. 1: 58-63.

Fleischman, J. 1997. Joining forces. *Earth* 6, no. 3: 34-39.

Jordon, T. H., and J. B. Minster. 1988. Measuring crustal deformation in the American west. *Scientific American* 259, no. 2: 48-59.

Kearey, P., and F. J. Vine. 1996. *Global tectonics.* 2d ed. Palo Alto, Calif: Blackwell Scientific Publishers.

Klein, G. D., ed. 1994. Pangaea! Paleoclimate, tectonics, and sedimentation during accretion, zenith, and breakup of a supercontinent. *Geological Society of America Special Paper 228.* Boulder, Colo.: Geological Society of America.

Luhman, J. G., J. B. Pollack, and L. Colin. 1994. The *Pioneer* mission to Venus. *Scientific American* 270, no. 4: 90-97.

Moores, E. 1996. The story of Earth. *Earth* 5, no. 6: 30-33.

Murphy, J. B., and R. D. Nance. 1992. Mountain belts and the supercontinent cycle. *Scientific American* 266, no. 4: 84-91.

Nance, R. D., T. R. Worsley, and J. B. Moody. 1988. The supercontinent cycle. *Scientific American* 259, no. 1: 72-79.

Parks, N. 1994. Loihi rumbles to life. *Earth* 6, no. 2: 42-49.

Pendick, O. 1997. Rocky Mountain why. *Earth* 6, no. 3: 26-33.

World Wide Web Activities

For these web site addresses, along with current updates and exercises, log on to

http://www.brookscole.com/geo/

▶ **ACTIVE TECTONICS**

This site functions mainly to disseminate information and links to other sites related to plate tectonics. Check out some of the other links listed.

▶ **PLATE TECTONICS**

This site contains pages of tutorial material from the University of Nevada's Seismological Laboratory at Reno. It contains the same information as found in this chapter but also has many images not seen in this chapter and is worth the visit.

▶ **TECTONIC PLATE MOTION**

Maintained by NASA, this site contains a wealth of information on how the various space geodetic technologies are used for calculating and tracking current plate movements.

1. What are the different ways plate movements can be determined and calculated?
2. Click on the various geographic sites listed to see what the average velocity and direction of movement of various plates are.

▶ **PLATE MOTION CALCULATOR**

This site allows you to calculate the direction and velocity of plate movements by comparing the "relatively moving plate" against the "relatively fixed plate." After you choose which two adjacent plates you want and select a latitude and longitude, the plate calculator then determines the direction and velocity of plate movement.

Explore the following *In-Terra-Active 2.0* CD-ROM module(s) and increase your understanding of key concepts and processes presented in this chapter.

Chapter Concept: **Evidence for Plate Tectonics**

Geologists consider plate tectonics a unifying theory because it provides a uniform explanation for (that is, unifies) a large body of observations that range from the geographic fit of the continents to the symmetrical pattern of magnetic fields on either side of mid-ocean ridges. This module introduces you to the evidence that scientists use to support plate tectonic theory and enables you to apply your skills to a hypothetical tectonic plate configuration.

Section: **Interior**
Module: **Plate Tectonics: Evidence**

Question: *If you were an astronaut landing on Venus, what evidence would you look for to confirm or deny the existence of plate tectonics there?*

Chapter Concept: **Plate Boundaries**

Current plate positions and the interactions at their boundaries are critical to recognizing and understanding ancient plate boundaries. This module illustrates the evolution of different plate boundaries and the deformation processes that accompany each type of plate boundary

Section: **Interior**
Module: **Plate Tectonics: Plate Boundaries**

The photograph shows a view of the Butte, Montana, copper deposits. Much of the copper minerals at Butte are found in porphyritic, silica-rich, igneous rock. These kinds of igneous rocks and associated copper depostits are typically formed in Arizona, the Andes Mountains, volcanic islands of the South Pacific, and the coast ranges of western Canada and the United States.

Question: *What do the copper deposits tell us about the plate tectonic setting of Butte, Montana, at the time this deposit formed, 50 to 60 million years ago?*

A History of the Universe, Solar System, and Planets

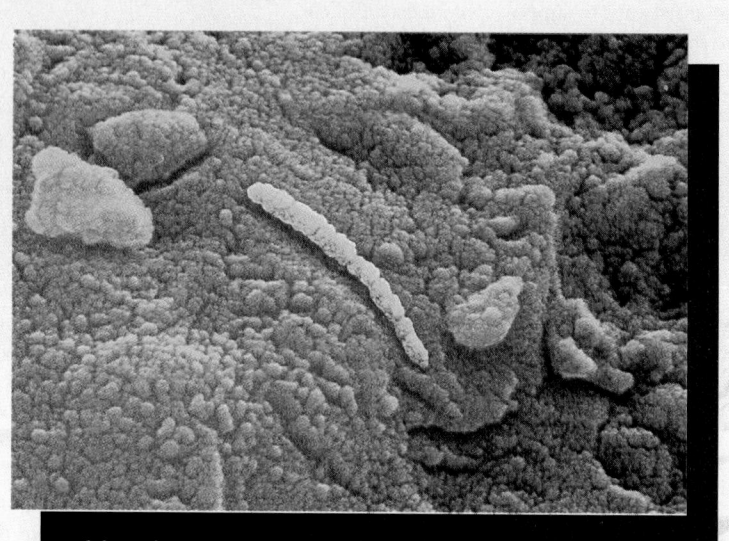

This microscopic tube-shaped structure was found within meteorite ALH84001, a meteorite discovered in Antarctica in 1984 and believed to have come from Mars. This microscopic structure is thought to possibly represent the fossil remains of primitive life that lived on Mars more than 3.6 billion years ago.

Prologue

Does life exist elsewhere in the universe, or is it unique to Earth? This question has intrigued people for centuries. The discovery of extraterrestrial life would certainly create a major scientific and theological upheaval as the fact that we are not alone sinks into the collective human consciousness.

The search for life elsewhere in the solar system and beyond goes back at least as far as Galileo. His discovery that other planets existed fueled people's imagination about whether life may also exist in these places. In 1894 the wealthy amateur astronomer Percival Lowell built his own observatory in Flagstaff, Arizona, to further the study of Mars, a planet in which he was especially interested. Inspired by earlier reports of dark lines on the Martian surface and his own observations, Lowell published a book in 1895 speculating that an advanced civilization had built canals (the dark crisscrossing lines some people observed) to bring water from the melting polar ice caps to the rest of the planet in response to a global drought.

Lowell's writings were instrumental in generating public interest in Mars and Martians and the possibility of extraterrestrial life forms. In 1898 H. G. Wells published his novel, *The War of the Worlds,* about an invasion of Earth by Martians. This theme was further amplified in a 1938 radio broadcast by Orson Welles in which Martians invaded New Jersey. Based on what we know presently about Mars from telescope observations and the various space probes sent to study the planet, scientists are convinced that Mars does not harbor any higher forms of life, yet the possibility remains that microorganisms may have lived on the Martian surface early in its history when water was present or are still living beneath its surface.

Do any other planets in the solar system or elsewhere in the universe have the conditions necessary for life to evolve? If they do, they would have to be located in a region from their sun where water is a liquid, rather than a solid or gas. Water is an ideal medium in which carbon-based organic compounds can dissolve and react with each other in the various ways necessary for life to exist.

Most scientists think that if extraterrestrial life exists, it will be carbon-based because carbon is found in so many different and complex compounds, as well as being the fourth most abundant element in the universe. Carbon compounds are found in meteorites and comets, as well as in huge molecular clouds floating throughout the cosmos. In fact, some scientists think that life evolved from organic compounds that fell to Earth from space early in its history. Whether life originated from organic compounds from space or near hot-water volcanic vents in the ocean as other scientists think, we know life evolved soon after Earth formed.

During this decade (1990s), several new planets have been discovered circling stars within the Milky Way Galaxy. What makes these discoveries so exciting is that, if these stars in the Milky Way Galaxy have planets circling them, the odds are greatly increased that other stars in the 100-billion-star Milky Way Galaxy have not only large, inhospitable planets circling them but also smaller, more life-friendly planets. Based on what we know about how life might have originated, we can say the probability is high that other planets in our galaxy also contain life.

At an August 1996 news conference, NASA announced the first tangible evidence that life may have existed on Mars. A 1.9-kg meteorite, named ALH84001, contains within it what appears to be chemical and fossil remains of Martian life. This potato-sized piece of rock was blasted from the surface of Mars 3.6 billion years ago and circled the Sun for millions of years, until finally crashing into an ice field in Antarctica's Allan Hills some 13,000 years ago. It remained there until it was discovered in 1984.

After studying the Martian meteorite for two years, the NASA-led team of scientists found organic molecules called polycyclic aromatic hydrocarbons, which on Earth are the byproducts of metabolic activity. In addition, traces of iron sulfides and magnetite were also found. These compounds are also associated with, but are not necessarily the result of, bacterial activity. Besides the chemical evidence, microscopic structures superficially resembling bacteria were also recovered from inside the meteorite. Although these segmented, tubelike objects look like bacteria, they are 100 times smaller than fossil bacteria found on Earth.

As startling as these discoveries are, many scientists urge caution before accepting this evidence as proof that life exists or has existed elsewhere in our solar system. Based on continued analysis of the ALH84001 meteorite, many scientists have recently concluded that the evidence cited for ancient Martian life can now be explained as either contamination or it is non-biogenic in origin.

Introduction

Of all the known planets and their moons only one, Earth, has life on it. This unique planet, revolving around the Sun every 365.25 days, is a dynamic and complex body. When viewed from the blackness of space, Earth is a brilliant, shimmering, bluish planet, wrapped in a veil of swirling white clouds. Beneath these clouds is a surface covered by oceans, seven continents, and numerous islands.

Earth has not always looked the way it does today. Based on various lines of evidence, many scientists think that it began as a homogeneous mass of rotating dust and gases that contracted, heated, and differentiated during its early history to form a medium-sized planet with a metallic core, a mantle composed of iron- and magnesium-rich rocks, and a thin crust. Overlying this crust is an atmosphere currently composed of 78% nitrogen and 21% oxygen.

As the third planet from the Sun, Earth seems to have formed at just the right distance from the Sun (150 million km) so that it is neither too hot nor too cold to support life as we know it. Furthermore, its size is just right to hold an atmosphere. If Earth were smaller, its gravity would be so weak that it could retain little, if any, atmosphere.

Origin of the Universe

Most scientists think that the universe originated approximately 15 billion years ago in what is popularly called the **Big Bang.** In a region infinitely smaller than an atom, both space and time were set at zero. Therefore, there is no "before the Big Bang," only what occurred after it. The reason for this is that space and time are unalterably linked to form a space–time continuum as demonstrated by Einstein's theory of relativity. Without space, there can be no time.

How do we know the Big Bang took place approximately 15 billion years ago? Why couldn't the universe have always existed as we know it today? Two fundamental phenomena indicate that the Big Bang occurred. The first is that the universe is expanding; the second is that it is permeated by background radiation.

That the universe is expanding was first recognized by Edwin Hubble in 1929. By measuring the optical spectra of distant galaxies, Hubble noted that the velocity at which a galaxy moves away from Earth increases proportionally to its distance from Earth. He observed that the spectral lines (wavelengths of light) of the galaxies are shifted toward the red end of the spectrum; that is, the lines are shifted toward longer wavelengths. Such a redshift would be produced by galaxies receding from each other at tremendous speeds (Figure 7.1). This is an example of the **Doppler effect,** which is the change in the frequency of a sound, light, or other wave caused by movement of the wave's source relative to the observer (Figure 7.2).

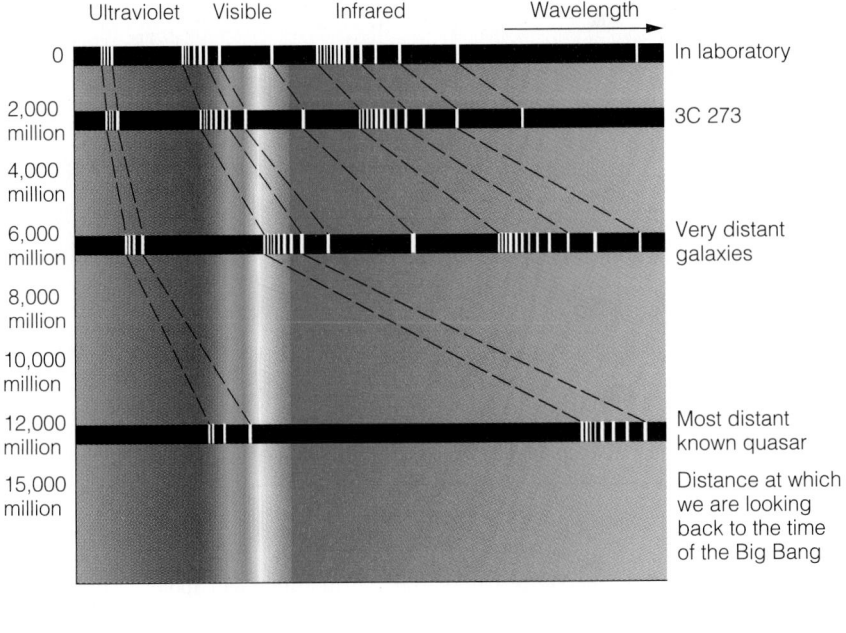

Ultraviolet   Visible   Infrared   Wavelength

0

2,000 million

4,000 million

6,000 million

8,000 million

10,000 million

12,000 million

15,000 million

Distance in light-years

In laboratory

3C 273

Very distant galaxies

Most distant known quasar

Distance at which we are looking back to the time of the Big Bang

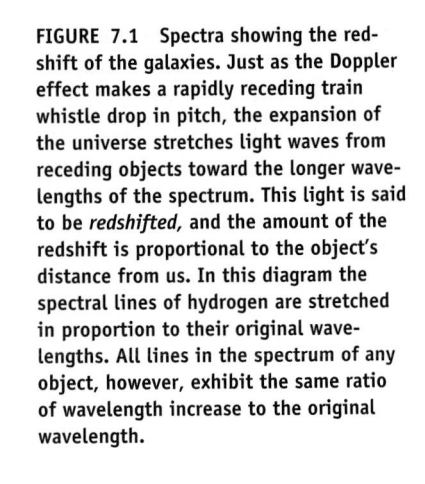

FIGURE 7.1 Spectra showing the redshift of the galaxies. Just as the Doppler effect makes a rapidly receding train whistle drop in pitch, the expansion of the universe stretches light waves from receding objects toward the longer wavelengths of the spectrum. This light is said to be *redshifted,* and the amount of the redshift is proportional to the object's distance from us. In this diagram the spectral lines of hydrogen are stretched in proportion to their original wavelengths. All lines in the spectrum of any object, however, exhibit the same ratio of wavelength increase to the original wavelength.

One way to envision how velocity increases with increasing distance is by reference to the popular analogy of a rising loaf of raisin bread in which the raisins are uniformly distributed throughout the loaf (Figure 7.3). Suppose that before the dough begins to rise, the raisins are 1 cm apart. After one hour, the dough has risen to the point where the raisins are 2 cm apart. Any raisin is now 2 cm from its nearest neighbor and 4 cm away from the next one, 6 cm from the next, and so on. From the perspective of any raisin, its nearest neighbor has moved away from it at a speed of 1 cm per hour (it originally was 1 cm away and is now 2 cm away), the next raisin over has moved away at a speed of 2 cm/h (it was 2 cm away, and is now 4 cm away), while the next one has moved away at 3 cm/h (it was originally 3 cm away and is now 6 cm away from the reference raisin), and so on. If the dough continues to rise until the distance between any two raisins is 4 cm, we would find that all the distances between raisin pairs are quadruple what they were to start with. The speeds needed to accomplish this are directly proportional to the distances between any two raisins: The farther away a given raisin is to begin with, the farther it must move to maintain the regular spacing during the expansion, and

hence the greater its velocity must be. In the same way that raisins move apart in a rising loaf of bread, galaxies are receding from each other at a rate proportional to the distance between them, which is exactly what astronomers see when they observe the universe.

A corollary to Hubble's expanding universe is that astronomers can use this expansion rate to calculate how long ago the galaxies were all together in a single point. Such calculations yield an estimated age for the universe of approximately 15 billion years.

The second important observation providing evidence of the Big Bang was made in 1965 by Arno Penzias and Robert Wilson of Bell Telephone Laboratories when they discovered that there is a pervasive background radiation of 2.7° above absolute zero (absolute zero equals −273° C) everywhere in the universe. This background radiation is thought to be the fading afterglow of the Big Bang.

EARLY HISTORY OF THE UNIVERSE

At the time of the Big Bang, matter as we know it did not exist, and the universe consisted of pure energy. Within the first second after the Big Bang, the four basic forces— **gravity, electromagnetic force, strong nuclear force,**

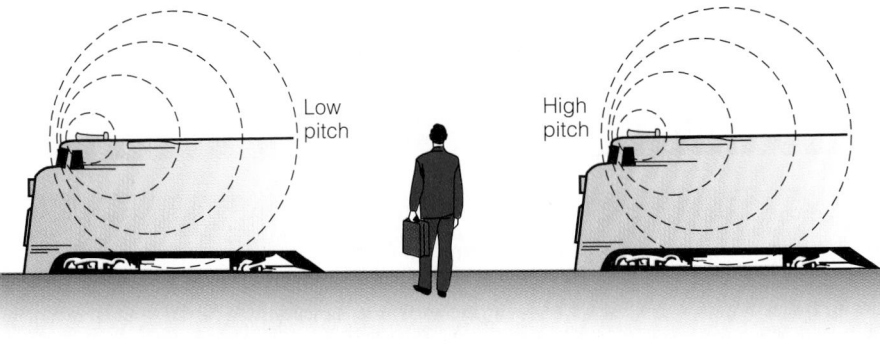

Low pitch

High pitch

FIGURE 7.2 The Doppler effect applied to sound waves. The sound waves of an approaching whistle are slightly compressed so that the individual hears a shorter-wavelength, higher-pitched sound. As the whistle passes and recedes from the individual, the sound waves are slightly spread out, and a longer-wavelength, lower-pitched sound is heard.

FIGURE 7.3 The motion of raisins in a rising loaf of raisin bread illustrates the relationship that exists between distance and speed and is analogous to an expanding universe. In this diagram, adjacent raisins are located 2 cm apart before the loaf rises. After one hour, any raisin is now 4 cm away from its nearest neighbor and 8 cm away from the next raisin over, and so on. Therefore, from the perspective of any raisin, its nearest neighbor has moved away from it at a speed of 2 cm per hour, and the next raisin over has moved away from it at a speed of 4 cm per hour. In the same way that raisins move apart in a rising loaf of bread, galaxies are receding from each other at a rate proportional to the distance between them.

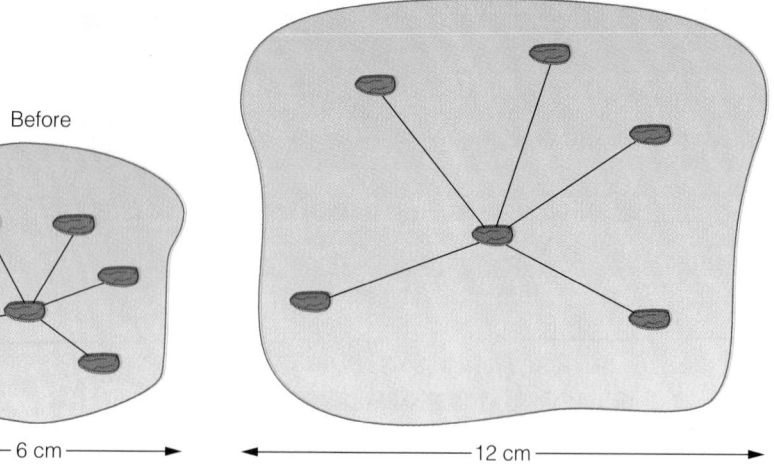

After

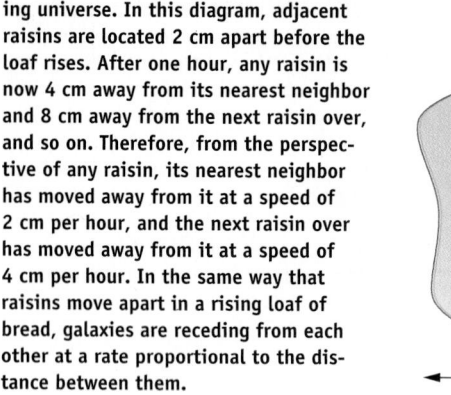

Before

6 cm

12 cm

and **weak nuclear force** (Table 7.1)—had all separated, and the universe experienced enormous expansion. Matter and antimatter collided and annihilated each other. Fortunately, a slight excess of leftover matter would become the universe. By the time the universe was three minutes old, temperatures were cool enough for protons and neutrons to fuse together to form the nuclei of hydrogen and helium atoms. Approximately 300,000 years later, electrons joined with the previously formed nuclei to make complete atoms of hydrogen and helium. At the same time, *photons* (the energetic particles of light) separated from matter, and light burst forth for the first time.

CHANGING COMPOSITION OF THE UNIVERSE

After about 200 million years, as the universe continued expanding and cooling, stars and galaxies began forming, and the chemical makeup of the universe changed. Early in its history, the universe was 100% hydrogen and helium, whereas today it is 98% hydrogen and helium by weight.

Over the course of their history, stars undergo many nuclear reactions whereby lighter elements are converted into heavier elements by nuclear fusion in which atomic nuclei combine to form more massive nuclei. Such reactions, which convert hydrogen to helium, occur in the cores of all stars. The subsequent conversion of helium to heavier elements, such as carbon, depends on the mass of the star. When a star dies, often explosively, the heavier elements that were formed in its core are returned to interstellar space and are available for inclusion in new stars.

When new stars form, they will have a small amount of these heavier elements, which may be converted to still heavier elements. In this way, the chemical composition of the galaxies, which are made up of billions of stars, is gradually enhanced in heavier elements.

Origin and Early Development of the Solar System

Having looked at the origin and brief history of the universe, we can now examine how our own solar system, which is part of the Milky Way Galaxy, formed. The solar system consists of the Sun, nine planets, 63 known moons, a tremendous number of asteroids—most of which orbit the Sun in a zone between Mars and Jupiter—and millions of comets and meteors, and interplanetary dust and gases (Table 7.2).

GENERAL CHARACTERISTICS OF THE SOLAR SYSTEM

Any theory that attempts to explain the origin and history of the solar system must take into account several general characteristics (Table 7.3). All planets revolve around the Sun in the same direction, in circular orbits, and in approximately the same plane (called the *plane of the ecliptic*), except for Pluto whose orbit is both highly elliptical and tilted 17 degrees to the orbital plane of the rest of the planets (Figure 7.4).

TABLE 7.1

The Four Basic Forces of the Universe

Four forces appear to be responsible for all interactions of matter:

1. **Gravity** is the attraction of one body toward another.
2. The **electromagnetic force** combines electricity and magnetism into the same force and binds atoms into molecules. It also transmits radiation across the various spectra at wavelengths ranging from gamma rays (shortest) to radio waves (longest) through massless particles called *photons*.

3. The **strong nuclear force** binds protons and neutrons together in the nucleus of an atom.
4. The **weak nuclear force** is responsible for the breakdown of an atom's nucleus, producing radioactive decay.

TABLE 7.2

Characteristics of the Sun, Planets, and Moon

OBJECT	MEAN DISTANCE TO SUN (km × 10⁶)	ORBITAL PERIOD (days)	ROTATIONAL PERIOD (days)	TILT OF AXIS	EQUATORIAL DIAMETER (km)	MASS (kg)	MEAN DENSITY (g/cm³)	NUMBER OF SATELLITES
Sun	—	—	25.5	—	1,391,400	1.99×10^{30}	1.41	—
Terrestrial planets								
Mercury	57.9	88.0	58.7	28°	4,880	3.33×10^{23}	5.43	0
Venus	108.2	224.7	243	3°	12,104	4.87×10^{24}	5.24	0
Earth	149.6	365.3	1	24°	12,760	5.97×10^{24}	5.52	1
Mars	227.9	687.0	1.03	24°	6,787	6.42×10^{23}	3.96	2
Jovian planets								
Jupiter	778.3	4,333	0.41	3°	142,796	1.90×10^{27}	1.33	16
Saturn	1,428.3	10,759	0.43	27°	120,660	5.69×10^{26}	0.69	18
Uranus	2,872.7	30,685	0.72	98°	51,200	8.69×10^{25}	1.27	17
Neptune	4,498.1	60,188	0.67	30°	49,500	1.03×10^{26}	1.76	8
Pluto	5,914.3	90,700	6.39	122°	2,300	1.20×10^{22}	2.03	1
Moon	0.38 (from Earth)	27.3	27.32	7°	3,476	7.35×10^{22}	3.34	—

All planets, except Venus and Uranus, and nearly all planetary moons rotate counterclockwise when viewed from a point high in space above Earth's North Pole. Furthermore, the axes of rotation of the planets, except for those of Uranus and Pluto, are nearly perpendicular to the plane of the ecliptic (Figure 7.4b).

The nine planets can be divided into two groups based on their chemical and physical properties. The four inner planets—Mercury, Venus, Earth, and Mars—are all small and have high mean densities (Table 7.3), indicating that they are composed of rock and metallic elements. They are known as the **terrestrial planets** because they are similar to *terra,* which is Latin for "earth."

The next four planets—Jupiter, Saturn, Uranus, and Neptune—are called the **Jovian planets** because they all resemble Jupiter. The Jovian planets are all large and have low mean densities, indicating they are composed of light-weight gases such as hydrogen and helium, as well as frozen compounds such as ammonia and methane. The outermost planet, Pluto, is small and has a low mean density of slightly more than 2.0 g/cm³.

The slow rotation of the Sun is another feature that must be accounted for in any comprehensive theory of the origin of the solar system. Finally, any theory of the origin of the solar system must accommodate the nature and distribution of the various types of interplanetary objects such as the asteroid belt, comets, and interplanetary gases and dust.

CURRENT THEORY OF THE ORIGIN AND EARLY HISTORY OF THE SOLAR SYSTEM

Various scientific theories of the origin of the solar system have been proposed, modified, and discarded since the French scientist and philosopher René Descartes first proposed in 1644 that the solar system formed from a gigantic whirlpool within a universal fluid. Most theories have involved an origin from a primordial rotating cloud of gas and dust. Through the forces of gravity and rotation, this cloud then shrank and collapsed into a rotating disk. Detached rings within the disk condensed into planets, and the Sun condensed in the center of the disk.

The problem with most of these theories is that they fail to explain the slow rotation of the Sun, which according to the laws of physics should be rotating rapidly. This problem was finally solved with the discovery of the solar wind, which is an outflow of ionized gases from the Sun that interacts with its magnetic field and slows its rotation through a magnetic braking process (Figure 7.5).

The currently accepted **solar nebula theory** for the origin of the solar system involves the condensation and

TABLE 7.3

General Characteristics of the Solar System

1. PLANETARY ORBITS AND ROTATION
   - Planetary and satellite orbits lie in a common plane.
   - Nearly all planetary and satellite orbital and spin motions are in the same direction.
   - The rotation axes of nearly all planets and satellites are roughly perpendicular to the plane of the ecliptic.

2. CHEMICAL AND PHYSICAL PROPERTIES OF THE PLANETS
   - The terrestrial planets are small, have high densities (4.0 to 5.5 g/cm³), and are composed of rock and metallic elements.
   - The Jovian planets are large, have low densities (0.7 to 1.7 g/cm³), and are composed of gases and frozen compounds.

3. THE SLOW ROTATION OF THE SUN

4. INTERPLANETARY MATERIAL
   - The existence and location of the asteroid belt.
   - The distribution of interplanetary dust.

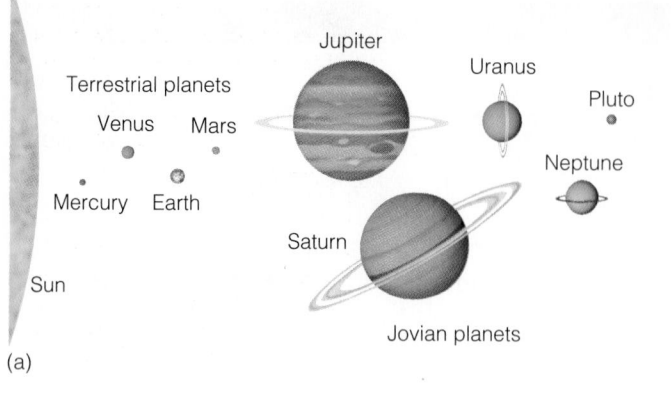

(a)

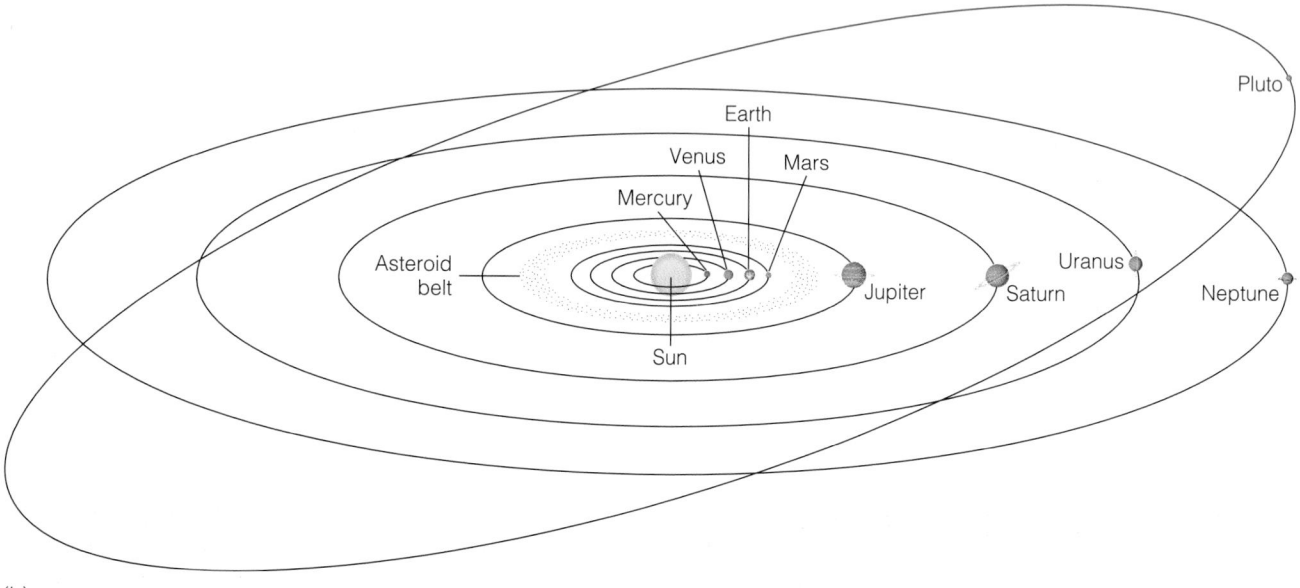

(b)

FIGURE 7.4 Diagrammatic representation of our solar system showing (a) the relative sizes of the planets and (b) their orbits around the Sun.

collapse of interstellar material in a spiral arm of the Milky Way Galaxy (Figure 7.6). As this cloud gradually collapsed under the influence of gravity, it flattened and began rotating counterclockwise, with about 90% of its mass concentrated in the central part of the cloud. The rotation and

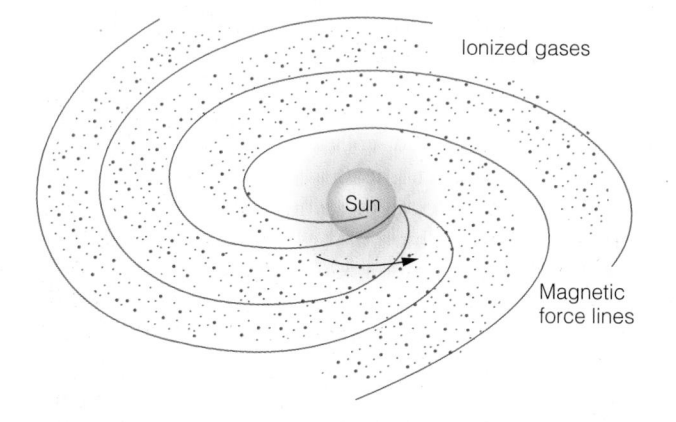

FIGURE 7.5 The slow rotation of the Sun is the result of the interaction of its magnetic force lines with ionized gases of the solar nebula. Thus, the rotation is slowed by a magnetic braking process.

concentration of material continued, and an embryonic Sun formed, surrounded by a turbulent, rotating cloud of material called a *solar nebula*. Within this solar nebula were localized eddies in which gas and solid particles condensed. Collisions between the various gases and particles resulted in accretion into *planetesimals* (Figure 7.7). As the planetesimals collided and grew in size and mass, they eventually became planets. According to various estimates, a planet the size of Earth could form from planetesimals in as little as 1 million years.

The composition and evolutionary history of the planets are determined in part by their distance from the Sun. The terrestrial planets are composed of rock and metallic elements that condensed at the high temperatures of the inner nebula. The Jovian planets, all of which have small central rocky cores compared to their overall size, are composed mostly of hydrogen, helium, ammonia, and methane, which condense at low temperatures. Thus, the farther away from the Sun that condensation took place, the lower the temperature and hence the higher the percentage of volatile elements, which condense at low temperatures, relative to refractory elements, which condense at high temperatures.

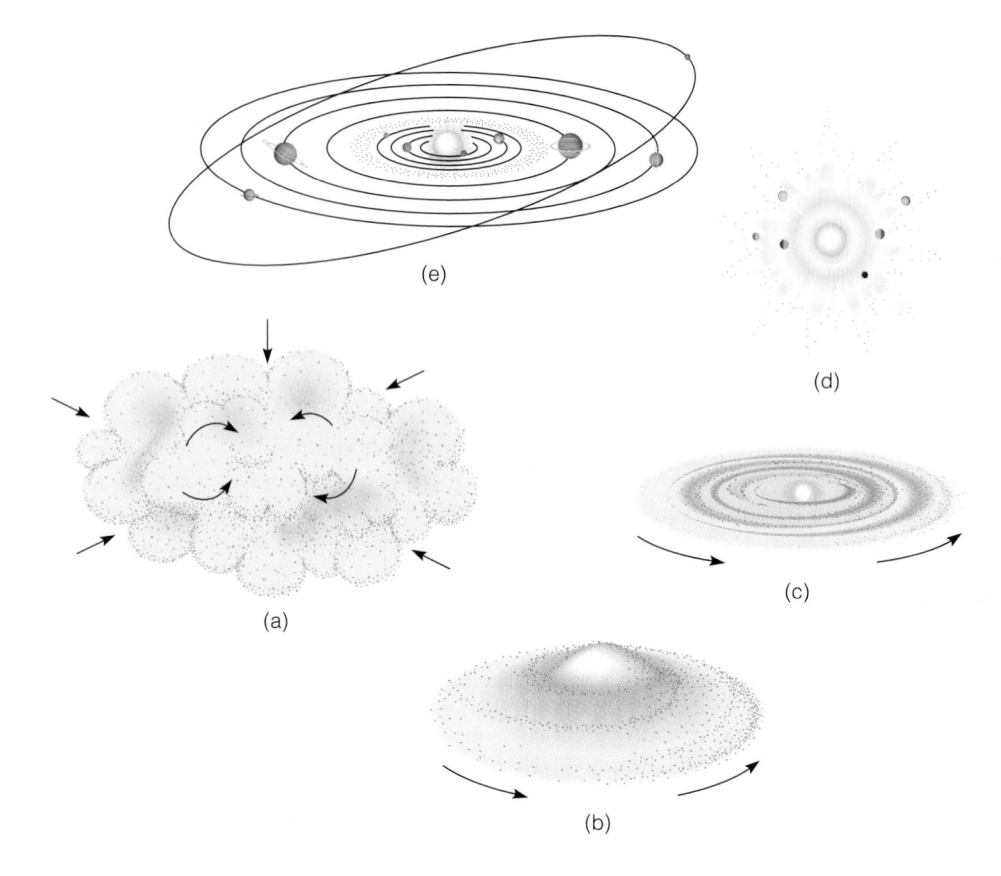

FIGURE 7.6 The currently accepted theory for the origin of our solar system involves (a) a huge nebula condensing under its own gravitational attraction, then (b) contracting, rotating, and (c) flattening into a disk, with the Sun forming in the center and eddies gathering up material to form planets. As the Sun contracted and began to visibly shine, (d) intense solar radiation blew away unaccreted gas and dust until finally, (e) the Sun began burning hydrogen and the planets completed their formation.

FIGURE 7.7 At the stage of development shown here, planetesimals have formed in the inner solar system, and large eddies of gas and dust remain at great distances from the protosun.

While the planets were accreting, material that had been pulled into the center of the nebula also condensed, collapsed, and was heated to several million degrees by gravitational compression. The result was the birth of a star, our Sun.

During the Sun's early history, it emitted a tremendous blast of energy that blew the solar system's unaccreted gases and dust into interstellar space. Such a blast is a normal phase in the evolution of a star and explains why the solar system is so free of extraneous debris. Also during the Sun's early history, its magnetic field interacted with the ionized gases of the solar nebula, slowing down its rotation (Figure 7.5). The discovery that the Sun's magnetic field exerted a force on the surrounding nebular gas solved the problem of why the Sun has such a slow rotation.

The asteroid belt is the final feature of the solar system that must be explained by the solar nebula theory. It is thought that *asteroids* probably formed as planetesimals in a localized eddy between what eventually became Mars and Jupiter in much the same way as other planetesimals formed the terrestrial planets. The tremendous gravitational field of Jupiter, however, prevented this material from ever accreting into a planet.

Comets, which are interplanetary bodies composed of loosely bound rocky and icy material, are thought to have condensed near the orbits of Uranus and Neptune. Each time comets pass by Jupiter and Saturn, the gravitational effect of those planets increases their speed, forcing comets farther out into the solar system.

The solar nebula theory of the formation of the solar system thus accounts for the similarities in orbits and rotation of the planets and their moons, the differences in composition between the terrestrial and Jovian planets, the slow rotation of the Sun, and the presence of the asteroid belt. Based on the available data, the solar nebula theory best explains the features of the solar system and provides a logical explanation for its evolutionary history.

Meteorites—Extraterrestrial Visitors

Meteorites are thought to be pieces of material that originated during the formation of the solar system 4.6 billion years ago. Early in the history of the solar system, a period of heavy meteorite bombardment occurred as the solar system cleared itself of the many pieces of material that had not yet accreted into planetary bodies or moons. Since then, meteorite activity has greatly diminished. Most meteorites that currently reach Earth are probably fragments resulting from collisions between asteroids.

STONES, IRONS, AND STONY-IRONS

Meteorites are classified into three broad groups based on their proportions of metals and silicate minerals (Figure 7.8). About 93% of all meteorites are composed of iron and magnesium silicate minerals and are thus known as **stones** (Figure 7.8b). Stony meteorites are not all the same and can be divided into three different types.

Ordinary chondrites are the most common type of stone meteorite and are composed of such high-temperature ferromagnesian silicate minerals as olivine and certain pyroxenes. Absolute dating reveals they are 4.6 billion years old and thus represent material that existed when the solar system formed. Most chondrites contain **chondrules,** which are small mineral bodies that formed by rapid cooling when the first solid material of the meteorite was condensing.

Carbonaceous chondrites have the same general composition as ordinary chondrites but also contain a small percentage of organic compounds. These organic molecules, which include some amino acids, are not biogenically produced, but may represent chemical precursors of organisms.

Acondrites, the third type of stone meteorite, do not contain chondrules. Their composition is similar to that of terrestrial basalts, and they are thought to result from collisions between larger asteroids.

Irons, the second group, accounting for about 6% of all meteorites are composed primarily of a combination of iron and nickel alloys (Figure 7.8c). Their large crystal size and chemical composition indicate that they cooled very slowly in large objects such as asteroids where the hot iron–nickel interior could be insulated from the cold of space. Collisions between such slowly cooling asteroids produced the iron meteorites we find today.

Stony-irons, the third group, are composed of nearly equal amounts of iron and nickel and silicate minerals; they make up less than 1% of all meteorites (Figure 7.8d). Stony-irons are generally thought to represent fragments from the zone between the silicate and metallic portions of a large differentiated asteroid.

Meteorites are important because their age—about 4.6 billion years—and composition provide scientists with information about the origin and history of the solar system. Furthermore, their collisions with Earth and the other planets (see Prologue) have affected the topography of the planets and, although there is no unanimity on this point, perhaps the evolution of life on Earth (see Perspective 7.1).

Origin and Differentiation of Early Earth

As matter was accreting in the various turbulent eddies that swirled around the early Sun, enough material eventually gathered together to form the planet Earth.

Earth consists of concentric layers of different composition and densities (Figure 7.9). This differentiation into concentric layers is a fundamental characteristic of all the terrestrial planets and presumably occurred very early in their history.

FIGURE 7.8 **(a)** Relative proportions of the three groups of meteorites. **(b)** Stony meteorite. **(c)** Iron meteorite. **(d)** Stony-iron meteorite.

Geologists know that Earth is 4.6 billion years old. The oldest known rocks, however, are 3.96-billion-year-old metamorphic rocks from Canada. Like the younger crustal rocks, these rocks are composed of relatively light silicate minerals. It appears that a crust, a heavier silicate mantle, and an iron–nickel core were already present 3.96 billion years ago, or 640 million years after Earth formed. Currently, most geologists accept the **homogeneous accretion theory** for how the core–mantle–crust density layering of Earth originated (Figure 7.10).

According to this theory, early Earth was probably cool, of generally uniform composition and density throughout, and composed mostly of silicate compounds, iron and magnesium oxides, and smaller amounts of all the other chemical elements (Figure 7.10a). For the iron and nickel to concentrate in the core, Earth must have heated up enough for them to melt and sink through the surrounding lighter silicate minerals.

This initial heating could have occurred in three ways. Some heat was no doubt generated by the impact of meteorites; most of this heat was radiated back into space, but some was probably retained by the accreting planet. Heat was also generated within early Earth as gravitational compression reduced it to a smaller volume. Rock is a poor conductor of heat, so this heat accumulated within Earth. The third cause of internal heating was the decay of radioactive elements such as uranium, thorium, and others. Even though these elements form only a very small portion

perspective 7.1

The Tunguska Event and Other Meteorite Impacts

On June 30, 1908, a bright object crossed the morning sky over Siberia, and seconds later a huge explosion occurred in the Tunguska River basin (Figure 1). The impact generated a shock wave that registered as an earthquake on seismographs around the world. The noise from the explosion was heard up to 1000 km away, trees were flattened for many kilometers around (Figure 2), and a column of incandescent matter rose to a height of about 20 km. Surprisingly, no evidence of a crater has ever been identified, even after several expeditions to the impact site.

The absence of a crater at the impact site has led to many strange hypotheses

regarding the cause of the explosion. These include a collision with a mini–black hole or piece of antimatter, a spontaneous nuclear explosion, and even the explosion of an alien spacecraft. Scientists at one time concluded that the Tunguska event was caused by a small, icy comet that exploded in the atmosphere. However, according to computer simulations, the nuclei of comets explode at too high an altitude to account for the pattern and magnitude of destruction found in the Tunguska River basin. Currently, most scientists think the Tunguska event was caused by a stony meteorite about 30 m in diameter, traveling at 15 km per

second and exploding in the atmosphere several kilometers above the ground, thus causing the observed damage, while not producing a crater.

The largest impact in recorded history to produce a crater also occurred in Siberia (Figure 1). On February 12, 1947, a large meteorite broke apart as it entered the atmosphere, and the ensuing pieces that impacted the ground produced more than 100 craters, some of which are up to 26 m in diameter.

Although collisions between meteorites and Earth may seem like a rare event, they actually are quite common. Recently declassified military satellite

FIGURE 1 Location of the 1908 Tunguska and 1947 Sikhote–Alin meteorite impacts.

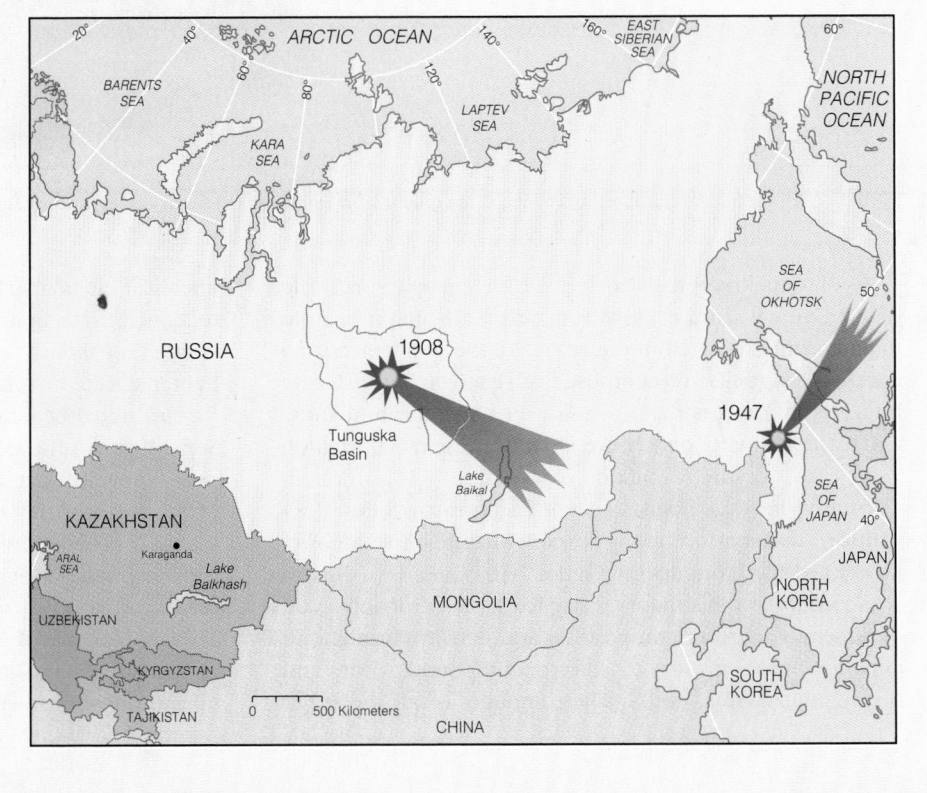

data show that meteorites with an energy equivalent to 1000 tons or more of high explosives strike Earth and vaporize in the atmosphere at an average rate of once a month. Furthermore, by examining the density of craters on the Moon and other planets, we know that the solar system experienced a period of heavy meteorite bombardment about 4 billion years ago. Since that time, meteorite impacts on Earth, though not as intense, have still been numerous. Most of these meteorites come from the asteroid belt, where there are uncounted millions of pieces of rock and small planetoids left over from the origin of the solar system.

Astronomers calculate that Earth can expect several collisions every million years with asteroids larger than a kilometer. One of the most famous results of an Earth–asteroid collision is Meteor Crater, Arizona, which was formed between 25,000 and 50,000 years ago and left a crater 1.2 km in diameter (Figure 3).

Collisions between large asteroids and Earth are not particularly common, but they do happen and can have devastating results. Geologists can identify more than 116 impact sites on Earth, many of which definitely are the result of a meteorite impact. It is thought that a collision with a meteorite about 10 km in diameter occurred 65 million years ago. Such a collision would have generated a tremendous amount of dust that would have blocked out the Sun, causing a significant reduction in photosynthesis and a lowering of global temperature, and perhaps resulting in the massive extinctions that occurred at the end of the Mesozoic Era (see Chapter 15).

FIGURE 3 Meteor Crater, Arizona, is the result of an Earth–asteroid collision that occurred between 25,000 and 50,000 years ago. It produced a crater 1.2 km in diameter and 180 m deep.

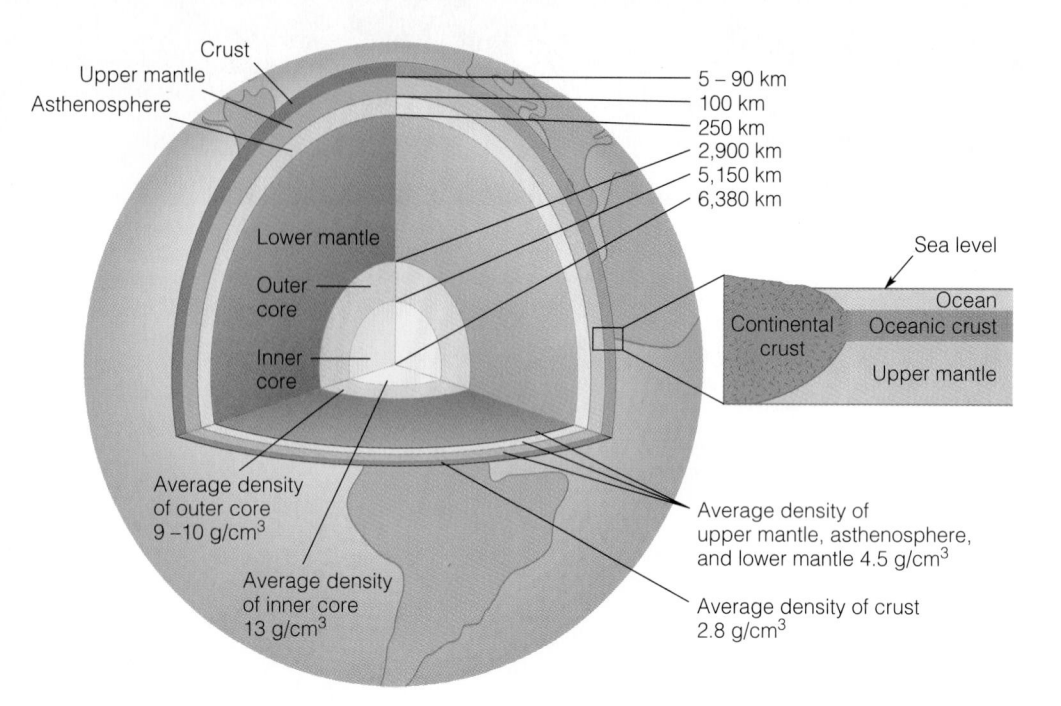

FIGURE 7.9 Cross section of Earth showing the various layers and their average density. The crust is divided into a continental and oceanic portion. Continental crust is 20 to 90 km thick; oceanic crust is 5 to 10 km thick.

Crust
Upper mantle
Asthenosphere

Lower mantle
Outer core
Inner core

5 – 90 km
100 km
250 km
2,900 km
5,150 km
6,380 km

Sea level
Ocean
Continental crust
Oceanic crust
Upper mantle

Average density of outer core 9 –10 g/cm³

Average density of inner core 13 g/cm³

Average density of upper mantle, asthenosphere, and lower mantle 4.5 g/cm³

Average density of crust 2.8 g/cm³

of Earth, the heat generated during radioactive decay was absorbed by the surrounding rock.

The combination of meteorite impacts, gravitational compression, and heat from radioactive decay increased the temperature of early Earth enough to melt iron and nickel, which being denser than silicate minerals, settled to Earth's center and formed the core (Figure 7.10b). Simultaneously, the lighter silicates slowly flowed upward, beginning the differentiation of the mantle from the core.

Calculations indicate that, with a uniform distribution of elements in an early solid Earth, enough heat could be generated to begin melting iron and nickel at depths between 400 and 850 km. Melting would have had to begin at shallow depths because the temperature at which melting begins increases with pressure; therefore, the melting point of any material increases toward Earth's center.

Most geologists currently favor the homogeneous accretion theory's explanation for Earth's internal zonation, but

another theory, the **inhomogeneous accretion** theory, has been proposed to explain Earth's internal differentiation. According to this theory, Earth's core, mantle, and crust condensed sequentially from the hot nebular gases forming early Earth.

Even though iron and nickel melt at lower temperatures than most silicates, they condense from the gaseous state at a slightly higher temperature. Calculations show that in a cooling cloud of hot nebular gases, iron and nickel would condense first, thus forming the core of early Earth. As the cloud continued to cool, iron and magnesium silicates would condense to form the mantle, while the lightest and most volatile elements that form the crust would condense last.

Both the homogeneous and inhomogeneous accretion theories provide a logical explanation for Earth's internal zonation, and both are based on the currently accepted solar nebula origin of the solar system. Regardless of which theory, if either, eventually proves correct, geologists do

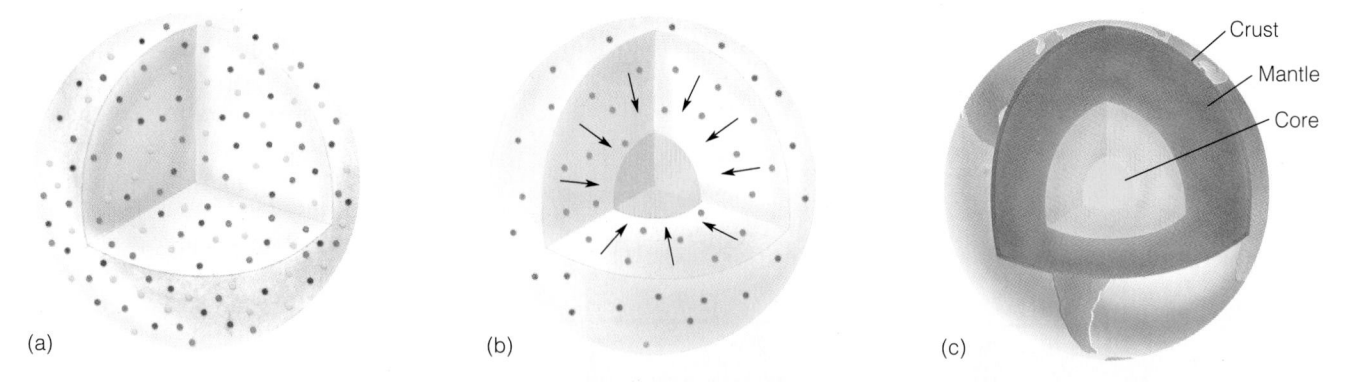

(a)   (b)   (c)

Crust
Mantle
Core

FIGURE 7.10 Homogeneous accretion theory for the formation of a differentiated Earth. (a) Early Earth was probably of uniform composition and density throughout. (b) Heating of early Earth reached the melting point of iron and nickel, which, being denser than silicate minerals, settled to Earth's center. At the same time, the lighter silicates flowed upward to form the mantle and the crust. (c) In this way, a differentiated Earth formed, consisting of a dense iron–nickel core, an iron-rich silicate mantle, and a silicate crust with continents and ocean basins.

know that the differentiation of Earth took place very early in its history. Not only did differentiation result in the formation of a crust and eventually continents, but it was also probably responsible for the outgassing of light volatile elements from the interior, which later resulted in the oceans and atmosphere.

Origin of the Earth–Moon System

We probably know more about our Moon than any other celestial object except Earth (Figure 7.11). Nevertheless, even though the Moon has been studied for centuries through telescopes and has been sampled directly, many questions remain unanswered.

The Moon is one-fourth the diameter of Earth, has a low density (3.3 g/cm³) relative to the terrestrial planets, and exhibits an unusual chemistry in that it is for the most part bone-dry, having been largely depleted of most volatile elements (Table 7.2). However, in 1995 a spacecraft circling the Moon in a vertical orbit discovered ice deposits in the south pole's Aitkin Basin. More recently in 1998, the *Lunar Prospector* spacecraft detected evidence of as much as 6 billion tons of ice near the lunar poles. While it is uncertain where the ice came from, it is thought that a collision with an icy meteor is the most likely source.

The Moon orbits Earth and rotates on its own axis at the same rate, so we always see the same side. Furthermore, the Earth–Moon system is unique among the terrestrial

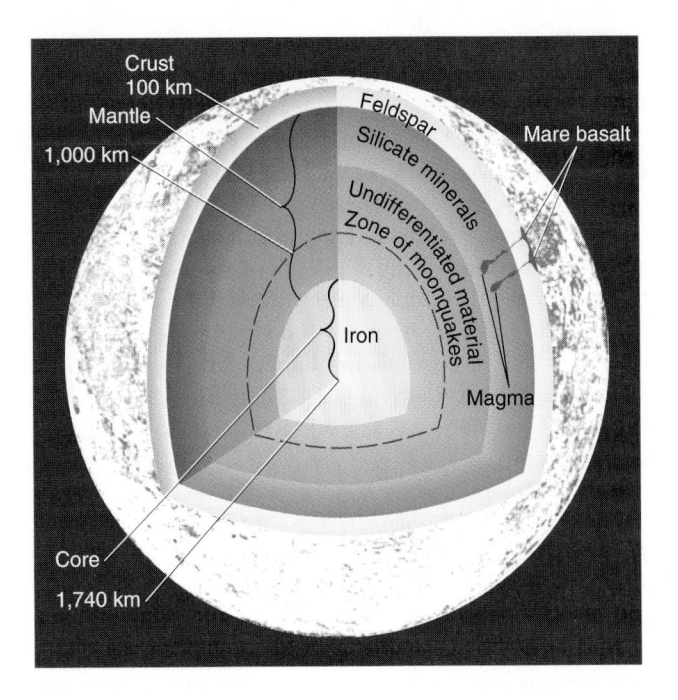

FIGURE 7.12 The internal structure of the Moon is different from that of Earth. The upper mantle is the source for the maria lavas. Moonquakes occur at a depth of 1000 km. Because seismic shear waves are not transmitted below this depth, it is thought that the innermost mantle is liquid. Below this layer is a small metallic core.

planets. Neither Mercury nor Venus has a moon, and the two small moons of Mars—Phobos and Deimos—are probably captured asteroids.

The surface of the Moon can be divided into two major parts: the low-lying dark-colored plains, called *maria,* and the light-colored *highlands* (Figure 7.11). The highlands are the oldest parts of the Moon and are heavily cratered, providing striking evidence of the massive meteorite bombardment that occurred in the solar system more than 4 billion years ago.

Study of the several hundred kilograms of rocks returned by the *Apollo* missions indicates that three kinds of materials dominate the lunar surface: igneous rocks, breccias, and dust. Basalt, a common dark-colored igneous rock on Earth, is one of the several different types of igneous rocks on the Moon and makes up the greater part of the maria. The presence of igneous rocks that are essentially the same as those on Earth shows that magmas similar to those on Earth were generated on the Moon long ago.

The lunar surface is covered with a regolith (or "soil") estimated to be 3 to 4 m thick. This gray covering, which is composed of breccia, glass spherules, and small particles of dust, is thought to be the result of debris formed by meteorite impacts.

The interior structure of the Moon is quite different from that of Earth, indicating a different evolutionary history (Figure 7.12). The highland crust is thick (65 to 100 km) and comprises about 12% of the Moon's volume. It was formed about 4.4 billion years ago, immediately following the Moon's accretion. The highlands are composed

FIGURE 7.11 The side of the Moon as seen from Earth. The light-colored areas are the lunar highlands, which were heavily cratered by meteorite impacts. The dark-colored areas are maria, which formed when lava flowed out onto the surface.

principally of the igneous rock anorthosite, which is made up of light-colored feldspar minerals that are responsible for their white appearance.

A thin covering (1 to 2 km thick) of basaltic lava fills the maria; lava covers about 17% of the lunar surface, mostly on the side facing Earth. These maria lavas came from partial melting of a thick underlying mantle of silicate composition. Moonquakes occur at a depth of about 1000 km, but below that depth seismic shear waves apparently are not transmitted. Because shear waves do not travel through liquid, their lack of transmission implies that the innermost mantle may be partially molten. There is increasing evidence that the Moon has a small (500 to 800 km diameter) solid iron core.

The origin and earliest history of the Moon are still unclear, but the basic stages in its subsequent development are well understood. It formed some 4.6 billion years ago and shortly thereafter was partially or wholly melted, yielding a silicate melt that cooled and crystallized to form the mineral anorthite. Because of the low density of the anorthite crystals and the lack of water in the silicate melt, the thick anorthosite highland crust formed. The remaining silicate melt cooled and crystallized to produce the zoned mantle, while the heavier metallic elements formed the core.

The formation of the lunar mantle was completed by about 4.4 to 4.3 billion years ago. The maria basalts, derived from partial melting of the upper mantle, were extruded during great lava floods between 3.8 and 3.2 billion years ago.

Numerous theories have been proposed for the origin of the Moon, including capture from an independent orbit, formation with Earth as part of an integrated two-planet system, breaking off from Earth during accretion, and formation resulting from a collision between Earth and a large planetesimal. The current theory that seems to account best for the Moon's particular composition and structure involves an impact by a large planetesimal with a young Earth (Figure 7.13). According to the *impact theory,* a giant planetesimal, the size of Mars or larger, struck Earth about 4.6 to 4.4 billion years ago, causing the ejection of a large quantity of hot material that formed the Moon. The material that was ejected was mostly in the liquid and vapor phase and came primarily from the mantle of the colliding planetesimal. As it cooled, the various lunar layers crystallized out in the order previously discussed.

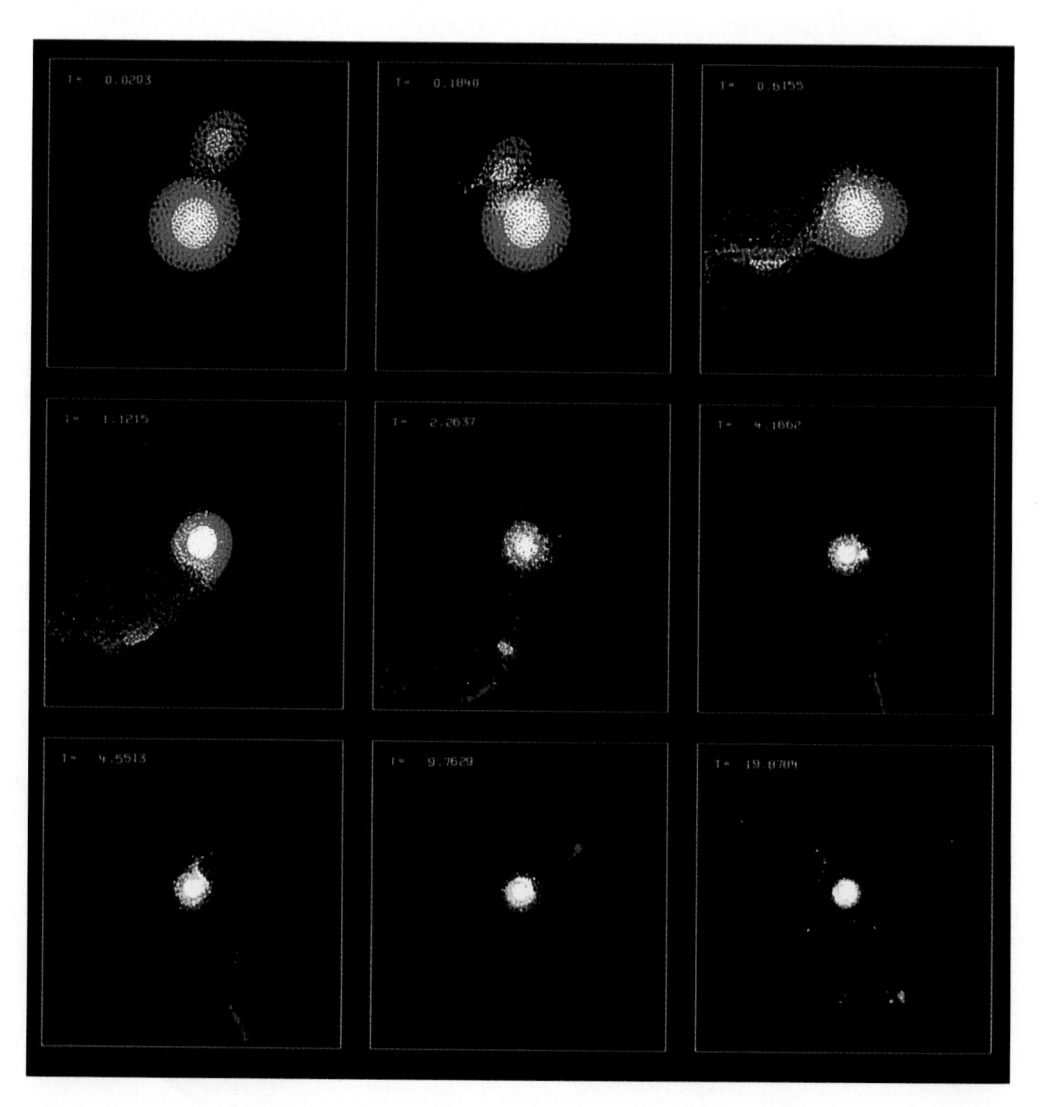

FIGURE 7.13 According to one hypothesis for the origin of the Moon, a large planetesimal the size of Mars crashed into Earth 4.6 to 4.4 billion years ago, causing the ejection of a mass of hot material that formed the Moon. This computer simulation shows the formation of the Moon as a result of an Earth–planetesimal collision.

The Planets

A tremendous amount of information about each planet in the solar system has been derived from Earth-based observations and measurements, as well as from the numerous space probes launched during the past 30 years. Such information about a planet's size, mass, density, composition, presence of a magnetic field, and atmospheric composition has allowed scientists to formulate hypotheses concerning the origin and history of the planets and their moons.

THE TERRESTRIAL PLANETS

All terrestrial planets appear to have experienced a similar early history during which volcanism and cratering from meteorite impacts were common. During the accretionary phase of formation and soon thereafter, each planet seems to have undergone differentiation as a result of heating by radioactive decay. The mass, density, and composition of the planets indicate that each formed a metallic core and a silicate mantle–crust during this phase. Images sent back by the various space probes also clearly show that volcanism and cratering by meteorites continued during the differentiation phase. Volcanic eruptions produced lava flows, and an atmosphere developed on each planet by *outgassing,* a process whereby light gases from the interior rise to the surface during volcanic eruptions.

Mercury   Mercury, the closest planet to the Sun, apparently has changed very little since it was heavily cratered during its early history (Figure 7.14a). Most of what we know about this small (4880-km diameter) planet comes from the measurements and observations of *Mariner 10* in 1974 and 1975 (Table 7.2). Its high overall density of 5.4g/cm³ indicates that it has a large metallic core. Further-more, Mercury has a weak magnetic field (about 1% as strong as Earth's), indicating the core is probably partially molten.

Images sent back by *Mariner 10* show a heavily cratered surface with the largest impact basins filled with what appear to be lava flows similar to the lava plains on the Moon. The lava plains are not deformed, however, indicating that there has been little or no tectonic activity. Another feature of Mercury's surface is a large number of scarps (Figure 7.14b). It is suggested that these scarps formed when Mercury cooled and contracted.

Because Mercury is so small, its gravitational attraction is insufficient to retain atmospheric gases; any atmosphere that it may have held when it formed probably escaped into space quickly. Nevertheless, *Mariner 10* detected small quantities of hydrogen and helium, thought to have originated from the solar winds that stream by Mercury.

Venus   Of all the planets, Venus is the most similar in size and mass to Earth (Table 7.2). It differs, though, in most other respects. Venus is searingly hot with a surface temperature of 475° C and an oppressively thick atmosphere composed of 96% carbon dioxide and 3.5% nitrogen, with traces of sulfur dioxide and sulfuric and hydrochloric acid.

Radar images from both orbiting spacecraft and the Venusian surface reveal a wide variety of terrains (Figure 7.15), some of which are unlike anything seen elsewhere in the solar system. Even though no active volcanism has been observed on Venus, the presence of volcanic domes, extensive lava flows, folded mountain ranges, a complex network of faults, and what appear to be deep trenches comparable to the oceanic trenches on Earth indicate that internal and surface activity has occurred during the past. There is, however, no evidence for active plate tectonics as on Earth.

(a)

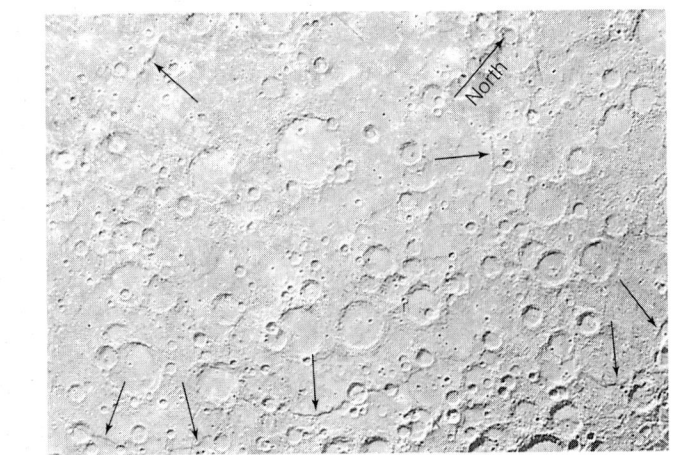

(b)

FIGURE 7.14 (a) Mercury has a heavily cratered surface that has changed very little since its early history. (b) Seven scarps (indicated by arrows) can clearly be seen in this image. It is thought that these scarps formed when Mercury cooled and contracted early in its history.

FIGURE 7.15 A nearly complete map of the northern hemisphere of Venus based on radar images beamed back to Earth from the *Magellan* space probe.

Mars Mars, the red planet, is differentiated, as are all the terrestrial planets, into a metallic core and a silicate mantle and crust (Figure 7.16). The thin Martian atmosphere consists of 95% carbon dioxide, 2.7% nitrogen, 1.7% argon, and traces of other gases. Mars has distinct seasons during

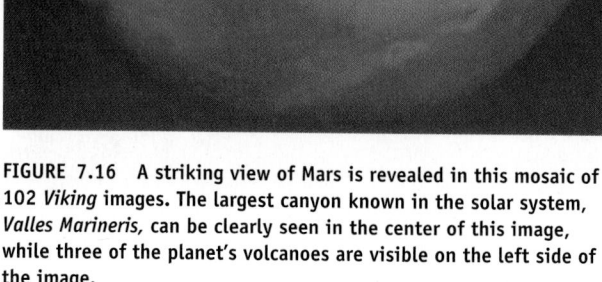

FIGURE 7.16 A striking view of Mars is revealed in this mosaic of 102 *Viking* images. The largest canyon known in the solar system, *Valles Marineris*, can be clearly seen in the center of this image, while three of the planet's volcanoes are visible on the left side of the image.

which its polar ice caps of frozen carbon dioxide expand and recede.

Perhaps the most striking aspect of Mars is its surface features, many of which have not yet been satisfactorily explained. Like the surfaces of Mercury and the Moon, the southern hemisphere is heavily cratered, attesting to a period of meteorite bombardment. *Hellas,* a crater with a diameter of 2000 km, is the largest known impact structure in the solar system and is found in the Martian southern hemisphere.

The northern hemisphere is much different, having large smooth plains, fewer craters, and evidence of extensive volcanism. The largest known volcano in the solar system, *Olympus Mons,* has a basal diameter of 600 km, rises 27 km above the surrounding plains, and is topped by a huge circular crater 80 km in diameter.

The northern hemisphere is also marked by huge canyons that are essentially parallel to the Martian equator. One of these canyons, *Valles Marineris,* is at least 4000 km long, 250 km wide, and 7 km deep and is the largest yet discovered in the solar system (Figure 7.16). If it were present on Earth, it would stretch from San Francisco to New York! It is not yet known how these vast canyons formed, although geologists postulate that they may have started as large rift zones that were subsequently modified by running water and wind erosion. Such hypotheses are based on comparison to rift structures found on Earth and topographic features formed by geologic agents of erosion such as water and wind.

Tremendous wind storms have strongly influenced the surface of Mars and led to dramatic dune formations. Even more stunning than the dunes, however, are the braided channels that appear to be the result of running water. Mars is currently too cold for surface water to exist, yet the channels strongly indicate that running water was on the planet during the past.

Further evidence strengthening the case for a warmer and wetter ancient Mars comes from the images and data radioed back by the *Mars Pathfinder,* which landed at *Ares Vallis* on Mars on July 4, 1997. This evidence includes the numerous rounded pebbles and possible conglomerates, which typically result from the action of running water, as well as abundant sand-sized particles, seen forming numerous small dunes.

The fresh-looking surfaces of its many volcanoes strongly suggest that Mars was tectonically active during the past and may still be. There is, however, no evidence that plate movement, such as occurs on Earth, has ever taken place.

THE JOVIAN PLANETS

The Jovian planets are completely unlike any of the terrestrial planets in size (Figure 7.17) or chemical composition and have had completely different evolutionary histories. While they all apparently contain a small core in relation to

FIGURE 7.17 A composite photograph, showing the Jovian planets and Earth to the same scale. The photographs of the Jovian planets were all obtained by the *Voyager* spacecraft.

their overall size, the bulk of a Jovian planet is composed of volatile elements and compounds, such as hydrogen, helium, methane, and ammonia, that condense at low temperatures (Figure 7.18).

Jupiter Jupiter is the largest of the Jovian planets (Table 7.2, Figure 7.17). With its moons, rings, and radiation belts, it is the most complex and varied planet in the solar system. Jupiter's density is only one-fourth that of Earth, but because it is so large, it has 318 times the mass (Table 7.2). It is an unusual planet in that it emits almost 2.5 times more energy than it receives from the Sun. One explanation is that most of the excess energy is left over from the time of its formation. When Jupiter

formed, it heated up because of gravitational contraction (as did all the planets) and is still cooling. Jupiter's massive size insulates its interior, and hence it has cooled very slowly.

Jupiter has a relatively small central rocky core (Figure 7.18). Above this core is a thick zone of liquid metallic hydrogen, followed by a thicker layer of liquid hydrogen; above that is a thin layer of clouds. Surrounding Jupiter are a strong magnetic field and an intense radiation belt.

Jupiter has a dense, hot and surprisingly dry atmosphere of hydrogen, helium, methane, and ammonia, which some think are the same gases that composed Earth's first atmosphere. Jupiter's cloudy and violent atmosphere (with winds up to 650 km/h) is divided into a series of different colored bands, as well as a variety of spots (the Great Red

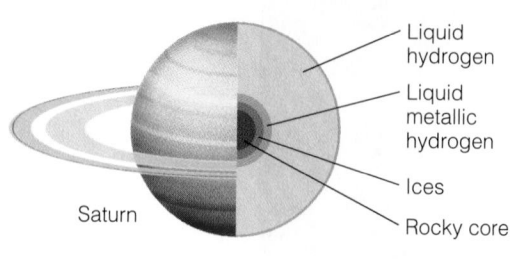

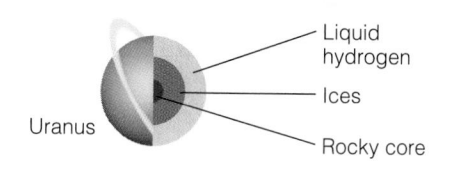

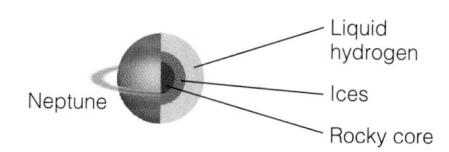

FIGURE 7.18 Probable interior makeup of the Jovian planets.

Even though the atmospheres of Saturn and Jupiter are similar, Saturn's atmosphere contains little ammonia because it is farther from the Sun and therefore is colder. The cloud layer on Saturn is thicker than on Jupiter, but it lacks the contrast between the different bands. Unlike Jupiter, Saturn has seasons because its axis tilts 27 degrees.

Uranus Uranus is much smaller than Jupiter (Figure 7.17), but their densities are about the same (Table 7.2). It is the only planet that lies on its side; that is, its axis of rotation nearly parallels the plane of the ecliptic. Some scientists think that a collision with an Earth-sized body early in its history may have knocked Uranus on its side.

Data gathered by the flyby of *Voyager 2* suggests that Uranus has a water zone beneath its cloud cover. Because the planet's density is greater than if it were composed entirely of hydrogen and helium, it is thought that Uranus must have a dense, rocky core, and this core may be surrounded by a deep global ocean of liquid water or ices (Figure 7.18).

The atmospheric composition of Uranus is similar to that of Jupiter and Saturn, with hydrogen being the dominant gas, followed by helium and some methane. Uranus also has a banded atmosphere and a circulation pattern much like those of Jupiter and Saturn. Surrounding Uranus is a huge corkscrew-shaped magnetic field that stretches for millions of kilometers into space. Uranus has at least nine thin, faint rings and 17 small moons circling it (Figure 7.20).

Neptune and Pluto The flyby of *Voyager 2* in August 1989 provided the first detailed look at Neptune and showed it to be a dynamic, stormy planet (Figure 7.17). Its atmosphere is similar to those of the other Jovian planets, and it exhibits a pattern of zonal winds and giant storm systems comparable to those of Jupiter. Neptune's internal structure is similar to that of Uranus in that it has a rocky core that is surrounded by a semifrozen slush of water and liquid hydrogen (Figure 7.18). Its atmosphere is composed of hydrogen and helium, with some methane. Winds up to 2000 km/h blow over the planet, creating tremendous storms—the largest of which, the Great Dark Spot, is in Neptune's southern hemisphere. It is nearly as big as Earth and is similar to the Great Red Spot on Jupiter. One of the mysteries raised by *Voyager 2*'s discovery is where Neptune gets the energy to drive such a storm system.

Encircling Neptune are three faint rings and eight moons, the most interesting of which is Triton, Neptune's largest moon (Figure 7.21). Triton's mottled surface of delicate pinks, reds, and blues consists primarily of water ice, with minor amounts of nitrogen and a methane frost. Geysers have been discovered erupting carbon-rich material and frozen nitrogen particles some 8 km above its surface, making it only the second place other than Earth undergoing active volcanism.

Spot) and other features, all interacting in incredibly complex motions.

Revolving around Jupiter are 16 moons varying greatly in geologic activity (Figure 7.19). Also surrounding Jupiter is a thin, faint ring, a feature shared by all the Jovian planets.

Saturn Saturn is slightly smaller than Jupiter, about one-third as massive and about one-half as dense, but has a similar internal structure and atmosphere (Table 7.2, Figure 7.17). Like Jupiter, Saturn gives off more energy (2.2 times as much) than it gets from the Sun. Saturn's most conspicuous feature is its ring system, consisting of thousands of rippling, spiraling bands of countless particles.

The composition of Saturn is similar to Jupiter's but consists of slightly more hydrogen and less helium. Saturn's core is not as dense as Jupiter's, but has a layer of ices overlying it, followed by a layer of liquid metallic hydrogen, a zone of liquid hydrogen and helium, and, lastly, a layer of clouds (Figure 7.18).

(a)

FIGURE 7.19 (a) In this view of Io, the active volcano Prometheus is visible in the right-center of the image. The black and red areas in the image are volcanic deposits no more than a few years old. (b) Europa, the second moon out from Jupiter, is covered by a thick surface layer of ice crisscrossed by numerous fractures. These fractures appear to be rifts where water has risen to the surface and frozen. This system of fractures and the lack of craters are evidence that Europa is a geologically active moon.

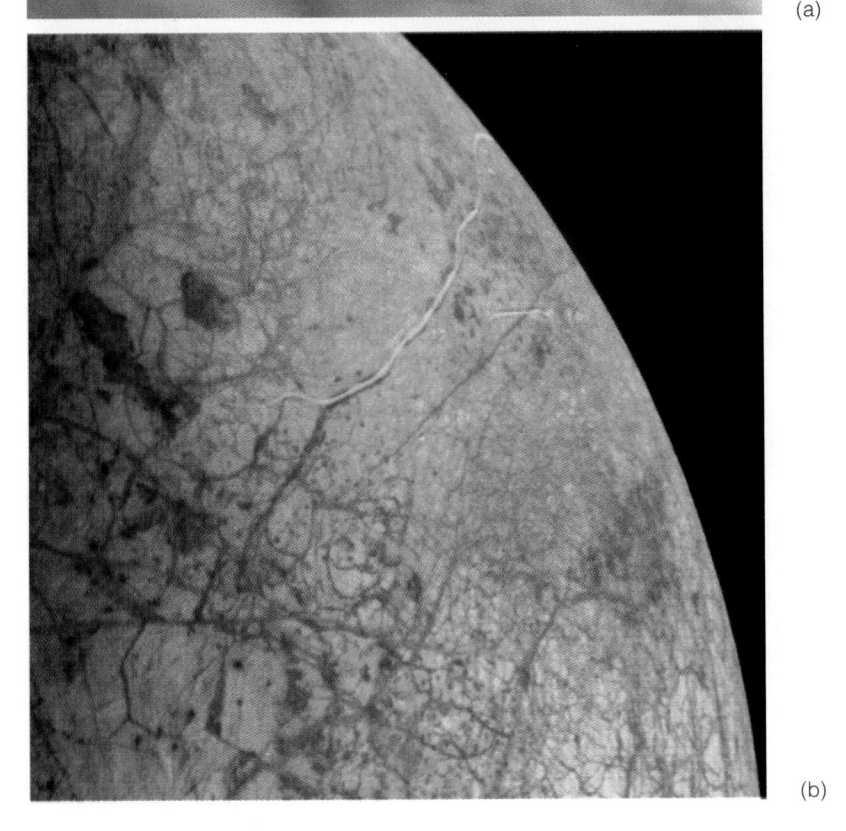

(b)

Some areas of Triton are smooth, whereas others have a very irregular appearance indicating numerous episodes of deformation. Heavily cratered areas bear witness to bombardment by meteorites or the collapse of its surface. Perhaps the most intriguing aspect of Triton is that it may have once been a planet—much like Pluto, which it resembles in size and possibly composition—that was captured by Neptune's gravitational field soon after the formation of the solar system.

With a diameter of only 2300 km, Pluto is the smallest

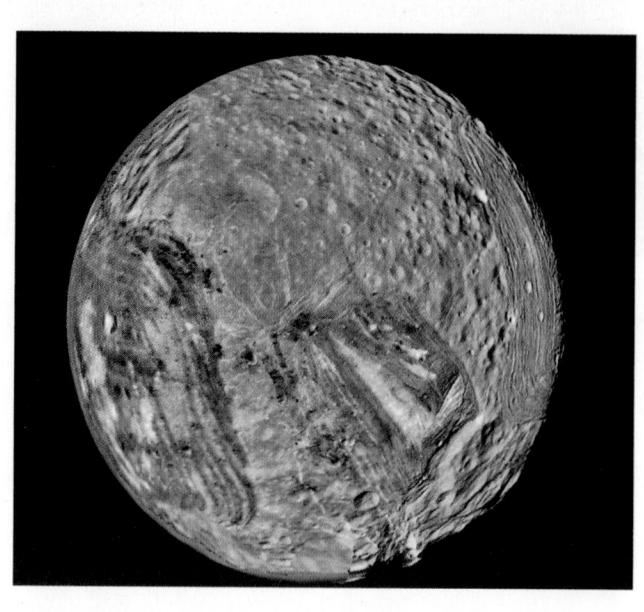

FIGURE 7.20 Miranda, Uranus's inner moon, shows a spectacular surface that may reflect a catastrophic early history.

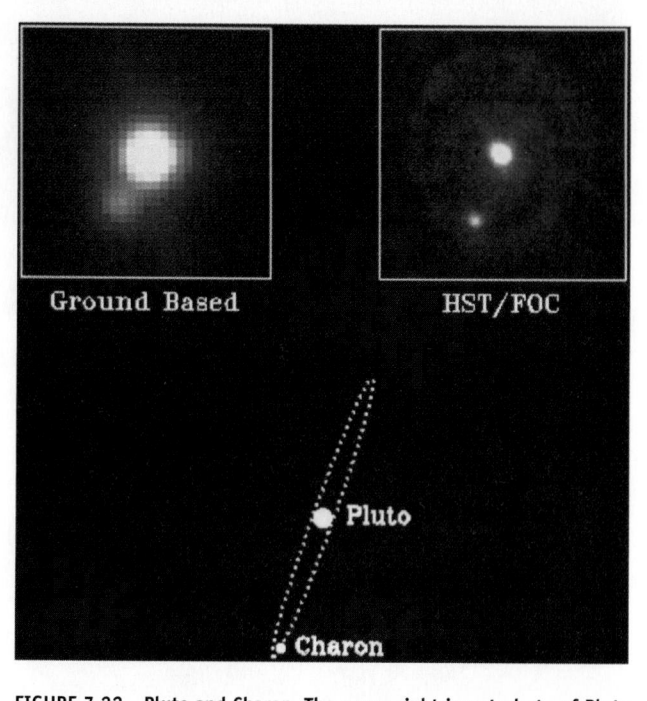

FIGURE 7.22 Pluto and Charon. The upper right insert photo of Pluto *(top)* and Charon *(bottom)* obtained by the Hubble Space Telescope is the best available image of these worlds.

planet and, strictly speaking, is not one of the Jovian planets (Table 7.2). Little is known about Pluto, but studies indicate it has a rocky core overlain by a mixture of frozen methane, nitrogen, and carbon monoxide. It has a thin, two-layer atmosphere with a clear upper layer overlying a more opaque lower layer.

Pluto differs from all other planets in that it has a highly elliptic orbit that is tilted with respect to the plane of the ecliptic. It has one known moon, Charon, that is nearly half its size (Figure 7.22).

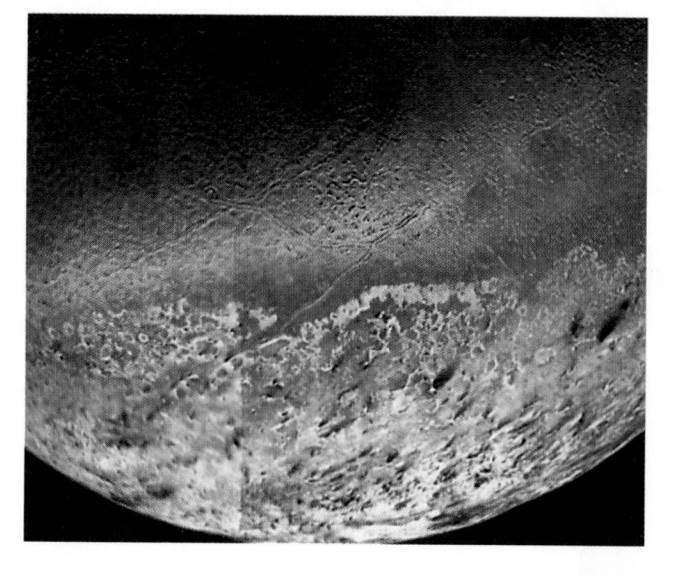

FIGURE 7.21 Neptune's moon Triton is described by scientists as "a world unlike any other." In this composite of numerous high-resolution images taken by *Voyager 2* during its August 1989 flyby, various features can be seen. The large south polar ice cap at the bottom consists mostly of frozen nitrogen that was deposited during the previous Tritonian winter and is slowly evaporating. The dark plumes in the lower right may be the result of volcanic activity. Smooth plains and fissures in the upper half are evidence of geologic activity in which the surface has been cracked and flooded by slushy ice that refroze.

Summary

1. The universe began with a Big Bang approximately 15 billion years ago. Astronomers have deduced this age from the fact that celestial objects are moving away from each other in what appears to be an ever-expanding universe. Furthermore, the universe has a background radiation of 2.7° above absolute zero, which is thought to be the faint afterglow of the Big Bang.

2. About 4.6 billion years ago, the solar system formed from a rotating cloud of interstellar matter. As this cloud condensed, it eventually collapsed under the influence of gravity and flattened into a counterclockwise rotating disk. Within this rotating disk, the Sun, planets, and moons formed from the turbulent eddies of nebular gases and solids.

3. Temperature as a function of distance from the Sun played a major role in the type of planets that evolved. The inner terrestrial planets are composed of rock and metallic elements that condense at high temperatures. The outer Jovian planets plus Pluto are composed mostly of hydrogen, helium, ammonia, and methane, all of which condense at lower temperatures.

4. During the solar system's early history, the Sun emitted a tremendous blast of energy that swept the solar system free of unaccreted gas and dust. During this time, most of the cratering of the planets and their satellites took place.

5. Meteorites provide vital information about the age and composition of the solar system. The three major groups of meteorites are stones, irons, and stony-irons. Each has a different composition, reflecting a different origin.

6. Earth formed from a swirling eddy of nebular material 4.6 billion years ago, and by at least 3.96 billion years ago, it had differentiated into its present-day structure. It probably accreted as a solid body, and then underwent differentiation during a period of internal heating.

7. The Moon probably formed as a result of a Mars-sized planetesimal crashing into Earth about 4.6 to 4.4 billion years ago and causing it to eject a large quantity of hot material. As the material cooled, the various lunar layers crystallized, forming a zoned body.

8. All terrestrial planets are differentiated into a core, mantle, and crust, and all seem to have had a similar early history during which volcanism and cratering from meteorite impacts were common.

9. The Jovian planets differ from the terrestrial planets in size and chemical composition and followed completely different evolutionary histories. All Jovian planets have a small core compared to their overall size; they are mainly composed of volatile elements and compounds.

Important Terms

acondrite

Big Bang

carbonaceous chondrite

chondrule

Doppler effect

electromagnetic force

gravity

homogeneous accretion

inhomogeneous accretion

irons

Jovian planets

meteorite

ordinary chondrite

solar nebula theory

stones

stony-irons

strong nuclear force

terrestrial planets

weak nuclear force

Review Questions

1. What two observations lead scientists to conclude that the Big Bang occurred approximately 15 billion years ago?
   a. _____ A steady-state universe and opaque background radiation
   b. _____ A steady-state universe and 2.7° above absolute zero background radiation
   c. _____ An expanding universe and opaque background radiation
   d. _____ An expanding universe and 2.7° above absolute zero background radiation
   e. _____ A shrinking universe and opaque background radiation

2. The change in frequency of a sound wave caused by movement of its source relative to an observer is known as the:
   a. _____ Curie point.
   b. _____ Hubble shift.
   c. _____ Doppler effect.
   d. _____ Quasar force.
   e. _____ None of these

3. Which of the following is not one of the basic forces in the universe?
   a. _____ Electromagnetic force
   b. _____ Gravity
   c. _____ Strong nuclear force
   d. _____ Quasar force
   e. _____ None of these

4. The current theory for the origin of the solar system is known as the:
   a. _____ nebular theory.
   b. _____ encounter theory.
   c. _____ evolutionary theory.
   d. _____ planetesimal theory.
   e. _____ solar nebula theory.

5. The most abundant meteorites are:
   a. _____ stones.
   b. _____ acondrites.
   c. _____ stony-irons.
   d. _____ irons.
   e. _____ carbonaceous chondrites.

6. The most widely accepted theory regarding the origin of the Moon involves:
   a. _____ capture from an independent orbit.
   b. _____ an independent origin from Earth.
   c. _____ breaking off from Earth during Earth's accretion.
   d. _____ formation resulting from a collision between Earth and a large planetesimal.
   e. _____ none of these.

7. The planets can be separated into terrestrial and Jovian primarily on the basis of which property?
   a. _____ Size
   b. _____ Atmosphere
   c. _____ Density
   d. _____ Color
   e. _____ None of these

8. The atmosphere of Venus is:
   a. _____ thick and composed of carbon dioxide.
   b. _____ similar to Earth's.
   c. _____ nonexistent.
   d. _____ thin, like that of Mars.
   e. _____ none of these.

9. The only planet whose axis of rotation nearly parallels the plane of the ecliptic is:
   a. _____ Venus.
   b. _____ Saturn.
   c. _____ Uranus.
   d. _____ Neptune.
   e. _____ Pluto.

10. The largest planet in the solar system is:
    a. _____ Neptune.
    b. _____ Uranus.
    c. _____ Saturn.
    d. _____ Jupiter.
    e. _____ Mars.

11. Which planets give off more energy than they receive?
    a. _____ Jupiter and Saturn
    b. _____ Saturn and Neptune
    c. _____ Neptune and Pluto
    d. _____ Jupiter and Neptune
    e. _____ Saturn and Pluto

12. Which of the following events did all of the terrestrial planets experience early in their history?
    a. _____ Accretion
    b. _____ Differentiation
    c. _____ Volcanism
    d. _____ Meteorite impacting
    e. _____ All of these

13. Discuss the origin and differentiation of Earth into its various internal layers.

14. Discuss the origin of the Earth–Moon system.

15. What two fundamental phenomena indicate that the Big Bang occurred?

16. If the oldest terrestrial rocks are 3.96 billion years old, why do geologists think Earth is 4.6 billion years old?

17. Why is the Doppler effect important to the theory of an expanding universe?

18. How does the solar nebula theory account for the general characteristics of the solar system?

19. What are the three groups of meteorites, and how do they aid geologists in determining the age and composition of the solar system?

20. How do the terrestrial planets differ from the Jovian planets?

Points to Ponder

1. The Sun is thought to have been much dimmer during the early history of the solar system. Because of this, it is thought that Venus had oceans early in its history, as did Mars. If the Sun was as bright during its early history as it is now, how might this have affected the evolution of the solar system, particularly the evolution of the terrestrial planets?

2. One of the theories proposed to account for the origin of the Moon is that it broke off from Earth during accretion.

What information about Earth and the Moon would you need to support this theory? How would you gather such information?

3. Based on our current knowledge about Mars, do you think it is possible for humans to colonize the planet effectively and become self-sufficient? What major problems must be overcome?

Additional Readings

Binzel, R. P., M. A. Barucci, and M. Fulchignoni. 1991. The origins of the asteroids. *Scientific American* 265, no. 4: 88-95.

Fernie, J. D. 1993. The Tunguska event. *American Scientist* 81, no. 5: 412-415.

Flamsteed, S. 1997. Impossible planets. *Discover* 18, no. 9: 78-83.

Freedman, D. H. 1998. The mysterious middle of the Milky Way. *Discover* 19, no. 11: 66-75.

Freedman, W. L. 1992. The expansion rate and size of the universe. *Scientific American* 267, no. 5: 54-61.

Golombek, M. P. 1998. The Mars *Pathfinder* mission. *Scientific American* 279, no. 1: 40-49.

Grieve, R. A. F. 1990. Impact cratering on the Earth. *Scientific American* 262, no. 4: 66-73.

Henry, J. P., U. G. Briel, and H. Böhringer. 1998. The evolution of galaxy clusters. *Scientific American* 279, no. 6: 52-57.

Kargel, J. S., and R. G. Strom. 1996. Global climatic change on Mars. *Scientific American* 275, no. 5: 80-88.

Kuhn, K. F. 1994. *In quest of the universe.* 2d ed. St. Paul: West.

Luhmann, J. G., J. B. Pollack, and L. Colin. 1994. The *Pioneer* mission to Venus. *Scientific American* 270, no. 4: 90-97.

Lunine, J. I. 1994. Does Titan have oceans? *American Scientist* 82, no. 2: 134-143.

McSween Jr., H. Y., and S. L. Murchie. 1999. Rocks at the Mars *Pathfinder* landing site. *American Scientist* 87, no. 1: 36-45.

Patrick, R. R., and R. C. Howe. 1994. Volcanism on the terrestrial planets. *Journal of Geological Education* 42, no. 3: 225-238.

Smith, B. A. 1994. New eyes on the universe. *National Geographic* 181, no. 1: 2-41.

Stone, R. 1996. The last great impact on Earth. *Discover* 17, no. 9: 60-71.

Taylor, G. J. 1994. The scientific legacy of *Apollo. Scientific American* 271, no. 1: 40-47.

Winters, J. 1998. A survey of ancient Mars. *Discover* 19, no. 7: 112-117.

For these web site addresses, along with current updates and exercises, log on to

`http://www.brookscole.com/geo/`

▶ WELCOME TO THE PLANETS

This site, maintained by the California Institute of Technology with U. S. government sponsorship, contains a collection of many of the best images of the planets from NASA's planetary exploration program. Each planet is listed, as well as the various spacecrafts. In addition, it provides links to other sites around the world and a What's New section. Click on any of the planets for a planet profile, as well as spectacular images of the planet and specific regions of it and information about those regions.

▶ NASA HOME PAGE

This site, maintained by NASA, describes the organization of NASA, what it does, and some of its projects and missions. It also contains extensive photo archives of images of Earth from space.

1. Click on the *Today@NASA* site. This will take you to the latest breaking news about space as well as sites for information and images from the Hubble Space Telescope, the current Shuttle mission, and much more.

2. Click on the *Mission to Planet Earth* site. This page is dedicated to understanding the various ways Earth is changing and how humans are influencing those changes.

▶ VIEWS OF THE SOLAR SYSTEM

This site is maintained by Calvin J. Hamilton and was created as an educational tour of the solar system. It contains hundreds of pages of information and images about the planets, asteroids, comets, meteoroids and meteorites, and the history of space exploration, as well as links to other astronomy web sites.

Explore the following *In-Terra-Active 2.0* CD-ROM module(s) and increase your understanding of key concepts and processes presented in this chapter.

Chapter Concept: **Geologic Processes on the Terrestrial Planets**

The solar nebula theory of how the solar system was formed accounts for the similarities in composition of the terrestrial planets. These planets also exhibit geomorphic features that suggest similar surficial and internal geologic processes. In this module you will explore the relationships between surface forms and processes on Earth's surface and apply these relationships to similar forms on the other terrestrial planets, as well as on Earth's moon.

Section: **Interior**

Module: **Planetary Geology**

The inner terrestrial planets (Mercury, Venus, Earth, and Mars) have many surface features in common. Thus, it appears that they share some of the same geologic processes. This image of the surface of Venus has been compiled from radar images from NASA. The actual surface is completely obscured by thick clouds.

Question: *What can we learn or determine about the geology and history of the inner terrestrial planets and moons of the solar system?*

8

Precambrian History: The Archean Eon

Archaen gneiss, schist, and granite are exposed in this view of the Teton Range in Grand Teton National Park, Wyoming. Younger rocks are also present, but not visible in this view.

Prologue

Imagine a barren, waterless, lifeless, hot planet with a poisonous atmosphere. Volcanoes erupt nearly continuously, meteorites and comets flash through the atmosphere, and cosmic radiation is intense. The planet's crust of dark-colored igneous rock is thin and unstable. Storms form in the turbulent atmosphere, and lightning discharges are common, but no rain falls. And because no oxygen is present in the atmosphere, nothing burns. Rivers and pools of molten rock emit a continuous reddish glow.

This may sound like a science fiction novel, but it is probably a reasonably accurate description of Earth shortly after it formed (Figure 8.1). We emphasize "probably" because no record exists for the earliest chapter of Earth history, the interval from 4.6 to 3.8 billion years ago, although one area of rocks 4 billion years old is now known in Canada. We can only speculate about what Earth was like during this time, based on our knowledge of how planets form and about other Earth-like planets.

When Earth formed, it had a tremendous reservoir of primordial heat, heat generated by colliding particles as Earth accreted, by compression, and by the decay of short-lived radioactive elements. Many geologists think that early Earth was so hot that it was partly or perhaps almost entirely molten. No one knows what Earth's surface temperature was during its earliest history, but it was almost certainly too hot for liquid water to exist or for any known organism to survive. Volcanism must have been ubiquitous and nearly continuous. Molten rock later solidified to form a thin, discontinuous, dark-colored crust, only to be disrupted by upwelling magmas.

Assuming that visitors to early Earth could tolerate the high temperatures, they would also have to contend with other factors. The atmosphere would be unbreathable by any of today's inhabitants. It probably contained considerable carbon dioxide and water vapor, but little or no oxygen. No ozone layer existed in the upper atmosphere, so our hypothetical visitors would receive a lethal dose of ultraviolet radiation, unless protected, and would be threatened constantly by comet and meteorite impacts. The view of the Moon would have been spectacular because it was much closer to Earth. However, its gravitational attraction would have caused massive Earth tides. And finally, our visitors would experience a much shorter day because Earth rotated on its axis in as little as ten hours.

Eventually, much of Earth's primordial heat was dissipated into space, and its surface cooled. As it cooled, water vapor began to condense, rain fell, and surface water began to accumulate. The bombardment by comets and meteorites slowed, and by 3.8 billion years ago, a few small areas of continental crust existed. The atmosphere still lacked oxygen and an ozone layer, but by as much as 3.5 billion years ago, life appeared. Some inconclusive evidence indicates that the most primitive lifeforms existed even earlier.

FIGURE 8.1   Earth as it is thought to have appeared about 4.6 billion years ago.

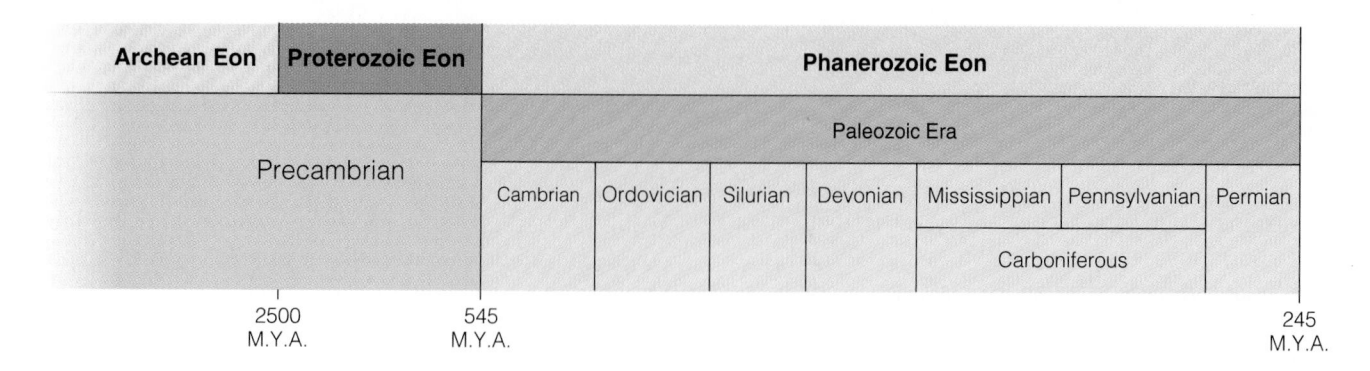

Archean Eon	Proterozoic Eon	Phanerozoic Eon							
Precambrian		Paleozoic Era							
		Cambrian	Ordovician	Silurian	Devonian	Mississippian	Pennsylvanian	Permian
						Carboniferous		

2500 M.Y.A. 　　545 M.Y.A. 　　　　　　　　　　　　　　　　　　245 M.Y.A.

Introduction

All crustal rocks lying below strata of the Cambrian System are referred to as **Precambrian.** The term is an informal one used to allude to both rocks and time. In reference to time, Precambrian includes all geologic time from Earth's origin 4.6 billion years ago to the beginning of the Phanerozoic Eon 545 million years ago. If all geologic time were represented by a 24-hour day, slightly more than 21 hours of it would be Precambrian (Figure 8.2). Accordingly, the Precambrian encompasses most of Earth history. Unfortunately for geologists, not all of this vast interval is recorded; only one area of rocks older than 3.8 billion years is known. Thus, practically no geologic record exists for the first 800 million years of Earth history.

Establishing formal, widely recognized subdivisions of the Precambrian is a difficult task. Precambrian rocks are exposed on all continents, but many have been complexly deformed and altered by metamorphism, and much of Precambrian Earth history is recorded by non-stratified rocks. Thus, the principle of superposition cannot be applied, particularly in older Precambrian rocks, making relative age determinations difficult. In addition, most correlations must be based on radiometric dates because Precambrian rocks contain few fossils useful for correlation. In short, the principle of fossil succession cannot be applied.

In 1982, in an effort to standardize terminology, the North American Commission on Stratigraphic Nomenclature proposed that two Precambrian eons be recognized, the *Archean* and *Proterozoic.* The **Archean Eon** began 4.0 billion years ago, the age of Earth's oldest known rocks, and ended 2.5 billion years ago, which marks the beginning of the **Proterozoic Eon** (see Table 8.1 on page 192). The interval from Earth's origin 4.6 billion years ago until the beginning of the Archean is informally referred to as the *Hadean.* This usage has gained wide acceptance and is followed in this book.

The Precambrian subdivisions in Table 8.1 are geochronologic, based on radiometric dates rather than time-stratigraphic ages. This departs from normal practice in which geologic systems based on stratotypes (see Chapter 3) are the basic time-stratigraphic units. An example will help clarify this point. The Cambrian Period is a geochronologic term, but it corresponds to the Cambrian System, a time-stratigraphic unit based on a body of rock with a stratotype in Wales. By contrast, Precambrian terminology is strictly geochronologic. There are no stratotypes for the subdivisions of the Precambrian.

An alternative scheme for designations of Precambrian time was adopted in 1971 by the U. S. Geological Survey (USGS) (Table 8.1). Instead of names, this scheme uses letter designations, which are also based on ages rather than stratotypes. Although this usage is not followed in this book, students will encounter it on recent USGS maps and in some books and articles.

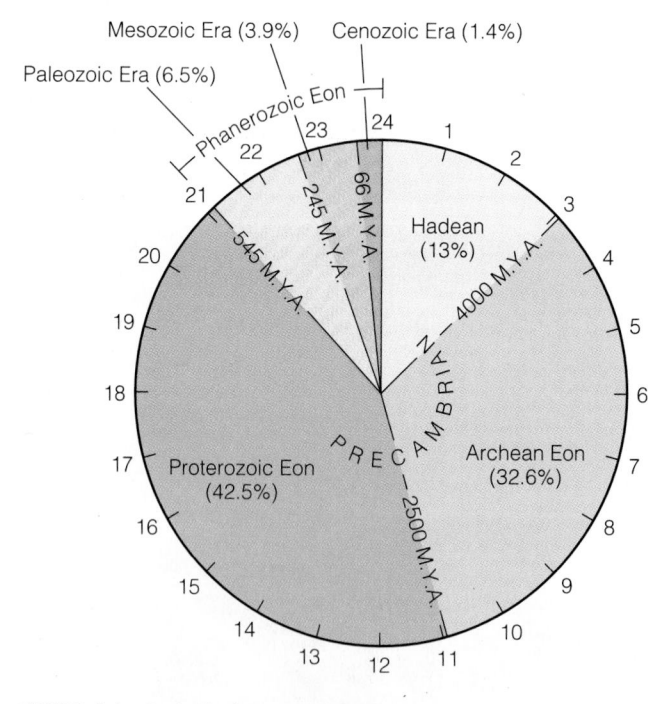

FIGURE 8.2 Geologic time represented on a 24-hour clock. Precambrian time includes more than 21 hours on this clock, or more than 88% of all geologic time.

Hadean Crustal Evolution

Geologists know that some continental crust existed at least 3.8 billion years ago because rocks this old are known

Phanerozoic Eon										
Mesozoic Era			Cenozoic Era							
Triassic	Jurassic	Cretaceous	Tertiary						Quaternary	
			Paleocene	Eocene	Oligocene	Miocene	Pliocene		Pleistocene	Holocene

245
M.Y.A.

66
M.Y.A.

from several areas, including Minnesota, Greenland, and South Africa. Moreover, many of these are metamorphic, which means they formed from even older rocks. In 1989 rocks were discovered in the Slave Province of Canada that date from 3.96 billion years ago. This rock unit, the Acasta Gneiss, is currently the oldest known crust, but Archean sedimentary rocks in Australia contain detrital zircons ($ZrSiO_4$) dated at 4.2 billion years, so source rocks at least that old must have been present.

Geologists generally agree that some kind of Hadean crust existed, but because no rocks this old are known, the origin and composition of the earliest crust must be inferred from geochemical considerations. Many investigators think that a crust formed during an early episode of partial melting, magma formation, and rising magmas. Whether this crust was worldwide or more restricted is debatable, as is the cause of the partial melting.

One crustal evolution model relies on internally generated radiogenic heat and decreasing heat production within Earth through time. An important aspect of this model is the evolution of sialic continental crust. Recall from Chapter 1 that sialic crust contains considerable silicon, oxygen, and aluminum. The earliest crust, however, was probably thin and unstable and was composed of ultramafic igneous rock. Such rock has a relatively low silica (SiO_2) content compared with other igneous rocks. According to the model, this early ultramafic crust was disrupted by upwelling basaltic magmas at ridges and consumed at subduction zones. Thus, due to its high density, which would have made recycling by subduction very likely, the ultramafic crust would have been destroyed. Apparently, only more sialic crust, because of its lower density, is immune to destruction by subduction.

With a decrease in Earth's production of radiogenic heat, a second stage of crustal evolution began. Partial melting of earlier-formed basaltic crust resulted in the formation of andesitic island arcs, and partial melting of lower crustal andesites yielded granitic magmas that were emplaced in the earlier-formed crust. Plate motions accompanied by subduction and collisions of island arcs formed several sialic continental nuclei by the Early Archean (Figure 8.3).

Shields and Cratons

Each continent is characterized by a **Precambrian shield** consisting of a vast area of exposed ancient rocks. Continuing outward from the shields are broad platforms of buried Precambrian rocks that underlie much of the continents. The shields and buried platforms are collectively called **cratons** (Figure 8.4). We can think of the cratons, which are composed of both Archean and Proterozoic rocks, as the ancient nuclei of the continents. The cratons are the relatively stable and immobile parts of the continents and form the foundations upon which Phanerozoic sediments were deposited. These cratons, including their exposed Precambrian shields, have been extremely stable since the beginning of the Phanerozoic Eon. Their stability during that time contrasts sharply with their Precambrian history of orogenic activity.

In North America the **Canadian Shield** includes most of northeastern Canada; a large part of Greenland; parts of the Lake Superior region in Minnesota, Wisconsin, and Michigan; and the Adirondack Mountains of New York (Figure 8.4). In general, the Canadian Shield is a vast area of subdued topography, numerous lakes, and exposed Precambrian rocks, thinly covered in places by Pleistocene glacial deposits. Both Archean and Proterozoic rocks are present, including intrusives, lava flows, various sedimentary rocks, and metamorphic equivalents of all of these (Figure 8.5).

Beyond the Canadian Shield, exposures of Precambrian rocks are limited to areas of uplift and deep erosion, as in the Appalachians, the southwestern United States, the Rocky Mountains, and the Black Hills of South Dakota (Figure 8.5). Geophysical evidence and deep drilling demonstrate that Precambrian rocks underlie most of North America.

The geologic history of the Canadian Shield is complex and not fully understood. Nevertheless, we can recognize several smaller cratons within the shield, each of which is delineated on the basis of radiometric ages and structural trends. These cratons and other buried cratons beyond the shield are the subunits that constitute the North American craton. Each of these subunits may have been independent minicontinents that were later assembled into the larger cratonic unit. The amalgamation of these small cratons occurred along deformation belts dur-

TABLE 8.1

Two Classification Schemes for Precambrian Rocks and Time in the United States

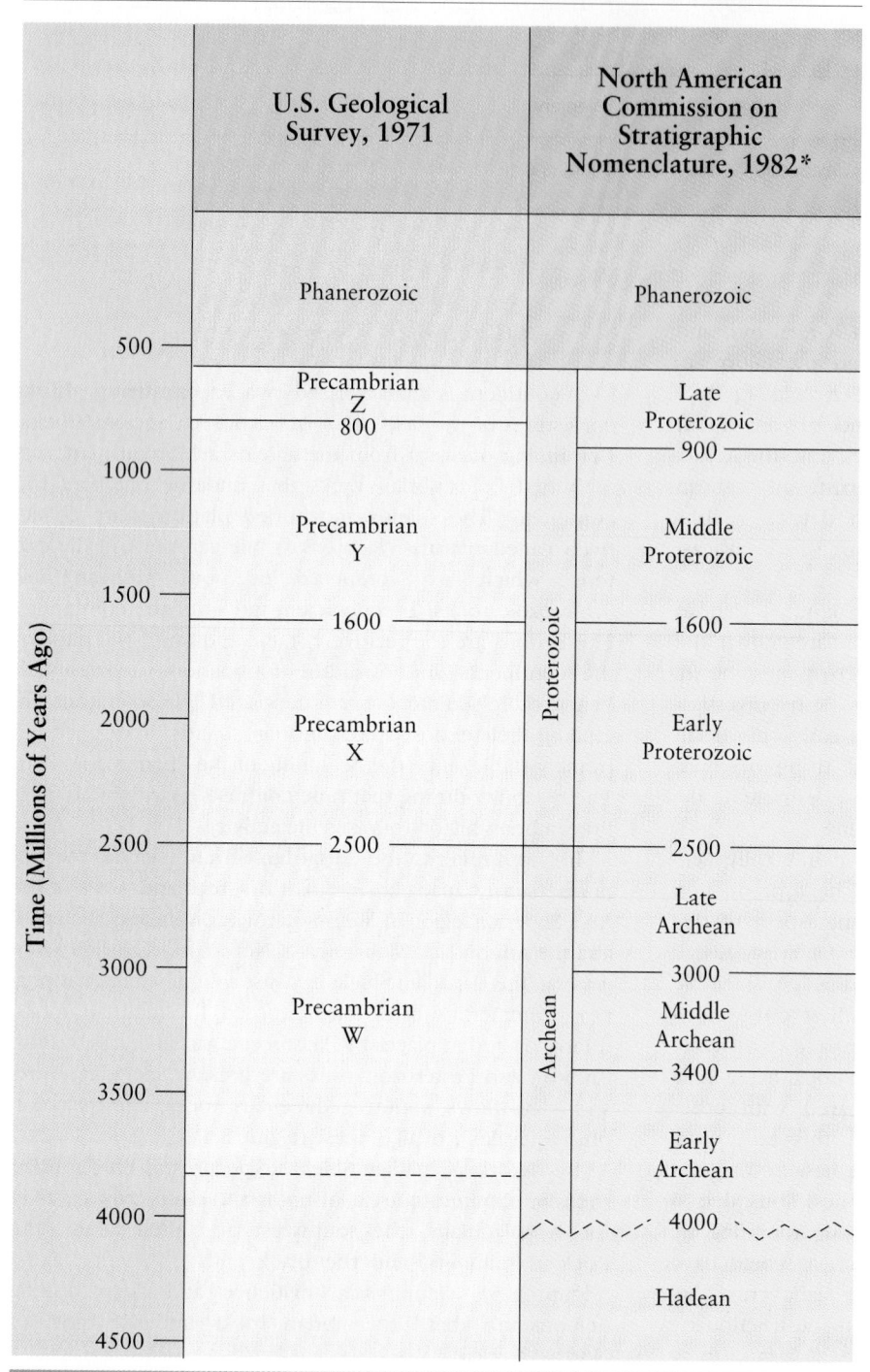

Time (Millions of Years Ago)	U.S. Geological Survey, 1971	North American Commission on Stratigraphic Nomenclature, 1982*
	Phanerozoic	Phanerozoic
500		
	Precambrian Z	Late Proterozoic
	—— 800 ——	—— 900 ——
1000		
	Precambrian Y	Middle Proterozoic
1500		
	—— 1600 ——	—— 1600 ——
2000	Precambrian X	Early Proterozoic
2500	—— 2500 ——	—— 2500 ——
		Late Archean
3000		—— 3000 ——
	Precambrian W	Middle Archean
		—— 3400 ——
3500		Early Archean
4000	- - - - - - - - -	~~~ 4000 ~~~
		Hadean
4500		

This scheme, which was proposed by Harrison and Peterman, is followed in this book..

ing the Early Proterozoic and will be considered more fully in Chapter 9.

Archean Rocks

Areas underlain by Archean rocks are characterized by two main types of rock bodies: **greenstone belts** and **granite–gneiss complexes.** By far the most abundant rocks are granites and gneisses. For example, the Rhodesian Province of southern Africa consists of about 83% gneiss and various granitic rocks; the remaining 17% is mostly greenstone belts.

GREENSTONE BELTS
The oldest large, well-preserved greenstone belts are those of South Africa, which date from 3.6 billion years ago.

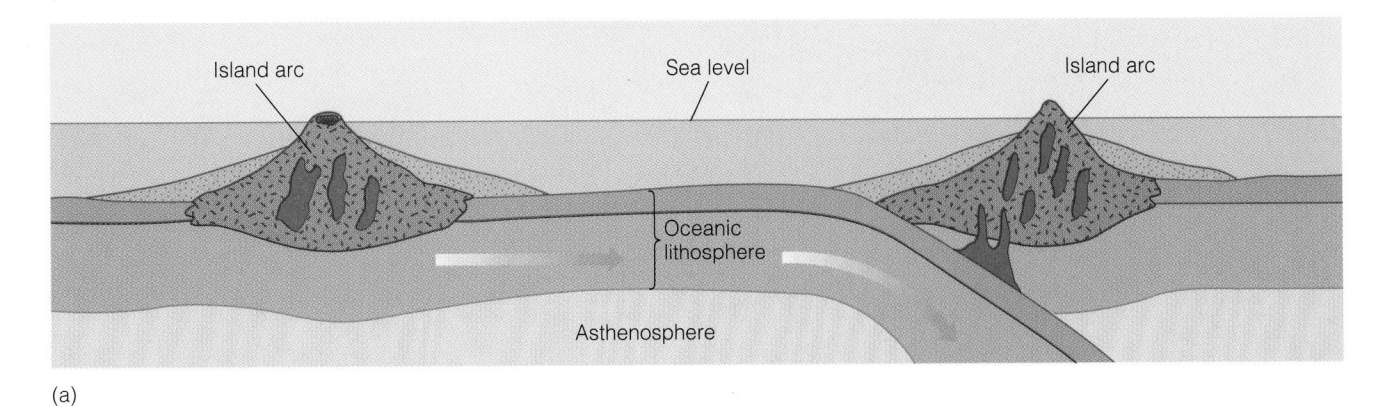

(a)

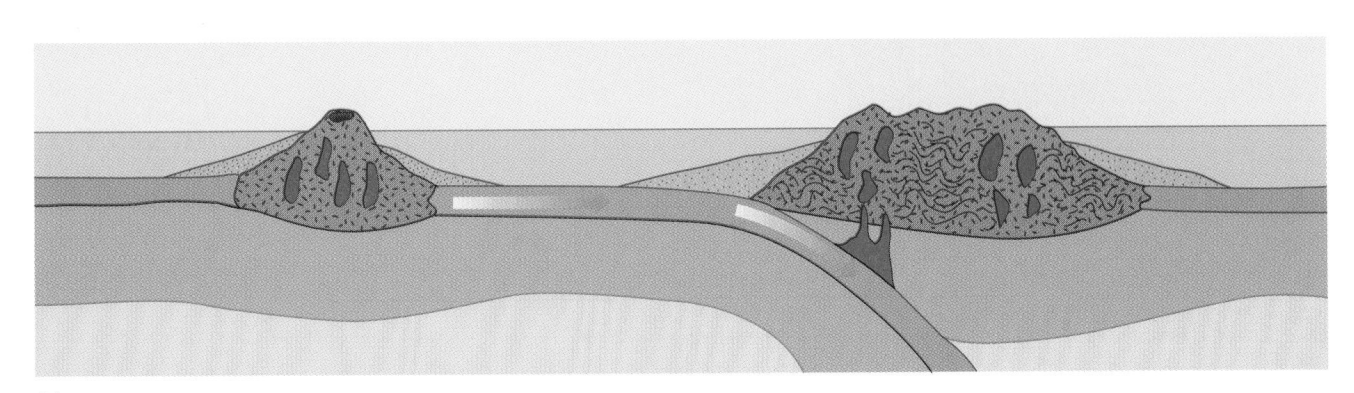

(b)

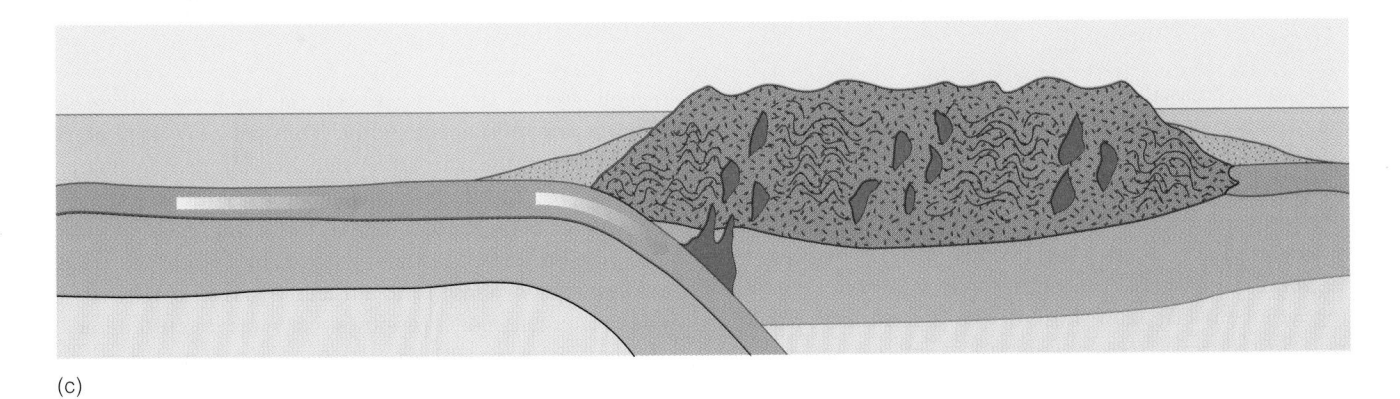

(c)

FIGURE 8.3 Origin of sialic continental crust—that is, crust with a granitic composition. Andesitic island arcs formed by partial melting of basaltic oceanic crust are intruded by granitic magmas. As a result of plate movements, island arcs collide and form larger units or cratons. **(a)** Two island arcs on separate plates move toward one another. **(b)** The island arcs shown in (a) collide. **(c)** The island arc shown in (b) collides with the craton.

In North America, greenstone belts are most common in the Superior and Slave cratons of the Canadian Shield (Figure 8.6), and most formed between 2.7 and 2.5 billion years ago.

An idealized greenstone belt consists of three major rock units; the lower and middle units are dominated by volcanic rocks, and the upper unit is sedimentary (Figure 8.7a). Most greenstone belts have a synclinal structure and are intruded by granitic magmas (Figure 8.7b), and many are complexly folded and cut by thrust faults. The volcanic rocks of greenstone belts are typically greenish due to the abundance of the mineral chlorite, which formed during low-grade metamorphism.

As the common occurrence of pillow basalts indicates,

much of the volcanism responsible for the igneous rocks of greenstone belts was subaqueous (Figure 8.7c). Shallow water and subaerial eruptions are indicated by pyroclastics, and in some areas large volcanic centers built up above sea level. Perhaps the most interesting igneous rocks in greenstone belts are ultramafic lava flows; such flows are rare in rocks younger than Archean. In order to erupt, an ultramafic magma requires near-surface magma temperatures over 1600°C; the highest recorded surface magma temperature for recent Hawaiian basalt lava flows is 1350°C. During Earth's early history, however, there was considerably more radiogenic heat; thus the mantle was hotter, perhaps 300°C hotter; and ultramafic magmas could be erupted onto the surface. Since the amount of radiogenic

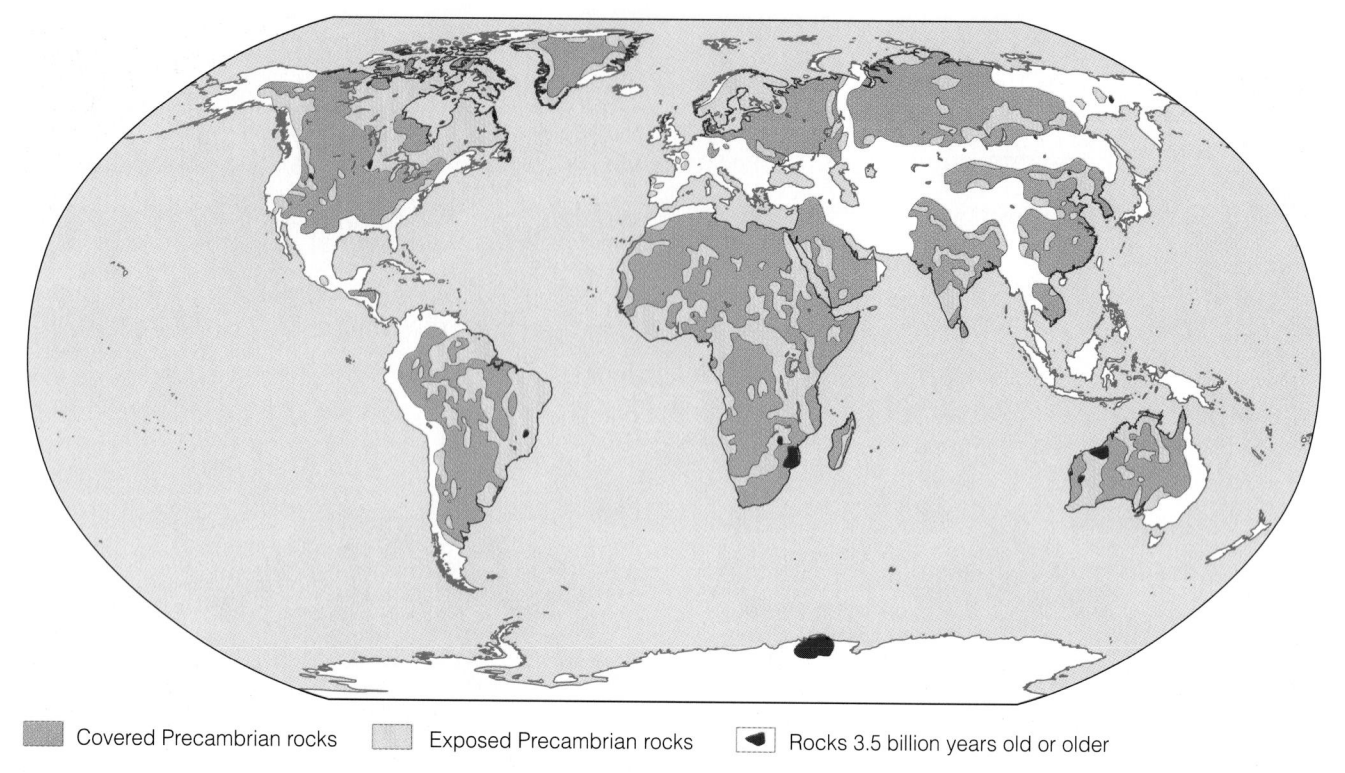

Covered Precambrian rocks   Exposed Precambrian rocks   Rocks 3.5 billion years old or older

FIGURE 8.4 Precambrian cratons of the world. The areas of exposed Precambrian rocks are the shields, while buried Precambrian rocks are the platforms. Shields and platforms collectively make up the cratons.

heat has decreased through time, Earth has cooled; and ultramafic lava flows have ceased to form.

Sedimentary rocks are a minor component in the lower parts of greenstone belts but become increasingly abundant toward the top (Figure 8.7a). The most common sedimentary rocks are successions of *graywacke* and *argillite*. Graywacke is a variety of sandstone containing abundant clay, and those in the greenstone belts are rich in volcanic rock fragments. Argillites are simply slightly metamorphosed mudrocks such as shale. Small-scale graded bedding and cross-bedding indicate that the graywacke–argillite successions were deposited by turbidity currents (see Figure 4.7). Some of the sedimentary rocks, such as quartz sandstones and shales, in the upper units show clear evidence of shallow-water deposition. These rocks were originally deposited in delta, tidal-flat, barrier-island, and shallow marine-shelf environments.

Other sedimentary rocks in greenstone belts include conglomerate, chert, carbonates, and banded iron formations. Some conglomerates are associated with graywackes and probably represent submarine slumps. Neither chert nor carbonates are very abundant, although there are local exceptions. Banded iron formations are present but are much more common in Proterozoic terranes and will therefore be discussed in the next chapter.

GREENSTONE BELT EVOLUTION

Most models for the development of greenstone belts rely on Archean plate movements. Figure 8.8 shows a model in which greenstone belts develop in **back-arc marginal basins** that subsequently close. Thus, an early stage of extension when the back-arc marginal basin opens, accompanied by volcanism and sedimentation, is followed by an episode of compression. During this compressional stage, the greenstone belt assumes its synclinal form (Figure 8.8) and is metamorphosed and intruded by granitic magmas.

An alternate model proposes that some greenstone belts formed in **intracontinental rifts.** This model assumes a preexisting sialic (silica- and aluminum-rich) crust and requires an ascending mantle plume (Figure 8.9). As the plume rises and spreads, it generates tensional forces that cause intracontinental rifting. The plume also serves as the source of the lower and middle volcanic units, and erosion of the rift flanks accounts for the upper sedimentary unit. And finally, there is an episode of subsidence, deformation, low-grade metamorphism, and plutonism (Figure 8.9).

The rift model has certain appealing aspects. For one, the ultramafic volcanics in some greenstone belts can be more easily accounted for by a mantle plume than by a back-arc marginal basin setting because subduction-related volcanics are more commonly andesitic. Also, the rift model has the advantage of explaining the variable sizes of greenstone belts because the size is related to how much the rift opens.

Both models may be reasonably used to explain greenstone belts. For belts with ultramafic rocks, the rift setting seems to be the best explanation. But those containing abundant andesites more likely formed in back-arc marginal basins.

(a)

(b)

(c)

(d)

FIGURE 8.5 Rocks of the Canadian Shield (a and b). (a) Outcrop of gneiss, Georgian Bay, Ontario, Canada. (b) Basalt (dark) and granite (light) along the banks of the Chippewa River, Ontario, Canada. Beyond the Canadian Shield, Precambrian rocks are exposed in areas of uplift (c) and deep erosion (d). (c) Archean metamorphic rocks exposed in the Rocky Mountains of Colorado. (d) The Archean Brama Schist is exposed in the deeper parts of the Grand Canyon, Arizona.

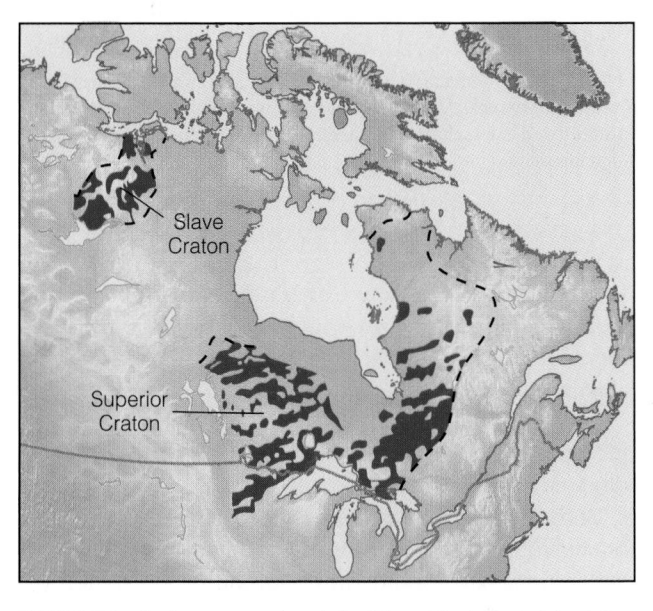

FIGURE 8.6 Archean greenstone belts (shown in dark green) of the Canadian Shield are mostly in the Superior and Slave cratons.

Development of Archean Cratons

We have already mentioned that by the beginning of the Archean several sialic continental nuclei or cratons had formed (Table 8.2). They may have been rather small, though, because rocks older than 3.0 billion years are of limited geographic extent, especially compared with those 3.0 to 2.5 billion years old (this latter interval seems to have been a time of rapid crustal evolution). A plate tectonic model has been proposed for the Archean crustal evolution of the southern Superior craton of Canada (Figure 8.10). This model incorporates sialic plutonism, greenstone belt formation, and collisions of microcontinents. It accounts for the origin of both greenstone belts and granite–gneiss complexes by the sequential accretion of island arcs (Figure 8.10).

The events leading to the origin of the southern Superior craton are part of a more extensive orogenic episode that took place near the end of the Archean. This

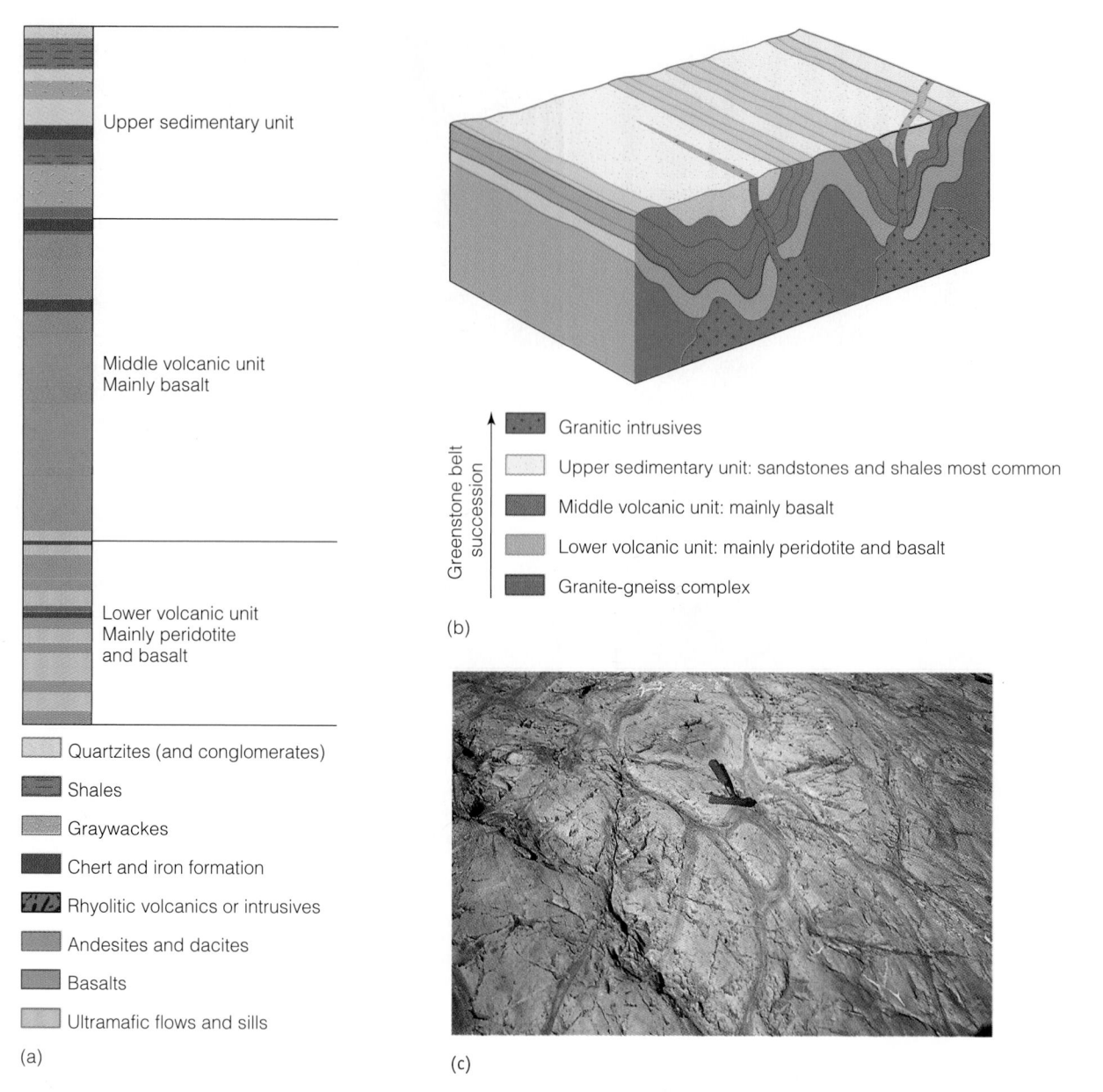

Quartzites (and conglomerates)

Shales

Graywackes

Chert and iron formation

Rhyolitic volcanics or intrusives

Andesites and dacites

Basalts

Ultramafic flows and sills

(a)

Greenstone belt succession

Granitic intrusives

Upper sedimentary unit: sandstones and shales most common

Middle volcanic unit: mainly basalt

Lower volcanic unit: mainly peridotite and basalt

Granite-gneiss complex

(b)

(c)

FIGURE 8.7 (a) Idealized stratigraphic column of an Archean greenstone belt. Older greenstone belts—those more than 2.8 billion years old—have an ultramafic lower unit succeeded upward by a basaltic unit, as shown here. In younger greenstone belts, the succession is a basaltic lower unit overlain by an andesite-rhyolite unit. The upper unit in both consists of sedimentary rocks. (b) Two adjacent greenstone belts showing their synclinal structure and relationship to granite-gneiss complexes. (c) Pillow structures in the Ispheming greenstone belt at Marquette, Michigan. These structures indicate a submarine eruption, probably at or near a spreading ridge.

episode of deformation was responsible for the formation of the Superior and Slave cratons as well as some Archean terranes within other parts of the Canadian Shield. It also affected Archean rocks in Wyoming, Montana, and the Minnesota River Valley.

Deformation during the Late Archean was the last major Archean event in North America. Several sizable cratons had formed, constituting the older parts of the Canadian Shield. These cratons, however, were independent units or microcontinents that were later assembled during the Early Proterozoic to form a larger craton. Large parts of the cratons of the other continents had also formed by the end of the Archean.

Archean Plate Tectonics

Undoubtedly, the present tectonic regime of opening and closing oceans has been a primary agent in Earth evolution for at least the last 2 billion years. In fact, this regime probably became established in the Early Proterozoic. Many geologists are becoming convinced that some sort of plate tectonics was operating in the Archean as well, but they disagree about the details.

Some Archean rocks appear to record plate movements as shown by deformation belts between presumed colliding cratons and island arcs. But ophiolite complexes, which mark younger convergent plate margins, are rare, al-

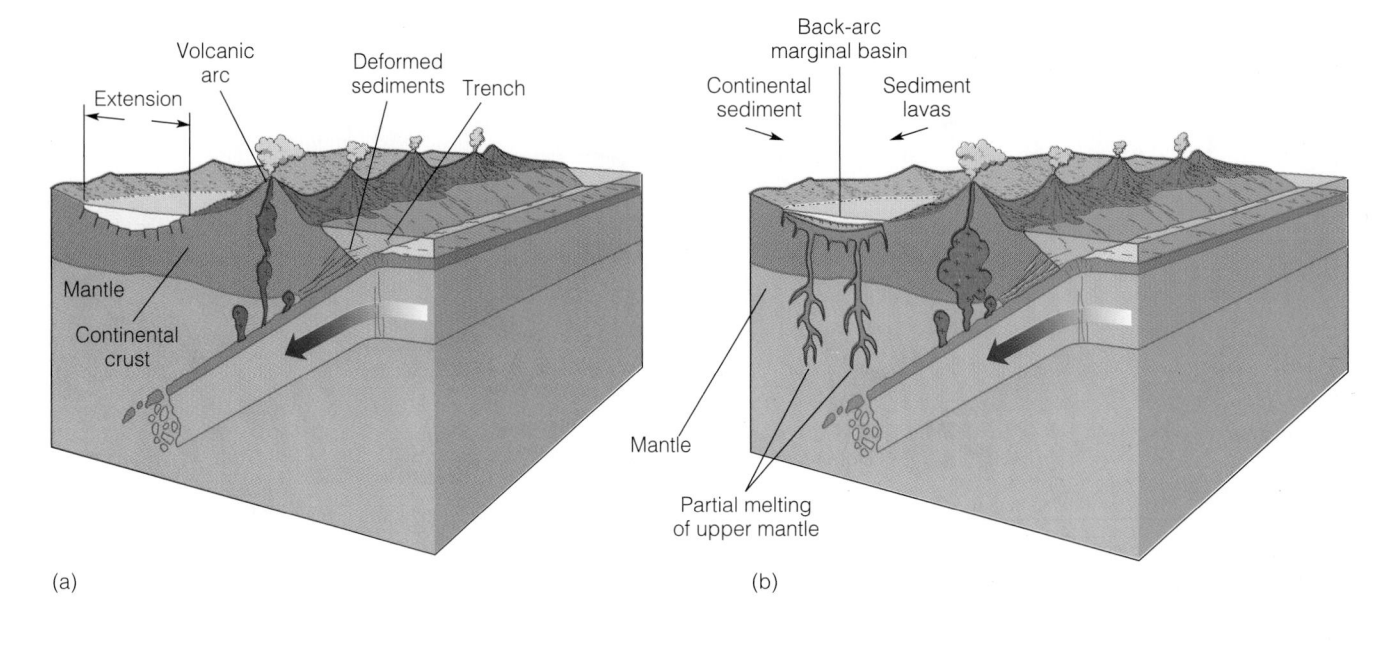

(a)

(b)

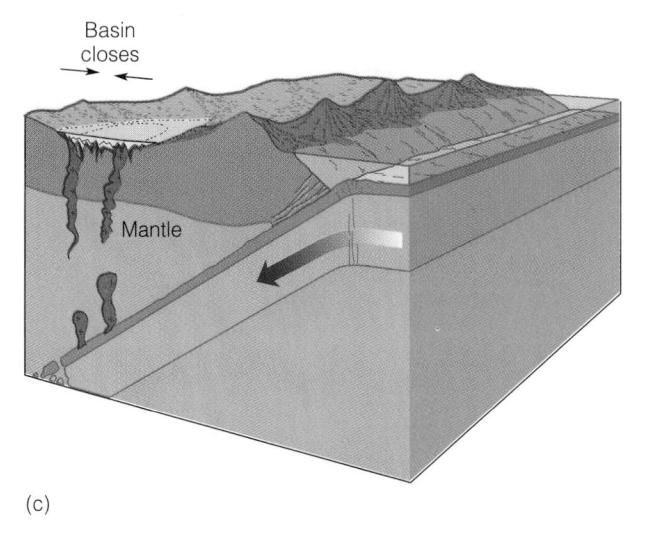

(c)

FIGURE 8.8 Formation of a greenstone belt in a back-arc marginal basin. **(a)** Rifting on the continent side of a volcanic island arc forms a back-arc marginal basin. Partial melting of subducted oceanic crust supplies andesitic and dioritic magmas to the island arc. **(b)** Basaltic lavas and sediments derived from the continent and island arc fill the back-arc marginal basin. **(c)** Closure of the back-arc marginal basin causes compression and deformation. The greenstone belt is deformed into a synclinal structure and is intruded by granitic magmas.

though Late Archean ophiolites have recently been reported from several areas.

Apparently, Earth's radiogenic heat production has diminished through time (Figure 8.11). Thus, during the Archean, when more heat was available, seafloor spreading and plate motions probably occurred faster, and magma was generated more rapidly. Nevertheless, Archean plates seem to have behaved differently from those in the Proterozoic. For example, sedimentary sequences typical of passive continental margins are uncommon in the Archean but quite common in the Proterozoic. Their near absence in the Archean indicates that continents with adjacent shelves and slopes were either not present or only poorly developed.

Another factor that favors some kind of Archean plate tectonics is the episode of rapid crustal growth 3.0 to 2.5 billion years ago. Like continents today, Archean continents probably grew by accretion at convergent plate margins. They probably grew more rapidly since plate motions were faster, thus accounting for an accelerated rate of accretion.

Today most geologists would probably agree that plate tectonics was operative in the Archean. Its details differed, however, from the present style of plate tectonics, which began during the Proterozoic when large, stable cratons were present.

The Atmosphere and Oceans

During its earliest history, Earth would have been particularly inhospitable for any of today's inhabitants. If we

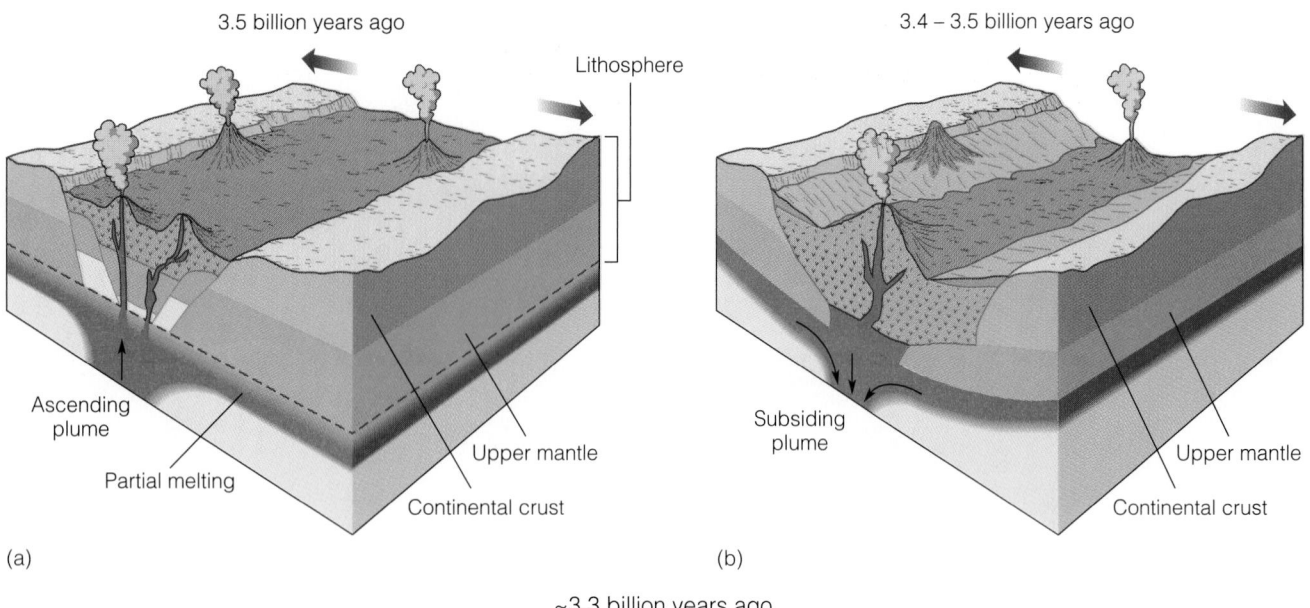

Lithosphere

Ascending
plume

Partial melting

Upper mantle

Continental crust

(a)

3.4 – 3.5 billion years ago

Subsiding
plume

Upper mantle

Continental crust

(b)

~3.3 billion years ago

Granitic
intrusion

Subsiding
plume

Upper mantle

Continental crust

(c)

FIGURE 8.9 **Model for the formation of a greenstone belt in an intracontinental rift. (a) An ascending mantle plume causes rifting and volcanism. (b) As the plume subsides, erosion of the rift flanks accounts for deposition of sediments. (c) Closure of the rift causes compression and deformation. Granitic magma intrudes the greenstone belt.**

could somehow go back and visit it, we would witness a barren, waterless surface, numerous meteorite impacts, be subjected to intense ultraviolet radiation, and be unable to breathe the atmosphere. Now the atmosphere is rich in nitrogen and oxygen and contains important trace amounts of carbon dioxide, water vapor, and other gases (Table 8.3). Ozone (O_3), although present in a minute quantity, is abundant enough in the upper atmosphere to block most of the Sun's ultraviolet radiation.

Earth's earliest atmosphere was probably composed of hydrogen and helium, because these gases are the most abundant elements in the universe. If so, Earth's gravity is insufficient to retain gases with such low molecular weight, and they would have escaped into space. In addition, before Earth had a differentiated core, it lacked a magnetic field and *magnetosphere*, the area around Earth within which the magnetic field is confined. The absence of a magnetosphere ensured that a strong solar wind, an outflow of ions from the Sun, would sweep away any gases that might have otherwise formed an atmosphere. Once the magnetosphere was established, gases derived from Earth's interior during volcanism began to accumulate by a process known as **outgassing** (Figure 8.12).

Gases emitted by present-day volcanoes are mostly water vapor with lesser amounts of carbon dioxide, sulfur dioxide, carbon monoxide, sulfur, chlorine, nitrogen, and hydrogen. Archean volcanoes probably emitted these same gases. The gases formed an early atmosphere but one no-

TABLE 8.2

Chronologic Summary of Events Important in the Archean Development of Cratons (Ages in Thousands of Millions of Years)

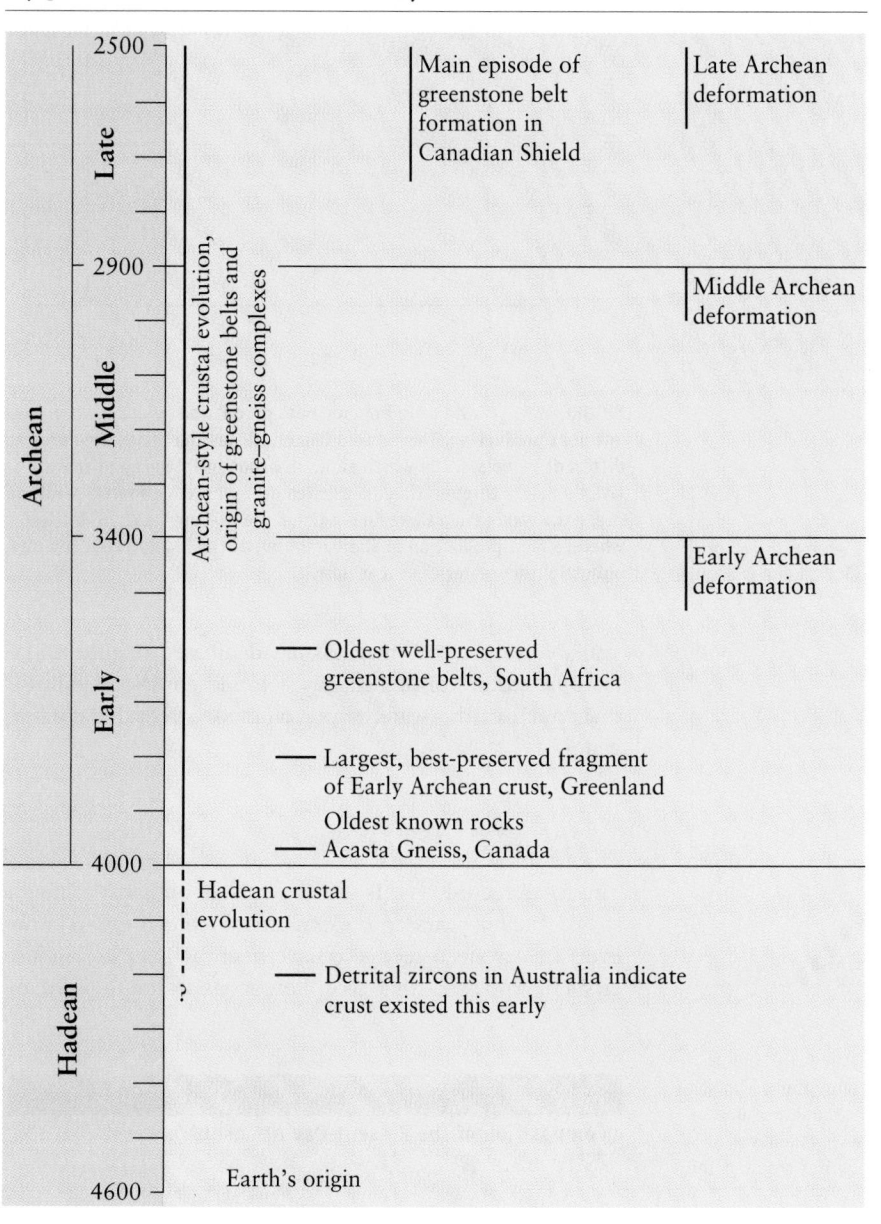

tably deficient in free oxygen; and without free oxygen, there could have been no ozone layer. Furthermore, the early atmosphere may have contained ammonia (NH_3) and methane (CH_4), both of which could have resulted from volcanic gases reacting chemically in the atmosphere.

An oxygen-deficient atmosphere appears to have persisted throughout the Archean as indicated by detrital deposits containing pyrite (FeS_2) and uraninite (UO_2). Both of these minerals are quickly oxidized in the presence of free oxygen. Iron in the oxidized state became quite common in the Proterozoic, indicating that at least some free oxygen was present by that time (see Chapter 9).

Two processes, both of which began in the Archean, can account for the introduction of free oxygen into the atmosphere. The first was **photochemical dissociation** of water vapor, in which water molecules were broken up by ultraviolet radiation in the upper atmosphere (Figure 8.13). This process eventually may have supplied up to 2% of present-day oxygen levels. At 2% free oxygen, ozone (O_3) will form, creating a barrier against incoming ultraviolet radiation and thus limiting the formation of more free oxygen by photochemical dissociation. Even more important were the activities of photosynthesizing organisms. During **photosynthesis** carbon dioxide and water combine into organic molecules, and oxygen is released as a waste product (Figure 8.13). Even so, the atmosphere at the end of

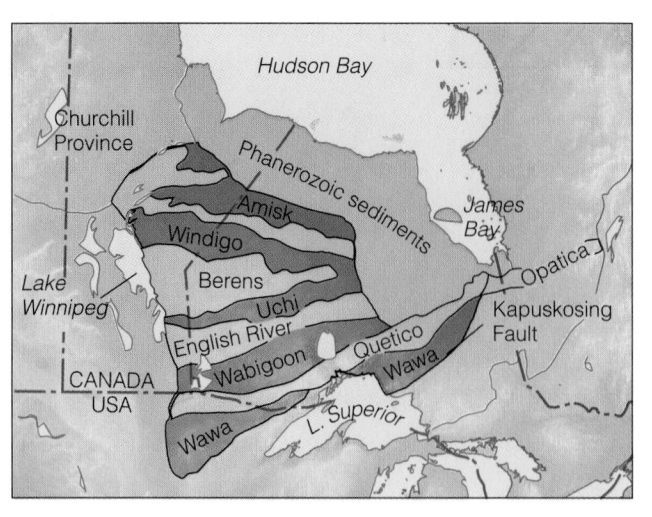

(a)

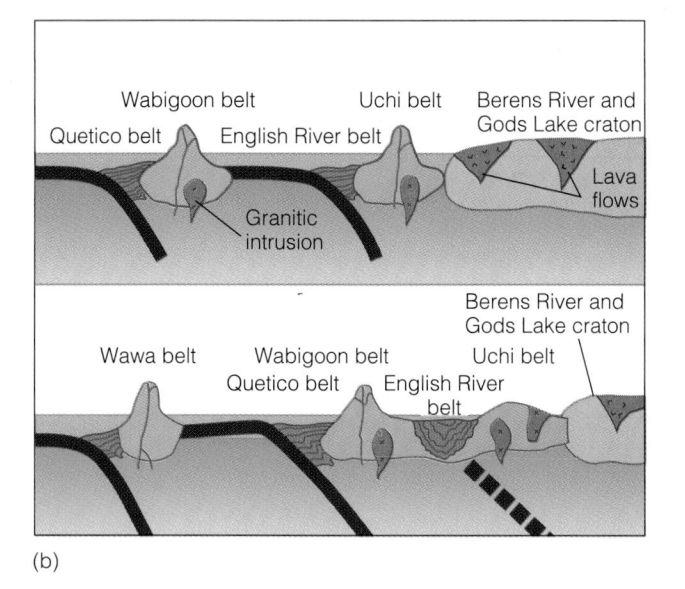

(b)

FIGURE 8.10 Origin of the southern Superior craton. (a) Geologic map showing greenstone belts (dark green areas) and granite–gneiss subprovinces (light green areas). (b) Plate tectonic model for development of the southern Superior craton. The figure represents a north–south section, and the upper diagram is an earlier stage of the lower diagram.

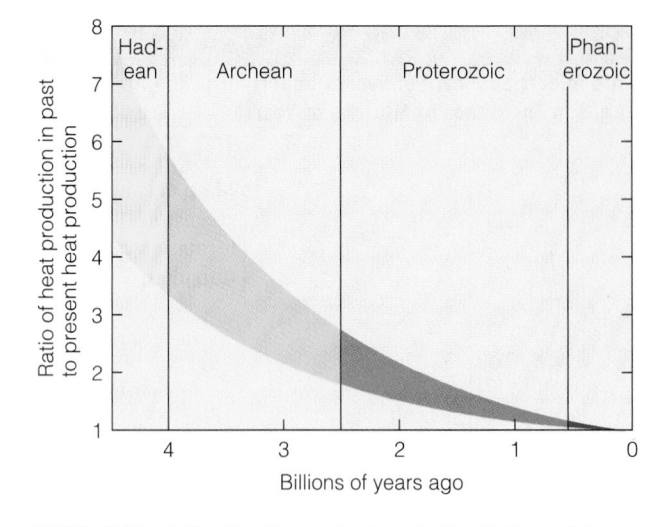

FIGURE 8.11 Ratio of radiogenic heat production in the past to present heat production. The colored band encloses the ratios according to different models, all of which show an exponential decay of radioactive elements through time. Estimates of heat production for 4 billion years ago range from three to six times the present heat production, whereas heat production at the beginning of the Phanerozoic Eon was only slightly greater than it is now.

available to generate magmas has decreased (Figure 8.11). Isotopic studies of present-day volcanic emissions indicate that much of the water vapor emitted is recycled surface water.

Archean Life

The fossil record reveals that life existed on Earth as much as 3.5 billion years ago. Compared to the present, however, the Archean seems to have been biologically impoverished. Today Earth's biosphere consists of millions of

the Archean may have contained no more than 1% of its present oxygen level.

Recall that the major gas emitted by volcanoes is water vapor, so the early atmosphere was also rich in this compound. Once Earth cooled sufficiently, water vapor condensed and surface waters began to accumulate. Good evidence exists for oceans in the Early Archean, although their volumes and extent are unknown. We can envision an early hot Earth with considerable volcanic activity and a rapid accumulation of surface waters. Is the volume of oceanic waters still increasing? Perhaps it is; but if so, the rate has slowed considerably because the amount of heat

TABLE 8.3		
Composition of the Present-Day Atmosphere		
	SYMBOL	PERCENTAGE BY VOLUME*
Nonvariable Gases		
Nitrogen	N_2	78.08
Oxygen	O_2	20.95
Argon	Ar	0.93
Neon	Ne	0.002
Others		0.001
Variable Gases and Particulates		
Water vapor	H_2O	0.1 to 4.0
Carbon dioxide	CO_2	0.034
Ozone	O_3	0.0006
Other gases		Trace
Particulates		Normally trace
Percentages, except for water vapor, are for dry air.		

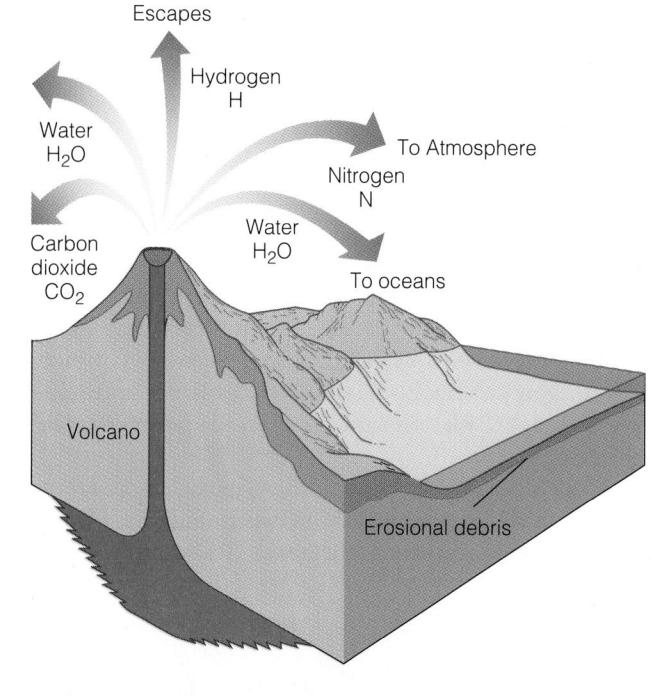

FIGURE 8.12 Outgassing released gases to form an early atmosphere composed of the gases shown. Chemical reactions in the atmosphere also probably yielded methane (CH_4) and ammonia (NH_3).

First, we must be very clear about what life is; that is, what is living and what is nonliving? Minimally, a living organism must reproduce and practice some kind of metabolism. Reproduction ensures the long-term survival of a group of organisms as a species; metabolism ensures short-term survival of an individual organism as a chemical system.

Using this reproduction-metabolism criterion, it would seem a simple matter to decide whether something is alive. Yet, this is not always the case. Viruses, for example, behave like living organisms when in an appropriate host cell, but outside a host cell they neither reproduce nor manufacture organic molecules. They are composed of a bit of genetic material, either deoxyribonucleic acid (DNA) or ribonucleic acid (RNA), enclosed in a protein capsule. So are they living or nonliving? Biologists can be found on both sides of this question, illustrating that defining life is not as clear-cut as it would seem to be.

Microspheres are simple organic molecules that form spontaneously and show greater organizational complexity than inorganic objects such as rocks. In fact, microspheres can even grow and divide, but in ways that are closer to random chemical processes than to organisms. Consequently, they are not considered living.

One may wonder what bearing viruses and microspheres have on the origin of life. First, they demonstrate that no matter how we define life, no absolute criterion exists for deciding whether or not some things are living. Second, if life originated on Earth by natural processes, it must have passed through a prebiotic stage, perhaps similar to microspheres.

species of animals, plants, and other organisms, all of which are thought to have evolved from one or a few primordial types. In Chapter 5 we considered the evolutionary processes whereby the diversification of life occurred, but here we are concerned with how life originated in the first place.

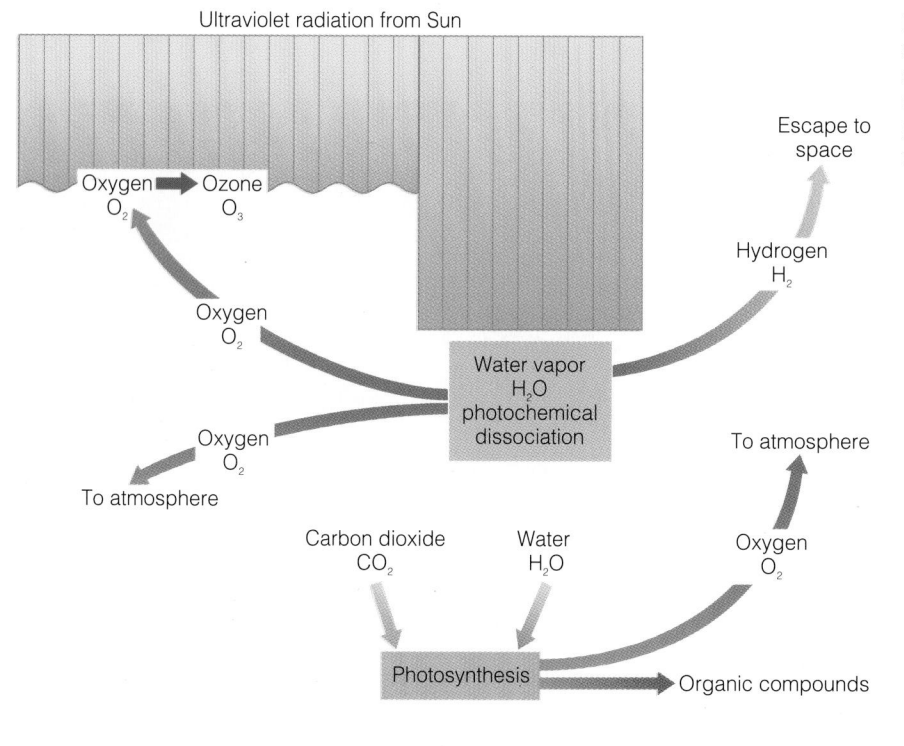

FIGURE 8.13 Photochemical dissociation and photosynthesis added free oxygen to the atmosphere. Once free oxygen was present, an ozone layer formed in the upper atmosphere and blocked most incoming ultraviolet radiation.

As early as 1924, the great Russian biochemist A. I. Oparin postulated that life originated when Earth's atmosphere contained little or no free oxygen. With no oxygen to destroy organic molecules and no ozone layer to block ultraviolet radiation, life could indeed have come into existence from nonliving matter.

THE ORIGIN OF LIFE

All investigators agree that two requirements were necessary for the origin of life: (1) a source of the appropriate elements from which organic molecules could have been synthesized and (2) an energy source to promote chemical reactions that synthesized organic molecules. All organisms are composed mostly of carbon, hydrogen, nitrogen, and oxygen, all of which were present in the early atmosphere in the form of carbon dioxide (CO_2), water vapor (H_2O), and nitrogen (N_2) and possibly methane (CH_4) and ammonia (NH_3). It is postulated that the elements necessary for life (C, H, N, and O) combined to form simple organic molecules called **monomers.** The energy sources that promoted these reactions were probably ultraviolet radiation and lightning. Typical monomers characteristic of organisms are amino acids.

Monomers are the basic building blocks of more complex organic molecules, but is it plausible that they originated in the manner postulated? Experimental evidence suggests so. In the 1950s, Stanley Miller synthesized several amino acids by circulating gases approximating the composition of Earth's early atmosphere in a closed glass vessel (Figure 8.14). This mixture of gases was subjected to an electric spark (simulating lightning), and in a few days the mixture became cloudy. Analysis showed that several amino acids typical of organisms had formed. In more recent experiments, all 20 amino acids common in organisms have been successfully synthesized.

Making monomers in a test tube is one thing, but the molecules of organisms are **polymers** such as proteins and nucleic acids (RNA and DNA), consisting of monomers linked together in a specific sequence. So how did this phenomenon known as polymerization take place? This is more difficult to answer, especially if polymerization occurred in an aqueous solution, which usually results in depolymerization. However, the noted researcher Sidney Fox has synthesized small molecules he calls *proteinoids,* some of which consist of more than 200 linked amino acids (Figure 8.15). He dehydrated concentrated amino acids and found that when heated they spontaneously polymerized to form proteinoids. Furthermore, it is not unreasonable to suppose that similar conditions for polymerization existed on early Earth.

At this stage we can refer to these molecules as *protobionts,* which are intermediate between inorganic chemical compounds and living organisms. These protobionts,

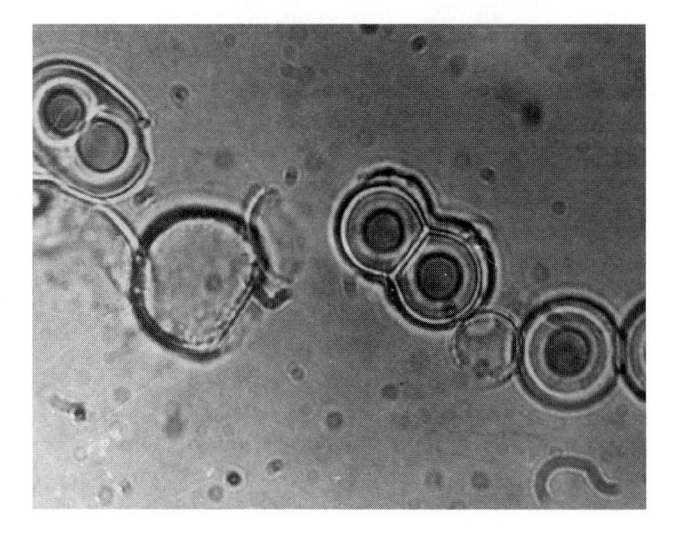

(a)

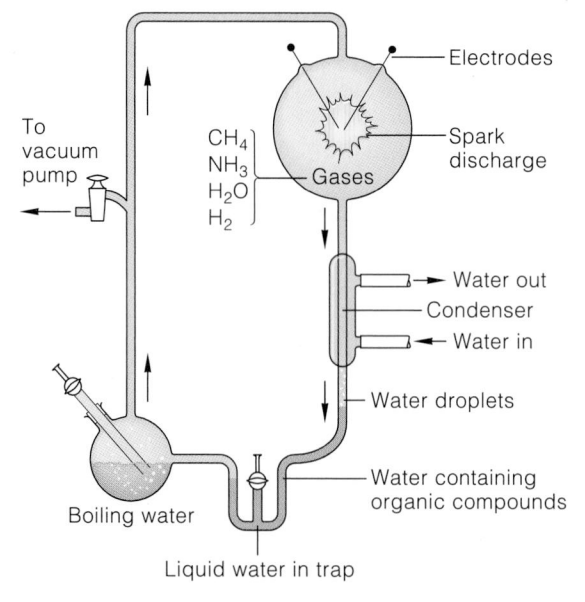

FIGURE 8.14 Experimental apparatus used by Stanley Miller. Several amino acids characteristic of organisms were artificially synthesized during Miller's experiments.

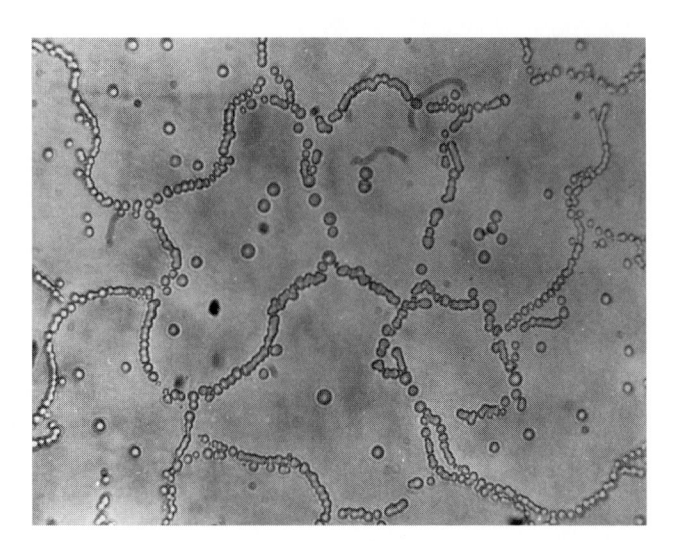

(b)

FIGURE 8.15 (a) Bacterium-like proteinoid. (b) Proteinoid microspheres.

however, would have been diluted and would have ceased to exist if some kind of outer membrane had not developed. In other words, they had to be self-contained as present-day cells are. Fox's experiments demonstrated that proteinoids will spontaneously aggregate into microspheres (Figure 8.15), which are bounded by a cell-like membrane and grow and divide much as bacteria do.

Fox's experimental results are interesting, but how can they be related to what may have taken place in the early history of life? Monomers likely formed continuously and in great abundance, accumulated in the oceans, and formed what the British biochemist J. B. S. Haldane characterized as a hot, dilute soup. According to Fox, the amino acids in this hot dilute soup may have washed up onto a beach or perhaps cinder cones, where they were concentrated by evaporation and polymerized by heat. The polymers were then washed back into the sea, where they reacted further.

Not much is known about the next step in the origin of life—the development of a reproductive mechanism. Fox's microspheres divide and are considered by some experts to represent a protoliving system. However, in present-day organisms nucleic acids, either as RNA or DNA, are necessary for reproduction. The problem is that nucleic acids cannot replicate without enzymes, and enzymes cannot be made without nucleic acids. Or so it seemed until recently.

Experimental evidence has demonstrated that small RNA molecules can replicate themselves without the aid of protein enzymes. In view of this evidence, it seems that the first replicating system may have been an RNA molecule. In fact, some researchers propose an early "RNA world" in which these molecules were intermediate between inorganic chemical compounds and the DNA-based molecules of organisms. Just how RNA molecules were naturally synthesized, however, remains a mystery, because they cannot easily be synthesized under the conditions that probably prevailed on early Earth.

It should be apparent from our discussion that much remains to be learned about the origin of life by natural processes. The steps involved and the significance of some experimental results are still debated. A common theme among many investigators, though, is that the earliest organic molecules were synthesized from atmospheric gases. But even this has been questioned by some who propose that life originated in hydrothermal vent systems on the seafloor (see Perspective 8.1).

THE EARLIEST ORGANISMS

Prior to the mid-1950s, we had very little knowledge of Precambrian life. Investigators had long assumed that the fossils so abundant in Cambrian strata must have had a long earlier history, but no such record was known. Some enigmatic Precambrian fossils had been reported, but these were mostly dismissed as inorganic in origin. In fact, the Precambrian was once referred to as the *Azoic,* meaning "without life."

In the early 1900s, Charles Walcott described layered moundlike structures from the Early Proterozoic Gunflint Iron Formation of Ontario. Walcott proposed that these structures, now called **stromatolites,** represented reefs constructed by algae (Figure 8.16a), but paleontologists did not demonstrate that stromatolites are the products of organic activity until 1954. Studies of present-day stromatolites show that they originate by the entrapment of sediment grains on sticky mats of photosynthesizing cyanobacteria or blue-green algae (Figure 8.16a). Morphologically distinct types of stromatolites may form, all of which are known from Precambrian rocks, some of which contain microfossils (Figure 8.16b).

Currently, the oldest known undisputed stromatolites are from 3.0-billion-year-old rocks in South Africa. Even older probable stromatolites have recently been discovered in the 3.3- to 3.5-billion-year-old Warrawoona Group near North Pole, Australia (Figure 8.17). Indirect evidence for even more ancient life comes from Archean rocks of western Greenland. These 3.8-billion-year-old rocks contain small carbon spheres that may be of biological origin, but the evidence is not conclusive.

The oldest known fossils are of photosynthesizing organisms, but photosynthesis is a complex metabolic process, and it seems reasonable that it was preceded by an even simpler process. In other words, nonphotosynthesizing organisms must have been present before cyanobacteria appeared.

No fossils are known of these earliest organisms, but they must have resembled tiny bacteria. And considering that the early atmosphere lacked free oxygen, they must have been **anaerobic,** meaning that they required no free oxygen. Also, very likely they were completely dependent on an external source of nutrients. That is, they must have been **heterotrophic** as opposed to **autotrophic** organisms, which make their own nutrients as in photosynthesis. Finally, they were **prokaryotic cells,** meaning they lacked a cell nucleus and other internal cell structures typical of more advanced *eukaryotic cells.* (We will discuss prokaryotic cells and the origin of eukaryotic cells more fully in Chapter 9.)

We can characterize these earliest organisms as anaerobic, heterotrophic prokaryotes. Their nutrient source was probably adenosine triphosphate (ATP), which was used to drive the energy-requiring reactions in cells. ATP can be synthesized from simple gases and phosphate, so it was no doubt available in the early Earth environment. The earliest lifeforms may have simply acquired their ATP from their surroundings. This situation could not have persisted for long, though, because as more and more cells competed for the same resources, the supply must have diminished. The first organisms to develop a more sophisticated metabolism probably used *fermentation* to meet their energy needs. Fermentation is an anaerobic process in which molecules such as sugars are split, releasing carbon dioxide, alcohol, and energy. In fact, most living prokaryotes ferment.

Other than the origin of life itself, the most significant biological event of the Archean was the development of

perspective 8.1

Submarine Hydrothermal Vents and the Origin of Life

*S*eawater seeps down into the oceanic crust through cracks and fissures at spreading ridges and becomes heated by magma. It then rises and discharges onto the seafloor as hydrothermal (hot water) vents (Figure 1). For many years geologists reasoned that a logical implication of plate tectonic theory was the presence of these vents, but none were observed until 1977. Hydrothermal vents were first observed from the submersible *Alvin* on the Galapagos Rift in the eastern Pacific Ocean. Since then, similar vents have been discovered in several other areas, including the Atlantic Ocean and the Sea of Japan.

Submarine hydrothermal vents are interesting to many scientists for several reasons. First, the rising hot water contains large quantities of dissolved minerals. Indeed, in many vents the water is black because it is saturated with dissolved minerals; hence, they are known as *black smokers* (Figure 1a). When the hot, mineral-rich water cools, iron, copper, and zinc sulfides and other minerals precipitate, thereby forming a chimneylike structure on the seafloor. Economic geologists are interested in this phenomenon because it may help them understand the origin of some ore deposits. Furthermore, the mineral deposits on the seafloor might someday become important resources.

FIGURE 1 **(a) A hydrothermal vent known as a black smoker. The plume of "smoke" is simply heated water saturated with dissolved minerals. (b) Bacteria that oxidize sulfur compounds form the base of a food chain for several types of organisms, including these tubeworms.**

(b)

(a)

Here we are mainly concerned with the biological aspects of these vents. Many organisms unknown to science have been discovered living near vents, and these organisms constitute the only known major communities that are not dependent directly on sunlight as an energy source. Bacteria that practice a metabolic process called *chemosynthesis* oxidize sulfur compounds from the hot vent waters and lie at the base of the food chain (Figure 1b). Thus, the ultimate source of energy for vent communities is radiogenic heat. Deposits similar to those forming adjacent to present-day black smokers in ancient rocks indicate that vents have been present since the Early Archean. If early Earth did indeed have more radiogenic heat than at present, it is reasonable to propose that these vents were much more common during the Archean.

Recently, investigators at the Oregon State University School of Oceanography have proposed that submarine hydrothermal vents were the sites at which life originated. The researchers point out that the more traditional hypotheses for the origin of life rely upon the synthesis of monomers such as amino acids from atmospheric gases, the formation of polymers, and the origin of some kind of self-replicating system. Recall from our earlier discussion that both monomers and polymers have been synthesized experimentally. The Oregon State investigators claim that

> The physical and chemical conditions of those experiments are found in a natural environment which has been present in the oceans since they first formed, in hot springs associated with submarine volcanism.*

_____
*J. B. Corliss, et al., *Oceanologica Acta* 4 (1981): 61.

The elements and energy source necessary for the synthesis of organic molecules were present in the hydrothermal vents. In fact, amino acids have been detected in hydrothermal vent solutions. Polymerization may have occurred on the surfaces of clay minerals, a process that has been proposed by other investigators. And finally, protocells would have been discharged with vent fluids and deposited on the seafloor. This scenario is questioned by some who point out that polymerization would be inhibited in this environment.

One of the most intriguing aspects of this hypothesis is that the process may be continuing at present! This does not mean that life is continually originating, because any polymers or protocells would be quickly devoured by existing vent community organisms. Nevertheless, this hypothesis can be tested, at least in part, in a natural environment.

(a)

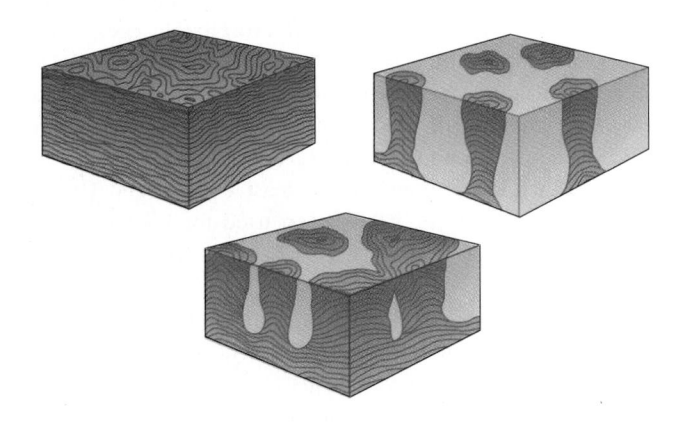

(b)

FIGURE 8.16   (a) Present-day stromatolites, Shark Bay, Australia. (b) Morphological types of stromatolites include irregular mats, columns, and columns linked by mats.

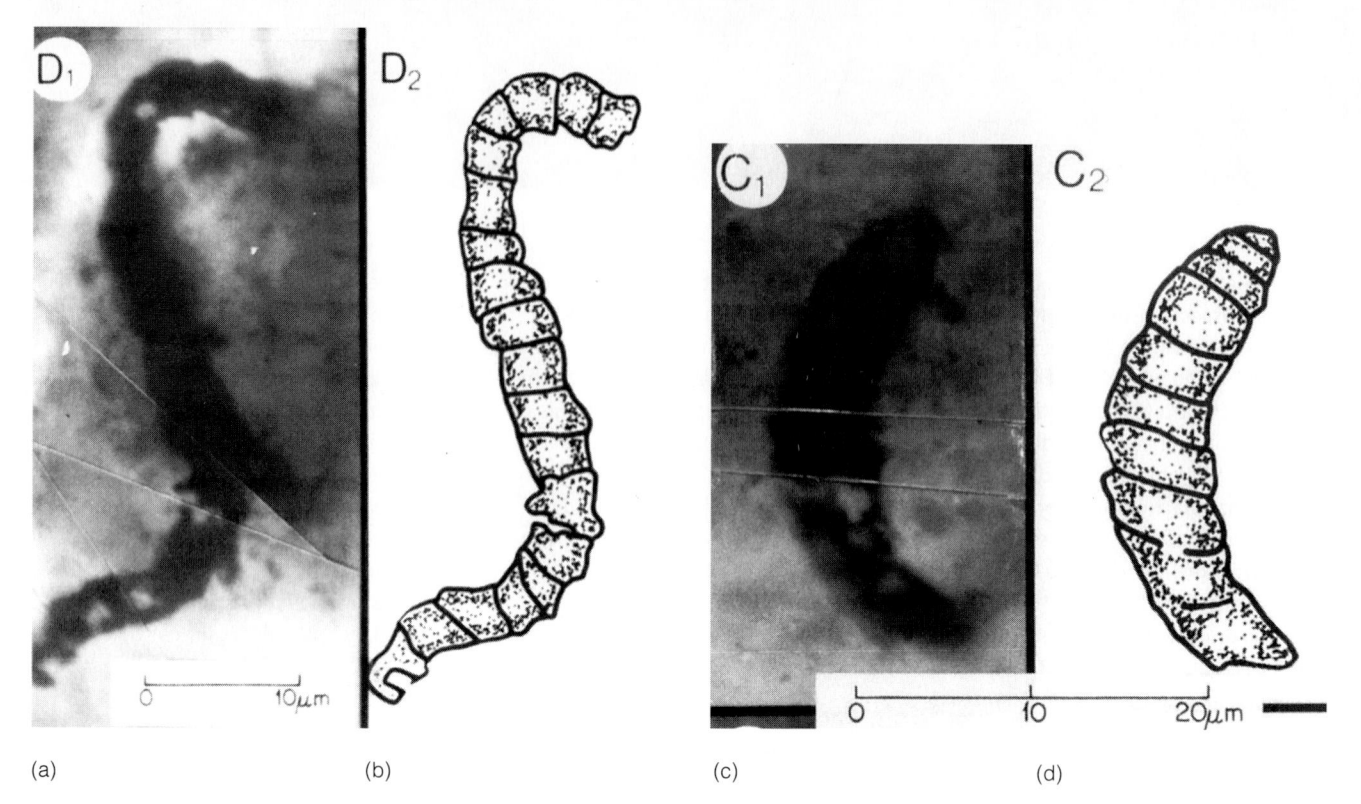

FIGURE 8.17 Photomicrographs (a) and (c) of fossil prokaryotes from the 3.3 to 3.5-billion-year-old Warrawoona Group, western Australia. Schematic restorations of (a) and (c) are shown in (b) and (d).

the autotrophic process of photosynthesis as much as 3.5 billion years ago. These cells were still anaerobic prokaryotes, but as autotrophs they were no longer completely dependent on an external source of preformed organic molecules as a source of nutrients. The Archean microfossils in Figure 8.17 are anaerobic, autotrophic prokaryotes. They belong to the kingdom Monera, which is represented today by bacteria and cyanobacteria or blue-green algae (Appendix B).

Archean Mineral Deposits

Although a variety of mineral resources are found in Archean rocks, the mineral most commonly associated with them is gold. Archean and Proterozoic rocks near Johannesburg, South Africa, have yielded more than 50% of the world's gold since 1886, much of it from conglomerate layers. A number of gold mining areas also occur within the Superior craton in Ontario, Canada.

Gold has been mined for at least 6000 years. During most of historic time, it has been used for jewelry, ornaments, and ritual objects and has served as a symbol of wealth and as a monetary standard. These traditional uses continue, but gold is now used for many other purposes, including gold plating, electrical circuitry, glass making, and the manufacture of chemicals.

A number of Archean-age deposits of massive sulfides of zinc, copper, and nickel are known from several areas, including western Australia, Zimbabwe, and the Abitibi greenstone belt of Ontario, Canada. Many of these deposits

probably formed as the result of hydrothermal activity associated with greenstone belt volcanism. Indeed, similar deposits are currently forming adjacent to black smokers, which are types of hydrothermal vents at or near spreading ridges (see Perspective 8.1).

About one-fourth of the world's chrome reserves are in Archean rocks, especially those in Zimbabwe. The deposits are in greenstone belt rocks and appear to have formed when crystals settled and became concentrated in the lower parts of mafic and ultramafic sills and other intrusive bodies. Chrome is used to produce hard chrome steel and for plating other metals. The United States has very few commercial chrome deposits so it must import most of what it needs.

One area of chrome deposits in the United States is in the Stillwater Complex in Montana. Low-grade ores from this area were mined during both World Wars as well as the Korean War, but they were simply stockpiled; that is, they were not refined for their chrome content. The Stillwater Complex is also a potential source of platinum, as are some other Archean rocks, but most platinum mined today comes from Proterozoic mafic rocks of the Bushveld Complex of South Africa.

Banded iron formations are sedimentary rocks consisting of alternating layers of silica (chert) and iron minerals. About 6% of the world's banded iron formations were deposited during the Archean Eon. Although Archean iron ores are mined in some areas, they are neither as thick nor as extensive as those of the Proterozoic Eon, which constitute the world's major iron deposits (see Chapter 9).

Pegmatites are very coarsely crystalline igneous rocks, commonly associated with plutons. Most are composed of quartz and feldspars and thus correspond to the composition of granite. Archean pegmatites are mostly granitic and of little economic value. Some, such as those of the Herb Lake district in Manitoba, Canada, and the Rhodesian Province in Africa, contain valuable minerals. In addition to minerals of gem quality, Archean pegmatites contain minerals mined for their lithium, beryllium, and rubidium. Furthermore, pegmatites are the only known commercial sources of some of the rare earth elements such as cesium.

Summary

1. Precambrian refers to any rock lying beneath Cambrian strata and to all geologic time from Earth's origin 4.6 billion years ago until the beginning of the Phanerozoic Eon 545 million years ago. Precambrian time is divided into two eons, the Archean and the Proterozoic.

2. The pre-Archean crust may have been composed of ultramafic rocks. None of this crust has been preserved, but by the beginning of the Archean, several sialic continental nuclei existed.

3. Each continent has an ancient, stable craton composed in part of Archean rocks. The exposed part of a craton is a Precambrian shield. The Canadian Shield of North America is made up of several cratons, which are delineated by ages and structural trends.

4. Archean rocks are predominantly greenstone belts and granite–gneiss complexes. Greenstone belts have a synclinal structure and occur as linear bodies within much more extensive areas of granite and gneiss.

5. Typical greenstone belts can be divided into three major rock sequences. The two lower units are dominated by volcanic rocks, some of which are ultramafic. The upper unit consists mostly of sedimentary rocks.

6. Most Archean sedimentary rocks are preserved in greenstone belts, and the most common are graywacke–argillite assemblages deposited by turbidity currents.

7. The evolution of greenstone belts is controversial, but most models rely on Archean plate motions. Those greenstone belts with ultramafic rocks may have formed in intracontinental rifts, while those rich in andesites probably formed in back-arc marginal basins.

8. During Late Archean deformation, the Superior, Slave, and parts of other cratons of the Canadian Shield were formed. This deformation was the last major Archean event in North America.

9. Plate tectonics similar to that of the present did not begin until the Early Proterozoic, but many geologists think some type of Archean plate tectonics occurred. Archean plates may have moved more rapidly, however, because Earth possessed more radiogenic heat.

10. The atmosphere and surface waters were derived from internally generated volcanic gases by a process called outgassing. The atmosphere so formed was deficient in free oxygen.

11. Most models for the origin of life require a nonoxidizing atmosphere. Atmospheric gases contained the elements necessary for simple organic molecules. Ultraviolet radiation and lightning probably provided the energy for synthesizing organic molecules.

12. The first naturally synthesized organic molecules were monomers such as amino acids. Monomers formed long chains, which gave rise to more complex molecules called polymers, such as proteins.

13. The method whereby a reproductive mechanism developed is unknown. Some investigators think RNA molecules may have been the first molecules capable of self-replication.

14. The Archean fossil record is very poor. A few localities contain unicellular, prokaryotic bacteria of the kingdom Monera. Stromatolites formed by photosynthesizing bacteria may date from 3.5 billion years ago.

15. The most important Archean mineral resources are gold, chrome, and massive sulfides of zinc, copper, and nickel.

Important Terms

anaerobic	craton	monomer	Precambrian
Archean Eon	granite–gneiss complex	outgassing	Precambrian shield
autotrophic	greenstone belt	photochemical dissociation	prokaryotic cell
back-arc marginal basin	heterotrophic	photosynthesis	Proterozoic Eon
Canadian Shield	intracontinental rift	polymer	stromatolite

Review Questions

1. The uppermost unit in a typical greenstone belt is composed mostly of:
   a. _____ sedimentary rocks.
   b. _____ ultramafic rocks.
   c. _____ andesite.
   d. _____ lava flows.
   e. _____ granite and gneiss.

2. The largest exposed area of the North American craton is the:
   a. _____ Wyoming craton.
   b. _____ Grand Canyon.
   c. _____ Canadian Shield.
   d. _____ American platform.
   e. _____ Minnesota lowlands.

3. Earth's atmosphere was probably derived from volcanic gases, a process known as:
   a. _____ dewatering.
   b. _____ photosynthesis.
   c. _____ autotrophism.
   d. _____ fermentation.
   e. _____ outgassing.

4. A polymer is an organic molecule:
   a. _____ capable of metabolism and reproduction.
   b. _____ made up of amino acids linked together.
   c. _____ characterized as heterotrophic and anaerobic.
   d. _____ that binds together loose sediment to make stromatolites.
   e. _____ found mostly in the lava flows of greenstone belts.

5. Most of the free oxygen was probably added to Earth's atmosphere by the process known as:
   a. _____ photosynthesis.
   b. _____ intracontinental rifting.
   c. _____ cratonization.
   d. _____ polymerization.
   e. _____ photochemical dissociation.

6. Greenstone belts with ultramafic rocks in their lower units probably formed in:
   a. _____ fluvial deposits.
   b. _____ granite-gneiss complexes.
   c. _____ intracontinental rifts.
   d. _____ fractures at spreading ridges.
   e. _____ back-arc marginal basins.

7. The ancient, stable nuclei of continents are known as:
   a. _____ cratons.
   b. _____ bedrock.
   c. _____ orogenic belts.
   d. _____ granite-gneiss belts.
   e. _____ pillow structures.

8. Deposition by turbidity currents was most likely responsible for _____ in greenstone belts.
   a. _____ granite-gneiss complexes
   b. _____ banded iron formations
   c. _____ extensive carbonates and evaporites
   d. _____ graywacke-argillite associations
   e. _____ stromatolites

9. Some scientists think the first self-replicating system might have been a(an):
   a. _____ stromatolite.
   b. _____ RNA molcule.
   c. _____ bacterium.
   d. _____ ATP cell.
   e. _____ proteinoid.

10. During the Archean, plates probably moved more rapidly because:
    a. _____ Earth possessed more heat.
    b. _____ Archean cratons were larger than those now present.
    c. _____ most of the Northern Hemisphere was covered by a vast ice sheet.
    d. _____ Earth was closer to the Sun.
    e. _____ the rate of Earth's rotation was much greater than at present.

11. The presence of detrital minerals such as pyrite (FeS_2) and uraninite (UO_2) in Archean sedimentary rocks indicate that the atmosphere:
    a. _____ was rich in nitrogen.
    b. _____ lacked free oxygen.

c. _____ developed by comet impacts.

d. _____ contained mostly water vapor and ozone.

e. _____ is where life originated.

12. The Archean Eon ended_____billion years ago.
    a. _____ 545   d. _____ 3.8
    b. _____ 1.3   e. _____ 3.96
    c. _____ 2.5

13. Why are Precambrian rocks so difficult to study compared with Phanerozoic rocks?

14. What is the evidence that some rocks were present before the beginning of the Archean Eon 4.0 billion years ago?

15. Why are ultramafic rocks so rare in rocks younger than Archean?

16. Summarize the experimental evidence that indicates both monomers and polymers could have formed by natural processes on early Earth.

17. Explain how photosynthesis and photochemical dissociation supplied oxygen to the Archean atmosphere.

18. What are Precambrian shields and cratons, and what are they composed of? Where would one go to visit the shield in North America?

19. Explain how greenstone belts might have formed in intra-continental rifts and back-arc marginal basins.

20. What are stromatolites, and how do they form?

21. What is meant by the terms *prokaryotic* and *heterotrophic*?

22. How and why did Archean plate tectonics differ from plate tectonics of later geologic time?

23. Explain how pre-Archean crust might have formed. Why is none of this ancient crust still present?

24. What are the similarities and differences between the hypotheses that life originated from atmospheric gases and that it originated near hydrothermal vents on the seafloor?

Points to Ponder

1. How is it that geologists have few difficulties correlating Phanerozoic rocks of the same age, but find it hard or impossible to do the same with Archean rocks? Also, why are even radiometric dates of limited use in this endeavor?

2. Explain fully why volcanism and the rate of plate movements has decreased through time. Speculate on the long-term effects this will have on Earth.

3. How can one use the principle of uniformitarianism to interpret the origin of greenstone belts?

Additional Readings

Bengtson, S., ed. 1994. *Early life on Earth: Nobel symposium no. 84.* New York: Columbia University Press.

Condie, K. C. 1994. *Archean crustal evolution.* Amsterdam: Elsevier.

———. 1997. *Plate tectonics and crustal evolution.* 4th ed. Boston: Butterworth Heinemann.

Eigen, M. 1992. *Steps toward life: A perspective on evolution.* New York: Oxford University Press.

Fox, S. W. 1991. Synthesis of life in the lab? Defining a protoliving system. *Quarterly Review of Biology* 66: 181–185.

Goodwin, A. M. 1996. *Principles of Precambrian geology.* London: Academic Press.

Horgan, J. 1991. In the beginning. *Scientific American* 164, no. 2: 116–125.

Margulis, L., and L. Olendzenski, eds. 1992. *Environmental evolution: Effects of the origin and evolution of life on planet Earth.* Cambridge. Mass.: MIT Press.

Naqvi, S. M., ed. 1990. *Precambrian continental crust and its economic resources.* Amsterdam: Elsevier.

Nisbet, E. G. 1987. *The young Earth: An introduction to Archean geology.* London: Allen & Unwin.

Reed, J. C., Jr., et al. 1993. Precambrian: Conterminous U.S. *The geology of North America, Volume C-2.* Decade of North American Geology. Boulder, Colo.: Geological Society of America.

Monastersky, R. 1998. The rise of life on Earth. *National Geographic* 193, no. 3: 54–81.

Windley, B. F. 1995. *The evolving continents.* 3d ed. New York: Wiley.

▶ UNIVERSITY OF CALIFORNIA MUSEUM OF PALEONTOLOGY

The Museum of Paleontology at the University of California, Berkeley, maintains this huge site with information on a variety of topics of interest to geologists, biologists, and earth scientists. Under *On-Line Exhibits* click *Time Periods,* which brings up a geologic time scale with clickable eras and periods. Click *Hadean Time* and note that the end of the Hadean at 3.8 billion years differs from the date given in the book. This age of 3.8 billion years corresponded to the ages of Earth's oldest known rocks until the discovery of the Atasca gneiss in Canada, which is nearly 4.0 billion years old. What kinds of rocks on Earth are known from Hadean time?

At the end of the section on Hadean Time click *Archean* and see what Earth's early atmosphere was likely composed of. Next, click *bacteria microfossils* and find out what the age of Earth's oldest fossils is. Read about cyanobacteria. What is another name for cyanobacteria?

▶ TRAVELS WITH GEOLOGY—GRAND TETON NATIONAL PARK

This is part of the Geology Gems site maintained by Winona State University, Winona, Minnesota. It has maps, images, and a discussion of the geologic history of the Teton Range of Wyoming. What kinds of Archean and Proterozoic rocks are found in the Teton Range? When did the present-day range form and what processes are responsible for its present rugged topography?

▶ SAVAGE EARTH: BLACK SMOKERS

This is a PBS online site. It is the web companion to a four-hour series on earthquakes, volcanoes, tsunamis, and other natural disasters. In addition to black smokers, this site has good animations of plate tectonics, mountain building, and volcanism. Read the article by Kathy Svitil and look at the accompanying images of black smokers, or hydrothermal vents on the seafloor.

1. What kinds of deposits have been found associated with black smokers?

2. Read about the organisms and fossils found in black smoker deposits.

3. Why do some scientists think that life might have originated near black smokers? See your text for help answering this question.

▶ ALGOMA-TYPE IRON FORMATION

The British Columbia Ministry of Energy and Mines maintains this site. The article on Algoma-Type Iron Formation is by G. A. Gross of the Geological Survey of Canada. What are the characteristics of Algoma-type iron formations? When and in what depositional environment did they form?

Explore the following *In-Terra-Active 2.0* CD-ROM module(s) and increase your understanding of key concepts and processes presented in this chapter.

Chapter Concept: Pressure-Temperature Conditions of Metamorphic Rocks

Precambrian-age cratonic areas are underlain by two main rock types, greenstone and granite gneiss, both metamorphic in origin. In this module, you will examine the textural and mineralogical changes associated with metamorphism and relate them to the pressure-temperature conditions that produce them.

Section: Materials
Module: Metamorphic Rocks

The coarse-grained garnet-mica schist shown on the left has essentially the same chemical composition as the fine-grained mudstone shown on the right. In the field, it is clear that these belong to the same stratigraphic unit.
Question: *How can this transformation take place in the solid state?*

Chapter Concept: Crustal Deformation of Cratons

The lithologies and structures present in Precambrian-age cratonic areas are the result of multiple periods of crustal deformation and igneous intrusion in a variety of plate tectonic settings. In this module, you explore the relation between crustal structures and the deformational processes that accompany plate tectonics.

Section: Interiors
Module: Crustal Deformation

The coarse-grained garnet-mica schist shown on the left has essentially the same chemical composition as the fine-grained mudstone shown on the right. In the field, it is clear that these belong to the same stratigraphic unit.
Question: *How can this transformation take place in the solid state?*

9

Precambrian History: The Proterozoic Eon

Proterozoic sedimentary rocks exposed in the Rocky Mountains of Montana.

Prologue

We began the last chapter with the statement "Imagine a barren, waterless, lifeless, hot planet with a poisonous atmosphere." Now as we begin the Proterozoic some 1.5 billion years later, the only life on land might have been some bacteria, but the planet was not waterless; extensive oceans were present, as well as lakes and streams. Meteorites no doubt still impacted Earth with greater frequency than at present but much less so than during the past. Furthermore, enough oxygen had accumulated in the atmosphere to at least reduce the amount of ultraviolet radiation. Still, the land was largely barren, and other than bacteria no land-dwelling organisms existed. But, the Proterozoic was much like the present in terms of geologic processes. Volcanism was probably still more intense, and tectonic plates moved more rapidly because of Earth's greater amount of internal heat (see Figure 8.11), but an essentially present-day style of plate tectonics was operating.

According to plate tectonic theory, oceanic crust is produced at spreading ridges, moves away from ridges by seafloor spreading, and is eventually consumed at subduction zones. This continuous destruction of oceanic crust explains the fact that none older than about 180 million years old is known except as fragments tectonically emplaced on continents. These fragments consisting of a sequence of rocks representing upper mantle, oceanic crust, and deep-sea sediments are known as **ophiolites.** Probably less than 0.001% of all oceanic crust that ever existed is now preserved as ophiolites. Nevertheless, ophiolites provide information on processes by which oceanic crust is generated at spreading ridges, and they are the primary features used to identify ancient convergent plate boundaries. Furthermore, the oldest ophiolites mark the time when a plate tectonic style similar to that of the present first began.

Ophiolites are well known from Phanerozoic rocks, and one with a probable age of 2.65 billion years has been reported from Wyoming. The oldest complete ophiolite so far identified, though, is the highly deformed 1.96-billion-year-old (Early Proterozoic) Jormua mafic–ultramafic complex of Finland (Figure 9.1). It compares closely with younger, well-documented ophiolites and represents a sequence of upper mantle, lower oceanic crust, submarine eruptions (pillow basalts), and metamorphosed deep-sea clay, oozes, and turbidites (Figure 9.1b).

According to Asko Kontinen of the Geological Survey of Finland, the Jormua complex formed at a spreading ridge during an extensional stage in the breakup of an Archean craton. As rifting of this craton proceeded, a large ocean basin formed in which deep-sea sediments and turbidites were deposited. About 1.96 billion years ago, a compressional stage began during which most of the previously formed oceanic crust was destroyed by subduction. The Jormua complex was tectonically emplaced on the craton and represents a small fragment of this ancient oceanic crust.

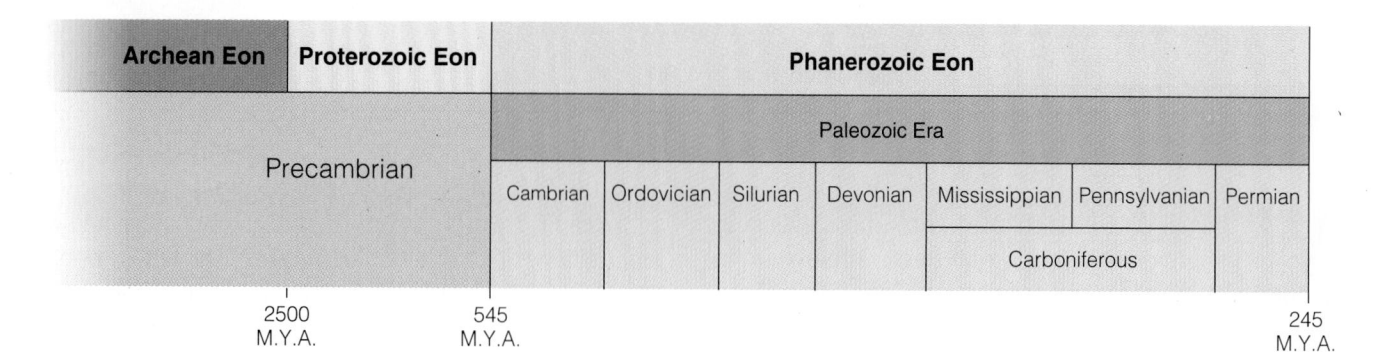

		Phanerozoic Eon						
Archean Eon	Proterozoic Eon							
		Paleozoic Era						
Precambrian		Cambrian	Ordovician	Silurian	Devonian	Mississippian	Pennsylvanian	Permian
						Carboniferous		

2500
M.Y.A.

545
M.Y.A.

245
M.Y.A.

Introduction

The Proterozoic Eon includes more than 42% of all geologic time (see Figure 8.2), yet we review this 1.955-billion-year episode of Earth history in a single chapter. A mere 11.8% of all geologic time, the Phanerozoic Eon, is the subject of the remainder of this book. This may seem disproportionate, but the study of the Proterozoic is subject to the same problems we encountered in studying Archean history, although to a lesser degree. In short, we know more about the Proterozoic than we do about the Archean, but Phanerozoic history is even better known, hence its greater coverage.

How does the Proterozoic differ from the Archean? The basic difference is the style of crustal evolution. Archean crust-forming processes generated greenstone belts and granite–gneiss complexes that were shaped into cratons; and although these rock assemblages continued to form during the Proterozoic, they did so at a considerably reduced rate. Many Archean rocks have been metamorphosed, although the degree of metamorphism varies considerably, and some are completely unaltered. In con-

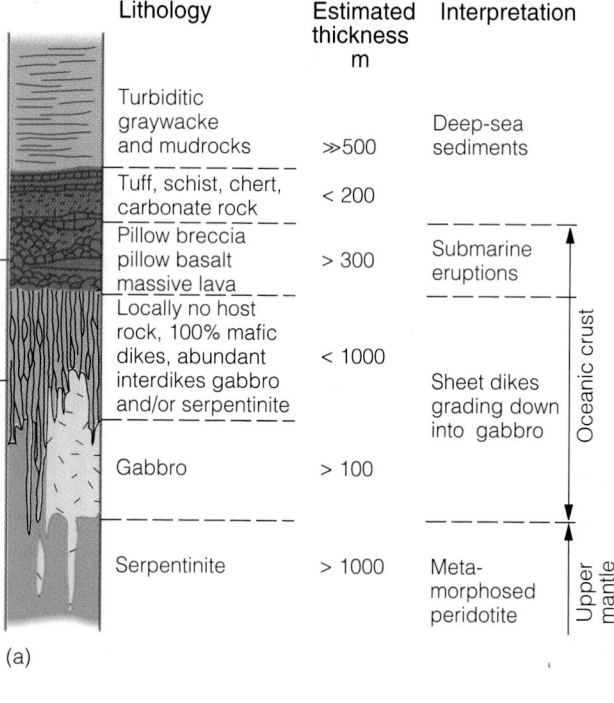

(a)

Lithology	Estimated thickness m	Interpretation		
Turbiditic graywacke and mudrocks	≫500	Deep-sea sediments		
Tuff, schist, chert, carbonate rock	< 200			
Pillow breccia pillow basalt massive lava	> 300	Submarine eruptions	Oceanic crust	
Locally no host rock, 100% mafic dikes, abundant interdikes gabbro and/or serpentinite	< 1000	Sheet dikes grading down into gabbro		
Gabbro	> 100			
Serpentinite	> 1000	Metamorphosed peridotite	Upper mantle	

(b)

(c)

FIGURE 9.1 (a) Reconstruction of the highly deformed Jormua mafic–ultramafic complex of Finland. This 1.96-billion-year-old sequence of rocks is the oldest known complete ophiolite. (b) A metamorphosed basaltic pillow lava. The code plate is 12 cm long. (c) Metamorphosed gabbro between mafic dikes. The hammer shaft is 65 cm long.

Phanerozoic Eon									
Mesozoic Era			Cenozoic Era						
Triassic	Jurassic	Cretaceous	Tertiary					Quaternary	
			Paleocene	Eocene	Oligocene	Miocene	Pliocene	Pleistocene	Holocene

245
M.Y.A.

66
M.Y.A.

trast, many Proterozoic rocks are unmetamorphosed and little deformed, and in many areas Proterozoic rocks are separated from Archean rocks by a profound unconformity. Finally, the Proterozoic is characterized by widespread assemblages of sedimentary rocks that are rare in the Archean and by a plate tectonic style that is essentially the same as that of the present.

Changes definitely occurred across the Archean-Proterozoic boundary, but these changes were not synchronous. For example, Archean crustal evolution in the Kaapvaal craton of South Africa was completed nearly 3.0 billion years ago when sedimentary deposits more typical of the Proterozoic began to accumulate. In North America, this same change did not occur until 2.6 billion years ago. Actually, the change in crust-forming processes occurred from 2.95 to 2.45 billion years ago, so placing the Archean–Proterozoic boundary at 2.5 billion years ago is rather arbitrary.

Proterozoic Crustal Evolution

In the last chapter we noted that Archean cratons assembled through a series of island arc and minicontinent collisions (see Figure 8.3). These provided the nuclei around which Proterozoic continental crust accreted, thereby forming much larger cratons. One large landmass, called **Laurentia,** consisted mostly of North America and Greenland, parts of northwestern Scotland, and perhaps parts of the Baltic Shield of Scandinavia. This chapter will emphasize the geologic evolution of Laurentia. Table 9.1 summarizes the crust-forming events discussed in the following sections.

EARLY PROTEROZOIC EVOLUTION OF LAURENTIA

The first major episode in the Proterozoic evolution of Laurentia took place between 2.0 and 1.8 billion years ago (Table 9.1). During this time, several major **orogens** developed, consisting of zones of deformed rocks, many of which have been metamorphosed and intruded by plutons. The Archean cratons were sutured along these deformed belts (Figure 9.2a). In other words, this was an episode of continental growth during which collisions between Archean cratons formed a larger craton, and new crust was accreted at the margins of the cratons so formed.

By 1.8 billion years ago, much of what is now Greenland, central Canada, and the north-central United States formed a large craton (Figure 9.2a).

One Early Proterozoic episode of colliding Archean cratons is recorded in rocks of the Thelon orogen in the northwestern Canadian Shield, where the Slave and Rae cratons collided (Figure 9.2). Here, the geologic record indicates subduction, granitic plutonism, volcanism, intense deformation, and regional metamorphism. The events leading to the formation of the Thelon orogen resulted in the collision and suturing of the Slave and Rae cratons.

Rocks of the Trans-Hudson orogen in northern Saskatchewan record initial rifting, extrusive volcanism, sedimentation, and the formation of oceanic crust, followed by ocean basin closure (Figure 9.3). As the ocean basin closed, a subduction zone formed resulting in island arc collisions, granitic plutonism, intense deformation, and regional metamorphism. The formation of the Trans-Hudson orogen resulted in the suturing of the Superior, Wyoming, and Hearne cratons.

Other notable events in the Early Proterozoic evolution of Laurentia include the development of the Wopmay, Mojave, and Penokean orogens (Figure 9.2a). Rocks of the Wopmay orogen, adjacent to the Slave craton in northwestern Canada, record the opening and closing of an ocean basin, or what is known as a *Wilson cycle.* Furthermore, some sedimentary rocks in this area are of particular interest because they form a **sandstone–carbonate–shale assemblage,** a suite of rocks typical of passive continental margins that first become abundant and widespread during the Proterozoic.

South of the Superior craton, the Penokean orogen accounts for the origin of another large segment of Laurentia (Figure 9.2a). Considerable disagreement exists on the tectonic processes responsible for this orogen. Some geologists think it formed in a manner similar to the Trans-Hudson orogen (Figure 9.3), but others appeal to intracratonic deformation.

Following the initial episode of amalgamation of Archean cratons, considerable accretion occurred along the southern margin of Laurentia. Between 1.8 and 1.6 billion years ago, growth continued in what is now the southwestern and central United States as successively younger belts were sutured to Laurentia, forming the Yavapai and Mazatzal–Pecos orogens (Figure 9.2b).

TABLE 9.1

Summary Chart Showing the Ages of Some of the Proterozoic Events Discussed in the Text

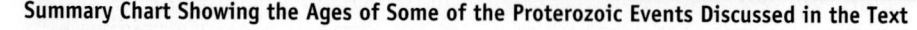

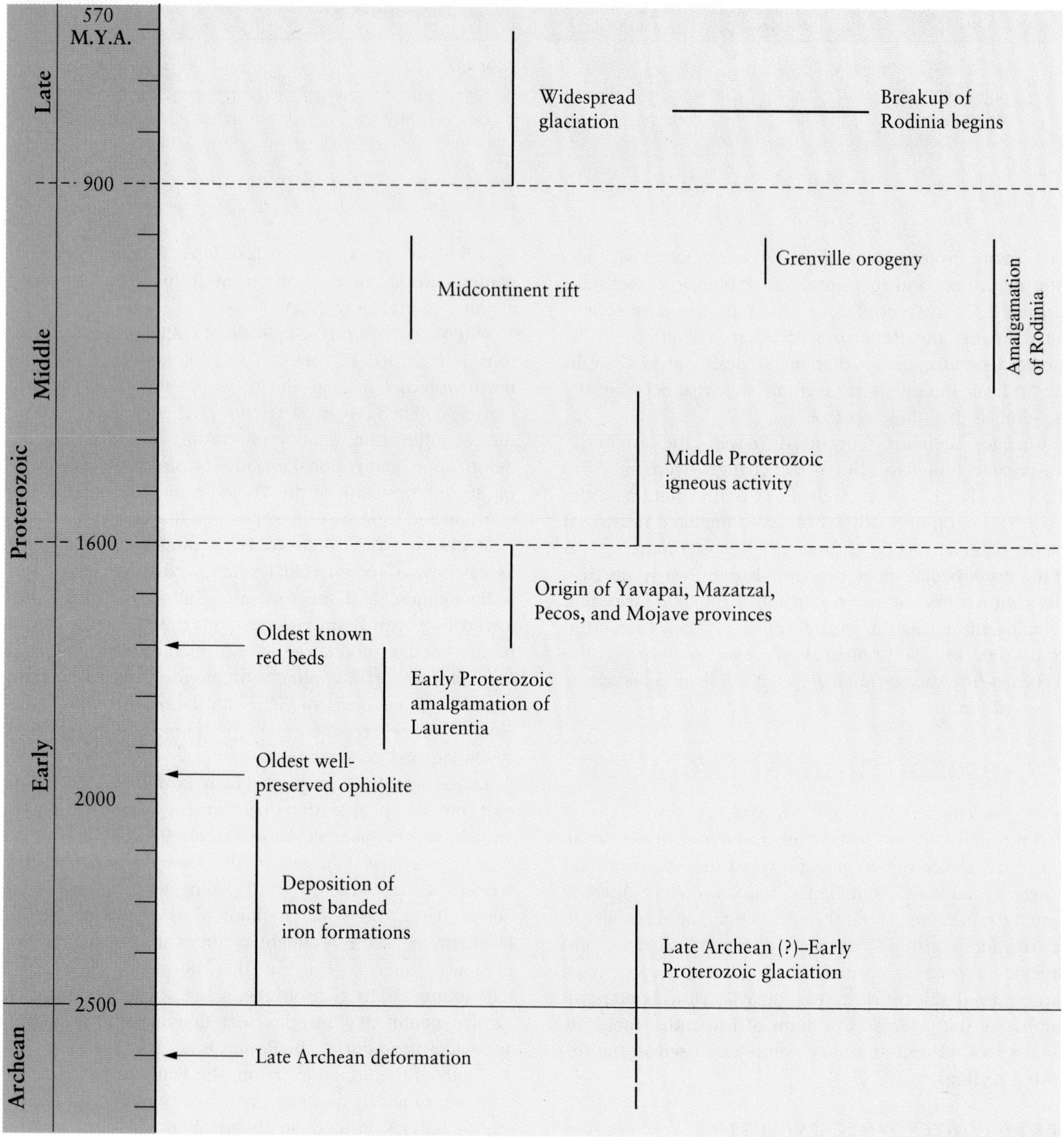

570 M.Y.A.	
Late	Widespread glaciation — Breakup of Rodinia begins
900	
Middle	Midcontinent rift
	Grenville orogeny
	Amalgamation of Rodinia
	Middle Proterozoic igneous activity
1600	
	Origin of Yavapai, Mazatzal, Pecos, and Mojave provinces
	Oldest known red beds
	Early Proterozoic amalgamation of Laurentia
Early	Oldest well-preserved ophiolite
2000	
	Deposition of most banded iron formations
	Late Archean (?)–Early Proterozoic glaciation
2500	
Archean	Late Archean deformation

The plate tectonic model accounting for the accretion of the Yavapai and Mazatzal–Pecos orogens contains both extensional and compressional phases. Greenstone belts formed in back-arc marginal basins (see Figure 8.8), whereas sandstone–carbonate–shale assemblages were deposited in fluvial, tidal, and shallow marine environments on a south-facing continental margin. Continental accretion occurred as northward-migrating island arcs collided with Laurentia, resulting in deformation, metamorphism, and the emplacement of granitic batholiths. The net effect was the accretion of more than 1000 km of continental crust along the southern margin of Laurentia (Figure 9.2b).

MIDDLE PROTEROZOIC IGNEOUS ACTIVITY

No major episode in the growth of Laurentia occurred between 1.6 and 1.3 billion years ago. Nevertheless, during this interval extensive igneous activity took place that was

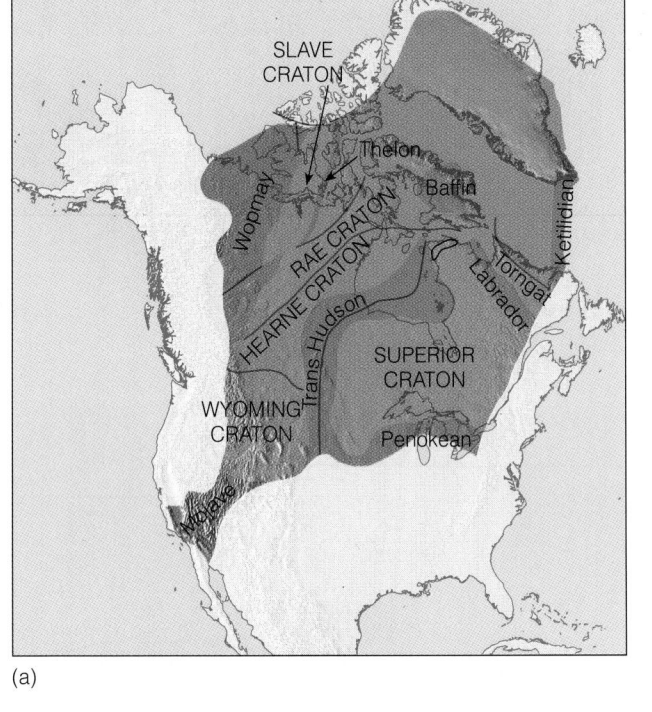

(a)

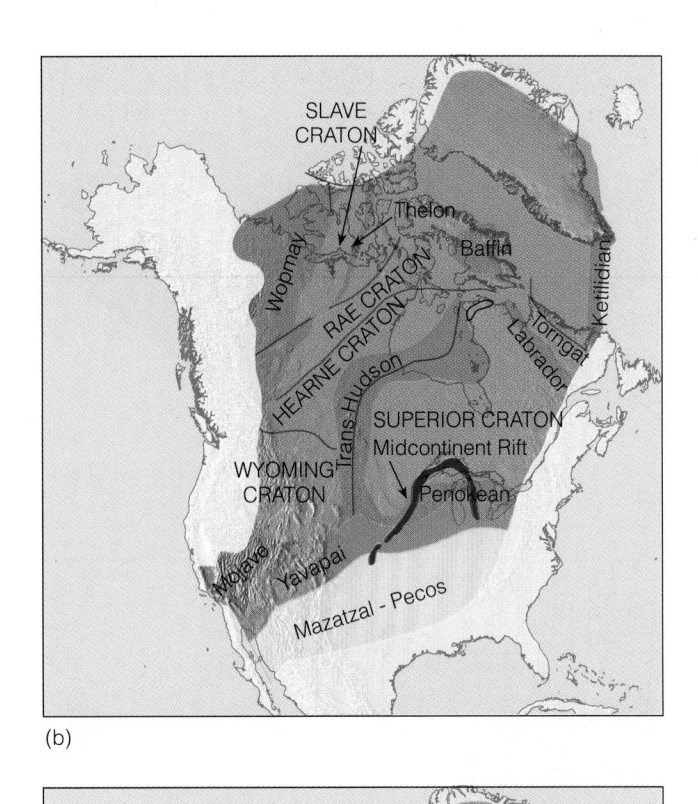

(b)

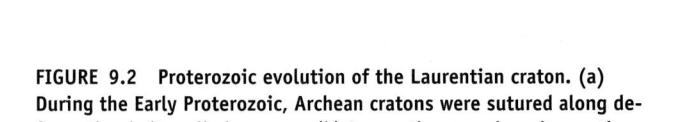

900 million - 1.2 billion

1.6 billion - 1.75 billion

1.75 billion - 1.8 billion

1.8 billion - 2.0 billion

2.5 billion - 3.0 billion

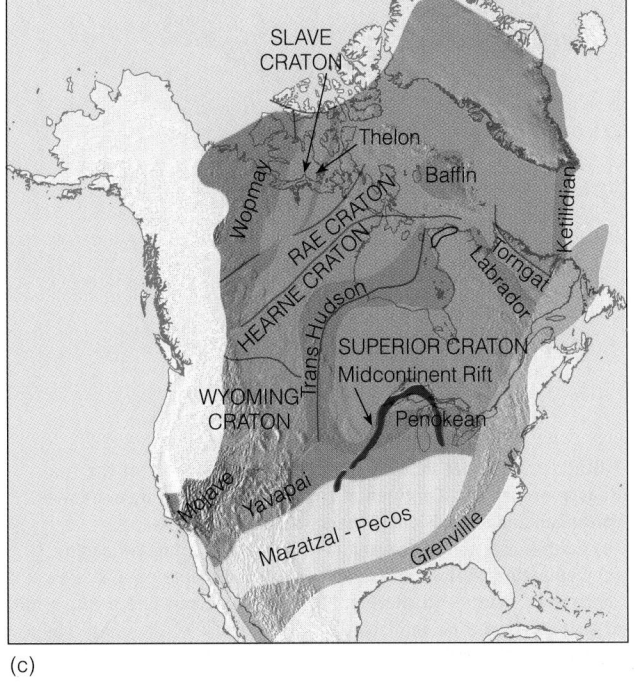

(c)

FIGURE 9.2 Proterozoic evolution of the Laurentian craton. (a) During the Early Proterozoic, Archean cratons were sutured along deformation belts called *orogens*. (b) Laurentia grew along its southern margin by accretion of the Central Plains, Yavapai, and Mazatzal orogens. (c) A final episode of Protoerozoic accretion occurred during the Grenville orogeny.

unrelated to orogenic activity (Figure 9.4). These rocks did not increase the area of Laurentia because they were intruded into or erupted upon already existing crust. These igneous rocks are buried beneath Phanerozoic strata in most areas, in the north–central and southwestern United States, for example. They are exposed in eastern Canada, extend across southern Greenland, and are also found in the Baltic Shield of Scandinavia. Recent investigations show that they are mostly granitic plutons, calderas and their fill, and vast sheets of rhyolite and ash flows, although gabbros and anorthosites (plutonic rocks composed mostly of plagioclase) are known as well.

The origin of these Middle Proterozoic igneous rocks continues to be the subject of debate. Paul Hoffmann of the Geological Survey of Canada postulates that large-scale mantle upwelling, or what he calls a superswell, rose beneath a Proterozoic supercontinent to produce these rocks. According to this hypothesis, the mantle temperature beneath a Proterozoic supercontinent would have been considerably higher than beneath later supercontinents because radiogenic heat production within Earth has decreased (see Figure 8.11). Accordingly, non-orogenic igneous activity would have occurred following the amalgamation of the first supercontinent.

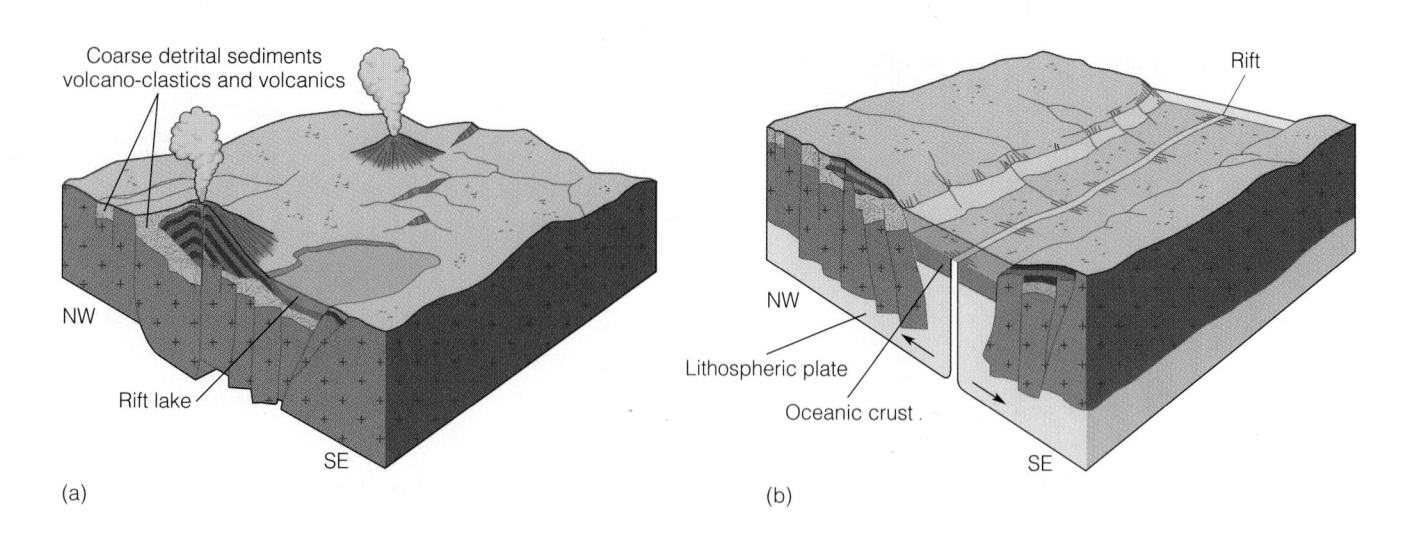

(a)

(b)

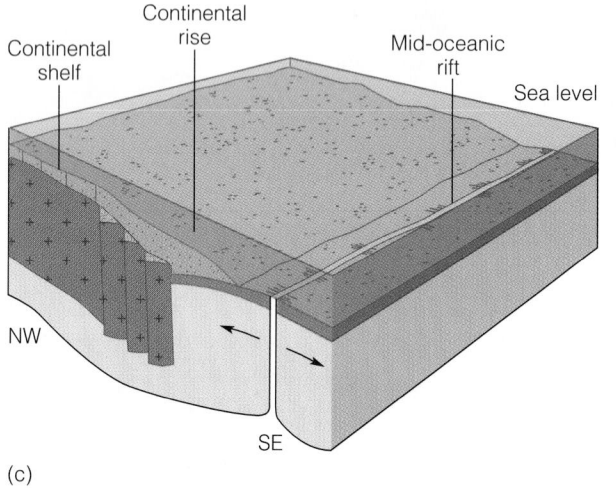

(c)

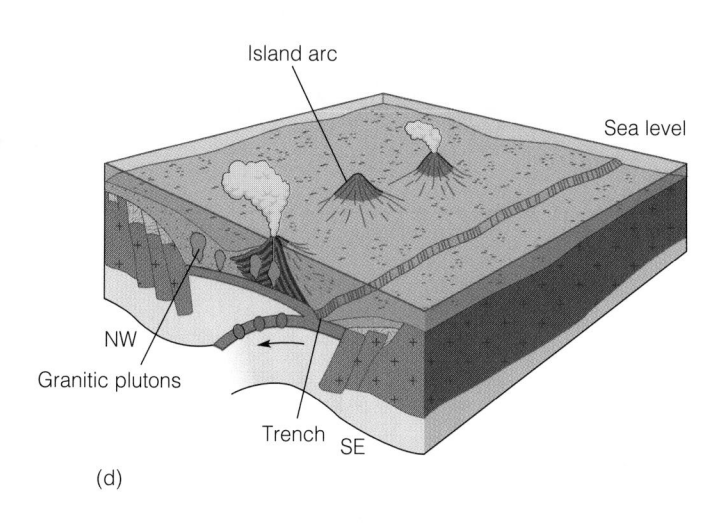

(d)

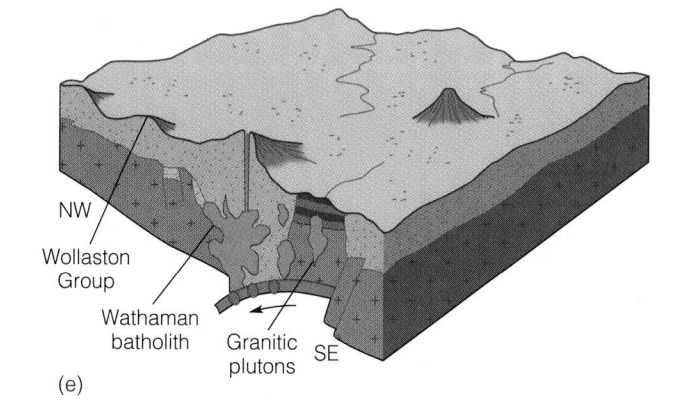

(e)

FIGURE 9.3 Model showing the development of part of the Trans-Hudson orogen. (a) Continental rifting accompanied by extrusive volcanism and deposition of coarse clastic sediments. (b) Continental separation and development of an ocean basin. (c) Deposition of shelf and deeper-water sediments on a passive continental margin. (d) Reversal of plate movement led to the origin of a subduction zone and island arc. (e) Closure of the ocean basin caused intrusive and extrusive volcanism and deformation of passive margin sediments.

MIDDLE PROTEROZOIC OROGENY AND RIFTING

Another major episode in the evolution of Laurentia, the **Grenville orogeny** in the eastern United States and Canada, occurred between 1.3 and 1.0 billion years ago (Figure 9.2c, Table 9.1). Rocks of the Grenville orogen are extensively exposed in southeastern Canada where they abut the Archean Superior Province. This belt of deformed rocks extends northeast into Greenland and continues into Scandinavia. It also continues southwest through the area of the present-day Appalachian Mountains.

The Grenville Supergroup, consisting of sandstones, shales, and carbonates, was highly deformed, metamorphosed, and intruded by various igneous bodies during the formation of the Grenville orogen (Figure 9.5). Many geologists think the Grenville orogen can be explained by the opening and then closing of an ocean basin. If so, the Grenville Supergroup may represent sediments deposited on a passive continental margin. However, some geologists think that Grenville deformation occurred in an intracontinental setting or was caused by major shearing. Whatever the cause, it represents the final episode of Proterozoic

FIGURE 9.7 Reconstructions of three possible Late Proterozoic supercontinents. Dashed lines indicate areas known or suspected to be underlain by Archean crustal rocks. The relative locations of Laurasia, East, and West Gondwana are unknown.

The components that ultimately formed Pangaea were evolving during the Precambrian, but few investigators agree on their size, shape, or associations. One group holds that a single Pangaea-like supercontinent persisted from the Late Archean to the end of the Proterozoic. Other investigators, pointing to uncertainties in the data, maintain that this reconstruction is not justified.

Three or more supercontinents existed in the Late Proterozoic, but their relative locations are unknown. As indicated in Figure 9.7, East Gondwana consisted of Australia, India, and Antarctica, while West Gondwana included Africa and South America. The third supercontinent, Laurasia, consisted of Laurentia, Europe, and most of Asia.

During the latest Proterozoic, all the Southern Hemisphere continents were strongly deformed. These events may mark the time during which East and West Gondwana were assembled into the larger supercontinent of Gondwana. In fact, some geologists think that Laurentia, Australia, and Antarctica were connected during the Late Proterozoic, forming a supercontinent (Figure 9.8). This supercontinent, known as **Rodinia,** formed between 1.3 and 1.0 billion years ago and then fragmented as an episode of rifting began about 750 million years ago. Although supercontinents might have existed prior to the assembly of Rodinia, this is the first one for which compelling geologic evidence exists.

FIGURE 9.8 Possible arrangement of the continents during the Late Proterozoic.

Proterozoic Rocks

Some Proterozoic rocks and rock associations deserve special attention. Greenstone belts, for example, are more typical of the Archean, but they continued to form during the Proterozoic, though they differed in detail. Sandstone–carbonate–shale assemblages also merit our attention because they closely resemble sedimentary rock associations of the Phanerozoic. Deposits attributed to two episodes of Proterozoic glaciation are recognized as well.

The onset of a present style of plate tectonics probably occurred in the Early Proterozoic, and the first well-preserved *ophiolites,* which mark convergent plate margins, are of this age (see the Prologue). Banded iron formations and red beds, too, are important Proterozoic rock types, but these are more conveniently discussed in the section on the atmosphere.

GREENSTONE BELTS

Greenstone belts are characteristic of the Archean, and most seem to have formed between 2.7 and 2.5 billion years ago. They are not, however, restricted to the Archean. Proterozoic greenstone belts are found on several continents, including North America, and at least one has been reported from the Cambrian of Australia. Several greenstone belts formed in the Yavapai and Mazatzal orogens as crustal accretion occurred south of the Wyoming craton (Figure 9.2).

Ultramafic rocks are rare in Proterozoic greenstone belts. The near-absence of Proterozoic and younger ultramafics is probably accounted for by the decreasing amount of radiogenic heat produced within Earth (see Figure 8.11). Otherwise, Proterozoic and Archean greenstone belts differ very little. Both appear to have formed in similar tectonic settings, intracratonic rifts and back-arc marginal basins (see Figures 8.8 and 8.9), although the latter setting became more prevalent during the Late Proterozoic.

SANDSTONE–CARBONATE–SHALE ASSEMBLAGES

Fully 60% of known Proterozoic rock successions are composed of *sandstone–carbonate–shale assemblages.* Widespread deposition of this assemblage occurred throughout the Proterozoic along rifted continental margins and in intracratonic basins. The most important aspect of these assemblages is that they represent suites of rocks similar to those of the Phanerozoic. Hence, their appearance indicates the widespread distribution of stable cratons with depositional environments much like those of the present.

Early Proterozoic sandstone-carbonate-shale assemblages are widely distributed in the Great Lakes region (Figure 9.9a). Both quartzite and shale are abundant, but conglomerates and graywackes are also present. The quartzites were derived from mature quartz sandstones, and many show wave-formed ripple marks and cross-

bedding (Figure 9.9b). Thick carbonates, mostly dolostone, containing abundant stromatolites, are also present (Figure 9.9c).

Many of these rocks were deposited on a shallow-water shelf, perhaps a rifted continental margin. Some intermittent deformation took place during their deposition, but all were further deformed and intruded by granite batholiths during the Early Proterozoic amalgamation of Laurentia. These rocks are now part of the Penokean orogen (Figure 9.2).

In the western United States and Canada, sandstone-carbonate-shale assemblages were deposited along the continental margin. These Middle to Late Proterozoic rocks are best preserved in three large basins (Figure 9.10; see Perspective 9.1). In the Belt Basin of the northwestern United States and adjacent parts of Canada, a thick sequence of sedimentary rocks was deposited between 1.45 billion and 850 million years ago. Like the deposits of the Great Lakes region, these rocks are predominantly sandstone, shale, and thick stromatolite-bearing carbonates (Figure 9.10b and c).

GLACIATION AND GLACIAL DEPOSITS

Glaciers deposit unsorted, unstratified sediment known as *till* that when lithified is called *tillite.* Tillites or till-like deposits have been reported from more than 300 Precambrian localities. Many of these may not be glacial deposits, but some almost certainly are. In fact, two major episodes of Proterozoic glaciation have been recognized—one occurred during the Early Proterozoic and the other during the Late Proterozoic (Table 9.1).

But how can we be sure there were any Proterozoic glaciers? Tillite is simply a type of conglomerate, but other processes, such as *mass wasting,* can produce very similar deposits. The deposits can be distinguished by their areal extent: mass-wasting deposits are very limited, whereas true tillites cover much larger areas. In addition, a suspected tillite can be established as a glacial deposit by its association with a striated bedrock pavement (Figure 9.11) and with finely laminated argillites (varved sediments) that contain large clasts dropped from floating ice (see Figure 4.15d).

Early Proterozoic glacial deposits are known from Canada, the United States, Australia, and South Africa. An example is the Huronian Supergroup of Ontario where the tillite of the Bruce Formation may date from 2.7 billion years ago, making it Late Archean in age. Tillites of the Gowganda Formation are widespread, overlie a striated pavement, and are associated with varved deposits.

Similar deposits thought to be the same age as the Gowganda tillite also occur in Michigan, the Medicine Bow Mountains of Wyoming, and Quebec, Canada. This distribution indicates that North America may have had an extensive Early Proterozoic ice sheet centered southwest of Hudson Bay (Figure 9.12). Tillites of about the same age also occur on other continents; the dating of these deposits is not precise enough to determine whether there

(b)

(a)

(c)

was a single, widespread glacial episode or a number of glacial events at different times in different places.

Widespread Late Proterozoic glaciation occurred between 900 and 600 million years ago (Table 9.1). Tillites and other glacial deposits have been recognized on all continents except Antarctica. Glaciation was not continuous during this entire interval but was episodic with four major glacial periods recognized. Figure 9.13 shows the approxi-

mate distribution of these Late Proterozoic glaciers, but we should emphasize that they are approximate because the geographic extent of glacial ice is unknown. In addition, the glaciers shown in Figure 9.13 were not all present at the same time. Nevertheless, the most extensive glaciation in Earth history was during the latest Proterozoic when ice sheets seem to have been present even in near-equatorial areas.

perspective 9.1

The Rocks of Waterton Lakes National Park, Canada, and Glacier National Park, Montana

Waterton Lakes National Park in Alberta, Canada, and Glacier National Park in Montana are adjacent to one another and in 1932 were designated an International Peace Park. Both parks have spectacular scenery, interesting wildlife such as mountain goats, bighorn sheep, and grizzly bears, and an impressive geologic history (Figure 1). The present-day topographic expression in the parks resulted from deformation and uplift from Cretaceous to Eocene times, followed by stream and glacial erosion. Visitors to the parks can see the results of the phenomenal forces at work during episodes of deformation; the Lewis overthrust fault is easily observed in several locations (see Chapter 16). And glacial landforms such as U-shaped glacial troughs, horns, arêtes, and cirques are some of the finest in North America. The present-day glaciers in the parks are small but still active.

Most of the rocks exposed in Glacier National Park belong to the Late Proterozoic Belt Supergroup, whereas those in Waterton Lakes National Park are assigned to the Purcell Surpergroup (see Figure 9.10a). The names are different north and south of the border, but the rocks are the same. Recall that a supergroup consists of two or more groups, whereas a group is made up of at least two formations. These Belt–Purcell rocks

in the parks are nearly 4000 m thick and were deposited between 1.45 billion and 850 million years ago.

Many of the formations are present in the parks and easily observed from the main roads. Some of the rocks are referred to as *argillite*, a particularly compact variety of mudrock lacking the fissility typical of shale. Many argillites have probably experienced low-grade metamorphism but have not developed slaty cleavage. Other features of these rocks are desiccation cracks and salt crystal casts, indicating

they were deposited in an alternately wet and dry environment such as a tidal flat or along the shores of an arid-region lake. The Grinnell Formation consisting of red (oxidized iron) and green (reduced iron)

FIGURE 1   View of Glacier National Park, Montana. The rocks exposed in the mountains belong to the Late Proterozoic Belt Supergroup. The topography of broad valleys and sharp, angular ridges and peaks resulted mostly from glacial erosion.

argillite is one of the most conspicuous rock units in Glacier National Park (see Figure 9.10b).

Several siltstone and sandstone formations are also present, and these show a variety of sedimentary structures. Cross-bedding and ripple marks are particularly common, but some units show graded bedding and other features thought to indicate deposition by turbidity currents. However, most of the siltstones and sandstones probably represent terrestrial depositional environments, perhaps in an arid region. In fact, some probably accumulated as braided stream and alluvial fan deposits.

Several thick limestone formations are also present. One of the most prominent is the 1210-m-thick Siyeh Limestone, which forms many of the high cliffs in the parks (Figure 2a). An interesting feature seen at Logan Pass in Glacier National Park is a dark band of igneous rocks within the Siyeh Limestone. This is the 1.08-billion-year-old Purcell sill, which was intruded along bedding planes and has metamorphosed the formation above and below to marble (Figure 2b).

Many of the park limestones contain structures produced by the activities of cyanobacteria, or what is commonly called blue-green algae. So-called molar tooth structures formed when bladelike algal structures were crumpled during compaction, and bulbous stromatolites can be seen in several areas (see Figure 9.10c). Present-day limestones form mostly in shallow, tropical seas, and it is enticing to make a similar interpretation for the Belt–Purcell limestones. However, some details of these rock units, as well as their overall composition and correlations, indicate that some may have been deposited in terrestrial environments, perhaps playa lakes.

The most impressive geologic structure in the area is the Lewis overthrust, a large fault along which Belt–Purcell rocks have moved at least 75 km eastward so they now rest upon younger rocks. At Marias Pass in Glacier National Park, the trace of the fault is clearly visible (see Figure 16.16). If one takes a trail to get a closer look at the trace of the fault, intense deformation of the rocks below the fault can be observed.

(a)

(b)

FIGURE 2 (a) The Siyeh Limestone is exposed here in a cliff at the upper end of a cirque. (b) This dark rock within the Siyeh Limestone is the Purcell sill.

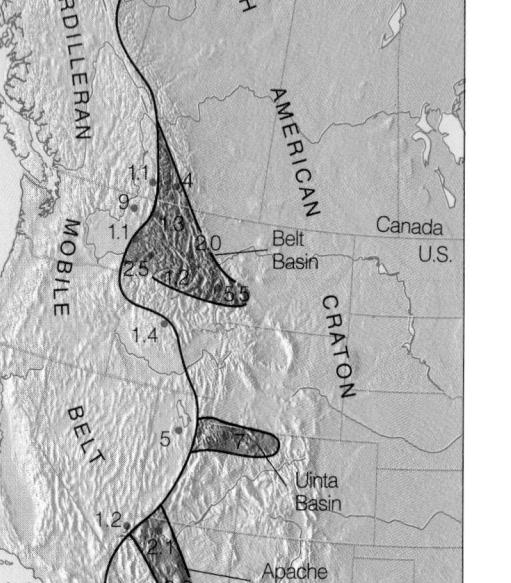

(b)

(c)

(a)

FIGURE 9.10 (a) Late Proterozoic sedimentary basins in the western United States and Canada. The numbers indicate the approximate thickness of rocks in kilometers. (b) and (c) Late Proterozoic rocks of the Belt Supergroup exposed in Glacier National Park, Montana. (b) Outcrop of red mudrock. (c) Limestone with stromatolites.

The Evolving Atmosphere

In the last chapter we noted that many geologists are convinced that the Archean atmosphere contained little or no free oxygen. In other words, the atmosphere was not strongly oxidizing as it is now. The late Preston Cloud, a specialist on the early history of life, estimated that the free oxygen content of the atmosphere increased from about 1 to 10% through the Proterozoic. It was not until about 400 million years ago that oxygen reached its present level. Most of the atmospheric oxygen was released as a waste product by photosynthesizing cyanobacteria (see Figure 8.13). As indicated by fossil stromatolites, cyanobacteria became common about 2.3 billion years ago.

BANDED IRON FORMATIONS

Banded iron formations (BIFs) are sedimentary rocks consisting of alternating thin layers of silica (chert) and iron minerals (Figure 9.14). The iron in these formations is mostly iron oxide [hematite (Fe_2O_3) and magnetite (Fe_3O_4)], but iron silicate, iron carbonate, and iron sulfide are also present.

Archean BIFs are relatively small lenticular (lens-shaped) bodies measuring a few kilometers across and a few meters thick; most appear to have been deposited in greenstone belt environments. Proterozoic BIFs are much more common, typically hundreds of meters thick and traceable for hundreds of kilometers. Most were deposited in shallow marine waters.

BIFs are found throughout the geologic column, but most formed during three main episodes of deposition (Table 9.2). These three depositional periods are bounded by major orogenies representing periods of increased global plate tectonic activity. The BIFs accumulated during the intervening quiescent intervals. The period from 2.5 to 2.0 billion years represents a unique time in Earth history, a time during which 92% of Earth's BIFs formed (Table

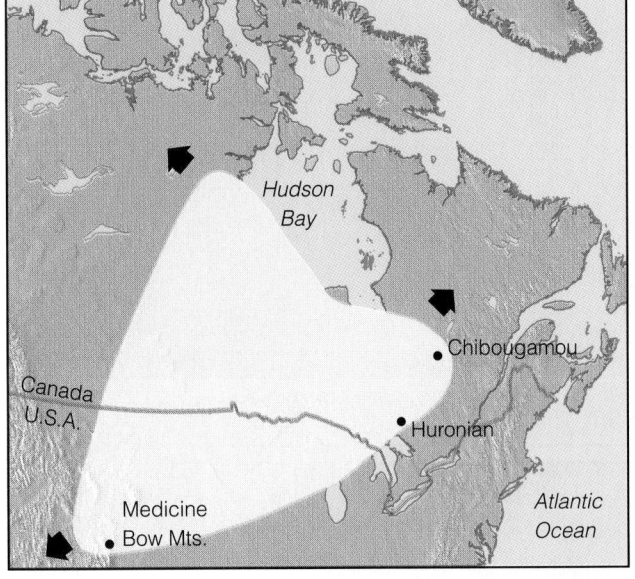

FIGURE 9.11 Evidence for Proterozoic glaciation in Norway. The Bigganjarga tillite overlies a striated bedrock surface on sandstone of the Veidnesbotn Formation.

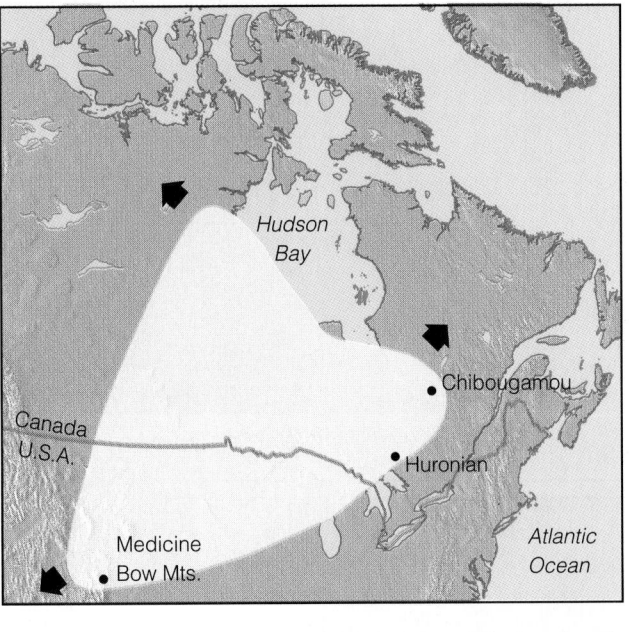

FIGURE 9.12 Glacial deposits of about the same age in Ontario and Quebec, Canada, and in Michigan and Wyoming suggest that an Early Proterozoic ice sheet was centered southwest of Hudson Bay.

9.2). Also remarkable are the consistent layering and extent of BIFs in shallow marine environments and their chemical purity; BIFs are composed mostly of iron oxides and silicon dioxide as chert.

Iron is a highly reactive element. In the presence of oxygen, it combines to form rustlike oxides that are not readily soluble in water. In the absence of free oxygen, however, iron is easily taken into solution and can accumulate in large quantities in the world's oceans. As we mentioned in Chapter 8, the Archean atmosphere was deficient in free oxygen; so it seems likely that little oxygen was dissolved in seawater. The increase in abundance of stromatolites about 2.3 billion years ago resulted in an increase in free oxygen in the oceans, because oxygen is a metabolic waste product of photosynthesizing cyanobacteria that form stromatolites. Apparently, this introduction of free oxygen into the world's oceans helped cause the precipitation of dissolved iron and silica and thus the formation of the BIFs.

One model for BIF precipitation involves a Precambrian ocean with an upper oxygenated layer overlying a large volume of oxygen-deficient water that contained reduced iron and silica. Upwelling of these deep, iron- and silica-rich waters onto newly created shallow continental shelves resulted in large-scale precipitation of iron oxides and silica as BIFs (Figure 9.15). A likely major source for this iron and silica was submarine volcanism, although weathering of the continents probably supplied iron as well. Recent research has shown that hydrothermal vents near spreading ridges (see Perspective 8.1) remove large quantities of iron and silica from the oceanic crust. The hy-

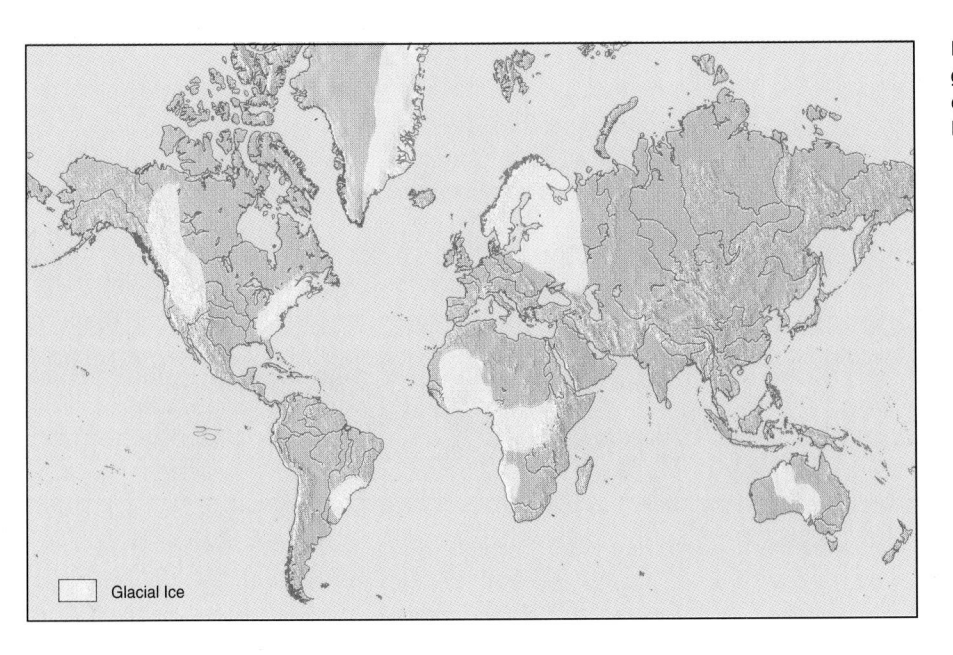

FIGURE 9.13 Major Late Proterozoic glacial centers shown on a map with the continents in their present positions. Extent of ice centers is approximate.

(a)

(b)

FIGURE 9.14 Early Proterozoic banded iron formation. (a) At this outcrop in Ishpeming, Michigan, the rocks are brilliantly colored alternating layers of red chert and silver iron minerals. (b) A more typical exposure of the formation near Negaunee, Michigan.

| TABLE 9.2 | |

Times of Deposition of Banded Iron Formations

TIME OF DEPOSITION (BILLIONS OF YEARS	TOTAL BIF DEPOSITED
3.5–3.0	6%
2.5–2.0	92
1.0–0.5	2

drothermally derived iron and silica combined with free oxygen released by photosynthesizing organisms and triggered the precipitation of massive quantities of BIFs between 2.5 and 2.0 billion years ago.

RED BEDS

Continental **red beds**—red sandstones and shales—first appeared about 1.8 billion years ago, following the deposition of the Proterozoic banded iron formations. The color of red beds is caused by the presence of ferric oxide, usually as the mineral hematite (Fe_2O_3), which forms under oxidizing conditions (Figure 9.10b). These deposits become increasingly abundant through the Proterozoic and are quite common in the Phanerozoic.

The appearance of continental red beds indicates an oxidizing atmosphere. But as we stated earlier, the Proterozoic atmosphere contained little free oxygen, perhaps as little as 1 to 2% in the early part of that eon. Is this sufficient to account for oxidized iron in sediments? Probably not; other attributes of this atmosphere must be considered.

Before abundant free oxygen was present, no ozone (O_3) layer existed in the upper atmosphere. Accordingly, as free oxygen (O_2) was released to the atmosphere as a waste product during photosynthesis, ultraviolet radiation

FIGURE 9.15 Depositional model for the origin of banded iron formations.

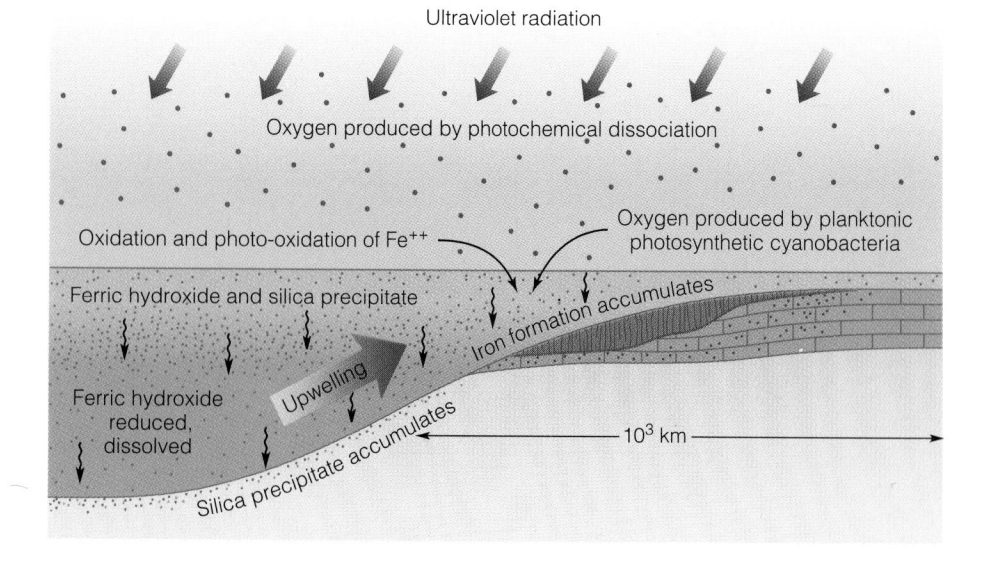

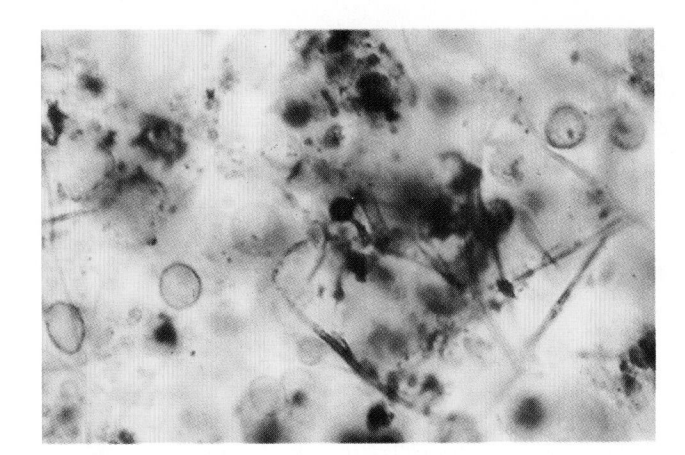

FIGURE 9.16 Photomicrographs of spheroidal and filamentous microfossils from stromatolitic chert of the Gunflint Iron Formation, Ontario.

Proterozoic Life

In the last chapter, we noted that Archean fossils are not common, and all are varieties of bacteria and cyanobacteria. The fossil record of the Early Proterozoic is likewise characterized by these same organisms. Apparently little diversification had taken place up to this time in the history of life. Thus, only unicellular prokaryotes existed during the time from the earliest known fossils about 3.5 billion years ago until the appearance of more complex eukaryotic cells about 2.1 billion years ago. Even the well-known 1.8- to 2.1-billion-year-old Gunflint Iron Formation of Canada, with 12 species of microorganisms, contains only bacteria and cyanobacteria (Figure 9.16).

Actually, the lack of organic diversity during the Archean and Early Proterozoic is not too surprising because prokaryotic cells reproduce asexually. Recall our discussion in Chapter 5 on the sources of variation in populations. Most variation results from sexual reproduction, during which the offspring receives half of its genes from each parent. Mutations may introduce new variations into any population, but the effects of mutations in prokaryotes are limited.

Consider this. Suppose a beneficial mutation occurred in a bacterium. How could this mutation spread through-

converted some of it into O and O_3, both of which oxidize surface materials more efficiently than O_2. Once an ozone layer became established in the upper atmosphere, most of the ultraviolet radiation failed to penetrate to the surface, and O_2 became the primary agent for oxidation of surface deposits.

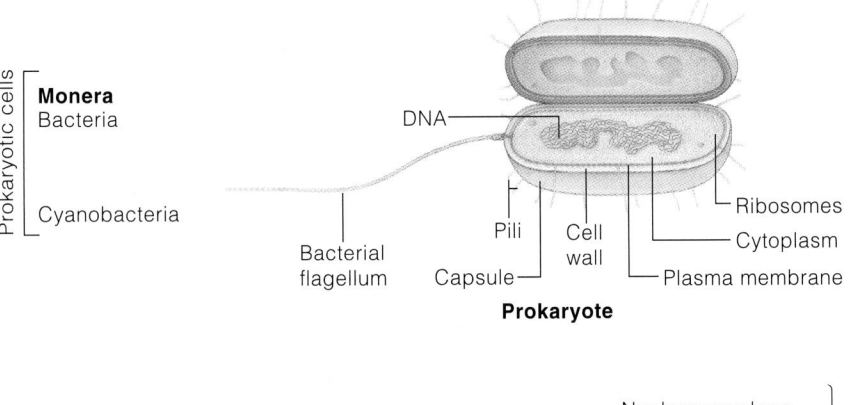

Prokaryotic cells

Monera
Bacteria

Cyanobacteria

DNA
Bacterial flagellum
Pili
Capsule
Cell wall
Plasma membrane
Ribosomes
Cytoplasm

Prokaryote

FIGURE 9.17 Prokaryotic and eukaryotic cells. Note that eukaryotic cells have a cell nucleus containing the genetic material and several organelles such as mitochondria and plastids, as well as chloroplasts in plant cells.

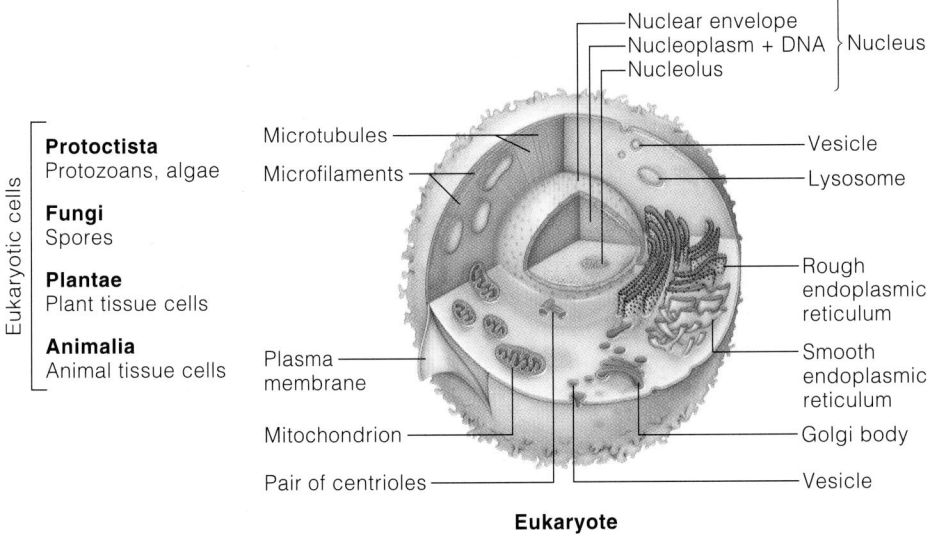

Eukaryotic cells

Protoctista
Protozoans, algae

Fungi
Spores

Plantae
Plant tissue cells

Animalia
Animal tissue cells

Nuclear envelope
Nucleoplasm + DNA — Nucleus
Nucleolus
Microtubules
Microfilaments
Vesicle
Lysosome
Rough endoplasmic reticulum
Smooth endoplasmic reticulum
Plasma membrane
Mitochondrion
Pair of centrioles
Golgi body
Vesicle

Eukaryote

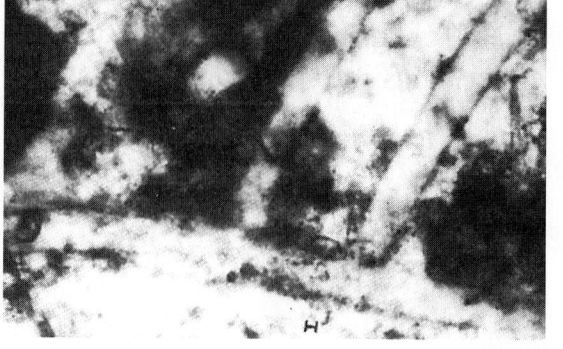
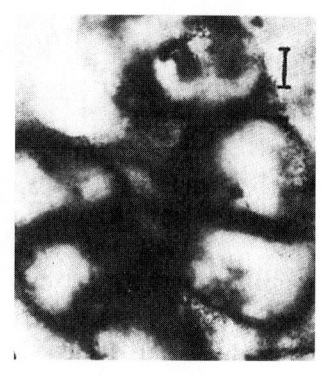

FIGURE 9.18 (a) Microfossils from the 750- to 800-million-year-old Beck Springs Dolomite of California. These are probably eukaryotic cells.

out the population? It could not, since the bacterium with the mutation does not share its genes with other bacteria.* By contrast, a beneficial mutation in a sexually reproducing organism could spread quickly and widely in an interbreeding population.

Before the appearance of cells capable of sexual reproduction, evolution was a comparatively slow process, accounting for the low organic diversity during the Archean and Early Proterozoic. This situation did not persist. By the Middle Proterozoic, cells appeared that reproduced sexually, and the tempo of evolution picked up markedly.

A NEW TYPE OF CELL APPEARS

The appearance of **eukaryotic cells** marks one of the most important events in life history. Eukaryotic cells are considerably more organizationally complex than prokaryotic cells and have a membrane-bounded nucleus that contains the genetic material (Figure 9.17). Organisms composed of eukaryotic cells are *eukaryotes* whereas those with prokaryotic cells are *prokaryotes.* Most eukaryotes are multicellular, although those of the kingdom Protoctista are unicellular (Appendix B). In marked contrast to prokaryotes, most eukaryotes reproduce sexually and most are aerobic; thus, eukaryotes could not have appeared until some free oxygen was present in the atmosphere.

A number of Proterozoic fossil localities have yielded unicellular eukaryotes, including the 750- to 800-million-year-old Beck Springs Dolomite of southeastern California (Figure 9.18a). And the 1-billion-year-old Bitter Springs Formation of Australia has fossils resembling unicellular algae that show evidence of both mitosis and meiosis, processes used only by eukaryotic cells. Recently, though, 2.1-billion-year-old fossils recovered from the Negaunee Iron Formation of Michigan are regarded as the oldest known eukaryotes (Figure 9.18b and Table 9.3).

The fossil evidence for the appearance of eukaryotic cells comes from size and relative complexity. Eukaryotic cells are larger, commonly much larger, than prokaryotic cells (Figure 9.17). Eukaryotic cells evolved more than 2

billion years ago, but cells larger than 60 microns appear in abundance about 1.4 billion years ago. As for relative complexity, prokaryotic cells are typically simple spherical or platelike structures. Proterozoic fossils of branched filaments, flask-shaped organisms, and some containing what appear to be internal, membrane-bounded structures all indicate a eukaryotic level of organization. Theories of how eukaryotic cells originated from prokaryotic cells are considered in Perspective 9.2.

Additional evidence for eukaryotes in the Proterozoic comes from a fossil group called *acritarchs.* These hollow fossils are probably the cysts of planktonic algae (Figure 9.19a and b), which became quite abundant during the Late Proterozoic but first appeared about 1.4 billion years ago. Numerous microfossils with vase-shaped skeletons have been recovered from Late Proterozoic rocks of the Grand Canyon (Figure 9.19c). These have been tentatively identified as cysts of some kind of algae.

MULTICELLULAR ORGANISMS

Multicellular organisms are not only composed of many cells, often billions, but also have cells specialized to perform specific functions such as reproduction and respiration. We know from fossils that multicellular organisms appeared in the Late Proterozoic, but we have no fossil evidence of the transition from their unicellular ancestors.

The study of present-day organisms gives some clues about how this transition may have occurred. Perhaps some unicellular organism divided and formed a group of cells that did not disperse but remained together as a colony. The cells in some colonies may have become somewhat specialized, similar to the situation in the living *colonial organisms* (Figure 9.20). Further specialization might have led to simple multicellular organisms such as sponges that consist of cells specialized for reproduction, respiration, and food gathering. In other words, specialization of cells might have led to the development of organs with specific functions.

Is there any particular advantage to being multicellular? After all, until the Late Proterozoic, life seems to have thrived even though all known organisms were unicellular. In fact, unicellular organisms are quite successful at what they do, but what they do is limited. For example, they

---

*Bacteria usually reproduce by binary fission and give rise to two cells having the same genetic makeup. Under some conditions, bacteria engage in conjugation during which some genetic material is transferred from one cell to another.

TABLE 9.3

Summary Chart for Proterozoic Fossils Discussed in the Text (dashed vertical lines indicate uncertainties in age ranges)

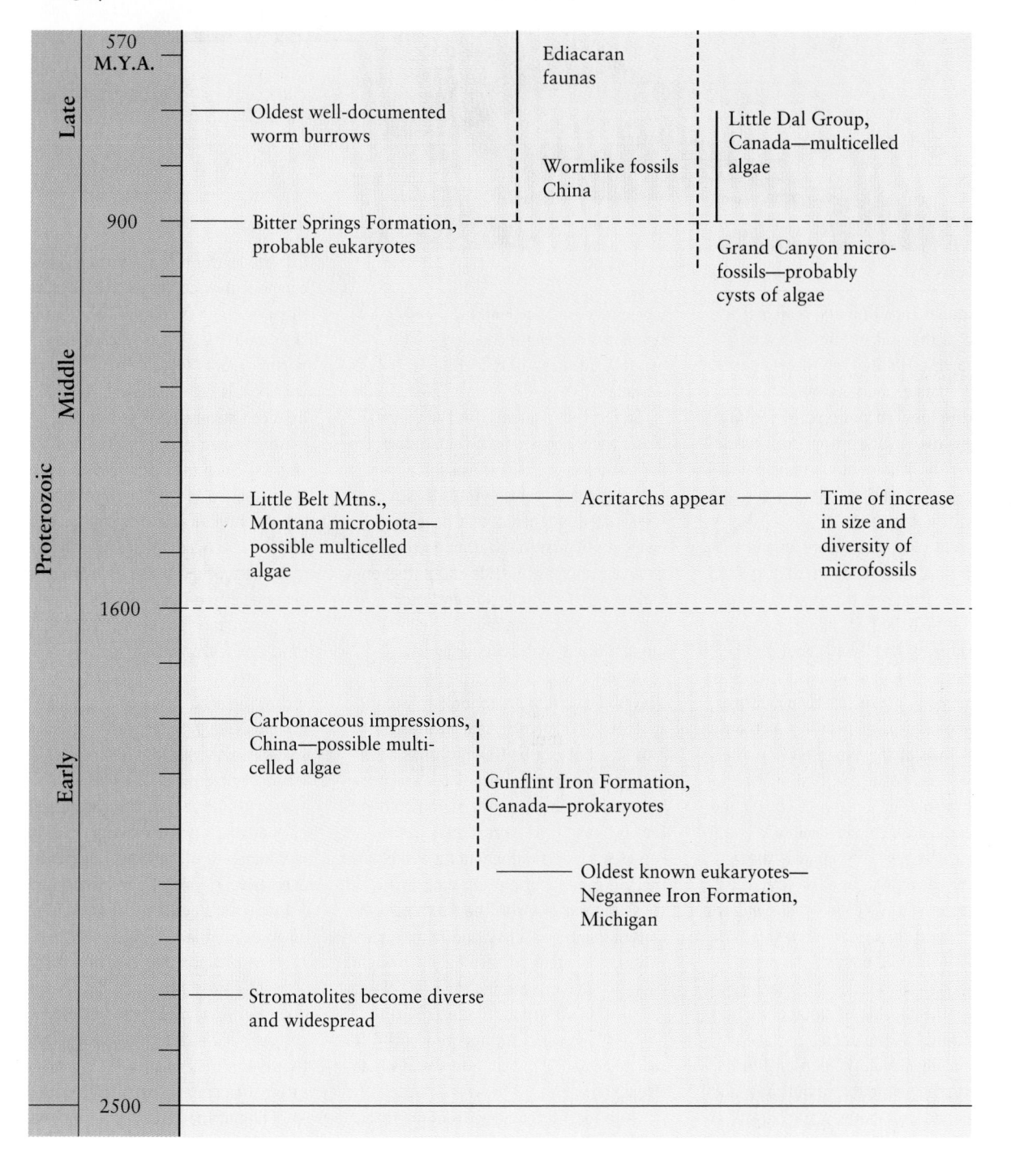

cannot become very large, because as size increases, proportionally less of the cell is exposed to the external environment in relation to its volume. In other words, as size increases, the amount of surface area compared to volume decreases; consequently, the process of transferring materials from the external environment into the cell becomes less efficient. Multicellular organisms live longer, since cells can be replaced; and more offspring can be produced, because some cells are specialized for reproduction. And finally, cells have increased functional efficiency as they become specialized into organs.

Multicellular Algae Well-preserved carbonaceous impressions of multicellular algae are known from rocks

The Origin of Eukaryotic Cells

*S*tratigraphic position and radiometric ages tell us that eukaryotic cells were present by the Middle Proterozoic. But although we know when eukaryotes appeared, the fossil record reveals nothing about how they evolved from prokaryotic ancestors. The study of living microorganisms, however, gives us some idea of how this may have occurred.

A currently popular theory among evolutionary biologists holds that eukaryotic cells formed from several prokaryotic cells that had established a symbiotic relationship. Symbiosis, the living together of two or more dissimilar organisms, is quite common among organisms today. It may take the form of parasitism in which one organism lives at the expense of another, or the two may coexist with mutual benefit. For example, lichens, once considered to be plants, are actually symbiotic fungi and algae. The type of symbiosis that might have given rise to eukaryotic cells is most appropriately called *endosymbiosis* because one of the symbionts resided permanently inside the body of the other.

In a symbiotic relationship, each symbiont must be capable of metabolism and reproduction, but the degree of dependence in many symbiotic relationships is such that the symbionts cannot grow and reproduce independently. Many parasites, for example, cannot exist outside a host organism. This may have been the case with Proterozoic prokaryotes: Two or more prokaryotes may have entered into a sym-

biotic relationship (Figure 1), and the symbionts became increasingly interdependent until the unit could exist only as a whole.

Supporting evidence for the endosymbiosis theory comes from the study of living eukaryotic cells. For example, eukaryotic cells contain internal structures called organelles that have their own genetic material. Although these organelles, such as plastids and mitochondia, cannot exist independently today, they apparently were once capable of reproduction as free-living organisms.

Another way of evaluating the endosymbiosis theory is to look at protein synthesis. Prokaryotic cells synthesize proteins, but can be thought of as a single system, whereas eukaryotes are a combination of protein-synthesizing systems; that is, some of the organelles within eukaryotes such as mitochondria and plastids are capable of protein synthesis. These organelles, with their own genetic material and protein-synthesizing capabilities, are thought to have been free-living bacteria that entered into a symbiotic relationship. With time, the interdependence of the various units grew until life was possible only as an integrated whole (Figure 1).

More recently, another theory for the origin of eukaryotic cells has been proposed. According to this theory, the sequence of events leading to the origin of eukaryotes began when early prokaryotes

lost the ability to synthesize an essential component of cell walls. One solution to this problem was to develop an internal skeleton consisting of microtubules and microfilaments (Figure 2a). This cytoskeleton, as it is called, allowed for an outer fluid cell membrane that infolded so that material coming into the cell could be enveloped. Such infolding is thought to have resulted in the origin of such intracellular structures as the cell nucleus (Figure 2b). According to this theory, the first eukaryotic cell was anaerobic; later it acquired a free-living aerobic organism by symbiosis, thus accounting for the origin of aerobic eukaryotes.

Although the fossil record does not record the acquisition of organelles or symbiosis, living eukaryotes can give some idea of what the first eukaryotes may have been like. The present-day giant amoeba *Pelomyxa,* which lives in the mud of ponds, lacks mitochondria. Two types and hundreds of individual bacteria, however, have a symbiotic relationship with *Pelomyxa* and perform the same function as mitochondria.

Pelomyxa provides evidence for the symbiotic theory for the origin of eukaryotes. However, recent studies of *Giardia,* a single-celled eukaryote, seem to indicate that eukaryotes may have acquired their internal membrane-bounded organelles by infolding of the cell wall. Although *Giardia* is a eukaryote, it shares many characteristics with prokaryotes and is capable of acquiring materials from outside by infolding of the cell wall.

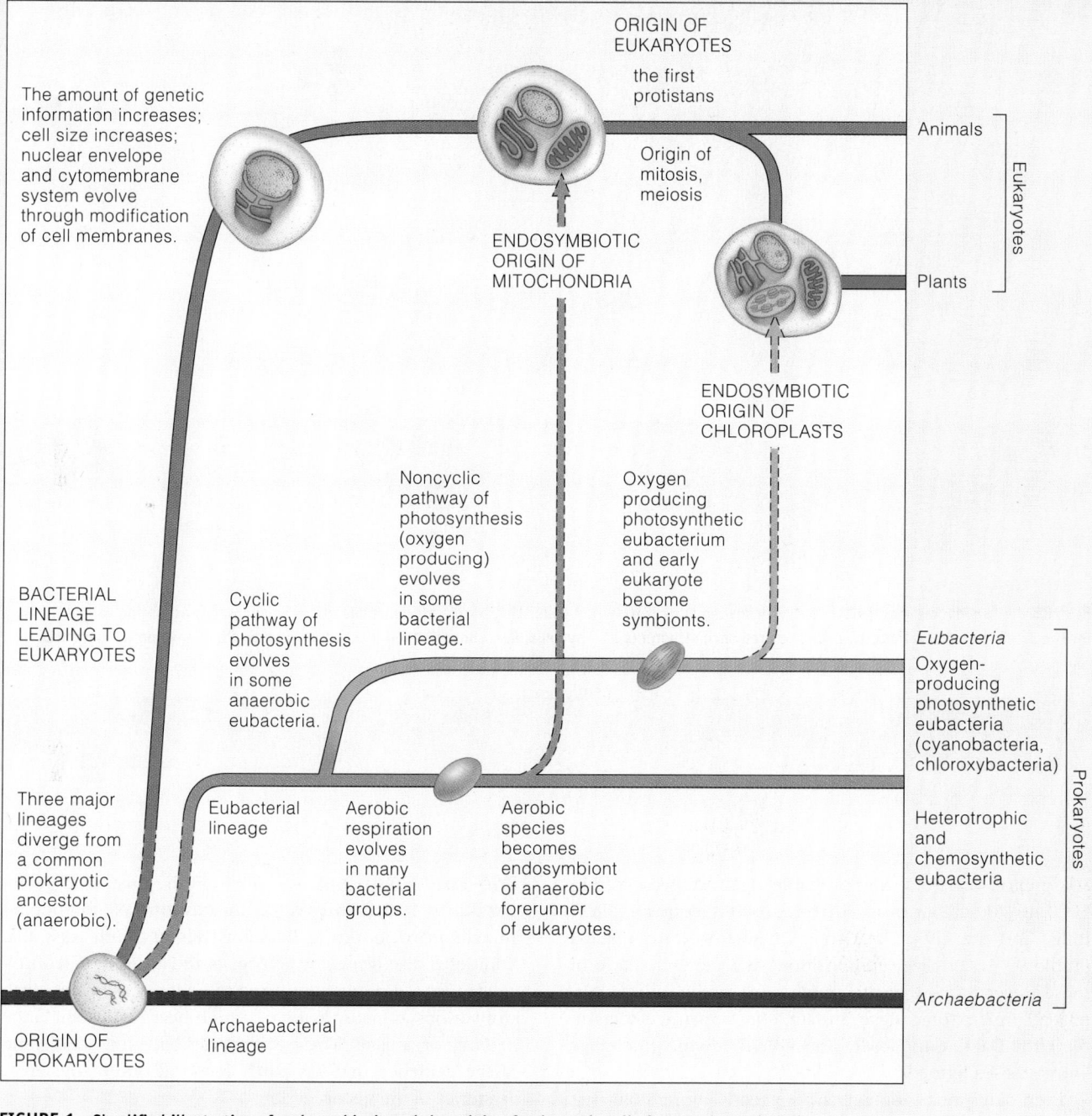

FIGURE 1 Simplified illustration of endosymbiosis and the origin of eukaryotic cells from prokaryotic cells.

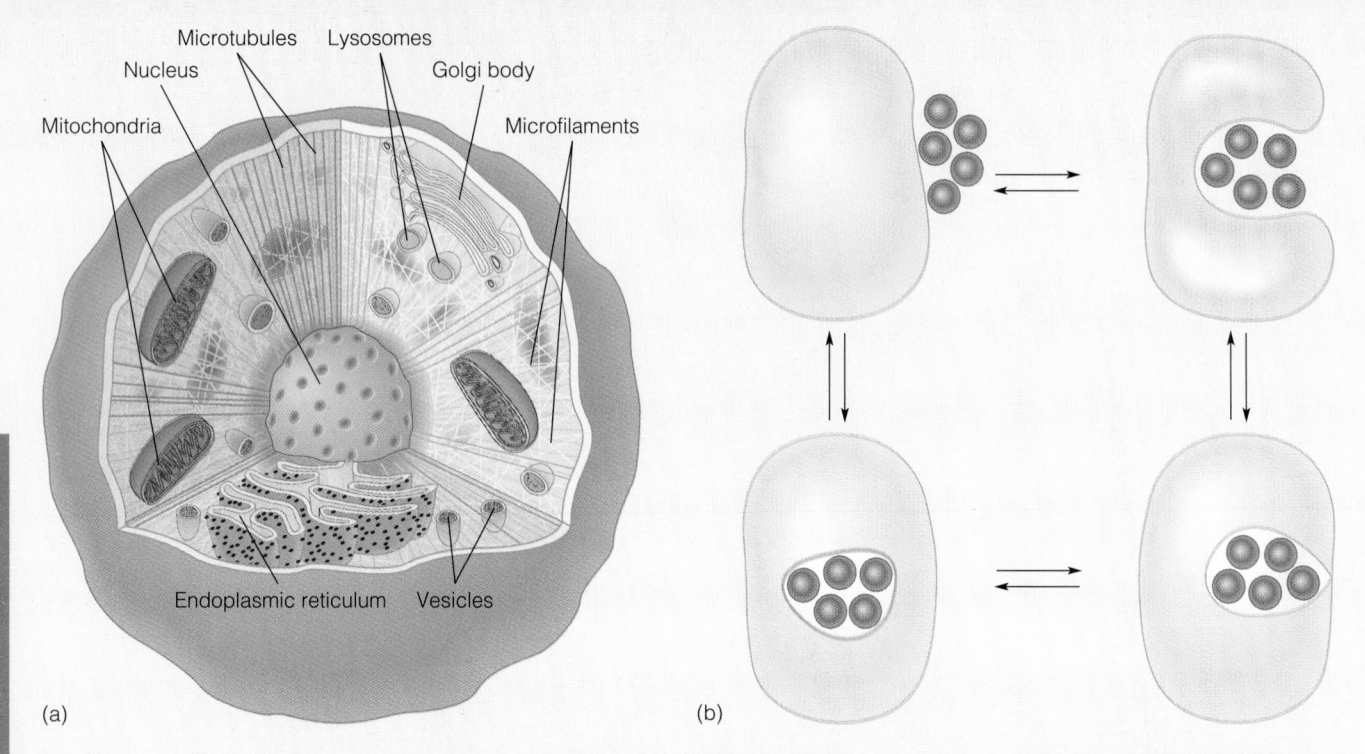

FIGURE 2 (a) Diagrammatic view of a cell showing the cytoskeleton consisting of microtubules and microfilaments. (b) According to one theory for the origin of eukaryotic cells, cells acquired such structures as a nucleus, mitochondria, and plastids by infolding and enveloping materials.

800 million years old. And probable algae are preserved in 1000- to 700-million-year-old rocks in Spitzbergen, China, India, and the Little Dal Group of northwestern Canada (Table 9.3). The size, composition, and general shape of these impressions indicate photosynthesizing eukaryotes, probably planktonic algae. In fact, some organic sheets in the Little Dal Group closely resemble the living green alga known as sea lettuce.

Even older rocks contain carbonaceous impressions that may be multicellular algae. For example, the 1.4-billion-year-old microbiota from the Little Belt Mountains of Montana contains filaments and spherical forms of uncertain affinities (Figure 9.21). Carbonaceous macroscopic filaments from 1.8-billion-year-old rocks of China are also suggestive of multicellular algae, and the 2.1-billion-year-old fossils from the Negaunee Iron Formation of Michigan are thought to be some kind of multicellular algae.

The Ediacaran Fauna In 1947 the Australian geologist R. C. Sprigg discovered impressions of soft-bodied animals in rocks of the Ediacara Hills of South Australia. Additional discoveries by geologists and amateur paleontologists turned up what appeared to be impressions of algae and various animals, many of which bear no resemblance to living organisms. These discoveries have provided some of the evidence that has partly resolved one of the great mysteries in the history of life.

Before the discovery of these Ediacaran fossils, geologists were perplexed by the apparent absence of animal fossils in strata older than the Cambrian. The shelly faunas of the Cambrian Period seemed to have appeared abruptly in the fossil record. In fact, the evidence for life before the Cambrian was so sparse that all pre-Paleozoic time was once referred to as the *Azoic,* meaning without life. Geologists assumed that the Cambrian shelly faunas must have been pre-

(a)

(b)

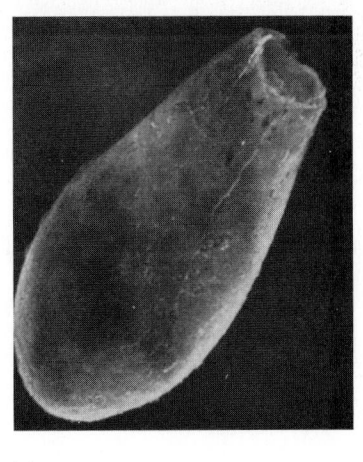
(c)

FIGURE 9.19 These common Late Proterozoic microfossils are thought to represent eukaryotic organisms. (a) and (b) Acritarchs are probably the cysts of algae. (c) Vase-shaped microfossil, probably a cyst of some kind of algae.

ceded by more primitive animals, but since none were known, a worldwide unconformity was proposed to account for the absence of such fossil-bearing rocks.

The rock unit in which the Ediacara Hills fossils were discovered, the Pound Quartzite, was initially thought to be Cambrian. However, a joint investigation by the South Australian Museum and the University of Adelaide demonstrated that the fossil-bearing strata lie more than 150 m below the oldest recognized Cambrian strata. Eventually, it became clear that what had been discovered was a unique assemblage of soft-bodied animals preserved as molds and casts on the undersides of sandstone layers (Figure 9.22). Martin Glaessner of the University of Adelaide thinks that these Ediacaran animals lived in a nearshore, shallow marine environment.

Some investigators, including Glaessner, are of the opinion that at least three present-day invertebrate phyla are represented: jellyfish and sea pens (phylum Cnidaria), segmented worms (phylum Annelida), and primitive members of the phylum Arthropoda, the phylum that includes insects, spiders, and crabs. One wormlike Ediacaran fossil, *Spriggina,* has been cited as a possible ancestor of trilobites, and another, *Tribachidium,* may be a primitive echinoderm (Figure 9.22b and c).

Researchers disagree, however, on exactly what these Ediacaran animals were and how they should be classified. Adolph Seilacher of Tübingen, Germany, for example, thinks that they represent an early evolutionary radiation quite distinct from the ancestry of the present-day invertebrate phyla. In other words, he thinks that the Ediacaran animals are not members of the phyla noted above and that they are not ancestral to the complex invertebrates that appeared in abundance during the Cambrian Period. In fact, Seilacher thinks they represent an evolutionary "dead end." He also thinks that these animals fed by passive absorption because there is no conclusive evidence of a mouth in the preserved specimens.

Ediacara-type faunas are now known on all continents except Antarctica. Collectively, these **Ediacaran faunas,** as they are commonly called, lived between 570 and 670 million years ago. These animals were widespread during this time, but their fossils are rare. Their scarcity should come as no surprise, though, because all lacked durable skeletons.

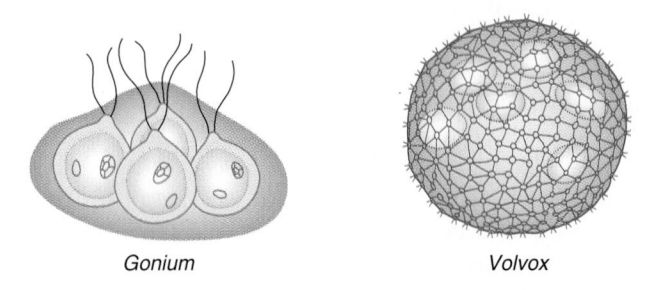

Gonium   *Volvox*

FIGURE 9.20 Living unicellular versus multicellular organisms. Although *Gonium* consists of as few as four cells, all cells are alike and can produce a new colony. *Volvox* has some cells specialized to perform different functions and has thus crossed the threshold that separates unicellular from multicellular organisms.

FIGURE 9.21 Carbonaceous impressions in Proterozoic rocks in the Little Belt Mountains, Montana. These may be impressions of multicellular algae, but this is uncertain.

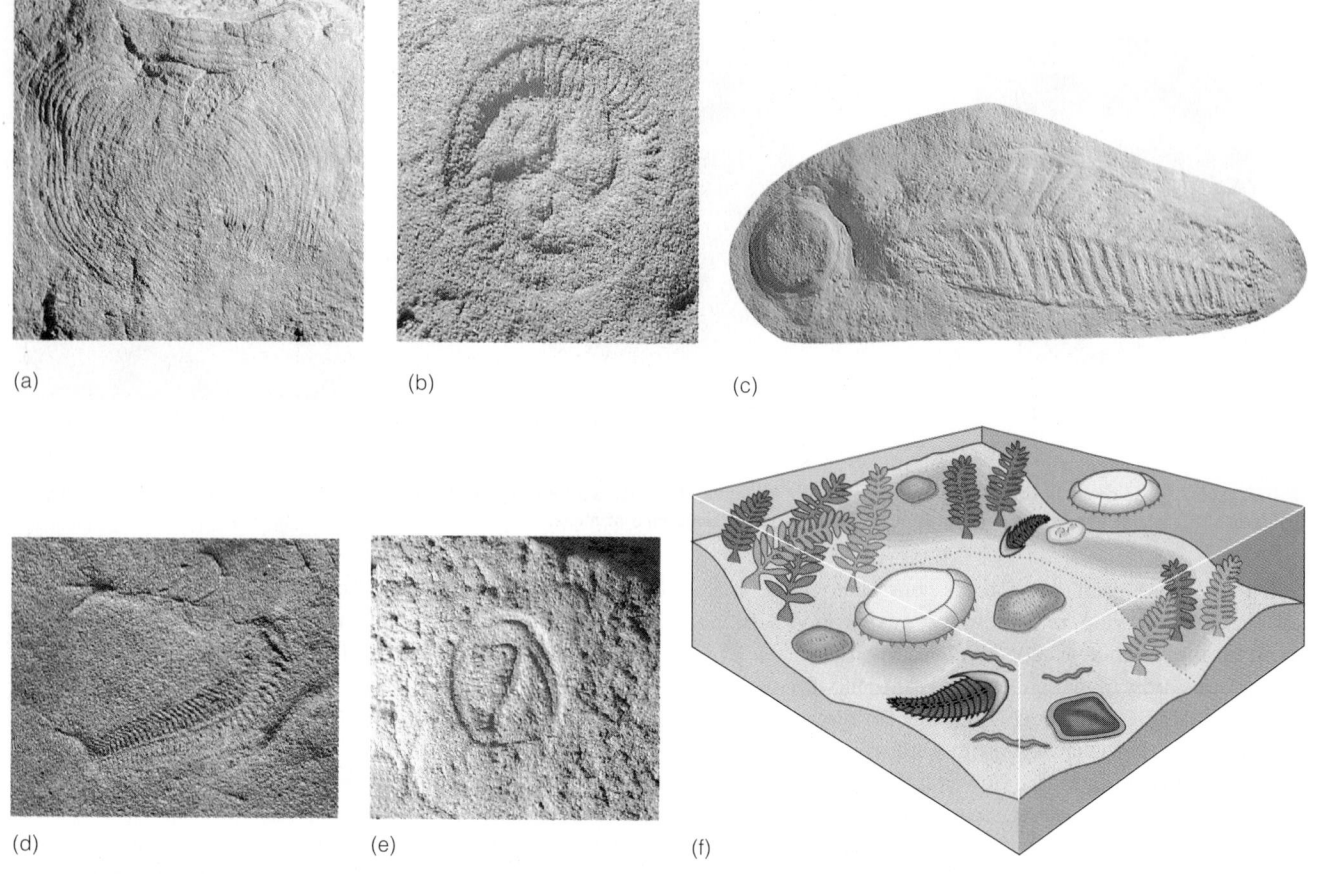

FIGURE 9.22   The Ediacaran fauna of Australia. Impressions of multicelled animals: (a) *Ovatoscutum concentricum;* **(b)** *Tribrachidium heraldicum,* a possible primitive echinoderm; **(c)** *Charniodiscus arboreus;* **(d)** *Spriggina floundersi,* a possible ancestor of trilobites; and **(e)** *Parvancorina minchami;* **(f)** Reconstruction of the Ediacaran environment.

The discovery of pre-Paleozoic fossil-bearing strata has prompted Preston Cloud, formerly of the University of California at Santa Barbara, and Glaessner to propose a new geologic period and system, the Ediacarian. According to this proposal, the Ediacarian Period constitutes the first period of the Paleozoic Era, and the stratotype for the Ediacarian System is in South Australia. Thus, strata of the Ediacarian System were deposited during the Ediacarian Period, which began 590 million years ago when multicelled organisms appeared in the fossil record and ended about 545 million years ago when shelly faunas of the Cambrian Period first appeared.

Cloud and Glaessner's proposal is a reasonable one and has been accepted by many geologists. Many other geologists, however, prefer the more traditional time scale, and in this book we consider the Ediacaran fauna to be latest Proterozoic in age.

Other Proterozoic Animal Fossils Although scarce, some evidence for pre-Ediacaran animals is known. For example, a jellyfish-like impression is present in rocks 2000 m below the Pound Quartzite. And in many areas, burrows—presumably made by worms—are found in rocks at least 700 million years old.

Wormlike fossils associated with fossil algae were recently reported from 700- to 900-million-year-old strata in China (Figure 9.23, Table 9.3). Perhaps these are worms, but both their biological affinities and their age have been questioned. For the present, we can consider these wormlike fossils as persuasive but not conclusive evidence for pre-Ediacaran animals.

All known Proterozoic animals were soft bodied; that is, they lacked the durable exoskeletons that characterize many Phanerozoic invertebrates. There is some evidence, however, that the earliest stages of skeletonization occurred during the Late Proterozoic. For example, some Ediacaran animals may have had chitin, possibly a chitinous carapace, and others may have possessed some calcareous skeletal elements. And the odd creature known as *Kimberella* from the latest Proterozoic of Russia had a tough outer covering similar to that of some present-day marine invertebrates. Paleontologists disagree on what *Kimberella* was; some think it was a sluglike creature, whereas others say it is mollusklike.

By the latest Proterozoic, several animals with skeletons probably existed. Evidence for this conclusion comes from minute scraps of shell-like material and denticles from larger animals, and spicules, presumably from sponges.

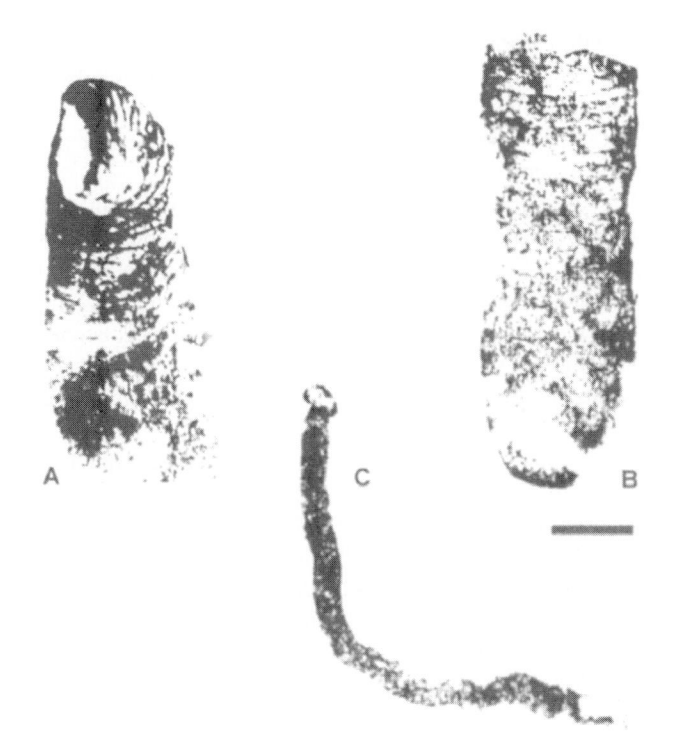

FIGURE 9.23 Wormlike body fossils from the Late Proterozoic of China.

Durable skeletons of chitin (a complex organic substance), silica, and calcium carbonate, however, began appearing in abundance at the beginning of the Phanerozoic Eon 545 million years ago.

Proterozoic Mineral Deposits

The most notable mineral deposits of Proterozoic age are banded iron formations (BIFs). BIFs are present in all Precambrian cratons; as noted earlier, 92% of all BIFs were deposited during the Late Proterozoic, although a few Archean and Phanerozoic examples are known. These deposits constitute the world's major iron ores. The largest producers of iron ores include Brazil, Australia, China, Ukraine, Sweden, South Africa, Canada, and the United States. Even though the United States is a major producer, it must still import about 30% of the iron ore used, mostly from Canada and Venezuela.

In North America, most of the large iron mines are in the Lake Superior region. Huge deposits of BIFs in Ontario, Canada, and adjacent states have been mined extensively for decades. Minnesota is the leading producer of iron ore in the United States, accounting for more than 70% of the total production, while Michigan produces most of the rest.

The richest ores—those containing up to 70% iron—in the Lake Superior region were depleted by the time of World War II. Continued mining has been possible because a method was developed to separate the iron ore from unusable rock of lower-grade ores and then shape the iron into pellets (Figure 9.24). These pellets contain about 65% iron and are easily shipped via the Great Lakes to the steel-producing centers.

The Sudbury mining district in Ontario, Canada, is an important area of nickel and platinum production. Nickel is essential in the production of nickel alloys such as stain-

(a)

(b)

FIGURE 9.24 (a) The Empire Mine at Palmer, Michigan. Iron ore from the Early Proterozoic Negaunee Iron Formation is mined and shaped into (b) pellets containing about 65% iron. The pellets measure about 1 cm in diameter.

less steel and Monel metal (nickel plus copper), which are valued for their strength and resistance to corrosion and heat. The United States must import more than 50% of all nickel used; most of these imports come from the Sudbury mining district.

Besides its economic importance, the Sudbury Basin, an elliptical area measuring more than 59 by 27 km, is interesting from the geologic perspective. One hypothesis for the concentration of ores is that they were mobilized from metal-rich rocks beneath the basin following a high-velocity meteorite impact.

Some platinum for jewelry, surgical instruments, and chemical and electrical equipment is also exported to the United States from Canada, but the major exporter is South Africa. The Bushveld Complex of South Africa is a layered complex of igneous rocks from which both platinum and chromite, the only ore of chromium, are mined. Much of the chromium used in the United States is imported from South Africa; it is used mostly in the manufacture of stainless steel.

Economically recoverable oil and gas have been discovered in Proterozoic rocks in China and Siberia, arousing some interest in the Midcontinent rift as a potential source of hydrocarbons (Figure 9.6a). So far, considerable land has been leased for exploration, and numerous geophysical studies have been done. However, even though some rock units within the rift are known to contain petroleum, no producing oil or gas wells are operating.

A number of Proterozoic pegmatites are important economically. The Dunton pegmatite in Maine, whose age is generally considered to be Late Proterozoic, has yielded magnificent gem-quality specimens of tourmaline and other minerals. Other pegmatites are mined for gemstones as well as for tin; industrial minerals, such as feldspars, micas, and quartz; and minerals containing such elements as cesium, rubidium, lithium, and beryllium.

More than 20,000 pegmatites have been identified in the country rocks adjacent to the Harney Peak Granite in the Black Hills of South Dakota. These pegmatites formed about 1.7 billion years ago when the granite was emplaced as a complex of dikes and sills. A few of these have been mined for gemstones, tin, lithium, and micas. In addition, some of the world's largest known mineral crystals have been discovered in these pegmatites.

Summary

1. The crust-forming processes that characterized the Archean continued into the Proterozoic but at a considerably reduced rate.

2. Archean cratons served as nuclei about which Proterozoic crust accreted. One large landmass formed by this process is called Laurentia. It consisted mostly of North America and Greenland.

3. The major events in the Proterozoic evolution of Laurentia were an Early Proterozoic episode of amalgamation of cratons, Middle Proterozoic igneous activity, and the Middle Proterozoic Grenville orogeny and Midcontinent rift.

4. By the Late Proterozoic, three or more supercontinents probably existed, Laurasia and East and West Gondwana. Laurasia consisted of Laurentia and what is now Europe and most of Asia. East and West Gondwana may have united in the latest Proterozoic to form Gondwana.

5. Greenstone belts formed during the Proterozoic but at a reduced rate, and they differ in detail from their Archean counterparts.

6. Ophiolite sequences, which mark convergent plate margins, are first well documented from the Early Proterozoic. A present-day style of plate tectonics seems to have been established during the Early Proterozoic.

7. Sandstone–carbonate–shale assemblages are known from the Late Archean but become common in Proterozoic rocks. These rock assemblages were deposited on passive continental margins and in intracratonic basins.

8. Widespread glaciation took place during the Early and Late Proterozoic.

9. The atmosphere became progressively richer in free oxygen through the Proterozoic. Photosynthesizing cyanobacteria were largely responsible for oxygena-

tion of the atmosphere, although photochemical dissociation of water vapor contributed some oxygen.

10. During the period from 2.5 to 2.0 billion years ago, most of the world's iron ores were deposited as banded iron formations.

11. The first continental red beds were deposited about 1.8 billion years ago. The widespread occurrence of oxidized iron in sedimentary rocks indicates an oxidizing atmosphere.

12. Unicellular prokaryotes are the only known lifeforms from the Early Proterozoic. Eukaryotic cells first appeared during the Middle Proterozoic.

13. The oldest fossils of multicellular organisms are carbonaceous impressions, probably of algae, in rocks between 1 billion and 700 million years old.

14. The Late Proterozoic Ediacaran faunas include the oldest well-documented animal fossils other than burrows. Animals were widespread at this time, but all were soft bodied, so fossils are not common.

15. Most of the world's iron ore production is from Proterozoic banded iron formations. Other important resources include nickel, platinum, and a variety of materials mined from pegmatites.

Important Terms

banded iron formation (BIF)

Ediacaran faunas

eukaryotic cell

Grenville orogeny

Laurentia

Midcontinent rift

multicellular organism

ophiolite

orogen

red beds

Rodinia

sandstone–carbonate–shale assemblage

Review Questions

1. Rocks of the Pound Quartzite in the Ediacara Hills of Australia:
   a. _____ contain vast deposits of platinum and chrome.
   b. _____ show evidence of Proterozoic glaciation.
   c. _____ have impressions of the oldest well-documented animal fossils.
   d. _____ were strongly deformed during the Grenville orogeny.
   e. _____ were deposited in the Midcontinent rift.

2. The Proterozoic Eon ended _____ million years ago.
   a. _____ 545
   b. _____ 2.5
   c. _____ 3800
   d. _____ 225
   e. _____ 4.2

3. Which one of the following rock associations is typical of passive continental margins?
   a. _____ Sandstone-granite-basalt
   b. _____ Basalt-andesite-ash fall
   c. _____ Granite-tillite-banded iron formation
   d. _____ Stromatolite-prokaryote-lava
   e. _____ Sandstone-carbonate-shale.

4. Most of the evidence for the origin of eukaryotic cells from prokaryotic cells comes from:

a. _____ chemicals preserved in Archean rocks.
b. _____ fossils in tillite.
c. _____ studies of present-day organisms.
d. _____ tracks and trails of trilobites.
e. _____ Early Proterozoic carbonate rocks.

5. The Middle Proterozoic of Laurentia was a time of:
   a. _____ the origin of most of Earth's greenstone belts.
   b. _____ igneous activity, rifting, and the Grenville orogeny.
   c. _____ appearance of the first stromatolites.
   d. _____ origin of animals with durable skeletons of chitin, calcium carbonate, and silicon dioxide.
   e. _____ widespread glaciation and meteorite impacts.

6. One indication of the presence of free oxygen in the Proterozoic atmosphere 1.8 billion years ago is:
   a. _____ continental red beds.
   b. _____ animals much like those of the present.
   c. _____ extinction of stromatolites.
   d. _____ evolution of prokaryotes from protobionts.
   e. _____ deposition of passive margin sediments in the Lake Superior region.

7. Which one of the following statements is correct?
   a. _____ 92% of all banded iron formations formed during the Late Archean.

b. _____ Eukaryotic cells have a membrane-bounded nucleus.

c. _____ Proterozoic plate tectonic processes were considerably different than those of the present.

d. _____ Photochemical dissociation accounts for the origin of multicellular organisms.

e. _____ Tillite indicates that the Proterozoic seas were chemically stratified.

8. The large landmass made up mostly of North America and Greenland that formed during the Proterozoic is:

a. _____ the Wyoming craton.

b. _____ Gowganda.

c. _____ Ediacara.

d. _____ Laurentia.

e. _____ the Trans-Hudson orogen.

9. The last major episode of accretion along the margin of Laurentia during the Proterozoic took place during the:

a. _____ Grenville orogeny.

b. _____ Midcontinent volcanic event.

c. _____ Wopmay orogeny.

d. _____ collison of the Slave and Superior cratons.

e. _____ origin of the Joruma mafic–ultramafic complex.

10. Probably the most important biological event to take place during the Proterozoic was the appearance of:

a. _____ ophiolites.

b. _____ eukaryotic cells.

c. _____ all present-day animal groups.

d. _____ ultramafic lava flows.

e. _____ trilobites and clams.

11. Archean and Proterozoic greenstone belts are similar except in the latter there is little or no:

a. _____ banded iron formation.

b. _____ tillite.

c. _____ quartzite and carbonate.

d. _____ basalt.

e. _____ ultramafic rock.

12. The associations of Late Proterozoic fossils similar to those in Australia are collectively known as the:

a. _____ Bitter Springs biota.

b. _____ Grenville microfossil complex.

c. _____ Ediacaran faunas.

d. _____ Pilbara greenstone fossils.

e. _____ multicellular prokaryote flora.

13. Explain how and when continental crust was accreted to Laurentia during the Proterozoic Eon.

14. Summarize the evidence indicating that eukaryotic cells appeared as much as 2.1 billion years ago.

15. What kind of evidence would confirm that a suspected Proterozoic tillite was actually a glacial deposit?

16. What is the Midcontinent rift, when did it form, and what kinds of rocks does it contain?

17. Of what significance are sandstone–carbonate–shale assemblages and ophiolites in deciphering Proterozoic Earth history?

18. Briefly discuss the Early Proterozoic accretion of continental crust along the margin of Laurentia.

19. How did the episode of Middle Proterozoic igneous activity affect the development of Laurentia? What kinds of rocks formed?

20. What are banded iron formations, and how are they thought to have formed? What do they indicate about the Proterozoic atmosphere?

21. What kinds of organisms are found in the Early Proterozoic fossil record, and why was organic diversity so low?

22. Discuss the endosymbiosis theory for the origin of eukaryotic cells.

23. The transition from unicellular to multicellular organisms is not recorded in the fossil record, so how can we account for this event?

24. Summarize the major differences between the Archean and Proterozoic. When did the Proterozoic begin and end?

Points to Ponder

1. Explain how continental red beds, banded iron formations, and the earliest eukaryotic cells give us some idea of the composition of the Proterozoic atmosphere.

2. The Belt-Purcell Supergroup of the northwestern United States and adjacent parts of Canada is 4000 m thick and was deposited between 1.45 billion and 850 million years ago. Calculate the average rate of sediment accumulation in millimeters per year. Why is this figure unlikely to represent the real rate of sedimentation?

3. Suppose you were to visit another planet that like Earth has oceans and continents. What evidence would indicate that this hypothetical planet's continents formed and evolved like those on Earth?

Additional Readings

Condie, K. C. 1992. *Proterozoic crustal evolution.* Amsterdam: Elsevier.

———.1997. *Plate tectonics and crustal evolution.* 4th ed. Boston: Butterworth Heinemann.

Goodwin, A. M. 1996. *Principles of Precambrian geology.* London: Academic Press.

Hanbrey, M. 1992. Secrets of a tropical ice age. *New Scientist* 133, no. 1806: 42–49.

Hoffmann, P. F. 1988. United plates of America, the birth of a craton: Early Proterozoic assembly and growth of Laurentia. *Annual Review of Planetary Sciences* 16: 543–603.

Kabnick, K. S., and D. A. Peattie. 1991. *Giardia:* A missing link between prokaryotes and eukaryotes. *American Scientist* 79, no. 1: 34–43.

McMenamin, M. A. S. 1998. *The garden of Ediacara.* New York: Columbia University Press.

Schopf, J. W., ed. 1992. *Major events in the history of life.* Boston: Jones & Bartlett.

Schopf, J. W., and C. Klein, eds. 1992. *The Proterozoic biosphere.* New York: Cambridge University Press.

Windley, B. F. 1995. *The evolving continents.* 3d ed. New York: Wiley.

► UNIVERSITY OF CALIFORNIA MUSEUM OF
PALEONTOLOGY

The Museum of Paleontology at the University of
California, Berkeley, maintains this huge web site with information on a variety of topics. Under *On-Line Exhibits*
click *Time Periods*, which brings up a geologic time scale
with clickable eras and periods. Click *Proterozoic.*

1. What is meant by oxygen build-up in the atmosphere?

2. How was the oxygen produced?

Next, click the *Localities* button and then click *Ediacara
Hills.* See the images and discussion about the fossils from
the Ediacara Hills, Australia. Look especially at the fossil
known as *Spriggina.* What kind of animal is it? How does it
compare with trilobites? Now click on *Mistaken Point,
Newfoundland* and see the fossil image and read the discussion.
Return to the *Introduction to the Proterozoic* and click
Vendian. What is the Vendian, and what localities shown
are Vendian age?

► SYMBIOTIC THEORY

This opens a site titled *Endosymbiotic Theory: Evolution
from Simple Prokaryotes to Complex Eukaryotes*, which
was designed, developed, and written by Joshua A. Bond, a
student at Dekalb College/Georgia State University. It is a
good two-page summary with figures about the endosymbiotic theory proposed by Lynn Margulis during the 1960s.
What was Margulis' prediction about the organelles of eukaryotic cells?

► STEEL—2000 MILLION YEARS IN THE MAKING

This site presented by the University of South Australia
Library has good images of banded iron formation rocks, as
well as a brief discussion of mining and refining iron ore to
produce iron and steel.

Explore the following *In-Terra-Active 2.0* CD-ROM module(s) and increase your understanding of key concepts and processes presented in this chapter.

Chapter Concept: **Pressure-Temperature Conditions of Metamorphic Rocks**

Precambrian-age cratonic areas are underlain by two main rock types, greenstone and granite gneiss, both metamorphic in origin. In this module, you will examine the textural and mineralogical changes associated with metamorphism and relate them to the pressure-temperature conditions that produce them.

Section: **Materials**

Module: **Metamorphic Rocks**

The coarse-grained garnet-mica schist shown on the left has essentially the same chemical composition as the fine-grained mudstone shown on the right. In the field, it is clear that these belong to the same stratigraphic unit.

Question: *How can this transformation take place in the solid state?*

10

Geology of the Early Paleozoic Era

Major John Wesley Powell, who led the first geologic expedition down the Colorado River through the Grand Canyon in Arizona.

Prologue

"The Grand Canyon is the one great sight which every American should see," declared President Theodore Roosevelt. "We must do nothing to mar its grandeur." And so, in 1908, he named the Grand Canyon a national monument to protect it from exploitation. In 1919 the Grand Canyon National Monument was upgraded to a national park primarily because its scenery and the geology exposed in the canyon are unparalleled.

When people visit the Grand Canyon, many are astonished by the seemingly limitless time represented by the rocks exposed in the walls. For most people, staring down 1.5 km at the rocks in the canyon is their only exposure to the concept of geologic time.

Major John Wesley Powell was the first geologist to explore the Grand Canyon region. Major Powell, a Civil War veteran who lost his right arm in the battle of Shiloh, led a group of hardy explorers down the uncharted Colorado River through the Grand Canyon in 1869. Without any maps or other information, Powell and his group ran the many rapids of the Colorado River in fragile wooden boats, hastily recording what they saw. Powell wrote in his diary that "all about me are interesting geologic records. The book is open and I read as I run."

From this initial reconnaissance, Powell led a second expedition down the Colorado River in 1871. This second trip included a photographer, a surveyor, and three topographers. This expedition made detailed topographic and geologic maps of the Grand Canyon area as well as the first photographic record of the region.

Probably no one has contributed as much to the understanding of the Grand Canyon as Major Powell. In recognition of his contributions, the Powell Memorial was erected on the South Rim of the Grand Canyon in 1969 to commemorate the hundredth anniversary of his first expedition.

When we stand on the rim and look down into the Grand Canyon, we are really looking far back in time, all the way back to the early history of our planet. More than 1 billion years of history are preserved in the rocks of the Grand Canyon, indicating episodes of mountain building as well as periods of transgressions and regressions of shallow seas.

The oldest rocks exposed in the Grand Canyon record two major mountain-building episodes during the Proterozoic Eon. The first episode, represented by the Vishnu and Brahma schists, records a time of uplift, deformation, and metamorphism. This mountain range was eroded to a rather subdued landscape and was followed by deposition of approximately 4000 m of sediments and lava flows of the Grand Canyon Supergroup. These rocks and the underlying Vishnu and Brahma schists were uplifted and formed a second Proterozoic mountain range. This mountain range was also eroded to a nearly flat surface by the end of the Proterozoic Eon.

The first sea of the Paleozoic Era transgressed over the region during the Cambrian Period, depositing sandstones, siltstones, and limestones. A major unconformity separates the Cambrian rocks from the Mississippian limestones exposed as the cliff-forming Redwall Limestone. Another unconformity separates the Redwall Limestone from the overlying Permian Kaibab Limestone, which forms the rim of the Grand Canyon.

The Grand Canyon in all its grandeur is a most appropriate place to start our discussion of the Paleozoic history of North America.

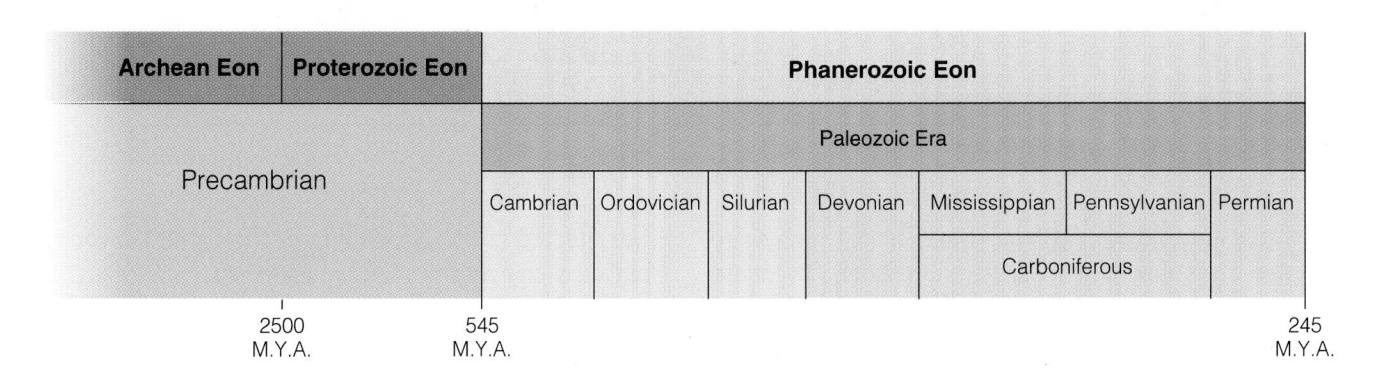

Archean Eon	Proterozoic Eon	Phanerozoic Eon							
Precambrian		Paleozoic Era							
		Cambrian	Ordovician	Silurian	Devonian	Mississippian	Pennsylvanian	Permian
						Carboniferous		

2500 M.Y.A.   545 M.Y.A.   245 M.Y.A.

Introduction

Having reviewed the geologic history of the Archean and Proterozoic eons, we now turn our attention to the Phanerozoic Eon, comprising the remaining 11.8% of geologic time. Recall that during the Archean, crust-forming processes generated greenstone belts and granite–gneiss complexes, which were shaped into cratons. The most common Archean sedimentary rocks are graywackes and argillites, typical of tectonically active regions. While a modern style of plate tectonics did not begin until the Early Proterozoic, many geologists think that the Archean style of plate tectonics involved numerous collisions of microcontinents, resulting in rapid crustal growth. The Proterozoic can be characterized as a time during which the sedimentary-rock assemblages and plate tectonic style were like those of the Phanerozoic. During the Proterozoic, plate movements, extensive plutonism, and regional metamorphism thickened, stabilized, and increased the area of continental crust.

At the beginning of the Phanerozoic, six major continental landmasses existed, four of which straddled the paleoequator. Plate movements during the Phanerozoic created a changing panorama of continents and ocean basins whose positions affected atmospheric and oceanic circulation patterns and created new environments for habitation by the rapidly evolving biota.

The Paleozoic history of most continents involves major mountain-building activity along their margins and numerous shallow-water marine transgressions and regressions over their interiors. These transgressions and regressions were caused by global changes in sea level probably related to plate activity and glaciation.

The following chapters present the geologic history of North America in terms of those major transgressions and regressions rather than a period-by-period chronology. While we focus on North American geologic history, we also place those events in a global context.

Continental Architecture: Cratons and Mobile Belts

During the Precambrian, continental accretion and orogenic activity led to the formation of sizable continents. At least three large continents existed during the Late Proterozoic, and some geologists think that these landmasses later collided to form a single Pangaea-like supercontinent (see Figure 9.8). This supercontinent began breaking apart sometime during the latest Proterozoic, and by the beginning of the Paleozoic Era, six major continents were present. Each continent can be divided into two major components: a craton and one or more mobile belts.

Cratons are the relatively stable and immobile parts of continents and form the foundation upon which Phanerozoic sediments were deposited. Cratons typically consist of two parts: a shield and a platform (Figure 10.1).

Shields are the exposed portion of the crystalline basement rocks of a continent and are composed of Precambrian metamorphic and igneous rocks that reveal a history of extensive orogenic activity during the Precambrian (see Chapters 8 and 9). During the Phanerozoic, however, shields were extremely stable and formed the foundation of the continents.

Extending outward from the shields are buried Precambrian rocks that constitute a *platform,* another part of the craton. Overlying the platform are flat-lying or gently dipping Phanerozoic detrital and chemical sedimentary rocks that were deposited in widespread shallow seas that transgressed and regressed over the craton. These seas, called **epeiric seas,** were a common feature of most Paleozoic cratonic histories. Changes in sea level caused primarily by continental glaciation as well as by plate movement were responsible for the advance and retreat of the seas.

While most of the Paleozoic platform rocks are still essentially flat lying, in some places they were gently folded into regional arches, domes, and basins (Figure 10.1). In many cases some of these structures stood out as low islands during the Paleozoic Era and supplied sediments to the surrounding epeiric seas.

Mobile belts are elongated areas of mountain-building activity. They are located along the margins of continents where sediments are deposited in the relatively shallow waters of the continental shelf and the deeper waters at the base of the continental slope. During plate convergence along these margins, the sediments are deformed and intruded by magma, creating mountain ranges.

Four mobile belts formed around the margin of the North American craton during the Paleozoic; these were the **Franklin, Cordilleran, Ouachita,** and **Appalachian mobile belts** (Figure 10.1). Each was the site of mountain

Phanerozoic Eon										
Mesozoic Era			Cenozoic Era							
Triassic	Jurassic	Cretaceous	Tertiary						Quaternary	
			Paleocene	Eocene	Oligocene	Miocene	Pliocene	Pleistocene		Holocene

245
M.Y.A.

66
M.Y.A.

building in response to compressional forces along a convergent plate boundary and formed such mountain ranges as the Appalachians and Ouachitas.

Paleozoic Paleogeography

One of the results of plate tectonics is that Earth's geography is constantly changing. The present-day configuration of the continents and ocean basins is merely a snapshot in time. As the plates move about, the location of continents and ocean basins constantly changes. One goal of historical geology is to provide paleogeographic reconstructions of the world for the geologic past. By synthesizing all the pertinent paleoclimatic, paleomagnetic, paleontologic, sedimentologic, stratigraphic, and tectonic data available, geologists can prepare paleogeographic maps of what the world looked like at a particular time in the geologic past (see Perspective 10.1).

The paleogeographic history of the Paleozoic Era, for example, is not as precisely known as for the Mesozoic and Cenozoic eras, in part because the magnetic anomaly patterns preserved in the oceanic crust were destroyed when much of the Paleozoic oceanic crust was subducted during the formation of Pangaea. Paleozoic paleogeographic reconstructions are therefore based primarily on structural relationships, climate-sensitive sediments such as

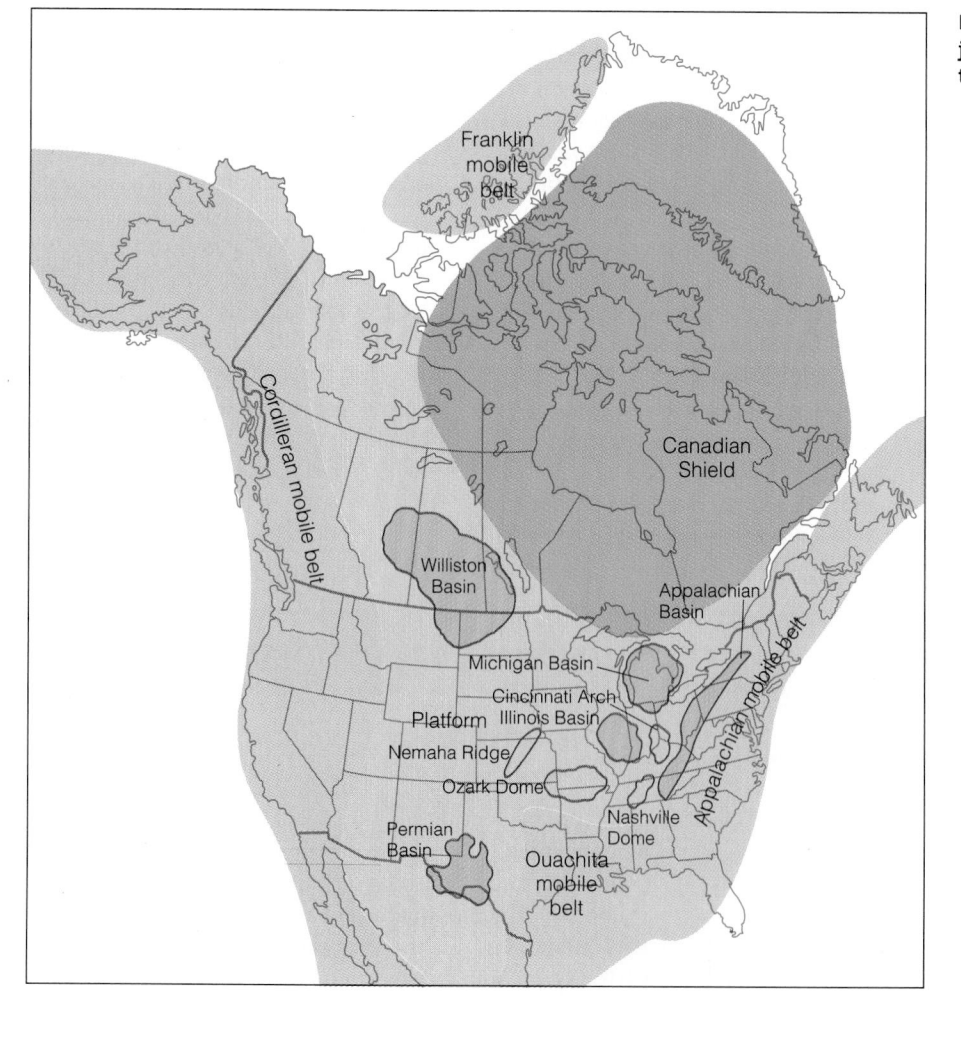

FIGURE 10.1 The mobile belts and major cratonic structures of North America that formed during the Paleozoic Era.

Paleogeographic Reconstructions and Maps

*T*he key to any reconstruction of world paleogeography is the correct positioning of the continents in terms of latitude and longitude as well as orientation of the paleocontinent relative to the paleonorth pole. The main criteria used for paleogeographic reconstructions are paleomagnetism, biogeography, tectonic patterns, and climatology.

Paleomagnetism provides the only source of quantitative data on the orientations of the continents. For the Paleozoic Era, the paleomagnetic data are often inconsistent and contradictory due to secondary magnetizations acquired through the effects of metamorphism or weathering.

The distribution of faunas and floras provides a useful check on the latitudes determined by paleomagnetism and can provide additional limits on longitudinal separation of continents. As is well known, the distribution of plants and animals is controlled by both climatic and geographic barriers. Such information can be used to position continents and ocean basins in a way that accounts for the biogeographic patterns indicated by fossil evidence.

Tectonic activity is indicated by deformed sediments associated with andesitic volcanics and ophiolites. Such features allow geologists to recognize ancient mountain ranges and zones of subduction. These mountain ranges may subsequently have been separated by plate movement, so the identification of large, continuous mountain ranges provides important information about continental positions in the geologic past.

Climate-sensitive sedimentary rocks are used to interpret past climatic conditions. Desert dunes are typically well sorted and cross-bedded on a large scale, and associated with other deposits, they indicate an arid environment. Coals form in freshwater swamps where climatic conditions promote abundant plant growth. Evaporites result when evaporation exceeds precipitation, such as in desert regions or along hot, dry shorelines. Tillites result from glacial activity and indicate cold, wet environments.

Paleogeographic features can be determined by associations of sedimentary rocks and sedimentary structures. For example, large-scale cross-beds may indicate windblown conditions such as in deserts.

Delta complexes and deep-sea fans have characteristic internal features and three-dimensional forms that can be recognized in the geologic record, just as coal and associated deposits usually follow a particular sequence. These features can be used to interpret such geographic features as lakes, streams, swamps, and shallow and deep marine areas.

Former mountain ranges can be recognized by folded and faulted sedimentary rocks associated with metamorphic and igneous rocks. We have already mentioned the association of andesites and ophiolites as evidence of former mountain building.

By combining all relevant geologic, paleontologic, and climatologic information, geologists can construct paleogeographic maps (see Figure 10.2). Such maps are simply interpretations of the geography of an area for a particular time in the geologic past. The majority of paleogeographic maps show the distribution of land and sea, probable climatic regimes, and such geographic features as mountain ranges, swamps, and glaciers.

red beds, evaporites, and coals, as well as the distribution of plants and animals.

Recall that six major continents were present at the beginning of the Paleozoic. Besides these large landmasses, geologists have also identified numerous small microcontinents and island arcs associated with various microplates that were present during the Paleozoic. We will be primarily concerned, however, with the history of the six major continents and their relationship to each other. The six major Paleozoic continents are **Baltica** (Russia west of the Ural Mountains and the major part of northern Europe), **China** (a complex area consisting of at least three

Paleozoic continents that were not widely separated and are here considered to include China, Indochina, and the Malay Peninsula), **Gondwana** (Africa, Antarctica, Australia, Florida, India, Madagascar, and parts of the Middle East and southern Europe), **Kazakhstania** (a triangular continent centered on Kazakhstan, but considered by some to be an extension of the Paleozoic Siberian continent), **Laurentia** (most of present North America, Greenland, northwestern Ireland, Scotland, and part of eastern Russia), and **Siberia** (Russia east of the Ural Mountains and Asia north of Kazakhstan and south of Mongolia). The paleogeographic reconstructions that follow (Figure 10.2) are based on the methods used to determine and interpret the location, geographic features, and environmental conditions on the paleocontinents.

EARLY PALEOZOIC GLOBAL HISTORY

In contrast to today's global geography, the Cambrian world consisted of six major continents dispersed around the globe at low tropical latitudes (Figure 10.2a). Water circulated freely among ocean basins, and the polar regions were apparently ice free. By the Late Cambrian, epeiric seas had covered large areas of Laurentia, Baltica, Siberia, Kazakhstania, and China, while major highlands were present in northeastern Gondwana, eastern Siberia, and central Kazakhstania.

During the Ordovician and Silurian periods, plate movement played a major role in the changing global geography (Figure 10.2b and c). Gondwana moved southward during the Ordovician and began to cross the South Pole as indicated by Upper Ordovician tillites found today in the Sahara Desert. In contrast to the passive continental margin Laurentia exhibited during the Cambrian, an active convergent plate boundary formed along its eastern margin during the Ordovician as indicated by the Late Ordovician Taconic orogeny that occurred in New England. During the Silurian, Baltica moved northwestward relative to Laurentia and collided with it to form the larger continent of Laurasia. This collision, which closed the northern Iapetus Ocean, is marked by the Caledonian orogeny. Following this orogeny, the southern part of the Iapetus Ocean still remained open between Laurentia and Gondwana (Figure 10.2c). Siberia and Kazakhstania moved from a southern equatorial position during the Cambrian to north temperate latitudes by the end of the Silurian Period.

With this plate tectonic overview in mind, we now focus our attention on North America (Laurentia) and the role it played in the Early Paleozoic geologic history of the world.

Early Paleozoic Evolution of North America

It is convenient to divide the geologic history of the North American craton into two parts: the first dealing with the relatively stable continental interior over which epeiric seas transgressed and regressed, and the other with the mobile belts where mountain building occurred.

In 1963 American geologist Laurence Sloss proposed that the sedimentary-rock record of North America could be subdivided into six cratonic sequences. A **cratonic sequence** is a large-scale (greater than supergroup) lithostratigraphic unit representing a major transgressive-regressive cycle bounded by cratonwide unconformities (Figure 10.3). The transgressive phase, which is usually covered by younger sediments, commonly is well preserved, whereas the regressive phase of each sequence is marked by an unconformity. Where rocks of the appropriate age are preserved, each of the six unconformities can be shown to extend across the various sedimentary basins of the North American craton and into the mobile belts along the cratonic margin.

Geologists have also recognized major unconformity bounded sequences in cratonic areas outside North America. Such global transgressive and regressive cycles are caused by sea-level changes and are thought to result from major tectonic and glacial events.

The realization that rock units can be divided into cratonic sequences and that these sequences can be further subdivided and correlated provides the foundation for an important concept in geology that allows high-resolution analysis of time and facies relationships within sedimentary rocks. **Sequence stratigraphy** is the study of rock relationships within a time–stratigraphic framework of related facies bounded by erosional or nondepositional surfaces. The basic unit of sequence stratigraphy is the *sequence,* which is a succession of rocks bounded by unconformities and their equivalent conformable strata. Sequence boundaries form as a result of a relative drop in sea level. Sequence stratigraphy is an important tool in geology because it allows geologists to subdivide sedimentary rocks into related units that are bounded by time–stratigraphically significant boundaries. Geologists use sequence stratigraphy for high-resolution correlation and mapping, as well as interpreting and predicting depositional environments.

The Sauk Sequence

Rocks of the **Sauk sequence** (Late Proterozoic–Early Ordovician) record the first major transgression onto the North American craton (Figure 10.3). During the Late Proterozoic and Early Cambrian, deposition of marine sediments was limited to the passive shelf areas of the Appalachian and Cordilleran borders of the craton. The craton itself was above sea level and experiencing extensive weathering and erosion. Because North America was located in a tropical climate at this time and there is no evidence of any terrestrial vegetation, weathering and erosion of the exposed Precambrian basement rocks must have proceeded rapidly. During the Middle Cambrian, the transgressive phase of the Sauk began with epeiric seas encroaching over the craton (see Perspective 10.2). By the Late Cambrian, the Sauk Sea had covered most of North America, leaving only a portion of the Canadian Shield and a few large islands above sea level (Figure 10.4). These is-

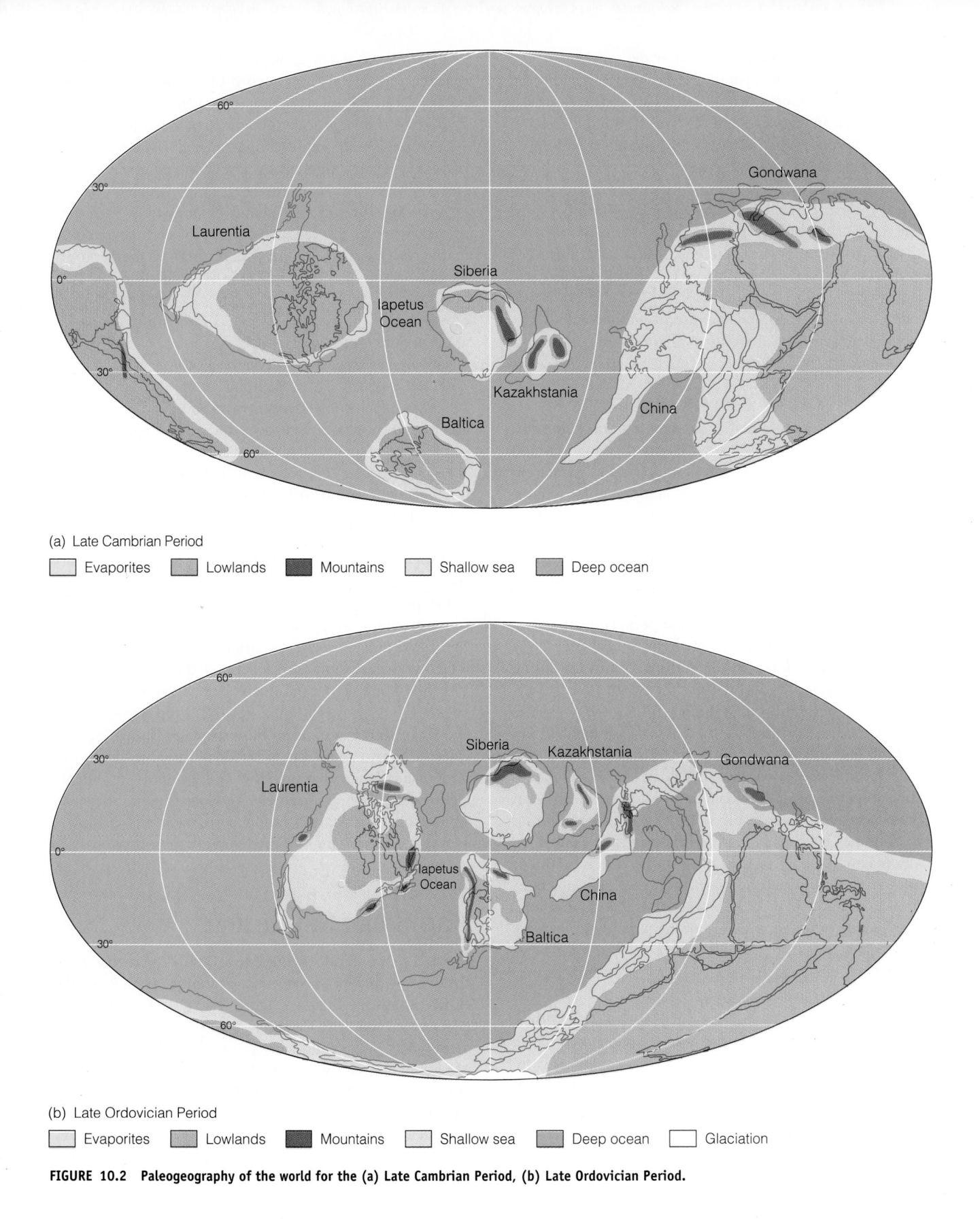

(a) Late Cambrian Period

☐ Evaporites ▨ Lowlands ■ Mountains ☐ Shallow sea ▨ Deep ocean

(b) Late Ordovician Period

☐ Evaporites ▨ Lowlands ■ Mountains ☐ Shallow sea ▨ Deep ocean ☐ Glaciation

FIGURE 10.2 Paleogeography of the world for the (a) Late Cambrian Period, (b) Late Ordovician Period.

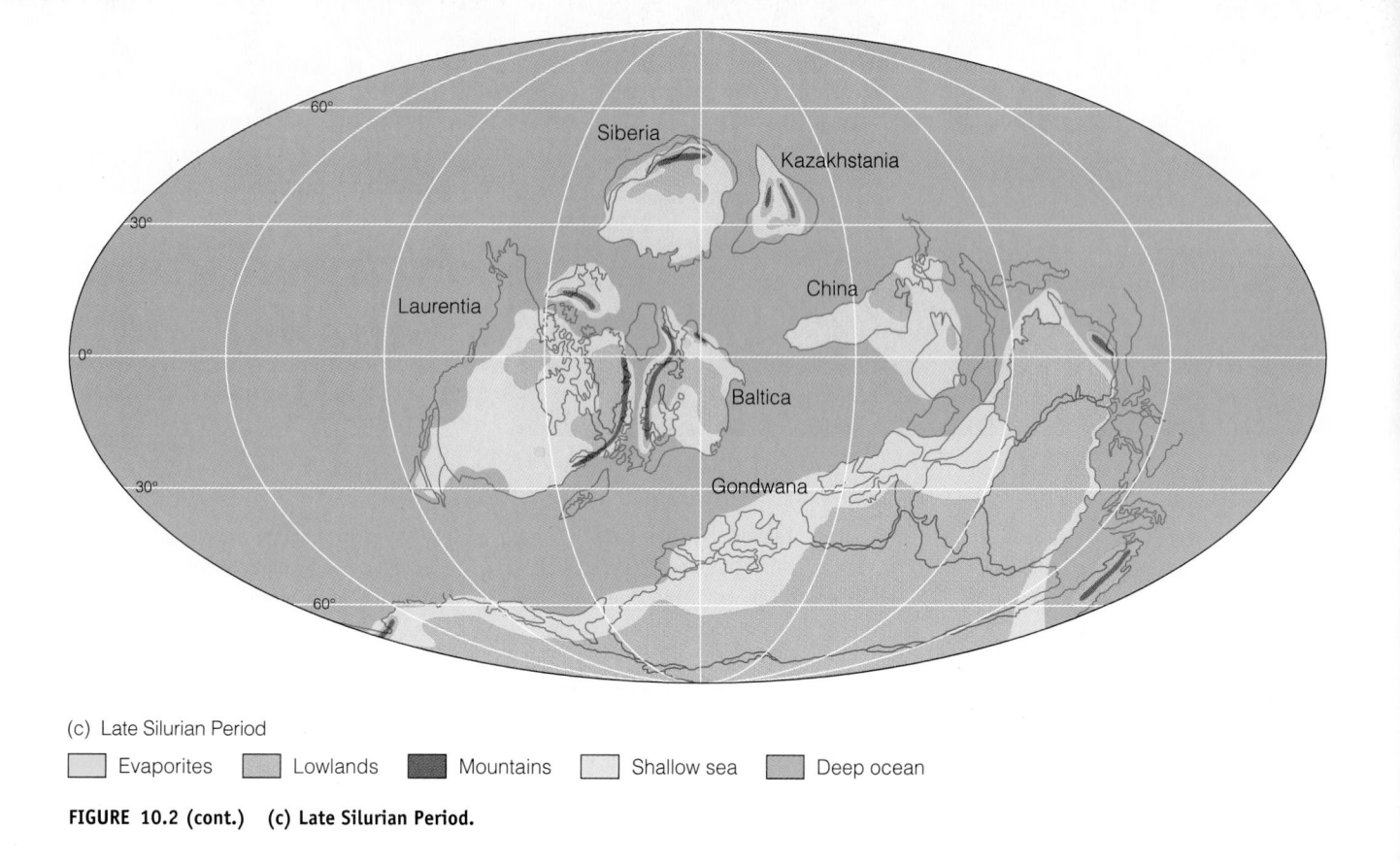

(c) Late Silurian Period

◻ Evaporites ◻ Lowlands ◼ Mountains ◻ Shallow sea ◻ Deep ocean

FIGURE 10.2 (cont.) (c) Late Silurian Period.

lands, collectively named the **Transcontinental Arch,** extended from New Mexico to Minnesota and the Lake Superior region.

The sediments deposited on both the craton and along the shelf area of the craton margin show abundant evidence of shallow-water deposition. The only difference between the shelf and craton deposits is that the shelf deposits are thicker. In both areas, the sands are generally clean and well sorted and commonly contain ripple marks and small-scale cross-bedding. Many of the carbonates are bioclastic (composed of fragments of organic remains), contain stromatolites, or have oolitic (small, spherical calcium carbonate grains) textures. Such sedimentary structures and textures indicate shallow-water deposition.

THE CAMBRIAN OF THE GRAND CANYON REGION: A TRANSGRESSIVE FACIES MODEL

Recall from Chapter 3 that sediments become increasingly finer the farther away from land one goes. Therefore, in a stable environment where sea level remains the same, coarse detrital sediments are typically deposited in the nearshore environment, and finer-grained sediments are deposited in the offshore environment. Carbonates form farthest from land in the area beyond the reach of detrital sediments. During a transgression, these facies (sediments that represent a particular environment) migrate in a landward direction (see Figure 3.12).

The Cambrian rocks of the Grand Canyon region (see the Prologue) provide an excellent model for illustrating the sedimentation patterns of a transgressing sea. The Grand Canyon region occupied the passive shelf and western margin of the craton during Sauk time. During the Late Proterozoic and Early Cambrian, most of the craton was above sea level, as the Sauk Sea was still largely restricted to the margins of the craton (continental shelves and slopes). In the Grand Canyon region, the Tapeats Sandstone represents the basal transgressive shoreline deposits that accumulated as marine waters transgressed across the shelf and just onto the western margin of the craton during the Early Cambrian (Figure 10.5). These sediments are clean, well-sorted sands of the type one would find on a beach today. As the transgression continued into the Middle Cambrian, muds of the Bright Angel Shale were deposited over the Tapeats Sandstone. By the Late Cambrian, the Sauk Sea had transgressed so far onto the craton that, in the Grand Canyon region, carbonates of the Muav Limestone were being deposited over the Bright Angel Shale. This vertical succession of sandstone (Tapeats), shale (Bright Angel), and limestone (Muav) forms a typical transgressive sequence and represents a progressive migration of offshore facies toward the craton through time (Figure 10.5).

Cambrian rocks of the Grand Canyon region also illustrate how many formations are time transgressive; that is, their age is not the same every place they are found.

Pictured Rocks National Lakeshore

*E*xposed along the south shore of Lake Superior between Au Sable Point and Munising in Michigan's Upper Peninsula is the beautiful and imposing wavecut sandstone called Pictured Rocks cliffs (Figure 1). The rocks exposed in this area, part of which is designated a national lakeshore, comprise the Upper Cambrian Munising Formation, which is divided into two members: the lower Chapel Rock Sandstone and the upper Miner's Castle Sandstone (Figure 1). The Munising Formation unconformably overlies the Upper Proterozoic Jacobsville Sandstone and is unconformably overlain by the Middle Ordovician Au Train Formation. The reddish brown, coarse-grained Jacobsville Sandstone was deposited in streams and lakes over an irregular erosion surface (Figure 1). Following deposition, the Jacobsville was slightly uplifted and tilted.

By the Late Cambrian, the transgressing Sauk Sea reached the Michigan area, and the Chapel Rock Sandstone was deposited. The principal source area for this unit was the Northern Michigan high-

FIGURE 1   Location of Pictured Rocks National Lakeshore and stratigraphy of rocks exposed in this area. The photograph of the Munising Formation shows the projection of the shoreline called Miner's Castle. The two turrets at the top of Miner's Castle formed by wave action when the water level of Lake Superior was higher. The contact between the Miner's Castle and Chapel Rock sandstone members is located just above lake level.

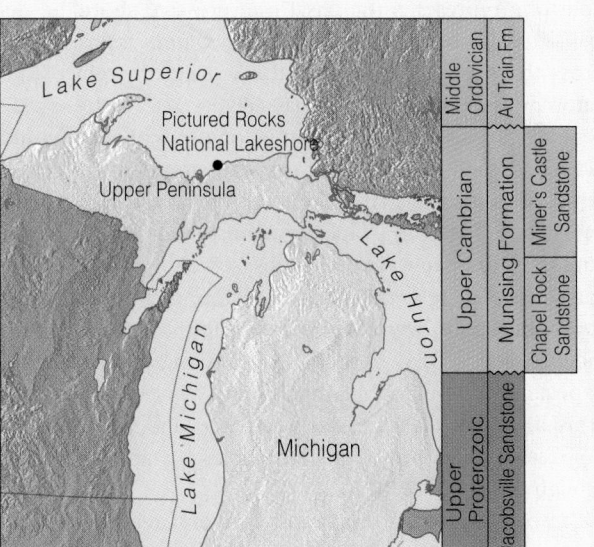

lands, an area that corresponds to the present Upper Peninsula. Following deposition of the Chapel Rock Sandstone, the Sauk Sea retreated from the area.

During a second transgression of the Sauk Sea in this area, the Miner's Castle Sandstone was deposited. This second transgression covered most of the Upper Peninsula of Michigan and drowned the highlands that were the source for the older Chapel Rock Sandstone.

The source area for the Miner's Castle Sandstone was the Precambrian Canadian Shield area to the north and northeast. The Miner's Castle Sandstone contains rounder, better sorted, and more abundant quartz grains than the Chapel Rock Sandstone, indicating a different source area. A major unconformity separates the Miner's Castle Sandstone from the overlying Middle Ordovician Au Train Formation.

One of the most prominent features of Pictured Rocks National Lakeshore is Miner's Castle, a wavecut projection along the shoreline (Figure 1). The lower sandstone unit at water level is the Chapel Rock Sandstone, while the rest of the feature is composed of the Miner's Castle Sandstone. The two turrets of the castle formed as sea stacks during a time following the Pleistocene when the water level of Lake Superior was much higher.

Mapping and correlations based on faunal evidence indicate that deposition of the Mauv Limestone had already started on the shelf before deposition of the Tapeats Sandstone was completed on the craton. Faunal analysis of the Bright Angel Shale indicates that it is Early Cambrian in age in California and Middle Cambrian in age in the Grand Canyon region, thus illustrating the time-transgressive nature of formations and facies.

This same facies relationship also occurred elsewhere on the craton as the seas encroached from the Appalachian and Ouachita mobile belts onto the craton interior (Figure 10.6). Carbonate deposition dominated on the craton as the Sauk transgression continued during the Early Ordovician, and the islands of the Transcontinental Arch were soon covered by the advancing Sauk Sea. By the end of Sauk time, the majority of the craton was submerged beneath a warm, equatorial epeiric sea (Figure 10.2a).

The Tippecanoe Sequence

As the Sauk Sea regressed from the craton during the Early Ordovician, it revealed a landscape of low relief. The rocks exposed were predominantly limestones and dolostones that experienced deep and extensive erosion because North America was still located in a tropical environment (Figure 10.7). The resulting cratonwide unconformity marks the boundary between the Sauk and Tippecanoe sequences.

Like the Sauk sequence, deposition of the **Tippecanoe sequence** (Middle Ordovician–Early Devonian) began with a major transgression onto the craton. This transgressing sea deposited clean, well-sorted quartz sands over most of the craton. The best known of the Tippecanoe basal sandstones is the St. Peter Sandstone, an almost pure quartz sandstone used in manufacturing glass. It occurs throughout much of the midcontinent and resulted from numerous cycles of weathering and erosion of Proterozoic and Cambrian sandstones deposited during the Sauk transgression (Figure 10.8).

The Tippecanoe basal sandstones were followed by widespread carbonate deposition (Figure 10.7). The limestones were generally the result of deposition by calcium carbonate–secreting organisms such as corals, brachiopods, stromatoporoids, and bryozoans. Besides the limestones, there were also many dolostones. Most of the dolostones formed as a result of magnesium replacing calcium in calcite, thus converting limestones into dolostones.

In the eastern portion of the craton, the carbonates grade laterally into shales. These shales mark the farthest extent of detrital sediments derived from weathering and erosion of the Taconic Highlands, a tectonic event we will discuss later.

TIPPECANOE REEFS AND EVAPORITES

Organic reefs are limestone structures constructed by living organisms, some of which contribute skeletal materials to the reef framework (Figure 10.9). Today, corals and calcareous algae are the most prominent reef builders, but in the geologic past other organisms played a major role. Regardless of the organisms dominating reef communities, reefs appear to have occupied the same ecologic niche in

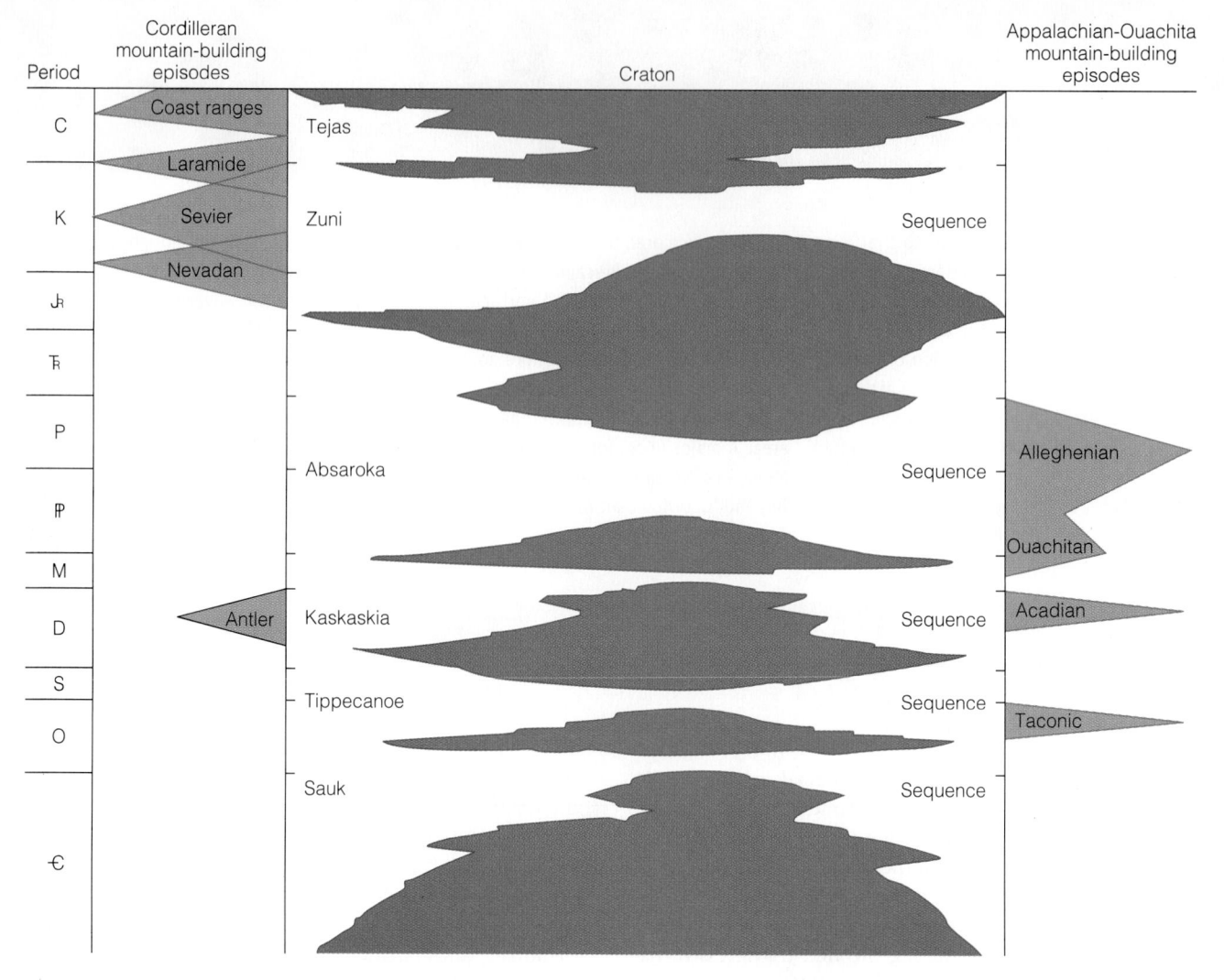

FIGURE 10.3 Cratonic sequences of North America. The white areas represent sequences of rocks that are separated by large scale unconformities shown as brown areas. The major Cordilleran orogenies are shown on the left side of the figure, and the major Appalachian orogenies are shown on the right side.

the geologic past as they do today. Because of the ecologic requirements of reef-building organisms, reefs today are confined to a narrow latitudinal belt between 30 degrees north and south of the equator. Corals, the major reef-building organisms today, require warm, clear, shallow water of normal salinity for optimal growth.

The size and shape of a reef are largely the result of the interaction between the reef-building organisms, the bottom topography, wind and wave action, and subsidence of the seafloor. Reefs also alter the area around them by forming barriers to water circulation or wave action.

Reefs are commonly long, linear masses that form a barrier between a shallow platform on one side and a comparatively deep marine basin on the other side. Such reefs are known as *barrier reefs* (Figure 10.9). Reefs create and maintain a steep seaward front that absorbs incoming wave energy. As skeletal material breaks off from the reef front, it accumulates along a fore-reef slope. The reef bar-

rier itself is porous and composed of reef-building organisms. The lagoon area is a low-energy, quiet-water zone where fragile, sediment-trapping organisms thrive. The lagoon area can also become the site of evaporitic deposits when circulation to the open sea is cut off. Modern examples of barrier reefs are the Florida Keys, Bahama Islands, and Great Barrier Reef of Australia.

Reefs have been common features since the Cambrian and have been built by a variety of organisms. The first skeletal builders of reeflike structures were archaeocyathids. These conical-shaped organisms lived during the Cambrian and had double, perforated, calcareous shell walls. Archaeocyathids built small mounds that have been found on all continents except South America (see Figure 12.6). Beginning in the Middle Ordovician, stromatoporoid–coral reefs became common in the low latitudes, and similar reefs remained so throughout the rest of the Phanerozoic Eon. The burst of reef building seen in the

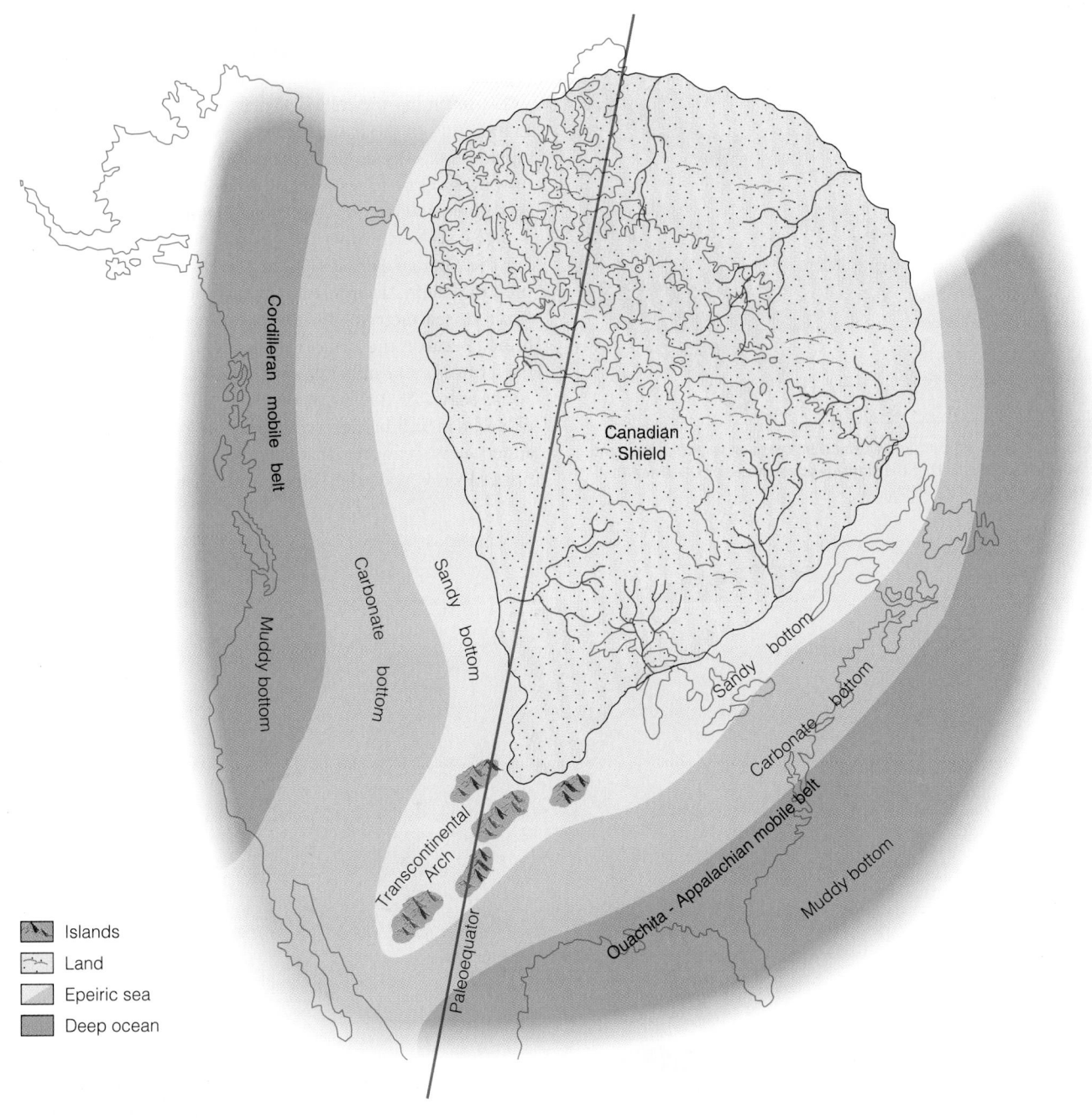

FIGURE 10.4 Paleogeography of North America during the Cambrian Period. Note the position of the Cambrian paleoequator. During this time North America straddled the equator as indicated in Figure 10.2a.

Late Ordovician through Devonian probably occurred in response to evolutionary changes triggered by the appearance of extensive carbonate seafloors and platforms beyond the influence of detrital sediments.

The Middle Silurian rocks (Tippecanoe sequence) of the present-day Great Lakes region are world famous for their reef and evaporite deposits and have been extensively studied (Figure 10.10). The most famous structure in the region, the Michigan Basin, is a broad, circular basin surrounded by large barrier reefs. No doubt these reefs contributed to increasingly restricted circulation and the precipitation of Upper Silurian evaporites within the basin (Figure 10.11).

Within the rapidly subsiding interior of the basin, other types of reefs are found. *Pinnacle reefs* are tall, spindly structures up to 100 m high. They reflect the rapid upward growth needed to maintain themselves near sea level during subsidence of the basin (Figure 10.11). Besides the pinnacle reefs, bedded carbonates and thick sequences of salt and anhydrite are also found in the Michigan Basin.

As the Tippecanoe Sea gradually regressed from the craton during the Late Silurian, precipitation of evaporite minerals occurred in the Appalachian, Ohio, and Michigan basins. In the Michigan Basin alone, approximately 1500 m of sediments were deposited, nearly half of which are halite and an-

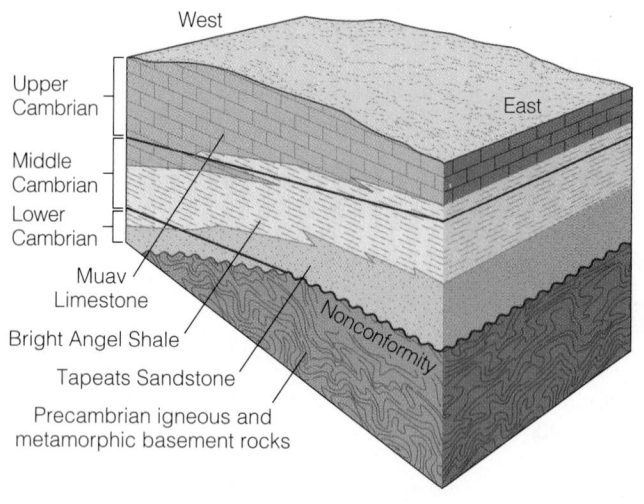

FIGURE 10.5 Cross section of Cambrian strata exposed in the Grand Canyon region illustrating the transgressive nature of the three formations.

West

East

Upper Cambrian

Middle Cambrian

Lower Cambrian

Muav Limestone

Bright Angel Shale

Tapeats Sandstone

Precambrian igneous and metamorphic basement rocks

Nonconformity

hydrite. How did such thick sequences of evaporites accumulate? One possibility is that when sea level dropped, the tops of the barrier reefs were as high as or above sea level, thus preventing the influx of new seawater into the basin. Evaporation of the basinal seawater would result in the precipitation of salts. A second possibility is that the reefs grew upward so close to sea level that they formed a sill or barrier that eliminated interior circulation (Figure 10.12).

Because North America was still near the equator during the Silurian Period (Figure 10.2c), temperatures were probably high. As circulation to the Michigan Basin was restricted, seawater within the basin evaporated, forming a brine. Because the brine was heavy, it concentrated near the bottom, and minerals precipitated on the basin floor. Some seawater flowed in over the sill and through channels cut in the barrier reefs, but this replenishment only added new seawater that later became concentrated as brine. In this way, the brine in the basin become increas-

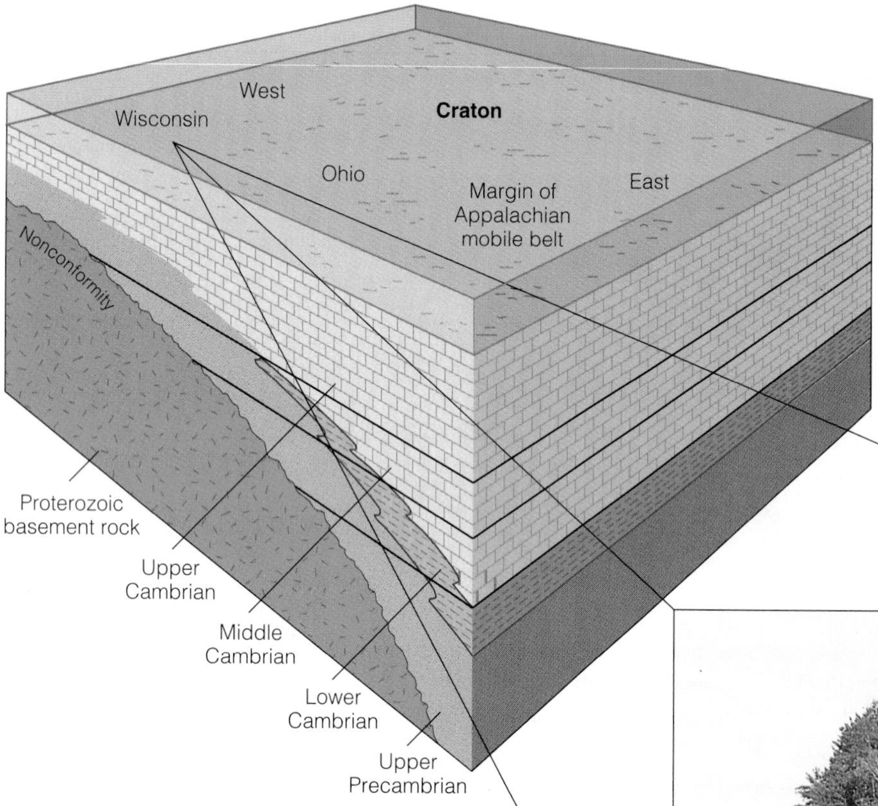

West

Craton

Wisconsin

Ohio

Margin of Appalachian mobile belt

East

Nonconformity

Proterozoic basement rock

Upper Cambrian

Middle Cambrian

Lower Cambrian

Upper Precambrian

(a)

FIGURE 10.6 (a) Cross section from the craton interior to the Appalachian mobile belt margin showing the three major Cambrian facies and the time-transgressive nature of the units. Note the progressive development of a carbonate facies due to the submergence of detrital source areas caused by the advancing Sauk Sea. (b) Outcrop of cross-bedded Upper Cambrian sandstone in the Dells area of Wisconsin.

(b)

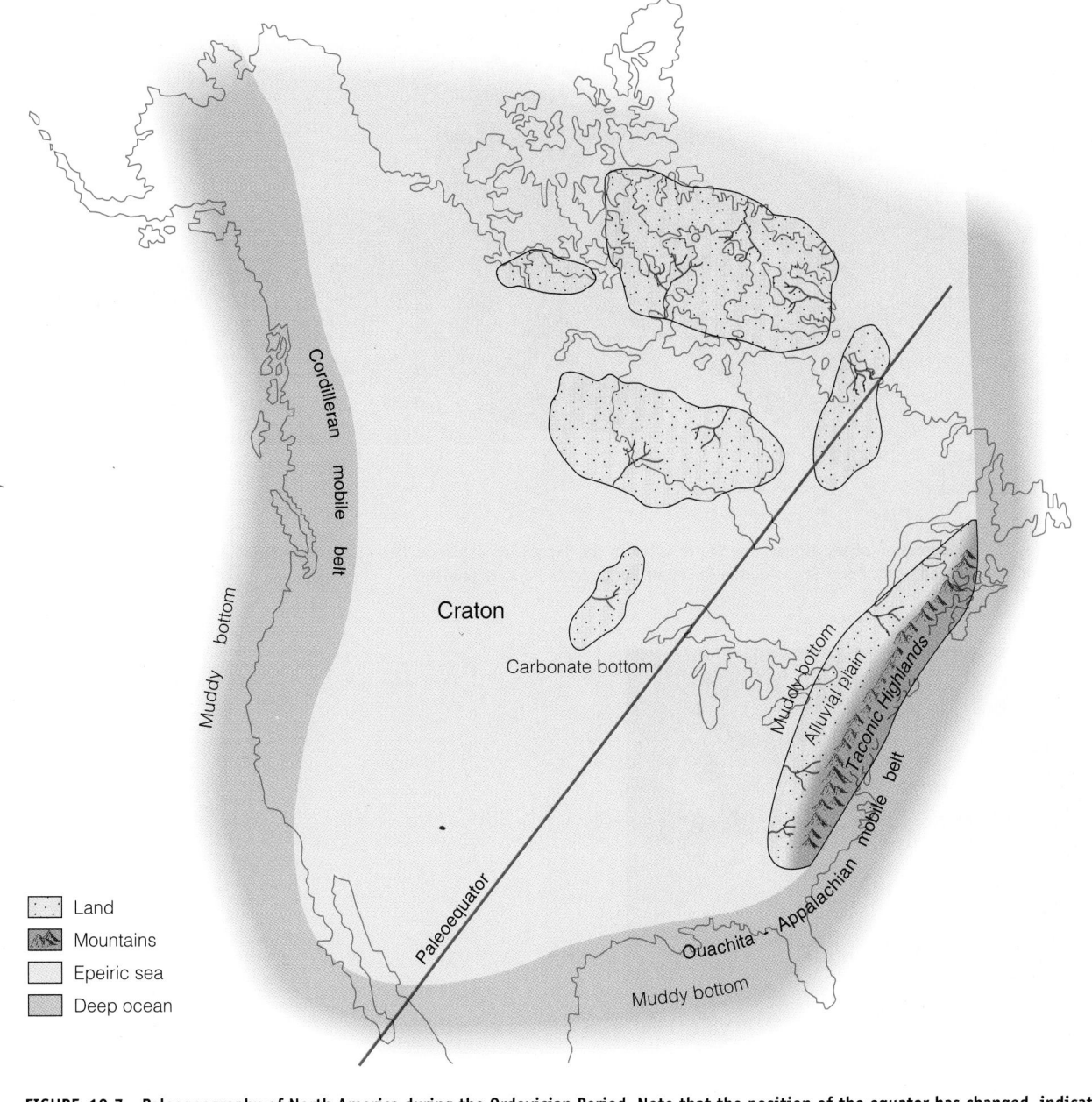

Land
Mountains
Epeiric sea
Deep ocean

Cordilleran mobile belt

Muddy bottom

Craton

Carbonate bottom

Paleoequator

Muddy bottom

Alluvial plain

Taconic Highlands

Appalachian mobile belt

Ouachita

Muddy bottom

FIGURE 10.7 Paleogeography of North America during the Ordovician Period. Note that the position of the equator has changed, indicating North America was rotating in a counterclockwise direction.

ingly concentrated until the salts no longer could stay in solution and thus precipitated to form evaporite minerals.

The order and type of salts that precipitate from seawater depend on their solubility, the original concentration of seawater, and local conditions of the basin. In general, salts precipitate in order beginning with the least soluble and ending with the most soluble. Therefore, calcium carbonate usually precipitates out first, followed by gypsum,* and lastly halite. Many lateral shifts and interfingering of the

limestone, anhydrite, and halite facies may occur, however, due to variations in the amount of seawater entering the basin and changing geologic conditions.

Thus, the periodic evaporation of seawater proposed by this model could account for the observed vertical and lateral distribution of evaporites in the Michigan Basin. Associated with those evaporites, however, are pinnacle reefs, and the organisms that constructed those reefs could not have lived in such a highly saline environment (Figure 10.11). How, then, can such contradictory features be explained? Numerous models have been proposed, ranging from cessation of reef growth followed by evaporite deposition, to alternation of reef growth and evaporite deposition. Although the Michigan Basin has been studied extensively for years, no

---

*Recall from Chapter 4 that gypsum (CaSO₄ · 2H₂O) is the common sulfate precipitated from seawater, but when deeply buried, gypsum loses its water and is converted to anhydrite (CaSO₄).

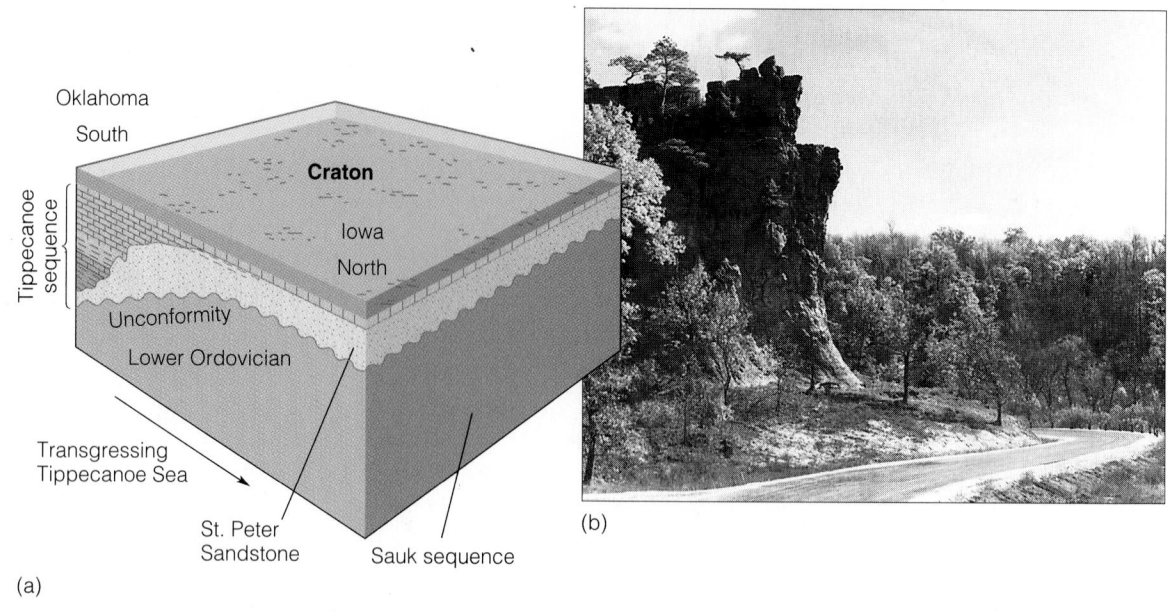

FIGURE 10.8 (a) The transgression of the Tippecanoe Sea resulted in the deposition of the St. Peter Sandstone (Middle Ordovician) over a large area of the craton. (b) Outcrop of St. Peter Sandstone in Governor Dodge State Park, Wisconsin.

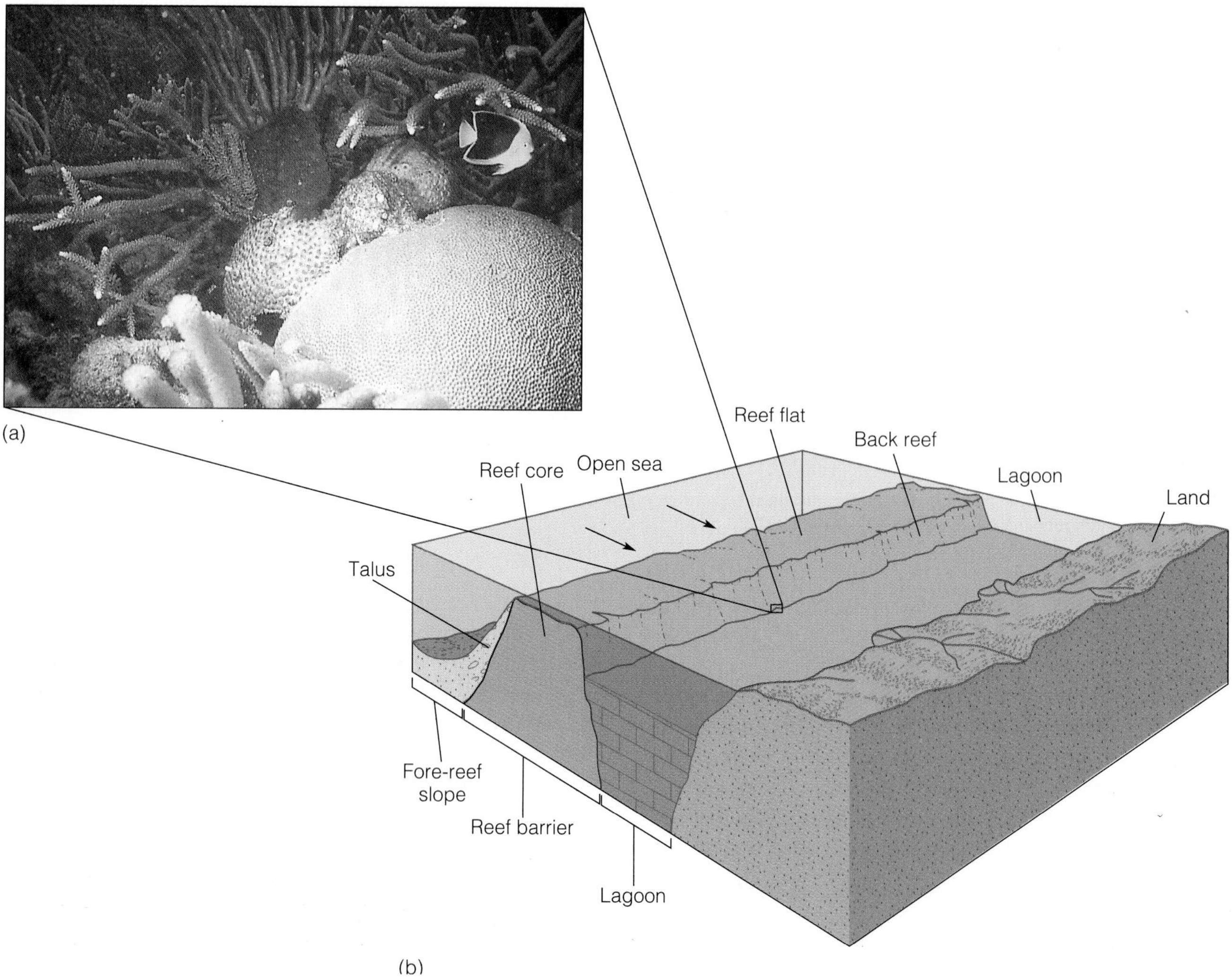

FIGURE 10.9 (a) Present-day reef community showing the various reef-building organisms. (b) Diagrammatic cross section of a reef showing the various environments within the reef complex.

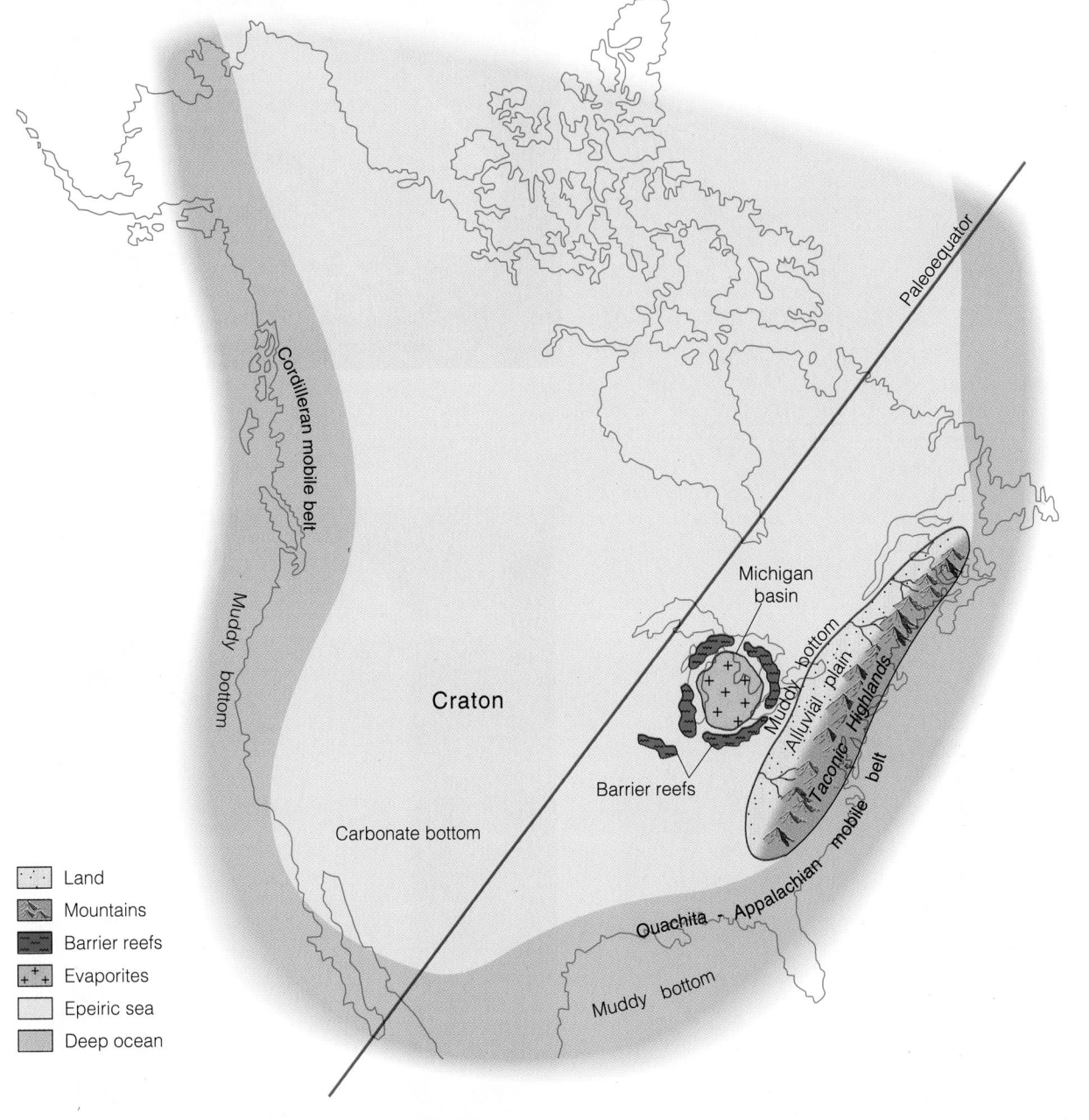

FIGURE 10.10 Paleogeography of North America during the Silurian Period. Note the development of reefs in the Michigan, Ohio, and Indiana–Illinois–Kentucky areas.

model yet proposed completely explains the genesis and relationship of its various reef, carbonate, and evaporite facies.

THE END OF THE TIPPECANOE SEQUENCE

By the Early Devonian, the regressing Tippecanoe Sea had retreated to the craton margin exposing an extensive lowland topography. During this regression, marine deposition was initially restricted to a few interconnected cratonic basins and, finally by the end of the Tippecanoe, to only the mobile belts surrounding the craton.

As the Tippecanoe Sea regressed during the Early Devonian the craton experienced mild deformation resulting in the formation of many domes, arches, and basins. These structures were mostly eroded during the time the craton was exposed so that they were eventually covered by deposits from the encroaching Kaskaskia Sea.

The Appalachian Mobile Belt and the Taconic Orogeny

Having examined the Sauk and Tippecanoe geologic history of the craton, we now turn our attention to the Appalachian mobile belt, where the first Phanerozoic orogeny began during the Middle Ordovician. The mountain building that occurred during the Paleozoic Era had a

FIGURE 10.11 (a) Generalized cross section of the northern Michigan Basin during the Silurian Period. (b) Stromatoporoid barrier-reef facies. (c) Evaporite facies. (d) Carbonate facies.

profound influence on the climate and sedimentary history of the craton. Additionally, it was part of the global tectonic regime that sutured the continents together, forming Pangaea by the end of the Paleozoic.

Throughout Sauk time, the Appalachian region was a broad, passive, continental margin. Sedimentation was closely balanced by subsidence as thick, shallow marine sands were succeeded by extensive carbonate deposits.

During this time, the **Iapetus Ocean** was widening as a result of movement along a divergent plate boundary (Figure 10.13a).

Beginning with the subduction of the Iapetus plate beneath Laurentia (an oceanic–continental convergent plate boundary), the Appalachian mobile belt was born (Figure 10.13b). The resulting **Taconic orogeny,** named after the present-day Taconic Mountains of eastern New York, cen-

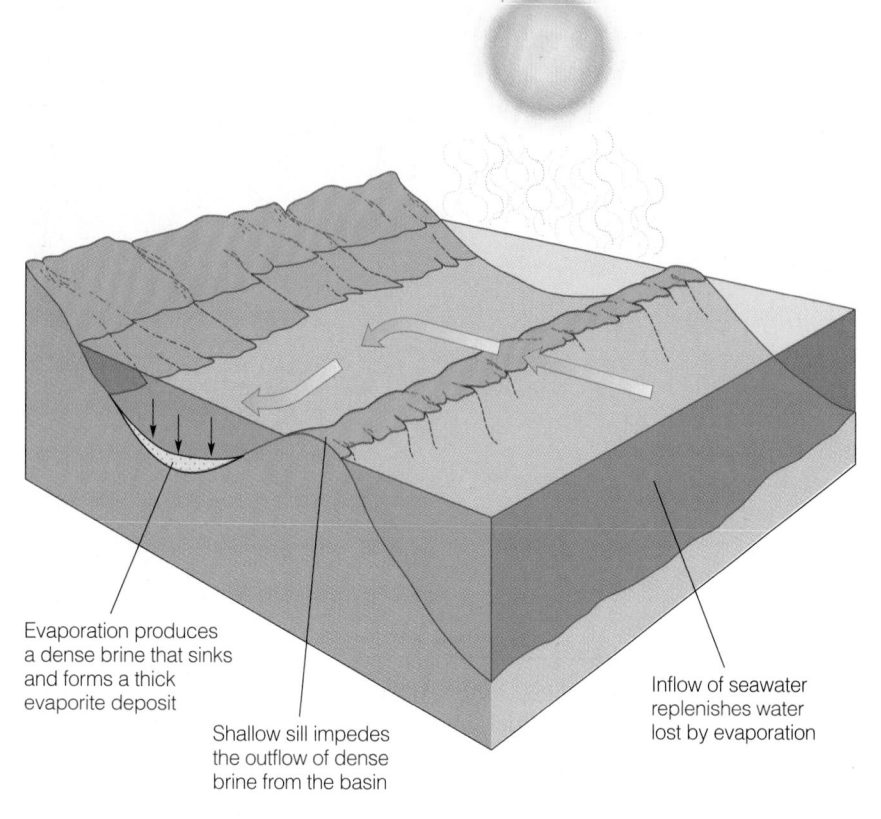

FIGURE 10.12 Silled basin model for evaporite sedimentation by direct precipitation from seawater. Vertical scale is greatly exaggerated.

Evaporation produces
a dense brine that sinks
and forms a thick
evaporite deposit

Shallow sill impedes
the outflow of dense
brine from the basin

Inflow of seawater
replenishes water
lost by evaporation

tral Massachusetts, and Vermont, was the first of several orogenies to affect the Appalachian region.

Obviously, any record of mountain building is complex, and we cannot go into all the details or subdivisions in a book such as this. We can, however, provide an overview of the event and explain its underlying causes in a plate tectonic framework as well as describe the evidence used by geologists to reconstruct such events.

The Appalachian mobile belt can be divided into two depositional environments. The first is the extensive, shallow-water carbonate platform that formed the broad eastern continental shelf and stretchedfrom Newfoundland to Alabama (Figure 10.13a). It formed during the Sauk Sea transgression onto the craton when carbonates were deposited in a vast shallow sea. The shallow-water depth on the platform is indicated by stromatolites, desiccation cracks, and other sedimentary structures.

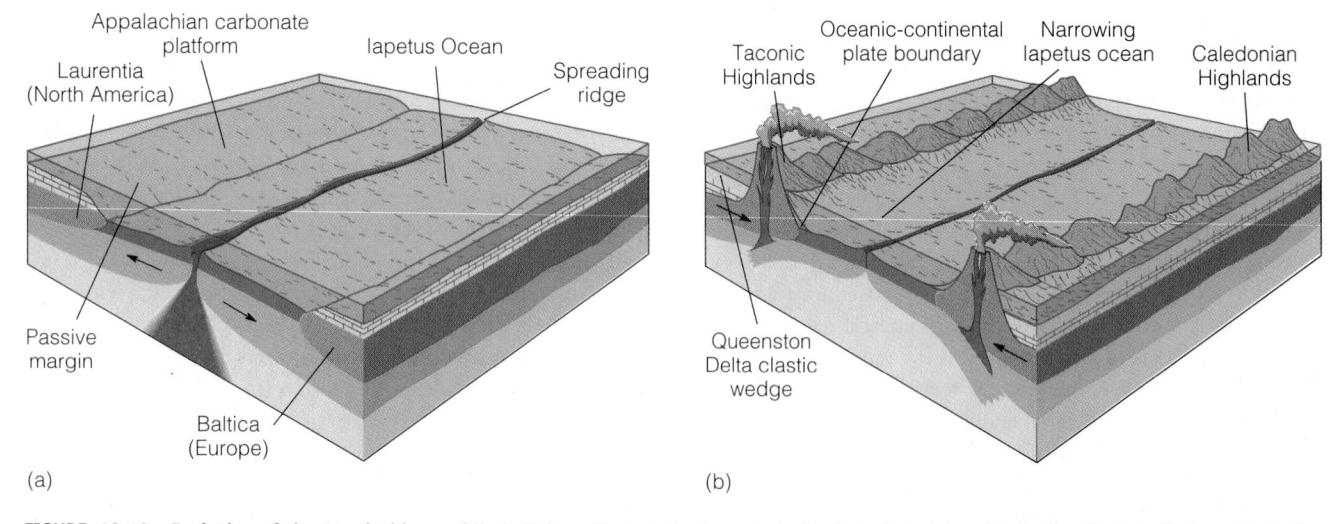

Appalachian carbonate
platform

Laurentia
(North America)

Iapetus Ocean

Spreading
ridge

Passive
margin

Baltica
(Europe)

(a)

Taconic
Highlands

Oceanic-continental
plate boundary

Narrowing
Iapetus ocean

Caledonian
Highlands

Queenston
Delta clastic
wedge

(b)

FIGURE 10.13 Evolution of the Appalachian mobile belt from the Late Proterozoic to the Late Ordovician. (a) During the Late Proterozoic to the Early Ordovician, the Iapetus Ocean was opening along a divergent plate boundary. Both the east coast of Laurentia and the west coast of Baltica were passive continental margins where large carbonate platforms existed. (b) Beginning in the Middle Ordovician, the passive margins of Laurentia and Baltica became oceanic–continental plate boundaries resulting in orogenic activity.

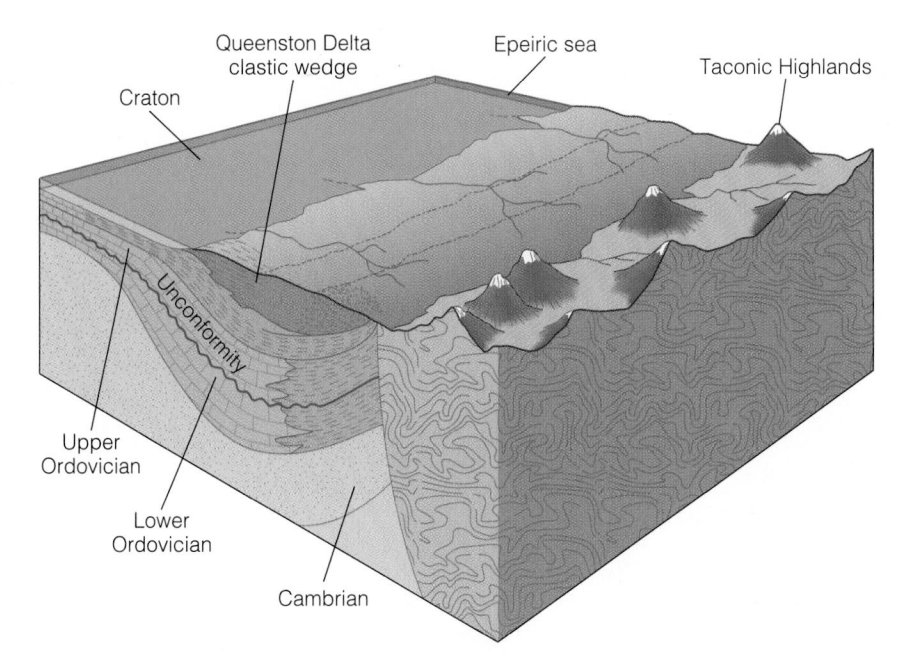

FIGURE 10.14 Reconstruction of the Taconic Highlands and Queenston Delta clastic wedge. The clastic wedge consists of thick, coarse-grained detrital sediments nearest the highlands and thins laterally into finer-grained sediments on the craton.

Carbonate deposition ceased along the East Coast during the Middle Ordovician and was replaced by deep-water deposits characterized by thinly bedded black shales, graded beds, coarse sandstones, graywackes, and associated volcanics. This suite of sediments marks the onset of mountain building, in this case, the Taconic orogeny. The subduction of the Iapetus plate beneath Laurentia resulted in volcanism and downwarping of the carbonate platform, forming an area where sediments accumulated (Figure 10.13). Throughout the Appalachian mobile belt, facies patterns, paleocurrents, and sedimentary structures all indicate that these deposits were derived from the east, where the Taconic Highlands and associated volcanoes were rising.

Additional structural, stratigraphic, petrologic, and sedimentologic evidence has provided much information on the timing and origin of this orogeny. For example, at many locations within the Taconic belt, pronounced angular unconformities occur where steeply dipping Lower Ordovician rocks are overlain by gently dipping or horizontal Silurian and younger rocks.

Other evidence includes volcanic activity in the form of deep-sea lava flows, volcanic ash layers, and intrusive bodies in the area from present-day Georgia to Newfoundland. These igneous rocks show a clustering of radiometric ages between 440 to 480 million years ago. In addition, regional metamorphism coincides with the radiometric dates.

The final piece of evidence for the Taconic orogeny is the development of a large **clastic wedge,** an extensive accumulation of mostly detrital sediments deposited adjacent to an uplifted area. These deposits are thickest and coarsest nearest the highland area and become thinner and finer grained away from the source area, eventually grading into the carbonate cratonic facies (Figure 10.14). The clastic wedge resulting from the erosion of the Taconic Highlands is referred to as the **Queenston Delta.**

The Taconic orogeny marked the first pulse of mountain building in the Appalachian mobile belt and was a response to the subduction taking place beneath the east coast of Laurentia. As the Iapetus Ocean narrowed and closed, another orogeny occurred in Europe during the Silurian. The *Caledonian orogeny* was essentially a mirror image of the Taconic orogeny and the Acadian orogeny (see Chapter 11) and was part of the global mountain-building episode that occurred during the Paleozoic Era. Even though the Caledonian orogeny occurred during Tippecanoe time, we will discuss it in the next chapter because it was intimately related to the Acadian orogeny.

Early Paleozoic Mineral Resources

Early Paleozoic–age rocks contain a variety of important mineral resources, including sand and gravel for construction, building stone, and limestone used in the manufacture of cement. Important sources of industrial or silica sand are the Upper Cambrian Jordan Sandstone of Minnesota and Wisconsin, the Lower Silurian Tuscarora Sandstone in Pennsylvania and Virginia, and the Middle Ordovician St. Peter Sandstone. The latter, the basal sandstone of the Tippecanoe sequence (Figure 10.8), occurs in several states, but the best-known area of production is in La Salle County, Illinois. Silica sand has a variety of uses including the manufacture of glass, refractory bricks for blast furnaces, and molds for casting iron, aluminum, and copper alloys. Some silica sands, called hydraulic fracturing sands, are pumped into wells to fracture oil- or gas-bearing rocks and provide permeable passageways for the oil or gas to migrate to the well.

Thick deposits of Silurian evaporites, mostly rock salt (NaCl) and rock gypsum ($CaSO_4 \cdot H_2O$) altered to rock an-

hydrite ($CaSO_4$), underlie parts of Michigan, Ohio, New York, and adjacent areas in Ontario, Canada. These rocks are important sources of various salts. In addition, barrier and pinnacle reefs in carbonate rocks associated with these evaporites are the reservoirs for oil and gas in Michigan and Ohio.

The host rocks for deposits of lead and zinc in southeast Missouri are Cambrian dolostones, although some Ordovician rocks contain these metals as well. These deposits have been mined since 1720 but have been largely depleted. Now most lead and zinc mined in Missouri come from Mississippian-aged sedimentary rocks.

The Silurian Clinton Formation crops out from Alabama north to New York, and equivalent rocks are found in Newfoundland. This formation has been mined for iron in many places. In the United States, the richest ores and most extensive mining occurred near Birmingham, Alabama, but only a small amount of ore is currently produced in that area.

Summary

Table 10.1 provides a summary of the geologic history of the North American craton and mobile belts as well as global events and sea level changes during the Early Paleozoic.

1. Six major continents existed at the beginning of the Paleozoic Era; four of them were located near the paleoequator.

2. During the Early Paleozoic (Cambrian-Silurian), Laurentia was moving northward and Gondwana moved to a south polar location, as indicated by tillite deposits.

3. Most continents consist of two major components: a relatively stable craton over which epeiric seas transgressed and regressed, surrounded by mobile belts in which mountain building took place.

4. The geologic history of North America can be divided into cratonic sequences that reflect cratonwide transgressions and regressions.

5. The Sauk Sea was the first major transgression onto the craton. At its maximum, it covered the craton except for parts of the Canadian Shield and the Transcontinental Arch, a series of large, northeast–southwest trending islands.

6. The Tippecanoe sequence began with deposition of an extensive sandstone over the exposed and eroded Sauk landscape. During Tippecanoe time, extensive carbonate deposition took place. In addition, large barrier reefs enclosed basins, resulting in evaporite deposition within these basins.

7. The eastern edge of North America was a stable carbonate platform during Sauk time. During Tippecanoe time an oceanic-continental convergent plate boundary formed, resulting in the Taconic orogeny, the first of several orogenies to affect the Appalachian mobile belt.

8. The newly formed Taconic Highlands shed sediments into the western epeiric sea, producing the Queenston Delta, a clastic wedge.

9. Early Paleozoic-age rocks contain a variety of mineral resources including building stone, limestone for cement, silica sand, hydrocarbons, evaporites, and iron ore.

TABLE 10.1

Summary of Early Paleozoic Geologic Events

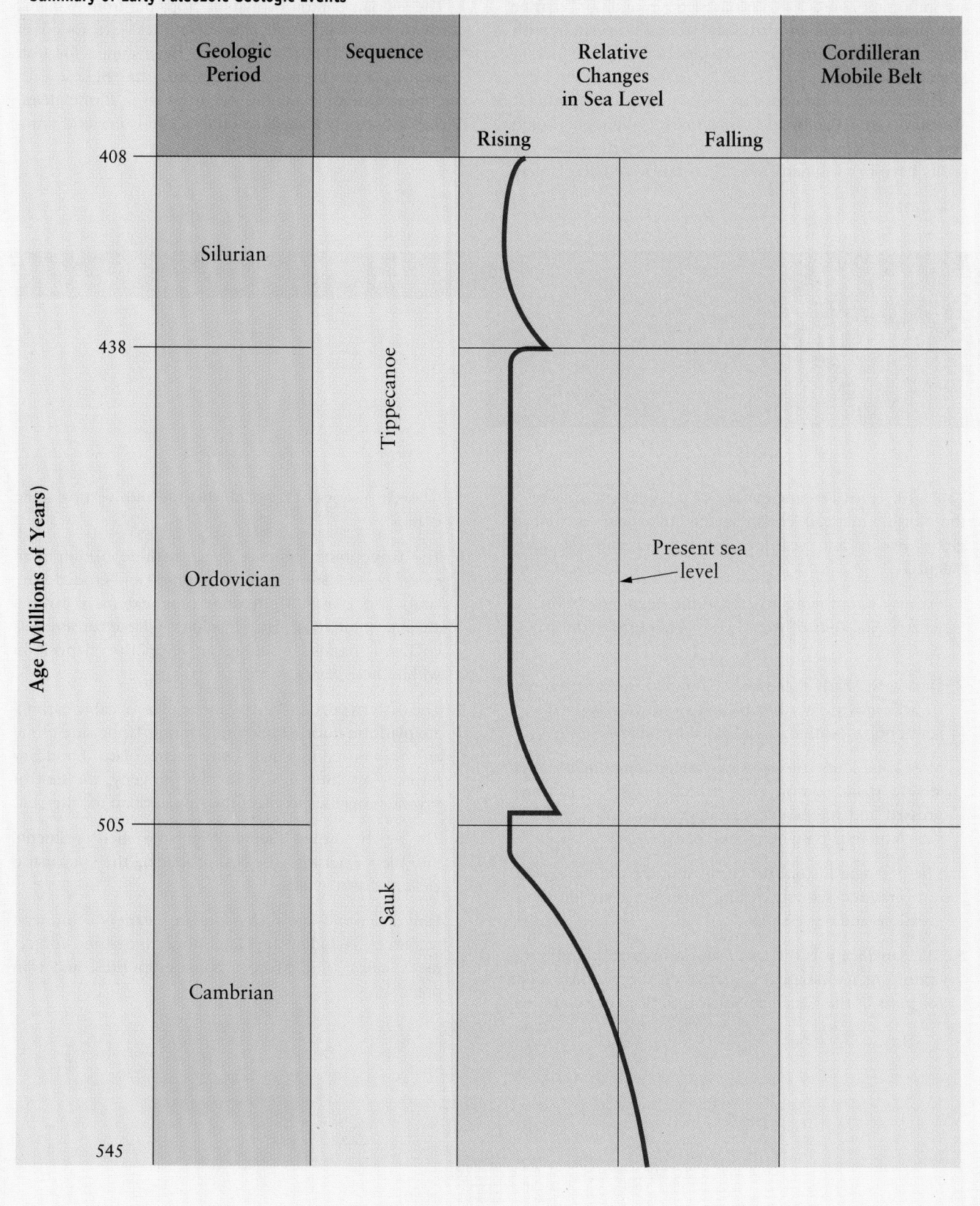

Geologic Period	Sequence	Relative Changes in Sea Level		Cordilleran Mobile Belt
		Rising	Falling	

Age (Millions of Years)

408

Silurian

438

Tippecanoe

Ordovician

Present sea level

505

Sauk

Cambrian

545

Craton	Ouachita Mobile Belt	Appalachian Mobile Belt	Major Events Outside North America
		Acadian orogeny	
Extensive barrier reefs and evaporites common.			Caledonian orogeny
Queenston Delta clastic wedge.		Taconic orogeny	Continental glaciation in Southern Hemisphere.
Transgression of Tippecanoe Sea.			
Regression exposing large areas to erosion.			
Canadian Shield and Transcontinental Arch only areas above sea level.			
Transgression of Sauk Sea.			

Important Terms

Appalachian mobile belt

Baltica

China

clastic wedge

Cordilleran mobile belt

craton

cratonic sequence

epeiric sea

Franklin mobile belt

Gondwana

Iapetus Ocean

Kazakhstania

Laurentia

mobile belt

organic reef

Ouachita mobile belt

Queenston Delta

Sauk sequence

sequence stratigraphy

Siberia

Taconic orogeny

Tippecanoe sequence

Transcontinental Arch

Review Questions

1. An elongated area marking the site of mountain building is a (an):
   a. _____ craton.
   b. _____ platform.
   c. _____ shield.
   d. _____ epeiric sea.
   e. _____ mobile belt.

2. Which of the following was not a Paleozoic continent?
   a. _____ Gondwana
   b. _____ Baltica
   c. _____ Kazakhstania
   d. _____ Eurasia
   e. _____ Laurentia

3. A major transgressive-regressive cycle bounded by craton-wide unconformities is (a)n:
   a. _____ cratonic sequence.
   b. _____ epeiric sea.
   c. _____ orogeny.
   d. _____ biostratigraphic unit.
   e. _____ cyclothem.

4. During deposition of the Sauk sequence, the only area above sea level besides a portion of the Canadian Shield was the:
   a. _____ Appalachian mobile belt.
   b. _____ Transcontinental Arch.
   c. _____ Taconic Highlands.
   d. _____ Queenston Delta.
   e. _____ cratonic margin.

5. The predominant cratonic rocks of the Tippecanoe sequence are:
   a. _____ coals.
   b. _____ sandstones and shales.
   c. _____ graywackes and cherts.
   d. _____ volcanics.
   e. _____ evaporites and reef carbonates.

6. During the Early Paleozoic Era, orogenic activity was confined to which mobile belt?
   a. _____ Cordilleran
   b. _____ Ouachita
   c. _____ Appalachian
   d. _____ Franklin
   e. _____ Answers (a) and (b)

7. The Taconic orogeny resulted from what type of plate movement?
   a. _____ Oceanic-oceanic convergent
   b. _____ Oceanic-continental convergent
   c. _____ Continental-continental convergent
   d. _____ Divergent
   e. _____ Transform

8. Which was the first major trangressive sequence onto the North American craton?
   a. _____ Sauk
   b. _____ Tippecanoe
   c. _____ Kaskaskia
   d. _____ Absaroka
   e. _____ Zuni

9. One of the major sources of Early Paleozoic-age iron ore is the _____ Formation:
   a. _____ St. Peter
   b. _____ Tuscarora
   c. _____ Jordan
   d. _____ Oriskany
   e. _____ Clinton

10. During which sequence was the eastern margin of Laurentia a passive plate margin?
    a. _____ Sauk
    b. _____ Tippecanoe
    c. _____ Kaskaskia
    d. _____ Absaroka
    e. _____ Zuni

11. Discuss how the Cambrian rocks of the Grand Canyon illustrate the sedimentation patterns of a transgressive sea.

12. Draw a diagrammatic cross section of a present-day reef, labeling and defining the various environments within the reef complex.

13. Briefly discuss the Early Paleozoic geologic history of the world.

14. What are some of the methods geologists can use to determine the locations of continents during the Paleozoic Era?

15. Why are cratonic sequences a convenient way to study the geologic history of the Paleozoic Era?

16. Where was the Transcontinental Arch? What evidence indicates that it existed?

17. Discuss how evaporites of the Michigan Basin may have formed during the Silurian Period.

18. What were the major differences between the Appalachian, Ouachita, and Cordilleran mobile belts during the Early Paleozoic?

19. What evidence in the geologic record indicates that the Taconic orogeny occurred?

20. What evidence indicates that the Iapetus Ocean began closing during the Middle Ordovician?

Points to Ponder

1. According to estimates made from mapping and correlation, the Queenston Delta contains more than 600,000 km³ of rock eroded from the Taconic Highlands. Based on this figure, geologists estimate the Taconic Highlands were at least 4000 m high. It is also estimated that the Catskill Delta (see Chapter 11) contains three times as much sediment as the Queenston Delta. From what you know about the geographic distribution of the Taconic Highlands and the Acadian Highlands (see Chapter 11), can you estimate how high the Acadian Highlands might have been?

2. Discuss how sequence stratigraphy can be used to make global correlations and why it is so useful in reconstructing past events.

3. Discuss the role plate tectonics plays in the formation and distribution of Early Paleozoic-age mineral resources.

Additional Readings

Bally, A. W., and A. R. Palmer, eds. 1989. *The geology of North America: An overview.* The Geology of North America. Vol. A. Boulder, Colo.: Geological Society of America.

Catacosinos, P. A., and P. A. Daniels Jr., eds. 1991. *Early sedimentary evolution of the Michigan Basin.* Geological Society of America Special Paper 256. Boulder, Colo.: Geological Society of America.

Dallmeyer, R. D., ed. 1989. *Terranes in the Circum-Atlantic Paleozoic orogens.* Geological Society of America Special Paper 230. Boulder, Colo.: Geological Society of America.

Harwood, D. S., and M. M. Miller, eds. 1990. *Paleozoic and Early Mesozoic paleogeographic relations: Sierra Nevada, Klamath Mountains, and related terranes.* Geological Society of America Special Paper 255. Boulder, Colo.: Geological Society of America.

Hatcher, R. D., Jr., W. A. Thomas, and G. W. Viele, eds. 1989. *The Appalachian-Ouachita orogen in the United States.* The Geology of North America. Vol. F-2. Boulder, Colo.: Geological Society of America.

McKerrow, W. S., and C. R. Scotese, eds. 1990. *Palaeozoic palaeogeography and biogeography.* Geological Society Memoir No. 12. London: Geological Society of London.

Moullade, M., and A. E. M. Nairn, eds. 1991. *The Phanerozoic geology of the world I: The Palaeozoic.* New York: Elsevier Science.

Osborne, R., and D. Tarling, eds. 1996. *The historical atlas of the Earth.* New York: Holt.

Sloss, L. L., ed. 1988. *Sedimentary cover—North American craton: U.S.* The Geology of North America. Vol. D-2. Boulder, Colo.: Geological Society of America.

Wilgus, C. K., B. S. Hastings, C. G. St. C. Kendall, H. Posamentier, C. A. Ross, and J. Van Wagoner, eds. 1988. *Sea-level changes: An integrated approach.* Society of Economic Paleontologists and Mineralogists Special Publication No. 42. Tulsa, Okla.: Society of Economic Paleontologists and Mineralogists.

World Wide Web Activities

▶ **STRATIGRAPHY OF THE BLACK HILLS**

This site contains information about the geology of the Black Hills, South Dakota. You can view a nice relief image of the Black Hills, as well as a geologic map of South Dakota. In addition, a stratigraphic column of the Black Hills is provided, with links to the various formations of the Black Hills. Under the Cambrian to Jurassic stratigraphic column, click on any of the Paleozoic formations to learn more about that particular formation.

▶ **UNION COLLEGE GEOLOGY DEPARTMENT**

This site is maintained by the Union College Geology Department, which is located in Schenectady, New York. This particular site provides information about the sediments that were deposited during the Middle and Late Ordovician Period in the present location of the Mohawk River Valley of New York. It also contains sites showing the paleogeography of the world and New York State during the Ordovician Period.

1. Click on the *New York* icon to see what the area looked like during the Late Ordovician. How does it compare to the paleogeographic map of North America during this time period?

2. Click on the *Paleocurrent indicators* icon. What was the general direction of the currents during this time? How was this determined?

▶ **GLOBAL EARTH HISTORY**

This site, maintained at the University of Northern Arizona, contains a series of plate tectonic reconstructions to show what Earth looked like at various times in the geologic past. Included are paleogeographic maps showing global, North American, and European reconstructions from the Cambrian to the present, as well as tectonic, sedimentation, and paleogeographic maps of the North Atlantic region.

▶ **NORTH AMERICAN OROGENIES**

This site is part of the *Global Earth History* site and shows simple, restored cross sections of the major orogenies taking place in North America from the Proterozoic through Mesozoic. Click on the various orogenies that occurred during the Early Paleozoic.

Explore the following *In-Terra-Active 2.0* CD-ROM module(s) and increase your understanding of key concepts and processes presented in this chapter.

Chapter Concept: **Crustal Deformation in Mobile Belts**

Multiple episodes of crustal deformation occurred during the Paleozoic era in the mobile belts of sediment adjacent to the stable shield areas of North America and other continents. In this module, you will explore the relationship between the folds and faults of mobile belts and the deformational processes that accompany plate tectonics.

Section: **Interior**

Module: **Crustal Deformation**

The coarse-grained garnet-mica schist shown on the left has essentially the same chemical composition as the fine-grained mudstone shown on the right. In the field, it is clear that these belong to the same stratigraphic unit.

Question: *How can this transformation take place in the solid state?*

11 Geology of the Late Paleozoic Era

Seedless vascular plant fossil in iron-carbonate concretion from Pennsylvanian-age deposits, Mazon Creek locality, Illinois.

Prologue

The unusual state of preservation of the Pennsylvanian-age Mazon Creek biota of northeastern Illinois provides us with significant insights about the soft-part anatomy of organisms rarely preserved in the fossil record. The biota is divided into marine and nonmarine components and contains the only known or oldest fossil representatives of several major animal groups.

The Mazon Creek fossils occur in spheroidal- to elliptical-shaped iron–carbonate concretions ranging from 1 to 30 cm long. Rapid burial and the formation of concretions around the organisms were primarily responsible for the excellent preservation of the fossils. Not only are the hard parts of organisms preserved, but also impressions and carbonaceous films of the soft-bodied animals and plants.

The environment in which these plants and animals lived was a large delta where sluggish southward-flowing rivers emptied into a subtropical epeiric sea that covered most of the present state of Illinois. Two major habitats are represented: a swampy forested lowland of the subaerial delta and the shallow-marine environment of the actively prograding delta.

A diverse marine fauna lived in the warm, shallow waters of the delta front and included cnadarians, mollusks, echinoderms, arthropods, worms, and fish. From this fauna have come the only known fossils of several animal groups, including the lamprey, whose gills and liver can be discerned in the impressions.

More than 350 plant species lived in the swampy lowlands surrounding the delta. Almost all the plants were seedless vascular plants, typical of the kinds that comprised the Pennsylvanian coal-forming swamps of North America. Also found in the swampy lowlands were numerous insects, including millipedes and centipedes as well as spiders and other animals such as scorpions and amphibians. In the ponds, lakes, and rivers were many fish, shrimp, and ostracodes.

Local farmers and townspeople began collecting the Mazon Creek biota in the mid-1800s. Their efforts were soon followed by scientific studies by such famous geologists and paleontologists as J. D. Dana, L. Lesquereux, and E. D. Cope. Study of the Mazon Creek fossils has been uneven through the years, with much of it concentrating on the descriptions of the many plant and animal species recovered. Today, research is still largely concerned with describing the complete assemblage, but researchers are also interested in determining the phylogenetic relationships of the organisms and the paleoecologic significance of the assemblage.

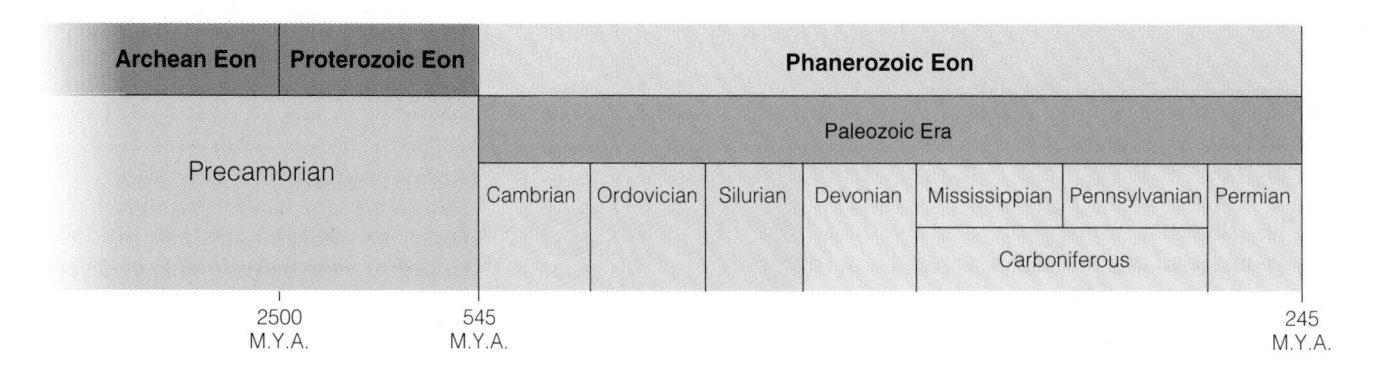

Archean Eon	Proterozoic Eon	Phanerozoic Eon							
		Paleozoic Era							
Precambrian		Cambrian	Ordovician	Silurian	Devonian	Mississippian	Pennsylvanian	Permian	
						Carboniferous			

2500 M.Y.A. 545 M.Y.A. 245 M.Y.A.

Introduction

The Late Paleozoic Era, composed of the Devonian, Carboniferous (Mississippian and Pennsylvanian in North American terminology), and Permian periods, witnessed the formation of the supercontinent Pangaea from the various landmasses that were present when the era began. Collisions between continents along convergent plate boundaries led to the assembly of Pangaea with the result that, by the end of the Permian, the world's geography consisted of one large landmass surrounded by a global ocean. During the Late Paleozoic, the various continents experienced a variety of climatic conditions as indicated by such sedimentary rocks as coals, evaporites, and tillites. Major glacial–interglacial intervals occurred throughout much of Gondwana as it continued moving over the South Pole during the Late Mississippian to Early Permian. The growth and retreat of these glaciers profoundly affected the world's biota and also contributed to global changes in sea level. The mountain building that resulted from plate convergence also strongly influenced oceanic and atmospheric circulation patterns. By the end of the Paleozoic, widespread arid and semiarid conditions prevailed over much of Pangaea.

Late Paleozoic Paleogeography

The Late Paleozoic was a time of continental collisions, mountain building, fluctuating sea levels, and varied climates. These events resulted from plate interactions and major continental ice-sheet advances and retreats. We will initially examine the changing paleogeography of the Late Paleozoic and then place North American events into this global framework.

THE DEVONIAN PERIOD

During the Silurian, Laurentia and Baltica collided along a convergent plate boundary to form the larger continent of **Laurasia.** This collision, which closed the northern Iapetus Ocean, is marked by the **Caledonian orogeny.** During the Devonian, as the southern Iapetus Ocean narrowed between Laurasia and Gondwana, mountain building continued along the eastern margin of Laurasia with

the **Acadian orogeny** (Figure 11.1a). The erosion of the resulting highlands provided vast amounts of reddish fluvial sediments that covered large areas of northern Europe (Old Red Sandstone) and eastern North America (the Catskill Delta). Other Devonian tectonic events, probably related to the collision of Laurentia and Baltica, include the Cordilleran **Antler orogeny,** the **Ellesmere orogeny** along the northern margin of Laurentia (which may reflect the collision of Laurentia with Siberia), and the change from a passive continental margin to an active convergent plate boundary in the Uralian mobile belt of eastern Baltica. The distribution of reefs, evaporites, and red beds, as well as the existence of similar floras throughout the world, suggests a rather uniform global climate during the Devonian Period.

THE CARBONIFEROUS PERIOD

During the Carboniferous Period, southern Gondwana moved over the South Pole, resulting in extensive continental glaciation (Figures 11.1b and 11.2a). The advance and retreat of these glaciers produced global changes in sea level that affected sedimentation patterns on the cratons. As Gondwana continued moving northward, it first collided with Laurasia during the Early Carboniferous and continued suturing with it during the rest of the Carboniferous (Figures 11.1b and 11.2a). Because Gondwana rotated clockwise relative to Laurasia, deformation generally progressed in a northeast-to-southwest direction along the Hercynian, Appalachian, and Ouachita mobile belts of the two continents. The final phase of collision between Gondwana and Laurasia is indicated by the Ouachita Mountains of Oklahoma, which were formed by thrusting during the Late Carboniferous and Early Permian.

Elsewhere, Siberia collided with Kazakhstania and moved toward the Uralian margin of Laurasia (Baltica), colliding with it during the Early Permian. It has been suggested that the northwestern margin of China collided with the southwestern margin of Siberia during the Late Carboniferous. By the end of the Carboniferous, the various continental landmasses were fairly close together as Pangaea began taking shape.

The Carboniferous coal basins of eastern North America, western Europe, and the Donets Basin of Ukraine all lay in the equatorial zone, where rainfall was high and temperatures were consistently warm. The absence of

Phanerozoic Eon												
Mesozoic Era			Cenozoic Era									
Triassic	Jurassic	Cretaceous	Tertiary						Quaternary			
			Paleocene	Eocene	Oligocene	Miocene	Pliocene		Pleistocene	Holocene		

245
M.Y.A.

66
M.Y.A.

strong seasonal growth rings in fossil plants from these coal basins is indicative of such a climate. The fossil plants found in the coals of Siberia and China, however, show well-developed growth rings, signifying seasonal growth with abundant rainfall and distinct seasons such as occur in the temperature zones (latitudes 40 degrees to 60 degrees north).

Glacial conditions and the movement of large continental ice sheets in the high southern latitudes are indicated by widespread tillites and glacial striations in southern Gondwana. These ice sheets spread toward the equator and, at their maximum growth, extended well into the middle temperate latitudes.

THE PERMIAN PERIOD

The assembly of Pangaea was essentially concluded during the Permian with the completion of many of the continental collisions that began during the Carboniferous (Figure 11.2b). Although geologists generally agree on the configuration and location of the western half of the supercontinent, there is no consensus on the number or configuration of the various terranes and continental blocks that composed the eastern half of Pangaea. Regardless of the exact configuration of the eastern portion, geologists know that the supercontinent was surrounded by various subduction zones and moved steadily northward during the Permian. Furthermore, an enormous single ocean, **Panthalassa,** surrounded Pangaea and spanned Earth from pole to pole (Figure 11.2b). Waters of this ocean probably circulated more freely than at present, resulting in more equable water temperatures.

The formation of a single large landmass had climatic consequences for the terrestrial environment as well. Terrestrial Permian sediments indicate that arid and semiarid conditions were widespread over Pangaea. The mountain ranges produced by the **Hercynian, Alleghenian,** and **Ouachita orogenies** were high enough to create rain shadows that blocked the moist, subtropical, easterly winds—much as the southern Andes Mountains do in western South America today. This produced very dry conditions in North America and Europe, as evident from the extensive Permian evaporites found in western North America, central Europe, and parts of Russia. Permian coals, indicative of abundant rainfall, were mostly limited to the northern temperate belts (latitude 40 degrees to

60 degrees north), while the last remnants of the Carboniferous ice sheets retreated to the mountainous regions of eastern Australia.

Late Paleozoic History of North America

The Late Paleozoic cratonic history of North America included periods of extensive shallow-marine carbonate deposition and large coal-forming swamps as well as dry, evaporite-forming terrestrial conditions. Cratonic events largely resulted from changes in sea level due to Gondwanan glaciation and tectonic events related to the assembly of Pangaea. Mountain building that began with the Ordovician Taconic orogeny continued with the Caledonian, Acadian, Alleghenian, and Ouachitia orogenies. These orogenies were part of the global tectonic process that resulted in the assembly of Pangaea by the end of the Paleozoic Era.

The Kaskaskia Sequence

The boundary between the Tippecanoe sequence and the overlying **Kaskaskia sequence** (Middle Devonian–Middle Mississippian) is marked by a major unconformity. As the Kaskaskia Sea transgressed over the low relief landscape of the craton, the majority of the basal beds deposited consisted of clean, well-sorted quartz sandstones. A good example is the Oriskany Sandstone of New York and Pennsylvania and its lateral equivalents (Figure 11.3). The Oriskany Sandstone, like the basal Tippecanoe St. Peter Sandstone, is an important glass sand as well as a good gas-reservoir rock.

The source areas for the basal Kaskaskia sandstones were primarily the eroding highlands of the Appalachian mobile belt area (Figure 11.4), exhumed Cambrian and Ordovician sandstones cropping out along the flanks of the Ozark Dome, and exposures of the Canadian Shield in the Wisconsin area. The lack of similar sands in the Silurian carbonate beds below the Tippecanoe–Kaskaskia unconformity indicates that the source areas of the basal Kaskaskia detrital rocks were submerged when the Tippecanoe sequence was deposited. Stratigraphic studies indicate that these source areas were uplifted and the

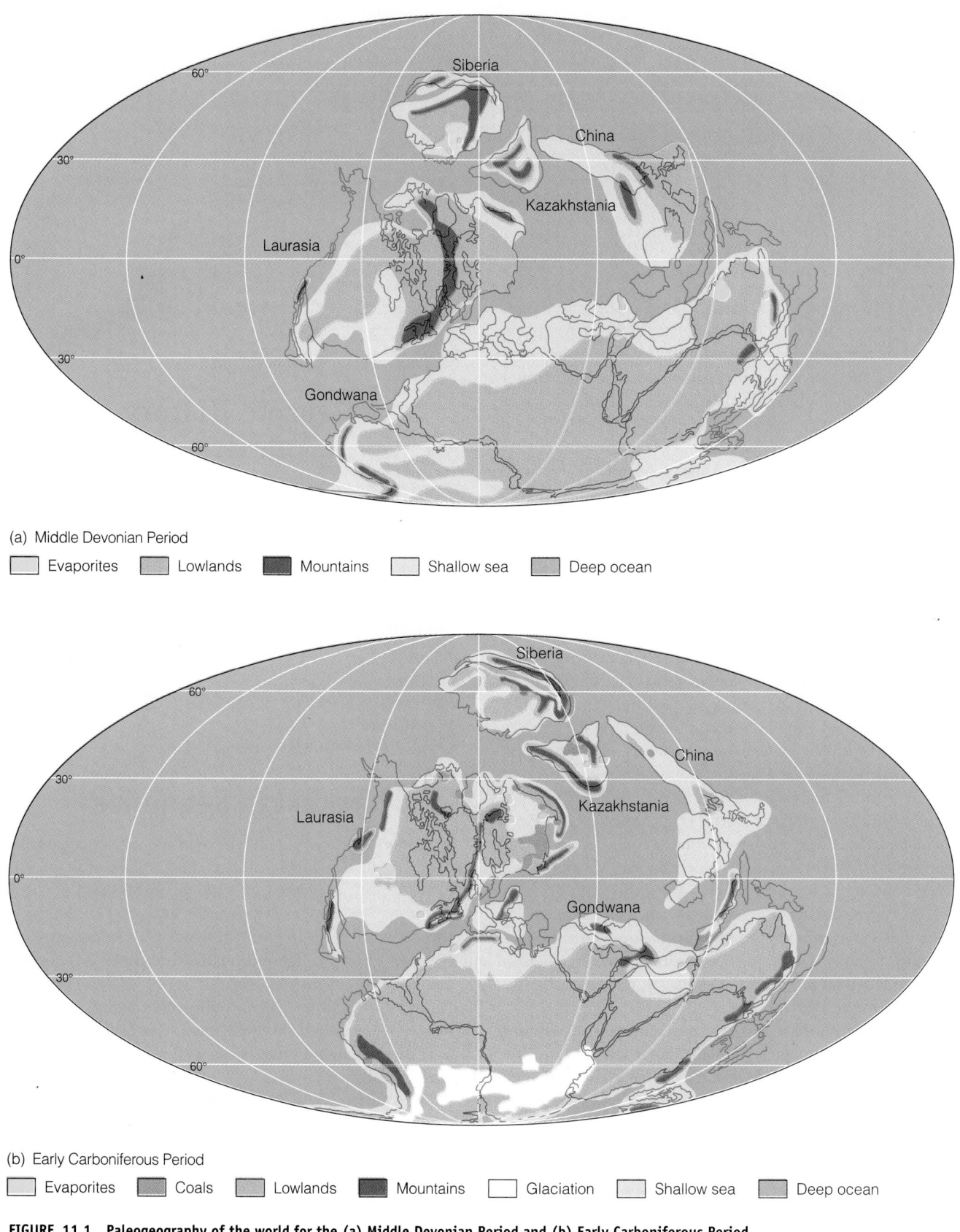

(a) Middle Devonian Period

□ Evaporites ■ Lowlands ■ Mountains □ Shallow sea ■ Deep ocean

(b) Early Carboniferous Period

□ Evaporites ■ Coals ■ Lowlands ■ Mountains □ Glaciation □ Shallow sea ■ Deep ocean

FIGURE 11.1 Paleogeography of the world for the (a) Middle Devonian Period and (b) Early Carboniferous Period.

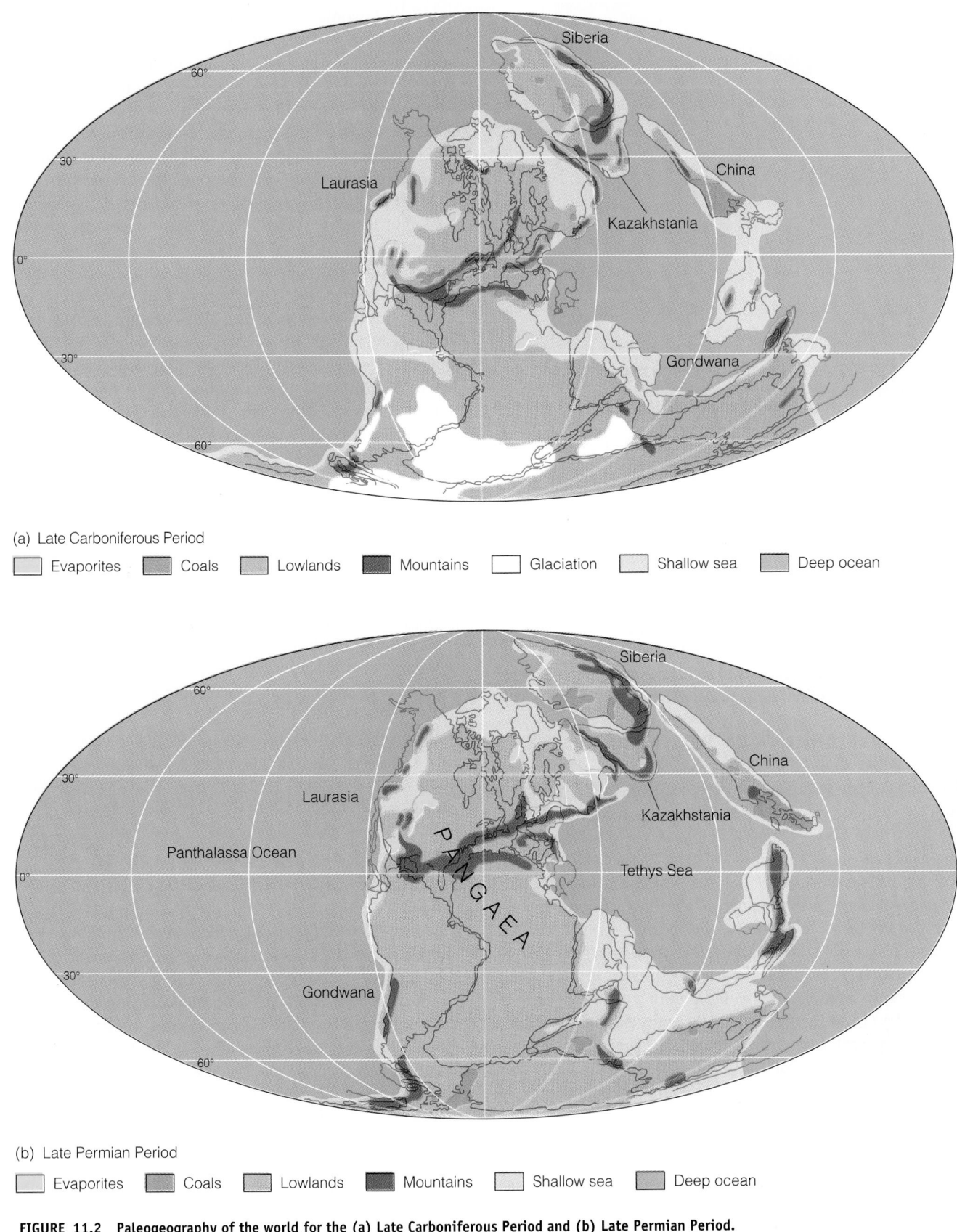

(a) Late Carboniferous Period

☐ Evaporites ☐ Coals ☐ Lowlands ■ Mountains ☐ Glaciation ☐ Shallow sea ☐ Deep ocean

(b) Late Permian Period

☐ Evaporites ☐ Coals ☐ Lowlands ■ Mountains ☐ Shallow sea ☐ Deep ocean

FIGURE 11.2 **Paleogeography of the world for the (a) Late Carboniferous Period and (b) Late Permian Period.**

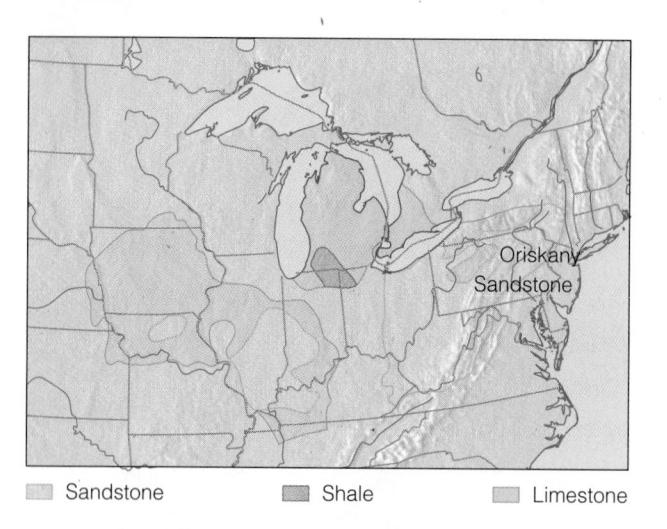

Sandstone   Shale   Limestone

FIGURE 11.3 Extent of the basal units of the Kaskaskia sequence in the eastern and north–central United States.

Tippecanoe carbonates removed by erosion prior to the Kaskaskia transgression. Kaskaskian basal rocks elsewhere on the craton consist of carbonates that are frequently difficult to differentiate from the underlying Tippecanoe carbonates unless they are fossiliferous.

Except for widespread Upper Devonian and Lower Mississippian black shales, the majority of Kaskaskian rocks are carbonates, including reefs, and associated evaporite deposits. In many other parts of the world, such as southern England, Belgium, central Europe, Australia, and Russia, the Middle and early Late Devonian epochs were times of major reef building (see Perspective 11.1).

REEF DEVELOPMENT IN WESTERN CANADA

The Middle and Late Devonian-age reefs of western Canada contain large reserves of petroleum and have been widely studied from outcrops and in the subsurface (Figure 11.5). These reefs began forming as the Kaskaskia Sea transgressed southward into western Canada. By the end of the Middle Devonian, they had coalesced into a large barrier-reef system that restricted the flow of oceanic water into the back-reef platform, thus creating conditions for evaporite precipitation (Figure 11.5). In the back reef area, up to 300 m of evaporites were precipitated in much the same way as in the Michigan Basin during the Silurian (see Figure 10.11). More than half of the world's potash, which is used in fertilizers, comes from these Devonian evaporites. By the middle of the Late Devonian, reef growth stopped in the western Canada region, although nonreef carbonate deposition continued.

BLACK SHALES

In North America, many areas of carbonate-evaporite deposition gave way to a greater proportion of shales and coarser detrital rocks beginning in the Middle Devonian and continuing into the Late Devonian. This change to detrital deposition resulted from the formation of new source areas brought on by the mountain-building activity associated with the Acadian orogeny in North America (Figure 11.4).

As the Devonian Period ended, a conspicuous change in sedimentation took place over the craton with the appearance of widespread black shales. In the eastern United States, these black shales are commonly called the Chattanooga Shale, but are known by a variety of local names elsewhere (for example, New Albany Shale and Antrim Shale). Although these black shales are best developed from the cratonic margins along the Appalachian mobile belt to the Mississippi Valley, correlative units can also be found in many western states and in western Canada (Figure 11.6).

The Upper Devonian-Lower Mississippian black shales of North America are typically noncalcareous, thinly bedded, and usually less than 10 m thick (Figure 11.6). Fossils are usually rare in these black shales, but some Upper Devonian black shales do contain rich conodont faunas. Because most black shales lack body fossils, they are difficult to date and to correlate. In places where they can be dated, usually by conodonts (microscopic animals), acritarchs (microscopic algae), or plant spores, the lower beds are Late Devonian, and the upper beds are Early Mississippian in age.

Although the origin of these extensive black shales is still being debated, the essential features required to produce them include undisturbed anaerobic bottom water, a reduced supply of coarser detrital sediment, and high organic productivity in the overlying oxygenated waters. High productivity in the surface waters leads to a shower of organic material, which decomposes on the undisturbed seafloor and depletes the dissolved oxygen at the sediment-water interface.

The wide extent of such apparently shallow-water black shales in North America remains puzzling. Nonetheless, these shales are rich in uranium and are an important source rock of oil and gas in the Appalachian region.

THE LATE KASKASKIA—A RETURN TO EXTENSIVE CARBONATE DEPOSITION

Following deposition of the widespread Upper Devonian-Lower Mississippian black shales, carbonate sedimentation on the craton dominated the remainder of the Mississippian Period (Figure 11.7). During this time, a variety of carbonate sediments were deposited in the epeiric sea as indicated by the extensive deposits of crinoidal limestones (rich in crinoid fragments), oolitic limestones, and various other limestones and dolostones (Figure 11.8). These Mississippian carbonates display cross-bedding, ripple marks, and well-sorted fossil fragments, all of which are indicative of a shallow-water environment. Analogous features can be observed on the present-day Bahama Banks. In addition, numerous small organic reefs occurred throughout the craton during the Mississippian. These were all much smaller than the large barrier-reef complexes that dominated the earlier Paleozoic seas.

During the Late Mississippian regression of the Kaskaskia Sea from the craton, carbonate deposition was replaced by vast quantities of detrital sediments. The re-

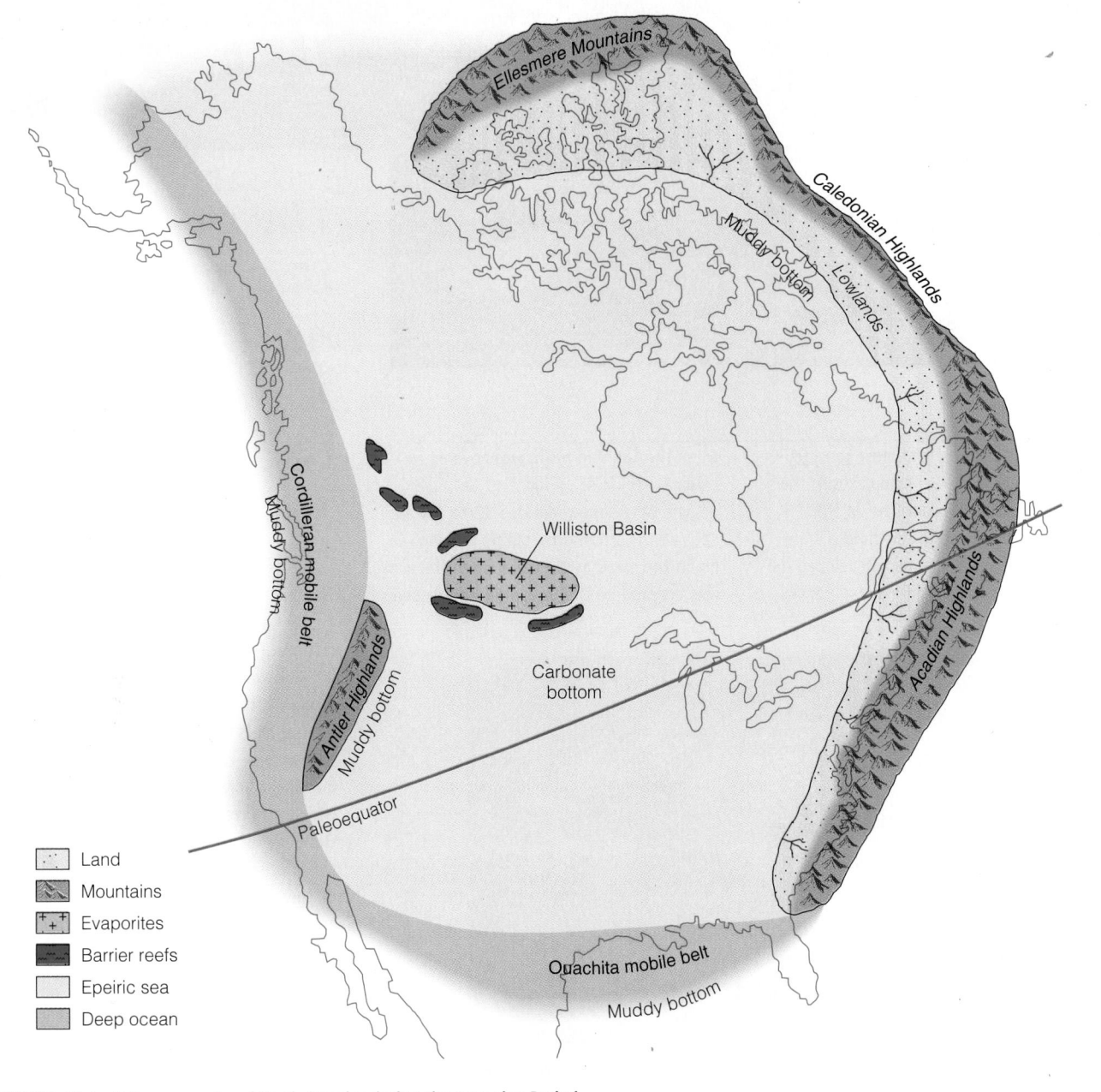

FIGURE 11.4 Paleogeography of North America during the Devonian Period.

Legend:
- Land
- Mountains
- Evaporites
- Barrier reefs
- Epeiric sea
- Deep ocean

Map labels: Ellesmere Mountains, Caledonian Highlands, Muddy bottom, Lowlands, Cordilleran mobile belt, Muddy bottom, Williston Basin, Antler Highlands, Muddy bottom, Carbonate bottom, Acadian Highlands, Paleoequator, Ouachita mobile belt, Muddy bottom

sulting sandstones, particularly in the Illinois Basin, have been studied in great detail because they are excellent petroleum reservoirs. Prior to the end of the Mississippian, the Kaskaskia Sea had retreated to the craton margin, once again exposing the craton to widespread weathering and erosion that resulted in a cratonwide unconformity when the Absaroka Sea began transgressing back over the craton.

The Absaroka Sequence

The **Absaroka sequence** includes rocks deposited during the latest Mississippian through Early Jurassic. In this chapter, however, we will be concerned only with the Paleozoic rocks of the Absaroka sequence. The extensive unconformity separating the Kaskaskia and Absaroka sequences essentially divides the strata into the North American Mississippian and Pennsylvanian systems. These two systems are equivalent to the European Lower and Upper Carboniferous systems, respectively. The rocks of the Absaroka sequence are not only different from those of the Kaskaskia sequence, but they are also the result of different tectonic regimes.

The lowermost sediments of the Absaroka sequence are confined to the margins of the craton. These deposits are generally thickest in the east and southeast, near the emerging highlands of the Appalachian and Ouachita mobile belts, and thin westward onto the craton. The lithologies also reveal lateral changes from nonmarine detrital rocks and coals in the east, through transitional marine–nonmarine beds, to largely marine detrital rocks and limestones farther west (Figure 11.9).

The Canning Basin, Australia— A Devonian Great Barrier Reef

One of the largest and most spectacularly exposed fossil-reef complexes in the world is the Great Barrier Reef of the Canning Basin, Western Australia (Figure 1). This barrier-reef complex developed during the Middle and Late Devonian Period when the Canning Basin was covered by a tropical epeiric sea (Figure 11.1a).

The limestone reefs rise 50 to 100 m above the surrounding plains looking much the same as they did when the area was covered by the Devonian epeiric sea.

FIGURE 1 Aerial view of Windjana Gorge showing the Devonian Great Barrier Reef exposed as a limestone ridge.

The shales and other soft sediments deposited on the open oceanside of the reef complex (see Figure 10.9) have been eroded away, leaving the resistant limestone reefs standing as ridges.

The reefs themselves were constructed primarily by calcareous algae, stromatoporoids, and various forms of corals, which also were the main components of the other major reef complexes in the world at that time. An interesting feature of these Canning Basin reefs is the presence of stromatolites, which are found in the reef, back-reef, and marginal-slope areas of the reef complex. Stromatolites are an unusual component because they ceased to be abundant by the end of the Proterozoic Eon. Throughout the Phanerozoic Eon, stromatolites typically formed only in areas generally inhospitable to other marine organisms.

CYCLOTHEMS

One of the characteristic features of Pennsylvanian rocks is their cyclical pattern of alternating marine and nonmarine strata. Such rhythmically repetitive sedimentary sequences are known as **cyclothems.** They result from repeated alternations of marine and nonmarine environments, usually in areas of low relief. Though seemingly simple, cyclothems reflect a delicate interplay between nonmarine deltaic and shallow-marine interdeltaic and shelf environments.

For purposes of illustration, we can look at a typical coal-bearing cyclothem from the Illinois Basin (Figure 11.10a). Such a cyclothem contains nonmarine units, capped by a coal and overlain by marine units. Figure 11.10c shows the depositional environments that produced the cyclothem. The initial units represent deltaic and fluvial deposits. Above them is an underclay that frequently contains root casts from the plants and trees that comprise the overlying coal. The coal bed results from ac-

FIGURE 2 Outcrop of the Devonian Great Barrier Reef along Windjana Gorge. The talus of the fore-reef can be seen on the left side of the picture sloping away from the reef core, which is unbedded. To the right of the reef core is the back-reef facies, which is horizontally bedded.

The outcrop along Windjana Gorge beautifully reveals the various features and facies of the Devonian Great Barrier Reef complex (Figure 2). The reef core consists of unbedded limestones composed predominantly of calcareous algae, stromatoporoids, and corals. The back-reef facies (see Figure 10.9) is bedded and makes up the major part of the total reef complex environment. A diverse and abundant fauna of calcareous algae, stromatoporoids, various corals, some bivalves, gastropods,

cephalopods, brachiopods, and crinoids lived in this lagoonal area behind the reef core.

On the seaward side of the reef core was the steep fore-reef slope (see Figure 10.9) where such organisms, as algae, sponges, and stromatoporoids lived. This facies contains considerable reef talus, an accumulation of debris eroded by waves from the reef front. The ocean-basin deposits contain the fossils of mainly swimming and floating organisms such as fish, radiolarians, and cephalopods.

Near the end of the Late Devonian, nearly all the reef-building organisms as well as much of the associated fauna of the Canning Basin Great Barrier Reef became extinct. As we will discuss in Chapter 12, few massive tabulate–rugose–stromatoporoid reefs are known from latest Devonian or younger rocks.

cumulations of plant material and is overlain by marine units of alternating limestones and shales, usually with an abundant marine invertebrate fauna. The marine cycle ends with an erosion surface. A new cyclothem begins with a nonmarine deltaic sandstone. All the beds illustrated in the idealized cyclothems are not always preserved because of abrupt changes from marine to nonmarine conditions or to removal of some units by erosion.

Cyclothems represent transgressive and regressive se-

quences with an erosional surface separating one cyclothem from another. Thus, an idealized cyclothem passes upward from fluvial–deltaic deposit, through coals, to detrital shallow-water marine sediments, and finally to limestones typical of an open marine environment.

Such places as the Mississippi delta, the Florida Everglades, and the Dutch lowlands represent modern coal-forming environments similar to those that existed during the Pennsylvanian (Figure 11.10d). By studying these mod-

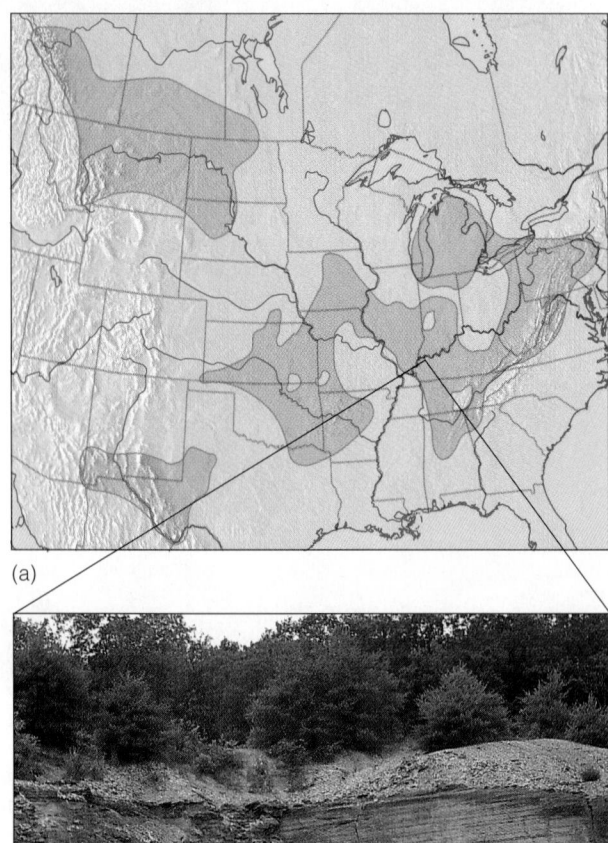

FIGURE 11.5 Reconstruction of the extensive Devonian Reef complex of western Canada. These extensive reefs controlled the regional facies of the Devonian epeiric seas.

The map in Figure 11.5 shows the following labels: Shallow basinal limestones, Northwest Territories, Canadian Shield, Yukon, Fore reef, Reef, Back reef, platform evaporites, Saskatchewan, Alberta, British Columbia.

(a)

(b)

FIGURE 11.6   (a) The extent of the Upper Devonian to Lower Mississippian Chattanooga Shale and its equivalent units in North America. (b) Upper Devonian New Albany Shale, Button Mold Knob Quarry, Kentucky.

ern analogues, geologists can make reasonable deductions about conditions that existed in the geologic past.

The coal swamps that existed during the Pennsylvanian Period must have been large lowland areas neighboring the sea (Figure 11.10c). In such cases, a very slight rise in sea level would have flooded large areas, while slight drops would have exposed large areas, resulting in alternating marine and nonmarine environments. The same result could have been caused by rising sea level and progradation of a large delta, such as occurs today in Louisiana.

Such regularity and cyclicity in sedimentation over a large area requires an explanation. In most cases, local cyclothems of limited extent can be explained by rapid but slight changes in sea level in a swamp–delta complex of low relief near the sea or by localized crustal movement. Explaining widespread cyclothems in more difficult.

The hypothesis currently favored by most geologists is a rise and fall of sea level related to advances and retreats of Gondwanan continental glaciers. When the Gondwanan ice sheets advanced, sea level dropped, and when they melted, sea level rose. Late Paleozoic cyclothem activity on all of the cratons closely corresponds to Gondwanan glacial–interglacial cycles.

CRATONIC UPLIFT—THE ANCESTRAL ROCKIES

Recall that cratons are stable areas, and when they do experience deformation, it is usually mild. The

Pennsylvanian Period, however, was a time of unusually severe cratonic deformation, resulting in uplifts of sufficient magnitude to expose Precambrian basement rocks. In addition to newly formed highlands and basins, many previously formed arches and domes, such as the Cincinnati Arch, Nashville Dome, and Ozark Dome, were also reactivated (see Figure 10.1).

During the Late Absaroka, the area of greatest deformation occurred in the southwestern part of the North American craton where a series of fault-bounded uplifted blocks formed the **Ancestral Rockies** (Figure 11.11a). Uplift of these mountains, some of which were elevated more than 2 km along near-vertical faults, resulted in the erosion of the

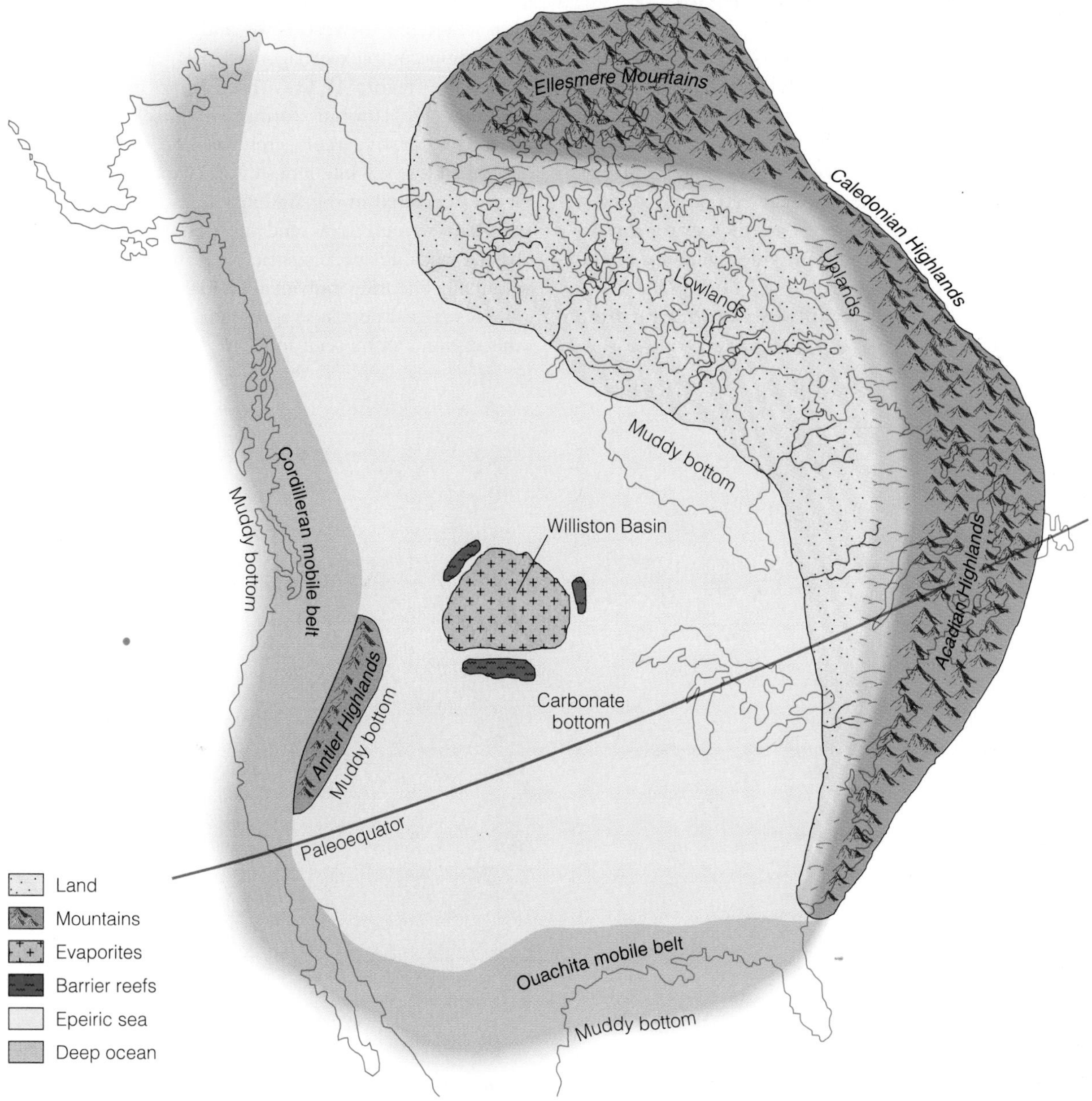

FIGURE 11.7 Paleogeography of North America during the Mississippian Period.

overlying Paleozoic sediments and exposure of the Precambrian igneous and metamorphic basement rocks (Figure 11.11b). As the mountains eroded, tremendous quantities of coarse, red arkosic sand and conglomerate were deposited in the surrounding basins. These sediments are preserved in many areas including the rocks of the Garden of the Gods near Colorado Springs (Figure 11.11c) and at the Red Rocks Amphitheatre near Morrison, Colorado.

Intracratonic mountain ranges are unusual, and their cause has long been debated. It is thought that the collision of Gondwana with Laurasia (Figure 11.2a) produced great stresses in the southwestern region of the North American craton. These crustal stresses were relieved by faulting that resulted in uplift of cratonic blocks and downwarp of adjacent basins, forming a series of ranges and basins.

FIGURE 11.8 Mississippian limestones exposed near Bowling Green, Kentucky.

THE LATE ABSAROKA—MORE EVAPORITE DEPOSITS AND REEFS

While the various intracratonic basins were filling with sediment during the Late Pennsylvanian, the Absaroka Sea slowly began retreating from the craton. During the Early Permian, the Absaroka Sea occupied a narrow region from Nebraska through west Texas (Figure 11.12). By the Middle Permian, the sea had retreated to west Texas and southern New Mexico. The thick evaporite deposits in Kansas and Oklahoma provide evidence of the restricted nature of the Absaroka Sea during the Early and Middle Permian and its southwestward retreat from the central craton.

During the Middle and Late Permian, the Absaroka Sea was restricted to west Texas and southern New Mexico, forming a complex of lagoonal, reef, and open-shelf environments (Figure 11.13). Three basins separated by two submerged platforms formed in this area during the Permian. Massive reefs grew around the basin margins (Figure 11.14), while limestones, evaporites, and red beds were deposited in the lagoonal areas behind the reefs. As the barrier reefs grew and the passageways between the basins became more restricted, Late Permian evaporites gradually filled the individual basins.

Spectacular deposits representing the geologic history of this region can be seen today in the Guadalupe Mountains

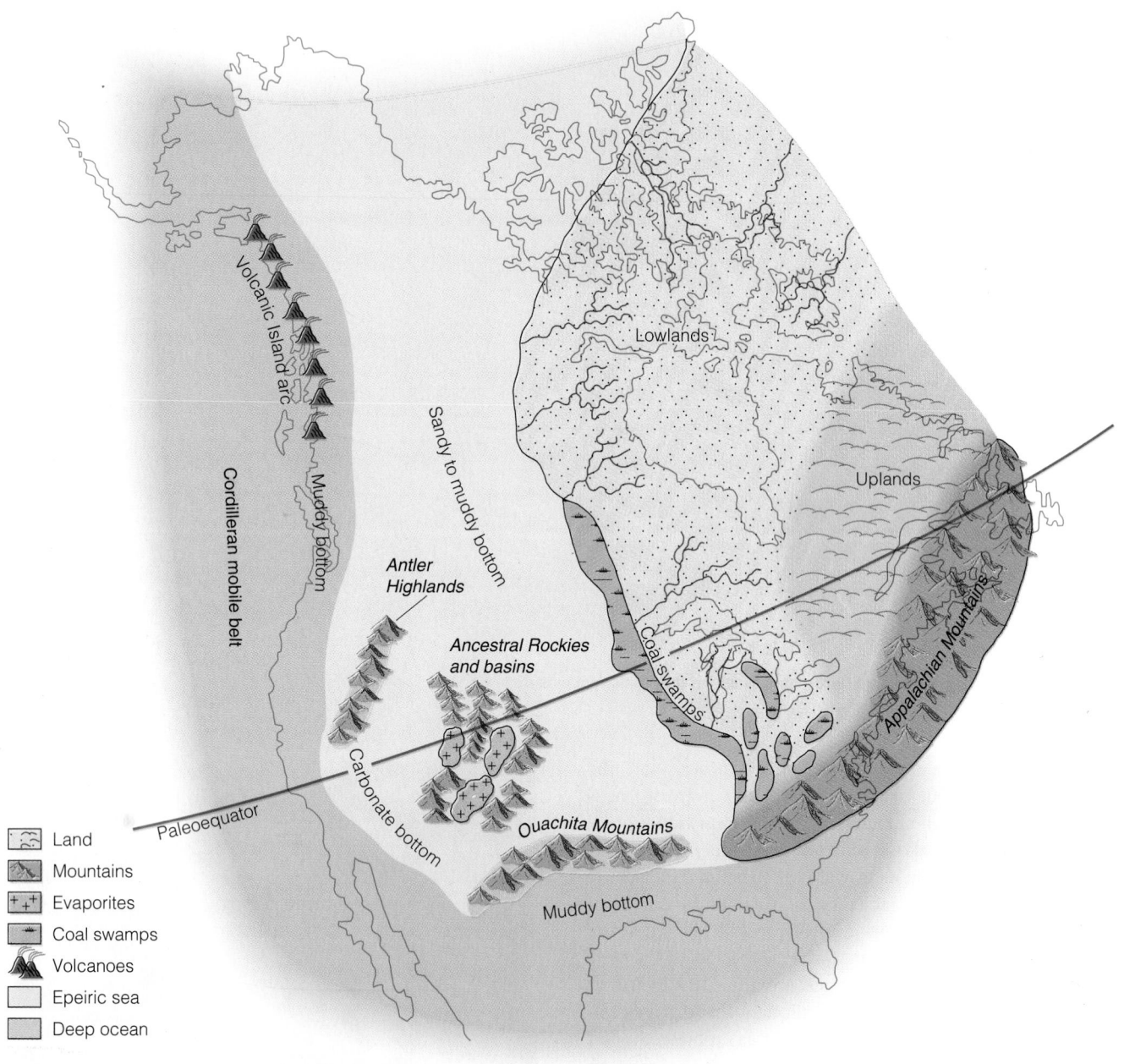

FIGURE 11.9 Paleogeography of North America during the Pennsylvanian Period.

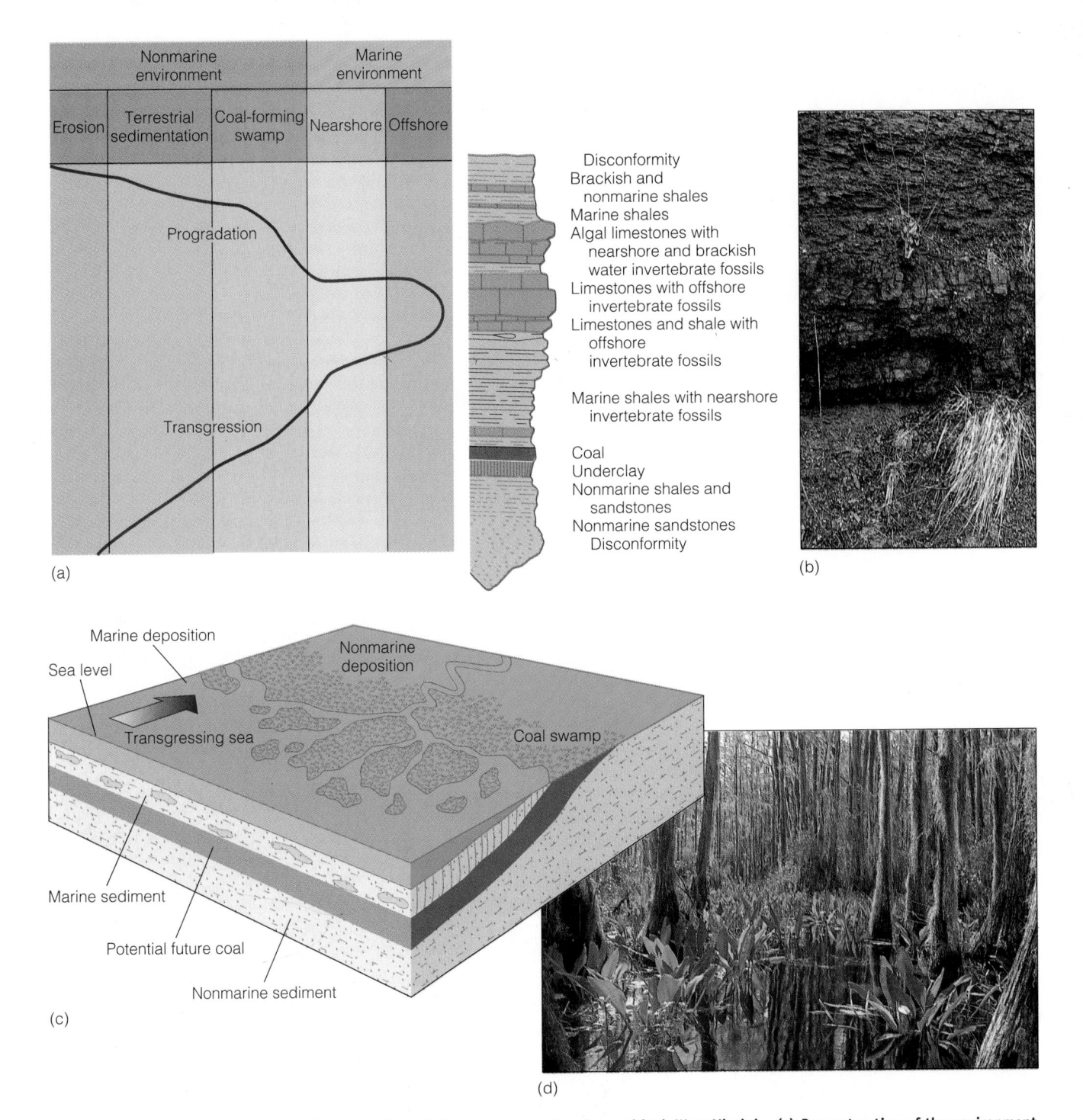

FIGURE 11.10 (a) Columnar section of a complete cyclothem. (b) Pennsylvanian coal bed, West Virginia. (c) Reconstruction of the environment of a Pennsylvanian coal-forming swamp. (d) The Okefenokee Swamp, Georgia, is a modern example of a coal-forming environment, similar to those occurring during the Pennsylvanian Period.

of Texas and New Mexico where the Capitan Limestone forms the caprock of these mountains (Figure 11.15). These reefs have been extensively studied because of the tremendous oil production that comes from this region.

By the end of the Permian Period, the Absaroka Sea had retreated from the craton exposing continental red beds over most of the southwestern and eastern region.

History of the Late Paleozoic Mobile Belts

Having examined the history of the craton for the Kaskaskia and Absaroka sequences, we now turn our attention to the orogenic activity in the mobile belts. The moun-

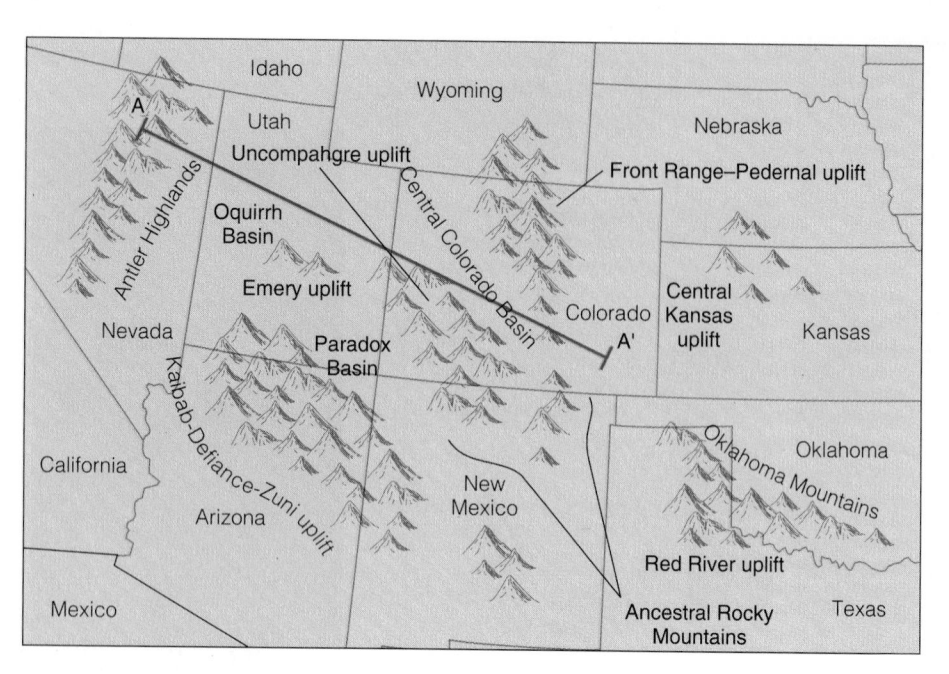

(a)

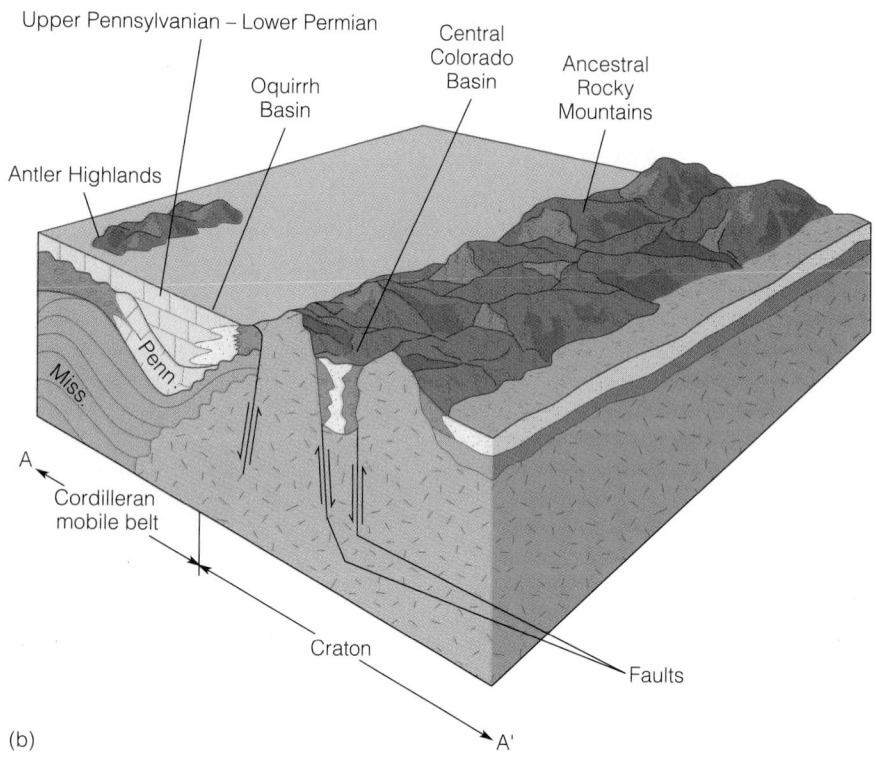

(b)

FIGURE 11.11 (a) Location of the principal Pennsylvanian highland areas and basins of the southwestern part of the craton. (b) Cross section of the Ancestral Rockies, which were elevated by faulting during the Pennsylvanian Period. Erosion of these mountains produced coarse, red-colored sediments that were deposited in the adjacent basins. (c) Garden of the Gods, storm sky view from Near Hidden Inn, Colorado Springs, Colorado.

tain building that occurred during this time had a profound influence on the climate and sedimentary history of the craton. In addition, it was part of the global tectonic regime that sutured the continents together, forming Pangaea by the end of the Paleozoic Era.

CORDILLERAN MOBILE BELT

During the Late Proterozoic and Early Paleozoic, the Cordilleran area was a passive continental margin along which extensive continental shelf sediments were deposited (see Figures 9.10 and 10.5). Thick sections of ma-

(c)

rine sediments graded laterally into thin cratonic units as the Sauk Sea transgressed onto the craton. Beginning in the Middle Paleozoic, an island arc formed off the western margin of the craton. A collision between this island arc and the western border of the craton took place during the Late Devonian and Early Mississippian, resulting in a highland area. Evidence for this collision can be seen along the Roberts Mountain thrust fault in central Nevada where deep-water continental slope deposits were thrust eastward over shallow-water continental shelf carbonates (Figure 11.16). This orogenic event, the *Antler orogeny*, was caused by subduction and resulted in the closing of the narrow ocean basin that separated the island arc from the craton. Erosion of the resulting Antler Highlands produced large quantities of sediment that were deposited to the east in the epeiric sea covering the craton and to the west in the deep sea. The Antler orogeny was the first in a series of orogenic events to affect the Cordilleran mobile belt. During the Mesozoic and Cenozoic, this area was the site of major tectonic activity caused by oceanic–continental convergence and accretion of various terranes.

OUACHITA MOBILE BELT

The Ouachita mobile belt extends for approximately 2100 km from the subsurface of Mississippi to the Marathon region of Texas. Approximately 80% of the former mobile belt is buried beneath a Mesozoic and Cenozoic sedimentary cover. The two major exposed areas in this region are the Ouachita Mountains of Oklahoma and Arkansas and the Marathon Mountains of Texas. Based on extensive study of the subsurface geology and the Ouachita and Marathon mountains, geologists have learned that this region had a complex geologic history (Figure 11.17).

During the Late Proterozoic to Early Mississippian, shal-low-water detrital and carbonate sediments were deposited on a broad continental shelf, while in the deeper-water portion of the adjoining mobile belt, bedded cherts and shales were also accumulating (Figure 11.17a). Beginning in the Mississippian Period, the rate of sedimentation increased dramatically as the region changed from a passive continental margin to an active convergent plate boundary (Figure 11.17b). Rapid deposition of sediments continued into the Pennsylvanian with the formation of a clastic wedge that thickened to the south. As much as 16,000 m of Mississippian- and Pennsylvanian-aged rocks crop out in the Ouachita Mountains, attesting to the rapid rate of sedimentation during this time. The formation of a clastic wedge marks the beginning of uplift of the area and formation of a mountain range during the Ouachita orogeny.

Thrusting of sediments continued throughout the Pennsylvanian and Early Permian as a result of the compressive forces generated along the zone of subduction as Gondwana collided with Laurasia (Figure 11.17c). The collision of Gondwana and Laurasia is marked by the formation of a large mountain range, most of which was eroded during the Mesozoic Era. Only the rejuvenated Ouachita and Marathon Mountains remain of this once lofty mountain range.

The Ouachita deformation was part of the general worldwide tectonic activity that occurred when Gondwana united with Laurasia. The Hercynian, Appalachian, and Ouachita mobile belts were continuous, and marked the southern boundary of Laurasia (Figure 11.2). The tectonic activity that resulted in the uplift in the Ouachita mobile belt was very complex and involved not only the collision of Laurasia and Gondwana but also several microplates between the continents that eventually became part of Central America. The compressive forces impinging on

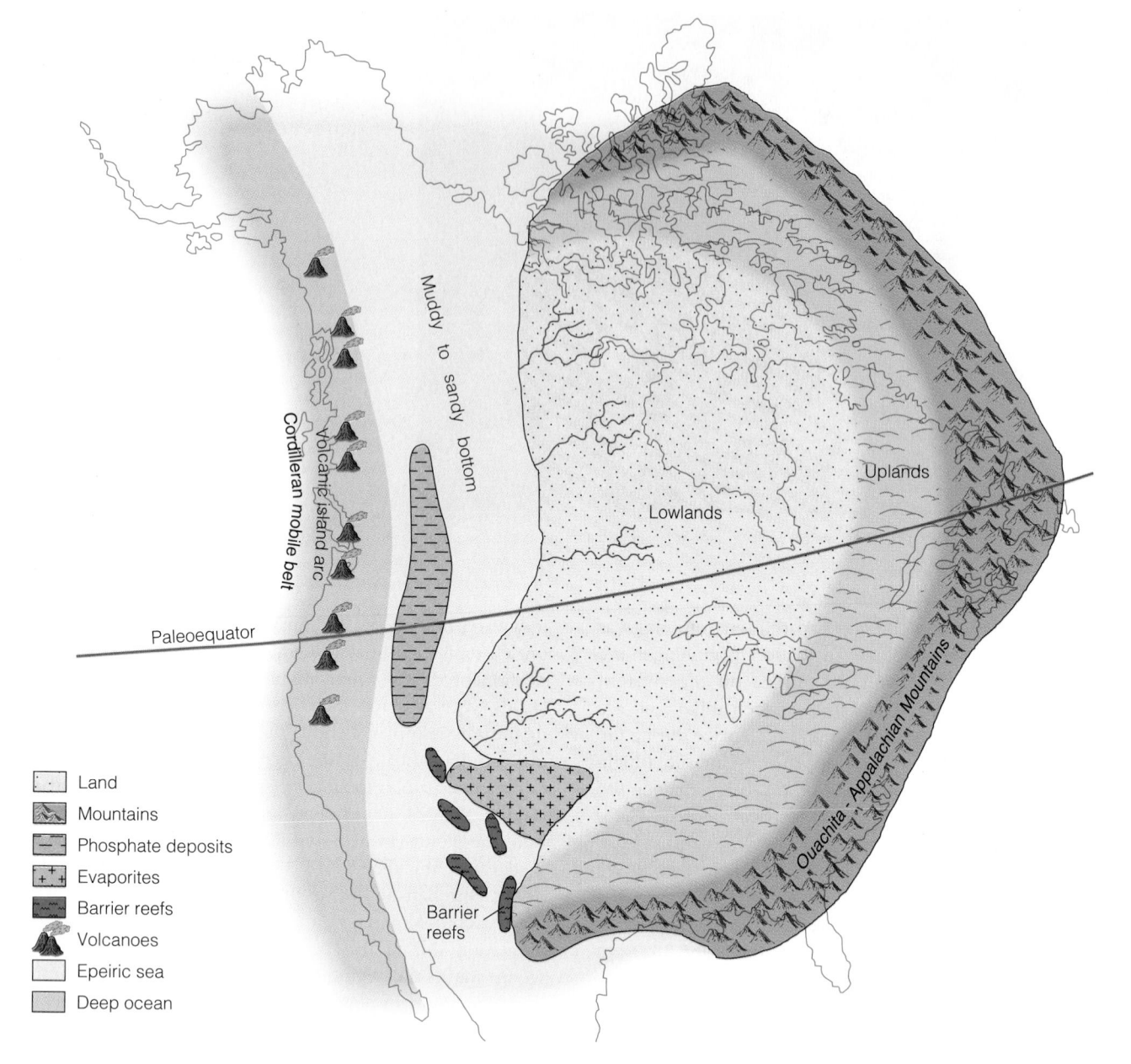

Land

Mountains

Phosphate deposits

Evaporites

Barrier reefs

Volcanoes

Epeiric sea

Deep ocean

FIGURE 11.12 Paleogeography of North America during the Permian Period.

the Ouachita mobile belt also affected the craton by causing broad uplift of the southwestern part of North America.

APPALACHIAN MOBILE BELT

Caledonian Orogeny The Caledonian mobile belt extends along the western border of Baltica and includes the present-day countries of Scotland, Ireland, and Norway (see Figure 10.2c). During the Middle Ordovician, subduction along the boundary between the Iapetus plate and Baltica (Europe) began, forming a mirror image of the convergent plate boundary off the east coast of Laurentia (North America).

The culmination of the *Caledonian orogeny* (men-

tioned earlier) occurred during the Late Silurian and Early Devonian with the formation of a mountain range along the margin of Baltica. Red-colored sediments deposited along the front of the Caledonian highlands formed a large clastic wedge known as the Old Red Sandstone.

Acadian Orogeny The third Paleozoic orogeny to affect Laurentia and Baltica began during the Late Silurian and concluded at the end of the Devonian Period. The *Acadian orogeny* (mentioned earlier) affected the Appalachian mobile belt from Newfoundland to Pennsylvania as sedimentary rocks were folded and thrust against the craton.

As with the preceding Taconic and Caledonian orogenies, the Acadian orogeny occurred along an oceanic–

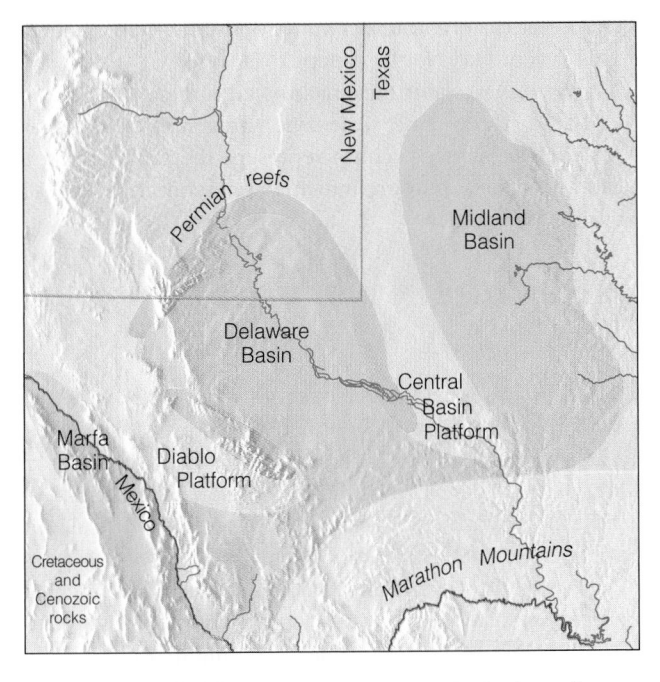

FIGURE 11.13 Location of the west Texas Permian basins and surrounding reefs.

continental convergent plate boundary. As the northern Iapetus Ocean continued to close during the Devonian, the plate carrying Baltica finally collided with Laurentia, forming a continental–continental convergent plate boundary along the zone of collision (Figure 11.1a).

As the increased metamorphic and igneous activity indicates, the Acadian orogeny was more intense and of longer duration than the Taconic orogeny. Radiometric dates from the metamorphic and igneous rocks associated with the Acadian orogeny cluster between 360 and 410 million years ago. And, just as with the Taconic orogeny, deepwater sediments were folded and thrust northwestward, producing angular unconformities separating Upper Silurian from Mississippian rocks.

Weathering and erosion of the Acadian Highlands produced the **Catskill Delta,** a thick clastic wedge named for the Catskill Mountains in upstate New York where it is well exposed. The Catskill Delta, composed of red, coarse conglomerates, sandstones, and shales, contains nearly three times as much sediment as the Queenston Delta (see Figure 10.14).

The Devonian rocks of New York are among the best studied on the continent. A cross section of the Devonian strata clearly reflects an eastern source (Acadian Highlands) for the Catskill facies (Figure 11.18). These clastic rocks can be traced from eastern Pennsylvania, where the coarse clastics are approximately 3 km thick, to Ohio, where the deltaic facies are only about 100 m thick and consist of cratonic shales and carbonates.

The red beds of the Catskill Delta derive their color from the hematite found in the sediments. Plant fossils and oxidation of the hematite indicate that the beds were deposited in a continental environment. Toward the west,

FIGURE 11.14 A reconstruction of the Middle Permian Capitan Limestone reef environment. Shown are brachiopods, corals, bryozoans, and large glass sponges.

the red beds grade laterally into gray sandstones and shales containing fossil tree trunks, which indicate a swamp or marsh environment.

The Old Red Sandstone The red beds of the Catskill Delta have a European counterpart in the Devonian Old Red Sandstone of the British Isles (Figure 11.18).

FIGURE 11.15 The prominent light-colored Capitan Limestone forms the caprock of the Guadalupe Mountains. The Capitan Limestones is rich in fossil corals and associated reef organisms.

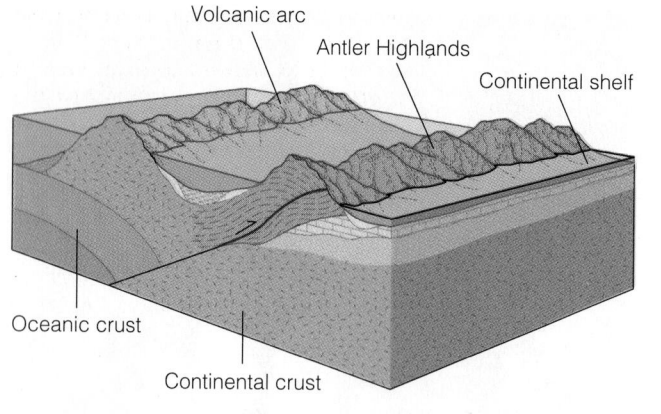

FIGURE 11.16 Reconstruction of the Cordilleran mobile belt during the Early Mississippian, showing the effects of the Antler orogeny.

The Old Red Sandstone was a Devonian clastic wedge that grew eastward from the Caledonian Highlands onto the Baltica craton. The Old Red Sandstone, just like its North American Catskill counterpart, contains numerous fossils of freshwater fish, early amphibians, and land plants.

By the end of the Devonian Period, Baltica and Laurentia were sutured together, forming Laurasia (Figure 11.18). The red beds of the Catskill Delta can be traced north, through Canada and Greenland, to the Old Red Sandstone of the British Isles and into northern Europe. These beds were deposited in similar environments along the flanks of developing mountain chains formed at convergent plate boundaries.

The Taconic, Caledonian, and Acadian orogenies are now thought to be part of the same major orogenic event related to the closing of the Iapetus Ocean (see Figures 10.13 and 11.18). This event began with paired oceanic-continental convergent plate boundaries during the Taconic and Caledonian orogenies and culminated along a continental–continental convergent plate boundary during the Acadian orogeny as Laurentia and Baltica collided. The Hercynian-Alleghenian orogeny then began, followed by orogenic activity in the Ouachita mobile belt.

Hercynian–Alleghenian Orogeny The Hercynian mobile belt of southern Europe and the Appalachian and Ouachita mobile belts of North America mark the zone along which Europe (part of Laurasia) collided with Gondwana (Figure 11.1). While Gondwana and southern Laurasia collided during the Pennsylvanian and Permian in the area of the Ouachita mobile belt, eastern Laurasia (Europe and southeastern North America) joined together with Gondwana (Africa) as part of the *Hercynian-Alleghenian orogeny* (Figure 11.2).

Initial contact between eastern Laurasia and Gondwana began during the Mississippian Period along the Hercynian mobile belt. The greatest deformation occurred during the Pennsylvanian and Permian periods and is referred to as the *Hercynian orogeny* (mentioned earlier). The central and southern parts of the Appalachian mobile belt (from New York to Alabama) were folded and thrust toward the

craton as eastern Laurasia and Gondwana were sutured. This event in North America is referred to as the *Alleghenian orogeny* (also mentioned earlier).

These three Late Paleozoic orogenies (Hercynian, Alleghenian, and Ouachita) represent the final joining of Laurasia and Gondwana into the supercontinent Pangaea during the Permian.

Paleozoic Microplates and the Formation of Pangaea

We have presented the geologic history of the mobile belts bordering the Paleozoic continents in terms of subduction along convergent plate boundaries. It is becoming increasingly clear, however, that accretion along the continental margins is more complicated than the somewhat simple, large-scale plate interactions that we have described. Geologists now recognize that numerous terranes or microplates existed during the Paleozoic and were involved in the orogenic events that occurred during that time. In this chapter and the previous one, we have been concerned only with the six major Paleozoic continents. A careful examination of the Paleozoic global paleogeographic maps (Figures 10.2, 11.1, and 11.2) shows numerous microplates, and their location and role during the formation of Pangaea must be taken into account. For example, the small continent of Avalonia is composed of some coastal parts of New England, southern New Brunswick, much of Nova Scotia, the Avalon Peninsula of eastern Newfoundland, southeastern Ireland, Wales, England, and parts of Belgium and northern France. This microplate existed as a separate continent during the Ordovician and collided with Baltica during the Silurian and Laurentia during the Devonian (see Figures 10.2 and 11.1).

Florida and parts of the eastern seaboard of North America make up the Piedmont microplate that was part of the larger Gondwana continent. This microplate became sutured to Laurasia during the Pennsylvanian Period. Numerous microplates occupied the region between Gondwana and Laurasia that eventually became part of Central America during the Pennsylvanian collision between these continents.

Thus, while the basic history of the formation of Pangaea during the Paleozoic remains the same, geologists now realize that microplates played an important role. Furthermore, the recognition of terranes within mobile belts helps explain some previously anomalous geologic situations.

Late Paleozoic Mineral Resources

Late Paleozoic-age rocks contain a variety of important mineral resources including energy resources and metallic and nonmetallic mineral deposits. Petroleum and natural gas are recovered in commercial quantities from rocks

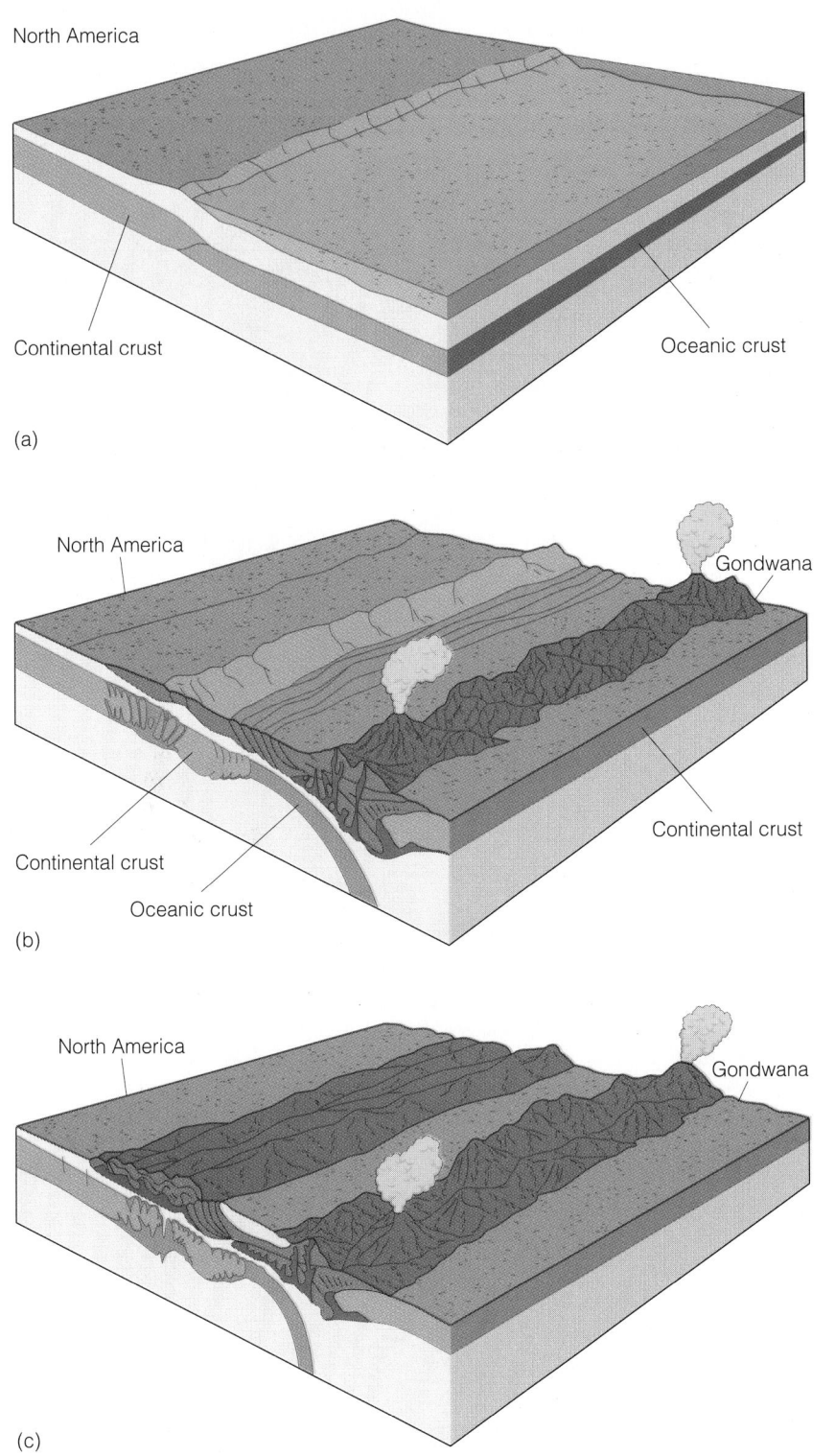

North America

Continental crust

Oceanic crust

(a)

North America

Gondwana

Continental crust

Continental crust

Oceanic crust

(b)

North America

Gondwana

(c)

FIGURE 11.17 Plate tectonic model for deformation of the Ouachita mobile belt. (a) Depositional environment prior to the beginning of orogenic activity. (b) Incipient continental collision between North America and Gondwana began during the Mississippian to Pennsylvanian. (c) Continental collision continued during the Pennsylvanian Period.

ranging from the Devonian through Permian. For example, Devonian-age rocks in the Michigan Basin, Illinois Basin, and the Williston Basin of Montana, South Dakota, and adjacent parts of Alberta, Canada, have yielded considerable amounts of hydrocarbons. Permian reefs and other strata in the western United States, particularly Texas, have also been prolific producers.

Although Permian-age coal beds are known from sev-

eral areas including Asia, Africa, and Australia, much of the coal in North America and Europe comes from Pennsylvanian (Late Carboniferous) deposits. Large areas in the Appalachian region and the midwestern United States are underlain by vast coal deposits (Figure 11.19). These coal deposits formed from the lush vegetation that flourished in Pennsylvanian coal-forming swamps (Figure 11.10).

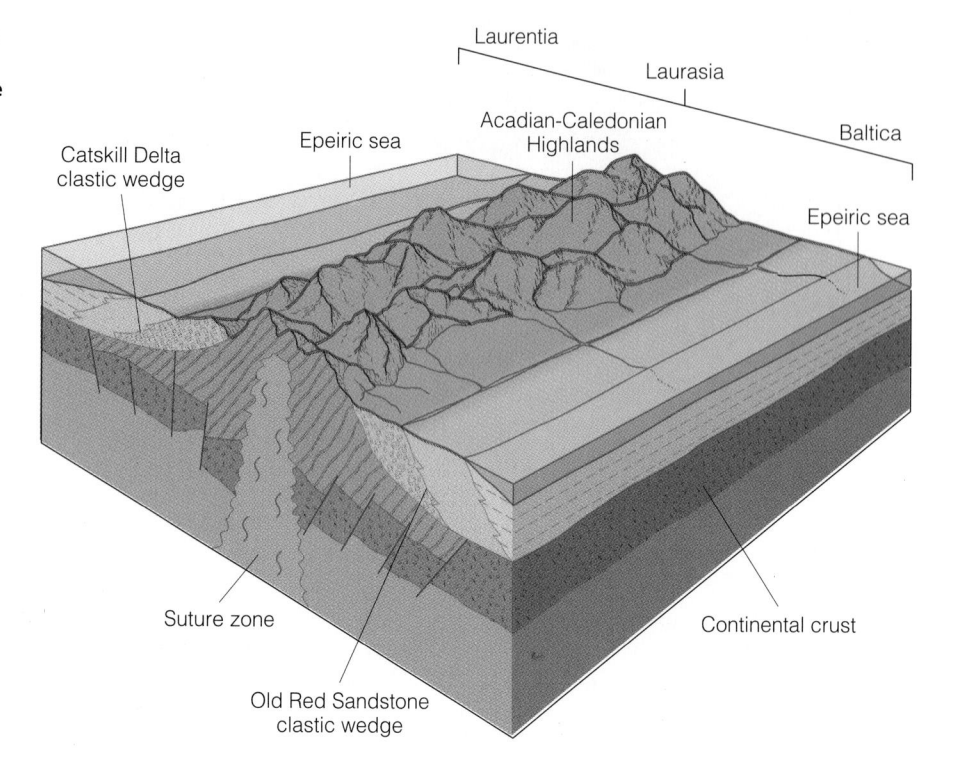

FIGURE 11.18 Cross section showing the area of collision between Laurentia and Baltica. Note the bilateral symmetry of the Catskill Delta clastic wedge and the Old Red Sandstone and their relationship to the Acadian and Caledonian Highlands.

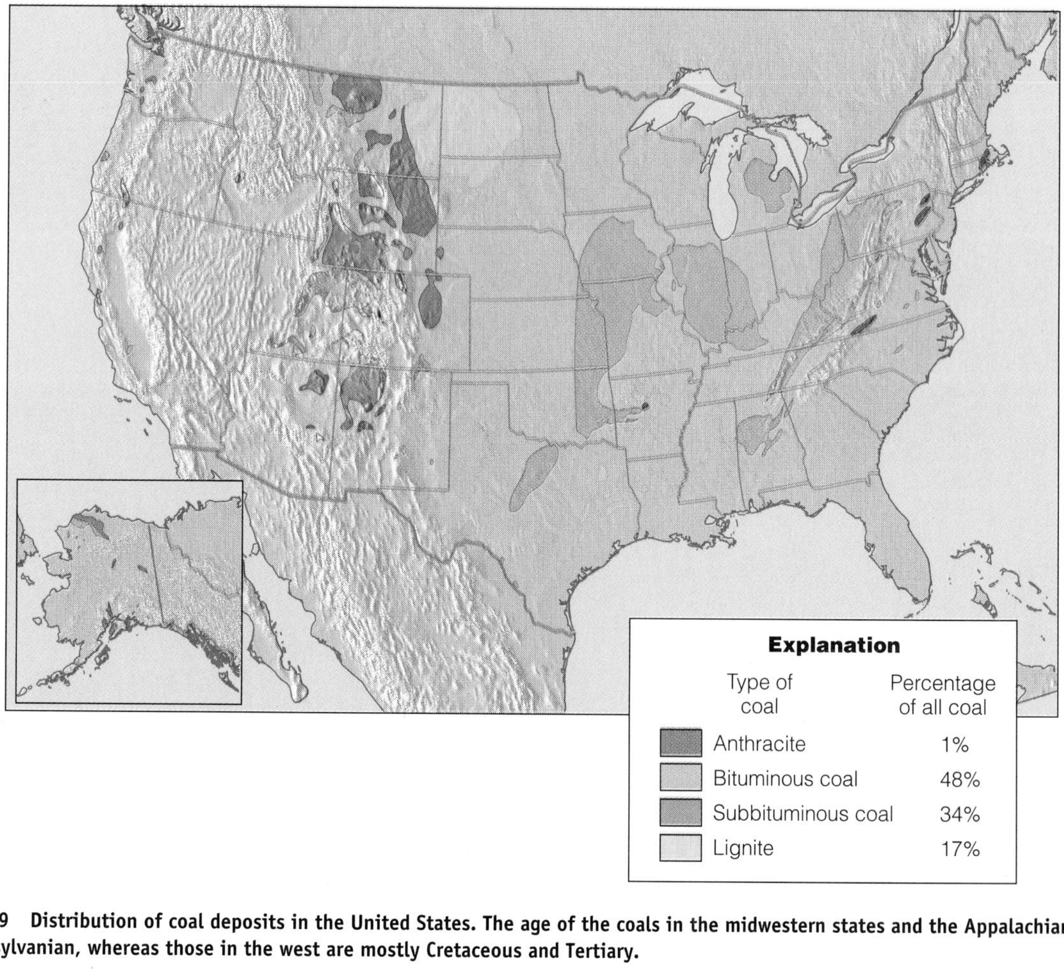

Explanation

	Type of coal	Percentage of all coal
	Anthracite	1%
	Bituminous coal	48%
	Subbituminous coal	34%
	Lignite	17%

FIGURE 11.19 Distribution of coal deposits in the United States. The age of the coals in the midwestern states and the Appalachian region are mostly Pennsylvanian, whereas those in the west are mostly Cretaceous and Tertiary.

Much of this coal is characterized as bituminous coal, which contains about 80% carbon. It is a dense, black coal that has been so thoroughly altered that plant remains can be seen only rarely. Bituminous coal is used to make *coke,* a hard, gray substance made up of the fused ash of bituminous coal. Coke is used to fire blast furnaces during the production of steel.

Some of the Pennsylvanian coal from North America is *anthracite,* a metamorphic type of coal containing up to 98% carbon. Most anthracite is in the Appalachian region (Figure 11.19). It is an especially desirable type of coal because it burns with a smokeless flame and it yields more heat per unit volume than other types of coal. Unfortunately, it is the least common type of coal, so much of the coal used in the United States is bituminous.

A variety of Late Paleozoic-age evaporite deposits are important nonmetallic mineral resources. The Zechstein evaporites of Europe extend from Great Britain across the North Sea and into Denmark, the Netherlands, Germany, and eastern Poland and Lithuania. Besides the evaporites themselves, Zechstein deposits form the caprock for the large reservoirs of the gas fields of the Netherlands and part of the North Sea region.

Other important evaporite mineral resources include those of the Permian Delaware Basin of west Texas and New Mexico and Devonian evaporites in the Elk Point basin of Canada. In Michigan, gypsum is mined and used in the construction of wallboard.

The majority of the silica sand mined in the United States comes from east of the Mississippi River, and much of this comes from Late Paleozoic-age rocks. For example, the Devonian Ridgeley Formation is mined in West Virginia, Maryland, and Pennsylvania, and the Devonian Sylvania Sandstone is mined near Detroit, Michigan. Recall from Chapter 10 that silica sand is used in the manufacture of glass; for refractory bricks in blast furnaces; for molds for casting aluminum, iron, and copper alloys; and for a variety of other uses.

Late Paleozoic-age limestones from many areas in North America are used in the manufacture of cement. Limestone is also mined and used in blast furnaces when steel is produced.

Metallic mineral resources including tin, copper, gold, and silver are also known from Late Paleozoic-age rocks, especially those that have been deformed during mountain building. Although the precise origin of the Missouri lead and zinc deposits remains unresolved, much of the ores of these metals come from Mississippian-age rocks. In fact, mines in Missouri account for a substantial amount of all domestic production of lead ores.

Summary

Table 11.1 provides a summary of the geologic history of the North American craton and mobile belts as well as global events and sea-level changes during the Late Paleozoic.

1. During the Late Paleozoic, Baltica and Laurentia collided, forming Laurasia. Siberia and Kazakhstania collided and finally were sutured to Laurasia. Gondwana moved over the South Pole and experienced several glacial–interglacial periods, resulting in global changes in sea level and transgressions and regressions along the low-lying craton margins.

2. Laurasia and Gondwana underwent a series of collisions beginning in the Carboniferous. During the Permian, the formation of Pangaea was completed. Surrounding the supercontinent was a global ocean, Panthalassa.

3. Much of the history of the North American craton can be deciphered from the rocks of the Kaskaskia and Absaroka sequences.

4. The basal beds of the Kaskaskia sequence that were deposited on the exposed Tippecanoe surface consisted of either sandstones, derived from the eroding Taconic Highlands, or carbonate rocks.

5. Most of the Kaskaskia sequence is dominated by carbonates and associated evaporites. The Devonian Period was a time of major reef building in western Canada, southern England, Belgium, Australia, and Russia.

6. A persistent and widespread black shale, the Chattanooga Shale and its equivalents, was deposited over a large area of the craton during the Late Devonian and Early Mississippian.

TABLE 11.1

Summary of Late Paleozoic Geologic Events

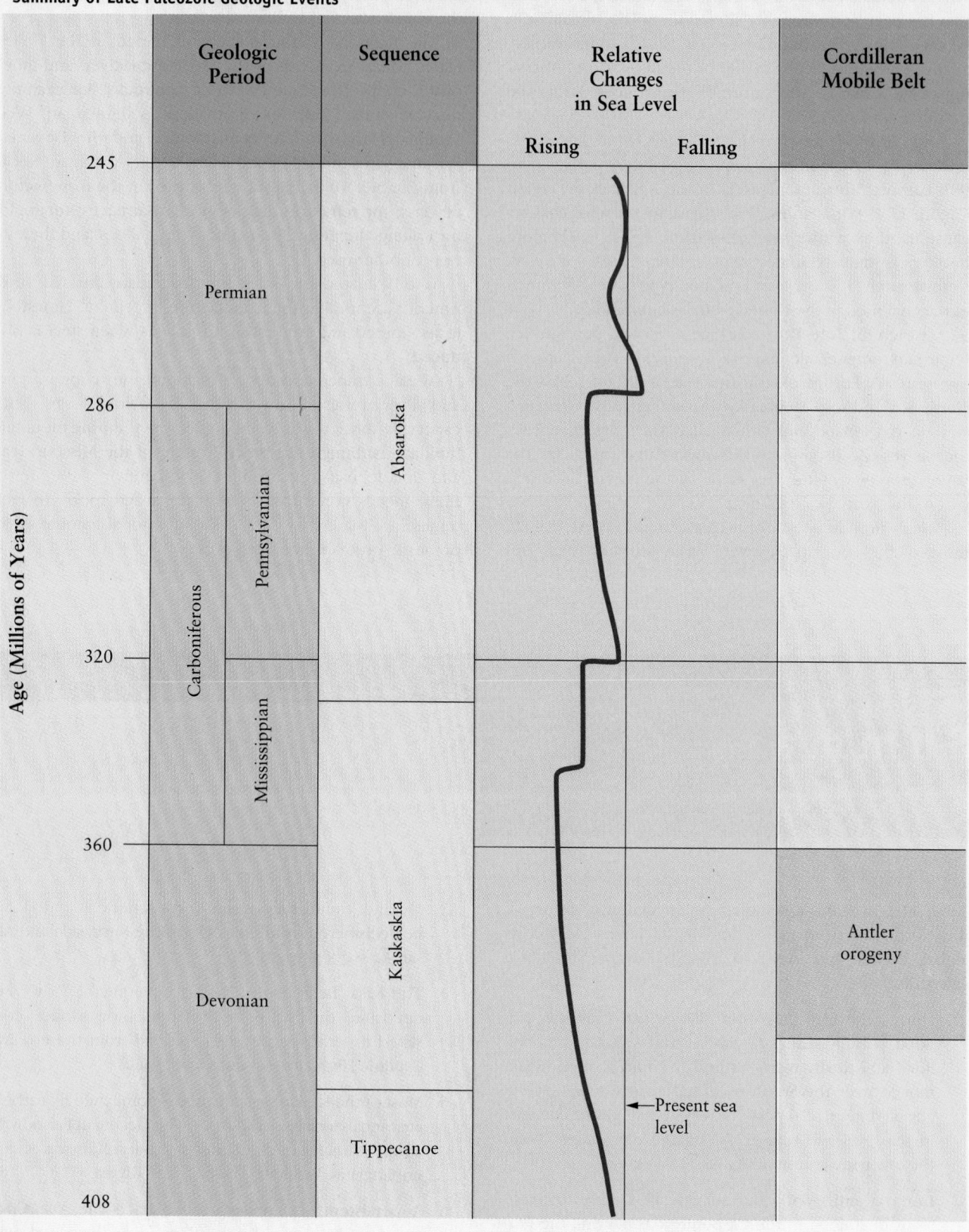

Craton	Ouachita Mobile Belt	Appalachian Mobile Belt	Major Events Outside North America
Deserts, evaporites, and continental red beds in southwestern United States. Extensive reefs in Texas area.		Formation of Pangaea.	
		Allegheny orogeny	Hercynian orogeny
Coal swamps common. Formation of Ancestral Rockies.	Ouachita orogeny		Continental glaciation in Southern Hemisphere.
Transgression of Absaroka Sea. Widespread black shales and limestones.			
Widespread black shales. Catskill Delta clastic wedge. Extensive barrier-reef formation in Western Canada. Transgression of Kaskaskia Sea.		Acadian orogeny	Old Red Sandstone clastic wedge in British Isles.
			Caledonian orogeny

7. The Mississippian Period was dominated for the most part by carbonate deposition.

8. Transgressions and regressions over the low-lying craton resulted in cyclothems and the formation of coals during the Pennsylvanian Period.

9. Cratonic mountain building occurred during the Pennsylvanian Period, and thick nonmarine detrital rocks and evaporites were deposited in the intervening basins.

10. By the Early Permian, the Absaroka Sea occupied a narrow zone of the south-central craton. Here, several large reefs and associated evaporites developed. By the end of the Permian Period, the Absaroka Sea had retreated from the craton.

11. The Cordilleran mobile belt was the site of the Antler orogeny, a minor Devonian orogeny during which deep-water sediments were thrust eastward over shallow-water sediments.

12. Mountain building occurred in the Ouachita mobile belt during the Pennsylvanian and Early Permian. This tectonic activity was partly responsible for the cratonic uplift that took place in the southwest, producing the Ancestral Rockies.

13. The Caledonian, Acadian, Hercynian, and Alleghenian orogenies were all part of the global tectonic activity resulting from the assembly of Pangaea.

14. During the Paleozoic Era, numerous microplates existed and played an important role in the formation of Pangaea.

15. Late Paleozoic-age rocks contain a variety of mineral resources including petroleum, coal, evaporites, silica sand, lead, zinc, and other metallic deposits.

Important Terms

Absaroka sequence	Antler orogeny
Acadian orogeny	Caledonian orogeny
Alleghenian orogeny	Catskill Delta
Ancestral Rockies	cyclothem

Ellesmere orogeny	Laurasia
Hercynian orogeny	Ouachita orogeny
Kaskaskia sequence	Panthalassa Ocean

Review Questions

1. The first Paleozoic orogeny to occur in the Cordilleran mobile belt was the:
   a. _____ Acadian.
   b. _____ Alleghenian.
   c. _____ Antler.
   d. _____ Caledonian.
   e. _____ Ellesmere.

2. Extensive continental glaciation of the Gondwana continent occurred during which period?
   a. _____ Devonian
   b. _____ Silurian
   c. _____ Carboniferous
   d. _____ Cambrian
   e. _____ Permian

3. Extensive cratonic black shales were deposited during which two periods?
   a. _____ Late Silurian-Early Devonian
   b. _____ Late Devonian-Early Mississippian
   c. _____ Late Mississippian-Early Pennsylvanian
   d. _____ Late Pennsylvanian-Early Permian
   e. _____ Late Permian-Early Triassic

4. Rhythmically repetitive sedimentary sequences are:
   a. _____ tillites.
   b. _____ cyclothems.
   c. _____ orogenies.
   d. _____ reefs.
   e. _____ evaporites.

5. Uplift in the southwestern part of the craton during the Late Absaroka resulted in which mountainous region?
   a. _____ Ancestral Rockies
   b. _____ Antler Highlands
   c. _____ Appalachians
   d. _____ Marathon
   e. _____ None of these

6. Weathering of which highlands or mountains produced the Catskill Delta?
   a. _____ Acadian
   b. _____ Alleghenian
   c. _____ Antler
   d. _____ Appalachian
   e. _____ Caledonian

7. The European counterpart to the Devonian Catskill Delta is the:
   a. _____ Phosphoria Formation.
   b. _____ Oriskany Sandstone.
   c. _____ Capitan Limestone.
   d. _____ Old Red Sandstone.
   e. _____ None of these

8. Which orogeny was not part of the closing of the Iapetus Ocean?
   a. _____ Acadian
   b. _____ Alleghenian
   c. _____ Antler
   d. _____ Caledonian
   e. _____ Taconic

9. During the Late Kaskaskia, what type of deposition predominated on the craton?
   a. _____ Black shales
   b. _____ Carbonates
   c. _____ Volcanics
   d. _____ Metamorphics
   e. _____ Cherts and graywackes

10. What is the main economic deposit of a cyclothem?
    a. _____ Evaporites
    b. _____ Carbonates
    c. _____ Gravels
    d. _____ Ores
    e. _____ Coals

11. Discuss how plate movement during the Late Paleozoic affected worldwide weather patterns.

12. Provide a geologic history for the Iapetus Ocean during the Paleozoic Era.

13. What are cyclothems? How do they form? Why are they economically important?

14. What was the relationship between the Ouachita orogeny and the cratonic uplifts that occurred on the craton during the Pennsylvanian Period?

15. What is the evidence for glaciation on Gondwana during the Late Paleozoic?

16. How are the Caledonian, Acadian, Ouachita, Hercynian, and Alleghenian orogenies related to modern concepts of plate tectonics?

17. Compare the Taconic, Caledonian, and Acadian orogenies in terms of the tectonic forces that caused them and the sedimentary features that resulted.

18. How does the origin of evaporite deposits of the Kaskaskia sequence compare with the origin of evaporites of the Tippecanoe sequence?

19. How did the sedimentary environments of the Kaskaskia sequence differ from those of the Absaroka sequence in North America?

20. How did the formation of Pangaea and Panthalassa affect the world's climate at the end of the Paleozoic Era?

Points to Ponder

1. Discuss how paleomagnetism and fossil evidence can be used in determining the distribution of microplates during the Paleozoic Era.

2. Based on the discussion of Milankovitch cycles and their role in causing glacial–interglacial cycles (Chapter 17), could these cycles be partly responsible for the transgressive-regressive cycles that resulted in cyclothems during the Pennsylvanian Period? Explain.

3. Discuss the role plate tectonics plays in the formation and distribution of Late Paleozoic–age mineral resources.

Additional Readings

Bally, A. W., and A. R. Palmer, eds. 1989. *The geology of North America: An overview.* The Geology of North America. Vol. A. Boulder, Colo.: Geological Society of America.

Dineley, D. L. 1984. *Aspects of a stratigraphic system—The Devonian.* New York: Wiley.

Harwood, D. S., and M. M. Miller, eds. 1990. *Paleozoic and Early Mesozoic paleogeographic relations: Sierra Nevada, Klamath Mountains, and related terranes.* Geological Society of America Special Paper 255. Boulder, Colo.: Geological Society of America.

Hatcher, R. D., Jr., W. A. Thomas, and G. W. Viele, eds. 1989. *The Appalachian-Ouachita orogen in the United States.* The Geology of North America. Vol. F-2. Boulder, Colo.: Geological Society of America.

McKerrow, W. S., and C. R. Scotese, eds. 1990. *Palaeozoic palaeogeography and biogeography.* The Geological Society Memoir No. 12. London: Geological Society of London.

Moullade, M., and A. E. M. Nairn, eds. 1991. *The Phanerozoic geology of the world I: The Paleozoic.* New York: Elsevier Science.

Osborne, R., and D. Tarling, eds. 1996. *The historical atlas of the Earth.* New York: Holt.

Shabica, C. W., and A. A. Hay, eds. 1997. *Richardson's guide to the fossil fauna of Mazon Creek.* Chicago: Northeastern Illinois University.

Sloss, L. L., ed. 1988. *Sedimentary cover—North American Craton: U.S.* The Geology of North America. Vol. D-2. Boulder, Colo.: Geological Society of America.

Woodrow, D. L., and W. D. Sevan. 1985. *The Catskill Delta.* Geological Society of America Special Paper 201. Boulder, Colo.: Geological Society of America.

Ziegler, P. A. 1989. *Evolution of Laurussia: A study in Late Palaeozoic plate tectonics.* Boston: Mass.: Kluwer Academic.

World Wide Web Activities

For these web site addresses, along with current updates and exercises, log on to

http://www.brookscole.com/geo/

▶ **GLOBAL EARTH HISTORY**

This site, maintained at the University of Northern Arizona, contains a series of plate tectonic reconstructions to show what Earth looked like at various times in the geologic past. Included are paleogeographic maps showing global, North American, and European reconstructions from the Cambrian to the present, as well as tectonic, sedimentation, and paleogeographic maps of the North Atlantic region.

▶ **NORTH AMERICAN OROGENIES**

This site is part of the *Global Earth History* site and shows simple restored cross sections of the major orogenies taking place in North America from the Proterozoic through Mesozoic. Click on the various orogenies that occurred during the Late Paleozoic.

Explore the following *In-Terra-Active 2.0* CD-ROM module(s) and increase your understanding of key concepts and processes presented in this chapter.

Chapter Concept: **Crustal Deformation in Mobile Belts**

Multiple episodes of crustal deformation occurred during the Paleozoic era in the mobile belts of sediment adjacent to the stable shield areas of North America and other continents. In this module, you will explore the relationship between the folds and faults of mobile belts and the deformational processes that accompany plate tectonics.

Section: **Interior**

Module: **Crustal Deformation**

The coarse-grained garnet-mica schist shown on the left has essentially the same chemical composition as the fine-grained mudstone shown on the right. In the field, it is clear that these belong to the same stratigraphic unit.

Question: *How can this transformation take place in the solid state?*

12 Life of the Paleozoic Era: Invertebrates

Diorama of the environment and biota of the Phyllopod bed of the Burgess Shale, British Columbia, Canada. In the background is a vertical wall of a submarine escarpment with algae growing on it. The large cylindrical ribbed organisms on the muddy bottom in the foreground are sponges.

Prologue

On August 30 and 31, 1909, near the end of the summer field season, Charles Walcott, geologist and head of the Smithsonian Institution, was searching for fossils along a trail on Burgess Ridge between Mount Field and Mount Wapta, near Field, British Columbia, Canada. On the west slope of this ridge, he discovered the first soft-bodied fossils from the Burgess Shale, a discovery of immense importance in deciphering the early history of life. During the following week, Walcott and his collecting party split open numerous blocks of shale, many of which yielded the impressions of a number of soft-bodied organisms beautifully preserved on bedding planes. Walcott returned to the site the following summer and located the shale stratum that was the source of his fossil-bearing rocks in the steep slope above the trail. He quarried the site and shipped back thousands of fossil specimens to the U.S. National Museum of Natural History, where he later cataloged and studied them.

The importance of Walcott's discovery is not that it was another collection of well-preserved Cambrian fossils, but rather that it allowed geologists a rare glimpse into a world previously almost unknown—that of the soft-bodied animals that lived some 530 million years ago. The beautifully preserved fossils from the Burgess Shale present a much more complete picture of a Middle Cambrian community than deposits containing only fossils of the hard parts of organisms. Specifically, the Burgess Shale contains species of trilobites, sponges, brachiopods, mollusks, and echinoderms, all of which have hard parts and are characteristic of Cambrian faunas throughout the world. But in addition to the diverse skeletonized fauna, a large and varied fossil assemblage of soft-bodied animals is also present. In fact, 60% of the total fossil assemblage is composed of soft-bodied animals, which usually are not preserved. In all, more than 100 genera of animals, at least 60 of which were soft bodied and preserved as impressions, have been recovered from the Burgess Shale. This proportion of soft-bodied animals to those with hard parts is comparable to present-day marine communities.

The Burgess Shale fauna reveals the evolutionary stage that marine life had reached by the Middle Cambrian: Highly advanced organisms comprised complex communities as diverse in structure and adaptation as many modern marine communities. The Burgess Shale fauna makes speculation about the early evolution of multicelled organisms difficult because it shows how little we know about early marine life. In fact, some of the organisms preserved may not represent any known phyla at all.

What conditions led to the remarkable preservation of the Burgess Shale fauna? When it was deposited, the Burgess Shale was located at the base of a steep submarine escarpment. The animals whose exquisitely preserved fossil remains are found in the Burgess Shale lived in and on mud banks that formed along the top of this escarpment. Periodically, this unstable area would slump and slide down the escarpment as a turbidity flow. At the base, the mud and animals carried with it were deposited in a deep-water anaerobic environment devoid of life. In such an environment, bacterial degradation did not destroy the buried animals, and they were compressed by the weight of the overlying sediments, eventually resulting in their preservation as carbonaceous impressions.

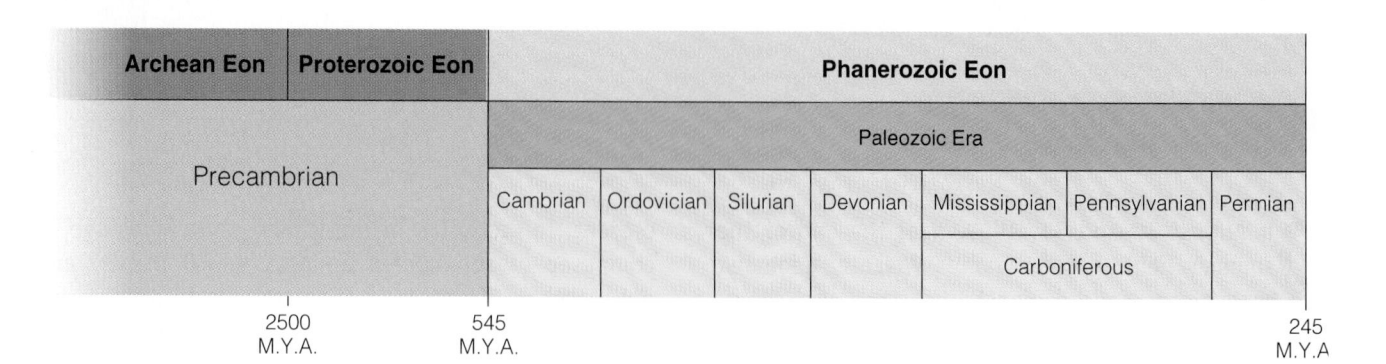

Archean Eon	Proterozoic Eon	Phanerozoic Eon						
Precambrian		Paleozoic Era						
		Cambrian	Ordovician	Silurian	Devonian	Mississippian	Pennsylvanian	Permian
						Carboniferous		

2500
M.Y.A.

545
M.Y.A.

245
M.Y.A

Introduction

In the following two chapters, we examine the history of Paleozoic life as a system in which its parts consist of a series of interconnected biologic and geologic events. The underlying processes of evolution and plate tectonics are the forces that drove this system. The opening and closing of ocean basins, transgressions and regressions of epeiric seas, the formation of mountain ranges, and the changing positions of the continents had a profound effect on the evolution of the marine and terrestrial communities.

A time of tremendous biologic change began with the appearance of skeletonized animals near the Precambrian–Cambrian boundary. Following this event, marine invertebrates began a period of adaptive radiation and evolution during which the Paleozoic marine invertebrate community greatly diversified. Indeed, the history of the Paleozoic marine invertebrate community was one of diversifications and extinctions.

Vertebrates also evolved during the Paleozoic. The earliest fossil records of vertebrates are of fish, which evolved during the Cambrian. One group of fish was ancestral to the first land animals, the amphibians, which evolved during the Devonian. Reptiles evolved from a group of amphibians during the Mississipian Period and were the dominant vertebrate animals on land by the end of the Paleozoic Era.

Plants preceded animals onto the land. Both plants and animals had to solve the same basic problems in making the transition from water to land. The method of reproduction proved to be the major barrier to expansion into new environments for both groups. With the evolution of the seed in plants and the amniote egg in animals, this limitation was removed, and both groups were able to move into all terrestrial environments.

The end of the Paleozoic Era was a time of major extinctions. The marine invertebrate community was greatly decimated, and many amphibians and reptiles on land also became extinct.

The Cambrian Explosion

At the beginning of the Paleozoic Era, animals with skeletons appeared rather abruptly in the fossil record. In fact, their appearance is described as an explosive development of new types of animals and is referred to as the "Cambrian explosion" by most scientists. This sudden and rapid appearance of new animals in the fossil record is, however, rapid only in the context of geologic time, having taken place over millions of years during the Early Cambrian Period.

This seemingly sudden appearance of animals in the fossil record is not a recent discovery. Early geologists observed that the remains of skeletonized animals appeared rather abruptly in the fossil record. Charles Darwin addressed this problem in *On the Origin of Species* and observed that, without a convincing explanation, such an event was difficult to reconcile with his newly expounded evolutionary theory.

The sudden appearance of shelled animals during the Early Cambrian contrasts sharply with the biota living during the preceding Proterozoic Eon. Up until the evolution of the Ediacaran fauna, Earth was populated primarily by single-celled organisms. Recall from Chapter 9 that the Ediacaran fauna, which is found on all continents except Antarctica, consists primarily of multicelled soft-bodied organisms. Microscopic calcareous tubes, presumably housing wormlike-suspension feeding organisms, have also been found at some localities. In addition, trails and burrows, which represent the activities of worms and other sluglike animals are also found associated with Ediacaran faunas throughout the world. The trails and burrows are similar to those made by present-day soft-bodied organisms.

Until recently, it appeared that there was a fairly long period of time between the extinction of the Ediacaran fauna and the first Cambrian fossils. That gap has been considerably narrowed in recent years with the discovery of new Proterozoic fossiliferous localities. Now, Proterozoic fossil assemblages continue right to the base of the Cambrian. Furthermore, recent work from Namibia indicates that Ediacaran-like fossils are even present above the first occurrence of Cambrian index fossils.

Nonetheless, the cause of the sudden appearance of so many different animal phyla during the Early Cambrian is still a hotly debated topic. Newly developed molecular techniques that allow evolutionary biologists to compare the similarity of molecular sequences of the same gene from different species is being applied to the phylogeny of many organisms. In addition, new fossil sites and detailed

Phanerozoic Eon										
Mesozoic Era			Cenozoic Era							
Triassic	Jurassic	Cretaceous	Tertiary						Quaternary	
			Paleocene	Eocene	Oligocene	Miocene	Pliocene		Pleistocene	Holocene

245
M.Y.A.

66
M.Y.A.

stratigraphic studies are shedding light on the early history and ancestry of the various invertebrate phyla.

It appears likely that the Cambrian explosion probably had its roots firmly planted in the Proterozoic. However, what the mechanism was that triggered this event is still unknown and was likely a combination of factors, both biologic and geologic. For example, geologic evidence indicates Earth was glaciated one or more times during the Proterozoic, followed by global warming during the Cambrian. These global environmental changes may have stimulated evolution and contributed to the Cambrian explosion. Recent work on *Hox* genes, which are sequences of genes that control the development of individual regions of the body, shows that the basic body plans for all animals was apparently established by the end of the Cambrian explosion, and only minor modifications have occurred since then. Whatever the ultimate cause of the Cambrian explosion was, the appearance of a skeletonized fauna and the rapid diversification of that fauna during the Early Cambrian was a major event in Earth's history.

The Emergence of a Shelly Fauna

The earliest organisms with hard parts are Proterozoic calcareous tubes found associated with Ediacaran faunas from several locations throughout the world. These are followed by other microscopic skeletonized fossils from the Early Cambrian (Figure 12.1) and the appearance of large skeletonized animals during the Cambrian explosion. Along with the question of why did animals appear so suddenly in the fossil record is the equally intriguing one of why they initially acquired skeletons and what selective advantage this provided.

A variety of explanations about why marine organisms evolved skeletons have been proposed, but none is completely satisfactory or universally accepted. One hypothesis is that the Precambrian oceans were deficient in calcium, carbonate, and phosphate ions, which are the materials found in mineralized skeletons. Without sufficient quantities of these ions, organisms could not precipitate hard shells. This explanation is now generally rejected because numerous Upper Precambrian and Lower Cambrian carbonate rocks are known from around the world.

Another hypothesis is that mineralized skeletons evolved as a response to an organism's need to eliminate mineral matter from its metabolic systems. One way to do this is to secrete a buildup of excess ions as a solid, thus producing a mineralized skeleton that would be advantageous to an organism in a variety of ways.

The formation of an exoskeleton confers many advantages on an organism: (1) It provides protection against ultraviolet radiation, allowing animals to move into shallower waters; (2) it helps prevent drying out in an intertidal environment; (3) it provides protection against predators. Recent evidence of actual fossils of predators and specimens of damaged prey, as well as antipredatory adaptations in some animals, indicates that the impact of predation during the Cambrian was great (Figure 12.2).

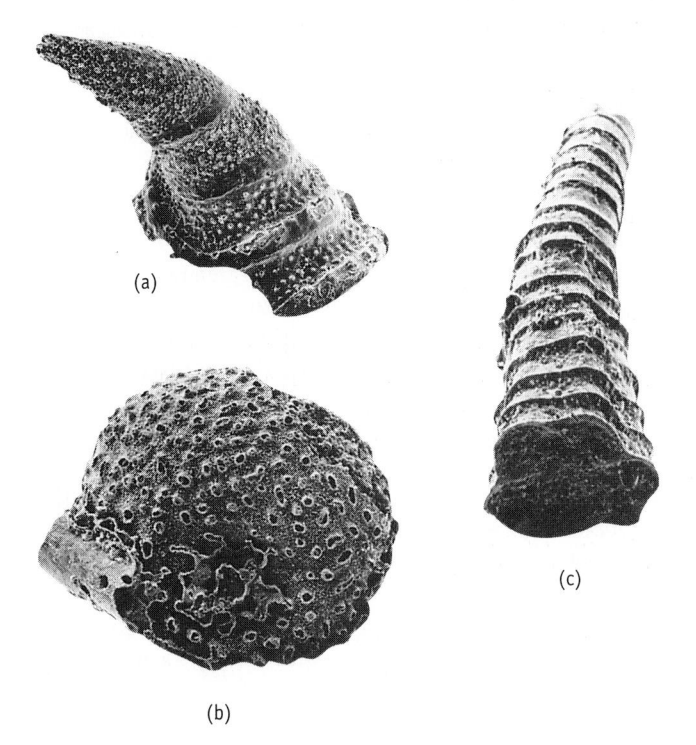

(a)

(b)

(c)

FIGURE 12.1 Three small Lower Cambrian shelly fossils. (a) A conical sclerite (a piece of the armor covering) of *Lapworthella* from Australia. (b) *Archaeooides*, an enigmatic spherical fossil from the Mackenzie Mountains, Northwest Territories, Canada. (c) The tube of an anabaritid from the Mackenzie Mountains, Northwest Territories, Canada.

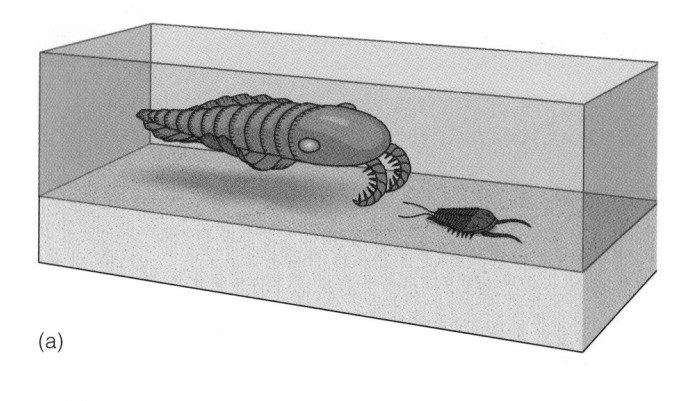

(a)

(b)

FIGURE 12.2 (a) Reconstruction of *Anomalocaris,* a predator from the Early and Middle Cambrian Period. It was about 45 cm long and probably fed on trilobites. Its gripping appendages presumably carried food to its mouth. (b) Wounds to the body of the trilobite *Olenellus robsonensis* (area just above the ruler). The wounds have healed, demonstrating that they occurred when the animal was alive and were not inflicted on an empty shell.

With predators playing an important role in the Cambrian marine ecosystem, any mechanism or feature that protected an animal would certainly be advantageous and confer an adaptive advantage to the organism. (4) A supporting skeleton, whether an exo- or endoskeleton, allows animals to increase their size and provides attachment sites for muscles.

There currently is no clear answer about why marine organisms evolved mineralized skeletons during the Cambrian explosion and shortly thereafter. They undoubtedly evolved because of a variety of biologic and environmental factors. Whatever the reason, the acquisition of a mineralized skeleton was a major evolutionary innovation allowing invertebrates to successfully occupy a wide variety of marine habitats.

Paleozoic Invertebrate Marine Life

Having considered the origin, differentiation, and evolution of the Precambrian–Cambrian marine biota, we now examine the changes that occurred in the marine invertebrate community during the Paleozoic Era. Rather than focusing on the history of each invertebrate phylum (Table 12.1), we will survey the evolution of the marine invertebrate communities through time, concentrating on the major features and changes that took place. To do that, we need to briefly examine the nature and structure of living marine communities so that we can make a reasonable interpretation of the fossil record.

THE PRESENT MARINE ECOSYSTEM

In analyzing the present-day marine ecosystem, we must look at where organisms live and how they get around, as well as how they feed (Figure 12.3). Organisms that live in the water column above the seafloor are called *pelagic.* They can be divided into two main groups: the floaters, or **plankton,** and the swimmers, or **nekton.**

Plankton are mostly passive and go where currents carry them. Plant plankton such as diatoms, dinoflagellates, and various algae, are called *phytoplankton* and are mostly microscopic. Animal plankton are called *zooplankton* and are also mostly microscopic. Examples of zooplankton include foraminifera, radiolarians, and jellyfish. The nekton are swimmers and are mainly vertebrates such as fish; the invertebrate nekton include cephalopods.

Organisms that live on or in the seafloor make up the **benthos.** They can be characterized as *epifauna* (animals) or *epiflora* (plants), for those that live on the seafloor, or as *infauna,* which are animals living in and moving through the sediments. The benthos can be further divided into those organisms that stay in one place, called *sessile,* and those that move around on or in the seafloor, called *mobile.*

The feeding strategies of organisms are also important in terms of their relationships with other organisms in the marine ecosystem. There are basically four feeding groups: **suspension feeders** remove or consume microscopic plants and animals as well as dissolved nutrients from the water; **herbivores** are plant eaters; **carnivore-scavengers** are meat eaters; and **sediment-deposit feeders** ingest sediment and extract the nutrients from it.

We can define an organism's place in the marine ecosystem by where it lives and how it eats. For example, an articulate brachiopod is a benthonic, epifaunal suspension feeder, whereas a cephalopod is a nektonic carnivore.

An ecosystem includes several **trophic levels,** which are tiers of food production and consumption within a feeding hierarchy. The feeding hierarchy and hence energy flow in an ecosystem comprise a food web of complex interrelationships among the producers, consumers, and decomposers (Figure 12.4). The **primary producers,** or *autotrophs,* are those organisms that manufacture their own food. Virtually all marine primary producers are phytoplankton. Feeding on the primary producers are the primary consumers, which are mostly suspension feeders. Secondary consumers feed on the primary consumers, and thus are predators, while tertiary consumers, which are

TABLE 12.1

The Major Invertebrate Groups and Their Stratigraphic Ranges

Phylum Protozoa	Cambrian-Recent	**Phylum Mollusca**	Cambrian-Recent
Class Sarcodina	Cambrian-Recent	Class Monoplacophora	Cambrian-Recent
Order Foraminifera	Cambrian-Recent	Class Gastropoda	Cambrian-Recent
Order Radiolaria	Cambrian-Recent	Class Bivalvia	Cambrian-Recent
Phylum Porifera	Cambrian-Recent	Class Cephalopoda	Cambrian-Recent
Class Demospongea	Cambrian-Recent	**Phylum Annelida**	Precambrian-Recent
Order Stromatoporoida	Cambrian-Oligocene	**Phylum Arthropoda**	Cambrian-Recent
Phylum Archaeocyatha	Cambrian	Class Trilobita	Cambrian-Permian
Phylum Cnidaria	Cambrian-Recent	Class Crustacea	Cambrian-Recent
Class Anthozoa	Ordovician-Recent	Class Insecta	Silurian-Recent
Order Tabulata	Ordovician-Permian	**Phylum Echinodermata**	Cambrian-Recent
Order Rugosa	Ordovician-Permian	Class Blastoidea	Ordovician-Permian
Order Scleractinia	Triassic-Recent	Class Crinoidea	Cambrian-Recent
Phylum Bryozoa	Ordovician-Recent	Class Echinoidea	Ordovician-Recent
Phylum Brachiopoda	Cambrian-Recent	Class Asteroidea	Ordovician-Recent
Class Inarticulata	Cambrian-Recent	**Phylum Hemichordata**	Cambrian-Recent
Class Articulata	Cambrian-Recent	Class Graptolithina	Cambrian-Mississippian

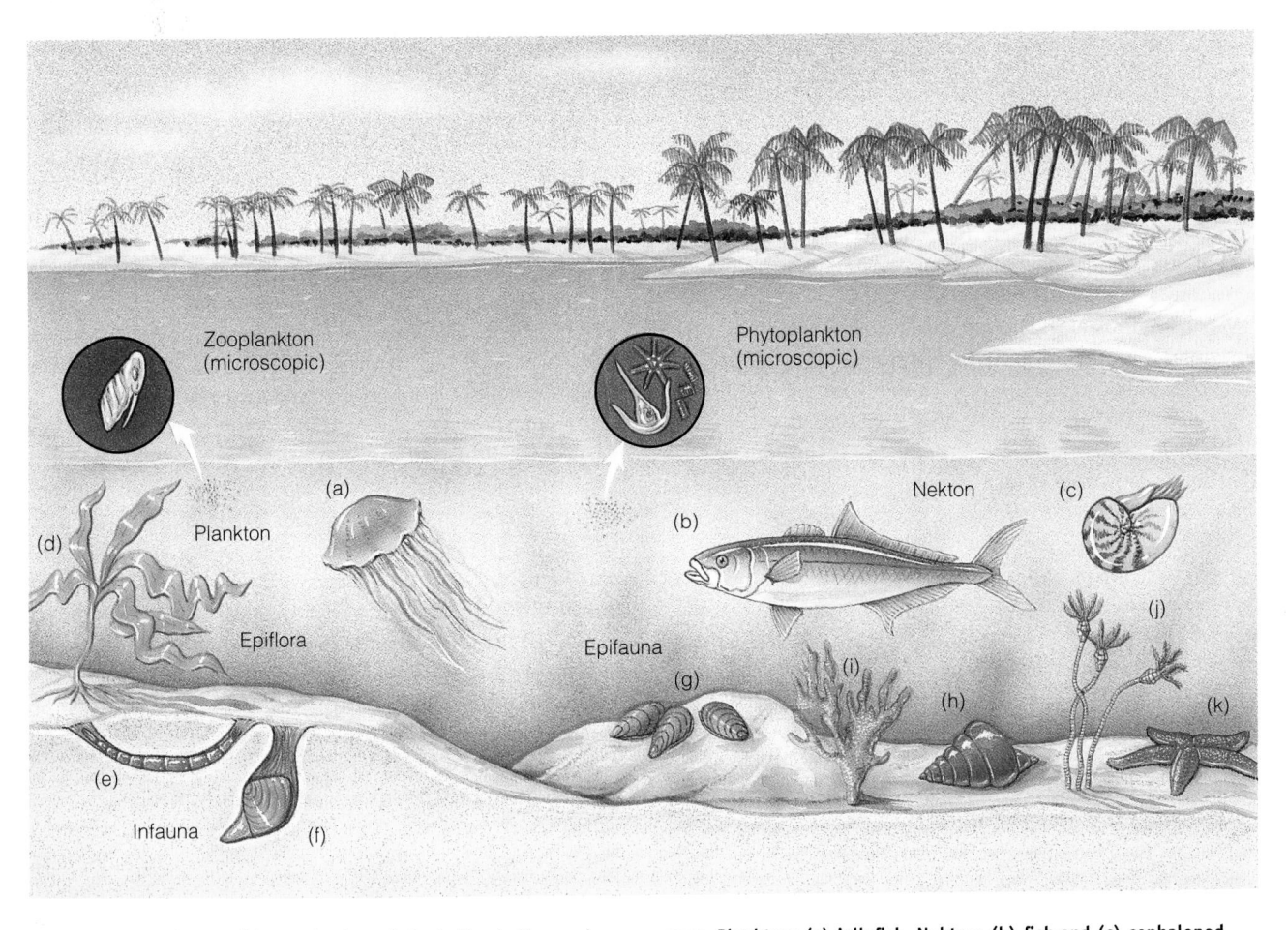

FIGURE 12.3 Where and how animals and plants live in the marine ecosystem. Plankton: (a) jellyfish. Nekton: (b) fish and (c) cephalopod. Benthos: (d) through (k). Sessile epiflora: (d) seaweed. Sessile epifauna: (g) bivalve, (i) coral, and (j) crinoid. Mobile epifauna: (k) starfish and (h) gastropod. Infauna: (e) worm and (f) bivalve. Suspension feeders: (g) bivalve, (i) coral, and (j) crinoid. Herbivores: (h) gastropod. Carnivores-scavengers: (k) starfish. Sediment-deposit feeders: (e) worm.

FIGURE 12.4 Marine food web showing the relationships among the producers, consumers, and decomposers.

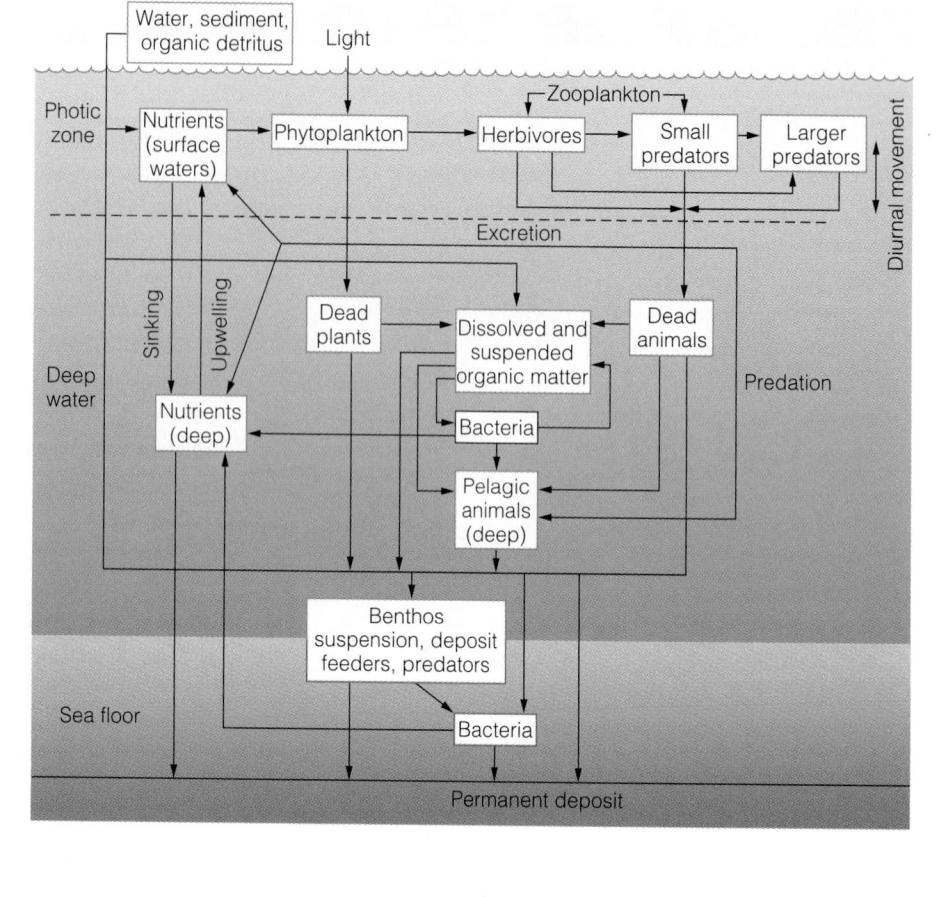

also predators, feed on the secondary consumers. Besides the producers and consumers, there are also transformers and decomposers. These are bacteria that break down the dead organisms that have not been consumed into organic compounds that are then recycled.

When we look at the marine realm today, we see a complex organization of organisms interrelated by trophic interactions and affected by changes in the physical environment. When one part of the system changes, the whole structure changes, sometimes almost insignificantly, other times catastrophically.

As we examine the evolution of the Paleozoic marine ecosystem, keep in mind how geologic and evolutionary changes can have a significant impact on its composition and structure. For example, the major transgressions onto the craton opened up vast areas of shallow seas that could be inhabited. The movement of continents affected oceanic circulation patterns as well as causing environmental changes.

CAMBRIAN MARINE COMMUNITY

The Cambrian Period was a time during which many new body plans evolved and animals moved into new niches. As might be expected, the Cambrian witnessed a higher percentage of such experiments than any other period of geologic history.

Although almost all the major invertebrate phyla evolved during the Cambrian Period (Table 12.1), many

were represented by only a few species. While trace fossils are common and echinoderms diverse, trilobites, inarticulate brachiopods, and archaeocyathids comprised the majority of Cambrian skeletonized life (Figure 12.5).

Trilobites were by far the most conspicuous element of the Cambrian marine invertebrate community and made up about half of the total fauna. Trilobites were benthonic mobile sediment-deposit feeders that crawled or swam along the seafloor. They first appeared in the Early Cambrian, rapidly diversified, reached their maximum diversity in the Late Cambrian, and then suffered mass extinctions near the end of the Cambrian from which they never fully recovered. As yet no consensus exists on what caused the trilobite extinctions, but a combination of factors were likely involved, including a reduction of shelf space, increased competition, and a rise in predators. It has also been suggested that a cooling of the seas may have played a role, particularly for the extinctions that took place at the end of the Ordovician Period.

Cambrian **brachiopods** were mostly primitive types called *inarticulates*. They secreted a chitinophosphate shell, composed of the organic compound chitin combined with calcium phosphate. Inarticulate brachiopods also lacked a tooth-and-socket-arrangement along the hinge line. The *articulate* brachiopods, which have a tooth-and-socket arrangement, were also present but did not become abundant until the Ordovician Period.

The third major group of Cambrian organisms were the **archaeocyathids** (Figure 12.6). These organisms were benthonic sessile suspension feeders that constructed reeflike structures. The rest of the Cambrian fauna consisted of representatives of the other major phyla, including many organisms that were short-lived evolutionary experiments (Figure 12.7).

THE BURGESS SHALE BIOTA

No discussion of Cambrian life would be complete without mentioning one of the best examples of a preserved soft-bodied fauna and flora, the Burgess Shale biota. As the Sauk Sea transgressed from the Cordilleran shelf onto the western edge of the craton, Early Cambrian sands were covered by Middle Cambrian black muds that allowed a diverse soft-bodied benthonic community to be preserved. As we discussed in the Prologue, these fossils were discovered in 1909 by Charles Walcott near Field, British Columbia. They represent one of the most significant fossil finds of the century because they consist of impressions of soft-bodied animals and plants (Figure 12.8), which are very rarely preserved in the fossil record. This discovery therefore allows us a valuable glimpse of rarely preserved organisms as well as the soft-part anatomy of many extinct groups.

In recent years, the reconstruction, classification, and interpretation of many of the Burgess Shale fossils have undergone a major change that has led to new theories and explanations of the Cambrian explosion of life. Recall that during the Late Proterozoic multicellular organisms evolved, and shortly thereafter animals with hard parts made their first appearance. These were followed by an explosion of invertebrate phyla during the Cambrian, some of which are now extinct. These Cambrian phyla represent the root stock and basic body plans from which all present-day invertebrates evolved. The question that paleon-

FIGURE 12.7 *Helicoplacus,* a primitive echinoderm that became extinct 20 million years after its first appearance about 510 million years ago. Such an organism (a representative of one of several short-lived echinoderm classes) illustrates the "experimental" nature of the Cambrian invertebrate fauna.

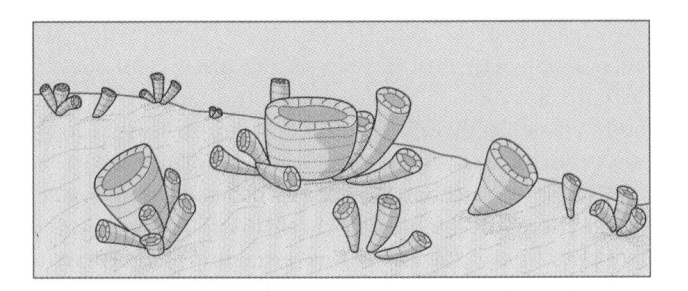

FIGURE 12.6 Restoration of a Cambrian reeflike structure built by archaeocyathids.

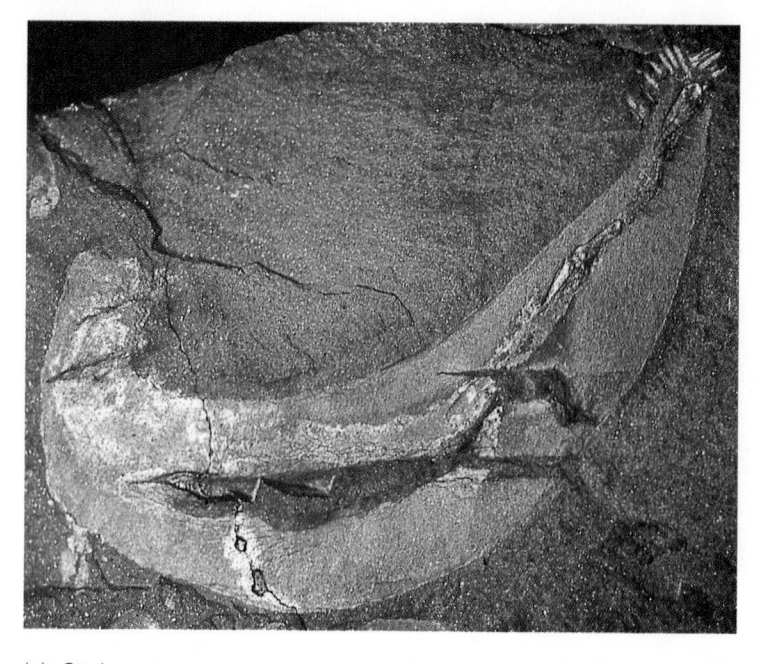

(a) *Ottoia*

(b) *Wiwaxia*

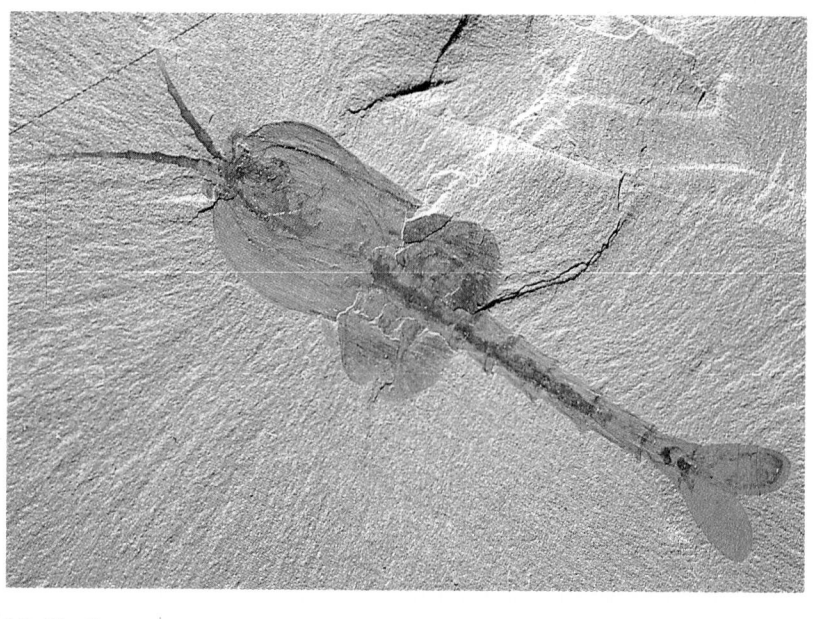

(c) *Hallucigenia*

(d) *Waptia*

FIGURE 12.8 Some of the fossil animals preserved in the Burgess Shale. (a) *Ottoia,* **a carnivorous worm. (b)** *Wiwaxia,* **a scaly armored sluglike creature whose affinities remain controversial. (c)** *Hallucigenia,* **a velvet worm. (d)** *Waptia,* **an anthropod.**

tologists are still debating is how many phyla arose during the Cambrian, and at the center of that debate are the Burgess Shale fossils. For years, most paleontologists placed the bulk of the Burgess Shale organisms into existing phyla, with only a few assigned to phyla that are now extinct. Thus, the phyla of the Cambrian world were viewed as being essentially the same in number as the phyla of the present-day world, but with fewer species in each phylum. According to this view, the history of life has

been simply a gradual increase in the diversity of species within each phylum through time. The number of basic body plans has therefore remained more or less constant since the initial radiation of multicellular organisms.

This view, however, has been challenged by other paleontologists who think that the initial explosion of varied lifeforms in the Cambrian was promptly followed by a short period of experimentation and then extinction of many phyla. The richness and diversity of modern life-

forms are the result of repeated variations of the basic body plans that survived the Cambrian extinctions. In other words, life was much more diverse in terms of phyla during the Cambrian than it is today. The reason members of the Burgess Shale biota look so strange to us is that no living organisms possess their basic body plan, and therefore many of them have been placed into new phyla.

Discoveries of new Cambrian fossils at localities such as Sirius Passet, Greenland, and Yunnan, China, have resulted in reassignment of some Burgess Shale specimens back into extant phyla. If these reassignments to known phyla prove to be correct, then no massive extinction event followed the Cambrian explosion, and life has gradually increased in diversity through time. Currently, there is no clear answer to this debate, and the outcome will probably be decided as more fossil discoveries are made.

ORDOVICIAN MARINE COMMUNITY

A major transgression that began during the Middle Ordovician (Tippecanoe sequence) resulted in the most widespread inundation of the craton. This vast epeiric sea, which experienced a uniformly warm climate during this time, opened numerous new marine habitats that were soon filled by a variety of organisms.

Not only did sedimentation patterns change dramatically from the Cambrian to the Ordovician, but the fauna underwent equally striking changes. Whereas the Cambrian invertebrate community was dominated by trilobites, inarticulate brachiopods, and archaeocyathids, the Ordovician was characterized by the adaptive radiation of many other animal phyla, (such as articulate brachiopods, bryozoans, and corals), with a consequent dramatic increase in the diversity of the total shelly fauna (Figure 12.9). The

Ordovician was also a time of increased diversity and abundance of the **acritarchs** (organic-walled phytoplankton of unknown affinity), which were the major phytoplankton group of the Paleozoic Era and the primary food source of the suspension feeders (Figure 12.10).

During the Cambrian, archaeocyathids were the main builders of reeflike structures, but bryozoans, stromatoporoids, and tabulate and rugose corals assumed that role beginning in the Middle Ordovician. Many of these reefs were small patch reefs similar in size to those of the Cambrian but of a different composition, whereas others were quite large. As with present-day reefs, Ordovician reefs exhibited a high diversity of organisms and were dominated by suspension feeders (Figure 12.11).

Three Ordovician fossil groups have proved to be particularly useful for biostratigraphic correlation—the *articulate brachiopods, graptolites,* and *conodonts* (Figure 12.12). The articulate brachiopods, present since the Cambrian, began a period of major diversification in the shallow-water marine environment during the Ordovician (Figure 12.12a). They became a conspicuous element of the invertebrate fauna during the Ordovician and in succeeding Paleozoic periods.

Most **graptolites** were planktonic animals carried about by ocean currents. Because most graptolites were planktonic and most individual species existed for less than a million years, graptolites are excellent guide fossils. They were especially abundant during the Ordovician and Silurian periods. Due to the fragile nature of their organic skeleton, graptolites are most commonly found in black shales (Figure 12.12b).

Conodonts are a group of well-known small toothlike fossils composed of the mineral apatite (calcium phos-

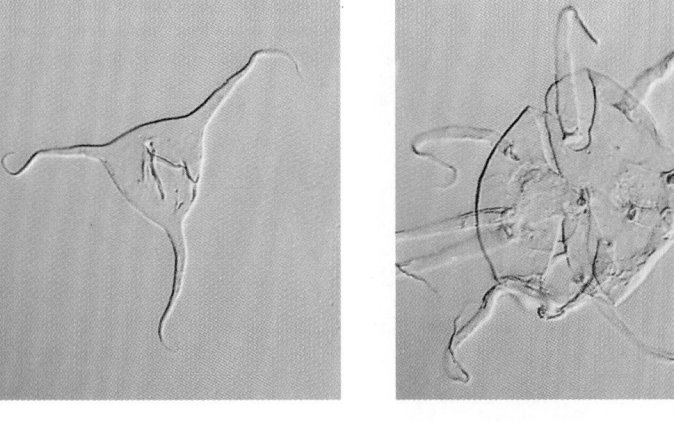
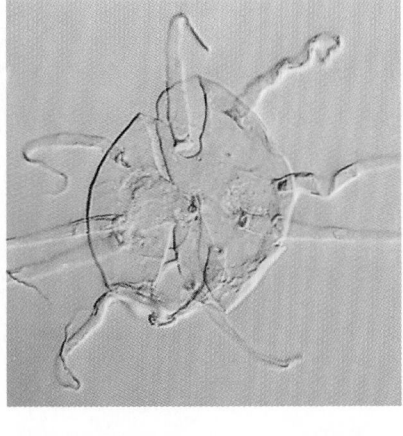
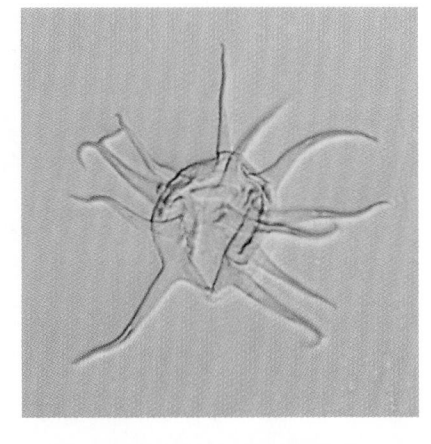

FIGURE 12.10 Acritarchs from the Upper Ordovician Sylvan Formation, Oklahoma. Acritarchs are organic-walled phytoplankton and were the primary food source of suspension feeders during the Paleozoic Era.

phate), the same mineral that composes bone (Figure 12.12c). Although conodonts have been known for more than 130 years, their affinity has been the subject of debate until the discovery of the conodont animal in 1983 (Figure 12.13). Several specimens of carbonized impressions of the conodont animal from Lower Carboniferous rocks of Scotland reveal that it is a member of a group of primitive jawless animals assigned to the phylum Chordata. Study of the specimens indicates that the conodont animal was probably an elongate swimming organism. The wide distri-

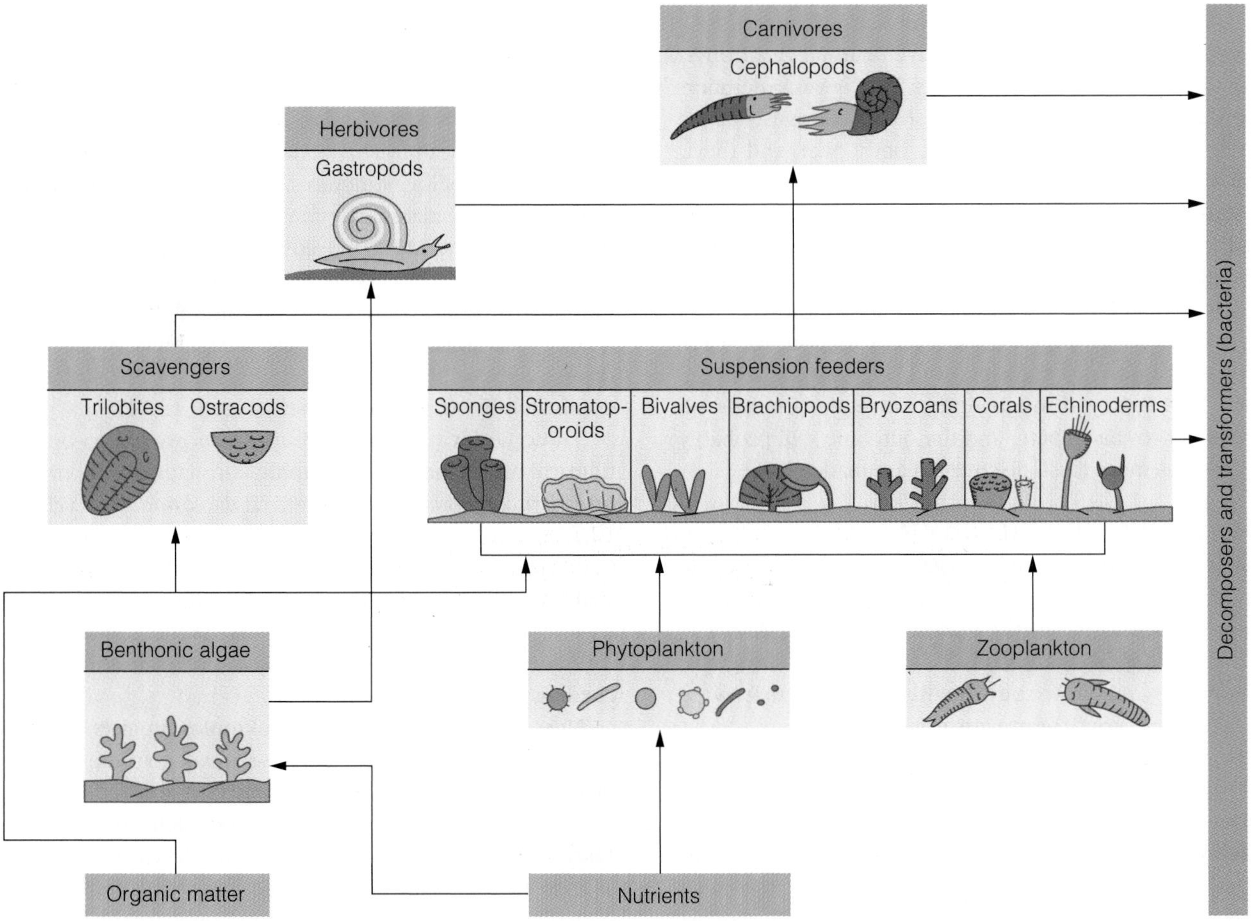

FIGURE 12.11 Trophic analysis of an Ordovician reef community showing the relationships among the various organisms of the community. The phytoplankton and benthonic algae are primary producers, occupying the lowest trophic level. The primary consumers are the suspension feeders, which make up the majority of the community. The highest trophic level is occupied by the carnivores, which in this community are the cephalopods.

(a)

(b)

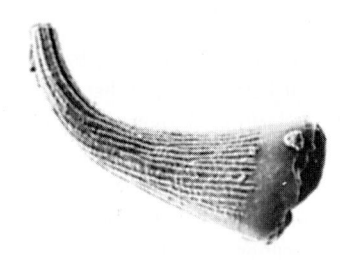

(c)

bution and short stratigraphic range of individual conodont species make them excellent fossils for biostratigraphic zonation and correlation.

The end of the Ordovician was a time of mass extinctions in the marine realm. More than 100 families of marine invertebrates became extinct, and in North America alone, approximately one-half of the brachiopods and bryozoans died out. What caused such an event? Many geologists think these extinctions were the result of extensive glaciation that occurred in Gondwana at the end of the Ordovician Period (see Chapter 10).

FIGURE 12.13 The conodont animal preserved as a carbonized impression in the Lower Carboniferous Granton Shrimp Bed in Edinburgh, Scotland. The animal measures about 40 mm long and 2 mm wide.

Mass extinctions, those geologically rapid events in which an unusually high percentage of the fauna and/or flora becomes extinct, have occurred throughout geologic time (at or near the end of the Ordovician, Devonian, Permian, and Cretaceous Periods) and are the focus of much research and debate (see Perspective 12.1).

SILURIAN AND DEVONIAN MARINE COMMUNITIES

The mass extinction at the end of the Ordovician was followed by rediversification and recovery of many of the decimated groups. Brachiopods, bryozoans, gastropods, bivalves, corals, crinoids, and graptolites were just some of the groups that rediversified again beginning during the Silurian.

As we discussed in Chapters 10 and 11, the Silurian and Devonian were times of major reef building. While most of the Silurian radiations of invertebrates represented repopulating of niches, organic reef builders diversified in new ways, building massive reefs larger than any produced during the Cambrian or Ordovician. This repopulation was probably due in part to renewed transgressions over the craton, and although a major drop in sea level occurred at the end of the Silurian, the Middle Paleozoic sea level was generally high (see Table 10.1).

The Silurian and Devonian reefs were dominated by tabulate and colonial rugose corals and stromatoporoids (Figure 12.14). While the fauna of these Silurian and Devonian reefs was somewhat different from that of earlier reefs and reeflike structures, the general composition and structure are the same as in present-day reefs.

Mass Extinctions and Their Possible Causes

*T*hroughout geologic history, various plant and animal species have become extinct. In fact, extinction is a common feature of the fossil record, and the rate of extinction through time has fluctuated only slightly. Just as new species evolve, others become extinct. There have, however, been brief intervals in the geologic past during which mass extinctions have eliminated large numbers of species. Extinctions of this magnitude could only occur due to radical changes in the environment on a regional or global scale.

When we look at the different mass extinction events that have occurred during the geologic past, several common themes stand out. The first is that mass extinctions typically have affected life both in the sea and on land. Second, tropical organisms, particularly in the marine realm, apparently are more affected than organisms from the temperate and high-latitude regions. Third, some animal groups repeatedly experience mass extinctions. During the first mass-extinction event, such groups are severely affected but not wiped out. Following the initial crisis, the survivors diversify, only to have their numbers reduced further by another mass extinction. Three marine invertebrate groups in particular display this characteristic: the trilobites, graptolites, and ammonoids. Each of these groups experienced high rates of extinction, followed by high rates of speciation.

When we examine the mass extinctions for the last 650 million years, we see that the first event involved only the acritarchs. Several extinction events occurred during the Cambrian, and these affected only marine invertebrates, particularly trilobites. Three other marine mass extinctions took place during the Paleozoic Era: one at the end of the Ordovician, involving many invertebrates; one near the end of the Devonian, affecting the major barrier reef-building organisms as well as the primitive armored fish; and the most severe at the end of the Permian, when about 90% of all marine species and more than 65% of all land animals became extinct.

The Mesozoic Era experienced several mass extinctions, the most devastating occurring at the end of the Cretaceous, when all large animals, including dinosaurs and seagoing animals such as plesiosaurs and ichthyosaurs, became extinct. Many scientists think that the terminal Cretaceous extinction event was caused by a meteorite impact.

Several mass-extinction events occurred during the Cenozoic Era. The most severe was near the end of the Eocene Epoch and is correlated with global cooling and climatic change.

Though many scientists think of the marine mass extinctions as sudden events from a geologic perspective, they were rather gradual from a human perspective, occurring over hundreds of thousands or even millions of years. Furthermore, many geologists think that climatic changes, rather than some catastrophe, were primarily responsible for the extinctions, particularly in the marine realm. Evidence of glacial episodes or other signs of climatic change, such as global warming, have been correlated with the extinction events recorded in the fossil record.

The Silurian and Devonian periods were also the time when *eurypterids* (arthropods with scorpion-like bodies and impressive pincers) were abundant, especially in brackish and freshwater habitats (Figure 12.15). *Ammonoids*, a subclass of the cephalopods, evolved from nautiloids during the Early Devonian and rapidly diversified. With their distinctive suture patterns, short stratigraphic ranges, and widespread distribution, ammonoids are excellent guide fossils for the Devonian through Cretaceous periods (Figure 12.16).

Another mass extinction occurred near the end of the Devonian and resulted in a worldwide near-total collapse of the massive reef communities. On land, however, the seedless vascular plants were seemingly

FIGURE 12.14 Reconstruction of a Middle Devonian reef from the Great Lakes area. Shown are corals, ammonoids, trilobites, and brachiopods.

unaffected, although the diversity of freshwater fish was greatly reduced. Thus, extinctions were most extensive in the marine realm, particularly in the reef and pelagic communities.

The demise of the Middle Paleozoic reef communities serves to highlight the geographic aspects of the Late Devonian mass-extinction event. The tropical groups were most severely affected; in contrast, the polar communities were seemingly little affected. Apparently, an episode of global cooling was largely responsible for the extinctions near the end of the Devonian. During such a cooling, the disappearance of tropical conditions would have had a severe effect on reef and other warm-water organisms. Cool-water species, on the other hand, could have simply migrated toward the equator. While cooling temperatures certainly played an important role in the Late Devonian extinctions, the closing of the Iapetus Ocean and the orogenic events of this time undoubtedly also played a role by

reducing the area of shallow shelf environments where many marine invertebrates lived.

CARBONIFEROUS AND PERMIAN MARINE COMMUNITIES

The Carboniferous invertebrate marine community responded to the Late Devonian extinctions in much the same way the Silurian invertebrate marine community responded to the Late Ordovician extinctions—that is, by renewed adaptive radiation and rediversification. The brachiopods and ammonoids quickly recovered and again assumed important ecologic roles, while other groups, such as the lacy bryozoans and crinoids, reached their greatest diversity during the Carboniferous. With the decline of the stromatoporoids and the tabulate and rugose corals, large organic reefs like those existing earlier in the Paleozoic virtually disappeared and were replaced by small patch reefs. These reefs were dominated by crinoids, blas-

FIGURE 12.15 Restoration of a Silurian brackish-marine bottom scene near Buffalo, New York. Shown are algae, eurypterids, worms, and shrimp.

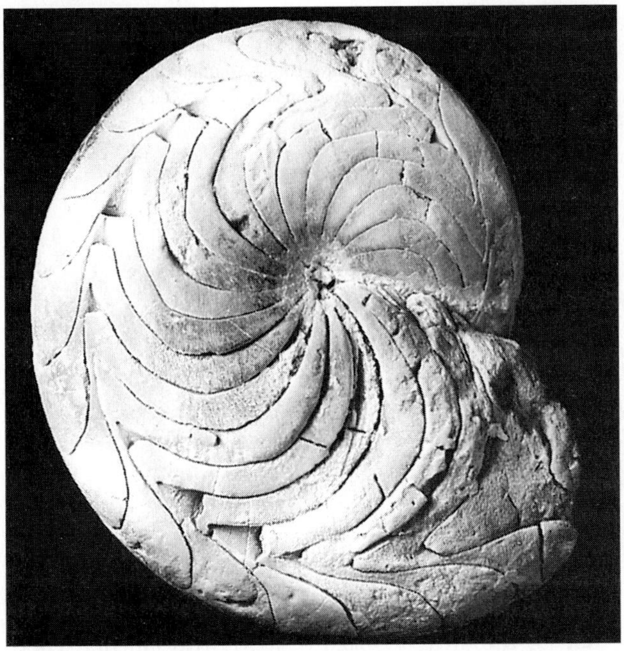

FIGURE 12.16 *Imitoceras rotatorium* (DeKoninck), a goniatitic ammonoid cephalopod from the Lower Mississippian Rockford Limestone, near Rockford, Indiana. The distinctive suture pattern, short stratigraphic range, and wide distribution make ammonoids excellent guide fossils.

toids, lacy bryozoans, brachiopods, and calcareous algae and flourished during the Late Paleozoic (Figure 12.17). In addition, bryozoans, crinoids, and fusulinids (spindle-shaped foraminifera) contributed large amounts of skeletal debris to the formation of the vast bedded limestones that constitute the majority of Mississippian sedimentary rocks.

The Permian invertebrate marine faunas resembled those of the Carboniferous, but were not as widely distributed because of the restricted size of the shallow seas on the cratons and the reduced shelf space along the continental margins (see Figure 11.12). The spiny and odd-shaped productids dominated the brachiopod assemblage and constituted an important part of the reef complexes that formed in the Texas region during the Permian (Figure 12.18). The fusulinids (Figure 12.19), which first evolved during the Late Mississippian and greatly diversified during the Pennsylvanian, experienced a further diversification during the Permian. Because of their abundance, diversity, and worldwide occurrence, fusulinids are important guide fossils for Pennsylvanian and Permian strata. Bryozoans, sponges, and some types of calcareous algae also were common elements of the Permian invertebrate fauna.

THE PERMIAN MARINE INVERTEBRATE EXTINCTION EVENT

The greatest recorded mass-extinction event to affect Earth occurred at the end of the Permian Period (Figure 12.20). Before the Permian ended, roughly 50% of all marine invertebrate families and about 90% of all marine invertebrate species became extinct. Fusulinids, rugose and tabulate corals, several bryozoan and brachiopod orders, as well as trilobites and blastoids did not survive the end of the Permian. All these groups had been very successful during the Paleozoic Era. In addition, more than 65% of all amphibians and reptiles, as well as nearly 33% of insects on land also became extinct.

What caused such a crisis for both marine and land-dwelling organisms? Various hypotheses have been proposed, but no completely satisfactory answer has yet been found. One hypothesis that can be immediately ruled out as causing the extinctions is a meteorite impact. No evidence, either direct or indirect, of a meteorite impact has been found; furthermore, it appears that the Permian mass-extinction event took place over an 8-million-year interval at the end of the Permian Period.

FIGURE 12.17 Marine life during the Mississippian based on an Upper Mississippian fossil site at Crawfordville, Indiana. Invertebrate animals shown include crinoids, blastoids, lacy bryozoans, and small corals.

Some hypotheses put forth to explain the extinctions include (1) a reduction of shelf space due to the suturing of the continents to form Pangaea, (2) a global drop in sea level due to glacial conditions, (3) a reduction of shelf space due to a large marine regression, and (4) climatic changes. The first two can be eliminated because most of the collisions of the continents had already taken place by the end of the Permian and the large-scale formation of glaciers took place during the Pennsylvanian Period. In addition, current evidence indicates a time of sudden warming at the end of the Permian.

Currently, many scientists think that a large-scale marine regression coupled with climatic changes in the form of global warming—due to an increase in carbon dioxide levels—may have been responsible for the mass extinctions recorded in the fossil record. In this scenario, a widespread lowering of sea level occurred near the end of the Permian, which greatly reduced the amount of shallow shelf space for marine organisms and exposed the shelf to erosion. Oxidation of the organic matter trapped in the sediments ensued, which reduced atmospheric oxygen levels as well as releasing large quantities of carbon dioxide into the atmosphere, resulting in increased global warming. During this time, widespread volcanic eruptions also took place, further releasing additional carbon dioxide into the atmosphere and contributing to increased climatic instability and ecologic collapse. At the end of the Permian, a rise in sea level flooded and destroyed the nearshore terres-

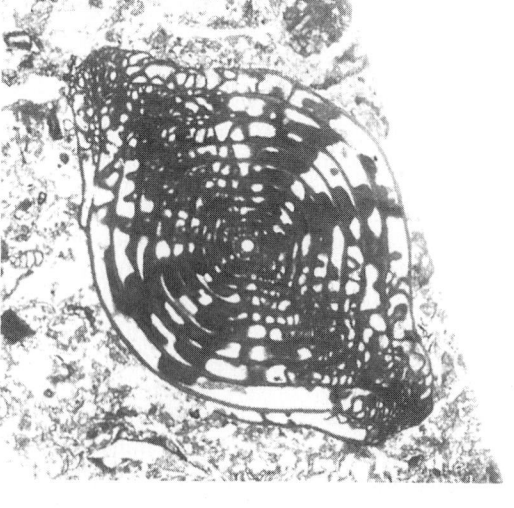

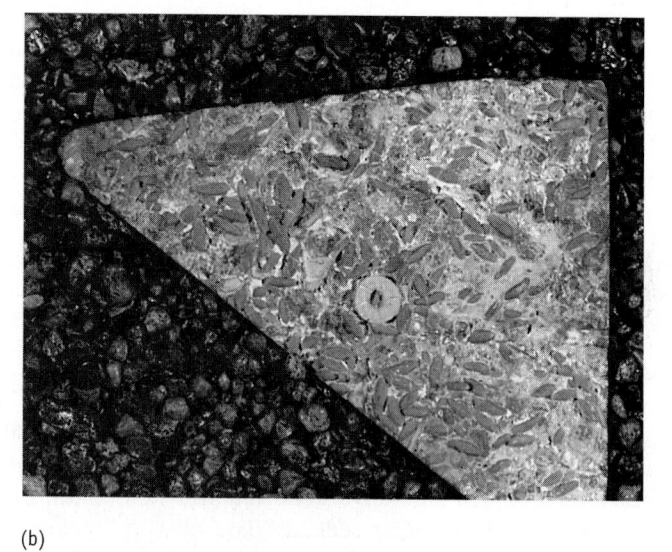

(a)   (b)

FIGURE 12.19 Fusulinids are large benthonic foraminifera that are excellent guide fossils for the Pennsylvanian and Permian periods. **(a)** *Leptotriticites tumidus,* Early Permian, Kansas. **(b)** *Triticites,* Pennsylvanian, Kansas.

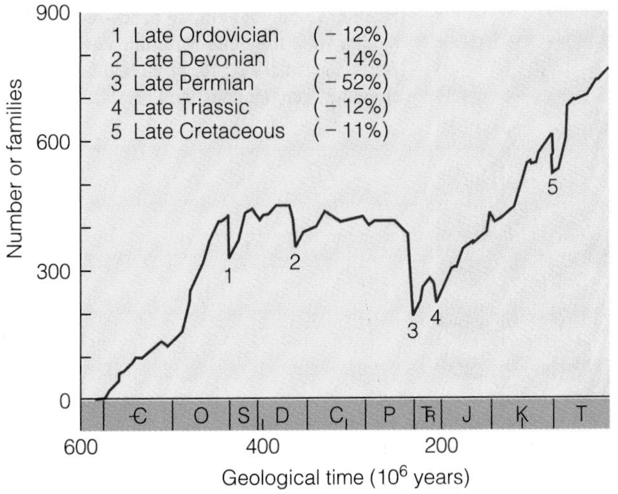

FIGURE 12.20 Phanerozoic diversity for marine invertebrate and vertebrate families. Note the three episodes of Paleozoic mass extinctions, with the greatest occurring at the end of the Permian Period.

trial habitats, thus causing the extinction of many terrestrial plants and animals.

In addition to the above scenario, scientists also suggest that a global turnover of deep-ocean waters would have increased the amount of carbon dioxide in the surface waters as well as releasing large amounts of carbon dioxide into the atmosphere. The end result would be the same: global warming and disruption of the marine and terrestrial ecosystem.

Regardless of the ultimate cause of the Permian mass extinctions, the fact is that Earth's biota was dramatically changed. The resulting Triassic marine faunas were of low diversity, but the surviving species tended to be abundant and widely distributed around the world. As we will see in Chapter 15, this fauna provided the root stock from which the world's oceans repopulated and evolved.

Summary

Table 12.2 summarizes the major evolutionary and geologic events of the Paleozoic Era and shows their relationships to each other.

1. Soft-bodied multicelled organisms presumably had a long Precambrian history during which they lacked hard parts. Invertebrates with hard parts suddenly appeared during the Early Cambrian in what is called the Cambrian explosion. Skeletons provided such advantages as protection against predators and support for muscles, enabling organisms to grow large and increase locomotor efficiency. Hard parts probably evolved as a result of various geologic and biologic factors rather than a single cause.

2. Marine organisms are classified as plankton if they are floaters, nekton if they swim, and benthos if they live on or in the seafloor.

3. Marine organisms can be divided into four basic feeding groups: suspension feeders, which remove or consume microscopic plants and animals as well as dissolved nutrients from water; herbivores, which are plant eaters; carnivore-scavengers, which are meat eaters; and sediment-deposit feeders, which ingest sediment and extract nutrients from it.

4. The marine ecosystem consists of various trophic levels of food production and consumption. At the base are primary producers, upon which all other organisms are dependent. Feeding on the primary producers are the primary consumers, which in turn can be fed upon by higher levels of consumers. The decomposers are bacteria that break down the complex organic compounds of dead organisms and recycle them within the ecosystem.

5. The Cambrian invertebrate community was dominated by three major groups, the trilobites, inarticulate brachiopods, and archaeocyathids. Little specialization existed among the invertebrates, and most phyla were represented by only a few species. The Middle Cambrian Burgess Shale contains one of the finest examples of a well-preserved soft-bodied biota in the world.

6. The Ordovician marine invertebrate community marked the beginning of the dominance by the shelly fauna and the start of large-scale reef building. The end of the Ordovician Period was a time of major extinctions for many invertebrate phyla.

7. The Silurian and Devonian periods were times of diverse faunas dominated by reef-building animals, while the Caroniferous and Permian periods saw a great decline in invertebrate diversity.

8. A major extinction occurred at the end of the Paleozoic Era, affecting the invertebrates as well as the vertebrates. Its cause is still being debated.

Important Terms

acritarch	carnivore-scavenger	nekton	suspension feeder
archaeocyathid	conodont	plankton	trilobite
benthos	graptolite	primary producer	trophic level
brachiopod	herbivore	sediment-deposit feeder	

Review Questions

1. The age of the Burgess Shale fauna is:
   a. _____ Cambrian.
   b. _____ Ordovician.
   c. _____ Silurian.
   d. _____ Devonian.
   e. _____ Mississippian.

2. Organisms living on or in the seafloor are:
   a. _____ epifauna.
   b. _____ epiflora.
   c. _____ infauna.
   d. _____ benthos.
   e. _____ All of these

3. An exoskeleton is advantageous because it:
   a. _____ provides protection against predators.
   b. _____ prevents drying out in an intertidal environment.
   c. _____ provides attachment sites for development of strong muscles.
   d. _____ provides protection against ultraviolet radiation.
   e. _____ All of these

4. The first organisms to construct reeflike structures were:
   a. _____ corals.
   b. _____ bryozoans.
   c. _____ archaeocyathids.
   d. _____ sponges.
   e. _____ mollusks.

5. The major phytoplankton group of the Paleozoic Era and the primary food source of the suspension feeders were:
   a. _____ dinoflagellates.
   b. _____ coccolithophorids.
   c. _____ diatoms.
   d. _____ acritarchs.
   e. _____ graptolites.

6. The greatest recorded mass-extinction event to affect the marine invertebrate community occurred at the end of which period?
   a. _____ Cambrian
   b. _____ Ordovician
   c. _____ Silurian
   d. _____ Devonian
   e. _____ Permian

7. What type of invertebrates dominated the Ordovician invertebrate community?
   a. _____ Epifaunal benthonic sessile suspension feeders
   b. _____ Infaunal benthonic sessile suspension feeders
   c. _____ Epifaunal benthonic mobile suspension feeders
   d. _____ Infaunal nektonic carnivores
   e. _____ Epifloral planktonic primary producers

8. A _____ is an example of an epifaunal benthonic suspension feeder.
   a. _____ trilobite
   b. _____ cephalopod
   c. _____ graptolite
   d. _____ articulate brachiopod
   e. _____ gastropod

9. Which group of invertebrates are excellent guide fossils for the Pennsylvanian and Permian periods?
   a. _____ Trilobites
   b. _____ Gastropods
   c. _____ Sponges
   d. _____ Bivalves
   e. _____ Fusulinids

10. The fossils of the Burgess Shale are significant because they provide a rare glimpse of _____.
   a. _____ the first shelled animals
   b. _____ the soft-part anatomy of extinct groups
   c. _____ soft-bodied animals
   d. _____ Answers (a) and (b)
   e. _____ Answers (b) and (c)

11. Discuss the possible causes for the Cambrian explosion.

12. Discuss the significance of the appearance of the first shelled animals and possible causes for the acquisition of a mineralized exoskeleton.

13. Discuss how Silurian and Devonian reefs differed from Cambrian reefs.

14. Draw a marine food web that shows the relationships among the producers, consumers, and decomposers.

15. Discuss the major differences between the Cambrian marine community and the Ordovician marine community.

TABLE 12.2

Major Evolutionary and Geologic Events of the Paleozoic Era

Age (Millions of Years)	Geologic Period		Invertebrates	Vertebrates
245	Permian		Largest mass extinction event to affect the invertebrates.	Acanthodians, placoderms, and pelycosaurs become extinct. Therapsids and pelycosaurs the most abundant reptiles.
286	Carboniferous	Pennsylvanian	Fusulinids diversify.	Amphibians abundant and diverse.
320	Carboniferous	Mississippian	Crinoids, lacy bryozans, blastoids become abundant. Renewed adaptive radiation following extinctions of many reef-builders.	Reptiles evolve.
360	Devonian		Extinctions of many reef-building invertebrates near end of Devonian. Reef building continues. Eurypterids abundant.	Amphibians evolve. All major groups of fish present—Age of Fish.
408	Silurian		Major reef building. Diversity of invertebrates remains high.	Ostracoderms common. Acanthodians, the first jawed fish, evolve.
438	Ordovician		Extinctions of a variety of marine invertebrates near end of Ordovician. Major adaptive radiation of all invertebrate groups. Suspension feeders dominant.	Ostracoderms diversify.
505	Cambrian		Many trilobites become extinct near end of Cambrian. Trilobites, brachiopods, and archaeocyathids are most abundant.	Earliest vertebrates—jawless fish called ostracoderms.
545				

Plants	Major Geologic Events
Gymnosperms diverse and abundant.	Formation of Pangaea. Alleghenian orogeny. Hercynian orogeny.
Coal swamps with flora of seedless vascular plants and gymnosperms.	Coal-forming swamps common. Formation of Ancestral Rockies. Continental glaciation in Gondwana.
Gymnosperms appear (may have evolved during Late Devonian).	Ouachita orogeny.
First seeds evolve. Seedless vascular plants diversify.	Widespread deposition of black shale. Antler orogeny. Acadian orogeny.
Early land plants—seedless vascular plants.	Caledonian orogeny. Extensive barrier reefs and evaporites.
Plants move to land?	Continental glaciation in Gondwana. Taconic orogeny.
	First Phanerozoic transgression (Sauk) onto North American craton.

16. How is the Carboniferous marine invertebrate community similar to the Silurian marine community?

17. Discuss the similarities and differences between the Carboniferous and Permian marine invertebrate communities.

18. Discuss some of the possible causes for the Permian mass-extinction event.

Points to Ponder

1. Discuss how the incompleteness of the fossil record may play a role in such theories as the Cambrian explosion of life.

2. If the Cambrian explosion of life was partly the result of the filling of unoccupied niches, why don't we see such rapid evolution following mass-extinction events such as occurred at the end of the Permian and Cretaceous periods?

3. Discuss how changing geologic conditions affected the evolution of life during the Paleozoic Era.

Additional Readings

Ausich, W. I., and N. G. Lane. 1999. *Life of the past.* 4th ed. Upper Saddle River, N.J.: Prentice Hall.

Conway Morris, S. 1998. *The crucible of creation—The Burgess Shale and the rise of animals.* New York: Oxford University Press.

Cowen, R. 1995. *History of life.* 2d ed. Cambridge, Mass.: Blackwell Scientific.

Donovan, S. K., ed. 1989. *Mass extinctions.* New York: Columbia University Press.

Doyle, P. 1996. *Understanding fossils.* Chichester, England: Wiley.

Droser, M. L., R. Fortey, and Xing Li. 1996. The Ordovician radiation. *American Scientist* 84, no. 2: 122–131.

Erwin, D. E. 1993. *The great Paleozoic crisis: Life and death in the Permian.* New York: Columbia University Press.

Erwin, D. E. 1996. The mother of mass extinctions. *Scientific American* 275, no. 1: 72–78.

Erwin, D., J. Valentine, and D. Jablonski. 1997. The origin of animal body plans. *American Scientist* 85, no. 2: 126–137.

Gore, R. 1993. The Cambrian Period: Explosion of life. *National Geographic* 184, no. 4: 38–45.

Gould, S. J. 1989. *Wonderful life.* New York: Norton.

Levinton, J. S. 1992. The big bang of animal evolution. *Scientific American* 267, no. 5: 84–93.

McGhee, G. R., Jr. 1995. *The Late Devonian mass extinction: The Frasnian/Famennian crisis.* New York: Columbia University Press.

McMenamin, M. A. S., and D. L. Schulte McMenamin. 1990. *The emergence of animals: The Cambrian breakthrough.* New York: Columbia University Press.

Prothero, D. R. 1998. *Bringing fossils to life.* New York: WCB McGraw-Hill.

Reid, M. 1992. Ghosts of the Burgess Shale. *Earth* 1, no. 5: 38–45.

Whittington, H. B. 1985. *The Burgess Shale.* New Haven, Conn.: Yale University Press.

World Wide Web Activities

► KEVIN'S TRILOBITE HOME PAGE

This site contains information, photos, and line drawings of trilobites. Along with a Table of Contents, it has links to other paleontology sites, a trilobite literature section, trilobite collectors and specialists, and trilobite classification—in other words, just about anything you would want to know about trilobites. It is maintained by Kevin Brett at the University of Alberta, Alberta, Canada.

1. Click on the *Trilobite Papers VII* site. What are the current areas of research concerning trilobites?

2. In scrolling through the images of trilobites from the various Paleozoic periods, are there any general evolutionary trends you can discern from these images?

► ILLINOIS STATE MUSEUM MAZON CREEK FOSSILS

This site contains some of the more interesting and dramatic types of fossils recovered from the Francis Creek Shale in Illinois.

1. Click on the *Where Are Mazon Creek Fossils Found* site. Where can Mazon Creek fossils be found? How did they form? What are the different types of fossils recovered from the Francis Creek Shale? Click on the names of some of these fossils to see what they looked like.

2. Click on *The Importance of Mazon Creek Fossils* site. Why are these fossils important?

► UNIVERSITY OF CALIFORNIA MUSEUM OF PALEONTOLOGY

This is an excellent site to visit for any aspect of geologic time, paleontology, and evolution.

1. Click on the *On-line Exhibits* to go to the *Paleontology Without Walls* home page, which is an introduction to the UCMP Virtual Exhibits. Click on the *Phylogeny* icon; this takes you to the *Phylogeny of Life* home page. Click on the *Metazoa, All Animals* icon; this takes you to the Introduction to the *Metazoa* home page. From here you can click on one of four sites to learn more about the life history and ecology of any animal group, its morphology or systematics, and its fossil record. Pick a fossil group and find out as much as you can about that group. For example, learn about trilobites and compare the information at this site with that presented in Kevin's Trilobite Home Page.

2. Click on *The Biosphere, All Life* icon on *The Phylogeny of Life* home page; this takes you to the *Three Domains of Life* home page. Click on the *Eucaryota* icon; this takes you to the *Introduction to the Eukaryota* home page. Click on the *Animals* site; this takes you to the *Introduction to the Metazoa* home page. Click on the *Fossil Record* icon; at this site you can then learn more about the fossil history of the various metazoans. Check out any group you want to learn more about.

13

Life of the Paleozoic Era: Vertebrates and Plants

Carboniferous sandstone casts of *Stigmaria* tree stumps are preserved in Glasgow's Victoria Park, Scotland. These casts were formed when an influx of sand filled the interior of the rotting trunks. The bark was later altered to thin coal layers.

Prologue

Palynology is the study of organic microfossils called *palynomorphs*. These include such familiar things as spores and pollen (both of which cause allergies for many people), but also such unfamiliar organisms as acritarchs (see Chapter 12), dinoflagellates (marine and freshwater single-celled phytoplankton, some species of which are responsible for red tides), chitinozoans (vase-shaped microfossils of unknown origin), scolecodonts (jaws of marine annelid worms), and microscopic colonial algae. Fossil palynomorphs are extremely resistant to decay and are extracted from sedimentary rocks by dissolving the rocks in various acids.

A specialty of palynology that attracts many biologists and geologists is the study of spores and pollen (Fig. 13.1). By examining the fossil spores and pollen preserved in sedimentary rocks, palynologists can tell when plants colonized Earth's surface, which, in turn, influenced weathering and erosion rates, soil formation, and changes in the composition of atmospheric gases. Furthermore, because plants are not particularly common as fossils, the study of spores and pollen can frequently reveal the time and region for the origin and extinction of various plant groups. As we will see in this chapter, the earliest known fossil plants are Middle Silurian in age, yet there is a palynologic record dating back to the Middle Ordovician.

Analysis of fossil spores and pollen can be used to solve many geologic and biologic problems. One of the more important uses of fossil spores and pollen is determining the geologic age of sedimentary rocks. Because spores and pollen are microscopic, resistant to decay, deposited in both marine and terrestrial environments, extremely abundant, and are part of the life cycle of plants (many of which evolve rapidly), they are very useful for dating. Many spore and pollen species have narrow time ranges that make them excellent guide fossils. Frequently, what are otherwise labeled unfossiliferous rocks by paleontologists looking only for megafossils, contain thousands, even millions, of fossil spores or pollen grains that allow palynologists to date the so-called unfossiliferous rock.

Fossil spores and pollen can also be used to correlate rocks both regionally and globally. This is particularly useful in trying to correlate marine and terrestrial sediments, which typically do not contain the same fossils. Because plants live on land, their spores and pollen are usually deposited on land. However, some spores and pollen may be dispersed by the wind over marine environments or carried by streams into the sea where they are deposited and may eventually become fossils, thus providing a link between the marine and terrestrial realms.

Fossil spores and pollen are also useful in determining the environment and climate in the past. Their presence in sedimentary rocks allows palynologists to determine what plants and trees were living at the time, even if the fossils of these plants and trees are not preserved. Plants are very sensitive to climatic changes, and by plotting the abundance and types of vegetation present, based on their preserved spores and pollen, palynologists can determine past climates and changes in climates through time.

The study of spores and pollen provides a tremendous amount of information about the vegetation in the past, its evolution, the type of climate, and changes in climate through time. In addition, spores and pollen are very useful for dating rocks and correlating marine and terrestrial rocks, both regionally and globally.

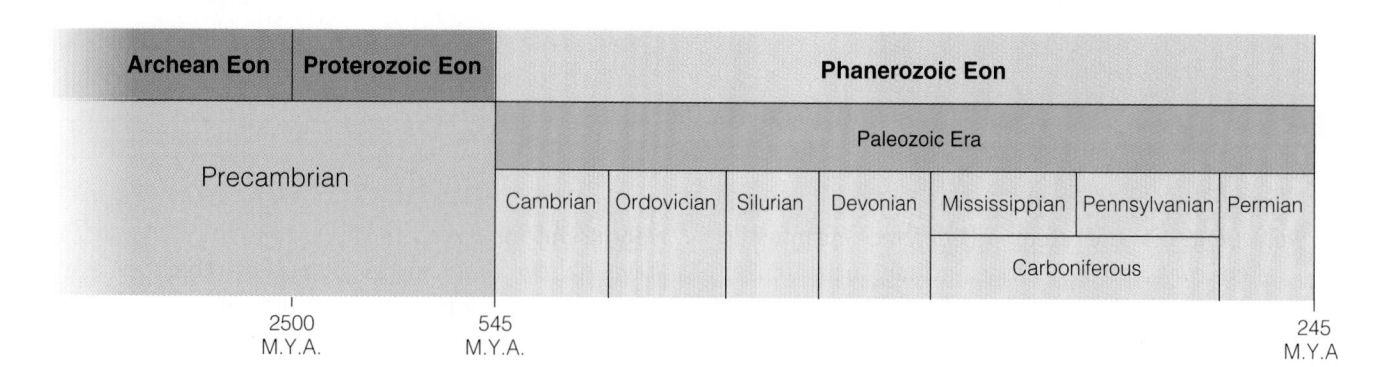

Archean Eon	Proterozoic Eon	Phanerozoic Eon						
Precambrian		Paleozoic Era						
		Cambrian	Ordovician	Silurian	Devonian	Mississippian	Pennsylvanian	Permian
						Carboniferous		

2500 M.Y.A.   545 M.Y.A.   245 M.Y.A

Introduction

In the previous chapter, we examined the Paleozoic history of invertebrates, beginning with the acquisition of hard parts and concluding with the massive Permian extinction event, in which about 90% of all invertebrates and more than 65% of all amphibians and reptiles died out. In this chapter we examine the Paleozoic evolutionary history of plants and vertebrates, which are those animals with a segmented vertebral column.

Vertebrates first evolved in the sea, and their earliest fossil record consists of phosphatic plate and scale fragments from Late Cambrian–age rocks. By the end of the Devonian, all major fish groups had evolved. During the Devonian, one group of fish evolved into amphibians, beginning the colonization of land by these animals. Before the end of the Paleozoic, reptiles had evolved from amphibians and had become the dominant group of terrestrial vertebrates.

Plants preceded animals onto the land, probably sometime during the Ordovician. In making the transition from water to land, both plants and animals had to solve the same basic problems. For both groups, the method of reproduction proved to be the major barrier to expansion into the various terrestrial environments. With the evolution of the seed in plants and the amniote egg in animals, this limitation was removed, and both groups were able to expand into all the terrestrial habitats.

The end of the Paleozoic Era was a time of major extinctions. Not only was the marine invertebrate community greatly decimated, but many amphibians and reptiles also became extinct.

Vertebrate Evolution

A **chordate** (phylum Chordata) is an animal that has, at least during part of its life cycle, a notochord, a dorsal hollow nerve cord, and gill slits (Figure 13.2). **Vertebrates,** which are animals with backbones, are simply a subphylum of chordates.

The ancestors and early members of the phylum Chordata were soft-bodied organisms that left few fossils. Recent discoveries from Chengjiang, Yunnan province, China, however, have pushed the evolution of the earliest chordates back to only a few million years after the Cambrian explosion (see Chapter 12). Until recently, *Yunnanozoon lividum*, a 5 cm-long animal was thought to be the oldest known chordate (Figure 13.3). That claim has been challenged by the discovery of yet another fossil *(Cathaymyrus diadexus)* from Chengjiang, a 2.2-cm-long animal that has several key chordate fea-

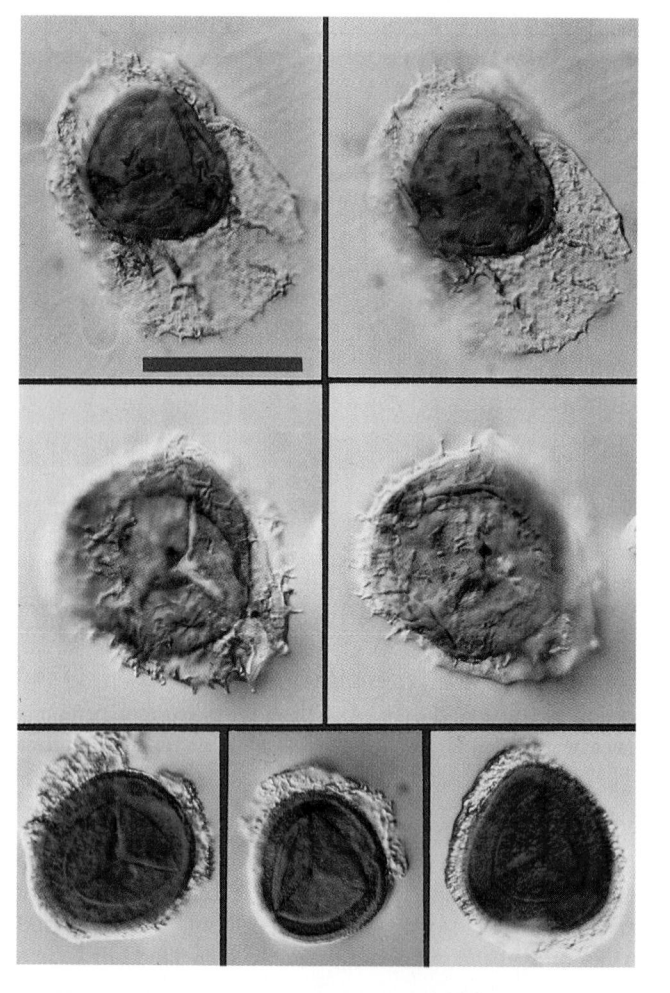

FIGURE 13.1 Three species of spores from the Middle Devonian Rapid Member of the Cedar Valley Formation, Iowa. These spores are characteristic of the Middle-Late Devonian and can be used for correlation with other rock units of the same age and for interpreting paleoenvironments.

Phanerozoic Eon											
Mesozoic Era			Cenozoic Era								
Triassic	Jurassic	Cretaceous	Tertiary						Quaternary		
			Paleocene	Eocene	Oligocene	Miocene	Pliocene		Pleistocene	Holocene	

245
M.Y.A.

66
M.Y.A.

tures. In fact, some scientists now claim that *Yunnanozoon lividum* is actually a hemichordate, a different but closely related phylum that includes the acorn worms.

Although the hemichordates and chordates appear quite different today, they share, along with the echinoderms (see Table 12.1), the same distinctive style of embryo development, suggesting a common ancestry (Figure 13.4). Furthermore, the biochemistry of muscle activity, blood proteins, and the larval stages are similar in all three phyla.

The evolutionary pathway to vertebrates thus appears to have taken place much earlier and more rapidly than many scientists have long thought. Based on fossil evidence and recent advances in molecular biology, one scenario suggests that vertebrates evolved shortly after an ancestral chordate, probably resembling *Cathaymyrus* or *Yunnanozoon,* acquired a second set of genes. According to this hypothesis, a random mutation produced a duplicate set of genes, allowing the ancestral vertebrate animal to evolve entirely new body structures, which proved to be evolutionarily advantageous. Not all scientists accept this hypothesis, and the evolution of vertebrates is still hotly debated.

Fish

The most primitive vertebrates are fish, and the oldest fish remains are found in the Upper Cambrian Deadwood Formation in northeastern Wyoming (Figure 13.5). Here phosphatic scales and plates of *Anatolepis,* a primitive

member of the class Agnatha (jawless fish) have been recovered from marine sediments. All known Cambrian and Ordovician fossil fish have been found in shallow, nearshore marine deposits, while the earliest nonmarine fish remains have been found in Silurian strata. This does not prove that fish originated in the oceans, but it does lend strong support to the idea.

As a group, fish range from the Late Cambrian to the present (Figure 13.6). The oldest and most primitive of the class Agnatha are the **ostracoderms,** whose name means "bony skin" (Table 13.1). These are armored jawless fish that first evolved during the Late Cambrian, reached their zenith during the Silurian and Devonian, and then became extinct.

The majority of ostracoderms lived on the seafloor. *Hemicyclaspis* is a good example of a bottom-dwelling ostracoderm (Figure 13.7a). Vertical scales allowed *Hemicyclaspis* to wiggle sideways, propelling itself along the seafloor, while the eyes on the top of its head allowed it to see such predators as cephalopods and jawed fish approaching from above. While moving along the sea bottom, it probably sucked up small bits of food and sediments through its jawless mouth. Another type of ostracoderm, represented by *Pteraspis,* was more elongated and probably an active swimmer, although it also seemingly fed on small pieces of food it could suck up.

The evolution of jaws was a major evolutionary advance among primitive vertebrates. While their jawless ancestors could only feed on detritus, jawed fish could chew food and become active predators, thus opening many new eco-

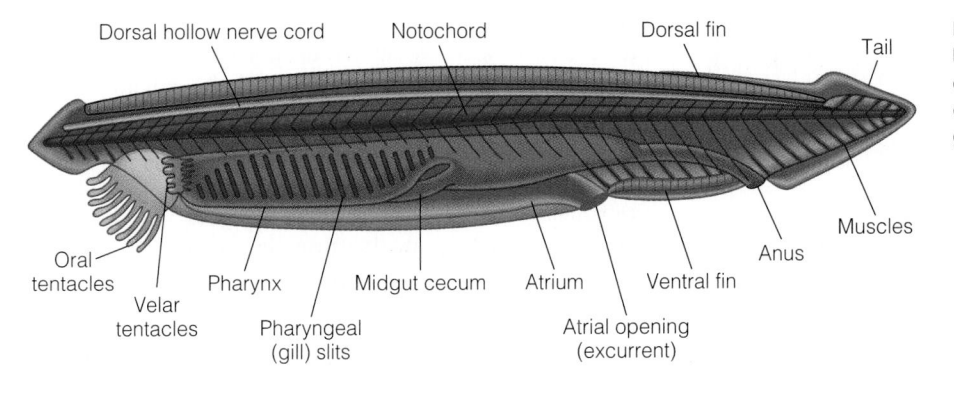

FIGURE 13.2 The structure of the lancelet *Amphioxus* illustrates the three characteristics of a chordate: a notochord, a dorsal hollow nerve cord, and gill slits.

Dorsal hollow nerve cord   Notochord   Dorsal fin   Tail

Oral tentacles   Velar tentacles   Pharynx   Pharyngeal (gill) slits   Midgut cecum   Atrium   Atrial opening (excurrent)   Ventral fin   Anus   Muscles

FIGURE 13.3 Found in 525-million-year-old rocks in Yunnan province, China, *Yunnanozoon lividum,* a 5 cm-long animal is one of the oldest known chordates.

FIGURE 13.5 A fragment of a plate from *Anatolepis* cf. *A. heintzi* from the Upper Cambrian Deadwood Formation of Wyoming. *Anatolepis* is the oldest known fish.

logic niches. The vertebrate jaw is an excellent example of evolutionary opportunism. Various studies suggest that the jaw originally evolved from the first three gill arches of jawless fish. Because the gills are soft, they are supported by gill arches composed of bone or cartilage. The evolution of the jaw may thus have been related to respiration rather than feeding (Figure 13.8). By evolving joints in the forward gill arches, jawless fish could open their mouths wider. Every time a fish opened and closed its mouth, it would pump more water past the gills, thereby increasing the oxygen intake. The modification from rigid to hinged forward gill arches enabled fish to increase both their food consumption and oxygen intake, and the evolution of the jaw as a feeding structure rapidly followed.

The remains of the first jawed fish, called **acanthodians** (Figure 13.7c, Table 13.1), are found in Lower Silurian nonmarine rocks. Acanthodians are an enigmatic group of fish characterized by large spines, scales covering much of the body, jaws, teeth, and reduced bony armor. Their relationship to other fish has not been well established. The

acanthodians were most abundant during the Devonian, declined in importance through the Carboniferous, and became extinct during the Permian.

While we do not know how the acanthodians were related to other, more complex fish, we do know that during the Devonian a major adaptive radiation of jawed fish took place. **Placoderms** (Table 13.1), whose name means "plate-skinned," evolved during the Late Silurian and, like acanthodians, reached their peak of abundance and diversity during the Devonian. Placoderms were heavily armored jawed fish that lived in both freshwater and the ocean. The placoderms exhibited considerable variety, including small bottom dwellers (Figure 13.7b) as well as large major predators such as *Dunkleosteus,* a Late Devonian fish that lived in the mid-continental North American epeiric seas (Figure 13.9a). It was by far the largest fish of the time, attaining a length of more than 12 m. It had a heavily armored head and shoulder region, a huge jaw lined with razor-sharp bony teeth, and a flexible tail, all features consistent with its status as a ferocious predator.

Besides the abundant acanthodians, placoderms, and ostracoderms, other fish groups, such as the cartilaginous and bony fish, also evolved during the Devonian Period. It is small wonder, then, that the Devonian is informally called the "Age of Fish," because all major fish groups were present during this time period.

Cartilaginous fish, class Chrondrichthyes (Table 13.1), represented today by sharks, rays, and skates, first evolved during the Middle Devonian, and by the Late Devonian, primitive marine sharks such as *Cladoselache* were quite abundant (Figure 13.9b). Cartilaginous fish have never been as numerous nor as diverse as their cousins, the bony fish, but they were, and still are, important members of the marine vertebrate fauna.

Along with cartilaginous fish, the **bony fish,** class Osteichthyes (Table 13.1), also first evolved during the Devonian. Because bony fish are the most varied and numerous of all fish and because the amphibians evolved from them, their evolutionary history is particularly important. There are two groups of bony fish: the common **ray-finned fish** (Figure 13.9d) and the less familiar **lobe-finned fish** (Table 13.1).

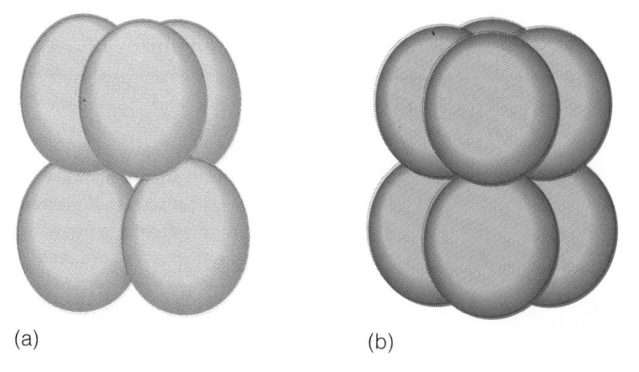

(a)   (b)

FIGURE 13.4 (a) Arrangement of cells resulting from spiral cleavage. In this arrangement, cells in successive rows are nested between each other. Spiral cleavage is characteristic of all invertebrates except echinoderms. (b) Arrangement of cells resulting from radial cleavage is characteristic of chordates and echinoderms. In this configuration, cells are directly above each other.

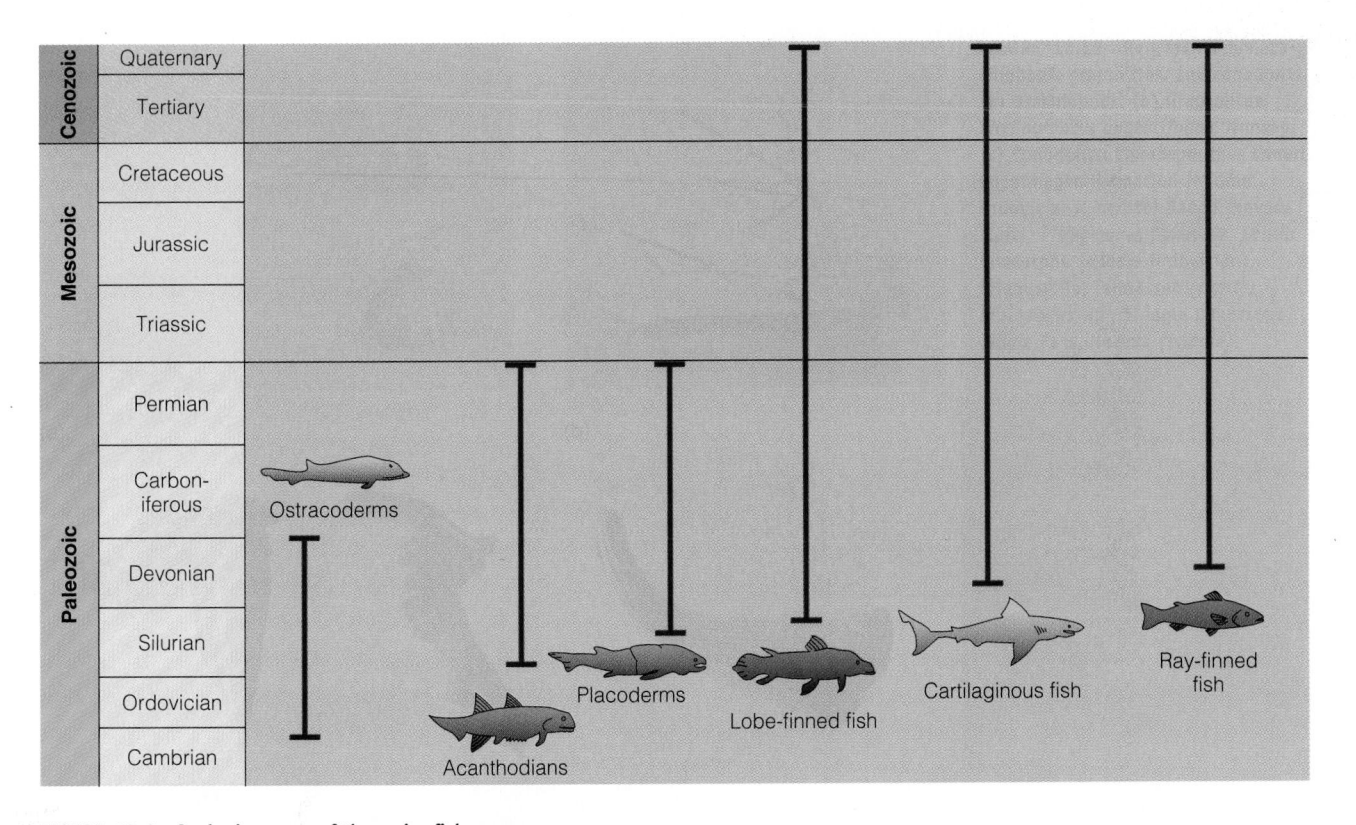

FIGURE 13.6 Geologic ranges of the major fish groups.

The term *ray-finned* refers to the way the fins are supported by thin bones that spread away from the body (Figure 13.10a). From a modest freshwater beginning during the Devonian, ray-finned fish, which include most of the familiar fish such as trout, bass, perch, salmon, and tuna, rapidly diversified to dominate the Mesozoic and Cenozoic seas.

Present-day lobe-finned fish are characterized by muscular fins. The fins do not have radiating bones but rather articulating bones with the fin attached to the body by a fleshy shaft (Figure 13.10b). Two major groups of lobe-

finned fish are recognized: *lungfish* and *crossopterygians* (Table 13.1). Lungfish were fairly abundant during the Devonian, but today only three freshwater genera exist, one each in South America, Africa, and Australia. Their present-day distribution presumably reflects the Mesozoic breakup of Gondwana.

Studies of present-day lung fish indicate that lungs evolved from saclike bodies on the ventral side of the esophagus. These saclike bodies enlarged and improved their capacity for oxygen extraction, eventually evolving into lungs. When the

---

TABLE 13.1

Brief Classification of Fish Showing the Groups Referred to in the Text

CLASSIFICATION	GEOLOGIC RANGE	LIVING EXAMPLE
Class Agnatha (jawless fish)	Late Cambrian-Recent	Lamprey, hagfish
Early members of the class are called ostracoderms		No living ostracoderms
Class Acanthodii (the first fish with jaws)	Early Silurian-Permian	None
Class Placodermii (armored jawed fish)	Late Silurian-Permian	None
Class Chondrichthyes (cartilagenous fish)	Devonian-Recent	Sharks, rays, skates
Class Osteichthyes (bony fish)	Devonian-Recent	Tuna, perch, bass, pike, catfish, trout, salmon, lungfish, *Latimeria*
Subclass Actinopterygii (ray-finned fish)	Devonian-Recent	Tuna, perch, bass, pike, catfish, trout, salmon
Subclass Sarcopterygii (lobe-finned fish)	Devonian-Recent	Lungfish, *Latimeria*
Order Dipnoi	Devonian-Recent	Lungfish
Order Crossopterygii	Devonian-Recent	*Latimeria*
Suborder Rhipidistia	Devonian-Permian	None

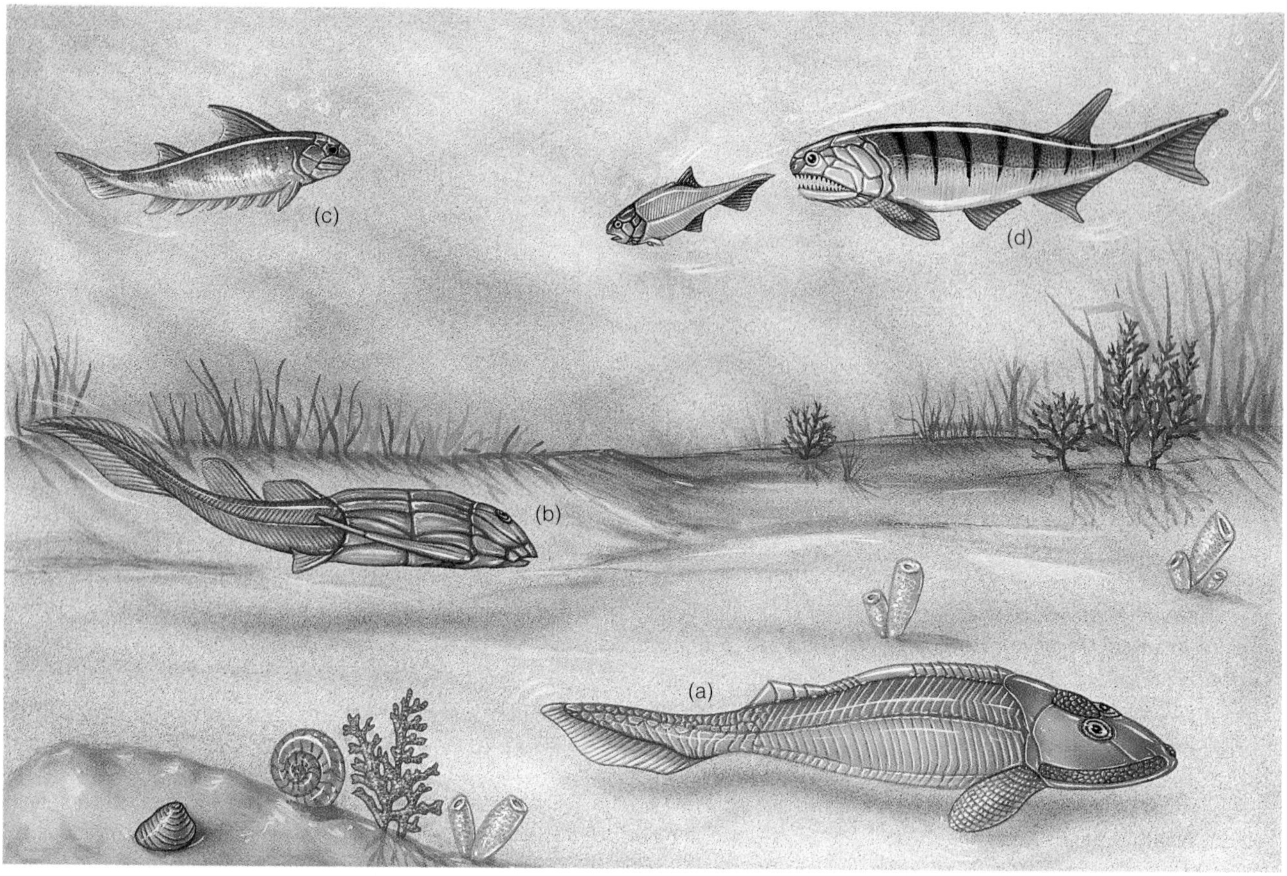

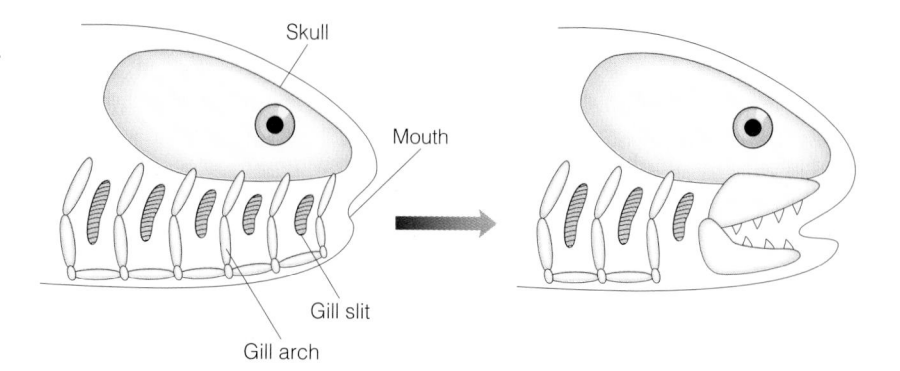

FIGURE 13.7 Recreation of a Devonian seafloor showing (a) an ostracoderm (*Hemicyclaspis*), (b) a placoderm (*Bothriolepis*), (c) an acanthodian (*Parexus*), and (d) a ray-finned fish (*Cheirolepis*).

lakes or streams in which lungfish live become stagnant and dry up, they breathe at the surface or burrow into the sediment to prevent dehydration. When the water is well oxygenated, however, lungfish rely upon gill respiration.

The **crossopterygians** are an important group of lobe-finned fish because amphibians evolved from them. During the Devonian, two separate branches of crossopterygians evolved. One led to the amphibians, while the other invaded the sea. This latter group, the *coelacanths,* were thought to have become extinct at the end of the Cretaceous. In 1938, however, fisherman caught a coelacanth in the deep waters off Madagascar (see Figure 5.19), and since then several dozen more have been caught.

The group of crossopterygians that is ancestral to amphibians are *rhipidistians* (Table 13.1). These fish, attaining lengths of over 2 m, were the dominant freshwater predators during the Late Paleozoic. *Eusthenopteron*, a good example of a rhipidistian crossopterygian, had an elongate body that enabled it to move swiftly in the water, as well as paired muscular fins that could be used for locomotion on land (Figure 13.11). The structural similarity between crossopterygian fish and the earliest amphibians is striking and one of the better documented transitions from one major group to another (Figure 13.12).

Before discussing this transition and the evolution of amphibians, it would be useful to place the evolutionary

FIGURE 13.8 The evolution of the vertebrate jaw is thought to have occurred from the modification of the first two or three anterior gill arches. This theory is based on the comparative anatomy of living vertebrates.

FIGURE 13.9 A Late Devonian marine scene from the midcontinent of North America. (a) The giant placoderm *Dunkleosteus* (length more than 12 m) is pursuing (b) the shark *Cladoselache* (length up to 1.2 m). Also shown are (c) the bottom-dwelling placoderm *Bothriolepis* and (d) the swimming ray-finned fish *Cheirolepis,* both of which attained a length of 40–50 cm.

history of Paleozoic fish in the larger context of Paleozoic evolutionary events. Certainly, the evolution and diversification of jawed fish as well as eurypterids and ammonoids had a profound effect on the marine ecosystem. Previously, defenseless organisms either evolved defensive mechanisms or suffered great losses, possibly even extinction. Recall from Chapter 12 that trilobites experienced major extinctions at the end of the Cambrian, recovered slightly during the Ordovician, then declined greatly from the end of the Ordovician to their ultimate demise at the end of the Permian. Perhaps their lightly calcified external covering made them easy prey for the rapidly evolving jawed fish and cephalopods. Ostracoderms, although armored, would also have been easy prey for the swifter jawed fishes. Ostracoderms became extinct by the end of the Devonian, a time that coincides with the rapid evolution of jawed fish. Placoderms also became extinct by the end of the Devonian, while acanthodians decreased in abundance after the Devonian and became extinct by the end of the Paleozoic Era. On the other hand, cartilaginous and ray-finned bony fish expanded during the Late Paleozoic, as did the ammonoid cephalopods, the other major predator of the Late Paleozoic seas.

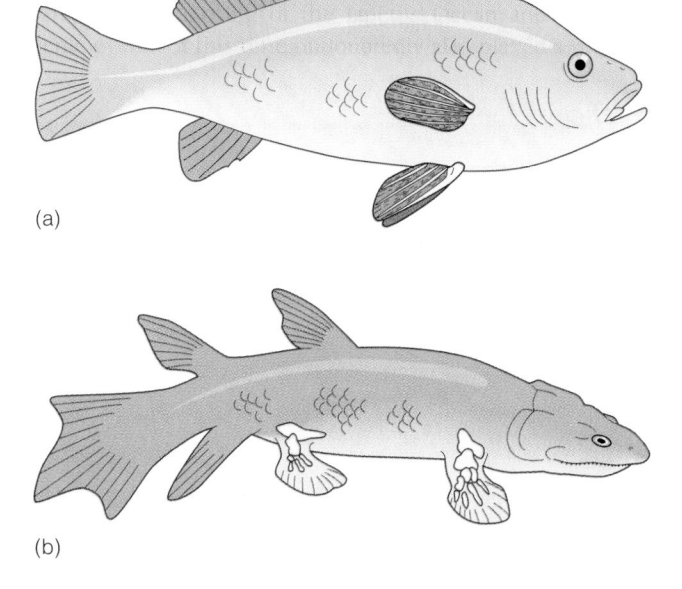

FIGURE 13.10 Arrangement of fin bones for (a) a typical ray-finned fish and (b) a lobe-finned fish. The muscles extend into the fin of the lobe-finned fish, allowing greater flexibility of movement than for the ray-finned fish.

Amphibians—Vertebrates Invade the Land

Although amphibians were the first vertebrates to live on land, they were not the first land-living organisms. Land plants, which probably evolved from green algae, first evolved during the Ordovician. Furthermore, insects, millipedes, spiders, and even snails invaded the land before amphibians (see Perspective 13.1). Fossil evidence indicates that such land-dwelling arthropods as scorpions and flightless insects had evolved by at least the Devonian.

FIGURE 13.11 *Eusthenopteron,* a member of the rhipidistian crossopterygians. The crossopterygians are the group from which the amphibians are thought to have evolved. *Eusthenopteron* had an elongate body and paired fins that could be used for moving about on land.

FIGURE 13.12 Similarities between the crossopterygian lobe-finned fish and the labyrinthodont amphibians. (a) Skeletal similarity. (b) Comparison of the limb bones of a crossopterygian *(left)* and amphibian *(right)*; color identifies the bones (u = ulna, shown in blue, r = radius, mauve, h = humerus, gold) that the two groups have in common. (c) Comparison of tooth cross sections shows the complex and distinctive structure found in both the crossopterygians *(left)* and amphibians *(right)*.

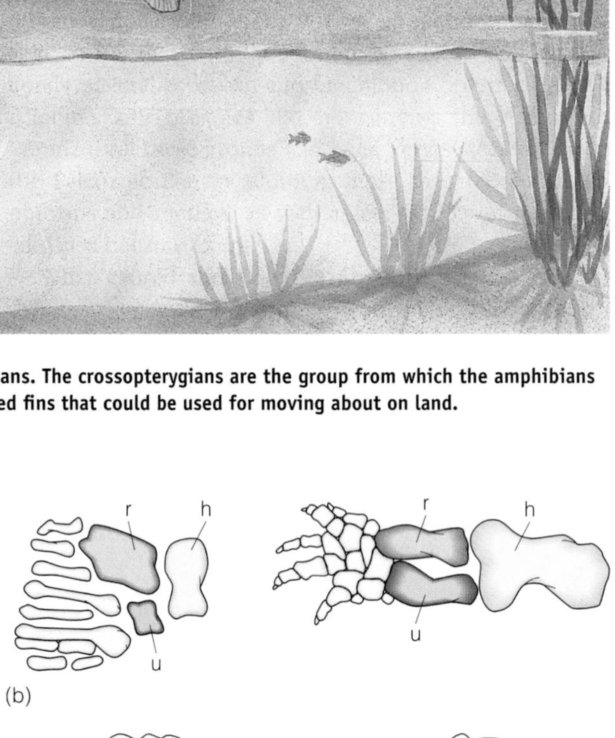

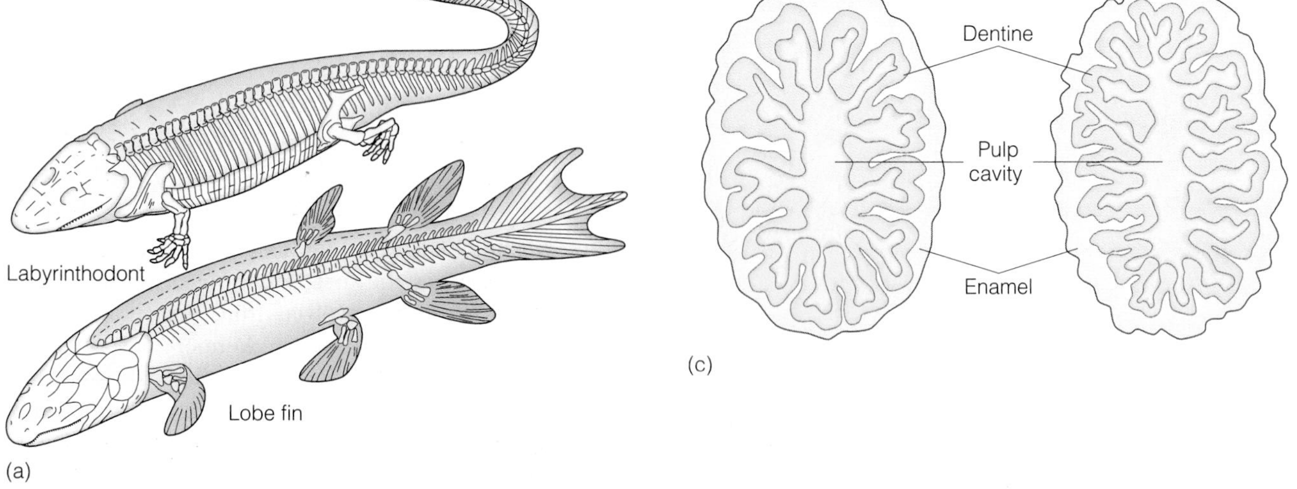

perspective 13.1

An Ordovician Invasion of Land

*T*he question of when land was first colonized has been much discussed. There is excellent fossil evidence that land was colonized by vascular plants during the Silurian Period. Fragments of plant remains have been found in Middle Silurian-age rocks, indicating primitive plants had invaded land by that time. These early plants, like their Devonian descendants, had to live near bodies of water due to their reproductive requirements.

While Silurian plant-fossil evidence is represented by a handful of geographically scattered fossils, the spore evidence for a Silurian land colonization is much more abundant and diverse. This may be due in part to the greater chance for preservation of spores, which have a resistant organic covering. In the North Atlantic region, Silurian vascular plant megafossil evidence is based primarily on specimens of *Cooksonia* (see Figure 13.20). Yet at least 14 spore genera are described for the Silurian Period.

Discoveries of probable vascular plant megafossils and characteristic spores indicate to many paleontologists that the evolution of vascular plants occurred well before the Middle Silurian. Sheets of cuticle-like cells—that is, the cells that cover the surface of modern land plants—and tetrahedral clusters that closely resemble the spore tetrahedrals of primitive land plants have been reported from Middle to Late Ordovician-age rocks from various areas of the world (Figure 1).

If land plants existed during the Ordovician, when did animals invade land? The first vertebrate animals made the transition from water to land during the Late Devonian. Terrestrial arthropods, however, evolved at least by the Late Silurian as indicated by their fossil record of arachnids and centipedes. Mites, arachnids, and small insects have been reported from the Lower Devonian Rhynie Chert of Scotland. This formation is also famous for its beautifully preserved fossil-plant flora. These earliest arthropods formed a terrestrial ecosystem consisting of herbivores, predators, and primitive early land plants.

The discovery of fossil burrows within a buried soil of the Upper Ordovician Juanita Formation in central Pennsylvania represents the oldest reported nonmarine trace fossils. While no hard parts or traces of organisms are associated with the burrows, their size, shape, and arrangement are consistent with a bilaterally symmetric burrower that was resistant to desiccation. It is hypothesized that these burrows may have been made by millipedes. Millipedes first appear in marine rocks of Early Silurian age, and some present-day millipedes are active burrowers.

The existence of sizable burrowing organisms on dry land during the Late Ordovician implies that some type of terrestrial vegetation existed for them to feed on. Burrowing animals could conceivably have fed on soil algae, now thought to have been in existence since the Late Precambrian. The discovery of probable primitive land plant spores from Middle to Upper Ordovician rocks means that these plants could have supported large populations of burrowing herbaceous arthropods as well as litter organisms. The discovery of Late Ordovician trace fossils in paleosoils adds to our increasing knowledge of Early Paleozoic terrestrial ecosystems.

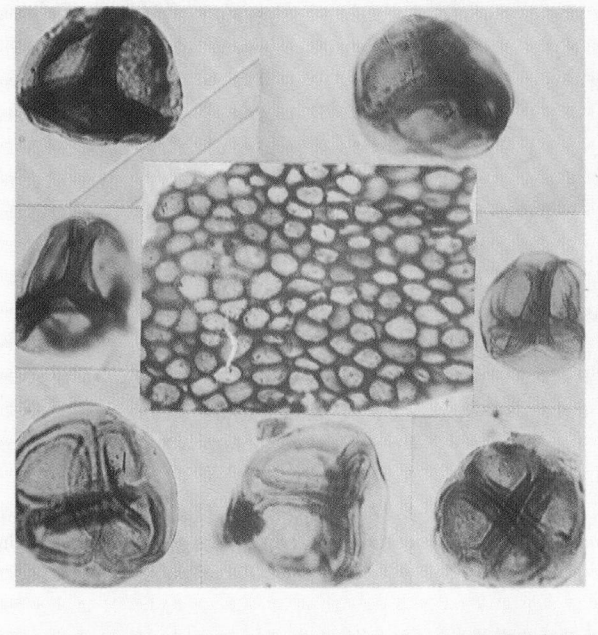

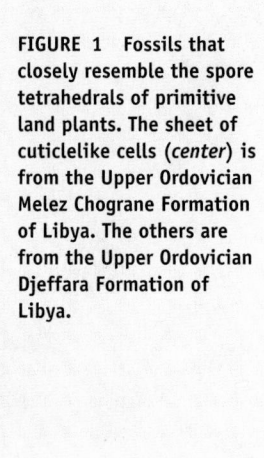

FIGURE 1 Fossils that closely resemble the spore tetrahedrals of primitive land plants. The sheet of cuticlelike cells (*center*) is from the Upper Ordovician Melez Chograne Formation of Libya. The others are from the Upper Ordovician Djeffara Formation of Libya.

FIGURE 13.13 A Late Devonian landscape in the eastern part of Greenland. Shown is *Ichthyostega,* an amphibian that grew to a length of about 1 m. The flora of the time was diverse, consisting of a variety of small and large seedless vascular plants.

The transition from water to land required that several barriers be surmounted. The most critical for animals were desiccation, reproduction, the effects of gravity, and the extraction of oxygen from the atmosphere by lungs rather than from water by gills. These problems were partly solved by the crossopterygians; they already had a backbone and limbs that could be used for walking and lungs that could extract oxygen (Figure 13.12).

The oldest amphibian fossils are found in the Upper Devonian Old Red Sandstone of eastern Greenland. These

FIGURE 13.14 Reconstruction of a Carboniferous coal swamp. The varied amphibian fauna of the time is shown, including the large labyrinthodont amphibian *Eryops (foreground),* the larval *Branchiosaurus (center),* and the serpentlike *Dolichosoma (background).*

amphibians, which belong to genera like *Ichthyostega,* had streamlined bodies, long tails, and fins. In addition, they had four legs, a strong backbone, a rib cage, and pelvic and pectoral girdles, all of which were structural adaptations for walking on land (Figure 13.13). The earliest amphibians appear to have had many characteristics that were inherited from the crossopterygians with little modification (Figure 13.12).

Because amphibians did not evolve until the Late Devonian, they were a minor element of the Devonian terrestrial ecosystem. Like other groups that moved into new and previously unoccupied niches, amphibians underwent rapid adaptive radiation and became abundant during the Carboniferous and Early Permian.

The Late Paleozoic amphibians did not at all resemble the familiar frogs, toads, newts, and salamanders that make up the modern amphibian fauna. Rather they displayed a broad spectrum of sizes, shapes, and modes of life (Figure 13.14). One group of amphibians were the **labyrinthodonts,** so named for the labyrinthine wrinkling and folding of the chewing surface of their teeth (Figure 13.12c). Most labyrinthodonts were large animals, as much as 2 m in length. These typically sluggish creatures lived in swamps and streams, eating fish, vegetation, insects, and other small amphibians (Figure 13.14).

Labyrinthodonts were abundant during the Carboniferous when swampy conditions were widespread (see Chapter 11), but soon declined in abundance during the Permian, perhaps in response to changing climatic conditions. Only a few species survived into the Triassic.

Evolution of the Reptiles— The Land Is Conquered

Amphibians were limited in colonizing the land because they had to return to water to lay their gelatinous eggs. The evolution of the **amniote egg** (Figure 13.15) freed reptiles from this constraint. In such an egg, the developing embryo is surrounded by a liquid-filled sac called the *amnion* and provided with both a yolk, or food sac, and an allantois, or waste sac. In this way the emerging reptile is in essence a miniature adult, bypassing the need for a larval stage in the water. The evolution of the amniote egg allowed vertebrates to colonize all parts of the land because they no longer had to return to the water as part of their reproductive cycle.

Many of the differences between amphibians and reptiles are physiological and are not preserved in the fossil record. Nevertheless, amphibians and reptiles differ sufficiently in skull structure, jawbones, ear location, and limb and vertebral construction to suggest that reptiles evolved from labyrinthodont ancestors by the Late Mississippian. This assessment is based on the discovery of a well-preserved skeleton of the oldest known reptile, *Westlothiana,* from Late Mississippian–age rocks in Scotland.

Some of the oldest known reptiles are from the Lower Pennsylvanian Joggins Formation in Nova Scotia, Canada.

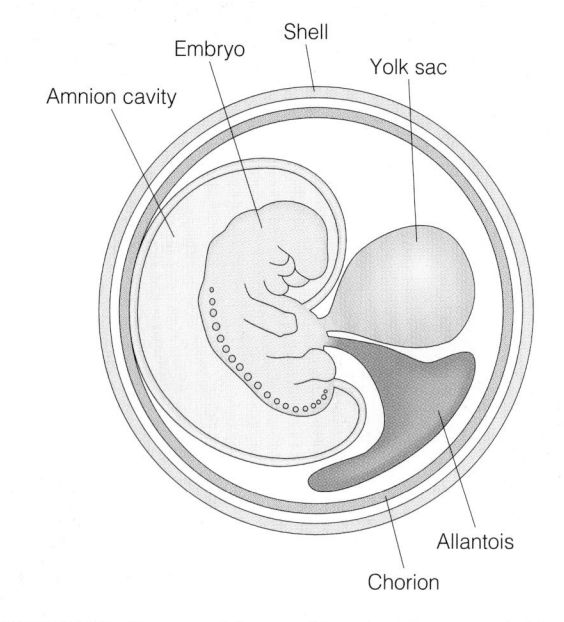

FIGURE 13.15 In an amniote egg, the embryo is surrounded by a liquid sac (amnion cavity) and provided with a food source (yolk sac) and waste sac (allantois). The evolution of the amniote egg freed reptiles from having to return to the water for reproduction and allowed them to inhabit all parts of the land.

Here, remains of *Hylonomus* are found in the sediments filling in tree trunks (Figure 13.16). These reptiles were small and agile and fed largely on grubs and insects. They are loosely grouped together as **protorothyrids,** whose members include the earliest reptiles (Figure 13.17). During the Permian Period, reptiles diversified and began displacing many amphibians. The success of the reptiles is due partly to their advanced method of reproduction and their more advanced jaws and teeth, as well as to their ability to move rapidly on land.

The **pelycosaurs,** or finback reptiles, evolved from the protorothyrids during the Pennsylvanian and were the

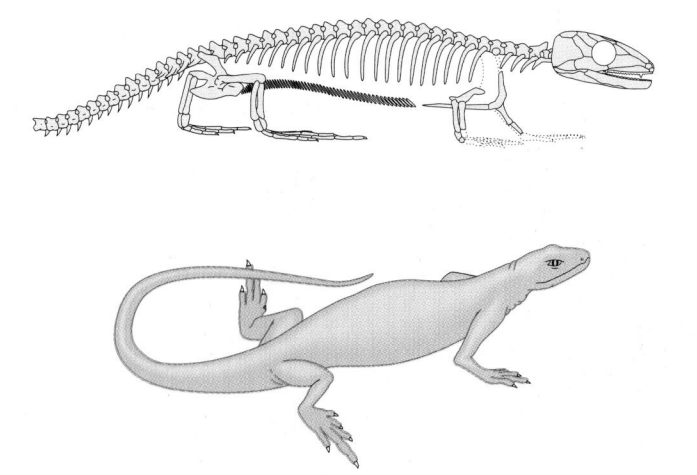

FIGURE 13.16 Reconstruction and skeleton of one of the oldest known reptiles, *Hylonomus lyelli* from the Pennsylvanian Period. Fossils of this animal have been collected from sediments that filled tree stumps. *Hylonomus lyelli* was about 30 cm long.

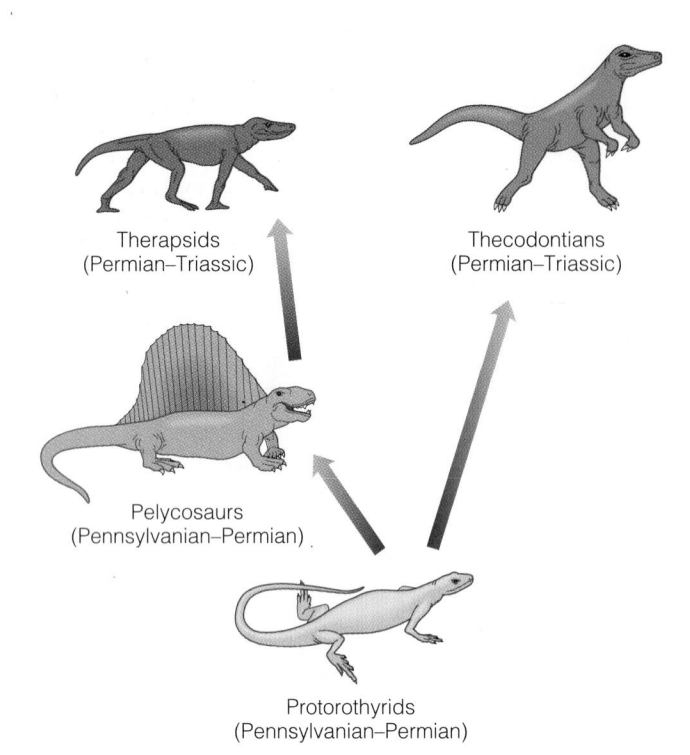

FIGURE 13.17 Evolutionary relationship among the Paleozoic reptiles.

Therapsids
(Permian–Triassic)

Thecodontians
(Permian–Triassic)

Pelycosaurs
(Pennsylvanian–Permian)

Protorothyrids
(Pennsylvanian–Permian)

dominant reptile group by the Early Permian. They evolved into a diverse assemblage of herbivores, exemplified by *Edaphosaurus,* and carnivores such as *Dimetrodon* (Figure 13.18). An interesting feature of the pelycosaurs is their sail. It was formed by vertebral spines that, in life, were covered with skin. The sail has been variously ex-

plained as a type of sexual display, a means of protection, and a display to look more ferocious. The current consensus seems to be that the sail served as some type of thermoregulatory device, raising the reptile's temperature by catching the sun's rays or cooling it by facing the wind. Because pelycosaurs are considered to be the group from which therapsids evolved, it is interesting that they may have had some sort of body-temperature control.

The pelycosaurs became extinct during the Permian and were succeeded by the **therapsids,** mammal-like reptiles that evolved from the carnivorous pelycosaur lineage and rapidly diversified into herbivorous and carnivorous lineages (Figure 13.19). Therapsids were small- to medium-sized animals displaying the beginnings of many mammalian features: fewer bones in the skull due to fusion of many of the small skull bones; enlargement of the lower jawbone; differentiation of the teeth for various functions such as nipping, tearing, and chewing food; and a more vertical position of the legs for greater flexibility, as opposed to the way the legs sprawled out to the side in primitive reptiles.

Furthermore, many paleontologists think therapsids were *endothermic,* or warm-blooded, enabling them to maintain a constant internal body temperature. This characteristic would have allowed them to expand into a variety of habitats, and indeed the Permian-age rocks in which their fossil remains are found have a wide latitudinal distribution.

As the Paleozoic Era came to an end, the therapsids constituted about 90% of the known reptile genera and occupied a wide range of ecologic niches. The mass extinctions that decimated the marine fauna at the close of the

FIGURE 13.18 Most pelycosaurs, or finback reptiles have a characteristic sail on their back. One hypothesis explains the sail as a type of thermoregulatory device. Other hypotheses are that it was a type of sexual display or a device to make the reptile look more intimidating. Shown here are (a) the carnivore *Dimetrodon* and (b) the herbivore *Edaphosaurus.*

FIGURE 13.19 A Late Permian scene in southern Africa showing various therapsids including *Dicynodon (left foreground)* and *Moschops (right)*. Many paleontologists think therapsids were endothermic and may have had a covering of fur as shown here.

TABLE 13.2

Major Events in the Evolution of Land Plants. The Devonian Period was a time of rapid evolution for the land plants. Major events were the appearance of leaves, heterospory, secondary growth, and the emergence of seeds.

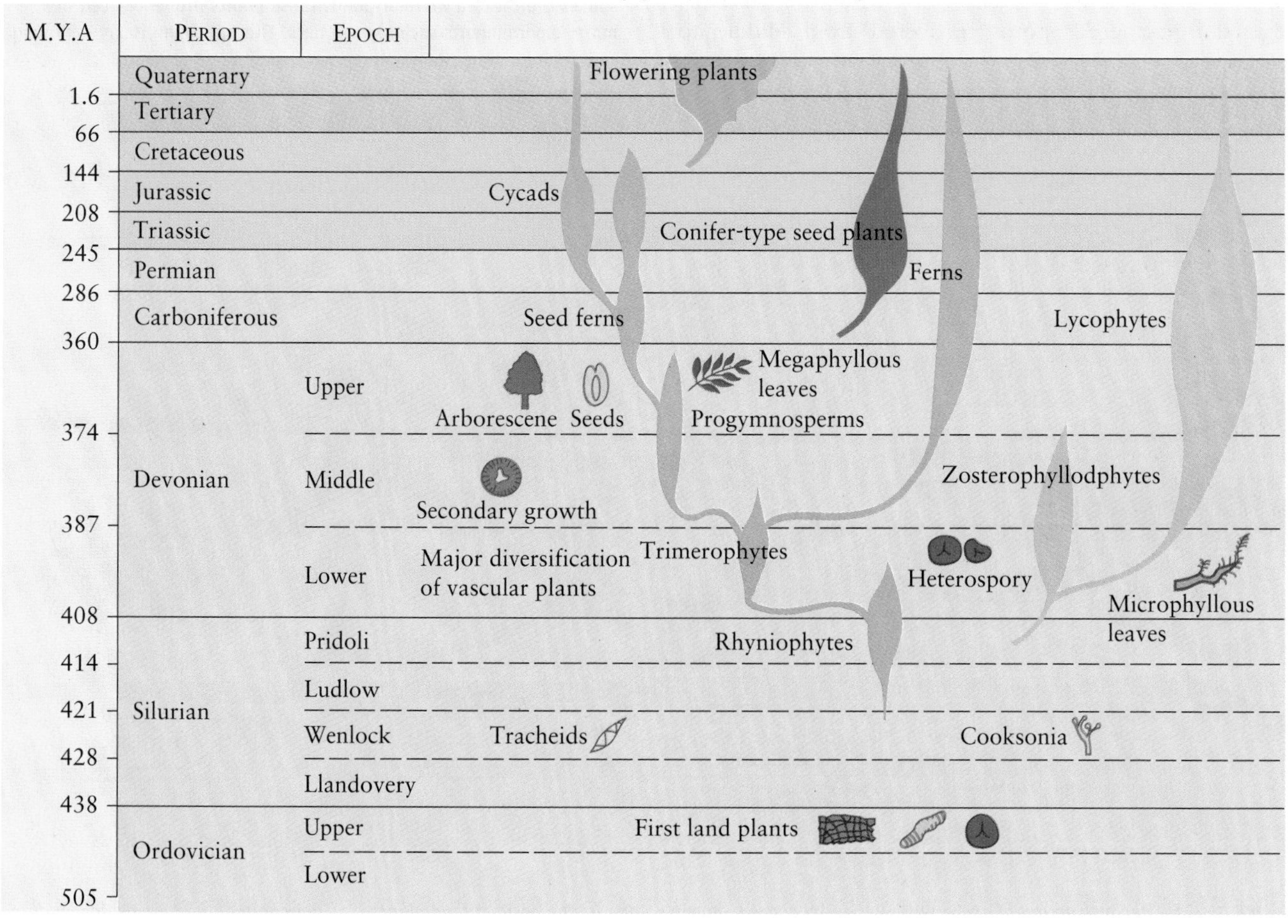

M.Y.A	PERIOD	EPOCH				
1.6	Quaternary			Flowering plants		
66	Tertiary					
144	Cretaceous					
208	Jurassic		Cycads			
245	Triassic			Conifer-type seed plants		
286	Permian				Ferns	
360	Carboniferous		Seed ferns			Lycophytes
374	Devonian	Upper	Arborescene Seeds	Megaphyllous leaves Progymnosperms		
387		Middle	Secondary growth			Zosterophyllodphytes
408		Lower	Major diversification of vascular plants	Trimerophytes	Heterospory	Microphyllous leaves
414	Silurian	Pridoli		Rhyniophytes		
421		Ludlow				
428		Wenlock	Tracheids		Cooksonia	
438		Llandovery				
505	Ordovician	Upper		First land plants		
		Lower				

Paleozoic had an equally great effect on the terrestrial population. By the end of the Permian, about 90% of all marine invertebrate species were extinct, compared with more than two-thirds of all amphibians and reptiles. Plants, on the other hand, apparently did not experience as great a turnover as invertebrates and animals.

Plant Evolution

When plants made the transition from water to land, they had to solve most of the same problems that animals did: desiccation, support, and the effects of gravity. Plants did so by evolving a variety of structural adaptations that were fundamental to the subsequent radiations and diversification that occurred during the Silurian, Devonian, and later periods (Table 13.2). Most experts agree that the ancestors of land plants first evolved in a marine environment, then moved into a freshwater environment and finally onto land. In this way, the differences in osmotic pressures between salt and freshwater were overcome while the plant was still in the water.

What was the land surface like before the plants made the transition to land? During the Early Paleozoic, the land was probably covered by scattered cyanobacterial microbial mats, which were followed later by various coccoid and filamentous green algae. Microfossil evidence indicates that sometime around the Middle Ordovician the terrestrial vegetation changed. Spores recovered from Middle to

FIGURE 13.20 The earliest known fertile land plant was *Cooksonia*, seen in this fossil from the Upper Silurian of South Wales. *Cooksonia* consisted of upright, branched stems terminating in sporangia (spore-producing structures). It also had a resistant cuticle and produced spores typical of a vascular plant. These plants probably lived in moist environments such as mud flats. This specimen is 1.49 cm long.

FIGURE 13.21 Reconstruction of an Early Devonian landscape showing some of the earliest land plants. (a) *Dawsonites.* (b) *Protolepidodendron.* (c) *Bucheria.*

Late Ordovician-age rocks from various locations suggest that land plants evolved well before the earliest known Middle Silurian fossil plant evidence indicates (see Perspective 13.1).

The higher land plants are composed of two major groups, the nonvascular and vascular plants. Most land plants are **vascular,** meaning they have a tissue system of specialized cells for the movement of water and nutrients. The **nonvascular** plants, such as bryophytes (liverworts, hornworts, and mosses) and fungi, do not have these specialized cells and are typically small and usually live in low, moist areas.

The earliest land plants from the Middle to Late Ordovician were probably small and bryophyte-like in their overall organization (but not necessarily related to bryophytes). The evolution of vascular tissue in plants was an important step as it allowed for the transport of food and water.

The ancestor of terrestrial vascular plants was probably some type of green algae. While no fossil record of the transition from green algae to terrestrial vascular plants exists, comparison of their physiology reveals a strong link. Primitive *seedless vascular plants* (discussed later in this chapter) such as ferns resemble green algae in their pigmentation, important metabolic enzymes, and type of reproductive cycle. Furthermore, the green algae are one of the few plant groups to have made the transition from salt water to freshwater.

The evolution of terrestrial vascular plants from an aquatic, probably green algal ancestry was accompanied by various modifications that allowed them to occupy this new and harsh environment. Besides the primary function of transporting water and nutrients throughout a plant, vascular tissue also provides some support for the plant body. Additional strength is derived from the organic compounds *lignin* and *cellulose,* which are found throughout a plant's walls.

The problem of desiccation was circumvented by the evolution of *cutin,* an organic compound found in the outer-wall layers of plants. Cutin also provides additional resistance to oxidation, the effects of ultraviolet light, and the entry of parasites.

Roots evolved in response to the need to collect water and nutrients from the soil and to help anchor the plant in the ground. The evolution of *leaves* from tiny outgrowths on the stem or from branch systems provided plants with an efficient light-gathering system for photosynthesis.

SILURIAN AND DEVONIAN FLORAS

The earliest known vascular land plants are small Y-shaped stems assigned to the genus *Cooksonia* from the Middle Silurian of Wales and Ireland. Together with Upper Silurian

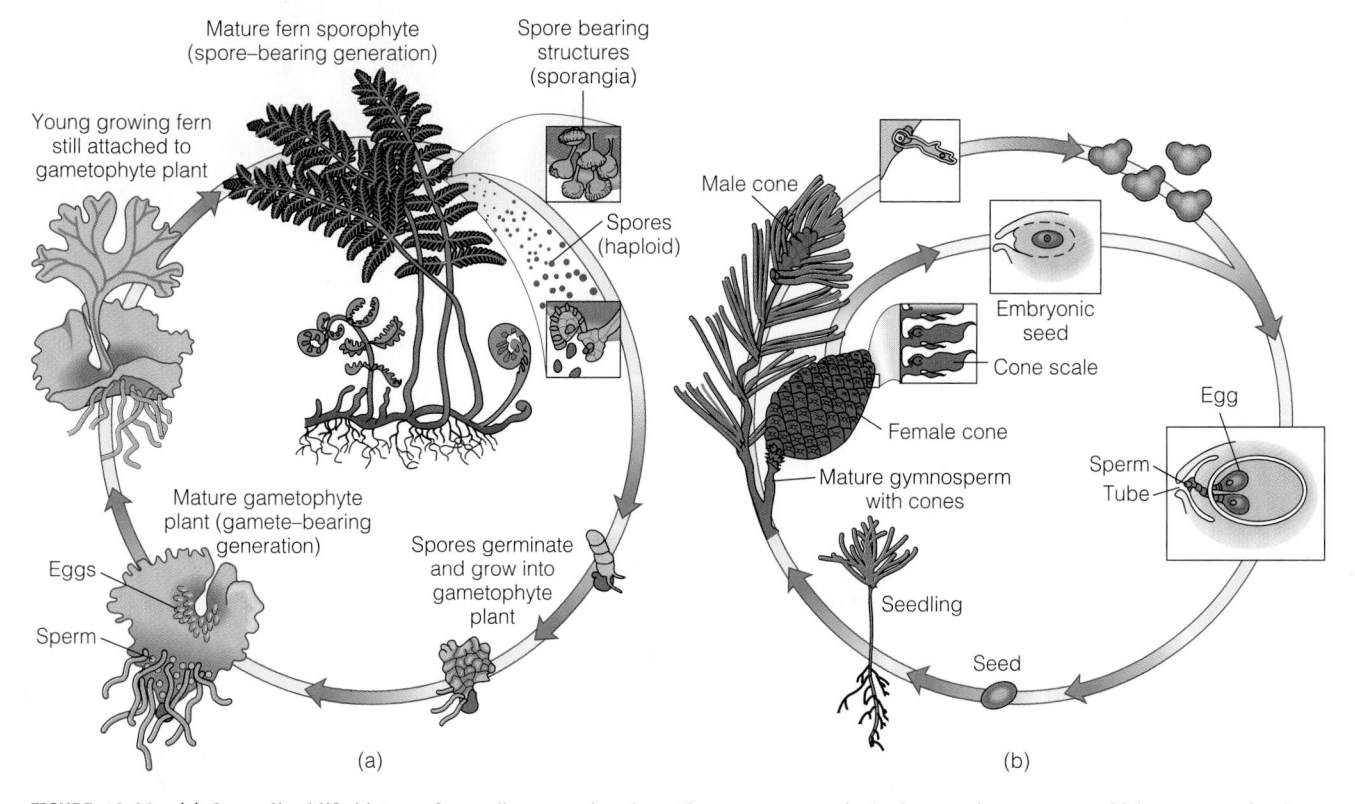

FIGURE 13.22 (a) Generalized life history of a seedless vascular plant. The mature sporophyte plant produces spores, which upon germination grow into small gametophyte plants that produce sperm and eggs. The fertilized eggs grow into the spore-producing mature plant, and the sporophyte–gametophyte life cycle begins again. (b) Generalized life history of a gymnosperm plant. The mature plant bears both male cones that produce sperm-bearing pollen grains and female cones that contain embryonic seeds. Pollen grains are transported to the female cones by the wind. Fertilization occurs when the sperm moves through a moist tube growing from the pollen grain and unites with the embryonic seed, which then grows into a cone-bearing mature plant.

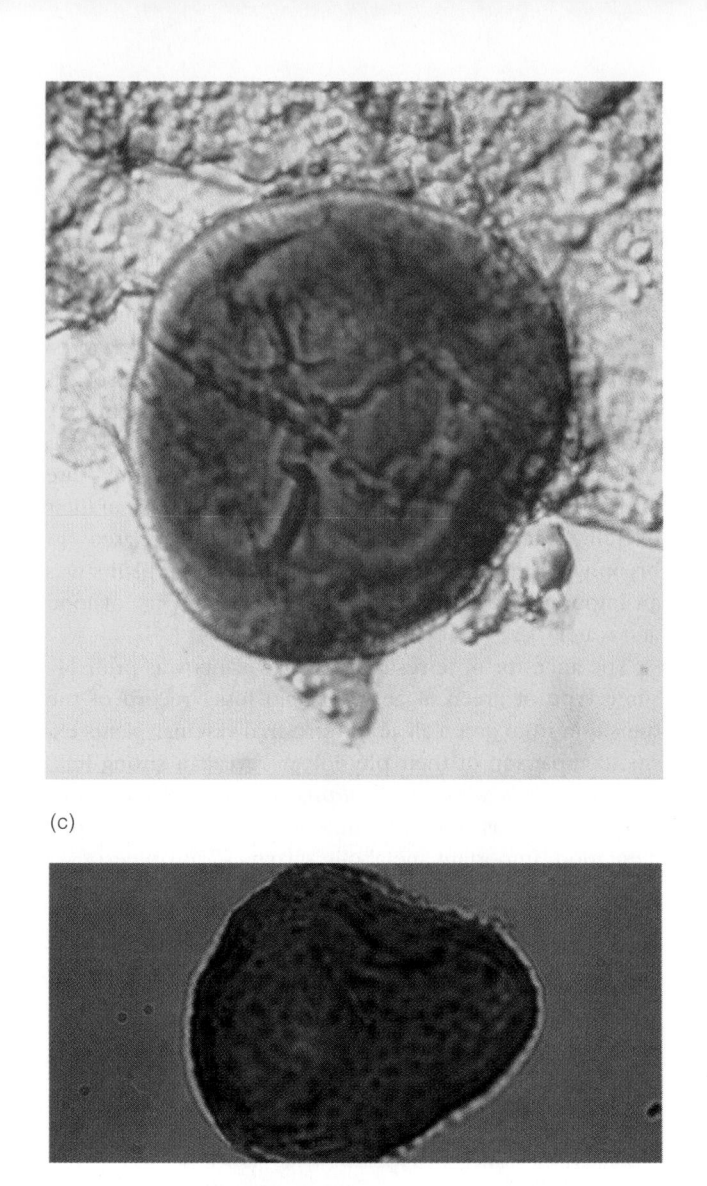

(a)

(c)

(b)

(d)

FIGURE 13.23 (a) Specimen and (b) reconstruction of the Early Devonian plant *Chaleuria cirrosa* from New Brunswick, Canada. This plant was heterosporous, meaning that it produced two sizes of spores. (c) The larger spores, which are thought to have produced female plants, range in size from 60 to 180 μm in diameter. (d) The smaller spores, which are believed to have produced male plants, have a size range of 30–40 μm in diameter.

and Lower Devonian species from Scotland, New York State, the Czech Republic, and the Commonwealth of Independent States, these earliest plants were small, simple, leafless stalks with a spore-producing structure at the tip (Figure 13.20); they are known as **seedless vascular plants** because they did not produce seeds. They also did not have a true root system. A *rhizome*, the underground part of the stem, transferred water from the soil to the plant and anchored the plant to the ground. The sedimentary rocks in which these plant fossils are found indicate that they lived in low, wet, marshy, freshwater environments.

An interesting parallel can be seen between seedless vascular plants and amphibians. When they made the transition from water to land, they had to overcome the problems such a transition involved. Both groups, while successful, nevertheless required a source of water in order to reproduce. In the case of amphibians, their gelatinous egg had to remain moist, while the seedless vascular plants required water for the sperm to travel through to reach the egg.

From this simple beginning, the seedless vascular plants evolved many of the major structural features characteristic of modern plants such as leaves, roots, and secondary growth. These features did not all evolve simultaneously but rather at different times, a pattern known as *mosaic evolution.* This diversification and adaptive radiation took place during the Late Silurian and Early Devonian and resulted in a tremendous increase in diversity (Figure 13.21). From the end of the Early Devonian to the end of the Devonian, the number of plant genera remained about the same, yet the composition of the flora changed. Whereas the Early Devonian landscape was dominated by relatively small, low-growing, bog-dwelling types of plants, the Late Devonian witnessed forests of large tree-sized plants up to 10 m tall.

In addition to the diverse seedless vascular plant flora of the Late Devonian, another significant floral event took place. The evolution of the seed at this time liberated land plants from their dependence on moist conditions and allowed them to spread over all parts of the land.

Seedless vascular plants require moisture for successful fertilization because the sperm must travel to the egg on the surface of the gamete-bearing plant (gametophyte) to produce a successful spore-generating plant (sporophyte). Without moisture, the sperm would dry out before reaching the egg (Figure 13.22a). In the seed method of reproduction, the spores are not released to the environment as they are in the seedless vascular plants but are retained on the spore-bearing plant, where they grow

FIGURE 13.24 Reconstruction of a Pennsylvanian coal swamp with its characteristic vegetation. The amphibian is *Eogyrinus*.

FIGURE 13.25 Living sphenopsids include the horsetail *Equisetum.*

spores of two distinct sizes (Figure 13.23). The appearance of heterospory was followed several million years later by the emergence of progymnosperms—Middle and Late Devonian plants with fernlike reproductive habits and a gymnosperm anatomy—which gave rise in the Late Devonian to such other gymnosperm groups as the seed ferns and conifer-type seed plants. While the seedless vascular plants dominated the flora of the Carboniferous coal-forming swamps, the gymnosperms made up an important element of the Late Paleozoic flora, particularly in the non-swampy areas.

LATE CARBONIFEROUS AND PERMIAN FLORAS

As discussed earlier, the rocks of the Pennsylvanian Period (Late Carboniferous) are the major source of the world's coal. Coal results from the alteration of plants living in low, swampy areas. The geologic and geographic conditions of the Pennsylvanian were ideal for the growth of seedless vascular plants, and consequently these coal swamps had a very diverse flora (Figure 13.24).

It is evident from the fossil record that while the Early Carboniferous flora was similar to its Late Devonian counterpart, a great deal of evolutionary experimentation was taking place that would lead to the highly successful Late Paleozoic flora of the coal swamps and adjacent habitats. Among the seedless vascular plants, the *lycopsids* and *sphenopsids* were the most important coal-forming groups of the Pennsylvanian Period.

The lycopsids were present during the Devonian, chiefly as small plants, but by the Pennsylvanian, they were the dominant element of the coal swamps, achieving heights up to 30 m in such genera as *Lepidodendron* and *Sigillaria.* The Pennsylvanian lycopsid trees are interesting because they lacked branches except at their top. The leaves were elongate and similar to the individual palm leaf of today. As the trees grew, the leaves were replaced from the top, leaving prominent and characteristic rows or spirals of scars on the trunk. Today, the lycopsids are represented by small temperate-forest ground pines.

The sphenopsids, the other important coal-forming plant group, are characterized by being jointed and having horizontal underground stem-bearing roots. Many of these plants, such as *Calamites,* average 5 to 6 m tall. Living sphenopsids include the horsetail *(Equisetum)* and scouring rushes (Figure 13.25). Small seedless vascular plants and seed ferns formed a thick undergrowth or ground cover beneath these treelike plants.

Not all plants were restricted to the coal-forming swamps. Among those plants occupying higher and drier ground were some of the *cordaites,* a group of tall gymnosperm trees that grew up to 50 m and probably formed vast forests (Figure 13.26). Another important non-swamp dweller was *Glossopteris,* the famous plant so abundant in Gondwana (see Figure 6.2), whose distribution is cited as critical evidence that the continents have moved through time.

into the male and female forms of the gamete-bearing generation. In the case of the **gymnosperms,** or flowerless seed plants, these are male and female cones (Figure 13.22b). The male cone produces pollen, which contains the sperm and has a waxy coating to prevent desiccation, while the egg, or embryonic seed, is contained in the female cone. After fertilization, the seed then develops into a mature, cone-bearing plant. In this way the need for a moist environment for the gametophyte generation is solved. The significance of this development is that seed plants, like reptiles, were no longer restricted to wet areas but were free to migrate into previously unoccupied dry environments.

Before seed plants evolved, an intermediate evolutionary step was necessary. This was the development of *heterospory,* whereby a species produces two types of spores: a large one (megaspore) that gives rise to the female gamete-bearing plant and a small one (microspore) that produces the male gamete-bearing plant. Prior to heterospory, plants were *homosporous;* that is, they produced spores that were all the same size and, upon germinating, grew into gametophyte plants that produced both male and female gametes.

The earliest evidence of heterospory is found in the Early Devonian plant *Chaleuria cirrosa,* which produced

FIGURE 13.26   A cordaite forest from the Late Carboniferous. Cordaites were a group of gymnosperm trees that grew up to 50 m tall.

The floras that were abundant during the Pennsylvanian persisted into the Permian, but due to climatic and geologic changes resulting from tectonic events (see Chapter 11), they declined in abundance and importance. By the end of the Permian, the cordaites became extinct, while the lycopsids and sphenopsids were reduced to mostly small, creeping forms. Those gymnosperms with lifestyles more suited to the warmer and drier Permian climates diversified and came to dominate the Permian, Triassic, and Jurassic landscapes.

Summary

Table 13.3 summarizes the major evolutionary and geologic events of the Paleozoic Era and shows their relationships to each other.

1. Chordates are characterized by a notochord, dorsal hollow nerve cord, and gill slits. The earliest chordates were soft-bodied organisms that were rarely fossilized. Vertebrates are a subphylum of the chordates.

2. Fish are the earliest known vertebrates with their first fossil occurrence in Upper Cambrian rocks. They have had a long and varied history, including jawless and jawed armored forms (ostracoderms and placoderms), cartilaginous forms, and bony forms. Crossopterygians, a group of lobe-finned fish, gave rise to the amphibians.

TABLE 13.3

Major Evolutionary and Geologic Events of the Paleozoic Era

Geologic Period			Invertebrates	Vertebrates
Permian		245	Largest mass extinction event to affect the invertebrates.	Acanthodians, placoderms, and pelycosaurs become extinct. Therapsids and pelycosaurs the most abundant reptiles.
Carboniferous	Pennsylvanian	286	Fusulinids diversify.	Amphibians abundant and diverse.
Carboniferous	Mississippian	320	Crinoids, lacy bryozans, blastoids become abundant. Renewed adaptive radiation following extinctions of many reef-builders.	Reptiles evolve.
Devonian		360	Extinctions of many reef-building invertebrates near end of Devonian. Reef building continues. Eurypterids abundant.	Amphibians evolve. All major groups of fish present—Age of Fish.
Silurian		408	Major reef building. Diversity of invertebrates remains high.	Ostracoderms common. Acanthodians, the first jawed fish, evolve.
Ordovician		438	Extinctions of a variety of marine invertebrates near end of Ordovician. Major adaptive radiation of all invertebrate groups. Suspension feeders dominant.	Ostracoderms diversify.
Cambrian		505	Many trilobites become extinct near end of Cambrian. Trilobites, brachiopods, and archaeocyathids are most abundant.	Earliest vertebrates—jawless fish called ostracoderms.
		545		

Age (Millions of Years)

Plants	Major Geologic Events
Gymnosperms diverse and abundant.	Formation of Pangaea. Alleghenian orogeny. Hercynian orogeny.
Coal swamps with flora of seedless vascular plants and gymnosperms.	Coal-forming swamps common. Formation of Ancestral Rockies. Continental glaciation in Gondwana.
Gymnosperms appear (may have evolved during Late Devonian).	Ouachita orogeny.
First seeds evolve. Seedless vascular plants diversify.	Widespread deposition of black shale. Antler orogeny. Acadian orogeny.
Early land plants—seedless vascular plants.	Caledonian orogeny. Extensive barrier reefs and evaporites.
Plants move to land?	
	Continental glaciation in Gondwana. Taconic orogeny.
	First Phanerozoic transgression (Sauk) onto North American craton.

3. The link between crossopterygians and the earliest amphibians is convincing and includes a close similiarity of bone and tooth structures. The transition from fish to amphibians occurred during the Devonian. During the Carboniferous, the labyrinthodont amphibians were the dominant terrestrial vertebrate animals.

4. The earliest fossil record of reptiles is from the Late Mississippian. The evolution of an amniote egg was the critical factor in the reptiles' ability to colonize all parts of the land.

5. Pelycosaurs were the dominant reptile group during the Early Permian, while therapsids dominated the landscape for the rest of the Permian Period.

6. Plants had to overcome the same basic problems as animals, namely desiccation, reproduction, and gravity in making the transition from water to land.

7. The earliest fossil record of land plants is from Middle to Late Ordovician–age rocks. These plants were probably small and bryophyte-like in their overall organization.

8. The evolution of vascular tissue was an important event in plant evolution as it allowed food and water to be transported throughout the plant and provided the plant with additional support.

9. The ancestor of terrestrial vascular plants was probably some type of green algae based on such similarities as pigmentation, metabolic enzymes, and the same type of reproductive cycle.

10. The earliest seedless vascular plants were small, leafless stalks with spore-producing structures on their tips. From this simple beginning, plants evolved many of the major structural features characteristic of today's plants.

11. By the end of the Devonian Period, forests with tree-sized plants up to 10 m had evolved. The Late Devonian also witnessed the evolution of the flowerless seed plants (gymnosperms) whose reproductive style freed them from having to stay near water.

12. The Carboniferous Period was a time of vast coal swamps, where conditions were ideal for the seedless vascular plants. With the onset of more arid conditions during the Permian, the gymnosperms became the dominant element of the world's flora.

Important Terms

acanthodian	crossopterygian	ostracoderm	seedless vascular plant
amniote egg	gymnosperm	pelycosaur	therapsid
bony fish	labyrinthodont	placoderm	vascular
cartilaginous fish	lobe-finned fish	protorothyrid	vertebrate
chordate	nonvascular	ray-finned fish	

Review Questions

1. Which of the following must an organism possess during at least part of its life cycle to be classified as a chordate?
   a. _____ Vertebrae, dorsal hollow nerve cord, gill slits
   b. _____ Notochord, ventral solid nerve cord, lungs
   c. _____ Notochord, dorsal hollow nerve cord, gill slits
   d. _____ Vertebrae, dorsal hollow nerve cord, lungs
   e. _____ Notochord, dorsal solid nerve cord, lungs

2. Based on similarity of embryo development, which invertebrate phylum is most closely allied with the chordates?
   a. _____ Arthropoda
   b. _____ Annelida
   c. _____ Porifera
   d. _____ Echinodermata
   e. _____ Mollusca

3. The "Age of Fish" refers to which period?
   a. _____ Cambrian
   b. _____ Ordovician
   c. _____ Silurian
   d. _____ Devonian
   e. _____ Mississippian

4. Placoderms are:
   a. _____ jawless fish.
   b. _____ primitive amphibians.
   c. _____ advanced reptiles.
   d. _____ plants.
   e. _____ jawed armored fish.

5. The ancestors of amphibians belong to which group?
   a. _____ Placoderms
   b. _____ Ostracoderms
   c. _____ Acanthodians
   d. _____ Crossopterygians
   e. _____ Cartilaginous

6. Labyrinthodonts are:
   a. _____ reptiles.
   b. _____ fish.
   c. _____ amphibians.
   d. _____ plants.
   e. _____ none of these.

7. In which period were amphibians most abundant?
   a. _____ Silurian
   b. _____ Devonian
   c. _____ Mississippian
   d. _____ Pennsylvanian
   e. _____ Permian

8. The most significant evolutionary change that allowed reptiles to colonize all parts of land was:
   a. _____ endothermy.
   b. _____ origin of limbs capable of supporting the animals on land.
   c. _____ evolution of the amniote egg.
   d. _____ evolution of a watertight skin.
   e. _____ evolution of tear ducts.

9. The dominant reptile group during the Permian, and the one that gave rise to the mammals, was the:
   a. _____ labyrinthodonts.
   b. _____ protorothyrids.
   c. _____ therapsids.
   d. _____ pelycosaurs.
   e. _____ acanthodians.

10. Which algal group was the probable ancestor of vascular plants?
    a. _____ Red
    b. _____ Blue-green
    c. _____ Green
    d. _____ Brown
    e. _____ Yellow

11. Which plant group was the first to successfully invade land?
    a. _____ Seedless vascular
    b. _____ Gymnosperms
    c. _____ Naked seed bearing
    d. _____ Angiosperms
    e. _____ Flowering

12. The first plant group that did not require a wet area for part of its life cycle was the:
    a. _____ seedless vascular plants.
    b. _____ gymnosperms.
    c. _____ angiosperms.
    d. _____ flowering plants.
    e. _____ labyrinthodonts.

13. Discuss the Paleozoic evolutionary history of the amphibians.

14. Discuss the significance and advantages of the pelycosaur sail.

15. Discuss the evidence for an Ordovician faunal and floral colonization of land.

16. Outline the evolutionary history of fish.

17. Describe the problems that had to be overcome before organisms could inhabit the land.

18. Why were the reptiles so much more successful at extending their habitat than the amphibians?

19. What are the major differences between seedless vascular plants and gymnosperms and why are these differences significant in terms of exploiting the terrestrial environment?

20. Discuss the current hypothesis for the evolution of vertebrates.

Points to Ponder

1. Why is it more likely that fish evolved in the seas and then migrated to freshwater environments? Could fish have evolved in freshwater and then migrated to the seas?

2. Discuss how the changing geologic conditions affected the evolution of life on land.

Additional Readings

Ausich, W. I., and N. G. Lane. 1999. *Life of the past.* 4th ed. Upper Saddle River, N.J.: Prentice Hall.

Carroll, R. L. 1992. The primary radiation of terrestrial vertebrates. *Annual Review of Earth and Planetary Sciences* 20: 45–84.

Colbert, E. H., and M. Morales. 1991. *Evolution of the vertebrates.* 4th ed. New York: Wiley.

Cowen, R. 1995. *History of life.* 2d ed. Cambridge, Mass.: Blackwell Scientific.

Forey, P. L. 1997. *History of the coelacanth fishes.* London: Chapman & Hall.

Forey, P., and P. Janvier. 1994. Evolution of the early vertebrates. *American Scientist* 82, no. 6: 554–565.

Gordon, M. S., and E. C. Olson. 1995. *Invasions of the land.* New York: Columbia University Press.

Gorr, T., and T. Kleinschmidt. 1993. Evolutionary relationships of the coelacanth. *American Scientist* 81, no. 1: 72–82.

Gray, J., and W. Shear. 1992. Early life on land. *American Scientist* 80, no. 5: 444–456.

Prothero, D. R. 1998. *Bringing fossils to life.* New York: WCB McGraw-Hill.

Schopf, J. W., ed. 1992. *Major events in the history of life.* Boston: Jones and Bartlett.

Shear, W. A., and W. D. Ian Rolfo. 1994. Lizzie the lizard. *Earth* 3, no. 7: 36–43.

Stewart, W. N., and G. W. Rothwell. 1993. *Paleobotany and the evolution of plants.* 2d ed. New York: Cambridge University Press.

Stokes, M. D., and N. D. Holland. 1998. The lancelet. *American Scientist* 86, no. 6: 552–560.

Thomson, K. W. 1991. Where did tetrapods come from? *American Scientist* 79, no. 6: 488–490.

White, M. E. 1986. *The greening of Gondwana.* Frenchs Forest, NSW, Australia: Reed Books Limited.

For these web site addresses, along with current updates and exercises, log on to

http://www.brookscole.com/geo/

▶ **ILLINOIS STATE MUSEUM MAZON CREEK FOSSILS**

This site contains some of the more interesting and dramatic types of fossils recovered from the Francis Creek Shale in Illinois.

1. Click on the *Where Are Mazon Creek Fossils Found* site. Where can Mazon Creek fossils be found? How did they form? What are the different types of fossils recovered from the Francis Creek Shale? Click on the names of some of these fossils to see what they looked like.

2. Click on *The Importance of Mazon Creek Fossils* site. Why are these fossils important?

▶ **UNIVERSITY OF CALIFORNIA MUSEUM OF PALEONTOLOGY**

This is an excellent site to visit for any aspect of geologic time, paleontology, and evolution.

1. Click on the *On-line Exhibits* to go to the *Paleontology Without Walls* home page, which is an introduction to the UCMP Virtual Exhibits. Click on the *Phylogeny* icon; this takes you to the *Phylogeny of Life* home page. Click on the *Vertebrates* icon; this takes you to the *Introduction to the Vertebrates* home page. From here check out the Paleozoic history of the vertebrates.

2. Click on *The Biosphere, All Life* icon on *The Phylogeny of Life* home page; this takes you to the *Three Domains of Life* home page. Click on the *Eucaryota* icon; this takes you to the *Introduction to the Eukaryota* home page. Click on the *Plants* site; this takes you to the *Introduction to the Plantae* home page. Click on the *Fossil Record* icon; at this site you can then learn more about the fossil history of plants. Check out any group you want to learn more about.

Geology of the Mesozoic Era

By 1852, mining operations were well under way on the American River near Sacramento.

Prologue

Approximately 150 to 210 million years after the emplacement of massive plutons created the Sierra Nevada (Nevadan orogeny), gold was discovered at Sutter's Mill on the South Fork of the American River at Coloma, California. On January 24, 1848, James Marshall, a carpenter building a sawmill for John Sutter, found bits of the glittering metal in the mill's tailrace. Soon, settlements throughout the state were completely abandoned as word of the chance for instant riches spread throughout California. Within a year after the news of the gold discovery reached the East Coast, the Sutter's Mill area was swarming with more than 80,000 prospectors, all hoping to make their fortune. In all, at least 250,000 gold seekers prospected the Sutter's Mill area, and though most were Americans, they came from all over the world, even as far away as China. Most of them thought the gold was simply waiting to be taken.

Of course, no one gave any thought to the consequences of so many people converging on the Sutter's Mill area, all intent on making easy money. In reality, only a small percentage of prospectors ever hit it big or were even moderately successful. The rest barely eked out a living until they eventually abandoned their dream and went home.

Although some prospectors dug $30,000 worth of gold dust a week out of a single claim and gold was found practically on the surface of the ground, most of this easy gold was recovered very early during the gold rush. Most prospectors made only a living wage working their claims. Nevertheless, during the five years from 1848 to 1853 that constituted the gold rush proper, more than $200 million in gold was extracted.

Would-be prospectors could follow three basic routes to the gold fields. The most popular was the overland journey by wagon train from the East to California—a route fraught with peril, including the threat of starvation, disease, and the crossing of the Sierra Nevada. Those who took this route and survived finally arrived at Sutter's Mill and dispersed from there to prospect. The Panama route to California began with a voyage by ship to Chagres on the Isthmus of Panama. Here the potential prospectors and their goods were transported to Panama City by boat and mules and then traveled by ship to California. For the most part, the prospectors who took this route to California were young, vigorous men without wives and children. They faced the usual hardships of steamy jungles, disease, and overcrowded ships, many of which were unseaworthy. The third route, by ship around the southern tip of South America, was the longest, but it avoided the overcrowding encountered on the Panama route and the perils of the overland journey. Those who went by ship arrived at San Francisco, where many decided to stay after seeing that there were opportunities to make money in this new town.

The earliest prospectors came from the West Coast and were, for the most part, honest and hardworking. Those who arrived from elsewhere in 1849 and later were also mostly honest and hardworking, but, unfortunately, some were thieves, outlaws, and murderers. Life in the mining camps was extremely hard and expensive. Frequently, the shopowners and traders made more money than the prospectors did. Gambling halls and saloons sprang up all over because the men had little to do with their time except gamble and drink. Family life was virtually nonexistent as women comprised only 2% of the population.

The gold these prospectors sought was mostly in the form of placer deposits. Weathering of gold-bearing igneous rocks and mechanical separation of minerals by density during stream transport forms placer deposits. Many prospectors searched for the mother lode, but all of the gold recovered during the gold rush came from placers. A common method of mining these deposits is by panning. A prospector dips a shallow pan into a streambed, swirls the material around, and pours off the lighter material. The gold, being about six times heavier than most sand grains and rock chips, concentrates on the bottom of the pan and can then be picked out.

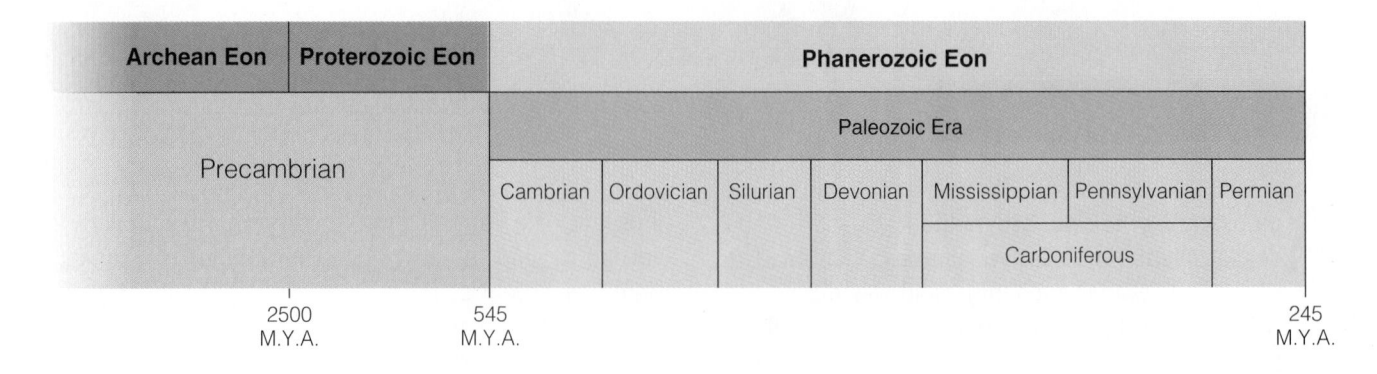

Archean Eon	Proterozoic Eon	Phanerozoic Eon							
Precambrian		Paleozoic Era							
		Cambrian	Ordovician	Silurian	Devonian	Mississippian	Pennsylvanian	Permian
						Carboniferous		

2500
M.Y.A.

545
M.Y.A.

245
M.Y.A.

Introduction

The Mesozoic Era (245 to 66 million years ago) is divided into three periods beginning with the Triassic, followed by the Jurassic, and finally the Cretaceous. The stratotypes for the systems from which these periods derive their names are in the Hercynian Mountains of Germany (Triassic), the Jura Mountains of Switzerland (Jurassic), and the Paris Basin of France (Cretaceous).

The Mesozoic Era is popularly known as the "Age of Reptiles," a phrase emphasizing the fact that reptiles, and particularly dinosaurs, were the dominant land-dwelling vertebrate animals. It also was the time during which birds, mammals, and angiosperms (flowering plants) first evolved and diversified (see Chapter 15).

The dawn of the Mesozoic ushered in a new era in Earth history. The major geologic event was the breakup of Pangaea, which affected oceanic and climatic circulation patterns and influenced the evolution of the terrestrial and marine biotas. Because most of the Mesozoic geologic history of North America involves the continental margins, we focus our attention on the eastern, Gulf, and western coastal regions. The transgressions and regressions of the final two epeiric seas and their depositional sequences (Absaroka and Zuni) will be incorporated into the geologic history of each of these regions.

The Breakup of Pangaea

Just as the formation of Pangaea influenced geologic and biologic events during the Paleozoic, the breakup of this supercontinent profoundly affected geologic and biologic events during the Mesozoic. The movement of continents affected the global climatic and oceanic regimes as well as the climates of the individual continents. Populations became isolated or were brought into contact with other populations, leading to evolutionary changes in the biota. So great was the effect of this breakup on the world, that it forms the central theme of this chapter.

Pangaea's breakup can be documented by the extensional structures (fault-block basins or rift valleys) that formed along the newly created continental margins. Modern-day analogues for the different stages of continental rifting can be found in various parts of the world such as the East African Rift valleys (see Figure 6.16). As a continent begins to rift, fault-block basins are formed (see Figure 6.15). Associated with these tensional structures are basaltic dikes, sills, and lava flows, which can be radiometrically dated to determine the time of rifting. As the continents continue separating and basaltic oceanic crust forms between them (see Figure 6.15), rates and direction of movement can be determined by radiometric and paleomagnetic dating of the oceanic crust (see Chapter 6).

THE FRAGMENTING OF PANGAEA

Geologic, paleontologic, and paleomagnetic data indicate that the breakup of Pangaea took place in four general stages. The first stage involved the rifting between Laurasia and Gondwana during the Late Triassic. By the end of the Triassic, the expanding Atlantic Ocean separated North America from Africa (Figure 14.1a). This change was followed by the rifting of North America from South America sometime during the Late Triassic and Early Jurassic.

Separation of the continents allowed water from the Tethys Sea to flow into the expanding central Atlantic Ocean, while Pacific Ocean waters flowed into the newly formed Gulf of Mexico, which at that time was little more than a restricted bay (Figure 14.2). During that time, these areas were located in the low tropical latitudes where high temperatures and high rates of evaporation were ideal for the formation of thick evaporite deposits.

The second stage in Pangaea's breakup involved rifting and movement of the various Gondwana continents during the Late Triassic and Jurassic periods. As early as the Late Triassic, Antarctica and Australia, which remained sutured together, separated from South America and Africa, while India split away from all four Gondwana continents and began moving northward (Figure 14.1a and b).

The third stage of breakup began during the Late Jurassic, when South America and Africa began separating (Figure 14.1b). The rifting and subsequent separation of these two continents formed a narrow basin where thick evaporite deposits accumulated from the evaporation of southern ocean waters (Figure 14.2). During this stage, the eastern end of the Tethys Sea began closing as a result of the clockwise rotation of Laurasia and the northward movement of Africa. This narrow Late Jurassic and Cretaceous seaway between Africa and Europe was the forerunner of the present Mediterranean Sea.

Phanerozoic Eon										
Mesozoic Era			Cenozoic Era							
Triassic	Jurassic	Cretaceous	Tertiary						Quaternary	
			Paleocene	Eocene	Oligocene	Miocene	Pliocene	Pleistocene		Holocene

245
M.Y.A.

66
M.Y.A.

By the end of the Cretaceous, Australia and Antarctica had separated, India had nearly reached the equator, South America and Africa were widely separated, and the eastern side of what is now Greenland had begun separating from Europe (Figure 14.1c).

A global rise in sea level during the Cretaceous resulted in worldwide transgressions onto the continents. These transgressions were caused by higher heat flow along the oceanic ridges due to increased rifting and the consequent expansion of oceanic crust (see Figure 3.14). By the Middle Cretaceous, sea level probably was as high as at any time since the Ordovician, and approximately one-third of the present land area was inundated by epeiric seas.

The final stage in Pangaea's breakup occurred during the Cenozoic. During this stage, Australia continued moving northward, and Greenland completely separated from Europe and rifted from North America to form a separate landmass.

THE EFFECTS OF THE BREAKUP OF PANGAEA ON GLOBAL CLIMATES AND OCEAN CIRCULATION PATTERNS

By the end of the Permian Period, Pangaea extended from pole to pole, covered about one-fourth of Earth's surface, and was surrounded by Panthalassa, a global ocean that encompassed about 300 degrees of longitude (see Fig. 11.2b). Such a configuration exerted tremendous influence on the world's climate and resulted in generally arid conditions over large parts of Pangaea's interior.

The world's climates result from the complex interaction between wind and ocean currents and the location and topography of the continents. In general, dry climates occur on large landmasses in areas remote from sources of moisture and where barriers to moist air, such as mountain ranges, exist. Wet climates occur near large bodies of water or where winds can carry moist air over land.

Past climatic conditions can be inferred from the distribution of climate-sensitive deposits. Evaporites are deposited where evaporation exceeds precipitation. While desert dunes and red beds may form locally in humid regions, they are characteristic of arid regions. Coal forms in both warm and cool humid climates. Vegetation that is eventually converted into coal requires at least a good seasonal water supply; thus, coal deposits are indicative of humid conditions.

Widespread Triassic evaporites, red beds, and desert dunes in the low and middle latitudes of North and South America, Europe, and Africa indicate dry climates in those regions, while coal deposits are found mainly in the high latitudes, indicating humid conditions. These high-latitude coals are analogous to today's Scottish peat bogs or Canadian muskeg. The lands bordering the Tethys Sea were probably dominated by seasonal monsoon rains resulting from the warm moist winds and warm oceanic currents impinging against the east-facing coast of Pangaea.

The temperature gradient between the tropics and the poles also affects oceanic and atmospheric circulation. The greater the temperature difference between the tropics and the poles, the steeper the temperature gradient and the faster the circulation of the oceans and atmosphere. Oceans absorb about 90% of the solar radiation they receive, while continents absorb only about 50%, even less if they are snow covered. The rest of the solar radiation is reflected back into space. Therefore, areas dominated by seas are warmer than those dominated by continents. By knowing the distribution of continents and ocean basins, geologists can calculate the average annual temperature for any region on Earth, as well as determining a temperature gradient.

The breakup of Pangaea during the Late Triassic caused the global temperature gradient to increase because the Northern Hemisphere continents moved farther northward, displacing higher-latitude ocean waters. Due to the steeper global temperature gradient caused by a decrease in temperature in the high latitudes and the changing positions of the continents, oceanic and atmospheric circulation patterns greatly accelerated during the Mesozoic (Figure 14.3). Though the temperature gradient and seasonality on land were increasing during the Jurassic and Cretaceous, the middle- and higher-latitude oceans were still quite warm, because warm waters from the Tethys Sea were circulating to the higher latitudes. The result was a relatively equable worldwide climate through the end of the Cretaceous.

The Mesozoic History of North America

The beginning of the Mesozoic Era was essentially the same in terms of tectonism and sedimentation as the preceding Permian Period in North America (see Figure

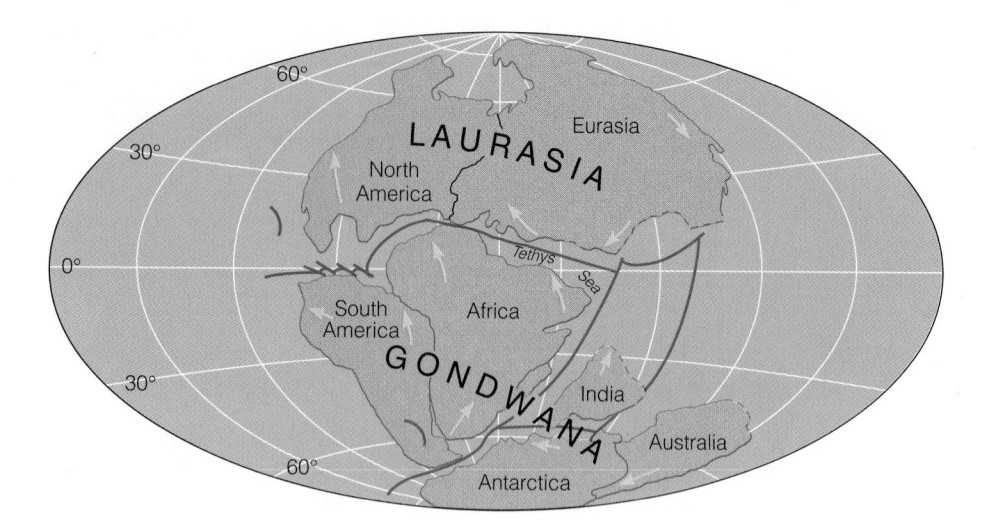

(a) Triassic Period (245–208 M.Y.A)

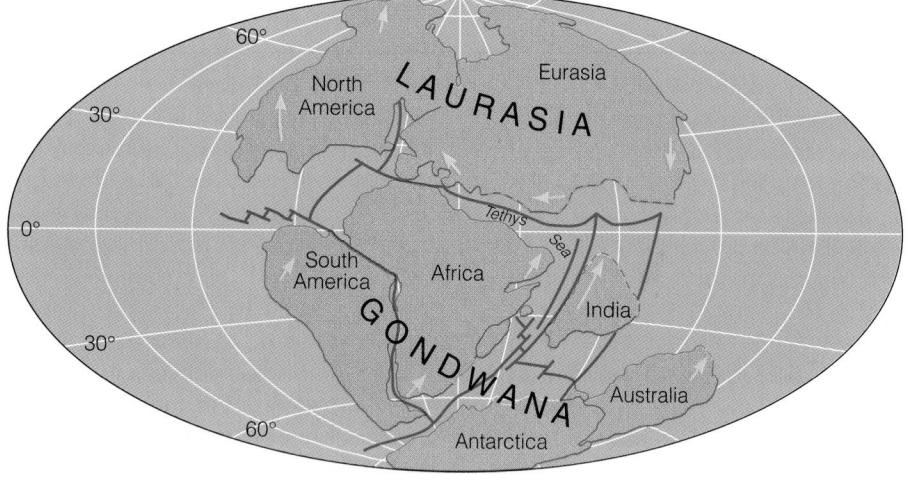

(b) Jurassic Period (208–144 M.Y.A)

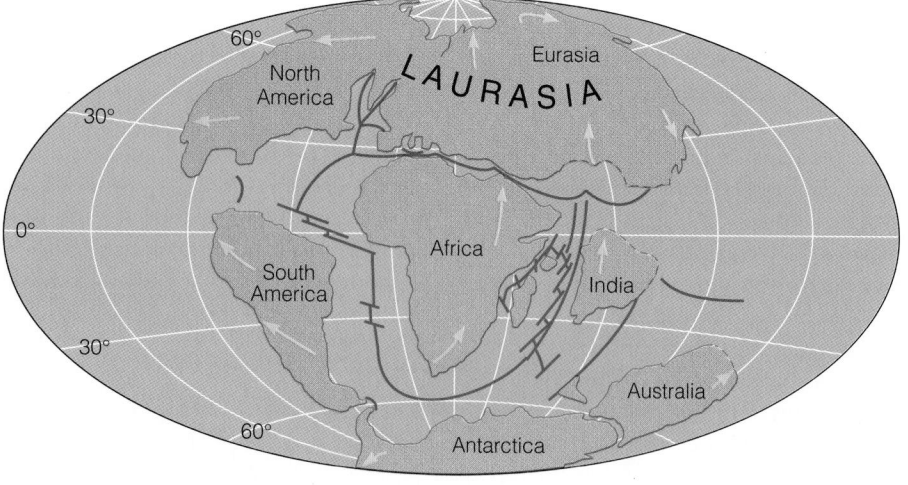

(c) Cretaceous Period (144–66 M.Y.A)

FIGURE 14.1 **Paleogeography of the world during the Mesozoic. Yellow arrows show the direction of movement for the continents. (a) The Triassic Period, (b) the Jurassic Period, and (c) the Cretaceous Period.**

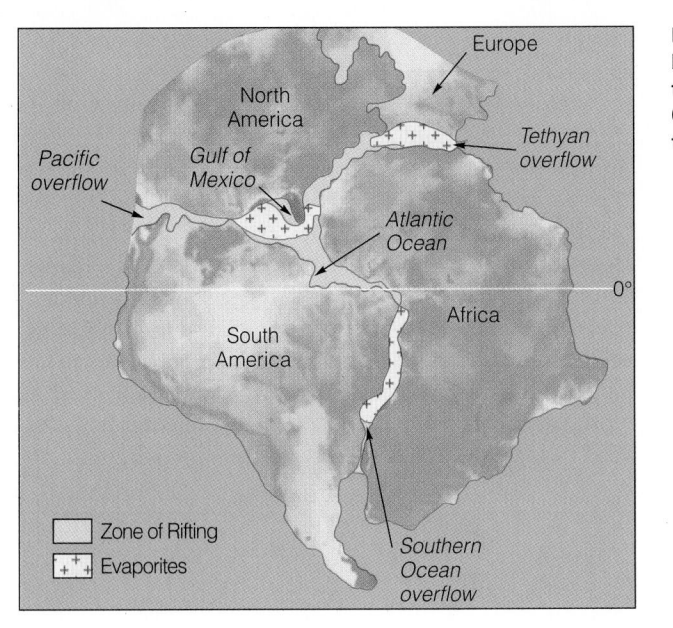

FIGURE 14.2 Evaporites accumulated in shallow basins as Pangaea broke apart during the Early Mesozoic. Water from the Tethys Sea flowed into the central Atlantic Ocean, while water from the Pacific Ocean flowed into the newly formed Gulf of Mexico. Marine water from the south flowed into the southern Atlantic Ocean.

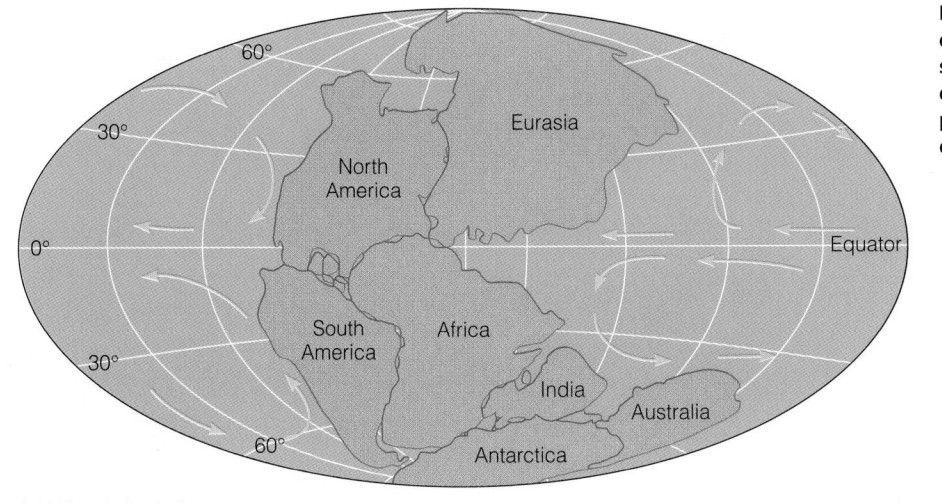

(a) Triassic Period

FIGURE 14.3 Oceanic circulation evolved from (a) a simple pattern in a single ocean (Panthalassa) with a single continent (Pangaea) to (b) a more complex pattern in the newly formed oceans of the Cretaceous Period.

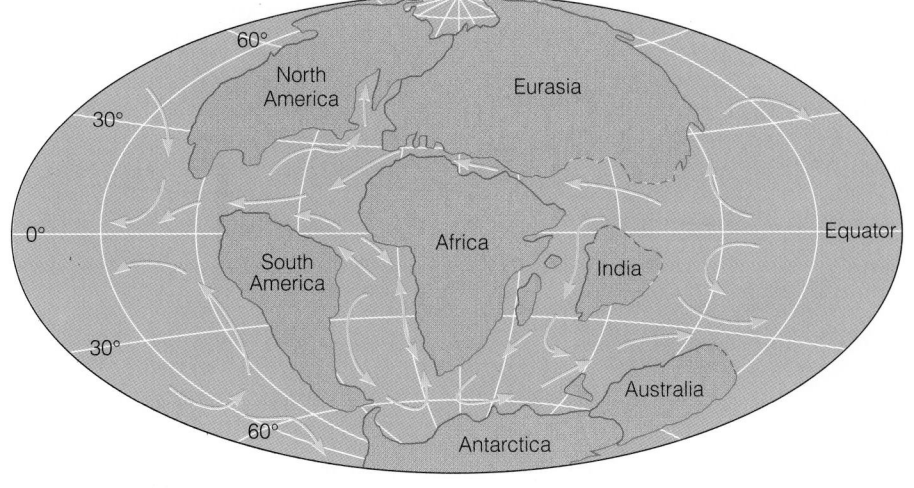

(b) Cretaceous Period

11.12). Terrestrial sedimentation continued over much of the craton, while block faulting and igneous activity began in the Appalachian region as North America and Africa began separating (Figure 14.4). The newly forming Gulf of Mexico experienced extensive evaporite deposition during the Late Triassic and Jurassic as North America separated from South America (Figure 14.5).

A global rise in sea level during the Cretaceous resulted in worldwide transgressions onto the continents (Figure 14.6). These transgressions were caused by higher heat flow along the oceanic ridges due to increased rifting and the consequent expansion of oceanic crust. By the Middle Cretaceous, sea level probably was as high as at any time since the Ordovician, and approxi-

mately one-third of the present land area was inundated by epeiric seas.

Marine deposition was continuous over much of the North American Cordillera. A volcanic island arc system that formed off the western edge of the craton during the Permian was sutured to North America sometime later during the Permian or Triassic. This event is referred to as the *Sonoma orogeny* and will be discussed later in the chapter. During the Jurassic, the entire Cordilleran area was involved in a series of major mountain-building episodes that resulted in the formation of the Sierra Nevada, the Rocky Mountains, and other lesser mountain ranges. While each orogenic episode has its own name, the entire mountain-building event is simply called the *Cordilleran orogeny* (also dis-

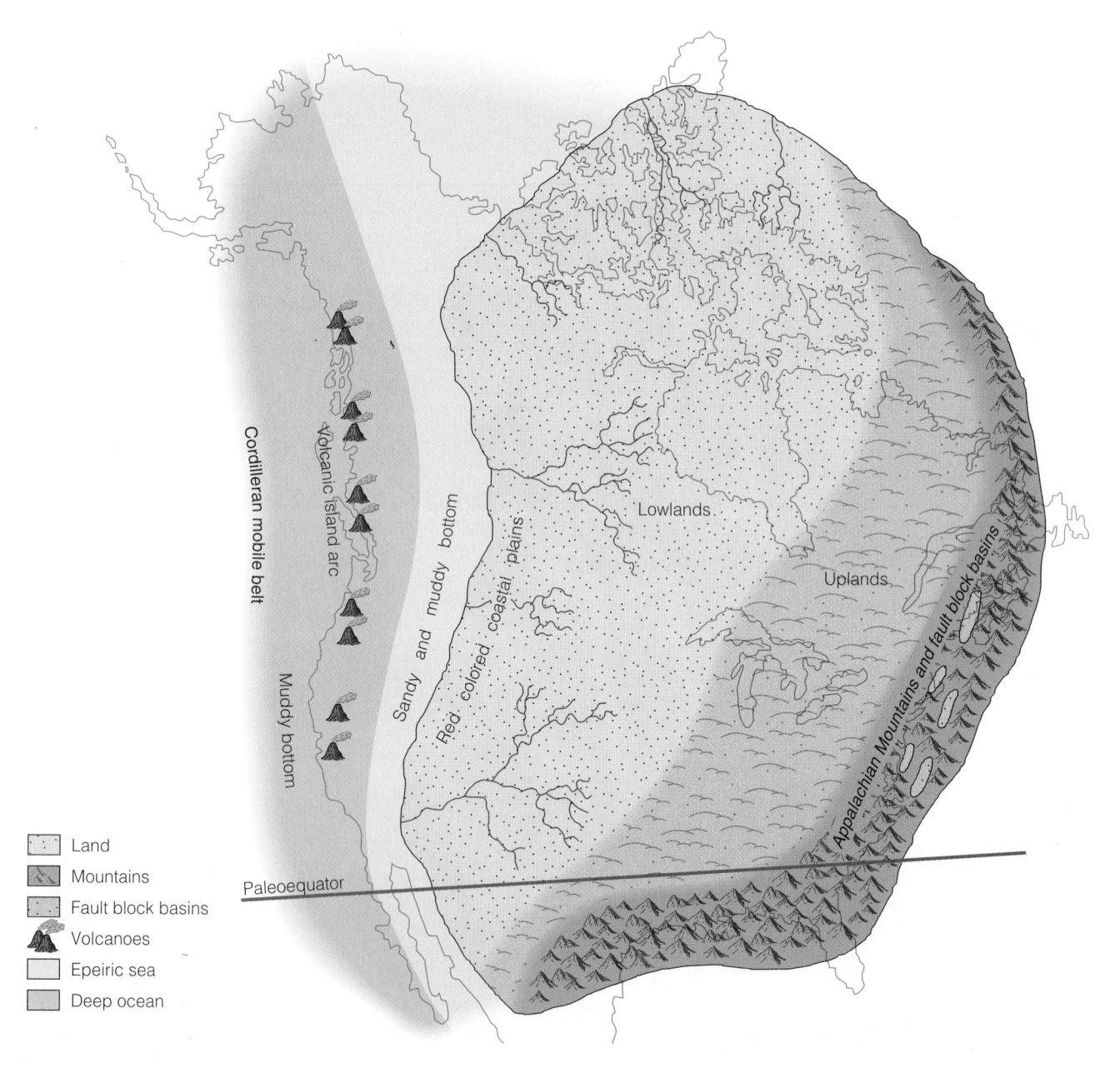

FIGURE 14.4 Paleogeography of North America during the Triassic Period.

Land
Mountains
Fault block basins
Volcanoes
Epeiric sea
Deep ocean

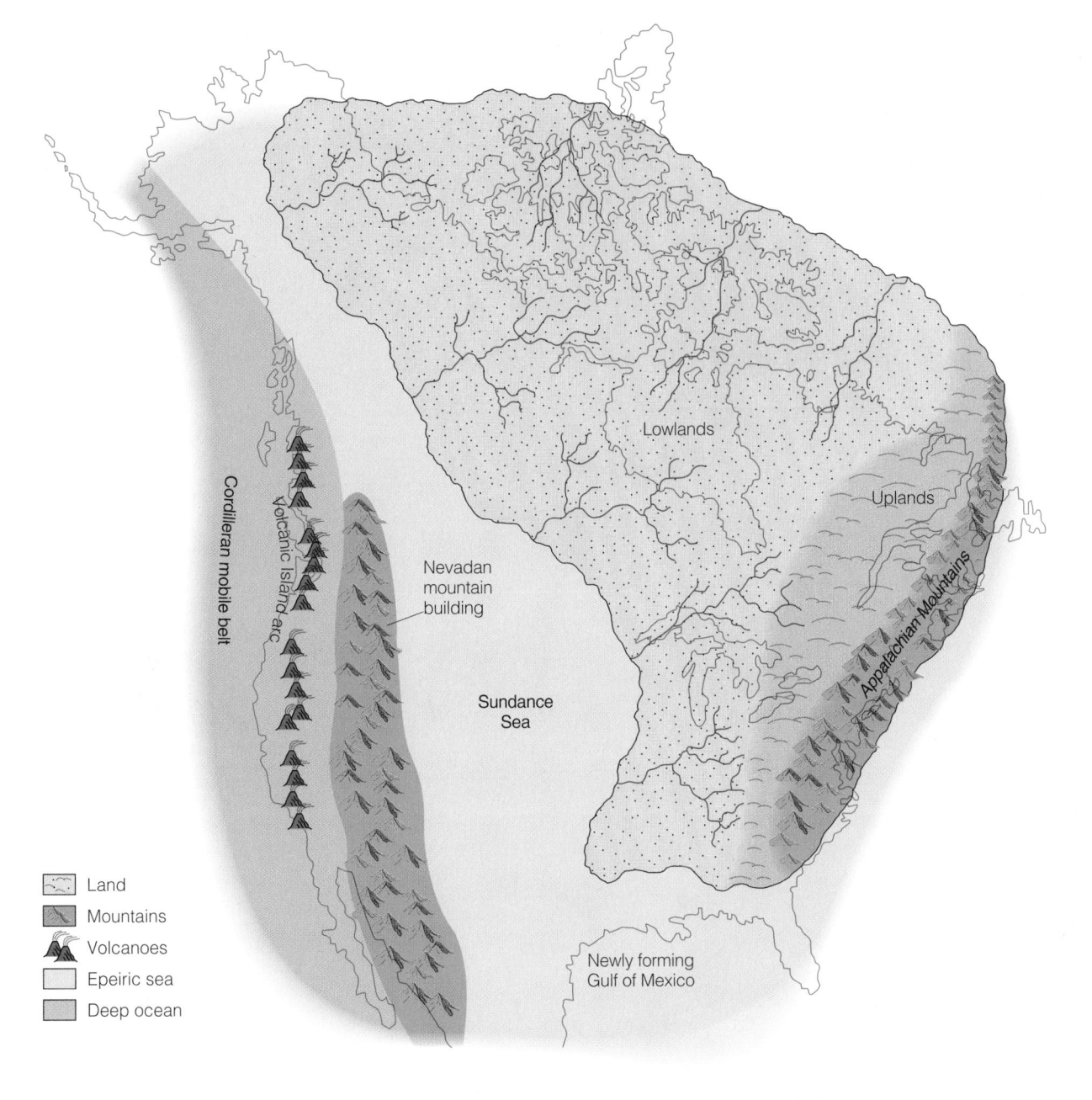

Land

Mountains

Volcanoes

Epeiric sea

Deep ocean

FIGURE 14.5 Paleogeography of North America during the Jurassic Period.

cussed later in this chapter). With this simplified overview of the Mesozoic history of North America in mind, we now examine the specific regions of the continent.

Continental Interior

Recall that the history of the North American craton can be divided into unconformity-bound sequences reflecting advances and retreats of epeiric seas over the craton (see Figure 10.3). While these transgressions and regressions played a major role in the Paleozoic geologic history of the continent, they were not as important during the Mesozoic. During the Mesozoic, most of the continental interior was well above sea level and did not experience epeiric sea inundation. Consequently, the two Mesozoic cratonic sequences, the Absaroka sequence (Late Mississippian to Early Jurassic) and **Zuni sequence** (Early Jurassic to Early Paleocene) (see Figure 10.3), are incorporated as part of the history of the three continental margin regions of North America.

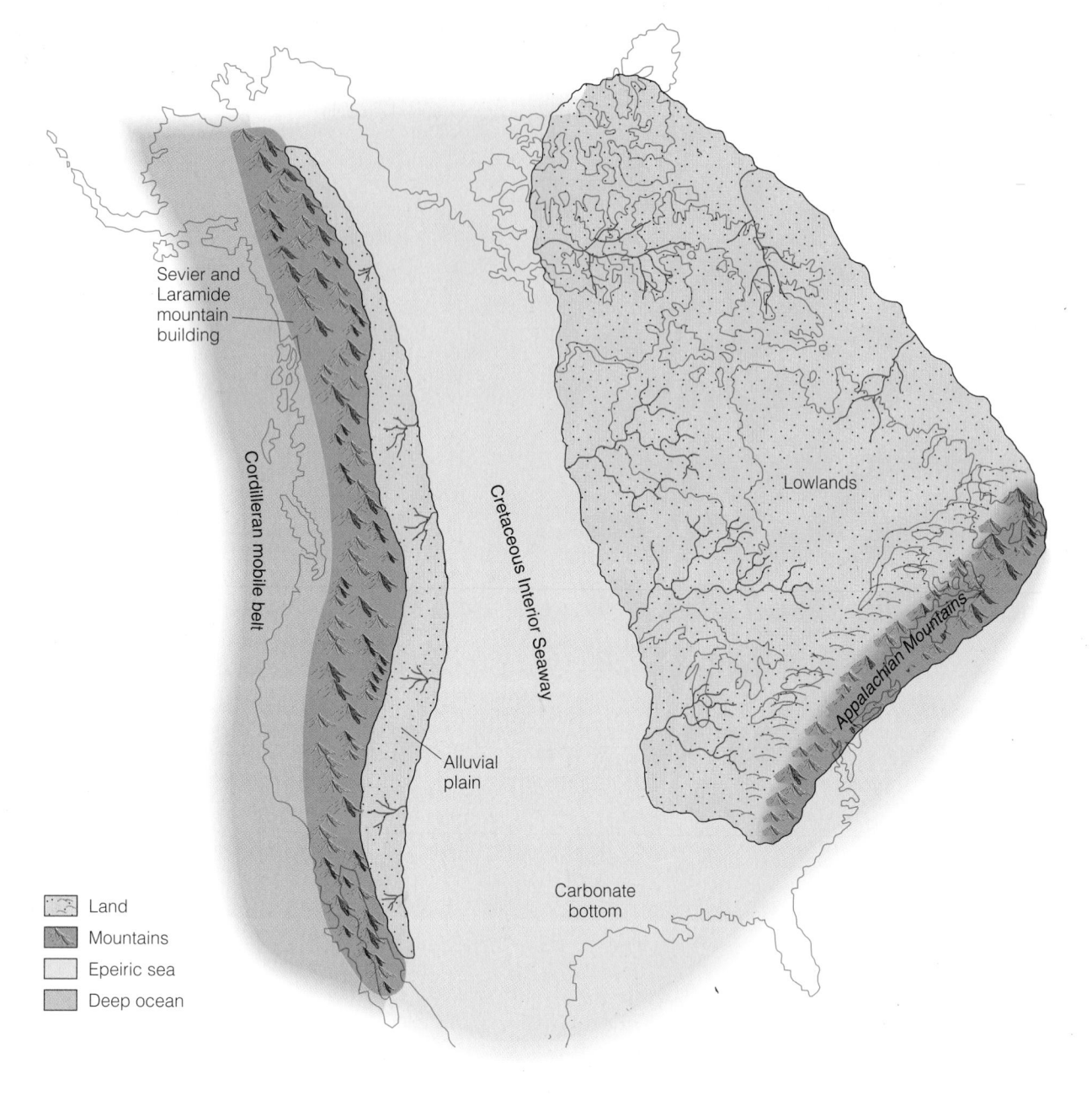

FIGURE 14.6 Paleogeography of North America during the Cretaceous Period.

Eastern Coastal Region

During the Early and Middle Triassic, coarse detrital sediments derived from the erosion of the recently uplifted Appalachians (Alleghenian orogeny) filled the various intermontane basins and spread over the surrounding areas. As erosion continued during the Mesozoic, this once lofty mountain system was reduced to a low-lying plain. During the Late Triassic, the first stage in the breakup of Pangaea began with North America separating from Africa. Fault-block basins developed in response to upwelling magma beneath Pangaea in a zone stretching from present-day Nova Scotia to North Carolina (Figure 14.7). Erosion of the adjacent fault-block mountains filled these basins with

great quantities (up to 6000 m) of poorly sorted red-colored nonmarine detrital sediments known as the *Newark Group*. Reptiles roamed along the margins of the various lakes and streams that formed in these basins, leaving their footprints and trackways in the soft sediments (Figure 14.8). Although the Newark rocks contain numerous dinosaur footprints, they are almost completely devoid of dinosaur bones. The Newark Group is mostly Late Triassic in age, but in some areas deposition began in the Early Jurassic.

Concurrent with sedimentation in the fault-block basins were extensive lava flows that blanketed the basin floors as well as intrusions of numerous dikes and sills. The most famous intrusion is the prominent Palisades sill along the

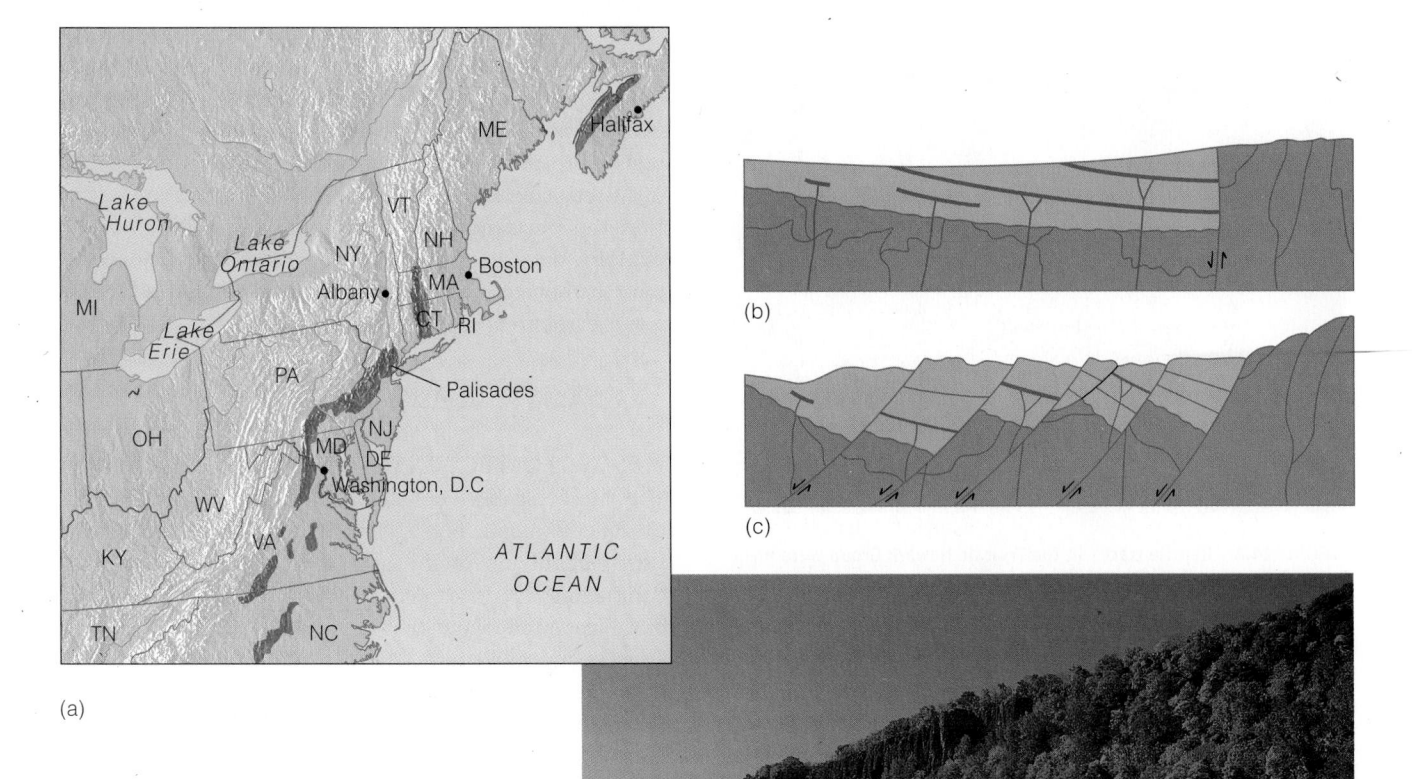

(a)

(b)

(c)

(d)

FIGURE 14.7 (a) Areas where Triassic fault-block basin deposits crop out in eastern North America. (b) After the Appalachians were eroded to a low-lying plain by the Middle Triassic, fault-block basins formed as a result of Late Triassic rifting between North America and Africa. (c) These valleys accumulated tremendous thickness of sediments and were themselves broken by a complex of normal faults during rifting. (d) Palisades of the Hudson River. This sill was one of many that were intruded into the Newark sediments during the Late Triassic rifting that marked the separation of North America from Africa.

Hudson River in the New York–New Jersey area (Figure 14.7d).

As the Atlantic Ocean grew, rifting ceased along the eastern margin of North America, and this once active plate margin became a passive, trailing continental margin. The fault-block mountains that were produced by this rifting continued eroding during the Jurassic and Early Cretaceous until all that was left was a large low-relief area. The sediments produced by this erosion contributed to the growing eastern continental shelf. During the Cretaceous Period, the Appalachian region was reelevated and once again shed sediments onto the continental shelf, forming a gently dipping, seaward-thickening wedge of rocks up to 3000 m thick. These rocks are currently exposed in a belt extending from Long Island, New York, to Georgia.

Gulf Coastal Region

The Gulf Coastal region was above sea level until the Late Triassic (Figure 14.4). As North America separated from South America during the Late Triassic, the Gulf of Mexico began to form (Figure 14.5). With oceanic waters flowing into this newly formed, shallow, restricted basin, conditions were ideal for evaporite formation. These Jurassic evaporites are thought to be the source for the Tertiary salt domes found today in the Gulf of Mexico and southern Louisiana (see Figure 16.30). The history of these salt domes and their associated petroleum accumulations will be discussed in Chapter 16.

By the Late Jurassic, circulation in the Gulf of Mexico was less restricted, and evaporite deposition ended.

FIGURE 14.8 Reptile tracks in the Triassic Newark Group were uncovered during the excavation for a new state building in Hartford, Connecticut. Because the tracks were so spectacular, the building site was moved, and the excavation was designated as a state park.

Normal marine conditions returned to the area with alternating transgressing and regressing seas. The resulting sediments were covered and buried by thousands of meters of Cretaceous and Cenozoic sediments.

During the Cretaceous, the Gulf Coastal region, like the rest of the continental margin, was inundated by northward-transgressing seas (Figure 14.6). As a result, nearshore sandstones are overlain by finer sediments characteristic of deeper waters. Following an extensive regression at the end of the Early Cretaceous, a major transgression began during which a wide seaway extended from the Arctic Ocean to the Gulf of Mexico (Figure 14.6). Sediments that were deposited in the Gulf Coastal region formed a seaward-thickening wedge.

Reefs were also widespread in the Gulf Coastal region during the Cretaceous. Bivalves called *rudists* were the main constituent of many of these reefs (see Figure 15.5). Because of their high porosity and permeability, rudistoid reefs make excellent petroleum reservoirs. A good example of a Cretaceous reef complex occurs in Texas (Figure 14.9). Here the reef trend had a strong influence on the carbonate platform deposition of the region. The facies patterns of these carbonate rocks are as complex as those found in the major barrier-reef systems of the Paleozoic Era.

Western Region

MESOZOIC TECTONICS

The Mesozoic geologic history of the North American Cordilleran mobile belt is very complex, involving the eastward subduction of the oceanic Pacific plate under the continental North American plate. Activity along this oceanic–continental convergent plate boundary resulted in an eastward movement of deformation. This orogenic activity progressively affected the trench and continental slope, the continental shelf, and the cratonic margin, causing a thickening of the continental crust.

Except for the Late Devonian–Early Mississippian Antler orogeny, the Cordilleran region of North America experienced little tectonism during the Paleozoic. During the Permian, however, an island arc and ocean basin formed off the western North American craton (Figure 14.4), followed by subduction of an oceanic plate beneath the island arc and the thrusting of oceanic and island arc rocks eastward against the craton margin (Figure 14.10). This event, known as the **Sonoma orogeny** occurred at or near the Permian–Triassic boundary.

Following the Late Paleozoic–Early Mesozoic destruction of the volcanic island arc during the Sonoma orogeny,

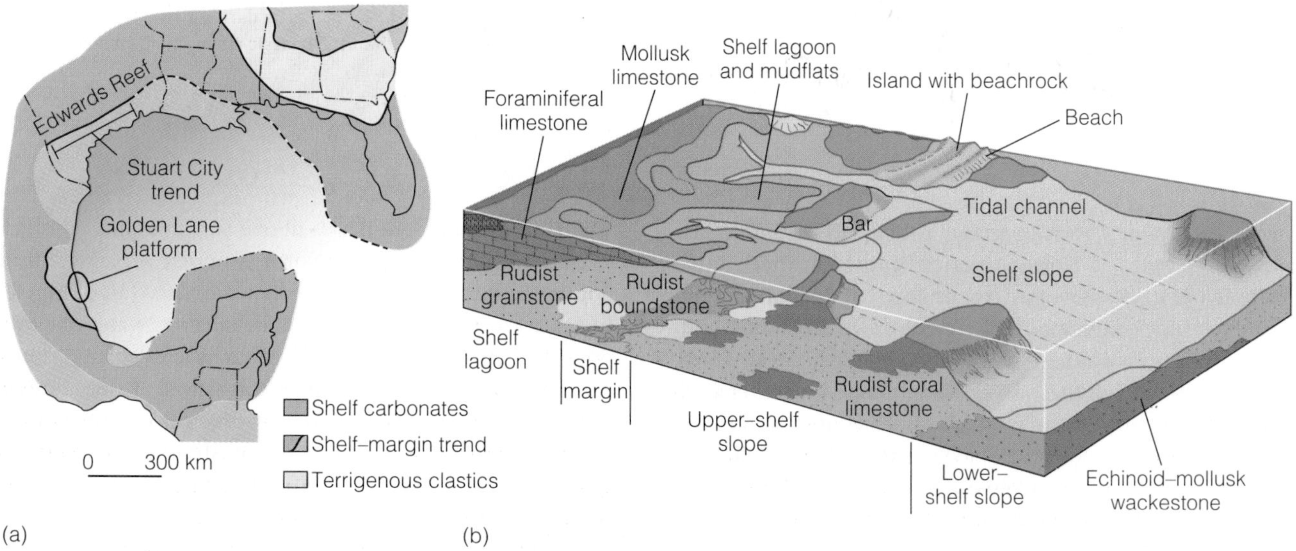

FIGURE 14.9 (a) Early Cretaceous shelf-margin facies around the Gulf of Mexico Basin. The reef trend is shown as a black line. (b) Reconstruction of the depositional environment and facies changes across the Stuart City reef trend, South Texas.

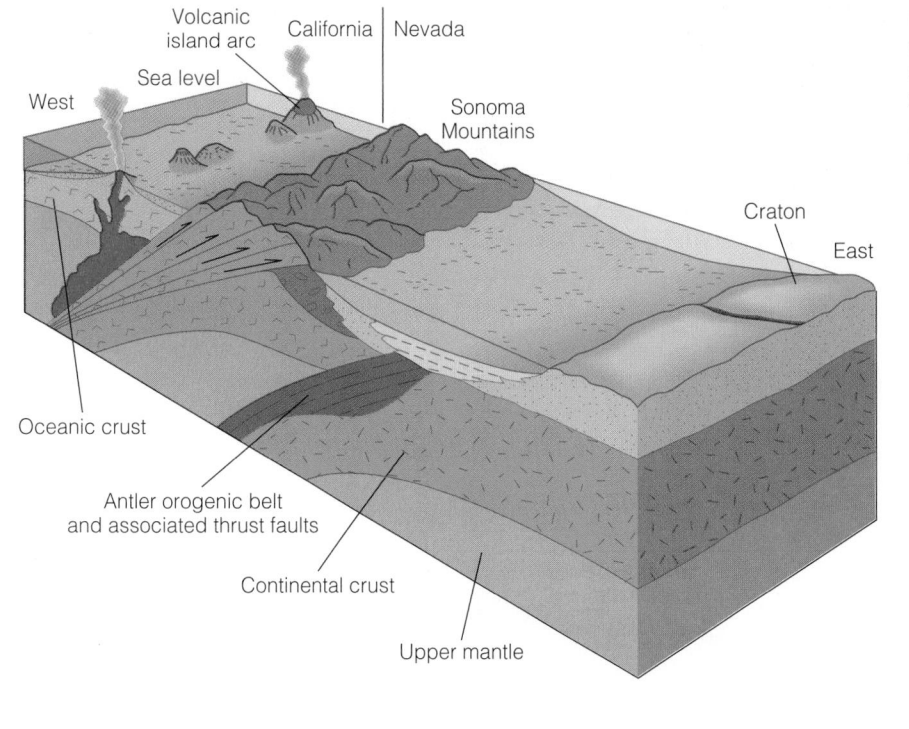

FIGURE 14.10 Tectonic activity that culminated in the Permian–Triassic Sonoma orogeny in western North America. The Sonoma orogeny was the result of a collision between the southwestern margin of North America and an island arc system.

the western margin of North America became an oceanic–continental convergent plate boundary. During the Late Triassic, a steeply dipping subduction zone developed along the western margin of North America in response to the westward movement of North America over the Pacific plate. This newly created oceanic–continental plate boundary controlled Cordilleran tectonics for the rest of the Mesozoic Era and for most of the Cenozoic Era; this subduction zone marks the beginning of the modern circum-Pacific orogenic system.

Two subduction zones, dipping in opposite directions from each other, formed off the west coast of North America during the Middle and early Late Jurassic (Figure 14.11). The more westerly subduction zone was eliminated by the westward-moving North American plate, which overrode the oceanic Pacific plate.

The *Franciscan Complex,* which is up to 7000 m thick, is an unusual rock unit consisting of a chaotic mixture of rocks that accumulated during the Late Jurassic and Cretaceous. The various rock types—graywacke, volcanic breccia, siltstone, black shale, chert, pillow basalt, and blueschist metamorphic rocks—suggest that continental-shelf, slope, and deep-sea environments were brought together in a submarine trench when North America overrode the subducting Pacific plate (Figure 14.12).

East of the Franciscan Complex and currently separated from it by a major thrust fault is the *Great Valley Group.* It consists of more than 16,000 m of Cretaceous conglomerates, sandstones, siltstones, and shales. These sediments were deposited on the continental shelf and slope at the same time the Franciscan deposits were accumulating in the submarine trench (Figure 14.12).

The general term **Cordilleran orogeny** is applied to the mountain-building activity that began during the

Jurassic and continued into the Cenozoic (Figure 14.13). The Cordilleran orogeny consisted of a series of individual mountain-building events that took place in different regions at different times. Most of this Cordilleran orogenic activity is related to the continued westward movement of the North American plate.

The first phase of the Cordilleran orogeny, the **Nevadan orogeny** (Figure 14.13), began during the Late Jurassic and continued into the Cretaceous as large volumes of granitic magma were generated at depth beneath the western edge of North America. These granitic masses ascended as huge batholiths that are now recognized as the Sierra Nevada, Southern California, Idaho, and Coast Range batholiths (Figure 14.14).

By the Late Cretaceous, most of the volcanic and plutonic activity had migrated eastward into Nevada and Idaho. This migration was probably caused by a change from high-angle to low-angle subduction, which resulted in the subducting oceanic plate reaching its melting depth farther east (Figure 14.15). Thrusting occurred progressively farther east so that by the Late Cretaceous, it extended all the way to the Idaho–Washington border.

The second phase of the Cordilleran orogeny, the **Sevier orogeny,** was mostly a Cretaceous event (Figure 14.13). Subduction of the Pacific plate beneath the North American plate continued during this time, resulting in numerous overlapping, low-angle thrust faults in which blocks of older strata were thrust eastward on top of younger strata (Figure 14.16). This deformation produced generally north-south–trending mountain ranges that stretch from Montana to western Canada.

During the Late Cretaceous to Early Cenozoic, the final pulse of the Cordilleran orogeny occurred (Figure 14.13). The **Laramide orogeny** developed east of the Sevier oro-

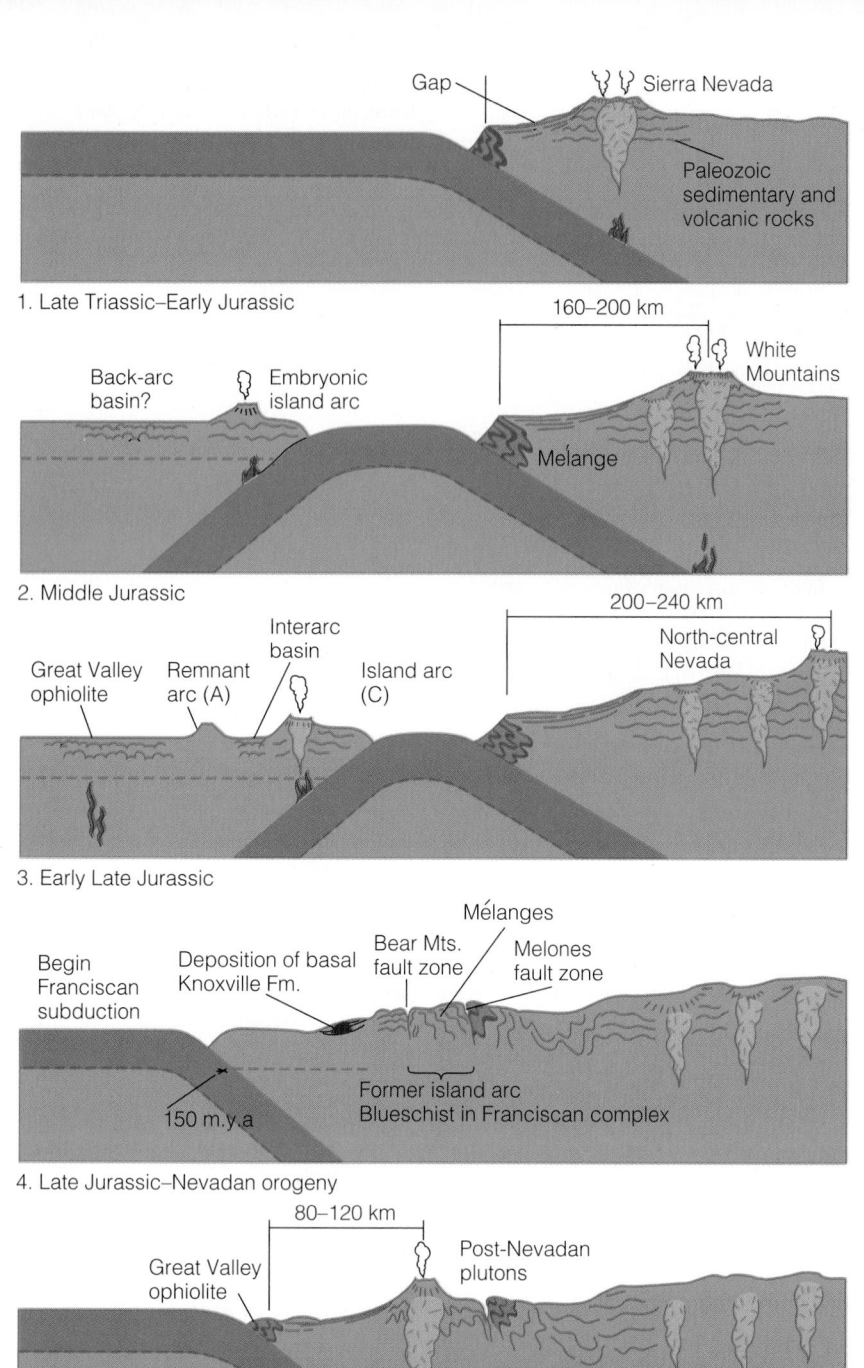

FIGURE 14.11 Interpretation of the tectonic evolution of the Sierra Nevada during the Mesozoic Era.

Gap — Sierra Nevada

Paleozoic sedimentary and volcanic rocks

1. Late Triassic–Early Jurassic

160–200 km

White Mountains

Back-arc basin? — Embryonic island arc

Mélange

2. Middle Jurassic

200–240 km

North-central Nevada

Great Valley ophiolite — Remnant arc (A) — Interarc basin — Island arc (C)

3. Early Late Jurassic

Mélanges

Bear Mts. fault zone — Melones fault zone

Begin Franciscan subduction — Deposition of basal Knoxville Fm.

150 m.y.a

Former island arc
Blueschist in Franciscan complex

4. Late Jurassic–Nevadan orogeny

80–120 km

Great Valley ophiolite — Post-Nevadan plutons

5. Latest Jurassic

genic belt in the present-day Rocky Mountain areas of New Mexico, Colorado, and Wyoming. Most of the features of the present-day Rocky Mountains resulted from the Cenozoic phase of the Laramide orogeny, and for that reason, it will be discussed in Chapter 16.

MESOZOIC SEDIMENTATION

Concurrent with the tectonism in the Cordilleran mobile belt, Early Triassic sedimentation on the western continental shelf consisted of shallow-water marine sandstones, shales, and limestones. During the Middle and Late

Triassic, the western shallow seas regressed farther west, exposing large areas of seafloor to erosion. Marginal marine and nonmarine Triassic rocks, particularly red beds, contribute to the spectacular and colorful scenery of the region.

The Lower Triassic *Moenkopi Formation* of the southwestern United States consists of a succession of brick-red and chocolate-colored mudstones (Figure 14.17). Such sedimentary structures as desiccation cracks and ripple marks, as well as fossil amphibians and reptiles and their tracks, indicate deposition in a variety of continental environ-

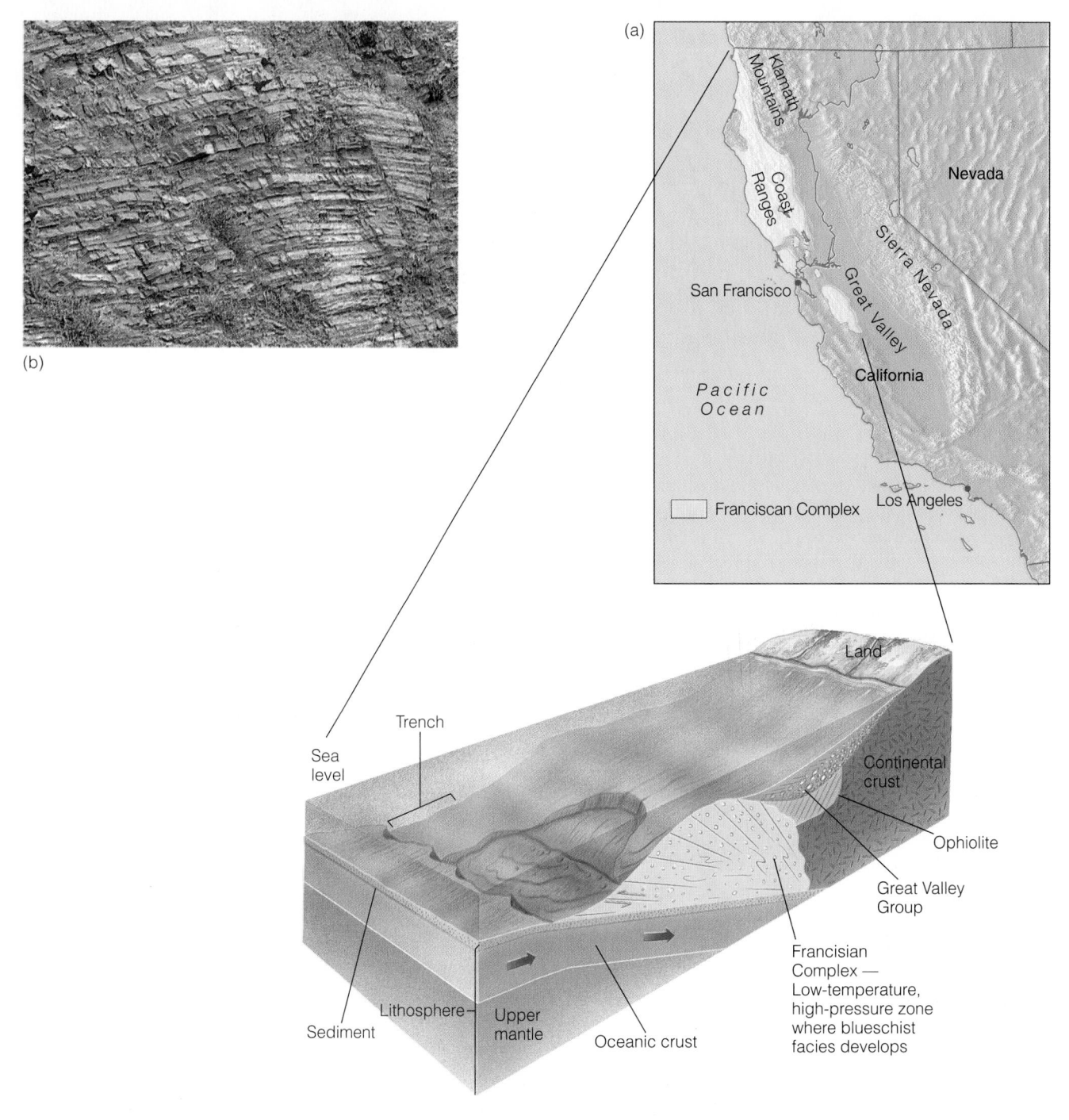

FIGURE 14.12 (a) Reconstruction of the depositional environment of the Franciscan Complex during the Late Jurassic and Cretaceous periods. (b) Exposures of the Franciscan Complex along the central California coast.

ments, including stream channels, floodplains, and fresh and brackish water ponds. Thin tongues of marine limestones indicate brief incursions of the sea, while local beds with gypsum and halite crystal casts attest to a rather arid climate.

Unconformably overlying the Moenkopi is the Upper Triassic *Shinarump Conglomerate,* a widespread unit generally less than 50 m thick.

Above the Shinarump are the multicolored shales, siltstones, and sandstones of the Upper Triassic *Chinle*

Formation (Figure 14.17). This formation is widely exposed over the Colorado Plateau and is probably most famous for its petrified wood, spectacularly exposed in Petrified Forest National Park, Arizona (see Perspective 14.1). While fossil ferns are found here, the park is best known for its abundant and beautifully preserved logs of gymnosperms, especially conifers and plants called *cycads* (see Figure 15.10a). Fossilization resulted from the silicification of the plant tissues. Weathering of volcanic ash beds interbedded with fluvial and deltaic Chinle sediments pro-

perspective 14.1

Petrified Forest National Park

Petrified Forest National Park is located in eastern Arizona about 42 km east of Holbrook. The park consists of two sections: the Painted Desert, which is north of Interstate 40, and the Petrified Forest, which is south of the Interstate.

The Painted Desert is a brilliantly colored landscape whose colors and hues change constantly throughout the day. The multicolored rocks of the Triassic Chinle Formation have been weathered and eroded to form a badlands topography of numerous gullies, valleys, ridges, mounds, and mesas. The Chinle Formation is composed predominantly of various-colored shale beds. These shales and associated volcanic ash layers are easily weathered and eroded. Interbedded locally with the shales are lenses of conglomerates, sandstones, and limestones, which are more resistant to weathering and erosion than the shales and form resistant ledges.

The Petrified Forest was originally set aside as a national monument to protect the large number of petrified logs that lay exposed in what is now the southern part of the park (Figure 1). When the transcontinental railroad constructed a coaling and watering stop in Adamana, Arizona, passengers were encouraged to take excursions to "Chalcedony Park," as the area was then called, to see the petrified forests. In a short time, collectors and souvenir hunters hauled off tons of petrified wood, quartz crystals, and Native American relics. It was not until a huge rock crusher was built to crush the logs for the manufacture of abrasives that the area

was declared a national monument and the petrified forests preserved and protected.

During the Triassic Period, the climate of the area was much wetter than today, with many rivers, streams, and lakes. About 40 different fossil plant species have been identified from the Chinle Formation. These include numerous seedless vascular plants such as rushes and ferns, as well as gymnosperms such as cycads and conifers. Such plants thrive in floodplains and marshes. Most logs are conifers and belong to the genus *Araucarioxylon*. Some of these trees were more than 60 m tall and up to 4 m in diameter. Apparently, most of the conifers grew on higher ground or riverbanks.

Although many trees were buried in place, most appear to have been uprooted and transported by raging streams during times of flooding. Burial of the logs was rapid, and groundwater saturated with silica from the ash of nearby volcanic eruptions quickly permineralized the trees.

Deposition continued in the Colorado Plateau region during the Jurassic and Cretaceous, further burying the Chinle Formation. During the Laramide orogeny, the Colorado Plateau area was uplifted and eroded, exposing the Chinle Formation. Because the Chinle is mostly shales, it was easily eroded, leaving the more resistant petrified logs and log fragments exposed on the surface—much as we see them today.

FIGURE 1 Petrified Forest National Park, Arizona. All the logs here are *Araucarioxylon*, which is the most abundant tree in the park. The petrified logs have been weathered from the Chinle Formation and are mostly in the position in which they were buried some 200 million years ago.

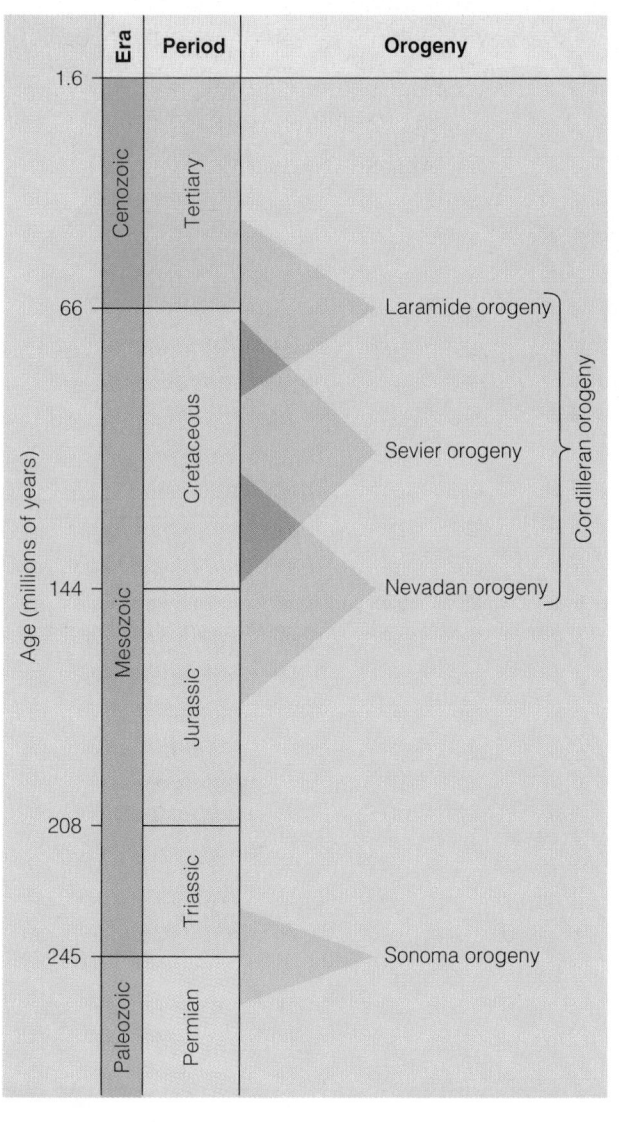

FIGURE 14.13 Mesozoic orogenies occurring in the Cordilleran mobile belt.

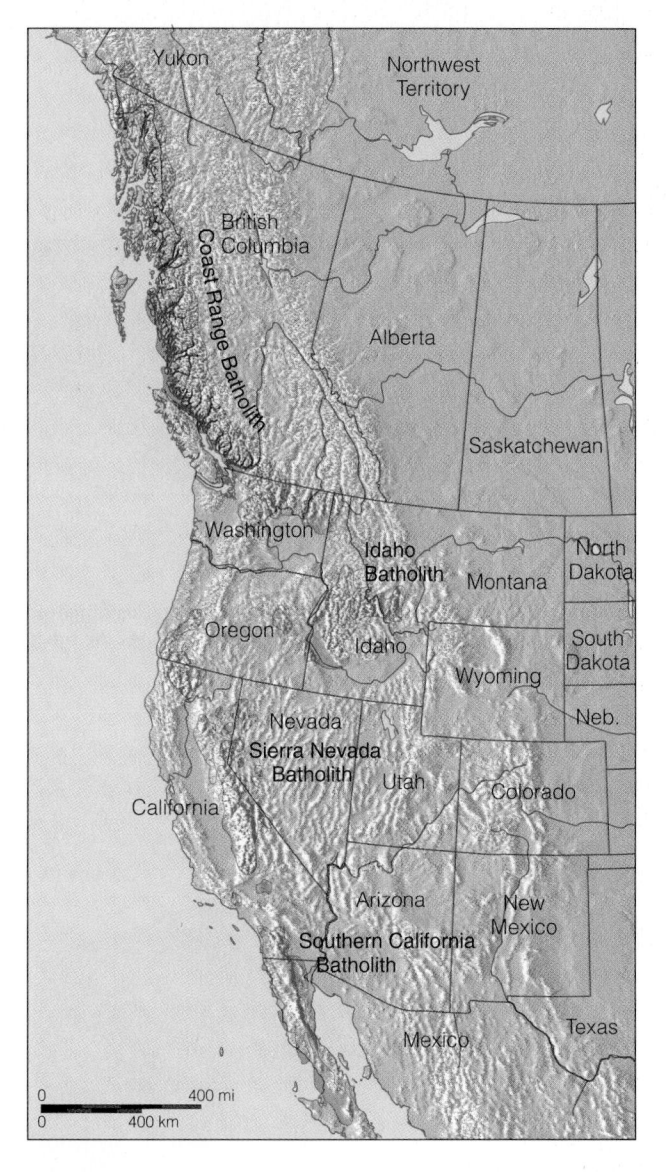

FIGURE 14.14 Location of Jurassic and Cretaceous batholiths in western North America.

vided most of the silica for silicification. Some trees were preserved in place, but most were transported during floods and deposited on sandbars and on floodplains, where fossilization took place. After burial, silica-rich groundwater percolated through the sediments and silicified the wood.

Though best known for its petrified wood, the Chinle Formation has also yielded fossils of labyrinthodont amphibians, phytosaurs, and small dinosaurs (see Chapter 15 for a discussion of the latter two animal groups).

The *Wingate Sandstone,* a desert dune deposit, and the *Kayenta Formation,* a stream and lake deposit, overlie the Chinle Formation. These two formations are well exposed in southwestern Utah and complete the Triassic stratigraphic succession in the Southwest (Figure 14.17).

Early Jurassic deposits in a large part of the western region consist mostly of clean, cross-bedded sandstones indicative of windblown deposits. The thickest and most prominent of these is the *Navajo Sandstone,* a widespread

cross-bedded sandstone that accumulated in a coastal dune environment along the southwestern margin of the craton. The sandstone's most distinguishing feature is its large-scale cross-beds, some of which are more than 25 m high (Figure 14.18). The upper part of the Navajo contains smaller cross-beds as well as dinosaur and crocodilian fossils.

Marine conditions returned to the region during the Middle Jurassic when a seaway called the Sundance Sea twice flooded the interior of western North America (Figure 14.5). The resulting deposits, the *Sundance Formation,* were produced from erosion of tectonic highlands to the west that paralleled the shoreline. These highlands resulted from intrusive igneous activity and associated volcanism that began during the Triassic.

During the Late Jurassic, a mountain chain formed in Nevada, Utah, and Idaho as a result of the deformation produced by the Nevadan orogeny. As the mountain chain grew and shed sediments eastward, the Sundance Sea be-

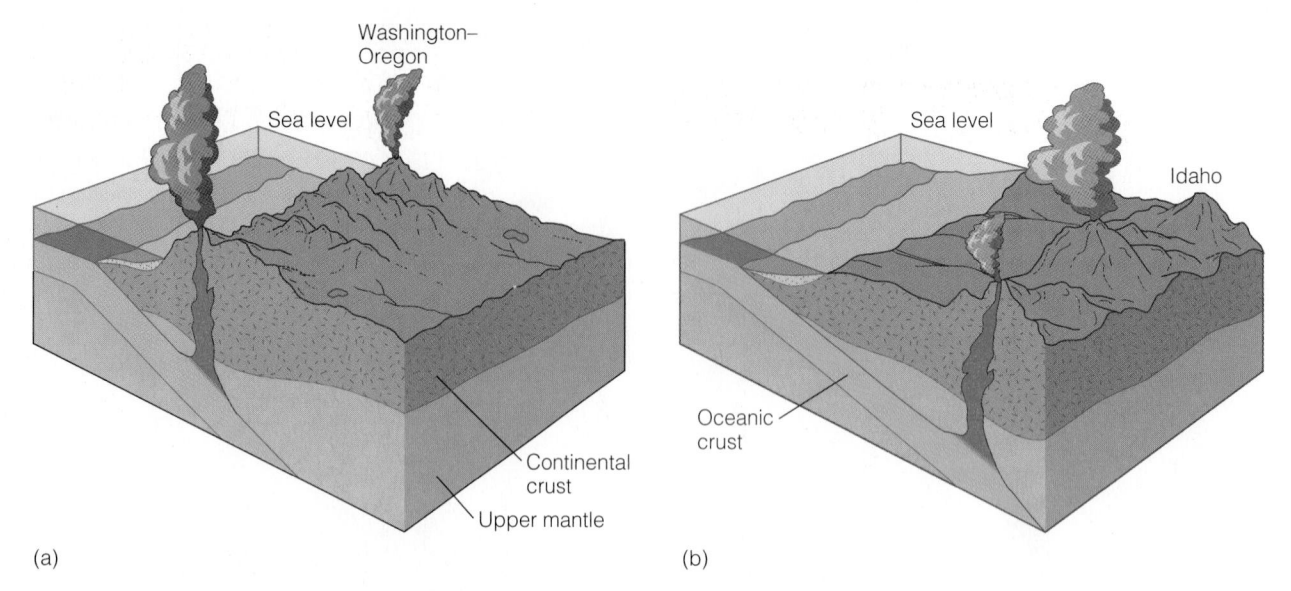

FIGURE 14.15 A possible cause for the eastward migration of igneous activity in the Cordilleran region during the Cretaceous Period was a change from (a) high-angle to (b) low-angle subduction. As the subducting plate moved downward at a lower angle, the depth of melting moved farther to the east.

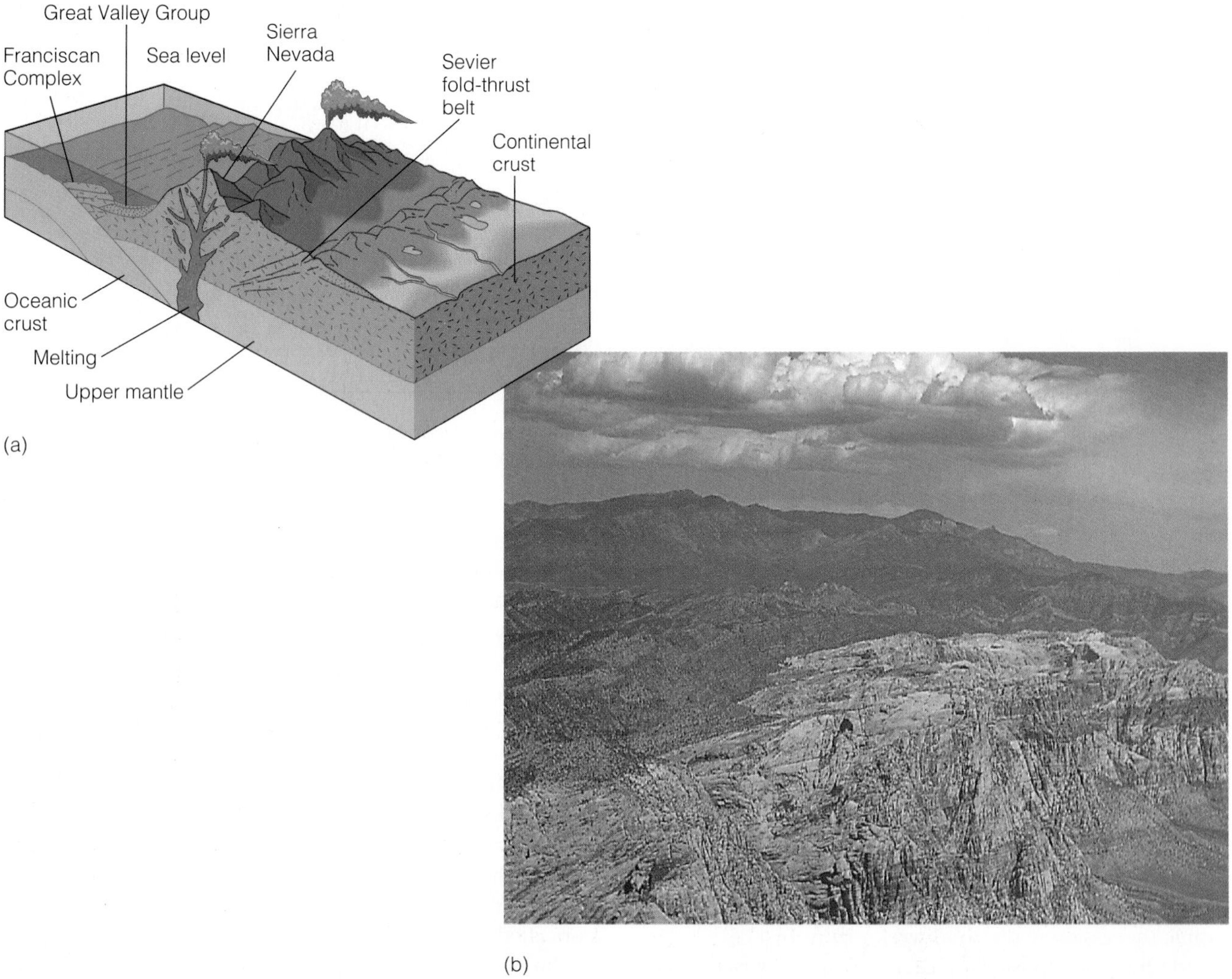

FIGURE 14.16 (a) Restoration showing the associated tectonic features of the Late Cretaceous Sevier orogeny due to subduction of the Pacific plate under the North American plate. (b) The Keystone thrust fault, a major fault in the Sevier overthrust belt is exposed west of Las Vegas, Nevada. The sharp boundary between the light-colored Mesozoic rocks and the overlying dark-colored Paleozoic rocks marks the trace of the Keystone thrust fault. *(Photo © John Shelton)*

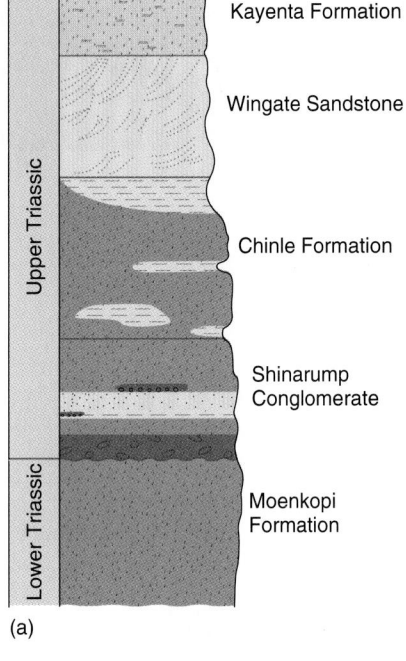

(a)

(b)

FIGURE 14.17 (a) Stratigraphic section of Triassic formations in the western United States. (b) View of East Entrance of Zion Canyon, Zion National Park, Utah. The light-colored massive rocks are the Jurassic Navajo Sandstone, while the slope-forming rocks below the Navajo are the Upper Triassic Kayenta Formation.

gan retreating northward. A large part of the area formerly occupied by the Sundance Sea was then covered by multi-colored sandstones, mudstones, shales, and occasional lenses of conglomerates that comprise the world-famous *Morrison Formation* (Figure 14-19).

The Morrison Formation contains the world's richest assemblage of Jurassic dinosaur remains. Although most of the dinosaur skeletons are broken up, as many as 50 individuals have been found together in a small area. Such a concentration indicates that the skeletons were brought together during times of flooding and deposited on sandbars in stream channels. Soils in the Morrison indicate that the climate was seasonally dry. The presence of turtle, crocodile, and fish fossils also indicates that the lakes at the time were probably very shallow and saline.

Although most major museums have either complete dinosaur skeletons or at least bones from the Morrison Formation, the best place to see the bones still embedded in the rocks is the visitors' center at Dinosaur National Monument near Vernal, Utah (see Perspective 14.2).

Shortly before the end of the Early Cretaceous, Arctic waters spread southward over the craton, forming a large inland sea in the Cordilleran foreland basin area. Mid-Cretaceous transgressions also occurred on other continents, and all were part of the global mid-Cretaceous rise in sea level that resulted from accelerated seafloor spreading as Pangaea continued to fragment.

FIGURE 14.18 Large cross-beds of the Jurassic Navajo Sandstone in Zion National Park, Utah.

FIGURE 14.19 Panoramic view of the Jurassic Morrison Formation as seen from the visitors' center at Dinosaur National Monument, Utah.

Dinosaur National Monument

*I*n 1909 Earl Douglass of the Carnegie Museum discovered "Dinosaur Ledge," a sandstone unit in the Upper Jurassic Morrison Formation that contained numerous dinosaur bones. After 13 years of excavating the layer, the Carnegie Museum had removed parts of 300 dinosaur specimens, two dozen of which were sufficiently complete to be reassembled (Figure 1). Ten different dinosaur species were represented, as well as many other reptiles. It was by far the finest collection of dinosaur remains in the world. In the years that followed, the Smithsonian Institution and the University of Utah also worked this rich quarry and recovered more fossil specimens.

Recognizing the scientific importance of this unit, the dinosaur quarry and 80 acres surrounding it were designated a national monument by President Wilson on October 4, 1915. Less than a year later, it was included in the newly created National Park System. In 1938 Dinosaur National Monument was further expanded to 200,000 acres.

As far back as 1915, Earl Douglass envisioned an exhibit in which the dinosaur bones would be exposed in relief in the tilted sandstone bed exactly where they came to rest 140 million years ago. In a letter to Dr. Charles Walcott, secretary of the Smithsonian Institution, Douglass wrote: "I hope that the Government, for the benefit of science and the people, will uncover a large area, leave the bones and skeletons in relief, and house them in

place. It would make one of the most astounding and instructive sights imaginable." It was not until 1953, however, that work begin on a museum constructed around the still-buried sandstone ledge. The sediment overlying the remaining tilted sandstone ledge was removed, and the dinosaur bones were carefully exposed in bas relief. This quarry wall now forms the north wall of the visitors' center at Dinosaur National Monument (Figure 2). The structure was completed and opened to the public in 1958.

What was the landscape like during the Jurassic when dinosaurs roamed the area that is now Dinosaur National Monument?

The land for kilometers around the present-day quarry was a low-lying desert during the Early Jurassic. This is indicated by the large cross-bedded sand dunes of the Navajo Sandstone. Following deposition of the Navajo, a shallow sea transgressed from the west, depositing sandstones, siltstones, and limestones. During the Late Jurassic, the area from Mexico to Canada and from central Utah to the Mississippi River was above sea level. To the west was a mountain chain. Streams flowing from these mountains deposited sand and silt in the adjacent plains. Small lakes were numerous, as well as some swamps. These stream, lake, and

FIGURE 1 *Camarasaurus lentus* is one of the best preserved and complete skeletons recovered from the Morrison Formation by the Carnegie Museum at Dinosaur National Monument.

(a)

(b)

FIGURE 2 (a) Visitors' center, Dinosaur National Monument. (b) North wall of visitors' center showing dinosaur bones in bas relief, just as they were deposited 140 million years ago.

swamp deposits of the Jurassic coastal plain comprise the Morrison Formation.

Semitropical conditions prevailed during the Late Jurassic in the area, and forests of ginkos, cycads, and tree-ferns covered the land. In this setting pterosaurs glided through the air while dinosaurs roamed the landscape below. Crocodiles, turtles, and small mammals were also present. Most of the bones of the dinosaurs and other animals were deposited in the streambeds. During Cretaceous through Eocene time, the area was uplifted during the Laramide orogeny, and later the Green and Yampa Rivers cut deep canyons in the area, exposing the rocks that make up Dinosaur National Monument.

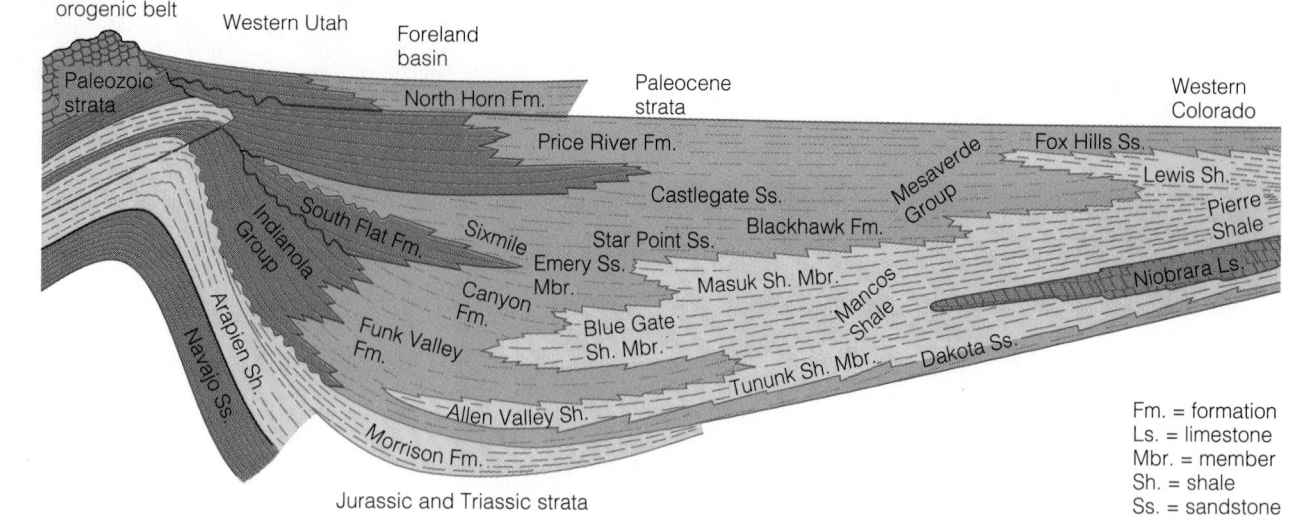

FIGURE 14.20 This restored west–east cross section of Cretaceous facies of the western Cretaceous Interior Seaway shows their relationship to the Sevier orogenic belt.

By the beginning of the Late Cretaceous, this incursion joined the northward-transgressing waters from the Gulf area to create an enormous **Cretaceous Interior Seaway** that occupied the area east of the Sevier orogenic belt. Extending from the Gulf of Mexico to the Arctic Ocean and more than 1500 km wide at its maximum extent, this seaway effectively divided North America into two large landmasses until just before the end of the Late Cretaceous (Figure 14.7).

Cretaceous deposits less than 100 m thick indicate that the eastern margin of the Cretaceous Interior Seaway subsided slowly and received little sediment from the emergent, low-relief craton to the east. The western shoreline, however, shifted back and forth, primarily in response to fluctuations in the supply of sediment from the Cordilleran Sevier orogenic belt to the west. The facies relationships show lateral changes from conglomerate and coarse sandstone adjacent to the mountain belt through finer sandstones, siltstones, shales, and even limestones and chalks in the east (Figure 14.20). During times of particularly active mountain building, these coarse clastic wedges of gravel and sand prograded even further east.

As the Mesozoic Era ended, the Cretaceous Interior Seaway withdrew from the craton. During this regression, marine waters retreated to the north and south, and marginal marine and continental deposition formed widespread coal-bearing deposits on the coastal plain.

Microplate Tectonics and the Growth of Western North America

In the preceding sections, we discussed orogenies along convergent plate boundaries resulting in continental accretion. Much of the material accreted to continents during such events is simply eroded older continental crust, but a significant amount of new material is added to continents as well—igneous rocks that formed as a consequence of subduction and partial melting, for example. While subduction is the predominant influence on the tectonic history in many regions of orogenesis, other processes are also involved in mountain building and continental accretion, especially the accretion of microplates.

Geologists now know that portions of many mountain systems are composed of small accreted lithospheric blocks that are clearly of foreign origin. These **microplates** differ completely in their fossil content, stratigraphy, structural trends, and paleomagnetic properties from the rocks of the surrounding mountain system and adjacent craton. In fact, these microplates are so different from adjacent rocks that most geologists think they formed elsewhere and were carried great distances as parts of other plates until they collided with other microplates or continents.

Geologic evidence indicates that more than 25% of the entire Pacific Coast from Alaska to Baja California consists of accreted microplates. The accreting microplates are composed of volcanic island arcs, oceanic ridges, seamounts, and small fragments of continents that were scraped off and accreted to the continent's margin as the oceanic plate with which they were carried was subducted under the continent. It is estimated that more than 100 different-sized microplates have been added to the western margin of North America during the last 200 million years (Figure 14.21).

The basic plate tectonic reconstruction of orogenies and continental accretion remains unchanged, but the details of such reconstructions are decidedly different in view of microplate tectonics. For example, growth along active continental margins is faster than along passive continental margins because of the accretion of microplates.

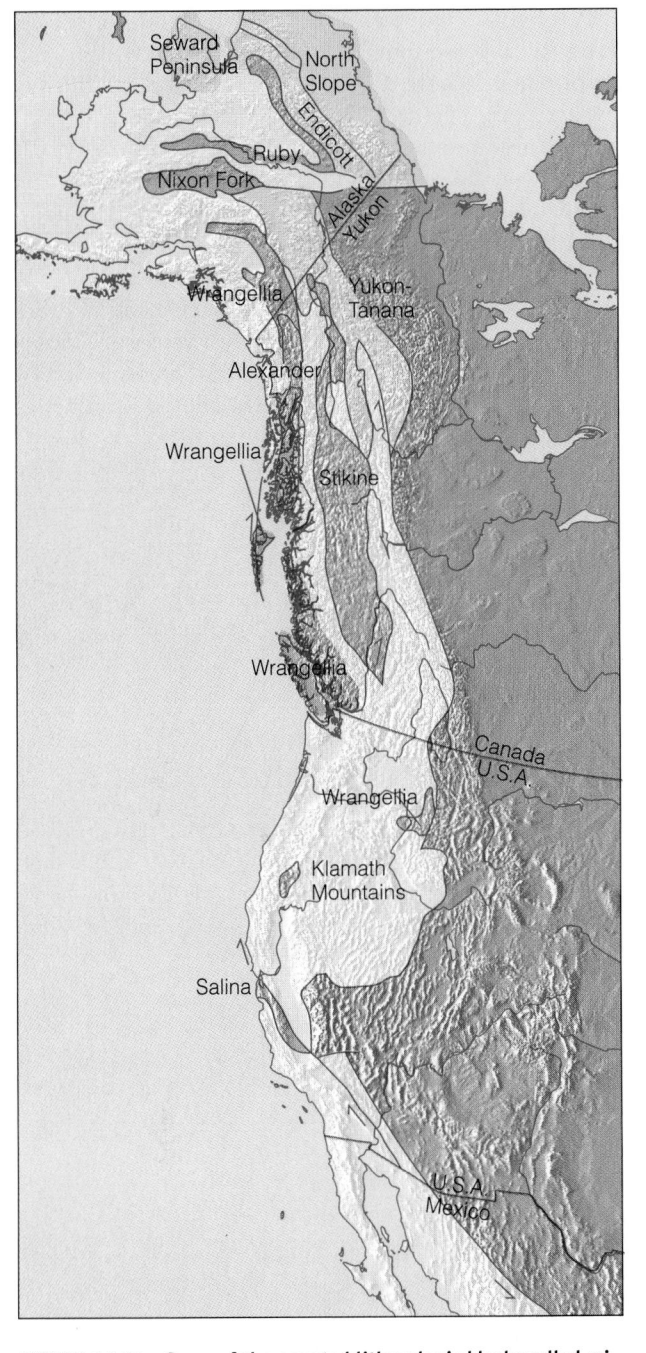

FIGURE 14.21 Some of the accreted lithospheric blocks called microplates that form the western margin of the North American Craton. The dark brown blocks probably originated as parts of continents other than North America. The light green blocks are possibly displaced parts of North America. The North American craton is shown in dark green.

Furthermore, these accreted microplates are often new additions to a continent, rather than reworked older continental material.

So far, most microplates have been identified in mountains of the North American Pacific Coast region, but a number of such plates are suspected to be present in other mountain systems as well. They are more difficult to recognize in older mountain systems, such as the Appalachians, however, because of greater deformation and erosion.

Thus, microplate tectonics provides another way of viewing Earth and gaining a better understanding of the geologic history of the continents.

Mesozoic Mineral Resources

Although much of the coal in North America is Pennsylvanian or Tertiary in age, important Mesozioc coals occur in the Rocky Mountains states. These are mostly lignite and bituminous coals, but some local anthracites are present as well. Particularly widespread in western North American are coals of Cretaceous age. Mesozoic coals are also known from Alberta and British Columbia, Canada, as well as from Australia, Russia, and China.

Large concentrations of petroleum occur in many areas of the world, but more than 50% of all proven reserves are in the Persian Gulf region. During the Mesozoic Era, what is now the Gulf region was a broad passive continental margin extending eastward from Africa. This continental margin lay near the equator where countless microorganisms lived in the surface waters, particularly during the Cretaceous Period when most of the petroleum formed. The remains of these organisms accumulated with the bottom sediments and were buried, beginning the complex processes of oil generation.

Similar conditions existed in what is now the Gulf Coast region of the United States and Central America. Here, petroleum and natural gas also formed on a broad shelf over which transgressions and regressions occurred. In this region, the hydrocarbons are largely in reservoir rocks that were deposited as distributary channels on deltas and as barrier-island and beach sands. Some of these hydrocarbons are associated with structures formed adjacent to rising salt domes. The salt, called the *Louann Salt*, initially formed in a long, narrow sea when North America separated from Europe and North Africa during the fragmentation of Pangaea (Figure 14.2).

The richest uranium ores in the United States are widespread in Mesozoic rocks of the Colorado Plateau area of Colorado and adjoining parts of Wyoming, Utah, Arizona, and New Mexico. These ores, consisting of fairly pure masses of a complex potassium-, uranium-, vanadium-bearing mineral called *carnotite*, are associated with plant remains in sandstones that were deposited in ancient stream channels.

As noted in Chapter 9, Proterozoic banded iron formations are the main sources of iron ores. There are, however, important exceptions. For example, the Jurassic-age "Minette" iron ores of Western Europe, composed of oolitic limeonite and hematite, are important ores in France, Germany, Belgium, and Luxembourg. In Great Britain, low-grade iron ores of Jurassic age consist of oolitic siderite, which is an iron carbonate. And in Spain, Cretaceous rocks are the host rocks for iron minerals.

South Africa, the world's leading producer of gem-quality diamonds and among the leaders in industrial diamond production, mines these minerals from conical igneous intrusions called kimberlite pipes. Kimberlite pipes are composed of dark gray or blue igneous rock known as kimberlite. Diamonds, which form at great depth where pressure and temperature are high, are brought to the surface during the explosive volcanism that forms kimberlite pipes. Although kimberlite pipes have formed throughout geologic time, the most intense episode of such activity in South Africa and adjacent countries was during the Cretaceous Period. Emplacement of Triassic and Jurassic diamond-bearing kimberlites also occurred in Siberia.

In the Prologue we noted that the mother lode or source for the placer deposits mined during the California gold rush is in Jurassic-age intrusive rocks of the Sierra Nevada. Gold placers are also known in Cretaceous-age conglomerates of the Klamath Mountains of California and Oregon.

Porphyry copper was originally named for copper deposits in the western United States mined from porphyritic granodiorite, but the term now applies to large, low-grade copper deposits disseminated in a variety of rocks. These prophyry copper deposits are an excellent example of the relationship between convergent plate boundaries and the distribution, concentration, and exploitation of valuable metallic ores. Magma generated by partial melting of a subducting plate rises toward the surface, and as it cools, it precipitates and concentrates various metallic ores. The world's largest copper deposits were formed during the Mesozoic and Tertiary in a belt along the western margins of North and South America (see Figure 6.30).

Summary

Table 14.1 provides a summary of the geologic history of North America, as well as global events and sea-level changes during the Mesozoic.

1. The breakup of Pangaea can be divided into four stages.

   a. The first stage involved the separation of North America from Africa during the Late Triassic, followed by the separation of North America from South America.

   b. The second stage involved the separation of Antarctica, India, and Australia from South America and Africa during the Jurassic. During this stage, India broke away from the still-united Antarctica and Australia landmass.

   c. During the third stage, which began in the Late Jurassic, South America separated from Africa, while Europe and Africa began converging.

   d. In the last stage, during the Cenozoic Era, Greenland separated from North America and Europe.

2. The breakup of Pangaea influenced global climatic and atmospheric circulation patterns. While the temperature gradient from the tropics to the poles gradually increased during the Mesozoic, overall global temperatures remained equable.

3. An increased rate of seafloor spreading during the Cretaceous Period caused sea level to rise and transgressions to occur.

4. Except for incursions along the continental margin and two major transgressions (the Sundance Sea and the Cretaceous Interior Seaway), the North American craton was above sea level during the Mesozoic Era.

5. The Eastern Coastal Plain was the initial site of the separation of North America from Africa that began during the Late Triassic. During the Cretaceous Period, it was inundated by marine transgressions.

6. The Gulf Coastal region was the site of major evaporite accumulation during the Jurassic as North America rifted from South America. During the Cretaceous, it was inundated by a transgressing sea, which, at its maximum, connected with a sea transgressing from the north to create the Cretaceous Interior Seaway.

7. Mesozoic rocks of the western region of North America were deposited in a variety of continental and marine environments. One of the major controls of sediment distribution patterns was tectonism.

8. Western North America was affected by four interrelated orogenies: the Sonoma, Nevadan, Sevier, and Laramide. Each involved igneous intrusions, as well as eastward thrust faulting and folding.

9. The cause of the Sonoma, Nevadan, Sevier, and Laramide orogenies was the changing angle of subduction of the oceanic Pacific plate under the continental North American plate. The timing, rate, and, to some degree, the direction of plate movement were related to seafloor spreading and the opening of the Atlantic Ocean.

10. Orogenic activity associated with the oceanic–continental convergent plate boundary in the Cordilleran mobile belt explains the structural features of the western margin of North America. It is thought, however, that more than 25% of the North American western margin originated from the accretion of microplates.

11. Mesozoic-age rocks contain a variety of mineral resources, including coal, petroleum, uranium, gold, and copper.

TABLE 14.1

Summary of Mesozoic Geologic Events

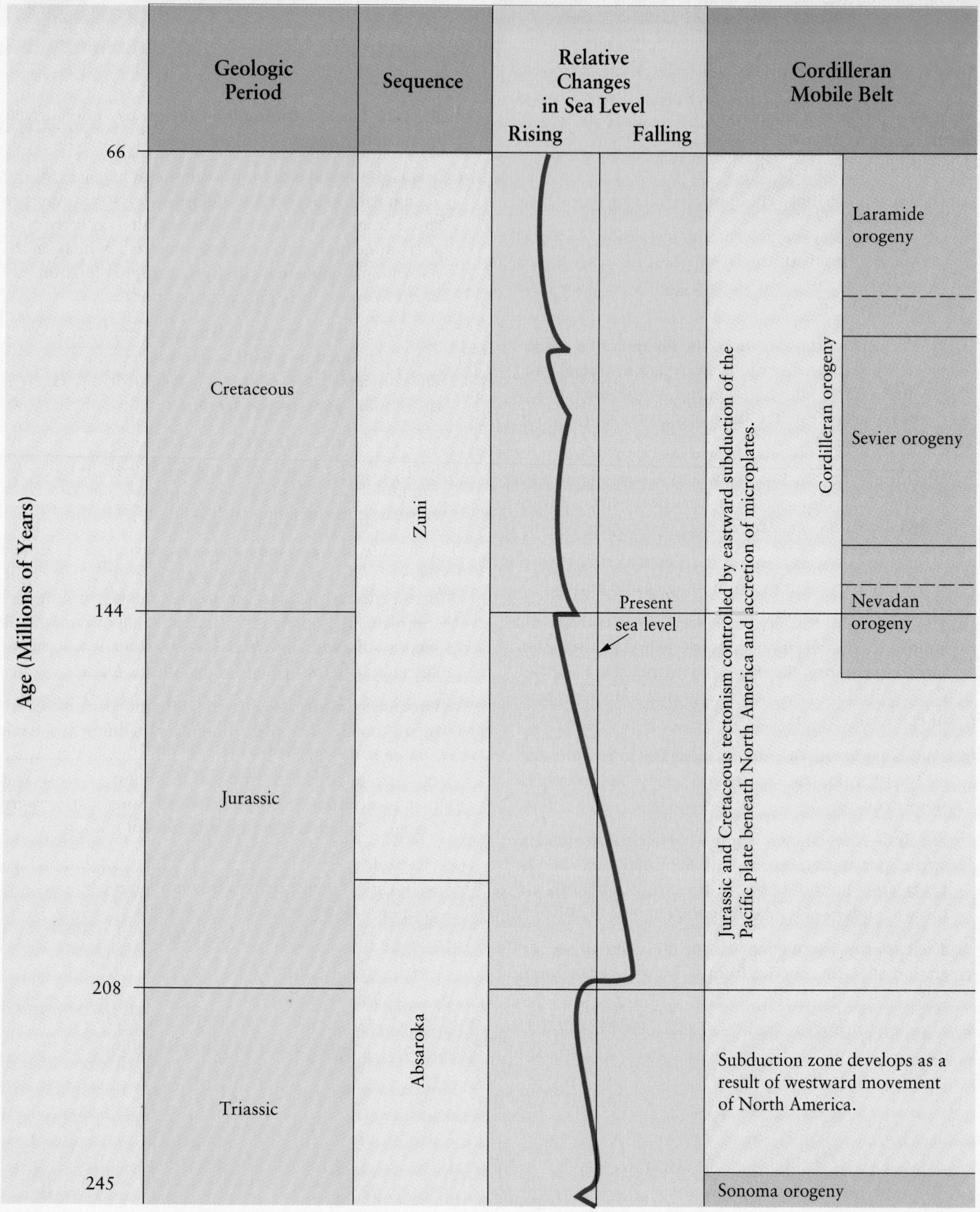

Geologic Period	Sequence	Relative Changes in Sea Level Rising / Falling	Cordilleran Mobile Belt
			Laramide orogeny
Cretaceous	Zuni	Present sea level	Sevier orogeny
			Nevadan orogeny
Jurassic			
Triassic	Absaroka		Subduction zone develops as a result of westward movement of North America.
			Sonoma orogeny

Age (Millions of Years): 66 — 144 — 208 — 245

Cordilleran orogeny

Jurassic and Cretaceous tectonism controlled by eastward subduction of the Pacific plate beneath North America and accretion of microplates.

North American Interior	Gulf Coastal Region	Eastern Coastal Region	Global Plate Tectonic Events
	Major Late Cretaceous transgression. Reefs particularly abundant.	Appalachian region uplifted.	South America and Africa are widely separated. Greenland begins separating from Europe.
	Regression at end of Early Cretaceous. Early Cretaceous transgression and marine sedimentation.	Erosion of fault-block mountains formed during the Late Triassic to Early Jurassic.	
	Sandstones, shales, and limestones are deposited in transgressing and regressing seas. Thick evaporites are deposited in newly formed Gulf of Mexico.		South America and Africa begin separating in the Late Jurassic.
		Fault-block mountains and basins develop in eastern North America.	
	Gulf of Mexico begins forming during Late Triassic.	Deposition of Newark Group; lava flows, sills, and dikes.	Breakup of Pangaea begins with rifting between Laurasia and Gondwana. Supercontinent Pangaea still in existence.

Important Terms

Cordilleran orogeny   microplate   Sevier orogeny   Sundance Sea

Cretaceous Interior Seaway   Nevadan orogeny   Sonoma orogeny   Zuni sequence

Laramide orogeny

Review Questions

1. Evidence for the breakup of Pangaea includes:
   a. _____ rift valleys.
   b. _____ dikes.
   c. _____ sills.
   d. _____ great quantities of poorly sorted nonmarine detrital sediments.
   e. _____ all of these.

2. The time of greatest post-Paleozoic inundation of the craton occurred during which geologic period?
   a. _____ Triassic
   b. _____ Jurassic
   c. _____ Cretaceous
   d. _____ Paleogene
   e. _____ Neogene

3. The first Mesozoic orogeny in the Cordilleran region was the:
   a. _____ Antler.
   b. _____ Laramide.
   c. _____ Nevadan.
   d. _____ Sevier.
   e. _____ Sonoma.

4. The breakup of Pangaea began with initial Triassic rifting between which two continental landmasses?
   a. _____ South America and Africa
   b. _____ Laurasia and Gondwana
   c. _____ North America and Eurasia
   d. _____ Antarctica and India
   e. _____ India and Australia

5. The Jurassic formation or complex famous for dinosaur fossils is the:
   a. _____ Navajo.
   b. _____ Chinle.
   c. _____ Sundance.
   d. _____ Morrison.
   e. _____ Franciscan.

6. The orogeny responsible for the present-day Rocky Mountains is the:
   a. _____ Sevier.
   b. _____ Sonoma.
   c. _____ Nevadan.
   d. _____ Antler.
   e. _____ Laramide.

7. The first major seaway to flood North America was the:
   a. _____ Sundance.
   b. _____ Newark.
   c. _____ Zuni.
   d. _____ Cordilleran.
   e. _____ Cretaceous Interior Seaway.

8. The Mesozoic tectonic history of the North American Cordilleran region is very complex and involves:
   a. _____ oceanic–continental convergence.
   b. _____ continental–continental convergence.
   c. _____ microplate accretion.
   d. _____ answers (a) and (b).
   e. _____ answers (a) and (c).

9. A possible cause for the eastward migration of igneous activity in the Cordilleran region during the Cretaceous was a change from:
   a. _____ oceanic–oceanic convergence to oceanic–continental convergence.
   b. _____ high-angle to low-angle subduction.
   c. _____ divergent to convergent plate margin activity.
   d. _____ divergent plate margin activity to subduction.
   e. _____ subduction to divergent plate margin activity.

10. The formation or complex responsible for the spectacular scenery of the Painted Desert and Petrified Forest is the:
    a. _____ Chinle.
    b. _____ Wingate.
    c. _____ Franciscan.
    d. _____ Navajo.
    e. _____ Morrison.

11. The Sierra Nevada, Southern California, Idaho, and Coast Range batholiths formed as a result of which orogeny?
    a. _____ Sonoma
    b. _____ Nevadan
    c. _____ Sevier
    d. _____ Laramide
    e. _____ None of these

12. Discuss the geologic evidence for the breakup of Pangaea.

13. Provide a general global history of the breakup of Pangaea.

14. Compare the tectonics of the Sonoma and Antler orogenies.

15. How did the breakup of Pangaea affect oceanic and climatic circulation patterns?

16. How did the Mesozoic rifting that took place on the East Coast of North America affect the tectonics in the Cordilleran mobile belt?

17. Discuss the tectonics of the Cordilleran mobile belt during the Mesozoic Era.

18. Using a diagram, explain how increased seafloor spreading can cause a rise in sea level along the continental margins.

19. Compare the tectonic setting and depositional environment of the Gulf of Mexico evaporites with the evaporite sequences of the Paleozoic Era.

20. How does microplate tectonics change our interpretations of the geologic history of the western margin of North America, and how does it relate to the Mesozoic orogenies that took place in that area?

Points to Ponder

1. The breakup of Pangaea influenced the distribution of continental landmasses, ocean basins, and oceanic and atmospheric circulation patterns, which in turn affected the distribution of natural resources, landforms, and the evolution of the world's biota. Reconstruct a hypothetical history of the world for a different breakup of Pangaea—one in which the continents separate in a different order or rift apart in a different configuration. How would such a scenario affect the distribution of natural resources? Would the distribution of coal and petroleum reserves be the same? How might evolution be affected? Would human history be different?

2. Discuss the similarities and differences between the orogenic activity that occurred in the Appalachian mobile belt during the Paleozoic Era and that which occurred in the Cordilleran mobile belt during the Mesozoic Era.

Additional Readings

Bally, A. W., and A. R. Palmer, eds. 1989. *The geology of North America: An overview.* The Geology of North America. Vol. A. Boulder, Colo.: Geological Society of America.

Bonatti, E. 1987. The rifting of continents. *Scientific American* 256, no. 3: 96–103.

Harwood, D. S., and M. M. Miller, eds. 1990. *Paleozoic and Early Mesozoic paleogeographic relations; Sierra Nevada, Klamath Mountains, and related terranes.* Geological Society of America Special Paper 255. Boulder, Colo.: Geological Society of America.

Howell, D. G. 1989. *Tectonics of suspect terranes.* New York: Chapman and Hall.

Jones, D. L., A. Cox, P. Coney, and M. Beck. 1982. The growth of western North America. *Scientific American* 247, no. 5: 70–128.

Murphy, J. B., G. L. Oppliger, G. H. Brimhall Jr., and A. Hynes. 1999. Mantle plumes and mountain building. *American Scientist* 87, no. 2: 146–153.

Pendick, D. 1997. Rocky Mountain why. *Earth* 6, no. 3: 26–33.

Nations, J. D., and J. G. Eaton, eds. 1991. *Stratigraphy, depositional environments, and sedimentary tectonics of the western margin, Cretaceous Western Interior Seaway.* Geological Society of America Special Paper 260. Boulder, Colo.: Geological Society of America.

Salvador, A., ed. 1991. *The Gulf of Mexico Basin.* The Geology of North America. Vol. J. Boulder, Colo.: Geological Society of America.

For these web site addresses, along with current updates and exercises, log on to

http://www.brookscole.com/geo/

▶ **MORRISON RESEARCH INITIATIVE**

The Morrison Research Initiative is a multidisciplinary study of the Upper Jurassic Morrison Formation, which is funded by the National Park Service and the USGS. What are the project objectives? Why is the Morrison Formation an important formation?

▶ **UNIVERSITY OF CALIFORNIA MUSEUM OF PALEONTOLOGY**

This is an excellent site to visit for any aspect of geologic time, paleontology, and evolution.

Click on the *Geology* icon under the On-Line Exhibits heading, which will take you to the *Geology and Geologic Time* home page. Click on the *Mesozoic* icon, which will take you to the *Introduction to the Mesozoic Era* home page. From here, click on any of the three periods of the Mesozoic and learn more about the stratigraphy, life, localities, and tectonics of that period. There are also links to other sites of interest for each geologic period.

Explore the following *In-Terra-Active 2.0* CD-ROM module(s) and increase your understanding of key concepts and processes presented in this chapter.

Chapter Concept: **The Breakup of Pangaea**

The breakup of Pangaea had a profound impact on geologic and biologic events during the Mesozoic Era. This module introduces you to the geologic and fossil evidence that scientists use to support the existence of the supercontinent Pangaea and its subsequent breakup.

Section: **Interior**

Module: **Plate Tectonics: Evidence**

Question: *If you were an astronaut landing on Venus, what evidence would you look for to confirm or deny the existence of plate tectonics there?*

Chapter Concept: **Spreading Ridges and the Creation of the Seafloor**

With the breakup of Pangaea, plate margin dynamics changed dramatically as passive margins became active, active margins became passive, and rifting initiated proto-oceans and the creation of new seafloor. In this module, you will study the different oceanic tools geologists use to determine margin type and composition and apply this knowledge to unknown margins.

Section: **Surface**

Module: **The Seafloor**

Mineralogists have it relatively easly when it comes to sampling objects for study—they can hike, climb rocks, and excavate to get their specimens. It's much harder for geologists studying the seafloor. The difficulties of penetrating the ocean depths and gathering solid information cannot be overestimated. The seafloor is wet and dark; it can be more than 9,000 m; and it is often covered with sediment that obscures the solid rock beneath.

Question: *Given these constraints, what do you think is the most effective way to examine the structure and composition of the seafloor?*

Chapter Concept: **Crustal Deformation in Mobile Belts**

Multiple episodes of crustal deformation occurred along the western margin of the North American plate during the Mesozoic Era. In this module, you will explore the relationship between the folds and faults of mobile belts and the deformational processes that accompany plate tectonics.

Section: **Interior**

Module: **Crustal Deformation**

The coarse-grained garnet-mica schist shown on the left has essentially the same chemical composition as the fine-grained mudstone shown on the right. In the field, it is clear that these belong to the same stratigraphic unit.

Question: *How can this transformation take place in the solid state?*

Life of the Mesozoic Era

Restoration of the Cretaceous dinosaur *Giganoto-saurus carolinii* from South America. Its length of nearly 13 m and weight of about 7.3 metric tons makes it one of the largest carnivorous dinosaurs. *Giganotosaurs* lived about 25 million years before the similar and better-known dinosaur *Tyrannosaurs rex.*

Prologue

Western European males dominated the early history of paleontology, a situation that no longer prevails. Now men and women from all continents are making important discoveries and contributions to our knowledge of life history. But probably the most notable early exception is Mary Anning (1799–1847), who began a remarkable career as a fossil collector when she was only 11 years old.

Mary Anning was born in Lyme Regis on England's south coast. When only 15 months old, she survived a lightning strike that, according to one report, killed three girls, and according to another, killed a nurse tending her. Of the ten or so children in the Anning family, only Mary and her brother Joseph reached maturity. In 1810 Mary's father, a cabinetmaker who also sold fossils part time, died leaving the family nearly destitute. Mary Anning (Figure 15.1) expanded the fossil business and became a professional fossil collector known to the paleontologists of her time, some of whom visited her shop to buy fossils or gather information. She collected fossils from the Dorset coast near Lyme Regis and is reported to have been the inspiration for the tongue twister, "She sells sea shells sitting on the sea shore."

Soon after her father's death, she made her first important discovery, a nearly complete skeleton of a Jurassic marine reptile known as an ichthyosaur, which was described in 1814 by Sir Everard Home. The sale of this fossil specimen provided considered financial relief for her family. In 1821 she made a second major discovery and excavated the remains of another Mesozoic marine reptile called a plesiosaur. And in 1818 she excavated the remains of the first Mesozoic flying reptile (pterosaur) found in England, which was sent to the eminent geologist William Buckland of Oxford University.

By 1830 Mary Anning's fortunes began declining as collectors and museums had less discretionary money to spend on fossils. Indeed, she may once again have become destitute if it had not been for her friend Henry Thomas de la Beche, also a resident of Lyme Regis. De la Beche drew a fanciful scene called *Duria antiquior*, meaning "An earlier Dorset," in which he brought to life the fossils Mary Anning had collected

FIGURE 15.1   Mary Anning (1799–1847), who lived in Lyme Regis on England's southern coast, began collecting and selling fossils when she was only 11 years old.

(Figure 15.2). The scene was printed and sold widely, the proceeds of which went directly to Mary Anning.

Mary Anning died of cancer in 1847, and although only 48 years old, she had a fossil-collecting career that spanned 36 years. Her contributions to paleontology are now widely recognized, but, unfortunately, soon after her death she was mostly forgotten. Apparently, people who purchased her fossils were credited with finding them. So even though Mary Anning became a respected fossil collector, many scientists of that time could not accept that an untutored girl could possess such knowledge and skill. "It didn't occur to them to credit a woman from the lower classes with such astonishing work. So an uneducated little girl, with a quick mind and an accurate eye, played a key role in setting the course of the 19th century geologic revolution. Then—we simply forgot about her."*

---

* John Lienhard, University of Houston.

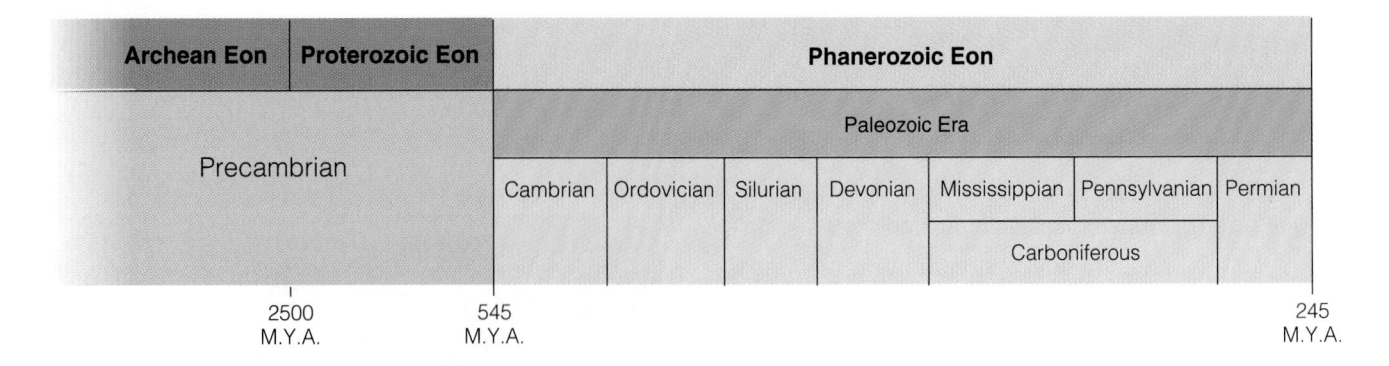

Archean Eon	Proterozoic Eon	Phanerozoic Eon							
		Paleozoic Era							
Precambrian		Cambrian	Ordovician	Silurian	Devonian	Mississippian	Pennsylvanian	Permian	
						Carboniferous			

2500 M.Y.A.   545 M.Y.A.   245 M.Y.A.

FIGURE 15.2 This fanciful scene, titled *Duria antiquior* by Henry De la Beche, features restorations of some of the fossils collected by Mary Anning. Note especially the ichthyosaurs, plesiosaurs, and pterosaurs.

Introduction

Many people find the Mesozoic Era the most interesting chapter in the history of life because dinosaurs, flying reptiles, and marine reptiles lived during this time. In fact, the Mesozoic is commonly referred to as the "Age of Reptiles," alluding to the fact that reptiles were the most diverse and abundant land-dwelling vertebrate animals. The fascination with dinosaurs and their close relatives overshadows the fact that, although Mesozoic reptile diversification was an important evolutionary event, other equally important events also took place. For instance, mammals evolved from mammal-like reptiles during the Triassic, and birds

evolved from reptiles, probably small carnivorous dinosaurs, during the Jurassic.

Vast changes took place in land-plant communities too. When the Mesozoic Era began, the dominant land plants were holdovers from the Paleozoic, but during the Cretaceous, the first flowering plants evolved. These plants diversified rapidly and soon became the most diverse and abundant land plants.

The Mesozoic was also a time of resurgence of marine invertebrates. Diversity among marine invertebrates had been drastically reduced as a result of the Permian extinction, but survivors rapidly diversified, giving rise to increasingly complex marine invertebrate communities. Ammonites

Phanerozoic Eon								
Mesozoic Era			Cenozoic Era					
Triassic	Jurassic	Cretaceous	Tertiary					Quaternary
			Paleocene	Eocene	Oligocene	Miocene	Pliocene	Pleistocene	Holocene

245
M.Y.A.

66
M.Y.A.

and planktonic microorganisms called *foraminifera* were particularly abundant and diverse.

The breakup of Pangaea that began during the Triassic continued throughout the Mesozoic. Nevertheless, the proximity of continents and mild Mesozoic climates allowed land animals and plants to occupy extensive geographic areas. As the fragmentation of Pangaea continued, however, some continents, Australia and South America especially, became isolated, and their faunas evolved independently.

Another mass extinction occurred at the end of the Mesozoic. Once again organic diversity declined markedly as dinosaurs and several other reptiles and some marine invertebrates died out. Because dinosaurs were victims of this extinction, it has received more publicity than any other, although the Permian extinctions were of greater impact (see Chapter 12).

Marine Invertebrates

Following the wave of extinctions at the end of the Paleozoic, the Mesozoic was a time when the marine invertebrates repopulated the seas. As the Atlantic and Indian Oceans formed, the epeiric seas once again inundated the low-lying continental margins, and invertebrate marine life returned with a flourish. Gone were many Paleozoic inver-

tebrates such as the fusulinid foraminifera, blastoids, rugose corals, and trilobites.

The Early Triassic invertebrate marine fauna, though widely distributed, was not very diverse. By the Late Triassic, the seas were once again richly populated with invertebrates. The mollusks became increasingly diverse and abundant throughout the Mesozoic. The brachiopods, however, never completely recovered from their near extinction at the end of the Paleozoic, and though they diversified during the Triassic and Jurassic, they declined in numbers during the Cretaceous, remaining a minor invertebrate phylum during the Cenozoic. In areas of warm, relatively clear, shallow marine waters, corals again proliferated, but they were of a new and familiar type, the *scleractinians* (Figure 15.3). Echinoids, which were rare during the Paleozoic, greatly diversified during the Mesozoic (Figure 15.4).

One of the major differences between Paleozoic and Mesozoic marine invertebrate communities was the increased abundance and diversity of burrowing organisms. Except for a few animals with hard parts, such as the inar-

FIGURE 15.3 Scleractinian corals evolved during the Triassic and proliferated in the warm, clear, shallow marine waters of the Mesozoic Era. Most living corals are scleractinians, represented here by the so-called staghorn coral.

FIGURE 15.4 Echinoids, or sea urchins, were rare during the Paleozoic but became diverse and abundant during the Mesozoic.

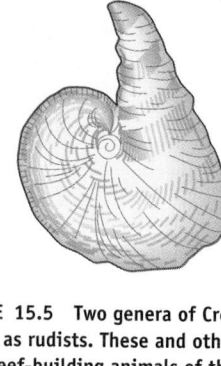
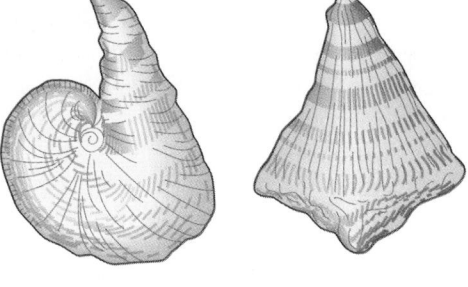

FIGURE 15.5 Two genera of Cretaceous reef-building bivalves known as rudists. These and other rudists displaced corals as the main reef-building animals of the Mesozoic.

FIGURE 15.6 (a) *Calcidiscus macintyrei*, a Miocene coccolith from the Gulf of Mexico *(left)*; *Discoaster variabilis*, a Pliocene-Miocene coccolith from the Gulf of Mexico *(right)*. (b) *Actinoptychus senarius*, an Upper Miocene diatom from Java *(left)*; *Cocconeis pellucida*, an Upper Miocene diatom from Java *(right)*. (c) *Deflandrea spinulosa*, an Eocene dinoflagellate from Alabama *(left)*; *Spiniferites mirabilis*, a Neogene dinoflagellate from the Gulf of Mexico *(right)*.

ticulate brachiopod *Lingula,* almost all Paleozoic burrowers were soft-bodied animals such as worms. The bivalves and echinoids, which were epifaunal elements during the Paleozoic, evolved various means of entering the infaunal habitats. This trend toward an infaunal existence may reflect an adaptive response to increasing predation from the rapidly evolving fish and cephalopods.

The end of the Mesozoic witnessed another major extinction, this time involving mostly large reptiles. The only extinctions of consequence in the marine invertebrate fauna were those affecting the ammonoid cephalopods, planktonic foraminifera and a group of reef-forming *rudist* bivalves (Figure 15.5).

Beginning with the primary producers, we now examine some of the major Mesozoic marine plant and invertebrate groups. *Coccolithophores* are an important group of living phytoplankton (Figure 15.6a). Their remains, which are constructed of microscopic calcareous plates, collect in tremendous numbers on the ocean floor when the organisms die. Coccolithophore remains comprise many Cretaceous chalk beds such as the White Cliffs of Dover in England and the Demopolis Chalk of the southwestern United States. Coccolithophores first appeared during the Jurassic and diversified tremendously during the Cretaceous. They continue to be abundant today and are one of the major primary producers supporting suspension feeders in present-day oceans. *Diatoms* first evolved during the Cretaceous but were more important as primary producers during the Cenozoic (Figure 15.6b). Diatoms construct their shells out of silica. They are presently most abundant in cooler waters. *Dinoflagellates* were common during the Mesozoic and today are the major primary producers in warm waters. Their remains are easily preserved as organic cysts (Figure 15.6c).

Mollusks, as previously noted, were the major invertebrate phylum of the Mesozoic. The mollusks include six classes, only three of which—the gastropods, bivalves, and cephalopods—are significant members of the marine invertebrate fauna. Gastropods increased in abundance and diversity during the Mesozoic, becoming most abundant during the Cretaceous, when carnivorous forms appeared. It was bivalves and cephalopods, however, that dominated the invertebrate community of the Mesozoic.

Mesozoic bivalves diversified to inhabit many epifaunal and infaunal niches. Oysters and other clams became particularly diverse and abundant epifaunal suspension feeders and, despite a reduction at the end of the Cretaceous, continued to be important throughout the Cenozoic to the present (Figure 15.7). Reef-forming rudists were a significant group of Mesozoic bivalves (Figure 15.5). These epifaunal bivalves were geologically short-lived and so are excellent guide fossils for the Late Jurassic through Cretaceous. Rudists are also significant because they

FIGURE 15.7 Bivalves, represented here by two Cretaceous forms, were particularly diverse and abundant during the Mesozoic.

FIGURE 15.8 Cephalopods such as the Late Cretaceous ammonoids *Baculites* (foreground) and *Helicoceros* were present throughout the Mesozoic Era, but they were particularly abundant during the Jurassic and Cretaceous Periods.

formed large tropical reefs, displacing corals as the main reef builders. Bivalves also expanded into the infaunal niche during the Mesozoic. By burrowing into the sediment, they escaped predation from cephalopods and fish.

Cephalopods were one of the most important Mesozoic invertebrate groups. Their rapid evolution and nektonic life-style make them excellent guide fossils (Figure 15.8). Recall that two orders of cephalopods arose during the Paleozoic: Nautiloidea, with simple sutures, and Ammonoidea, with wrinkled sutures. Ammonoidea are divided into three groups: goniatites, ceratites, and ammonites. Ammonites, which are characterized by extremely complex suture patterns, were present during all three Mesozoic periods but were most prolific during the Jurassic and Cretaceous. Their tremendous success was a reflection of their ability to adapt to a wide range of marine environments. While most ammonites were coiled, some attaining diameters of 2 m, others were uncoiled and led a near benthonic existence.

Despite their successful adaptations, ammonites became extinct at the end of the Cretaceous; explanations for their demise have ranged from competition with fish to

extinction caused by a meteorite impact. Although ammonites became extinct at or near the end of the Cretaceous, two other groups of cephalopods survived into the Cenozoic—nautiloids and belemnoids, a group of squidlike cephalopods that was highly successful during the Jurassic and Cretaceous (Figure 15.8).

Although mollusks dominated the Mesozoic, other phyla were also important. The stalked echinoderms, which were abundant during the Paleozoic, became minor members of the Mesozoic marine invertebrate community. However, echinoids, which were exclusively epifaunal during the Paleozoic, branched out into the infaunal habitat and became very diverse and abundant (Figure 15.4). Bryozoans, although rare in Triassic strata, diversified and expanded during the Jurassic and Cretaceous.

As is true today, where shallow marine waters were warm and clear, coral reefs proliferated. Mesozoic corals belong to the order Scleractinia (Figure 15.3). Scleractinians use aragonite rather than calcite in constructing their skeletons, and most today have a symbiotic relationship with certain dinoflagellates. Because dinoflagellates require sufficient sunlight to function, they can live only in shallow water. Whether scleractinian corals evolved from the rugose order or from an as yet unknown soft-bodied group that left no fossil record is still unresolved.

Lastly, the foraminifera underwent an explosive radiation during the Jurassic and Cretaceous that continued to the present. The planktonic forms (Figure 15.9) in particular underwent rapid diversification, but they were also affected by the extinction at the end of the Cretaceous. Most of the planktonic genera became extinct at the end of the Cretaceous, and only a few genera survived into the Cenozoic.

In general terms, we can think of the Mesozoic as a time of increasing complexity of the marine invertebrate com-

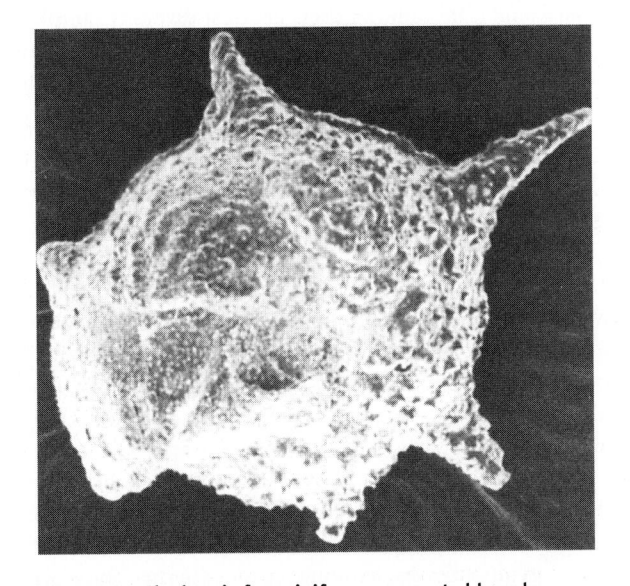

FIGURE 15.9 Planktonic foraminifera, represented here by *Globotruncana calcarata* from the Cretaceous Pecan Gap Chalk of Texas, became diverse during the Jurassic and Cretaceous, but were affected by the extinction at the end of the Cretaceous.

munity as it evolved from a simple one with low diversity and short food chains at the beginning of the Triassic to a highly complex community with interrelated food chains near the end of the Cretaceous. This evolutionary history reflects the change in geologic conditions influenced by plate tectonic activity that we discussed in Chapter 14.

Fish and Amphibians

The cartilaginous fish, which include sharks and their relatives, increased in abundance through the Mesozoic, but even so they never came close to matching the diversity of the bony fishes. Nevertheless, sharks were, and still are, important elements of the marine fauna.

Among the bony fish, lungfish and crossopterygians are represented by few Mesozoic genera and species. In fact, crossopterygians declined and were almost extinct by the close of the era; only one living species is known (see Figure 5.19b), and the group has no known Cenozoic fossil record.

All other bony fish belong to three groups, which for convenience we can call primitive, intermediate, and advanced. The most primitive of these bony fish existed mostly during the Paleozoic but persisted throughout the Mesozoic when most of them became extinct. Living members of this group include sturgeon and paddlefish. By Middle Mesozoic time, the most common fish were members of the intermediate group, which included a variety of now-extinct oceanic species and the ancestors of the present-day freshwater-dwelling gar pikes and dogfish, or mudfish. By the Cretaceous, the advanced group, also known as *teleosts,* was the dominant marine and freshwater fish. Their dominance is indicated by the fact that at least 20,000 species exist, making them the most numerous of all vertebrates.

A few labyrinthodont amphibians persisted into the Mesozoic but died out by the end of the Triassic. Since the Pennsylvanian, the time of their greatest diversity, amphibians have made up only a small part of the total vertebrate fauna. Frogs and salamanders appeared during the Mesozoic, but their fossil records are poor.

Primary Producers on Land— Plants

As a prelude to our discussion of Mesozoic land animals, we consider land-plant communities. After all, plants as photosynthesizers lie at the base of the food chain. In other words, they are the primary producers on land, while the animals, as consumers, are dependent upon plants.

Triassic and Jurassic land-plant communities, like those of the Late Paleozoic, were composed of seedless vascular plants such as ferns, horsetail rushes, and club mosses and various gymnosperms. Among gymnosperms, however, the large seed ferns became extinct by the end of the Triassic. *Ginkgos* remained abundant throughout the Mesozoic Era; ginkgos nearly disappeared during the

Cenozoic, and only one species survives today. *Conifers* also persisted from the Late Paleozoic, continued to diversify, and now are the most common land plants at high elevations and at the margins of the Arctic. A new type of gymnosperm, the *cycads,* evolved during the Triassic. Cycads superficially resemble palm trees, and several varieties still exist in tropical and subtropical areas (Figure 15.10a).

Marked changes in the composition of land-plant communities occurred at the beginning of the Late Jurassic. The long dominance of seedless plants and gymnosperms ended as many were replaced by **angiosperms,** or flowering plants (Figure 15.10b and c). The earliest angiosperms evolved from a specialized group of seed ferns. Since the angiosperms evolved, they have adapted to nearly every terrestrial habitat from mountains to deserts, and some have adapted to shallow coastal waters. A few varieties are known as "carnivorous" plants. But rather than digesting prey, mostly insects, as animals do, they dissolve the exoskeleton and absorb nitrogen molecules because they live in nitrogen-deficient soils.

(a)

(b)

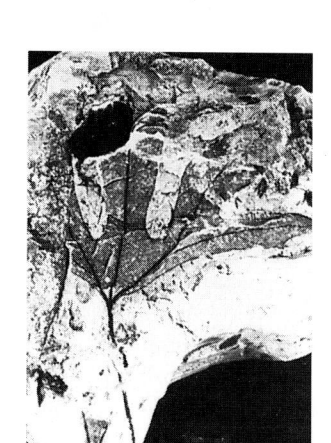

(c)

FIGURE 15.10 (a) Living cycads, southern California. (b) and (c) Fossil angiosperms from the Lower Cretaceous Potomac Group of the eastern United States. (b) *Sapindopsis,* Cecil County, Maryland. (c) *Aralia* from New Jersey.

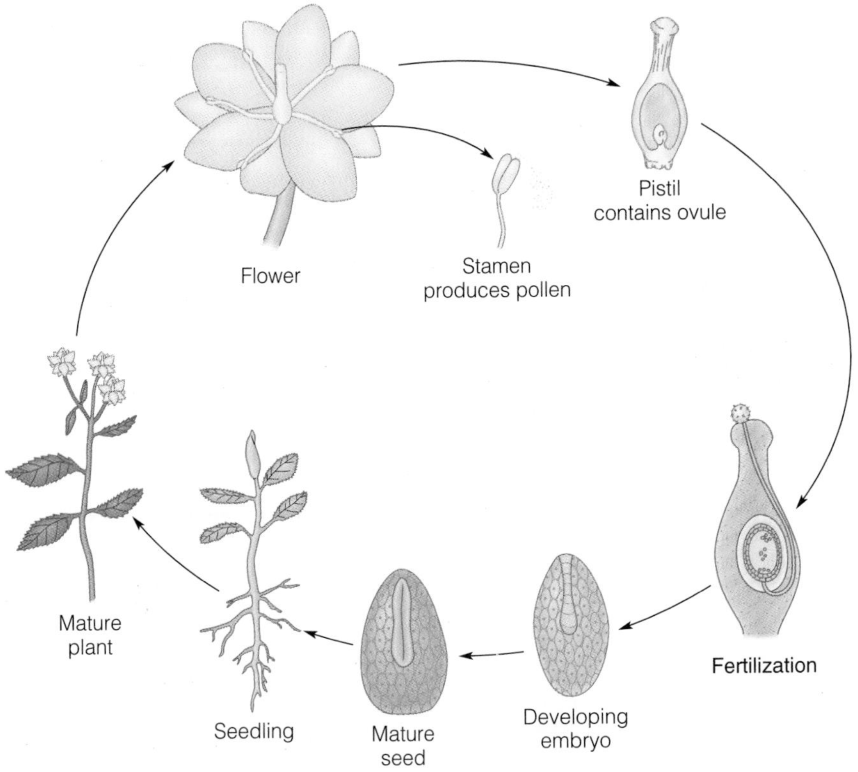

FIGURE 15.11 The reproductive cycle in angiosperms.

Several factors account for the phenomenal success of flowering plants, but chief among them is their method of reproduction (Figure 15.11). Two developments were particularly important: the evolution of flowers, which attract animal pollinators, especially insects, and the evolution of enclosed seeds. Angiosperm seeds are dispersed widely in several ways—blown by the wind, embedded in a fleshy fruit, or carried in burrs that become attached to animal fur.

Seedless plants and gymnosperms are important and still flourish in many environments; in fact, many botanists regard ferns and conifers as emerging groups. Nevertheless, a measure of the angiosperms' success is that today they account for more than 90% of all land-plant species.

Reptiles

The evolutionary relationships among the groups of fossil and living reptiles are summarized in Figure 15.12. Reptile diversification began during Mississippian time with the evolution of the protorothyrids, apparently the first animals to lay amniote eggs (see Chapter 13). From this basic stock of so-called *stem reptiles,* all other reptiles evolved. Birds and mammals, too, have their ancestors among the reptiles, so they are a part of this major evolutionary diversification.

Recall from Chapter 13 that pelycosaurs were the dominant land vertebrates of the Pennsylvanian and Permian periods. Here we continue our story of reptile diversification with a group called *archosaurs.*

ARCHOSAURS AND THE ORIGIN OF DINOSAURS

A group of reptiles know as **archosaurs** (*archo* meaning "ruling," and *sauros* meaning a "lizard") includes crocodiles, pterosaurs (flying reptiles), dinosaurs, and birds. The inclusion of such diverse animals in a single group implies they share a common ancestor, and indeed they possess several characteristics that unite them. For instance, teeth set in individual sockets are found in all these animals except present-day birds, but even the earliest birds possessed this feature. We will have more to say about pterosaurs and birds later in this chapter.

Dinosaurs are characterized by many shared characteristics, but two orders, the **Saurischia** and **Ornithischia,** are recognized based on their pelvic structure (Figure 15.13). Saurischian dinosaurs had a lizardlike pelvis and are therefore referred to as lizard-hipped dinosaurs. Ornithischians had a birdlike pelvis; hence, they are called bird-hipped dinosaurs. The traditional interpretation is that each order evolved independently during the Late Triassic, but it is now agreed that they had a single common ancestor much like archosaurs from Middle Triassic rocks of Argentina. These animals were small (less than 1 m long), long-legged, carnivores that walked and ran on their hind limbs, so they were **bipedal,** as opposed to **quadrupedal** animals that move on all four limbs.

DINOSAURS

More than any other type of animal, dinosaurs have inspired awe and have thoroughly captured the public imagination. As often as not, unfortunately, their

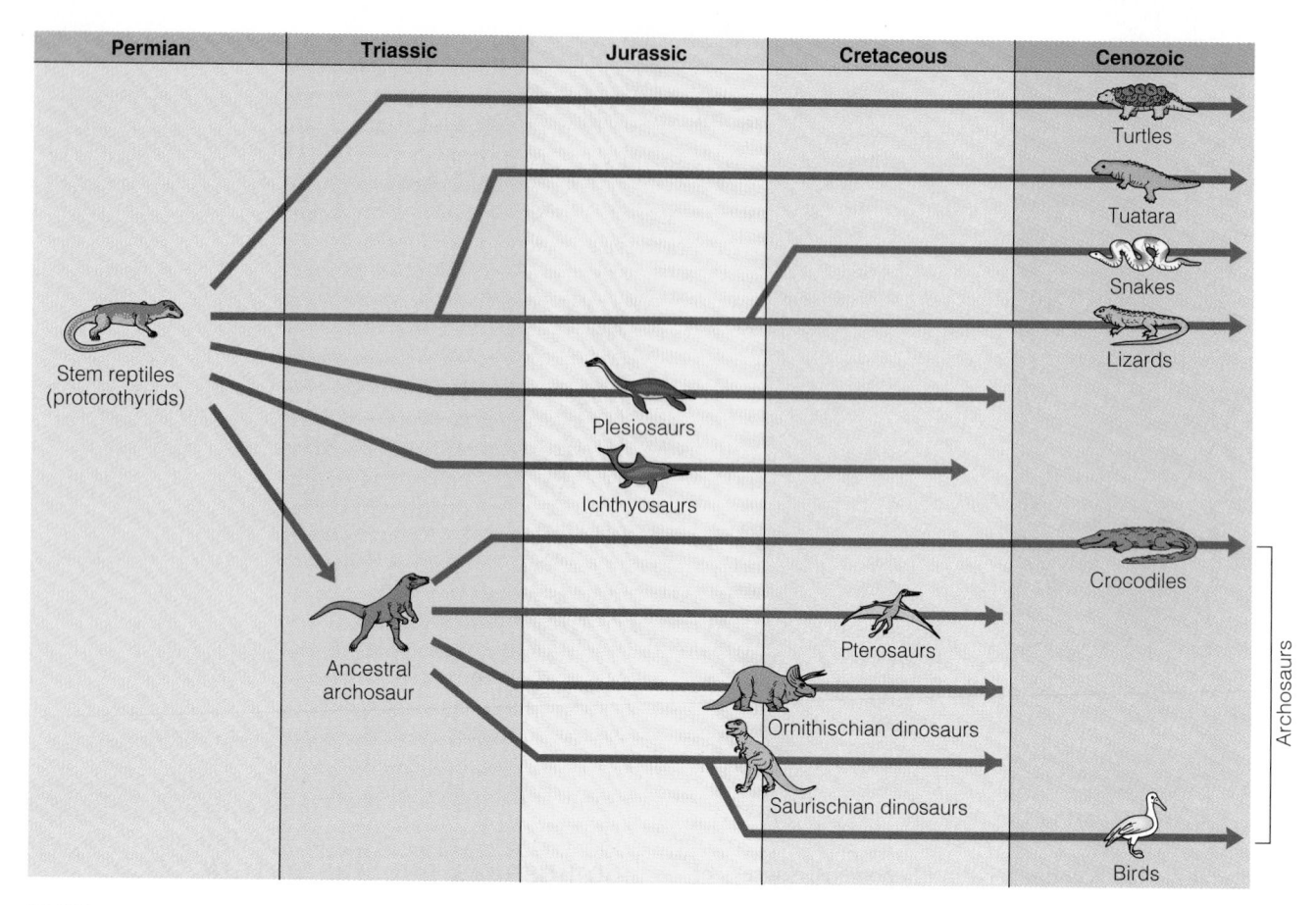

FIGURE 15.12 Relationships among fossil and living reptiles and birds.

popularization in cartoons, movies, and many books has led to misunderstandings about dinosaurs in general. It is true that many were large, indeed the largest animals ever to live on land. But not all were large. In fact, dinosaurs varied from giants to those no larger than a chicken (Figure 15.14).

A common but erroneous perception of dinosaurs is that they were poorly adapted animals that had trouble surviving. True, they became extinct, but to consider this a failure is to ignore the fact that for more than 140 million years they were the dominant land vertebrates. During their existence, they diversified into numerous types and adapted to a wide variety of environments. Eventually, dinosaurs did die out, an event that then enabled mammals to become the predominant land vertebrates.

Nor were dinosaurs the lethargic beasts often portrayed in various media. Evidence now available indicates that at least some dinosaurs may have been very active and possibly warm-blooded. It also appears that some species cared for their young long after hatching, a behavioral characteristic most often associated with birds and mammals.

Saurischian Dinosaurs Two groups of saurischians are recognized, theropods and sauropods (Figure 15.13). All theropods were carnivorous bipeds that ranged in size from the tiny *Compsognathus* (Figure 15.14a), which probably weighed only 2 or 3 kg, to the 3- to 5-ton *Tyran-*

nosaurus, which until recently was the largest terrestrial carnivore known. In 1995, though, dinosaurs similar to *Tyrannosaurus* that might have weighed 7 or 8 tons were discovered in Africa and Argentina. A 1-m-long theropod known as *Eoraptor* was also recently discovered in Late Triassic rocks of Argentina. Although this dinosaur is too specialized to be the common ancestor of all dinosaurs, details of it structure indicate that it was close to that ancestry.

The movie *Jurassic Park* and its sequel *The Lost World* popularized some of the smaller theropods such as *Velociraptor,* a 1.8-m long predator with a large sickle-shaped claw on each hind foot. *Velociraptor* and other raptors such as the larger *Deinonychus* are thought to have used this claw in a slashing type of attack (Figure 15.14b). A particularly interesting fossil from Mongolia shows a *Velociraptor* grasping a herbivorous dinosaur known as *Protoceratops.* It appears that both animals died when the former attacked the latter.

Included among the sauropods are the giant, quadrupedal herbivores such as *Apatosaurus, Diplodocus,* and *Brachiosaurus,* the largest-known land animals of any kind. According to one estimate, *Brachiosaurus,* a giant even by sauropod standards, weighed more than 75 tons! Discoveries of partial remains in Colorado, New Mexico, and elsewhere indicate that even larger sauropods existed, but just how large they were is not resolved (see

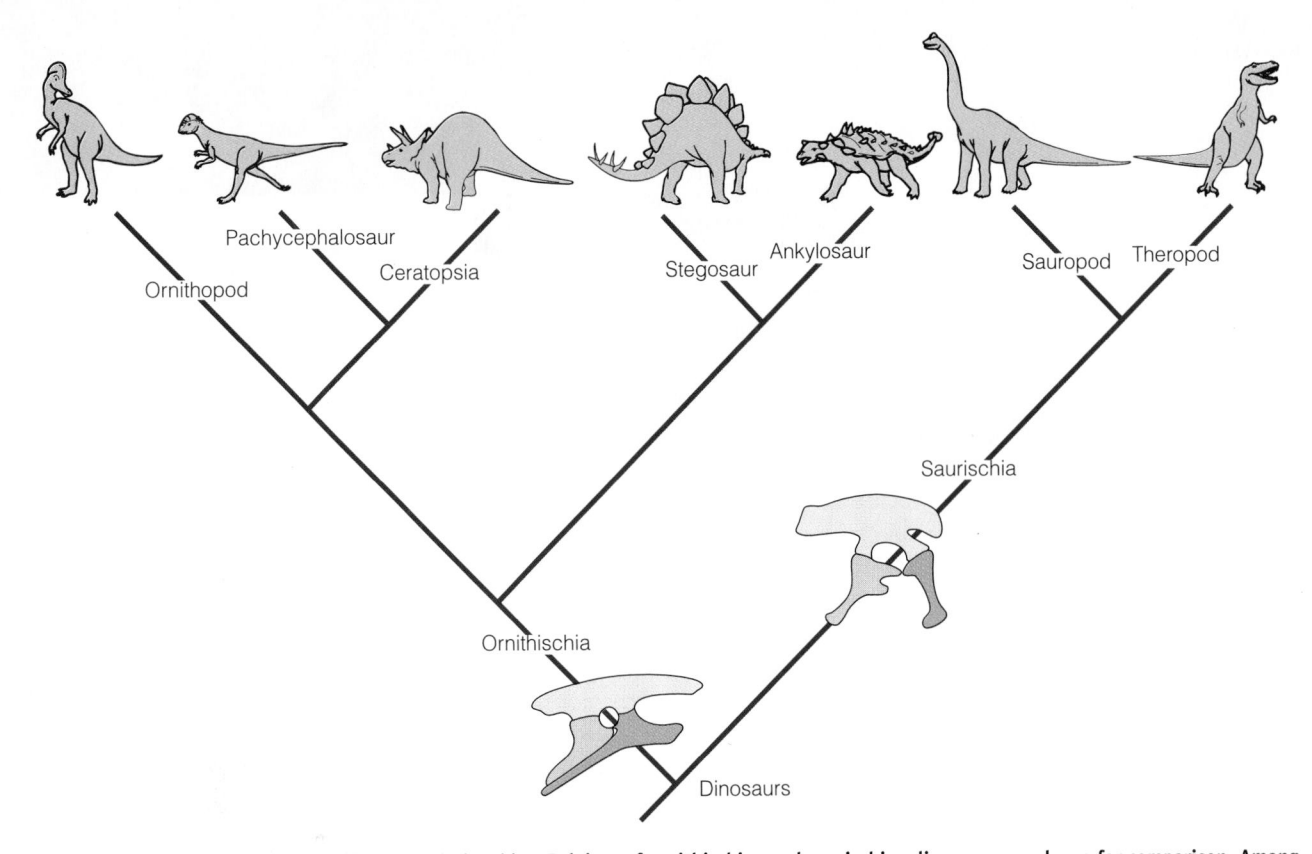

FIGURE 15.13 Cladogram showing dinosaur relationships. Pelvises of ornithischian and saurischian dinosaurs are shown for comparison. Among the seven subgroups of dinosaurs theropods were carnivorous and all others were herbivores.

Perspective 15.1). Sauropod trackways provide evidence for the idea that these animals moved around in herds. They probably depended on their size and herding behavior rather than speed for protection from predators.

Sauropods were preceded in the fossil record by a group of considerably smaller dinosaurs known as *prosauropods.* These Late Triassic and Early Jurassic animals were closely related to sauropods but probably not their ancestors. Sauropods were especially diverse and abundant during the Jurassic, but only a few genera are known from Cretaceous rocks.

Ornithischian Dinosaurs

The great diversity of ornithischians is manifested by the fact that five distinct groups are recognized: ornithopods, pachycephalosaurs, ankylosaurs, stegosaurs, and ceratopsians (Figure 15.13). Ornithopods include the duck-billed dinosaurs, which had flattened, bill-like mouths (Figure 15.15). These dinosaurs were particularly varied and abundant during the Cretaceous, and some species were characterized by head crests (Figure 15.15a). Such crests may have functioned as a means of species recognition, as display devices to attract mates, or as resonating chambers to amplify bellowing. All ornithopod genera were herbivores and primarily bipedal, but their well-developed forelimbs allowed them to walk in a quadrupedal fashion, too.

The pachycephalosaurs constitute a most peculiar group of ornithischian dinosaurs. The most distinctive feature of these bipedal herbivores is the dome-shaped skull that resulted from thickening of the bones. According to one hypothesis, these domed skulls were used in intraspecific butting contests for dominance and mates. The few known genera of pachycephalosaurs lived mostly during the Late Cretaceous.

Ankylosaurs were heavily armored, quadrupedal herbivores, and some were quite large. Bony armor protected the back, flanks, and top of the head, and the tail ended in a large, bony clublike growth. No doubt a blow delivered by the powerful tail could seriously injure an attacking predator.

The stegosaurs, represented by the familiar genus *Stegosaurus* (Figure 15.13), were quadrupedal herbivores with bony spikes on the tail, which were undoubtedly used for defense, and bony plates on the back. The exact arrangement of these plates is debated, but many paleontologists think they functioned as a device to absorb and dissipate heat.

The final group of ornithischian dinosaurs is the ceratopsian, or horned, dinosaurs (Figure 15.13). A rather good fossil record indicates that large, Late Cretaceous genera such as *Triceratops* evolved from small, Early Cretaceous ancestors (Figure 15.16). The later ceratopsians were characterized by huge heads, a large bony frill over the top of the neck, and a large horn or horns on the skull. Fossil trackways indicate that these large, quadrupedal herbivores moved in herds.

How Large Was the Largest Dinosaur?

When speaking of an animal's size, one usually means mass or simply weight rather than height or length. With this in mind, we can say that the largest living land animals are African elephants, some of which are 4 m tall and weigh 6.5 metric tons. Adult weight of elephants varies considerably, but by any measure they are large animals, and in fact weigh more than many dinosaurs. But at least half of all known dinosaurs weighed more than 2 metric tons, whereas only about 2% of all mammals attain this weight, the most notable land-dwelling ones being elephants, hippopotamuses, and rhinoceroses. Also, the 2% of mammals weighing more than 2 metric tons includes whales, which are aquatic, and their apparent weight is reduced by the buoyant effects of water.

Many dinosaurs were large, but even among dinosaurs the real giants were the sauropods. *Diplodocus* at 27 m long and 10.5 metric tons was a lightweight compared to much-more-heavily-built *Brachiosaurus,* which measured 23 to 27 m long and probably weighed 45 to 55 tons. Weighing a large living animal such as an elephant is difficult, so how does one estimate the weight of an animal known only from its bones? Two methods are used. The first involves determining the cross-sectional area of a limb bone such as the femur, then using this data in an equation developed for estimating the weights of living vertebrate animals. The calculated value gives only the weight borne by the limb bone, so an adjustment must be made to estimate the weight of the entire animal. The second method involves making a scale model and immersing it in water to

determine its volume. From this information an estimate can be made of a full-sized dinosaur's weight (Figure. 1). Unfortunately, both methods are subject to error, so estimates vary depending on the accuracy of a scale model, for instance, so dinosaur weights are simply reasonable approximations of the actual weight.

Given that several sauropod dinosaurs weighed several tens of metric tons, it was thought for many years that they were too large to walk on land; hence, they were depicted partly submerged in swamps and lakes (Figure 2). This interpretation is no longer accepted because sauropod bones and trackways clearly show that they lived on dry land. Furthermore, if brachiosaurs had lived as indicated in Figure 2, their lungs would have been more than 6 m underwater, making breathing difficult because water pressure would prevent the chest from expanding.

Currently, *Brachiosaurus* is the largest dinosaur whose weight has been estimated from rather complete remains, but even larger ones might have existed. Unfortunately, only partial remains are available for dinosaurs known by such common names as supersaurus, ultrasaurus, and seismosaurus. According to one estimate, ultrasaurus was more than 30 m long and weighed as much as 136 metric tons. Seismosaurus might have been between 39 and 52 m and even heavier. Estimating dinosaur weight from rather complete remains is difficult, so these estimates based on fragmentary remains must be taken with caution.

It may seem odd that ultrasaurus's estimated length is not much greater than that of *Brachiosaurus,* yet its estimated weight is more than 1.7 times as much. What this means is that ultrasaurus was not much longer but considerably heavier. The reason

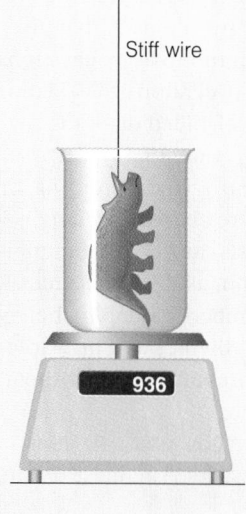

Water

Balance

840

Stiff wire

936

FIGURE 1 Using a scale model to determine a dinosaur's weight. Submerging the model shows that it displaces a volume of water with a mass of 96 g as indicated by the different readings on the balance. Because the model is 1-to-40 scale, we can find the volume of the living dinosaur by multiplying $40 \times 40 \times 40 = 64{,}000$. If we assume the dinosaur's mass equaled that of water, 1 g/cm^3, we multiply $96 \times 64{,}000 = 6{,}144{,}000$, giving the dinosaur's weight in grams. In other words, the dinosaur weighed just over 6 metric tons.

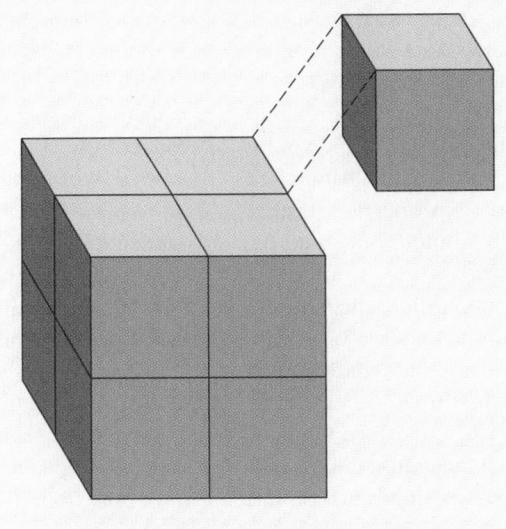

FIGURE 2 It was once thought that large dinosaurs, such as these sauropods, were too heavy to walk on dry land and must have been semi-aquatic. Evidence from fossil footprints indicates that sauropods were quite capable of walking on land. Furthermore, if they had submerged themselves, as in this illustration, breathing would have been difficult because water pressure would prevent the chest from expanding.

is that the weight of an animal, or any object regardless of its shape or density, increases by a factor of 8 if the linear dimensions are doubled. In other words, if the length, height, and width of an object are doubled, its weight increases eightfold. To understand this concept more easily, think of a cube weighing 2 kg. If you were to double all linear dimensions you would end up with eight cubes, and the weight would be 16 kg (Figure 3). Animals are of course not cubic, but the same mathematical relationship holds if all their dimensions are doubled. Thus, if your pet dog is twice as long, tall, and wide as your neighbor's, yours will weigh eight times as much.

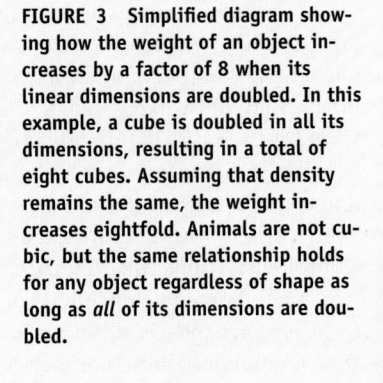

FIGURE 3 Simplified diagram showing how the weight of an object increases by a factor of 8 when its linear dimensions are doubled. In this example, a cube is doubled in all its dimensions, resulting in a total of eight cubes. Assuming that density remains the same, the weight increases eightfold. Animals are not cubic, but the same relationship holds for any object regardless of shape as long as *all* of its dimensions are doubled.

(a)

FIGURE 15.14 (a) The small theropod *Compsognathus* measured only 60 cm long and weighed 2 to 3 kg. Bones found inside its rib cage indicate this carnivore ate lizards. (b) This lifelike restoration shows the theropod *Deinonychus* in its probable attack posture. Large, curved claws on its back feet were probably used in a slashing type of attack. Whether *Deinonychus* had scaly skin or some other kind of body covering is unknown.

(b)

Dinosaur Behavior Typically, carnivorous dinosaurs are portrayed in movies as ferocious, aggressive beasts. Titanic battles between *Tyrannosaurus* and some large armored dinosaur is common fare, but, just as carnivores do today, they most likely avoided large dangerous prey and went for the easy kill. Preying on the old, weak, and young or simply dining on carrion when it was available is a much more likely type of behavior. No one doubts that many carnivorous dinosaurs actively hunted, but some paleontologists think that *Tyrannosaurus* was primarily a scavenger. Some evidence indicates that dinosaurs such as *Deinonychus* and *Velociraptor* hunted in packs.

Undoubtedly, some dinosaurs were solitary animals, but the available evidence indicates that the large herbivores were gregarious, congregating in vast herds. Two kinds of evidence support this conclusion. One consists of fossil trackways indicating that a number of dinosaurs of the same type moved together. The other kind of evidence is so-called bone beds in which only one species of dinosaur is found. These concentrations of fossils indicate that large numbers of animals perished quickly as the result of some catastrophe such as drowning while crossing a river or suffocation by volcanic ash. In Perspective 3.1 we noted that several sauropods were trapped in mud at what is now known as Howe Quarry in Wyoming and that several thousand duck-billed dinosaurs died en masse during the Late Cretaceous in northwestern Montana. Bone beds of ceratopsian dinosaurs in Alberta, Canada, indicate that these dinosaurs also congregated in herds.

In preceding sections we mentioned several dinosaur features that might have served some behavioral function. The crests of ornithopods are the most obvious, but horns, frills, and spikes in, for instance, ceratopsians might also have functioned in species recognition or displays to at-

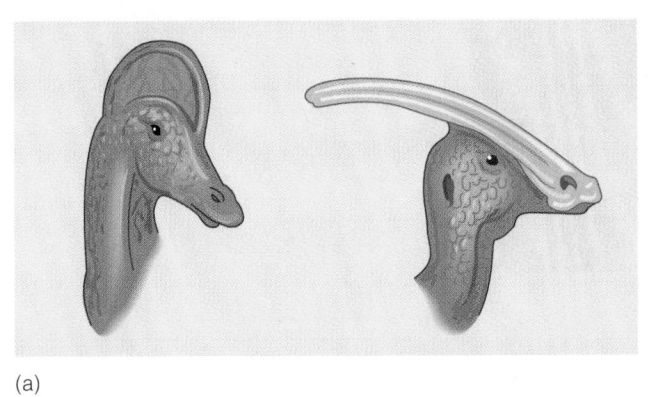

(a)

(b)

FIGURE 15.15 (a) Two duck-billed dinosaurs from the Late Cretaceous of western North America. *Corythosaurus (left)* and *Parasaurolophus (right)* were two of several duck-billed dinosaurs with head crests. (b) Scene from the Late Cretaceous of northern Montana. A female *Maiasaura* leads her young to a feeding area.

tract mates. Paleontologists think stegosaur plates were some kind of devise to absorb and dissipate heat, but they might also have been for display and species recognition.

Nesting sites discovered in Montana clearly demonstrate that at least three types of dinosaurs not only nested in colonies but also used the same site repeatedly. One of these dinosaurs, named *Maiasaura* (good mother dinosaur), laid eggs in 2-m-diameter nests each spaced about 7 m apart, or about the average length of an adult maiasaur. It seems that these dinosaurs nested in colonies much as some living birds do. Some of the nests contain juveniles up to 1 m long, which is much greater than their length at the time of hatching. This evidence indicates that the young remained in the nest area for some time after hatching and that adults cared for them. It seems likely that adults protected and fed the young (Figure 15.15b).

FIGURE 15.16 Skeleton of the ceratopsian dinosaur *Triceratops* in the Natural History Museum, London, England. This rhinoceros-sized dinosaur was particularly common during the Late Cretaceous in western North America.

Nests and eggs of the ostrichlike dinosaur *Oviraptor* are well known in Late Cretaceous rocks of Mongolia. Unfortunately, it has not yet been determined whether they nested in colonies, but one remarkable discovery in 1993 shows a 2.4-m-long *Oviraptor* that might have been incubating about 15 eggs. The dinosaur's forelimbs encircle the eggs and the back legs are folded underneath the animal's body, but this posture may have simply been to protect the eggs from a predator.

In November 1998 American and Argentine scientists reported the first eggs and embryos of sauropod dinosaurs, probably titanosaurs. A Late Cretaceous nesting ground with thousands of eggs, some with unhatched embryos, was discovered in the badlands of Patagonia in Argentina. Apparently, adult titanosaurs gathered by the thousands and nested on floodplains where some of their eggs were buried during floods. The abundance of eggs and egg-shell fragments shows that these dinosaurs nested in colonies, but it is not known if they cared for their young.

Warm-Blooded Dinosaurs? All living reptiles are **ectotherms,** that is, cold-blooded animals whose body temperature varies in response to the outside temperature. **Endotherms,** warm-blooded animals, such as birds and mammals, are capable of maintaining a rather constant body temperature regardless of the outside temperature. Some investigators think that dinosaurs, or at least some dinosaurs, were endotherms.

Proponents of dinosaur endothermy note that dinosaur bones are penetrated by numerous passageways that, when the animals were living, contained blood vessels. Bones of endotherms typically have this structure, but considerably fewer of these passageways are found in bones of ectotherms. Living crocodiles and turtles have this so-called endothermic bone structure, yet they are ec-

What Did Dinosaurs Have on the Outside?

*T*he skeletal structure of many dinosaurs is well known from fossils representing all or nearly all the bones of the body. Actually, a complete skeleton is not necessary to know what the entire skeleton was like, because many bones are mirror images of one another. For example, if one finds the right front forelimb of a dinosaur, or any other vertebrate animal, the left forelimb looks the same except reversed. Accordingly, what dinosaur skeletons looked like is rather well known. However, the body covering of vertebrate animals consisting of skin, scales, hair, fur, or feathers is rarely preserved. We discussed the fact that feather impressions of *Archaeopteryx* are found on several fossils, otherwise these fossils might have been called dinosaurs.

Although rare, skin impression and preserved skins are known from several types of dinosaurs. For some, the skin is made up of tubercle-like scales (Figure 1). That is, the skin has noblike projections giving it a pebbly appearance, whereas today's snakes and most lizards have overlapping scales. However, recently discovered sauropod embryos with patches of preserved skin in Argentina show that these dinosaurs had scaly skin

similar to that of present-day lizards. And at least one species of sauropod similar to *Diplodocus* had a row of triangular spines along the back from head to tail.

Restorations of dinosaurs in museums

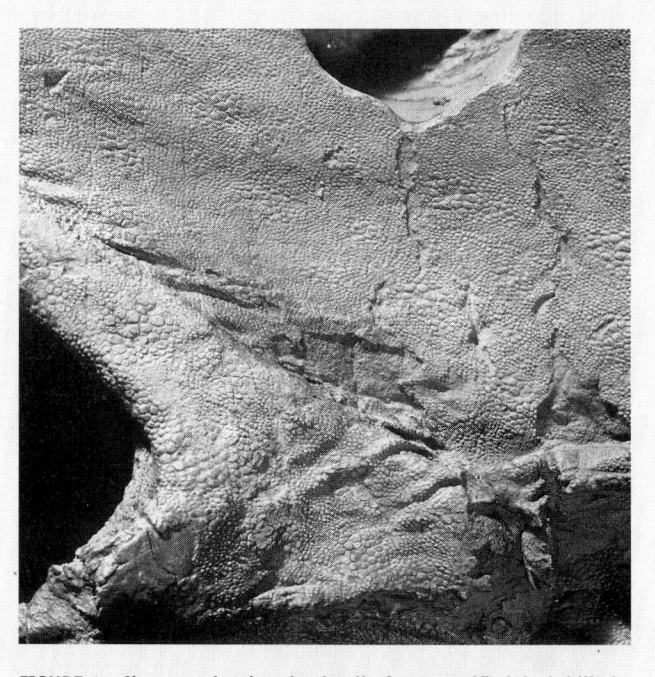

FIGURE 1   Close-up showing the detail of a mummified duck-billed dinosaur skin. See Figure 3.19a for a more complete view of this dinosaur mummy.

and in illustrations are reasonably accurate as far as skin texture is concerned, but the flamboyant colors often used are purely conjectural. Some dinosaur skins had different patterns, and perhaps this is related to color differences; but color

FIGURE 2 Restoration of the Early Cretaceous feathered dinosaur *Caudipteryx* from China. Feathers were preserved on the arms (wings) and tail. The fact that *Caudipteryx* had short forelimbs, symmetric feathers, and was larger than *Archaeopteryx* indicates that it was flightless.

had a coat of downy feathers. Several experts have since examined the specimen and concluded that the body covering is not true feathers but downy bristles or the forerunners of feathers. One critic, however, claims it is nothing more than a reptilian frill.

There is little doubt, though, that Early Cretaceous fossils from northeast China represent three genera of feathered dinosaurs. All of these—*Protarchaeopteryx, Caudipteryx* (Figure 2), and *Confuciusornis*— possess a curious mix of dinosaur and bird characteristics. And even though all were feathered, only *Confuciusornis* appears to have been a true flyer; the others were apparently ground dwellers, as indicated by relatively short forelimbs and other anatomical characteristics. All were carnivorous bipeds with typical theropod teeth, and while they probably ate a variety of small animals, one fossil specimen contains the jawbones of a mammal.

The fact that these dinosaurs had feathers implies that they were endotherms. After all, living endotherms— birds and mammals—have an outer body covering to preserve heat. Furthermore, feathered theropods provide compelling evidence for the hypothesis that some small theropod is the ancestor of birds. But why, if these fossils have feathers, are they not birds rather than dinosaurs? According to the Chinese paleontologists who discovered the fossils, they are more primitive than *Archaeopteryx,* the oldest known bird, and they possess many theropod characteristics. In fact, according to cladistic analyses, these feathered dinosaurs and *Archeaopteryx* constitute a group of theropod dinosaurs.

does not fossilize, so we can only make some educated guesses about the color of the animals when living. For instance, large terrestrial mammals such as elephants and rhinoceroses are typically gray, so perhaps the large sauropods were similarly colored. It is likely that many di-nosaurs had various color patterns for concealment or bright colors to attract mates. Whatever these colors might have been, though, is not known.

In 1996 Chinese paleontologists discovered the remains of a dinosaurs, later named *Sinosauropteryx,* that they claimed totherms. And in some small mammals the bone structure is more typical of ectotherms, yet we know that they are capable of maintaining a constant body temperature. It may be that bone structure is more related to body size and growth patterns than to endothermy.

Because endotherms have high metabolic rates, they must eat more than ectotherms of comparable size. Consequently, endothermic predators require large prey populations. They would therefore constitute a much smaller proportion of the total animal population than their prey. In contrast, the proportion of ectothermic predators to their prey population is much greater. Where data are sufficient to allow an estimate, dinosaur predators appear to have made up 3 to 5% of the total population. These figures are comparable to present-day mammalian populations. However, a number of uncertainties about the composition of fossil communities make this argument for endothermy unconvincing to many paleontologists.

Living endotherms have a large brain in relation to body size. A relatively large brain is not necessary for endothermy, but endothermy does seem to be a prerequisite for having a large brain because a complex nervous system requires a rather constant body temperature. Some dinosaurs, particularly the small carnivores, did have a large brain in relation to their body, but many did not. That the small carnivorous ones usually had a large brain seems to be a good argument for endothermy, but there is an even more compelling argument. The relationship of birds to small carnivores such as *Compsognathus* (Figure 15.14a) implies that these small dinosaurs were endothermic or at least trending in that direction. And recent discoveries in China of dinosaurs with a covering of feathers not only supports the idea of endothermic theropods but also provides compelling evidence for the relationship of theropods to birds (see Perspective 15.2).

The large sauropods were probably not endothermic but nevertheless may have been able to maintain their body temperatures within narrow limits as endotherms do. A large animal heats up and cools down slowly because it has a small surface area compared to its volume. With proportionately less surface area to allow heat loss, sauropods probably retained body heat more efficiently than smaller dinosaurs.

Another point on endothermy in dinosaurs is that the flying reptiles, the *pterosaurs,* evolved from ancestral ardrosaurs as did the dinosaurs (Figure 15.12). At least one species of pterosaur had hair or hairlike feathers. This is interesting because an insulating covering of hair or feathers is known only in endotherms. Furthermore, the physiology of active flight requires endothermy. Such evidence indicates that perhaps both ancestral ardrosaurs and dinosaurs were endothermic.

Obviously, considerable disagreement exists on dinosaur endothermy. In general, a fairly good case can be made for endothermic, small, carnivorous dinosaurs and pterosaurs, but for the others the question is still open.

FLYING REPTILES

The first flying animals were Paleozoic insects, but the first vertebrates to accomplish flight were Mesozoic **pterosaurs,** or what are commonly called flying reptiles. Pterosarus evolved from ancestral archosaurs during the Triassic and were numerous until the end of the Mesozoic when they became extinct. The earliest pterosaurs were small animals with long tails; the largest had a wingspan of about 60 cm. Their descendants were tailless and generally larger (Figure 15.17). Modifications for flight include a wing membrane supported by an elongate fourth finger (Figure 15.17c), light hollow bones, a bony structure on the breast bone for flight-muscle attachment, and development of those parts of the brain associated with muscular coordination and sight. Among the large pterosaurs were *Pteranodon* (Figure 15.17b), and *Quetzalcoatlus* from the Cretaceous of Texas had a wingspan of at least 12 m. The fact that at least one pterosaur had a coat of hair or hairlike feathers and was a flyer indicates that it, and perhaps all pterosaurs, were endotherms.

Were pterosaurs, particularly the larger ones, active, wing-flapping fliers or simply gliders? Experiments with scale models indicate that *Pteranodon* (Figure 15.17b) could actively fly, but once airborne probably took advantage of thermal updrafts to stay aloft and ranged far out over the open sea, mostly by soaring. Perhaps much like the present-day frigate bird, *Pteranodon* glided close to the surface and plucked fish from the water with its long beak. Most paleontologists agree that the small pterosaurs were wing-flapping fliers.

The function of the large crest on the head of *Pteranodon* is uncertain. It may have functioned aerodynamically in maneuvering or braking, or it may have served as a counterbalance for the long beak. Some specimens seem to lack this feature, indicating that it may have been a sexual characteristic of some sort. If so, whether males or females had crests is unknown.

MARINE REPTILES

Ichthyosaurs are undoubtedly the most familiar of the Mesozoic marine reptiles (Figure 15.18a). Many species of these completely aquatic animals were about 3 m long, but some of gigantic proportions are known. One of these is well represented at Berlin-Ichthyosaur State Park, Nevada, where individuals up to 15 m long were fossilized. Ichthyosaur aquatic adaptations include a streamlined, somewhat fishlike body, flipperlike forelimbs for maneuvering, and a powerful tail for propulsion. Numerous sharp teeth indicate that they ate fish. Reptile eggs cannot be laid in water, so females probably retained the eggs within their bodies and gave birth to live young. Fossil ichthyosaurs with young in the appropriate part of the body cavity support this interpretation. Ichthyosaurs superficially resemble dolphins and probably lived much like dolphins do, thus providing a good example of convergent evolution (see Figure 5.22).

A second group of Mesozoic marine reptiles known as

(a)

(b)

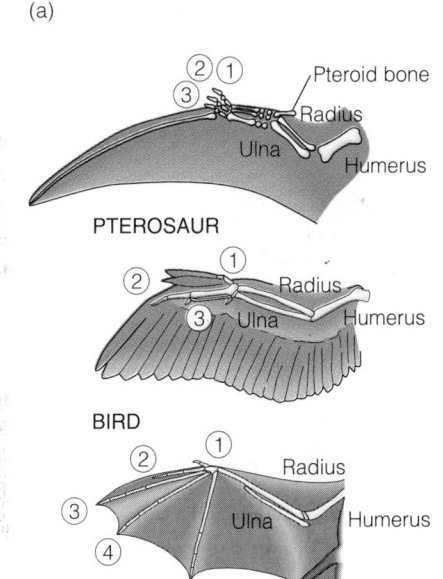

(c)

FIGURE 15.17 (a) *Pterodactylus,* a long-tailed pterosaur, is well known from the Late Jurassic Solnhofen Limestone of Germany. Among the several known species, wing span ranged from 50 cm to 2.5 m. (b) The short-tailed pterosaur *Pteranodon* was a large, Cretaceous animal with a wing span up to 9 m. (c) In all flying vertebrates, the forelimb has been modified into a wing. In pterosaurs, a long fourth finger provides support for the wing, the wing support in birds is fused second and third fingers, and in bats fingers two through five support the wing.

plesiosaurs is also quite familiar. Two distinct varieties are represented by short-necked and long-necked plesiosaurs (Figure 15.18b). Most plesiosaurs were 3.6 to 6 m long, but one species from Antarctica measures 15 m. Long-necked plesiosaurs were heavy-bodied animals with mouthfuls of sharp teeth and limbs modified into oarlike paddles. They might have used their long necks in snakelike fashion to capture fish. Plesiosaurs probably came ashore to lay eggs. On land they were likely rather awkward but no more so than present-day walruses or seals.

Mosasaurs were a group of Late Cretaceous marine lizards related to the living Komodo dragon, or monitor lizard. These animals had a worldwide distribution, but they are particularly well known from the Niobrara Chalk of Kansas. Mosasaurs varied from small species only 2.5 m long to giants such as *Tylosaurus* measuring more than 9 m (Figure 15.19). Mosasaur limbs resembled paddles and were probably used mostly for maneuvering, whereas their long tail provided propulsion. All were predators. The preserved stomach contents of one specimen include a diving bird, a marine bony fish, a shark, and part of a smaller mosasaur. They also ate a variety of marine invertebrates including ammonites. Fossils of deep-diving mosasaurs show signs of the pathological condition known as avascular necrosis, or what is more commonly called "the bends."

CROCODILES, TURTLES, LIZARDS, AND SNAKES

By Jurassic time, crocodiles had become the dominant freshwater predators. All crocodiles are amphibious,

(a)

(b)

FIGURE 15.18. Mesozoic marine reptiles: (a) ichthyosaurs and (b) a long-necked plesiosaur.

spending much of their time in water, but they are also well equipped for walking on land. Overall, crocodile evolution has been conservative, involving changes mostly in size from a meter or so in Jurassic forms to 15 m in some Cretaceous species.

Turtles, too, have been evolutionarily conservative since their appearance during the Triassic. The most remarkable feature of turtles is their heavy, bony armor; turtles are more thoroughly armored than any other vertebrate animal, living or fossil. Turtle ancestry is uncertain. One Permian animal had eight broadly expanded ribs, which may represent the first stages in the development of turtle armor.

Lizards and snakes are closely related, and lizards were in fact ancestral to snakes. The limbless condition in snakes (some lizards are limbless, too) and skull modifications that allow snakes to open their mouths very wide are the main differences between these two groups. Lizards

are known from Upper Permian strata, but they did not become abundant until the Late Cretaceous. Recall that mosasaurs are also lizards, but because they were thoroughly aquatic, we discussed them in the section on marine reptiles.

Snakes first appear during the Cretaceous, but the families to which most living snakes belong differentiated since the Early Miocene. Although their Cretaceous and Early Tertiary fossil record is poor, one Early Cretaceous genus from Israel appears to show characteristics intermediate between snakes and their lizard ancestors.

Birds

In Chapter 5 we briefly discussed the origin of birds to illustrate how fossils provide evidence for evolution. We noted that *Archaeopteryx*, from Late Jurassic rocks in Germany, has several avian features such as feathers and a

FIGURE 15.19 *Tylosaurus* was a large mosasaur from the Late Cretaceous. It measured up to 9 m long.

wishbone, so accordingly it was a bird (Figure 15.20). In most other characteristics—including reptilelike teeth, a long tail, and forelimb structure—*Archaeopteryx* more closely resembles theropod dinosaurs. Most paleontologists now think that some kind of small theropod was the ancestor of birds. Even the wishbone, consisting of fused clavicles, is present in many theropods, and the recent discovery of theropods with some kind of feathery covering is further evidence of theropod–bird relationships (see Perspective 15.2).

Two hypotheses are currently being debated about how bird flight evolved. No one doubts that *Archaeopteryx* could fly, but it probably did so rather poorly compared with today's birds. However, it had asymmetric feathers, meaning that the feathers were not equally as large on each side of a central shaft, just as in flying birds now. In contrast, flightless birds have symmetric feathers. In any case, some think that bird flight evolved from the ground up; that is, fast running theropods leaped into the air with arms outstretched to catch insects or avoid predators. According to the arboreal or from-the-trees-down hypothesis, birds, or their immediate ancestors, climbed trees and their feathered forelimbs assisted them in gliding down.

The fossil record of birds is not nearly as good as that of many other animals, but it is continually getting better. For instance, in recent years several discoveries have been made that shed light on bird evolution. One fossil specimen from China is slightly younger than *Archaeopteryx* and possesses both primitive and advanced characteristics. It retains abdominal ribs quite similar to those of *Archaeopteryx* and theropod dinosaurs, but it has a reduced tail typical of present-day birds. A fossil bird from Spain is 20 to 30 million years younger than *Archaeopteryx,* and it too possesses a mix of primitive and advanced features, but it does appear to lack abdominal ribs. And a Late Jurassic to Early Cretaceous bird from Asia has a toothless beak similar to that of today's birds. Recent finds of Cretaceous bird fossils in Madagascar and Argentina further support the idea that theropods gave rise to birds.

Archaeopteryx's status as the oldest known bird has been challenged. Two crow-sized fossils named *Protoavis* from Triassic rocks has convinced one paleontologist that birds originated earlier than previously thought. According to this claim, *Protoavis* was more birdlike than *Archaeopteryx*, but the remains are extremely fragmentary and no feather impressions were found. In view of these facts, many paleontologists think that *Protoavis* was actually a small carnivorous dinosaur.

Two Late Cretaceous bird fossils from rocks in Kansas are interesting. *Hesperornis* was a swimming bird with

FIGURE 15.20 *Archaeopteryx,* a Jurassic-age animal from Germany, had feathers and a wishbone and is therefore classified as a bird. In most other anatomic features, though, it more closely resembles small theropod dinosaurs. For instance, it had reptilelike teeth, claws on the wings, and a long tail—none of which are found in present-day birds.

powerful kicking feet and vestigial wings (Figure 15.21). *Ichthyornis,* on the other hand, was an active flyer who seems to have lived much like terns. Neither bird had a long tail, but both appear to have had teeth. No Cenozoic bird is known to have had teeth, so toothless birds must have been present by the end of the Mesozoic.

From Reptile to Mammal

Therapsids, or the advanced mammal-like reptiles, were briefly described in Chapter 13. These reptiles diversified into numerous species of herbivores and carnivores, and during the Permian they were the most diverse and numerous terrestrial vertebrates. One particular group of carnivorous therapsids called *cynodonts* was the most mammal-like of all and by the Late Triassic gave rise to mammals.

CYNODONTS AND THE ORIGIN OF MAMMALS

The transition from **cynodonts** to mammals is well documented by fossils and is so gradational that classification of some fossils as either reptile or mammal is difficult. We can easily recognize living mammals as warm-blooded animals with hair or fur that have mammary glands and, except for the platypus and spiny ant-eater, give birth to live young (Figure 15.22).

Obviously, most criteria for recognizing living mammals are inadequate for classifying fossils. For them, we must use skeletal structure only. Several skeletal modifications characterize the transition from mammal-like reptiles to mammals, but distinctions between the two groups are based largely on details of the middle ear, the lower jaw, and the teeth (Table 15.1).

Reptiles have only one small bone in the middle ear—the stapes—while mammals have three—the incus, the malleus, and the stapes. Also, the lower jaw of a mammal is composed of a single bone called the *dentary,* but a reptile's jaw is composed of several bones (Figure 15.23). In addition, a reptile's jaw is hinged to the skull at a contact between the articular and quadrate bones, while in mammals the dentary contacts the squamosal bone of the skull (Figure 15.23).

During the transition from cynodonts to mammals, the quadrate and articular bones that had formed the joint between the jaw and skull in reptiles were modified into the incus and malleus of the mammalian middle ear (Figure 15.23). Fossils clearly document the progressive enlargement of the dentary until it became the only element in the mammalian jaw (Figure 15.24). Likewise, a progressive change from the reptile to mammal jaw joint is documented by fossil evidence. In fact, some of the most advanced cynodonts were truly transitional because they had a compound jaw joint consisting of (1) the articular and quadrate bones typical of reptiles and (2) the dentary and squamosal bones as in mammals (Table 15.1).

In Chapter 5 we noted that the study of embryos provides evidence for evolution. Opossum embryos clearly show that the middle-ear bones of mammals were originally part of the jaw. In fact, even when opossums are born, the middle-ear elements are still attached to the dentary (Figure 15.23d), but as they develop further, these elements migrate to the middle ear, and a typical mammal jaw joint develops.

Several other aspects of cynodonts also indicate they were ancestral to mammals. Their teeth were somewhat differentiated into distinct types in order to perform specific functions. In mammals the teeth are fully differentiated into incisors, canines, and chewing teeth, but typical reptiles do not have differentiated teeth. In addition, mammals have only two sets of teeth during their lifetimes—a set of baby teeth and the permanent adult teeth. Typical reptiles have teeth replaced continuously throughout their lives. The notable exception is seen in some cynodonts who in mammalian fashion had only two sets of teeth. Another important feature of mammal teeth is occlusion; that is, the chewing teeth (premolars and molars) meet surface to surface to allow grinding. Thus, mammals chew their food, but typical reptiles, amphibians, and fish do not. However, tooth occlusion is known in some advanced cynodonts (Table 15.1).

Another mammalian feature, the secondary palate, was partially developed in advanced cynodonts (Figure 15.25). This secondary palate is a bony shelf above the mouth that separates the nasal passages from the mouth cavity. It is an adaptation for eating and breathing at the same time, a necessary requirement for endotherms with their high demands for oxygen.

Mesozoic Mammals

Even though mammals evolved during the Late Triassic, their diversity remained low during the rest of the Mesozoic (Figure 15.26). A few Mesozoic mammals may have been as large as raccoons, but most were quite small—about the size of mice or rats. Their fossil record, which consists of

FIGURE 15.21 A Cretaceous bird from Kansas. Restoration of *Hesperornis,* a toothed, aquatic bird with vestigial wings.

TABLE 15.1

Summary Chart Showing Some Characteristics and How They Changed During the Transition from Reptiles to Mammals

	TYPICAL REPTILE	CYNODONT	MAMMAL
Lower Jaw	Dentary and several other bones	Dentary enlarged, other bones reduced	Dentary bone only, except in earliest mammals
Jaw–Skull Joint	Articular–quadrate	Articular–quadrate; some advanced cynodonts had both the reptile jaw–skull joint and the mammal jaw–skull joint	Dentary–squamosal
Middle-Ear Bones	Stapes	Stapes	Stapes, incus, malleus
Secondary Palate	Absent	Partly developed	Well developed
Teeth	No differentiation	Some differentiation	Fully differentiated into incisors, canines, and chewing teeth (premolars and molars)
Tooth replacement	Teeth replaced continuously	Only two sets of teeth in some advanced cynodonts	Two sets of teeth
Occlusion (chewing teeth meet surface to surface to allow grinding)	No occlusion	Occlusion in some advance cynodonts	Occlusion
Endothermic vs. ectothermic	Ectothermic	Probably endothermic	Endothermic
Body covering	Scales	One fossil shows it had skin similar to that of mammals	Skin with hair or fur

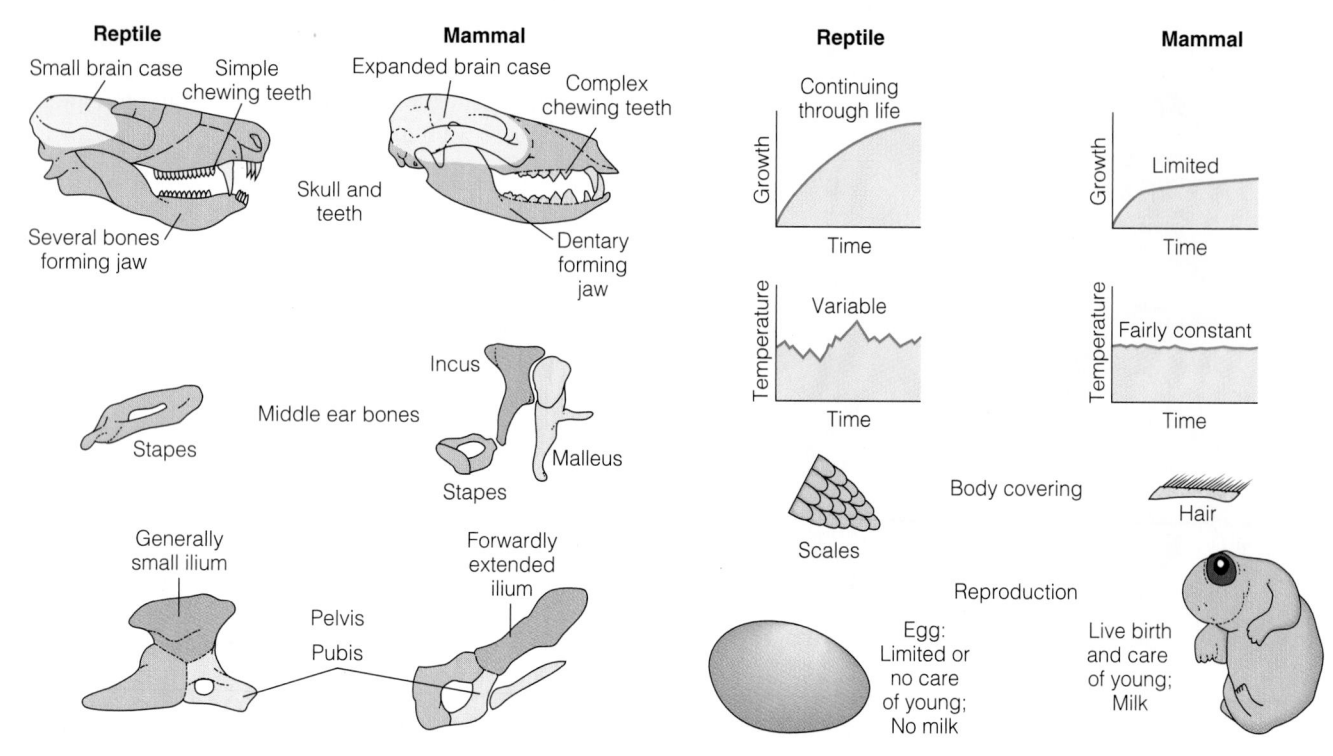

FIGURE 15.22 Some contrasting characteristics of reptiles and mammals. Most living mammals can be easily identified by these characteristics. The platypus and spiny anteater, however, both lay eggs, but in most other respects are mammals.

isolated teeth, jaw fragments, and fragmentary bones, is not particularly good, although a few skulls and even some fairly complete skeletons are known. Nevertheless, these fossils are important because two-thirds of mammalian history is recorded by Mesozoic fossils.

The first mammals retained several reptilian characteristics but had mammalian features as well. For example, the Triassic triconodonts (Figure 15.27) had the fully differentiated teeth typical of mammals, but they had both the reptile and mammal types of jaw joints. The earliest symmetrodonts retained several reptilian bones in the lower jaw, but by the Cretaceous only the dentary was present. In short, some mammalian features evolved more rapidly than others, thereby accounting for animals that possessed characteristics of both reptiles and mammals. Recall from Chapter 5 that evolution in different features of the same organism at such varying rates is known as *mosaic evolution*.

The best-known early mammal is *Morganucodon* from Late Triassic to Early Jurassic rocks of Europe and China. It was a tiny shrewlike animal with a skull measuring only 20 to 30 mm long. Small openings on its snout were likely the

FIGURE 15.23   (a) The skull of an opossum showing the typical mammalian dentary–squamosal jaw joint. (b) The skull of a cynodont shows the articular–quadrate jaw joint of reptiles. (c) Enlarged view of an adult opossum's middle-ear bones. (d) View of the inside of a young opossum's jaw showing that the elements of the middle ear are attached to the dentary during early development. This is the same arrangement of bones that is found in the adults of the ancestral mammals.

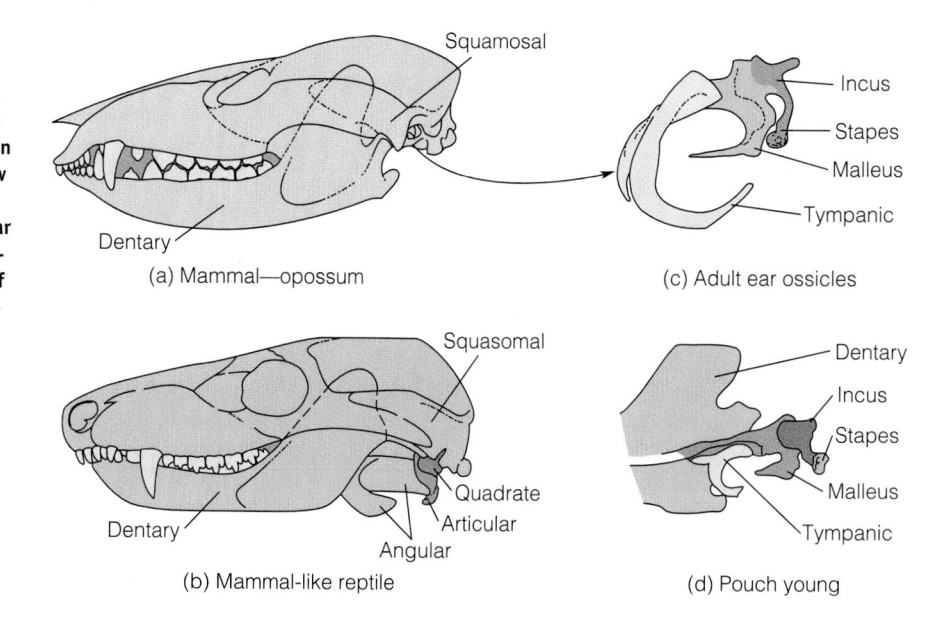

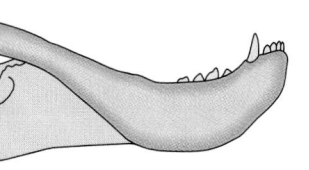

Pelycosaur
(Primitive mammal-like reptile)

Advanced therapsid, a cynodont

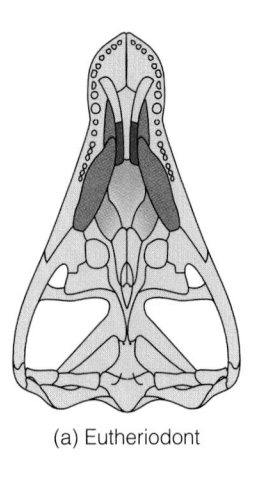

Primitive therapsid

Mammal

FIGURE 15.24 During the transition from reptiles to mammals, the dentary bone (brown) became progressively larger until it became the only bone in the lower jaw.

sites of sensory whiskers; thus, it probably had hair. In typical mammal fashion, it had only two sets of teeth during its life, and when born it appears to have lacked teeth. This being the case, it might have been nursed, further indicating its mammalian nature. *Morganucodon* was almost certainly endothermic, yet this tiny animal still possessed some reptile features such as some small bones in the lower jaw, in addition to the dentary.

Figure 15.26 shows that early mammals diverged into two distinct branches. One branch includes the triconodonts and their probable evolutionary descendants, the **monotremes,** or egg-laying mammals. Living monotremes are the platypus and spiny anteater of the Australian region. Also included in this branch are the *multituberculates,* the first mammalian herbivores; all other Mesozoic mammals probably preyed on insects, worms, and grubs. Multituberculates seem to have lived much like present-day rodents and, in fact, were replaced by rodents during the Early Cenozoic.

The second evolutionary branch shown in Figure 15.26 includes the symmetrodonts and the **eupantotheres** and their descendants—the **marsupial** (pouched) **mammals** and the **placental mammals.** All living mammals except

monotremes have ancestries that can be traced back through this branch.

Eupantotheres were mouse-sized animals with a poor fossil record, but details of their teeth indicate they were ancestral to both marsupial and placental mammals. Furthermore, one eupantothere fossil possesses bones projecting from the pelvis that could have supported a pouch, so in this respect it was similar to living marsupials. The divergence of marsupials and placentals from a common ancestor probably occurred during the Early Cretaceous, but undoubtedly both were present by the Late Cretaceous. The earliest known placental mammals were members of the order *Insectivora* (Figure 15.28), an order represented today by shrews, moles, and hedgehogs.

Mesozoic Climates and Paleogeography

The present continental positions and climatic patterns largely restrict the distribution of organisms. Land plants and animals have little opportunity to colonize distant areas because of physical barriers (especially the ocean

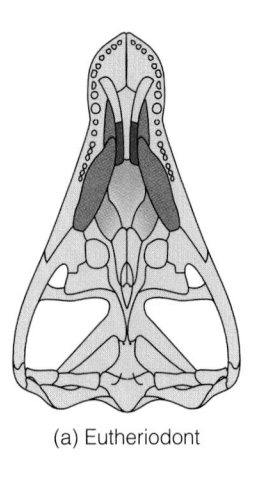

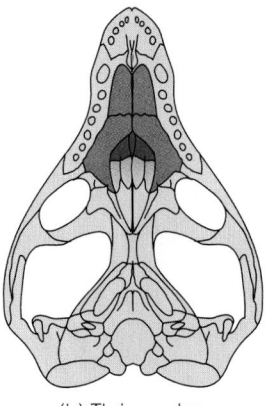

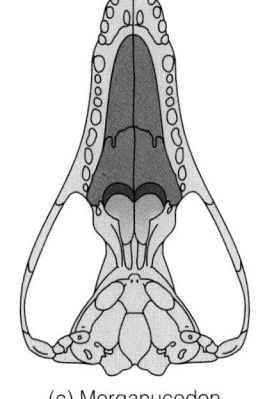

FIGURE 15.25 Views of the bottoms of skulls of (a) an early therapsid, (b) a cynodont, and (c) an early mammal, showing the progressive development of the bony secondary palate (brown).

(a) Eutheriodont

(b) Thrinaxodon

(c) Morganucodon

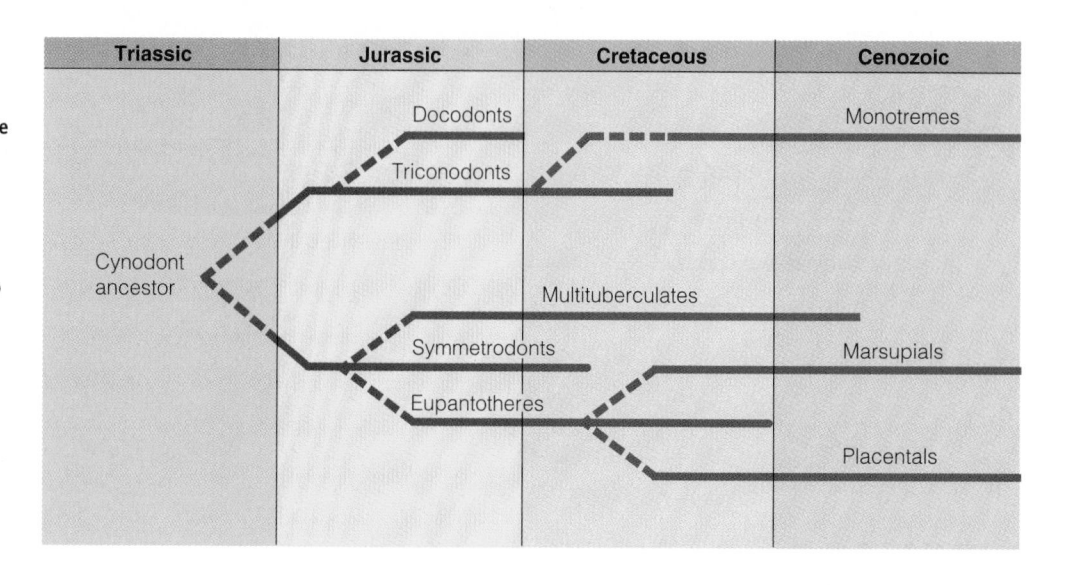

FIGURE 15.26 Possible relationships among the early mammals and their descendants. Several relationships are uncertain, but apparently mammalian evolution proceeded along two branches. One branch led to the monotremes, or egg-laying mammals, and the other led to marsupial and placental mammals.

Triassic	Jurassic	Cretaceous	Cenozoic

Cynodont ancestor

Docodonts — Monotremes

Triconodonts

Multituberculates

Symmetrodonts — Marsupials

Eupantotheres

Placentals

basins) and climatic barriers. Mesozoic barriers to migration were apparently not as effective as they are now, because some Mesozoic organisms are known from areas that are now widely separated.

Fragmentation of the supercontinent Pangaea began by the Late Triassic and continues to the present, but during much of the Mesozoic, close connections existed between the various landmasses. The proximity of these landmasses, however, is not sufficient to explain Mesozoic biogeographic distributions because climates are also effective barriers to wide dispersal. During much of the Mesozoic, though, climates were more equable and lacked the strong north and south zonation characteristic of the present. In short, Mesozoic plants and animals had greater opportunities to occupy much more extensive geographic ranges.

Pangaea persisted as a single unit through most of the

Triassic. The Triassic climate was warm-temperate to tropical, although some areas, such as the present southwestern United States, were arid. Mild temperatures extended 50 degrees north and south of the equator, and even the polar regions may have been temperate. The fauna was truly worldwide in its distribution. Archosaurs known as phytosaurs were present in North America, Europe, and Madagascar. Some dinosaurs had continuous ranges across Laurasia and Gondwana. The peculiar gliding lizards were in New Jersey and England.

By the Late Jurassic, Laurasia had become partly fragmented by the opening North Atlantic, but a connection still existed. The South Atlantic had begun to open so that a long, narrow sea separated the southern parts of Africa and South America. Otherwise the southern continents were still close together.

FIGURE 15.27 One of the earliest mammals, the triconodont *Triconodon*.

The mild Triassic climate persisted into the Jurassic. Ferns, whose living relatives are now restricted to the tropics of southeast Asia, are known from areas as far as 63 degrees south latitude and 75 degrees north latitude. Dinosaurs roamed widely across Laurasia and Gondwana. Specimens from the Morrison Formation in western North America and those in the Tendagura beds of eastern Africa are quite similar. For example, the giant dinosaur *Brachiosaurus* is known from both areas. Stegosaurs and some families of carnivorous dinosaurs lived throughout Laurasia and in Africa.

By the Late Cretaceous, the North Atlantic had opened further, and Africa and South America were completely separated. South America remained an island continent until late in the Cenozoic. Its fauna, evolving in isolation, became increasingly different from faunas of the other continents. Marsupial mammals reached Australia from South America via Antarctica, but the South American connection was eventually severed. Placentals, other than bats and a few rodents, never reached Australia. This explains why the marsupials continue to dominate the continent's fauna even today.

Cretaceous climates were more strongly zoned by latitude, but they remained warm and equable until the close of that period. Climates then became more seasonal and cooler, a trend that persisted into the Cenozoic. Dinosaur and mammal fossils demonstrate that interchange was still possible, especially between the various components of Laurasia.

As one might expect, the paleontology of Antarctica is poorly known. Marine reptiles, including plesiosaurs 15 m long have been found on Seymour Island near the tip of the Antarctic Peninsula, but since they were marine reptiles, their presence tells us little about continental connections. Nevertheless, Antarctica remained close to South America well into Cenozoic time, as indicated by the fact that a land-dwelling marsupial of Eocene age has been found there. An Early Tertiary coal bed on Seymour Island indicates that the climate was temperate.

Mass Extinctions—A Crisis in the History of Life

Given that most organisms that ever existed are now extinct, it would seem that extinctions are common in the history of life. This so-called *background extinction* differs in degree from **mass extinctions,** which are times of higher extinction rates during which organic diversity is sharply reduced. The greatest mass extinction took place at the end of the Paleozoic Era (see Chapter 12), but the one at the close of the Mesozoic has attracted more attention because among its casualties were dinosaurs, flying reptiles, marine reptiles, and several kinds of marine invertebrates. Among the latter were ammonites, which had been so abundant through the Mesozoic, rudistid clams, and some planktonic organisms.

Numerous hypotheses have been proposed to explain Mesozoic extinctions, but most have been dismissed as improbable, untestable, or inconsistent with the available data. A proposal that has become popular since 1980 was based on a discovery at the Cretaceous-Tertiary boundary in Italy—a 2.5-cm-thick clay layer with a remarkably high concentration of the platinum group element iridium (Figure 15.29). High iridium concentrations have now been identified at many other Cretaceous-Tertiary boundary sites.

The significance of this *iridium anomaly,* as it is now called, lies in the fact that iridium is rare in crustal rocks but is found in much higher concentrations in some meteorites. Accordingly, some investigators have proposed a meteorite impact to explain the iridium anomaly, and they further postulate that the impact of a meteorite perhaps 10 km in diameter set in motion a chain of events leading to extinctions (Figure 15.30a). Some Cretaceous-Tertiary boundary sites also contain soot and shock-metamorphosed quartz grains, both of which are cited as additional evidence of an impact.

According to the impact hypothesis, about 60 times the mass of the meteorite was blasted from the crust high into the atmosphere, and the heat generated at impact started raging forest fires that added more particulate matter to the atmosphere. Sunlight was blocked for several months, causing a temporary cessation of photosynthesis, food chains collapsed, and extinctions followed. Furthermore, with sunlight greatly diminished, Earth's surface temperatures were drastically reduced, adding to the biologic stress. Another proposed consequence of an impact is that sulfuric acid (H_2SO_4) and nitric acid (HNO_3) would have resulted from

vaporized rock and atmospheric gases. Both would have contributed to strongly acid rain that might have had devastating effects on vegetation and marine organisms.

The iridium anomaly is real, but its origin and significance are debatable. We know very little about the distribution of iridium in crustal rocks or how it may be distributed and concentrated. Some geologists suggest that the iridium was derived from within Earth by volcanism, but it is not conclusively supported by evidence.

Some investigators now claim that they have found the probable impact site centered on the town of Chicxulub on the Yucatán Peninsula of Mexico (Figure 15.30b). The structure lies beneath layers of sedimentary rock, so it has been detected only in drill holes and by geophysical work. For example, magnetic data indicate that the structure is symmetric and measures 180 km in diameter. Furthermore, it appears to be the right age. Evidence supporting the conclusion that the Chicxulub structure is an impact crater includes shocked quartz,

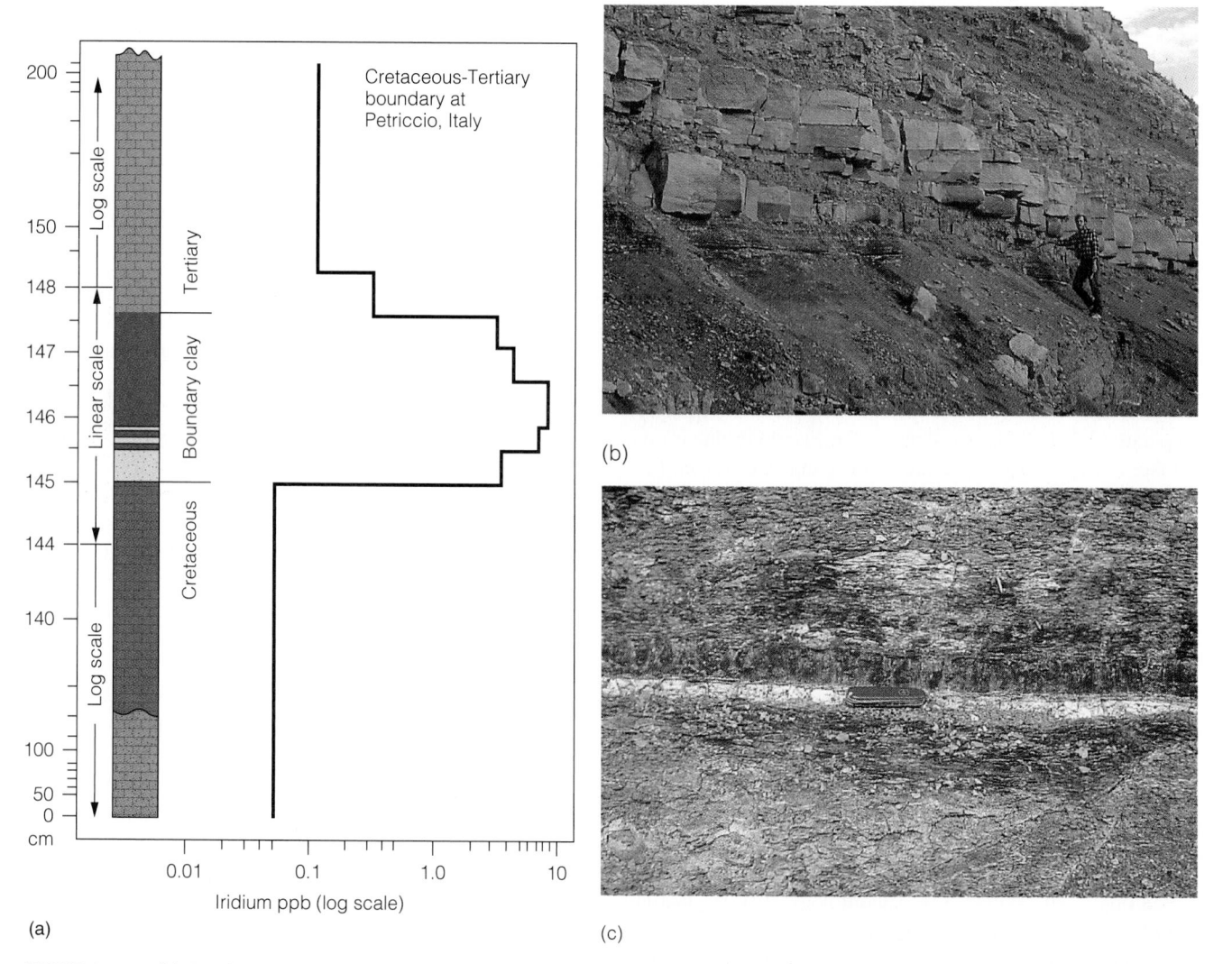

(a)

(b)

(c)

FIGURE 15.29 (a) Stratigraphy of the Cretaceous–Tertiary boundary site in Italy. The 2.5-cm-thick boundary clay shows an abundance of iridium. Iridium is present in crustal rocks in quantities measured in parts per billion (ppb), but its concentration increases markedly at the Cretaceous–Tertiary boundary. (b) In the Raton Basin, New Mexico, the boundary clay is the thin, white layer at knee level of geologist R. Farley Fleming. (c) Close-up of the boundary clay.

*20 mm=320 kilometers

(b)

Belize

Mexico

Guatemala   Honduras

0   200 km   El Salvador

Nic.

(a)

FIGURE 15.30 (a) Proposed meteorite impact at the end of the Cretaceous. (b) A large circular structure centered on Chixculub on the Yucatan Peninsula of Mexico is thought to be a meteorite impact site. It dates from about the end of the Cretaceous.

what appear to be the deposits of huge waves, and tektites, which are small pieces of rock that were melted during the proposed impact and hurled into the atmosphere. Although an impact origin for this structure is gaining acceptance, some geologists think it is some kind of volcanic feature.

Even if a meteorite did hit Earth, did it lead to these extinctions? If so, both terrestrial and marine extinctions must have occurred at the same time. To date, strict time equivalence between terrestrial and marine extinctions has not been demonstrated. The selective nature of the extinctions is also a problem. In the terrestrial realm, large animals were the most drastically affected, but not all dinosaurs were large, and crocodiles, close relatives of dinosaurs, were unaffected. Some paleontologists think that dinosaurs, some marine invertebrates, and many plants were already on the decline and headed for extinction before the end of the Cretaceous. A meteorite impact, if one actually occurred, may have simply hastened the process. Some evidence even indicates that dinosaurs survived into the Early Cenozoic, several tens of thousands of years after the proposed impact.

Investigators at the University of Chicago have proposed that the Mesozoic extinction was only one of several to occur at 26-million-year intervals during the last 250 million years. One possible cause of these cyclic extinctions is periodic meteorite showers. Some investigators have suggested that a companion star to the Sun with a highly eccentric orbit could provide a mechanism for periodic meteor showers. In this scenario, when this star is close to the Sun, it perturbs cometary orbits, thereby causing terrestrial impacts and mass extinctions.

Since attributing one mass extinction to a meteorite impact has generated controversy, one can imagine just how controversial this proposal is. The evidence for cyclic mass extinctions seems compelling, but not all paleontologists are convinced. The critics of cyclic mass extinctions think that some minor extinctions have been overemphasized. Furthermore, they point out that if mass extinctions are caused by periodic meteorite impacts, iridium anomalies corresponding with extinctions should be present, but few such anomalies have been identified.

In the final analysis, we have no widely accepted explanation for Mesozoic extinctions. We do know that vast shallow seas occupied large parts of the continents during the Cretaceous and that by the latest Cretaceous they had largely withdrawn. We also know that the mild, equable climates of the Mesozoic became harsher and more seasonal by the end of the Mesozoic. Changes such as these seem adequate to many paleontologists to explain Mesozoic mass extinctions. But the fact remains that this extinction was very selective, and no explanation accounts for all aspects of this crisis in the history of life.

Summary

Table 15.2 is a summary of biological and physical events for the Mesozoic Era.

1. Among the marine invertebrates, survivors of the Permian extinction diversified and gave rise to increasingly complex Mesozoic marine invertebrate communities.

2. Some of the most abundant and diverse invertebrates were ammonites and foraminifera.

3. Triassic and Jurassic land-plant communities were composed of seedless plants and gymnosperms. Angiosperms, or flowering plants, appeared during the Early Cretaceous, diversified rapidly, and soon became the dominant land plants.

4. The Triassic archosaurs included small bipedal carnivores that were ancestral to dinosaurs.

5. Dinosaurs appeared during the Late Triassic, but were most abundant and diverse during the Jurassic and Cretaceous. Based on pelvic structure, two distinct orders of dinosaurs are recognized—Saurischia (lizard-hipped) and Ornithischia (bird-hipped).

6. The carnivores and the giant quadrupeds were saurischian dinosaurs. Ornithischians, some of which were heavily armored, were more diverse than saurischians. All ornithischians were herbivores.

7. Bone structure, brain size, and predator–prey relationships have been cited as evidence for endothermy in dinosaurs. Considerable disagreement on dinosaur endothermy exists, but a fairly good case can be made for endothermic, small, carnivorous dinosaurs.

8. Pterosaurs were the first flying vertebrate animals. Small pterosaurs were probably active, wing-flapping fliers, while large ones may have depended more on thermal updrafts and soaring to stay aloft. At least one pterosaur species had hair or feathers, so it was very likely endothermic.

9. The fish-eating, porpoiselike ichthyosaurs were thoroughly adapted to an aquatic life. Female ichthyosaurs probably retained eggs within their bodies and gave birth to live young. Plesiosaurs were heavy-bodied marine reptiles that probably came ashore to lay eggs. Marine lizards called mosasaurs were present in Cretaceous seas.

10. During the Jurassic, crocodiles became the dominant freshwater predators. Turtles and lizards were present through most of the Mesozoic. Snakes appeared during the Cretaceous.

11. Birds probably evolved from small carnivorous dinosaurs during the Jurassic. The oldest known fossil bird is *Archaeopteryx,* which had feathers and a wishbone but retained reptilelike teeth and a long tail. Recent finds of other Mesozoic bird fossils provide additional evidence for bird–dinosaur relationships.

12. Among the mammal-like reptiles, the cynodonts show a transformation from the reptilian to the mammalian condition. The earliest mammals evolved during the Late Triassic, but they are difficult to distinguish from advanced cynodonts. Details of the teeth, the middle ear, and lower jaw are used to distinguish the two.

13. Several types of Mesozoic mammals existed, but all were small and their diversity was low. A group of Mesozoic mammals called eupantotheres gave rise to both marsupials and placentals during the Cretaceous.

14. The same types of some Mesozoic animals are found as fossils in areas that are now widely separated. Because the continents were close together during much of the Mesozoic and climates were mild even at high latitudes, animals and plants dispersed very widely.

15. Mesozoic mass extinctions account for the disappearance of dinosaurs, several other groups of reptiles, and a number of marine invertebrates. One hypothesis holds that the extinctions were caused by the impact of a large meteorite. Many paleontologists reject the meteorite proposal and claim that withdrawal of shallow seas and climatic changes can account for this extinction.

TABLE 15.2

Summary of Biologic and Physical Events for the Mesozoic Era

Age (Millions of Years)	Geologic Period	Invertebrates	Vertebrates	
66 —				
	Cretaceous	Continued diversification of ammonites and belemnoids. Rudist become major reef-builders. Extinction of ammonites, rudists, and most planktonic foraminifera at end of Creataceous.	Extinctions of dinosaurs, flying reptiles, and marine reptiles. Placental and marsupial mammals diverge.	GREATEST DIVERSITY OF DINOSAURS
144 —				
	Jurassic	Ammonites and belemnoid cephalopods increase in diversity. Scleractinian coral reefs common. Appearance of rudist bivalves.	First birds. Time of giant sauropod dinosaurs.	
208 —				
	Triassic	The seas are repopulated by invertebrates that survived the Permian extinction event. Bivalves and echinoids expand into the infaunal niche.	Mammals evolve from cynodonts. Cynodonts become extinct. Ancestral archosaurs give rise to dinosaurs. Flying reptiles and marine reptiles evolve.	
245				

Plants	Climate	Plate Tectonics
Angiosperms evolve and diversify rapidly. Seedless plants and gymnosperms still common but less varied and abundant.	North-south zonation of climates more marked, but remains equable. Climate becomes more seasonal and cooler at end of Cretaceous.	Further fragmentation of Pangaea. South America and Africa have separated. Australia separated from South America but remains connected to Antarctica. North Atlantic continues to open.
Seedless vascular plants and gymnosperms only.	Much like Triassic. Ferns with living relatives restricted to tropics live at high latitudes, indicating mild climates.	Fragmentation of Pangaea continues, but close connections exist among all continents.
Land flora of seedless vascular plants and gymnosperms as in Late Paleozoic.	Warm-temperate to tropical. Mild temperatures extend to high latitudes; polar regions may have been temperate. Local areas of aridity.	Fragmentation of Pangaea begins in Late Triassic.

Important Terms

angiosperm
archosaur
bipedal
cynodont
ectotherm

endotherm
eupantothere
ichthyosaur
marsupial mammal
mass extinction

monotreme
mosasaur
Ornithischia
placental mammal
plesiosaur

pterosaur
quadrupedal
Saurischia
therapsid

Review Questions

1. A typical mammal's jaw–skull joint is between which of these pairs of bones?
   a. _____ Stapes–quadrate
   b. _____ Dentary–squamosal
   c. _____ Articular–malleus
   d. _____ Ethmoid–jugular
   e. _____ Occipital–incus

2. Because of their rapid evolution and nektonic life style, the _____ are excellent guide fossils.
   a. _____ angiosperms
   b. _____ trilobites
   c. _____ bivalves
   d. _____ therapsid
   e. _____ cephalopods

3. All carnivorous dinosaurs belong to the group known as:
   a. _____ theropods.
   b. _____ ankylosaurs.
   c. _____ mosasaurs.
   d. _____ ammonites.
   e. _____ gymnosperms.

4. Which of the following pairs of animals were marine reptiles?
   a. _____ Angiosperm–ginkgo
   b. _____ Ichthyosaur–plesiosaur
   c. _____ Cephalopod–rudist
   d. _____ Cynodont–marsupial
   e. _____ Monotreme–endotherm

5. An important group of marine phytoplankton, the _____, first evolved during the Jurassic and remain numerous today.
   a. _____ oysters
   b. _____ burrowing worms
   c. _____ coccolithophores
   d. _____ belemnoids
   e. _____ teleosts

6. Dinosaurs, crocodiles, pterosaurs, and birds are collectively known as:
   a. _____ placentals.
   b. _____ archosaurs.

c. _____ mosasaurs.
   d. _____ invertebrates.
   e. _____ stem reptiles.

7. Animals that walk on two legs are characterized as:
   a. _____ endotherms.
   b. _____ bipeds.
   c. _____ tree dwellers.
   d. _____ aquatic.
   d. _____ sauropods.

8. Which following statement is correct?
   a. _____ *Archaeopteryx* had feathers only on its wings.
   b. _____ The most mammal-like of therapsids were the docodonts.
   c. _____ Monotremes were the most common mammals during the Cretaceous.
   d. _____ Multituberculates were the first mammalian herbivores.
   e. _____ The largest dinosaurs were pachycephalosaurs.

9. One argument for endothermy in dinosaurs is:
   a. _____ size.
   b. _____ body proportions.
   c. _____ upright posture.
   d. _____ that predators were more numerous than herbivores.
   e. _____ color.

10. An important feature of mammal teeth is occlusion, which means that:
    a. _____ the canine teeth are very large.
    b. _____ incisors are modified into tusks.
    c. _____ they have continuous tooth replacement throughout their lives.
    d. _____ two middle-ear bones were derived from jaw bones.
    e. _____ the chewing teeth meet surface to surface to allow chewing.

11. Much higher than expected levels of iridium at a number of Cretaceous-Tertiary boundary sites is cited as evidence for a(an):
    a. _____ marine regression.

b. _____ meteorite impact.

c. _____ change in vegetation.

d. _____ episode of mountain building.

e. _____ Mesozoic ice age.

12. An important Mesozoic event in the history of land plants was the:

a. _____ origin of ferns and horsetail rushes.

b. _____ extinction of cycads and conifers.

c. _____ first appearance and diversification of angiosperms.

d. _____ dominance of seedless vascular plants.

e. _____ prevalance of ginkgos and ceratopsians.

13. Discuss the changes that took place in the jaw and middle ear when cynodonts gave rise to mammals.

14. How does the breakup of Pangaea help us better understand the Mesozoic distribution of plants and animals?

15. Briefly summarize the evidence for and against endothermy in dinosaurs.

16. Why do paleontologists think that reptiles and birds are more closely related to one another than to other groups of vertebrates? In what specific group of reptiles is the probable ancestor of birds?

17. How do the Mesozoic marine invertebrate communities differ from those of the Paleozoic?

18. Mesozoic fossil cephalopods, especially ammonites, and foraminifera are very good guide fossils. Explain why.

19. Summarize the important events that took place among fish during the Mesozoic.

20. What are the two main groups of dinosaurs, and how do they differ?

21. Discuss the modifications that occurred for flight in pterosaurs and for an aquatic life in ichthyosaurs.

22. What are the three main groups of mammals, and how do they differ from one another?

23. How did land plant communities of the Triassic and Jurassic differ from those of the Cretaceous?

24. Briefly summarize the evidence for and against the hypothesis that a meteorite impact caused Mesozoic mass extinctions.

Points to Ponder

1. At least some carnivorous dinosaurs appear to have had a covering of feathers. Speculate on what function these feathers might have served. Also, construct two cladograms showing the relationships among dinosaurs, other reptiles, and birds. Which of these cladograms shows the most probable relationships among these animals?

2. Discuss how fossils of Mesozoic mammals illustrate the concept of mosaic evolution. Do present-day mammals possess any characteristics that indicate a reptile ancestry? If so, what?

3. Devise a testable hypothesis for Mesozoic extinctions other than the possible meteorite/comet impact. Remember, your hypothesis must not only be testable but also account for the selective nature of this extinction and the fact that both land-dwelling and marine organisms went extinct.

Additional Readings

Ackerman, J. 1998. Dinosaurs take wing. *National Geographic* 194, no. 1: 74–99.

Archibald, J. D. 1996. *Dinosaur extinction and the end of an era: What the fossils say.* New York: Columbia University Press.

Bakker, R. 1993. Jurassic sea monsters. *Discover* 14, no. 9: 78–85.

Benton, M. J. 1991. *The rise of the mammals.* New York: Crescent Books.

Callaway, J. M., and E. L. Nicholls (eds.). 1997. *Ancient marine reptiles,* San Diego: Academic Press.

Currie, P. J. 1996. The great dinosaur egg hunt. *National Geographic* 189, no. 5: 96–111.

Dingus, L., and T. Rowe. 1998. *The mistaken extinction: Dinosaur evolution and the origin of birds.* New York: Freeman.

Farlow, J. O., and M. K. Brett-Surman (eds.). 1997. *The complete dinosaur.* Bloomington, Ind.: Indiana University Press.

Fastovsky, D. E., and D. B. Weishampel. 1996. *The evolution and extinction of dinosaurs.* New Haven, Conn.: Yale University Press.

Gore, R. 1993. Dinosaurs. *National Geographic* 183, no. 1: 2–53.

Horner, J. R., and E. Dobb. 1997. *Dinosaur lives: Unearthing an evolutionary saga.* New York: HarperCollins.

———, and D. Lessem. 1993. *The complete* T. rex. New York: Simon & Schuster.

Lucas, S. G. 1997. *Dinosaurs: The textbook.* 2d ed. Dubuque, Iowa: Brown.

McGowan, C. 1991. *Dinosaurs, spitfires, and sea dragons.* Cambridge, Mass.: Harvard University Press.

Morell, V. 1997. The origin of birds: The dinosaur debate. *Audubon* 99, no. 2: 36–45.

Padian, K., and L. M. Chiappe. 1998. The origin of birds and their flight. *Scientific American* 278, no. 2: 38–47.

Schaal, S., and W. Ziegler (eds.). 1992. *Messel: An insight* *into the history of life and of the Earth.* Oxford: Clarendon Press.

Sereno, P. C. 1996. Africa's dinosaur castaways: *National Geographic* 189, no. 6: 106–119.

Shipman, P. 1998. *Taking wing:* Archaeopteryx *and the evolution of bird flight.* New York: Simon & Schuster.

Vickers-Rich, P., and T. H. Rich. 1993. Australia's polar dinosaurs. *Scientific American* 269, no. 1: 42–49.

Wellnhofer, P. 1990. *Archaeopteryx. Scientific American.* 262, no. 5: 70–77.

———. 1991. *The illustrated encyclopedia of pterosaurs.* New York: Crescent Books.

World Wide Web Activities

For these web site addresses, along with current updates and exercises, log on to

http://www.brookscole.com/geo/

▶ ILLINOIS STATE MUSEUM MAZON CREEK FOSSILS

This site contains some of the more interesting and dramatic types of fossils recovered from the Francis Creek Shale in Illinois.

1. What is the age of the Francis Creek Shale and in what enviroment(s) was it deposited?

2. Where can Mazon Creek fossils be found, and what kinds of different plant fossils have been recovered? Click on the names of some of these fossils and see what they look like.

3. Click on *Importance of Mazon Creek Fossils.* Why are these fossils important?

4. What kinds of fossil animals are found in the Francis Creek Shale?

▶ UNIVERSITY OF CALIFORNIA MUSEUM OF PALEONTOLOGY

This is an excellent site to visit for any aspect of geologic time, paleontology, evolution, and many other subjects.

1. Click on the *Phylogeny* icon. This will take you to the Phylogeny of Life home page. Click on the *Vertebrates* icon. This will take you to the Introduction to the Vertebrates home page. Click on the *Systematics* icon. At this site, click on the *Tetrapods* icon, then go to the *Dinosaur* site. This page gives a lot of information about dinosaurs, including a list of other web sites worth visiting.

2. While on the *Vertebrates* home page, click the icons for *diapsids* and *birds.* What is a diapsid, and why are birds considered diapsids? Scroll down and see *Protoavis.* Why is there some doubt that *Protoavis* is actually a bird?

3. Under *On-Line Exhibits* go to vertebrates and find *Pterosaurs.* What is the common name for pterosaurs, which one was largest, and how do they differ from birds and bats?

4. Also under *Vertebrates* find *multituberculates.* What was the time of existence of this group, and why are they referred to as the "Lost Tribe of Mammals"?

▶ ROYAL TYRRELL MUSEUM WEB SITE

This site contains a tremendous amount of information about dinosaurs and other Mesozoic animals. It is supported by the Royal Tyrrell Musem Cooperating Society, "a non-profit society funded to support the scientific, educational, recreational and public operations of the Royal Tyrrell Museum of Paleontology," at Drumheller, Alberta, Canada. The home page contains icons for Directions to the Museum, Admissions, Virtual Tours, Explorers Programmes, and Educational Programming.

1. Click on the *Ichthyosaurs* site. What are Ichthyosaurs? Where have they been found in Canada?

2. Click on the *Virtual Tour* icon. It takes you to a map of the museum as well as an index of the various museum exhibits and locations. Click on *The Origin of Dinosaurs* site. What is the origin of dinosaurs? Where is the oldest dinosaur found?

3. Check out the various sites and the What's New section to find out about life of the Mesozoic Era.

4. Go to the *Dinosaur Hall* and see what types of dinosaurs *Albertosaurs* and *Ornitholetes* were.

▶ ILLINOIS STATE GEOLOGICAL SURVEY—DINOSAURS AND VERTEBRATE PALEONTOLOGY LINKS

This is a great place to access dinosaur information on the Internet. Maintained by Russell J. Jacobson—also known as

Dino Russ, keeper of the Lair—this site lists hundreds of links to all manner of dinosaur information. The site is broken down into categories such as Dinosaur Art, Dinosaur Digs, Dinosaur Eggs, Dinosaur Information, and Dinosaur News and Commentaries, to name a few. Each category has highlighted links for you to click which will take you to that particular site. Each site listed has a short description of what will be found at that site.

▶ DINOSAUR PROVINCIAL PARK

This site regarding Dinosaur Provincial Park, Alberta, Canada, is brief but has good descriptions of the fossils and the park, along with several photographs. What was the climate like when the dinosaur fossils were preserved in this area? Also, how many types of dinosaurs have been found here and what other types of animal fossils have been recovered?

▶ BERI'S DINOSAUR WORLD

This is a "vertebrate paleontology" site, designed, written, painted, animated, and edited by the international paleolife illustrator Berislaw Krzic. At the site click *Enter* and go to the Table of Contents. Notice that discussions of many topics about a variety of fossil vertebrates are available. Click *Feathered Dinosaurs* and see the illustrations and read the discussion about the recent finds of feathered dinosaurs and birdlike dinosaurs. Also, be sure to see the page on *The Origin of Feathers and the Mystery of Evolution of Bird Flight.*

▶ YALE PEABODY MUSEUM

A site maintained by the Peabody Museum of Natural History at Yale University. On the main page click *China's Feathered Dinosaurs* and then click *Online Exhibits* and see the images of feathered dinosaurs and find out about their discovery, age, and relationships. Click on the names *Sinosauropteryx, Protarchaeopteryx,* and *Caudipteryx* to see images and discussions of the genera of feathered dinosaurs or those with protofeathers.

Return to the main page and click *The Age of Reptiles,* then click each of the Mesozoic periods, *Triassic, Jurassic,* and *Cretaceous,* for discussions of dinosaurs from these times.

▶ OTHER SITES

Many web sites have information on dinosaurs and other Mesozoic animals and plants. A search using "dinosaurs" will yield numerous results, but one can find reliable information and good images at the following sites, to name a few; American Museum of Natural History (New York), Field Museum of Natural History (Chicago), Museum of the Rockies (Bozeman, Montana), Los Angeles County Museum (Los Angeles), Natural History Museum (London), Carnegie Museum of Natural History (Pittsburgh), Canadian Museum of Nature (Ottawa, Canada), Denver Museum of Natural History (Denver), University of Michigan Museum of Paleontology (Ann Arbor), Florida Museum of Natural History (Gainesville), and University of Kansas Natural History Museum (Lawrence).

Cenozoic Geologic History: Tertiary Period

Cenozoic sedimentary rocks, mostly siltstone and sandstone, and volcanic ash layers exposed at Scott's Bluff National Monument, Nebraska.

Only 1.4% of all geologic time is represented by the Cenozoic Era, yet rocks of this age are those most commonly encountered because they are at or near the surface, burying older rocks in many areas. They are easily observed in large parts of North America, especially the West. Eastern North America was an area of active erosion during much of the Cenozoic and accordingly has few rocks of this age. The notable exceptions are along the Atlantic Coastal Plain and the Gulf Coastal Plain including Florida. In contrast, tectonism in the western part of the continent yielded a number of basins that became sites of sediment accumulation, and a vast sheet of sediment was also deposited on the Great Plains east of the present-day Rocky Mountains. Coastal California was tectonically active and the site of a number of basins that subsided below sea level and are now filled with Cenozoic marine sedimentary rocks, many of which contain oil and natural gas.

Vast exposures of sedimentary rocks are found in such areas as Badlands National Park, South Dakota (see Perspective 4.2), John Day Fossil Beds National Monument, Oregon (see Prologue to Chapter 18), Bryce Canyon National Park, Utah, several areas in Florida, and in the Calvert Cliffs of Maryland. In addition, many interesting volcanic rocks of Cenozoic age are also present. For instance, the Cascade Range of California, Oregon, Washington, and southern British Columbia, Canada, is composed of Cenozoic volcanic rocks, and volcanism continues in that area (see Perspective 16.2). Several picturesque Cenozoic intrusive igneous bodies are also present.

The exposed sedimentary rocks at Bryce Canyon National Park, Utah, are among the most brilliantly colored anywhere. Actually, Bryce Canyon is not a canyon but the eroded margin of a fault-bounded plateau. Nevertheless, its Eocene rocks constitute the Wasatch Formation, which is composed mostly of limestone deposited in lakes along with some sandstones and conglomerates. Differential weathering and erosion of massive limestone beds has yielded the fantastic spires, sharp angular ridges, and mazes of closely spaced gullies of the famous Pink Cliffs (Figure 16.1a).

The Wasatch Formation and many others in the West were deposited in continental depositional environments. In contrast, Cenozoic marine sedimentary rocks are exposed in the Calvert Cliffs along the shoreline of Chesapeake Bay in Maryland (Figure 16.1b). These rocks are composed of diatomite (sedimentary rock composed of the siliceous shells of diatoms), marl (a term usually referring to earthy mixtures of carbonate minerals and clay and other fine-grained particles), clay layers, and fine-grained sandstone. The Calvert Cliffs are not only picturesque but their rocks also contain numerous fossils of marine microorganisms, invertebrates, sharks, and marine mammals.

Cenozoic igneous rocks are found only in the West. Vast outpourings of lava took place in several areas, but

(a)

(b)

FIGURE 16.1 (a) Differential weathering and erosion of the Eocene Wasatch Formation has yielded the scenic Pink Cliffs of Bryce Canyon National Park, Utah. (b) Miocene sedimentary rocks exposed in the Calvert Cliffs of Maryland.

two especially interesting intrusive bodies are present at Devil's Tower National Monument, Wyoming, and at Shiprock, New Mexico. According to one legend, Devil's Tower (Figure 16.2a) formed when the Great Spirit caused it to rise from the ground, carrying with it several Native American children who were trying to escape from a gigantic grizzly bear. Another legend tells of six brothers who were also pursued by a grizzly bear. The youngest brother carried a small rock and when he sang a song, the rock grew to the present size of Devil's Tower. In both legends, the bear's attempts to reach the humans left deep scratch marks in the tower's rocks. Geologists have a less dramatic explanation for the tower's origin. It is a small igneous intrusion that was emplaced in Earth's crust 45 to 50 million years ago. As it cooled and contracted, columnar joints formed, thus accounting for the near vertical lines.

Shiprock is a conical monolith rising nearly 550 m above the surrounding plain in northwestern New Mexico, with three wall-like ridges of igneous rock radiat-ing outward from it (Figure 16.2b). Navajo legend holds that Shiprock represents a giant bird and the wall-like ridges are snakes that turned to stone. Once again, geologists have a rather mundane explanation for the origin of Shiprock. It is a feature known as a volcanic neck, an eroded volcanic conduit in which magma cooled and solidified, and the wall-like ridges are dikes. Shiprock and its associated dikes were intruded about 27 million years ago.

Shiprock is but one of several volcanic necks in this area that formed by explosive eruptions, during which volcanic debris as well as pieces of preexisting rocks were hurled high into the atmosphere. The dikes, however, came later and appear to have resulted when magma rose rather quietly. Shiprock now stands nearly 550 m above the surrounding plain, so at least that much erosion must have taken place to expose it in its present form. We can only speculate on how much higher and larger it might have been when it was part of an active volcano.

(a)

(b)

FIGURE 16.2 (a) Devil's Tower in northeastern Wyoming is a small igneous intrusion. It rises about 650 m above its base and can be seen from 48 km away. (b) Shiprock, a volcanic neck in northwestern New Mexico, rises nearly 550 m above the surrounding plain. The wall-like ridge in the foreground is a dike.

Archean Eon	Proterozoic Eon	Phanerozoic Eon							
		Paleozoic Era							
Precambrian		Cambrian	Ordovician	Silurian	Devonian	Mississippian	Pennsylvanian	Permian	
						Carboniferous			

2500
M.Y.A.

545
M.Y.A.

245
M.Y.A.

Introduction

Geologists have traditionally divided the Cenozoic Era into two periods, the **Tertiary** (66 to 1.6 million years ago) and the **Quaternary** (1.6 million years ago to the present), and each period is further divided into epochs (Figure 16.3). Tertiary and Quaternary are widely used, but a different scheme for designating Cenozoic time is becoming increasingly popular. This scheme retains the Quaternary Period for the most recent part of geologic time, but the Tertiary is replaced by the *Paleogene Period* (66 to 24 million years ago) and *Neogene Period* (24 to 1.6 million years ago) (Figure 16.3). In this book we follow the traditional usage of Tertiary and Quaternary.

In Chapter 8 we noted that Precambrian time encompasses more than 88% of all geologic time. Thus, if all geologic time were represented by a 24-hour day, slightly more than 21 hours would be the duration of the Precambrian (see Figure 8.2). The Cenozoic Era at 66 million years long is comparatively brief, accounting for only about 20 minutes of our 24-hour day. Nevertheless, 66 million years is an extremely long period of time by any measure, certainly long enough for significant evolution of Earth and its biosphere to have taken place. Furthermore, Cenozoic rocks are at or near the surface and have been little altered, thereby making access and interpretation easier than for rocks of previous eras.

Many of Earth's features have long histories, but the present distribution of land and sea and the topographic expression of the continents and their landforms all resulted from processes operating during the Cenozoic Era. For instance, the Appalachian Mountain region began its evolution during the Precambrian, but its present expression is largely the product of Cenozoic uplift and erosion. Likewise, the uplift of the Sierra Nevada of California and Nevada and the origin of the Himalaya of Asia took place during the Cenozoic. And many of the distinctive landscapes such as badlands topography, glacial valleys, and deep canyons of our national parks developed during the past few thousands to several millions of years. In other words, Earth's existing distinctive aspects formed very recently in the context of geologic time.

Cenozoic Plate Tectonics— An Overview

The Late Triassic fragmentation of the supercontinent Pangaea began an episode of plate movements that continues even now (see Figures 14.1 and 14.3). Spreading ridges such as the Mid-Atlantic Ridge and East Pacific Rise formed, and it is along these features that new oceanic crust is continuously generated. The age distribution of oceanic crust in the Pacific, though, is strongly asymmetric because

Phanerozoic Eon									
Mesozoic Era			Cenozoic Era						
Triassic	Jurassic	Cretaceous	Tertiary					Quaternary	
			Paleocene	Eocene	Oligocene	Miocene	Pliocene	Pleistocene	Holocene

245
M.Y.A.

66
M.Y.A.

Era	Period		Epoch	Duration, millions of years (approx.)	Millions of years ago (approx.)
			Holocene/Recent .01		
Cenozoic	Quarternary		Pleistocene	1.59	
	Tertiary	Neogene	Pliocene	3.7	1.6
			Miocene	18.7	5.3
			Oligocene	13	24
		Paleogene	Eocene	21	37
			Paleocene	8	58
					66

FIGURE 16.3 The geologic time scale for the Cenozoic Era.

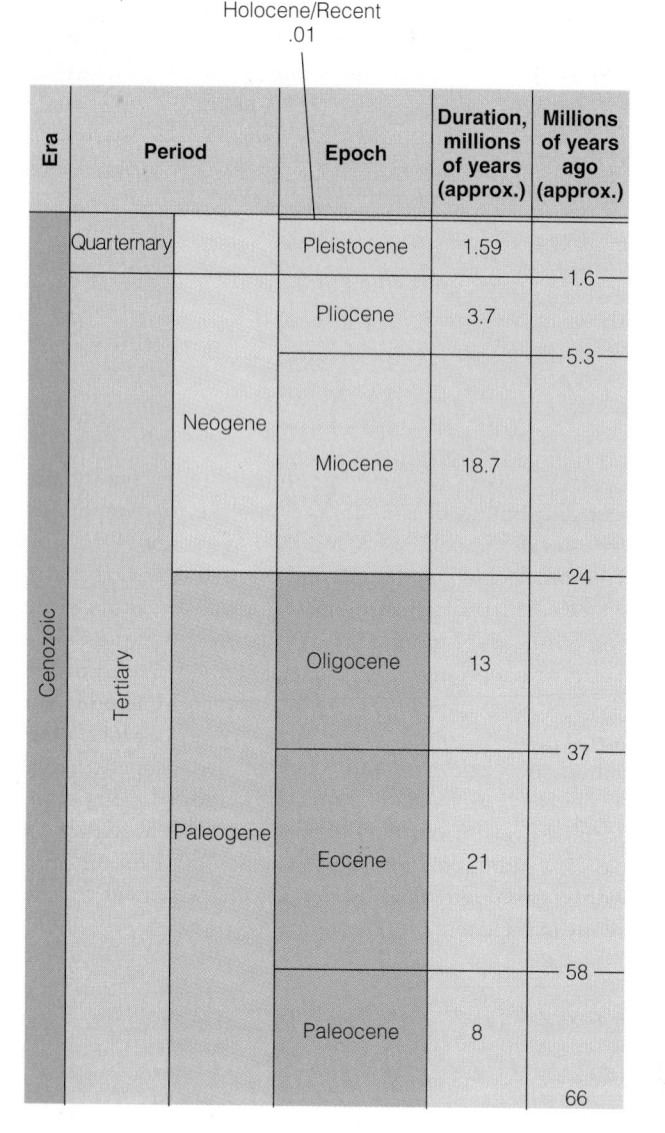

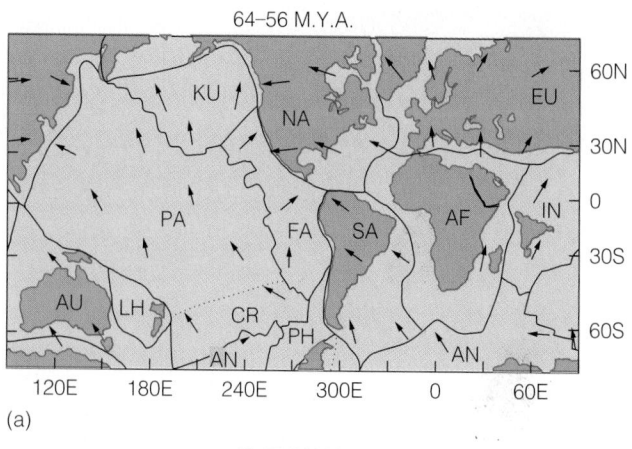

(a)

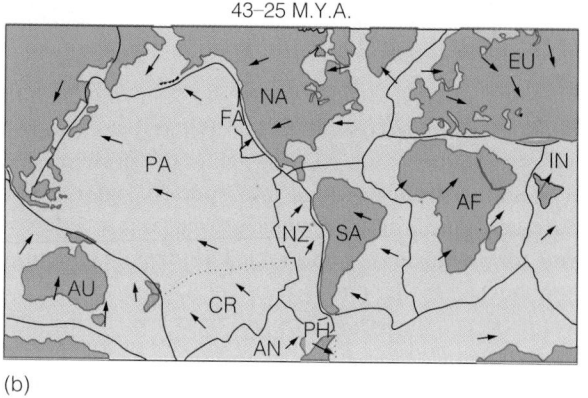

(b)

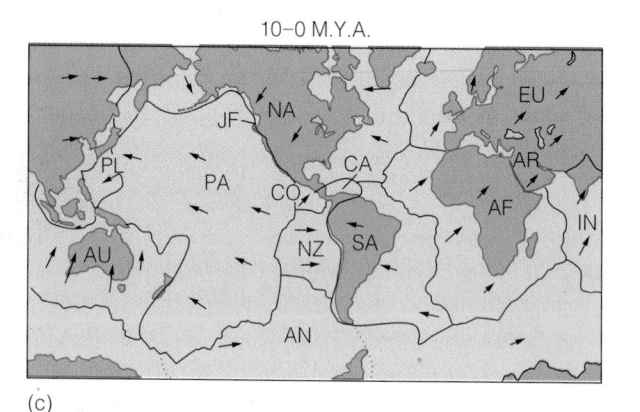

(c)

FIGURE 16.4 Cenozoic plate movements. The Atlantic Ocean basin opened as the Americas moved westward from Europe and Africa, while the Pacific Ocean basin decreased in size. The Farallon plate was mostly consumed beneath the Americas; the Nazca, Cocos, and Juan de Fuca plates are remnants of this plate. Africa moved northward and partly closed the Tethys Sea. India also moved northward and collided with Asia. Australia moved northward to its present position. Abbreviations for plates: AF, African; AN, Antarctic; AR, Arabian; AU, Australian; CA, Caribbean; CO, Cocos; CR, Chatham Rise; EU, Eurasian; FA, Farallon; IN, Indian; JF, Juan de Fuca; KU, Kula; LH, Lord Howe; NA, North American; NZ, Nazca; PA, Pacific; PL, Philippine; PH, Phoenix; SA, South American.

much of the crust in the eastern Pacific has been consumed at subduction zones along the western Americas (see Figure 1.10). In the Atlantic Ocean basin, separation at the Mid-Atlantic Ridge resulted in the present-day positions of the Americas with respect to Europe and Africa (Figure 16.4).

Another important event was the northward movement of the Indian plate and its collision with southern Asia. India's long journey began during the Cretaceous when it separated from Gondwana and moved progressively north. In doing so, India, along with the northward-moving African plate, caused the closure of the Tethys Sea (Figure 16.4). During the Early Tertiary, Australia separated from Antarctica and moved north to its present position.

The breakup of Pangaea and the drift of its various fragments account for the present geographic distribution of continents and oceans. But continental rifting is not restricted to the Late Triassic. A triple junction is currently located at the junction of the East African Rift System, the

rift in the Red Sea, and the rift in the Gulf of Aden (Figure 16.5). Rifting in this region began during the Late Tertiary and continues at present.

Rifting in East Africa seems to be in its early stage, because the continental crust has not yet stretched and

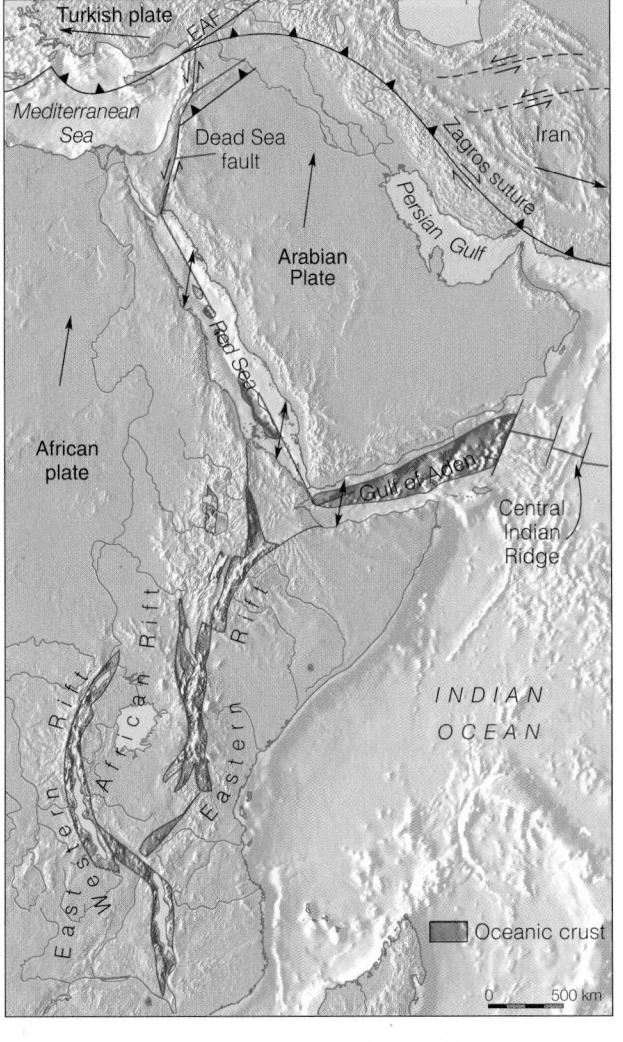

FIGURE 16.5 A triple junction exists at the junction of the East African Rift System and the rifts in the Red Sea and Gulf of Aden. Oceanic crust began forming in the Gulf of Aden about 10 million years ago. Rifting began later in the Red Sea, and oceanic crust is now forming. In East Africa, the continental crust has not yet stretched and thinned enough for oceanic crust to form from below. Much of the tectonic activity in the Middle East is caused by the Arabian plate moving northward against Eurasia.

thinned enough for oceanic crust to form from below. In the Red Sea region, rifting was preceded by vast eruptions of basalt. In the succeeding rifting stage, a long, narrow sea formed, and by the Late Pliocene, oceanic crust began forming along the rift axis (Figure 16.5). Rifting began even earlier in the Gulf of Aden than in the Red Sea. By the Late Miocene, the continental crust had stretched and thinned, and upwelling basaltic magma was sufficient to form oceanic crust.

Studies of this Late Cenozoic rifting in Africa and adjacent regions are important for two reasons. First, they allow us to better understand the Late Triassic rifting of Pangaea and the origin of the Atlantic Ocean basin. Second, all three branches of these rifts remain active, al-

though rifting in East Africa is very slow. As a consequence of rifting in the Red Sea and Gulf of Aden, the Arabian plate has separated from Africa. Arabia's northward motion is responsible for a major north–south trending shear zone, the Dead Sea fault, and much of the present tectonic activity in the Middle East (Figure 16.5).

Cenozoic Orogenic Belts

Cenozoic orogenic activity was largely concentrated in two major zones or belts, the **Alpine–Himalayan belt** and the **circum-Pacific belt** (Figure 16.6). Each belt consists of a number of **orogens,** or zones of deformed rocks, many of which have been metamorphosed and intruded by plutons. In many of these orogens, deformation dates well back into the Mesozoic but continued into the Cenozoic, and some, such as the Himalayan orogen, are active today.

THE ALPINE–HIMALAYAN OROGENIC BELT

The Alpine-Himalayan orogenic belt includes the mountainous regions of the Mediterranean and extends eastward through the Middle East and India and into southeast Asia (Figure 16.6). Remember that the Tethys Sea separated much of Gondwana from Eurasia during Mesozoic time (Figure 16.4). Plate motions beginning during the Mesozoic culminated in the Cenozoic with the closure of the Tethys Sea as Africa and India moved northward and collided with Eurasia.

The Alps   The **Alpine orogeny** produced a zone of deformation in southern Europe extending from the Atlantic Ocean eastward to Greece and Turkey. Concurrent deformation also occurred south of the Mediterranean basin along the northwest coast of Africa (Figure 16.6). Unraveling the complexities of this deformational event has been a long and arduous task, but the broad picture is now becoming clear, even though many details are still poorly understood.

Although events leading to the Alpine orogeny began during the Mesozoic, major deformation took place during the Eocene to Late Miocene. Deformation was caused by the northward movements of the African and Arabian plates against Eurasia, but the story is not quite so simple. The complexities of Alpine geology are compounded by the fact that several small plates collided with Europe and were subsequently deformed as the larger Eurasian and African plates converged.

Deformation resulting from plate convergence in this region formed the Pyrenees Mountains between Spain and France, the Alps of mainland Europe, and the Apennines of Italy, and other mountain ranges (Figure 16.6). The compressional forces generated by plate collisions produced complex thrust faults and large overturned folds called *nappes* (Figure 16.7). Another result of plate convergence in this region was the formation of an isolated sea in the Mediterranean basin, which had formerly been a part of the Tethys Sea. Evaporites up to 2 km thick were de-

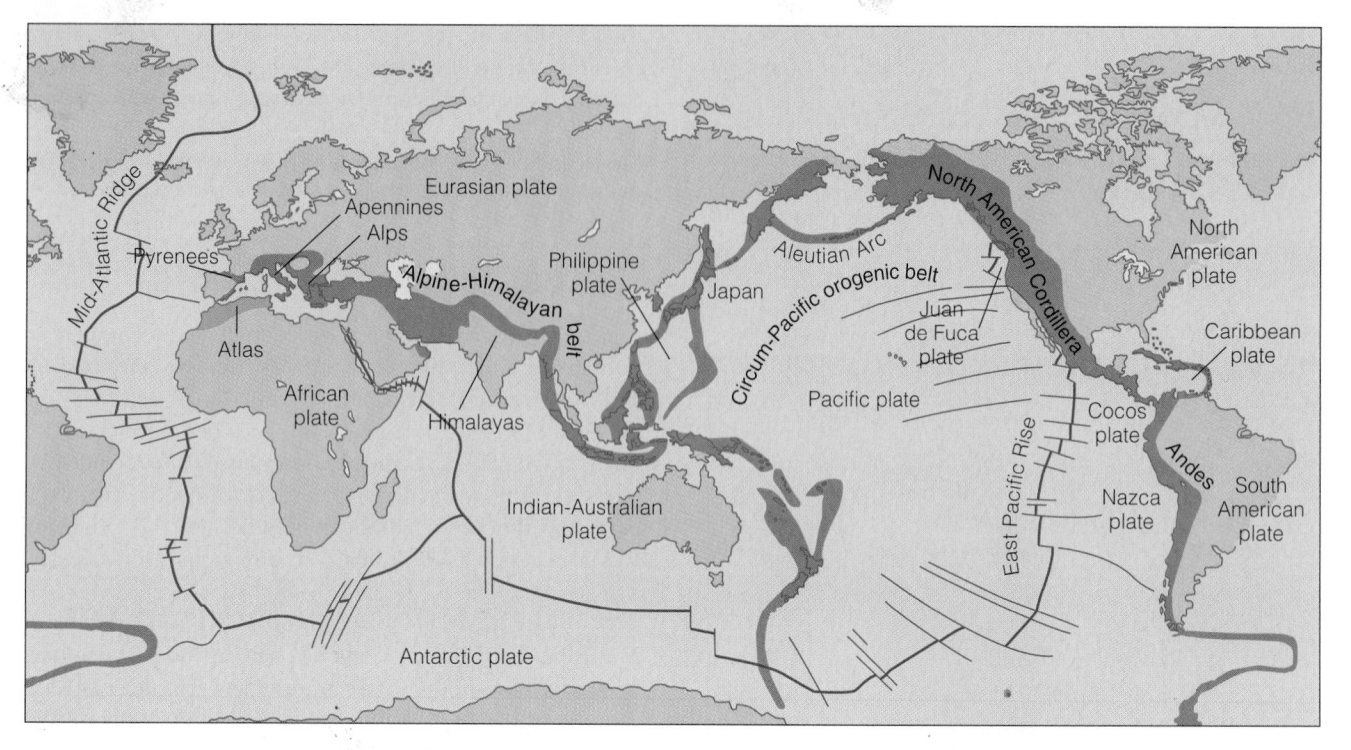

FIGURE 16.6 Areas of Cenozoic orogenic activity. Orogenesis occurred mainly in two major belts: the Alpine–Himalayan belt and the circum-Pacific belt.

posited in this isolated sea during the Late Miocene (see Perspective 16.1).

The Atlas Mountains of northwestern Africa (Figure 16.6) also formed as the African plate collided with Eurasia. Farther east in the Mediterranean basin, Africa is still forcing oceanic lithosphere northward beneath Greece and Turkey. Active volcanoes in Italy and seismic activity in much of southern Europe and the Middle East indicate that the Mediterranean basin remains geologically active. In 1990, for example, an earthquake measuring 7.3 on the Richter scale killed 40,000 people in Iran, and an earthquake in Turkey in 1992 resulted in at least 570 fatalities. And since eruptions of Mount Vesuvius destroyed the cities of Pompeii and Herculaneum in A.D. 79, it has erupted 80 times, most violently in 1631 and 1906; it last erupted in 1944.

Roof of the World—The Himalaya   During Early Cretaceous time, India broke away from Gondwana and began moving north (Figure 16.4). At the same time, a subduction zone formed along the south-facing margin of Asia where oceanic lithosphere was consumed (Figure 16.8a). Partial melting of this descending oceanic lithosphere generated magma, which rose to form a volcanic chain and large granitic intrusions in what is now Tibet. India eventually approached this chain and destroyed it as it collided with Asia to form a **collision orogen.** As a result, two continental plates became sutured—India and Asia were now one.

The exact time of India's collision with Asia is uncertain, but sometime between 40 and 50 million years ago India's northward drift rate decreased abruptly, from between 15 and 20 cm per year to about 5 cm per year. Because continental lithosphere is not dense enough to be subducted, this decrease in rate seems to mark the time of collision and India's resistance to subduction (Figure 16.8b).

Because of its low density and resistance to subduction, the leading margin of India was underthrust beneath Asia, causing crustal thickening, thrusting, and uplift. Sedimentary rocks that were deposited in the sea south of Asia were thrust northward into Tibet, and two major thrust faults carried Paleozoic and Mesozoic rocks of Asian origin onto the Indian plate (Figure 16.8c). Rocks that were deposited in the shallow seas along India's northern margin now form the higher parts of the Himalaya.

As the Himalaya were uplifted, they were also eroded, but at a rate insufficient to match the uplift. Much of the debris shed from the rising mountains was transported to the south and deposited as a vast blanket on the Ganges Plain and as huge submarine fans in the Arabian Sea and in the Bay of Bengal (Figure 16.9). Continued movement on the Main Boundary fault has folded and faulted rocks of the Ganges Plain along the southern margin of the Himalayas (Figure 16.8d).

Since its collision with Asia, India has been underthrust about 2000 km beneath Asia! Currently, India continues

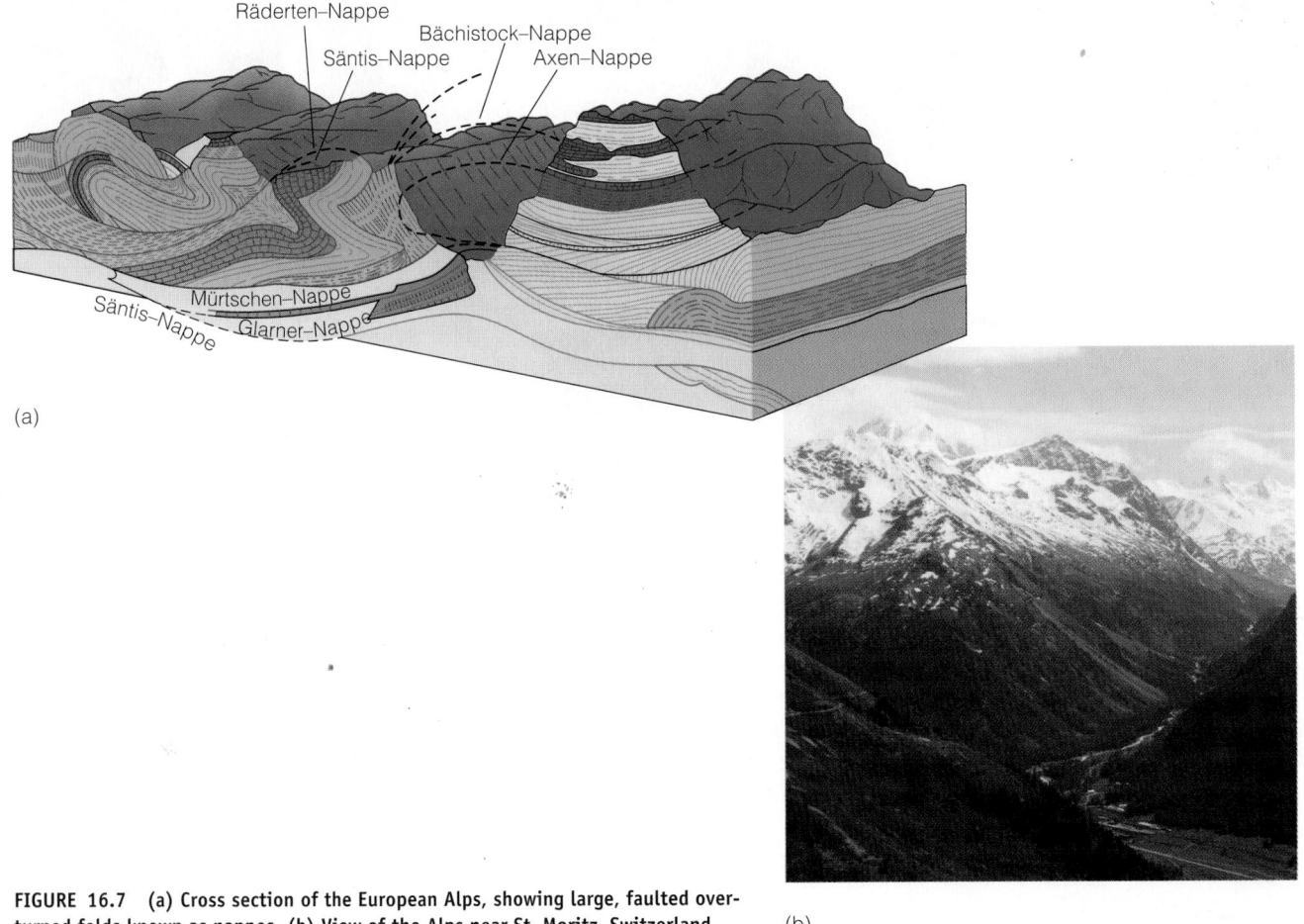

Räderten–Nappe
Säntis–Nappe
Bächistock–Nappe
Axen–Nappe

(a)

Mürtschen–Nappe
Säntis–Nappe
Glarner–Nappe

(b)

FIGURE 16.7 (a) Cross section of the European Alps, showing large, faulted over-turned folds known as nappes. (b) View of the Alps near St. Moritz, Switzerland.

moving north at about 5 cm per year. In other words, the Himalaya are still forming.

THE CIRCUM-PACIFIC OROGENIC BELT

The Pacific plate is being consumed at subduction zones along the western and northern margins of the Pacific Ocean basin (Figure 16.4). This process has continued throughout the Cenozoic, giving rise to orogens in the Aleutians, the Philippines, Japan, and several other areas in the southwestern Pacific Ocean basin (Figure 16.6).

Orogens of the western and northern Pacific are **arc orogens** characterized by subduction of oceanic lithosphere, deformation, and igneous activity. Japan, for example, is bounded on the east by the Japan Trench, where the Pacific plate is subducted. The Sea of Japan, a **back-arc marginal basin,** lies between Japan and the mainland of Asia. The origin of back-arc marginal basins is not fully understood, but according to one theory, Japan was once part of mainland Asia and was separated as back-arc spreading occurred (Figure 16.10). When separation began sometime during the Cretaceous, Japan moved eastward over the Pacific plate, and oceanic crust formed in the Sea of Japan. Japan's geology is complex, and much of its deformation predates the Cenozoic. Nevertheless, consider-

able deformation, metamorphism, and volcanism occurred during the Cenozoic and, in fact, continues to the present.

Spreading at the East Pacific Rise is carrying the Cocos and Nazca plates eastward, where they are being subducted beneath Central and South America, respectively (Figure 16.4). Volcanism and seismic activity indicate that the orogens in Central and South America remain active. We are reminded of this activity by events such as the tragic 1985 earthquake that struck Mexico City.

The Andes Mountains in South America, with more than 49 peaks higher than 6000 m, are the highest in the Western Hemisphere. They formed as Mesozoic-Cenozoic plate convergence caused crustal thickening as sedimentary rocks were deformed, uplifted, and intruded by granitic plutons. A large thrust fault carries the mountains over the Brazilian Shield, further increasing the crustal thickness of South America's western margin (Figure 16.11). Although compression continues in this area, parts of the high Andes are subjected to tensional forces, resulting in displacement of large blocks along normal faults.

Following its separation from Africa during the Mesozoic, South America was an island continent until the Late Tertiary. The land connection between North and South America formed as a result of subduction at the Middle

perspective 16.1

History of the Mediterranean

*I*f it were not for the connection between the Mediterranean Sea and the Atlantic Ocean at the Strait of Gibraltar, the Mediterranean would eventually dry up and become a vast desert lying far below sea level. The reason for this is that the Mediterranean is in an arid region where it loses more water to evaporation than it receives from rainfall and stream runoff (Figure 1). Inflow of seawater from the Atlantic, however, keeps the Mediterranean at a constant level. Studies of 2-km-thick evaporite beds beneath the floor of the present-day Mediterranean indicate that they were deposited in shallow water rather than a deep-ocean basin as is now present. Such observations have led some geologists to hypothesize that during the Late Miocene the Mediterranean lost its Atlantic connection and dried up. With its supply of oceanic water cut off, the sea evaporated to near dryness in as little as 1000 years and became a vast desert basin lying 3000 m below sea level.

The lowest elevation in today's world is about 395 m below sea level at the Dead Sea on the Israel–Jordan border.

More evidence indicating that the Mediterranean dried up comes from studies of southern European and northern African rivers that flow into the sea. The Mediterranean serves as base level for these rivers; thus, they can erode no lower than present sea level. However, beneath these rivers are buried valleys deeply incised into bedrock far below sea

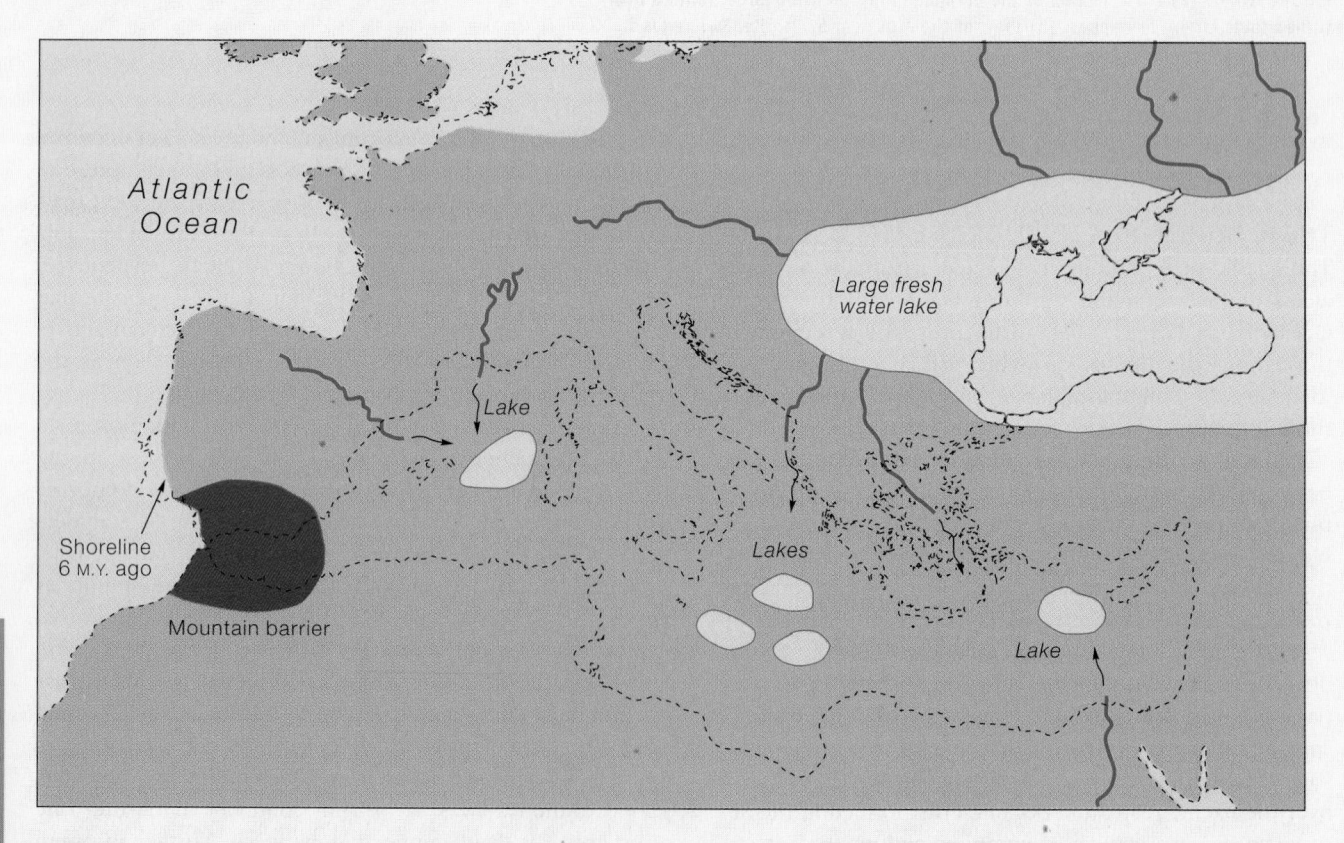

FIGURE 1 About 6 million years ago, Spain and Africa collided to create a barrier that blocked off the Mediterranean, which soon dried out. Rivers fed small playa lakes on the dry seafloor. A large freshwater lake sprawled across much of southeastern Europe and covered the present Black, Caspian, and Aral Seas. This lake overflowed into the Mediterranean Basin and also fed small playa lakes.

level. For instance, a buried channel lies more than 900 m below the surface of the Rhone delta in southern Europe. And at the Aswan High Dam in Egypt, some 465 km upstream on the Nile River, a buried river valley has incised more than 200 m below present sea level. Apparently, these valleys were eroded when sea level in the Mediterranean was much lower, and they were later filled when sea level rose.

Simply evaporating a body of water as deep as the present Mediterranean, whose average depth is about 1500 m, would yield only 25 m or so of evaporite deposits. Yet the evaporites beneath the sea are more than 2 km thick! It seems, though, that periods of isolation of the Mediterranean basin alternated with periods when an oceanic connection was reestablished. When an oceanic connection existed, gravel and silt were deposited around the basin margin, and deep-sea deposition took place near the basin center. At times when the oceanic connection was lost, though, carbonates followed by gypsum and halite were deposited in a body of water that became shallower and increasingly saline.

Several episodes of filling followed by evaporation could yield a considerable thickness of evaporites, but unless these episodes occurred many times they still could not account for 2 km of evaporites. However, shallow-water deposition in a deep basin could yield deposits limited in thickness only by the basin's initial depth, which in this case is proposed to be about 3 km.

The model just described for Mediterranean evaporites is widely accepted among geologists, but it does not have unanimous support. Another idea is that these thick evaporites formed in a deep, water-filled basin. According to this model, the Mediterranean and Atlantic were continuously connected so that inflow from the Atlantic replenished losses to evaporation. But because it was in an arid environment, the Mediterranean eventually reached the saturation point and precipitation of evaporite minerals began.

America Trench, together with arc magmatism. We will have more to say about this land connection between the Americas in Chapter 17 because it played an important role in the intercontinental migrations and extinctions of animals.

The North American Cordillera

The **North American Cordillera** is a complex mountainous region extending from Alaska into central Mexico (Figure 16.12). In the western United States, the Cordillera widens to about 1200 km where it had a complicated tectonic evolution during the Cenozoic. Evolution of the North American Cordillera actually began during the Late Proterozoic when thick sequences of sediments were deposited in continental environments and along a west-facing continental margin (see Figure 9.10 and Perspective 9.1). Deposition continued during the Paleozoic, and during the Devonian part of the region was deformed during the Antler orogeny (see Chapter 11). Another episode of deformation continuing from the Late Jurassic into the Early Tertiary took place as the Nevadan, Sevier, and Laramide orogenies progressively affected areas from west to east (see Figure 14.13). The Nevadan and Sevier orogenies were discussed in Chapter 14, and the *Laramide orogeny,* a Late-Cretaceous-to-Eocene event, will be considered in the following section.

Following the Laramide orogeny, several other events that contributed to the Cordillera's evolution took place. For instance, the block-faulted mountains and basins of the Basin and Range Province developed when the area was subjected to tensional forces. Vast outpourings of rhyolitic and especially basaltic volcanics occurred in the Pacific Northwest, and volcanism continues in the Cascade Range of California, Oregon, Washington, and British Columbia. The North American Cordillera continues to evolve, especially along its western margin where the *San Andreas transform fault* cuts through coastal California (Figure 16.12).

THE LARAMIDE OROGENY

The **Laramide orogeny** was the culmination of a series of deformational events that affected the Cordilleran mobile belt beginning in the Jurassic (see Figure 14.13). However, this orogeny differed from previous ones in that it was much farther inland from a convergent plate boundary and was not accompanied by significant volcanism or batholith

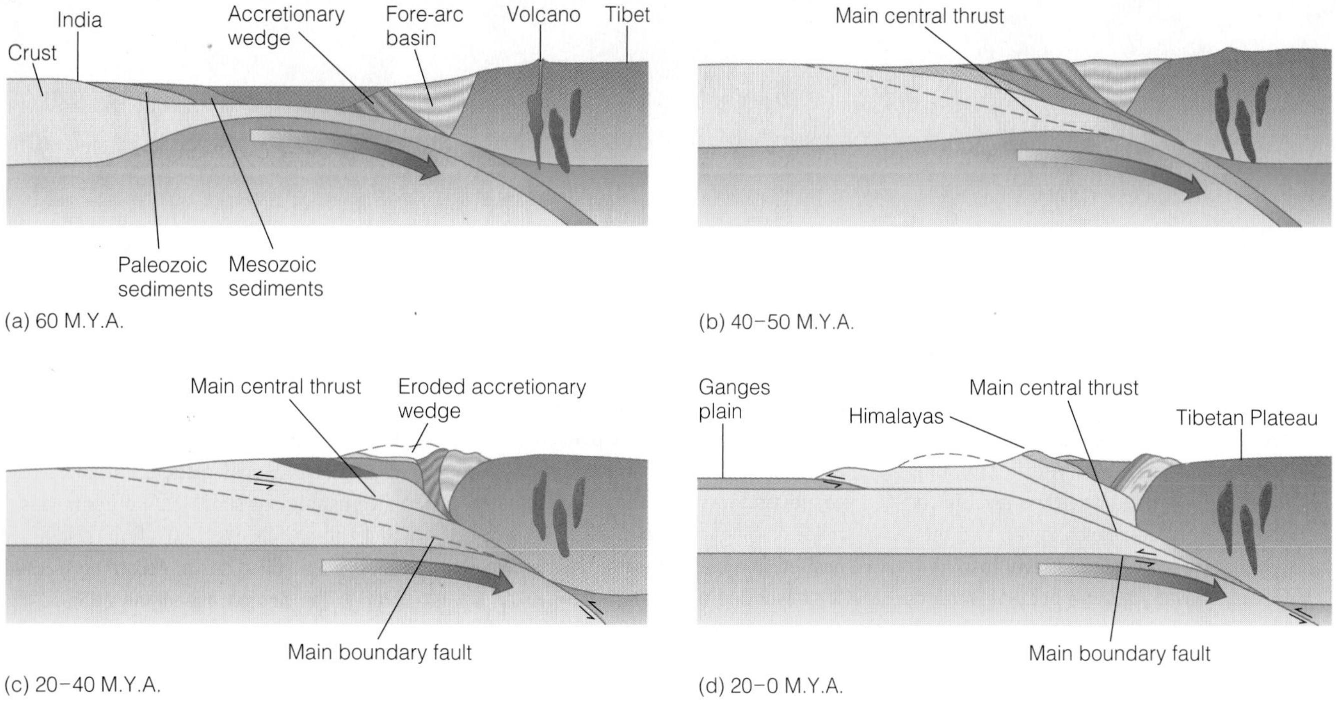

(a) 60 M.Y.A.

India
Crust
Accretionary wedge
Fore-arc basin
Volcano
Tibet
Paleozoic sediments
Mesozoic sediments

(b) 40–50 M.Y.A.

Main central thrust

(c) 20–40 M.Y.A.

Main central thrust
Eroded accretionary wedge
Main boundary fault

(d) 20–0 M.Y.A.

Ganges plain
Himalayas
Main central thrust
Tibetan Plateau
Main boundary fault

FIGURE 16.8 Simplified cross sections showing the collision of India with Asia and the origin of the Himalaya. (a) The northern margin of India before its collision with Asia. Subduction of oceanic lithosphere beneath southern Tibet as India approached Asia. (b) About 40 to 50 million years ago, India collided with Asia, but because India was too light to be subducted, it was underthrust beneath Asia. (c) Continued convergence accompanied by thrusting of rocks of Asian origin onto the Indian subcontinent. (d) Since about 10 million years ago, India has moved beneath Asia along the main boundary fault. Shallow marine sedimentary rocks that were deposited along India's northern margin now form the higher parts of the Himalaya.

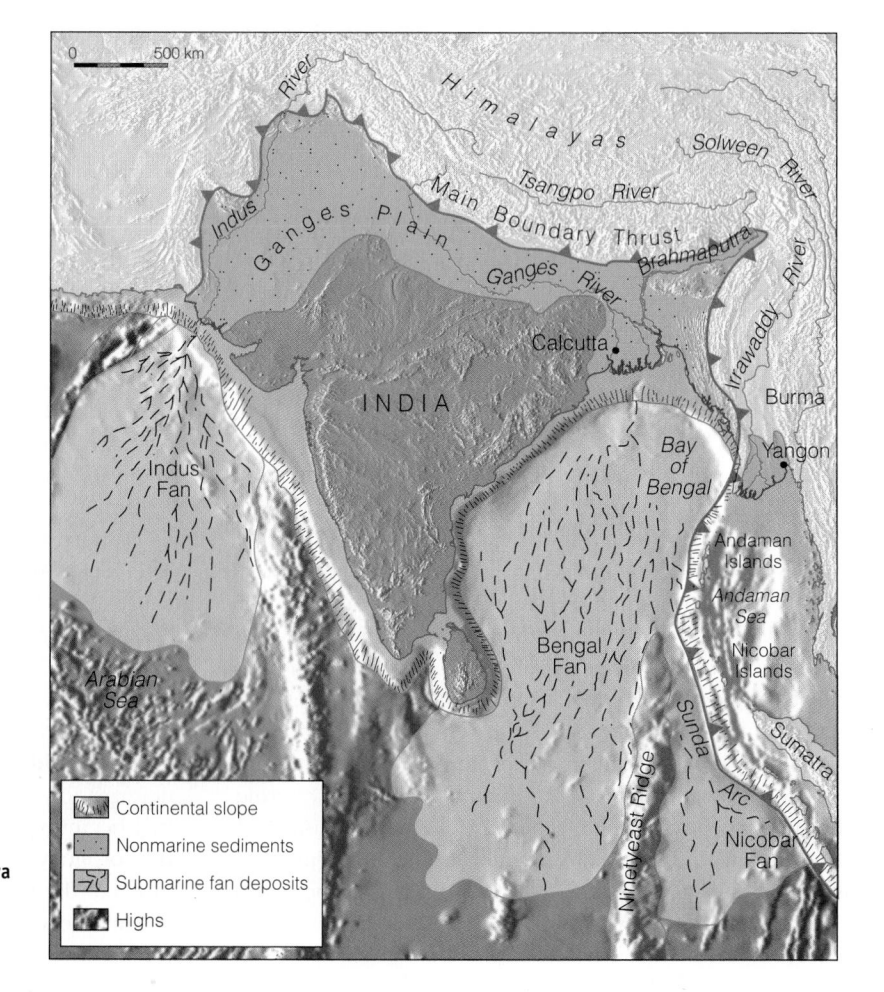

FIGURE 16.9 Sediment eroded from the Himalaya has been deposited as a vast blanket on the Ganges Plain and as large submarine fans in the Arabian Sea and the Bay of Bengal.

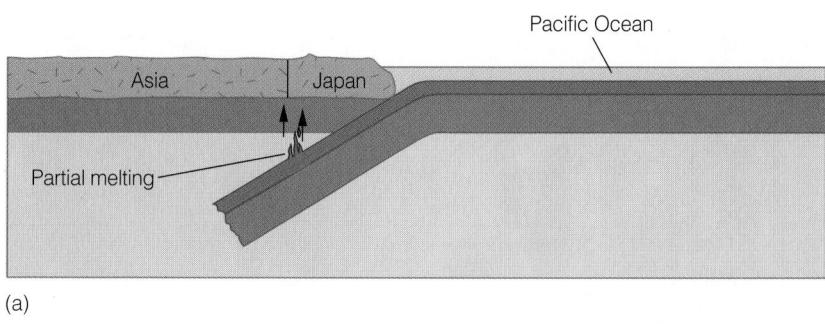

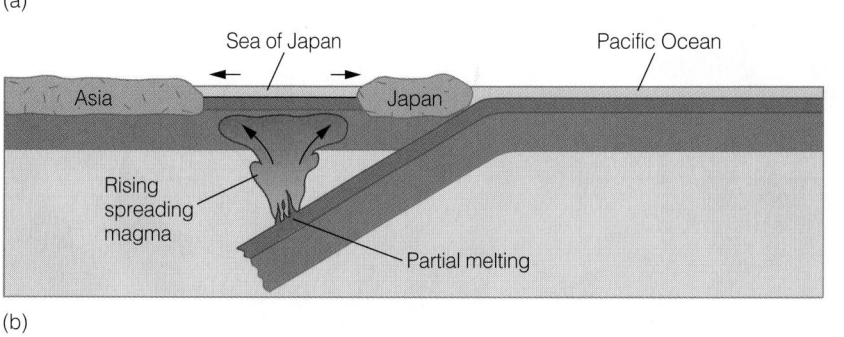

FIGURE 16.10 The back-arc marginal basin occupied by the Sea of Japan is thought to have formed by back-arc spreading. This model shows an initial stage (a) in the opening of a back-arc marginal basin, and a more advanced stage (b).

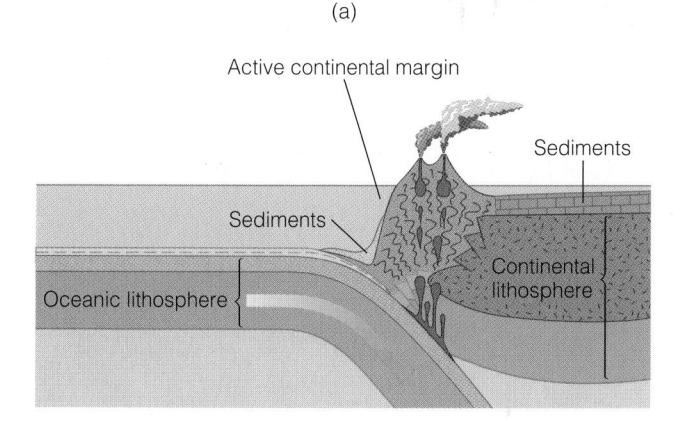

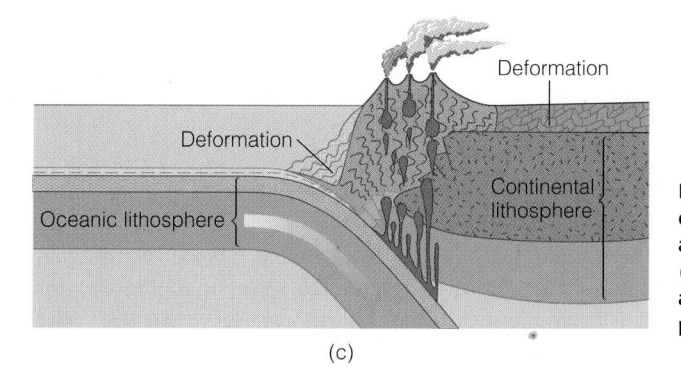

FIGURE 16.11 Generalized diagrams showing three stages in the development of the Andes of South America. (a) Prior to 200 million years ago, the west coast of South America was a passive continental margin. (b) Orogenesis began when the west coast of South America became an active continental margin. (c) Continued deformation, volcanism, and plutonism.

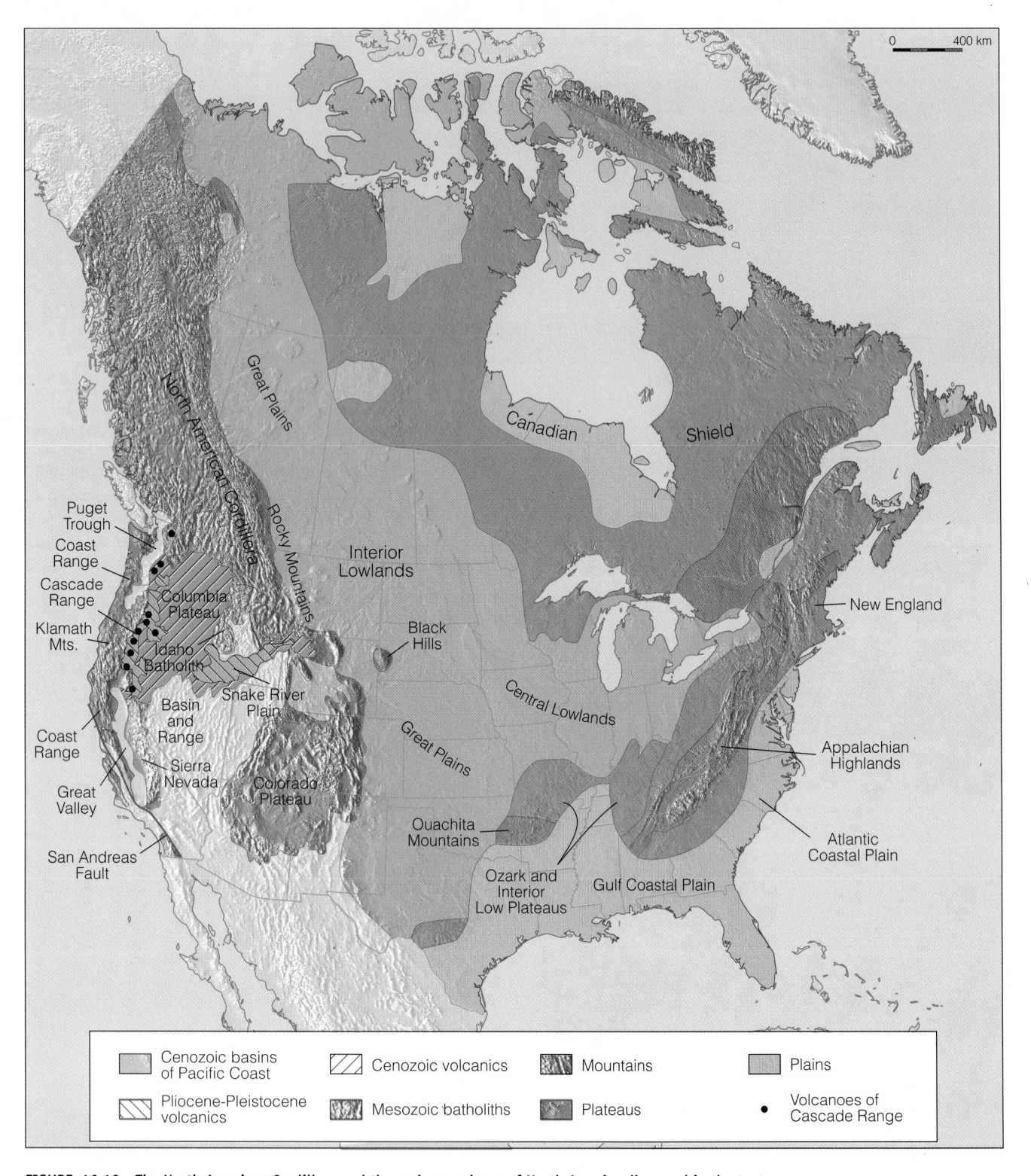

FIGURE 16.12 The North American Cordillera and the major provinces of North America discussed in the text.

emplacement. Furthermore, deformation was mostly in the form of vertical uplifts rather than the compression-induced folding and thrust faulting typical of arc orogens. To account for these observations, geologists have modified the classic arc orogen model in which oceanic lithosphere is subducted beneath continental lithosphere at an angle of 30° or more, arc magmatism occurs inland from the trench, and sedimentary rocks of the continental margin are deformed (Figure 16.13a). In the model for the Laramide-style of deformation (Figure 16.13b), the subducted oceanic plate descends at a lower angle and moves nearly horizontally beneath the continental lithosphere, thus accounting for deformation far inland from the continental margin. Shallow subduction would also account for the lack of magmatism because it occurs only when a subducted plate penetrates as deep as the mantle.

This model for Laramide deformation has been widely accepted even though no one could explain why subduction should change from steep to shallow. Recently, though, a further modification of the Laramide model has been proposed to account for shallow subduction (Figure 16.14). During the Late Cretaceous, a subduction zone existed along the entire West Coast of North America, where the Farallon plate was consumed. The Farallon plate descended at about a 50° angle, and arc magmatism took place 150 to 200 km inland from the trench (Figure 16.14a). During the Early Tertiary Period, the westward-moving North America plate had overridden part of the Farralon plate beneath which was the deflected head of a mantle plume. The lithosphere immediately above the plume was buoyed up, thus accounting for the change from steep to shallow subduction (Figure 16.14b). As a result, arc magmatism shifted farther inland and eventually ceased, because the descending Farallon plate no longer penetrated to the mantle.

Another consequence of the decreasing angle of subduction was a change in tectonic style. The fold-thrust deformation of the preceding Sevier orogeny gave way to large-scale buckling and fracturing, which produced fault-bounded, vertical uplifts. Uplift along these faults yielded mountain ranges with intervening intermontane basins, which became sites of Early Tertiary deposition of sediments eroded from the uplifted blocks (Figure 16.15).

The Laramide orogen is centered mostly in the middle and southern Rocky Mountains of Wyoming and Colorado, but deformation of the Cordillera also took place far to the north and south. In the northern Rocky Mountains of Montana and Alberta, Canada, large slabs of pre-Laramide strata were transported to the east along large overthrust faults. A large slab of Precambrian rocks moved at least 75 km eastward along the Lewis overthrust in Montana (Figure 16.16). In the Canadian Rocky Mountains of Alberta, thrust sheets piled one upon another in shinglelike fashion, resulting in complex structural relationships. Far to the south of the main Laramide orogen, carbonates and detrital rocks in the Sierra Madre Oriental of east-central Mexico are now part of a major fold-thrust belt.

By about Middle Eocene time, Laramide deformation ceased and intrusive and extrusive magmatism resumed in the Cordillera when the mantle plume beneath the lithosphere disrupted the overlying oceanic plate (Figure 16.14c). The uplifted blocks of the Laramide orogen continued to erode, and the debris filled the adjacent intermontane basins. By Late Tertiary time, the rugged, eroded mountains had been nearly buried in their own erosional debris, forming a vast plain across which streams flowed (Figure 16.17). During a renewed cycle of erosion, these streams removed much of the basin-fill sediments and incised their valleys into the uplift blocks. The deep canyons cutting through the ranges formed during this cycle of erosion (Figure 16.17). Late Tertiary uplift accounts for the elevations of the present-day ranges; in some areas uplift continues to the present.

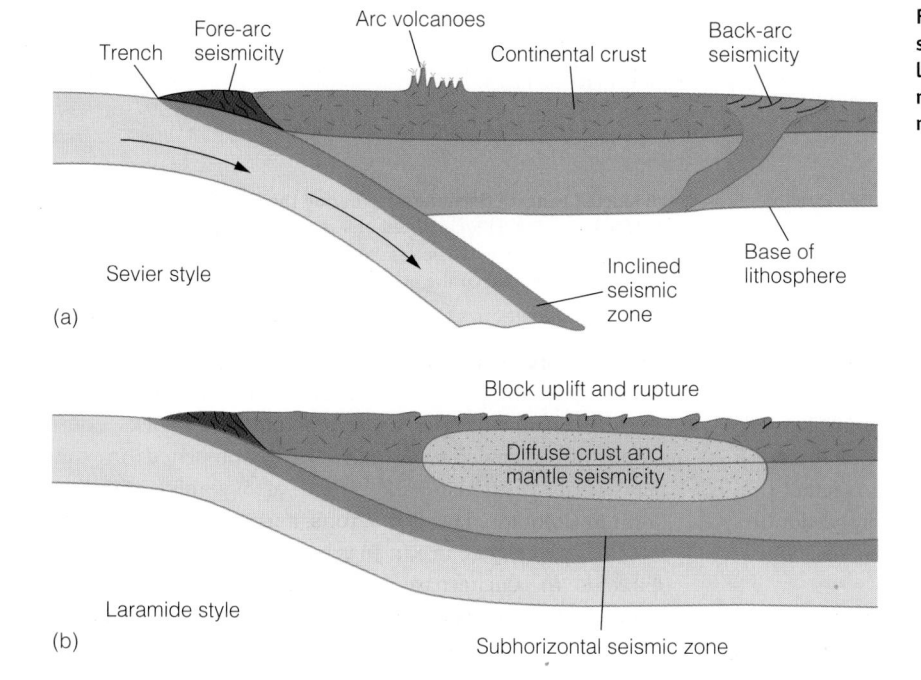

FIGURE 16.13 Arc orogens resulting from (a) steep and (b) shallow subduction. In the shallow-subduction model, the subducted slab moves nearly horizontally beneath the continent, and arc volcanism ceases.

Trench

Fore-arc seismicity

Arc volcanoes

Continental crust

Back-arc seismicity

Sevier style

Inclined seismic zone

Base of lithosphere

(a)

Block uplift and rupture

Diffuse crust and mantle seismicity

Laramide style

Subhorizontal seismic zone

(b)

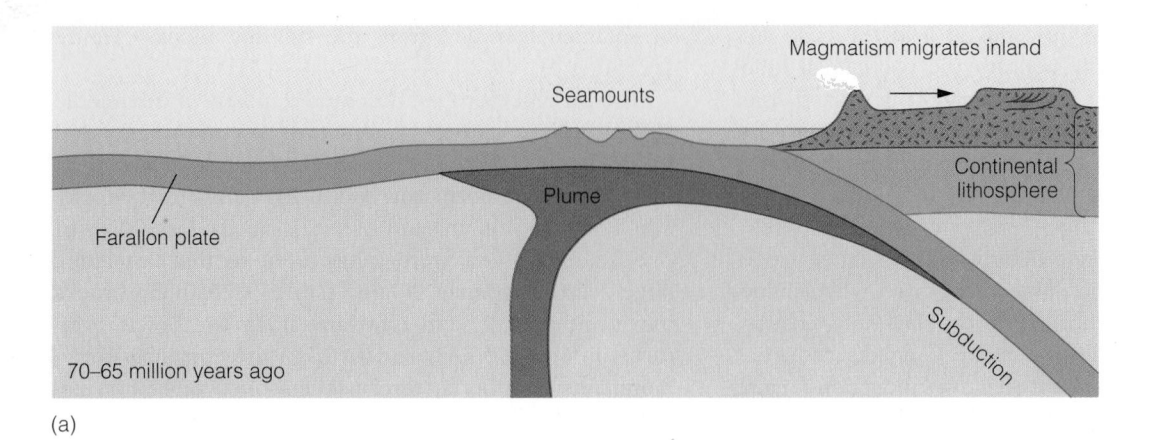

(a)

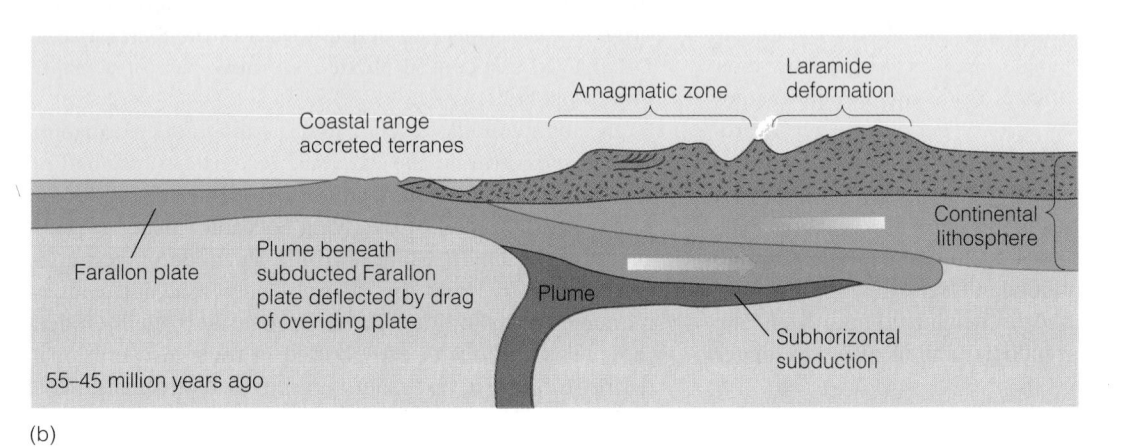

(b)

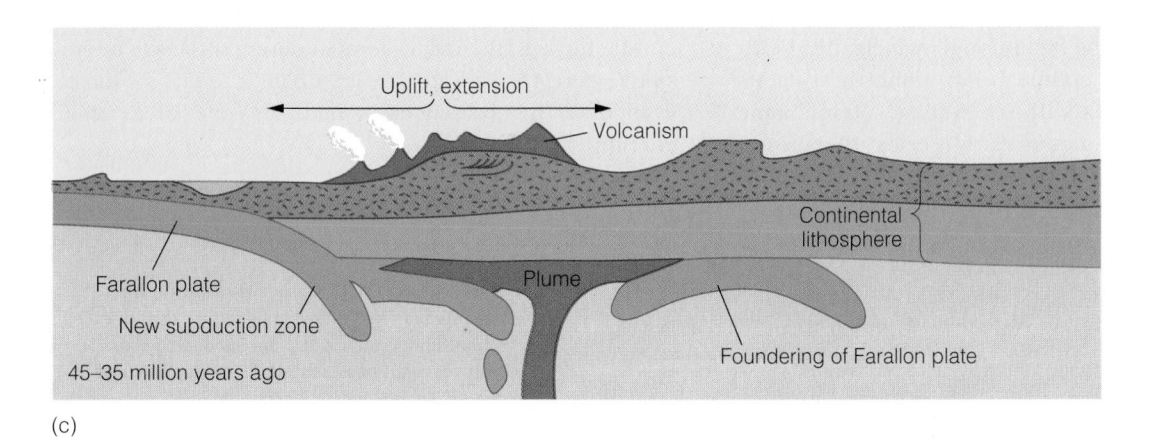

(c)

FIGURE 16.14. The Laramide orogeny resulted when the Farallon plate was subducted beneath North America during the Late Cretaceous to Eocene. **(a)** As North America moved westward over the Farallon plate, beneath which was the deflected head of a mantle plume, the angle of subduction decreased and magmatism shifted inland. **(b)** With nearly horizontal subduction, magmatism ceased and the continental crust was deformed mostly by vertical forces. **(c)** Disruption of the oceanic plate by the mantle plume marked the onset of renewed volcanism.

CORDILLERAN VOLCANISM

The vast batholiths of the Sierra Nevada in California, Idaho, and in the Coast Ranges of British Columbia were largely emplaced during the Mesozoic Era, but intrusive activity continued into the Tertiary. Numerous small plutons were emplaced, including copper and molybdenum-bearing stocks in Utah, Nevada, Arizona, and New Mexico.

Tertiary volcanism took place more or less continuously in the Cordillera, although it varied in eruptive style and lo-

cation (Figure 16.18), and it ceased for some time but later resumed in the area of the Laramide orogen (Figure 16.14b). Eocene lava flows and sedimentary rocks composed of volcanic rock particles accumulated in the Yellowstone National Park region of Wyoming. Further south in Colorado, lava flows, tuffs, and large calderas characterize the Oligocene San Juan volcanic field. In Arizona, Pliocene to Quaternary volcanism built up the San Francisco Mountains (Figure 16.18c). Here, volcanism may have ceased as recently as 1200 years ago.

In the Pacific Northwest, the Columbia Plateau (Figure 16.12) is underlain by about 200,000 km³ of Miocene lava flows known as the Columbia River basalts. These flows issued from long fissures and resulted in overlapping flows that are now well exposed in the walls of the deep gorges cut by the Snake and Columbia Rivers (Figure 16.19a). The magnitude of some of these eruptive events is difficult to imagine. For example, the single Roza flow covers 40,000 km² and has been traced more than 300 km west from its source. Most of the Columbia River basalts were erupted during a span of 3.5 million years and filled a basin with an aggregate thickness of 2500 m at the center.

Despite considerable study, though, the relationship of this phenomenal eruptive event to plate tectonics remains unclear, but some geologists think that it is related to the presence of a mantle plume beneath western North America (Figure 16.14c).

The Snake River Plain (Figure 16.12) is actually a depression in crust that has been filled mostly by Miocene and younger rhyolite, ash, and basalt (Figure 16.19b). The volcanics are oldest in the southwestern part of the Snake River Plain and become progressively younger toward the northeast, leading some geologists to propose that North America has migrated over a mantle plume (Figure 16.14). This plume,

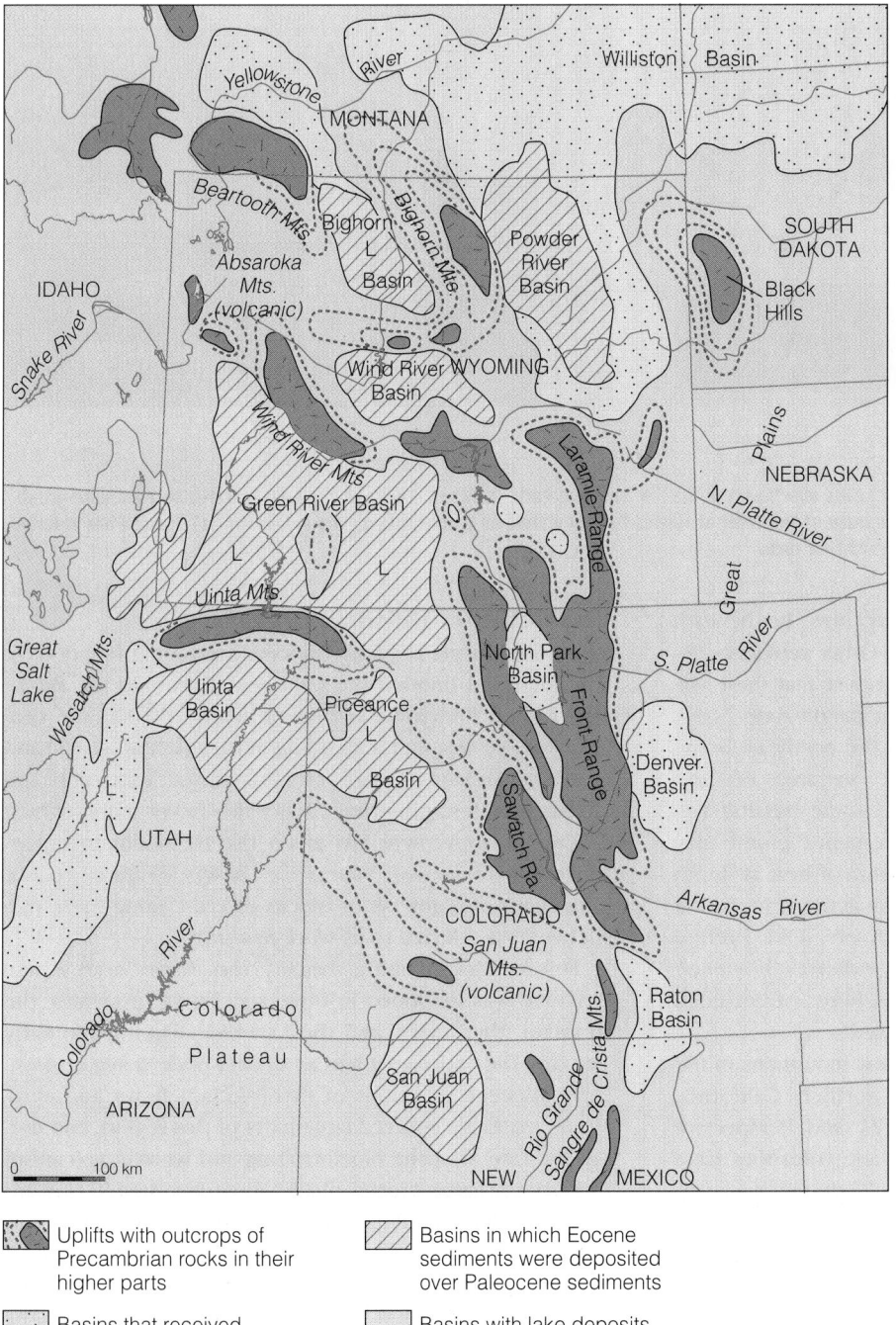

FIGURE 16.15 Map of Laramide uplifts and basins.

Uplifts with outcrops of Precambrian rocks in their higher parts

Basins that received Paleocene sediments

Basins in which Eocene sediments were deposited over Paleocene sediments

Basins with lake deposits (mostly Eocene)

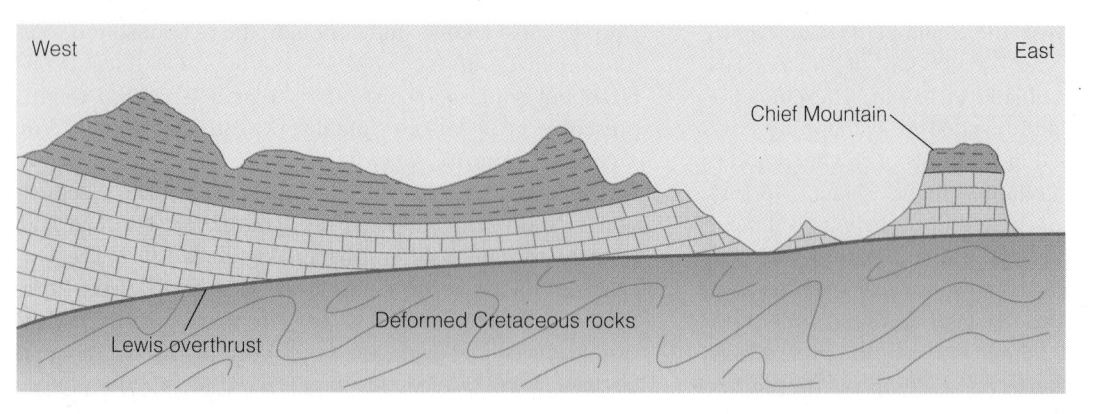

(a)

(b)

(c)

FIGURE 16.16 (a) Diagrammatic view of the Lewis overthurst in Glacier National Park, Montana. Rocks of the Late Proterozoic Belt Supergroup rest upon deformed Cretaceous strata. (b) The trace of the fault at Marias Pass is visible as a light line on the mountain. (c) Erosion has isolated Chief Mountain from the rest of the slab of overthrust rock.

although its existence is debated, may now lie beneath Yellowstone National Park in Wyoming. Other geologists disagree with the plume hypothesis and propose that these volcanics were erupted along an intracontinental rift zone.

Bordering the Snake River Plain on the northeast is the Yellowstone Plateau (Figure 16.18), an area of Late Pliocene and Quaternary rhyolitic and some basaltic volcanism (Figure 16.20). As just noted, a mantle plume may be located beneath this area. Some source of heat at depth is indicated by the current hydrothermal activity, including one of the world's most famous geysers, Old Faithful (Figure 16.20c). Geophysical evidence indicates, however, that the heat source may be only a body of intruded magma that has not yet completely cooled.

Some of the most majestic and highest mountains in the Cordillera are in the Cascade Range of northern California, Oregon, and Washington (Figure 16.21 and Perspective 16.2). Twice during this century, Cascade volcanoes have erupted. Mount Lassen, in northern California, was active from 1914 to 1917, and more recently, Mount St. Helens, in Washington, devastated a large area when it erupted in May 1980. Cascade volcanism continues as a result of subduction of the Juan de Fuca plate.

BASIN AND RANGE PROVINCE

Beginning during the Late Miocene, tensional forces have stretched and thinned the continental crust in the **Basin and Range Province,** an area of nearly 780,000 km² centered in Nevada but extending into adjacent states and northern Mexico (Figure 16.12). Crustal extension has yielded north-south–oriented normal faults along which differential movement has given rise to mountain ranges separated by broad valleys or basins (Figure 16.22). Movement of segments or blocks of crust along these parallel faults is referred to as *block-faulting.*

Before block-faulting began, the Basin and Range Province was deformed in three successive orogenies: the Nevadan, the Sevier, and the Laramide. During the Early Tertiary, the entire area was an upland undergoing erosion; Early Miocene eruptions of rhyolitic lava flows and pyroclastic materials covered large parts of this region, and during the Late Miocene block-faulting and basaltic volcanism had began. Large crustal blocks moved relatively upward along parallel normal faults and other crustal blocks moved downward, forming the ranges and basins characteristic of this region (Figure 16.22a). As the ranges were uplifted, they were eroded, and the debris was transported into the

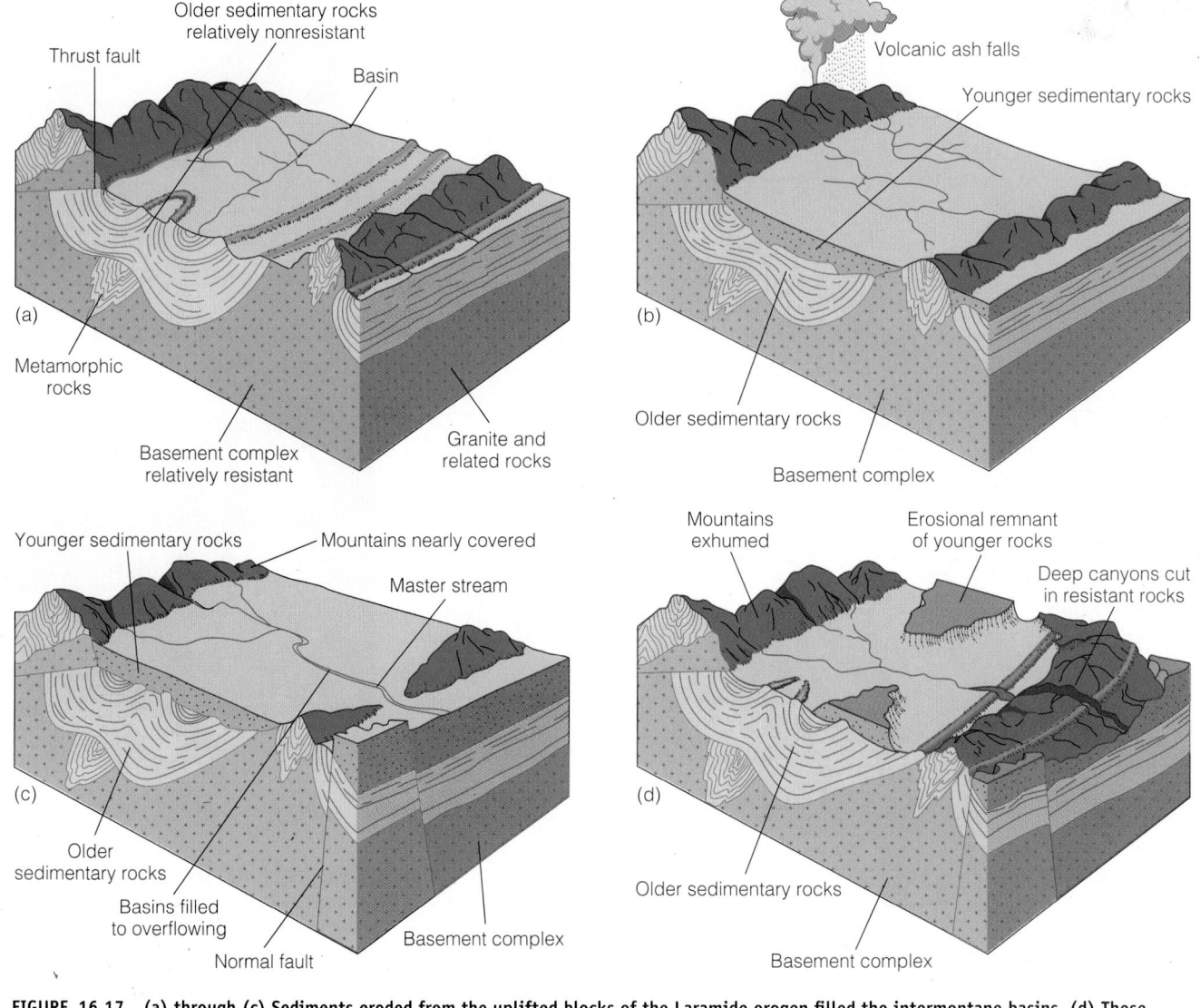

FIGURE 16.17 **(a) through (c) Sediments eroded from the uplifted blocks of the Laramide orogen filled the intermontane basins. (d) These basin sediments were partially excavated, and streams eroded deep canyons into the uplifted blocks.**

adjacent basins where it was deposited as thick alluvial fans and playa-lake deposits.

At its western margin, the Basin and Range Province is bounded by a large escarpment that forms the east face of the Sierra Nevada (Figure 16.22b). The Sierra Nevada block was uplifted along normal faults during the Pliocene and Pleistocene and tilted to the west. Its crest now stands 3000 m above the basins immediately to the east. Prior to uplift, the Basin and Range was in a subtropical climatic regime, but the rising Sierra Nevada created a rain shadow that caused the climate to become increasingly arid.

Several models have been proposed to account for basin-and-range structure. Among these are back-arc spreading, spreading at the East Pacific Rise (which is thought to now lie beneath North America), a mantle plume beneath the area, and deformation related to the San Andreas transform fault (Figure 16.23). There is little

agreement on the mechanism, but we can be sure that whatever it is, it is still active.

COLORADO PLATEAU

The Colorado Plateau (Figure 16.12) is a vast area of deep canyons, broad mesas, volcanic mountains, and brilliantly colored rocks (Figure 16.24). Exposed strata range in age from Archean to Cenozoic. Remember from our discussions in Chapters 11 and 14 that the Colorado Plateau region was the site of deposition of extensive red beds during the Permian and Triassic. Many of these formations are now exposed in the uplifts and walls of canyons.

At the end of the Cretaceous, the Colorado Plateau was near sea level, as indicated by marine sedimentary rocks of that age. During the Early Tertiary, Laramide deformation produced broad anticlines and arches and basins. These basins became the depositional sites for sediments shed from adjacent highlands. A number of large normal faults

(a)

Legend:
- Predominantly andesitic
- Basalt and andesite
- Predominantly basaltic

(b)

(c)

FIGURE 16.18 (a) The distribution of Cenozoic volcanic rocks in the western United States. (b) Snake River Plain basalt flow with spatter cones on its surface in Craters of the Moon National Monument, Idaho. (c) Volcanic rocks in the San Francisco Mountains, Arizona.

also cut the area, but overall deformation was far less intense than it was elsewhere in the Cordillera. During its Early Tertiary stage of development, the Colorado Plateau had no deep canyons, and the region was nearly encircled by mountains, making it a basin of internal drainage in which sediments accumulated.

Late Tertiary uplift elevated the region from near sea level to the 1200–1800 m elevations seen today. As uplift proceeded, deposition ceased, and streams began eroding deep canyons. Geologists disagree on the details of canyon cutting. For example, were the streams antecedent or superposed? If antecedent, the streams were already established in their present courses before the existing topography developed. Thus, they simply eroded downward as uplift proceeded. If superposed, the implication is that the entire area was buried by younger strata upon which streams flowed. With uplift, the streams stripped

away these strata and eroded downward into the underlying rocks.

PACIFIC COAST

The present plate tectonic elements of the Pacific Coast developed as a result of the westward drift of North America, the partial consumption of the **Farallon plate,** and the collision of North America with the Pacific-Farallon ridge. Prior to the Eocene, the entire Pacific Coast was bounded by a subduction zone that stretched from Mexico to Alaska. Most of the Farallon plate was consumed at this subduction zone, and now only two small remnants exist—the Juan de Fuca and Cocos plates (Figure 16.25). As discussed earlier, continuing subduction of these small plates accounts for seismicity and volcanism in the Pacific Northwest and Central America, respectively.

(a)

(a)

(b)

(b)

FIGURE 16.19 (a) Several lava flows of the Columbia River basalts in Washington. (b) Volcanic rocks of the Snake River Plain at Twin Falls, Idaho.

(c)

FIGURE 16.20 Rocks and processes in Yellowstone National Park, Wyoming. (a) Two basalt flows separated by the Tower Creek Conglomerate. (b) The Grand Canyon of the Yellowstone River has been eroded into hydrothermally altered Yellowstone Tuff. (c) Old Faithful geyser and other hydrothermal features indicate that a heat source is still present beneath the park.

perspective 16.2

Volcanoes of the Cascade Range

*T*he volcanoes to erupt most recently in the continental United States—Mount St. Helens, Washington, and Lassen Peak, California—are 2 of 15 large volcanoes in the 970-km-long Cascade Range of northern California, Oregon, Washington, and southern British Columbia (Figure 1, Table 1). Most are composite volcanoes or stratovolcanoes, but shield volcanoes—such as Newbery Volcano, Oregon, and Medicine Lake Volcano, California—are also present, and Lassen Peak is a lava dome. (Newberry and Medicine Lake Volcanoes are just east of the main Cascade Range but are usually included in discussions of the range.) In addition, many cinder cones are found throughout the range. At least 7 of these volcanoes have erupted during the last 200 years, and several show signs that eruptions will occur again (Figure 1, Table 1).

Lassen Peak's eruptions beginning in 1914 culminated on May 22, 1915, with the "Great Hot Blast," a huge steam explosion that devastated a large area of forest on the volcano's eastern and northeastern flanks (Figure 2a). The area was only sparsely settled, so no injuries or deaths resulted. A mudflow resulting from viscous lava flowing onto snow and melting it moved nearly 30 km from the peak. Boiling mud pots and gas vents known as fumeroles remind us that a source of heat is still present beneath the Lassen Peak area.

On May 18, 1980, after two months of intermittent activity, a tremendous northward-directed lateral blast blew out the north side of Mount St. Helens, Washington (Figure 2b). The blast accelerated to more than 1000 km/h and obliterated everything in its path. Some 600 km² of forest

were destroyed, trees were snapped off at their bases and strewn around the countryside, and trees as much as 30 km away were seared by the intense heat. Tens of

thousands of animals were killed; roads, bridges, and buildings were destroyed; and 61 people perished in the greatest volcanic disaster of U.S. history.

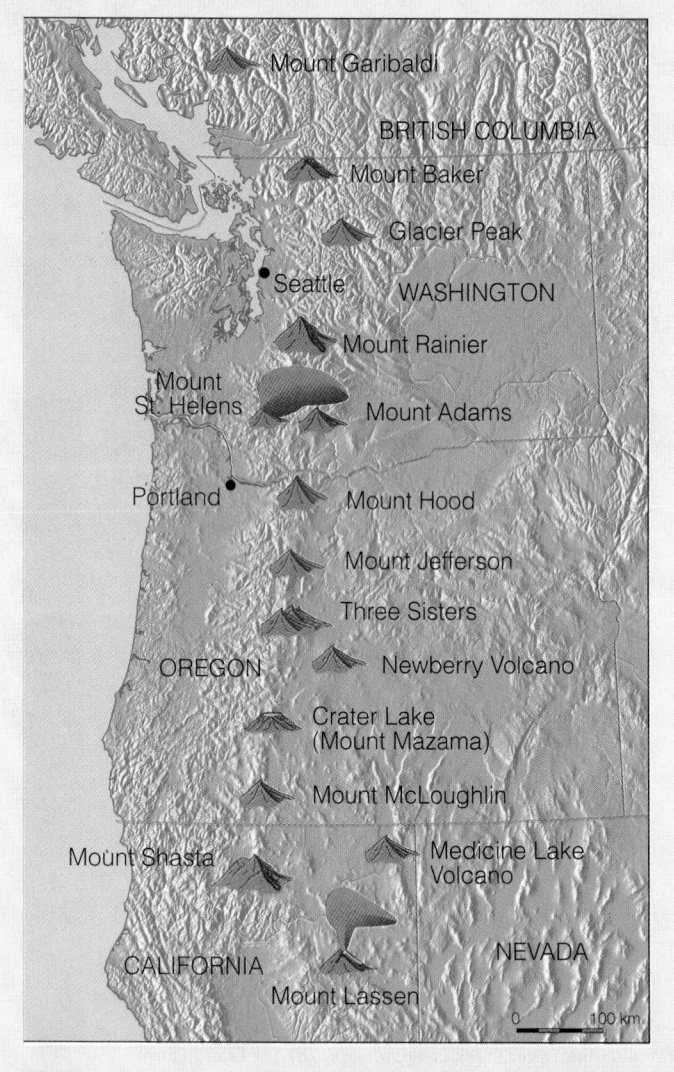

FIGURE 1 The large volcanoes of the Cascade Range. Lassen Peak and Mount St. Helens have erupted during this century, and several others have been active during the last 200 years. Medicine Lake Volcano, California, and Newberry Volcano, Oregon, are just east of the main Cascade Range but are usually considered in discussions of the range.

During the last 4000 years, Mount St. Helens has been the Cascade Range's most active volcano (Figure 1, Table 1), so its eruption in 1980 was no surprise to geologists. As a matter of fact, geologists of the U.S. Geological Survey warned in 1978 that Mount St. Helens was a particularly dangerous volcano. Several other Cascade volcanoes also pose threats to nearby populated areas. Mount Shasta in northern California (see Figure 16.21) has erupted about once every 600 years during the last 4500 years, and future eruptions will cause damage and perhaps fatalities in several nearby communities.

Mount Hood, Oregon, is less than 65 km from the densely populated Portland area. These and several other Cascade volcanoes will no doubt erupt in the future, but probably the most dangerous one is Mount Rainier, Washington.

Mount Rainier is the highest peak in the Cascade Range and is second only to Mount Shasta in volume. (Medicine Lake Volcano in California is more voluminous than both Rainier and Shasta, but it is a shield volcano [Table 1].) Volcanic debris flows and mudflows are the greatest danger from Mount Rainier, about 60 of which have occurred during the last

10,000 years. The largest flow consisting of about 4 km^3 of debris covers an area now occupied by more than 120,000 people.

The preceding discussion has focused on recent eruptions of Cascade Range volcanoes, but the range has a history going back into the Oligocene Epoch, some 36 million years ago. Older volcanic rocks are present in the range, but these probably formed during a volcanic episode that was rather widespread in the Northwest rather than confined to a relatively narrow belt. Most of the rocks composing the range are andesite and pyroclastic materials, but

TABLE 1

History of Eruptions of Some Cascade Range Volcanoes

LOCATION	VOLCANO	LAST ERUPTED/COMMENTS	LOCATION	VOLCANO	LAST ERUPTED/COMMENTS
British Columbia	Mount Garibaldi (composite volcano)	10,000 years ago. Might be extinct.	Oregon	Newberry Volcano (shield volcano)	600 years ago. Large summit caldera. Obsidian flow about 1600 years ago.
Washington	Mount Baker (composite volcano)	Erupted five or six times between 1820 and 1870. Has erupted extensive pyroclastic flows. Fumeroles on flanks and in crater.		Crater Lake (Mount Mazama) (composite volcano)	Explosive eruptions and origin of caldera 6600–7000 years ago. Cinder cone formed within caldera 800 to 900 years ago.
	Glacier Peak (composite volcano)	Between 240–300 years ago. Typically erupts large quantities of ash and pumice.	California	Medicine Lake Volcano (shield volcano)	A.D. 1065. Largest volcano in Cascade Range (600 km^3).
	Mount Rainier (composite volcano)	About 5 times between 1820 and 1882. Highest Cascade volcano. History of massive debris flows and mudflows.		Mount Shasta (composite volcano)	Probably erupted in 1786. Erupted about once every 600 years during last 4500 years. Largest composite volcano in Cascade Range (350 km^3).
	Mount St. Helens (composite volcano)	1980 to present. Most active volcano in the Cascade Range. Erupted 14 or 15 times during last 4000 years.		Lassen Peak (lava dome)	1914–1917. Hydrothermal activity in several areas. Eruptions at Cinder Cone near Lassen Peak in the 1650s.
Oregon	Mount Hood (composite volcano)	Minor explosive eruptions 1859, 1865, and possibly 1907. Fumeroles in crater area.			

basalt, rhyolite, and several other rock types are present as well. For instance, during the 1650s a basalt lava flow was erupted from Cinder Cone near Lassen Peak, California (Figure 3).

Apparently, volcanism began as a result of subduction of the Juan de Fuca plate beneath North America, a phenomenon that continues at about 4 cm/yr. Most authorities agree that volcanism in the Cascade Range has been episodic rather than continuous, the most recent episode beginning during the Late Miocene or Early Pliocene about 5 million years ago. Volcanism during this time has yielded more than 3000 vents, but the most impressive ones are the 15 large volcanoes shown in Figure 1.

Before the Cascade Range was present, warm, moist air from the Pacific Ocean moved far inland and supported subtropical forests in, for example, what is now central Oregon (see the Prologue to Chapter 18). However, once the range began forming, this flow of moist air was reduced and the climate inland became drier. In fact, the area east of the range is called high desert. To fully appreciate the climatic impact the range has had, consider that the west slopes of the Cascades receive on the order of 40 cm of rain annually, and in the higher parts of the range yearly snowfalls of

(a) (b)

FIGURE 2 Volcanism in the Cascade Range of northern California, Oregon, and Washington. (a) Lassen Peak in northern California, shown erupting in 1915, erupted from 1914 to 1917. (b) The eruption of Mount St. Helens, Washington, on May 18, 1980.

more than 15 m are not unusual. A number of the higher peaks support glaciers. In contrast, precipitation in the high desert amounts to only a few centimeters per year.

Besides its interest as a recent volcanically active area, the Cascade Range is also noted for its spectacular scenery. For instance, Lassen Volcanic National Park, California, Crater Lake National Park, Oregon, and Mount Rainier and North Cascades National Parks, Washington, all have much to offer in terms of geologic phenomenon, wildlife, and impressive scenery.

(a)

(b)

FIGURE 3 Although andesite and rocks composed of pyroclastic materials are most common in the Cascade Range, other rock types are present as well. (a) Basalt lava flows make up the Fantastic Lava Beds of Lassen Volcanic National Park, California. The most recent flow took place in the 1650s. The light-colored material seen at the right center is ash that fell on a still-hot lava flow. (b) The lava beds in (a) were erupted from this volcano known as Cinder Cone. It is nearly 230 m high.

(a)

(b)

FIGURE 16.21. Volcanism in the Cascade Range of northern California, Oregon, Washington, and southern British Columbia, Canada. (a) Mount Shasta in California probably last erupted in 1786. (b) Crater Lake, Oregon, is a caldera that formed when the summit of Mount Mazama collapsed about 6600 years ago. The small cone within the caldera is a cinder cone known as Wizard Island.

Westward drift of North America also resulted in its collision with the Pacific–Farallon ridge and the origin of the Queen Charlotte and San Andreas transform faults (Figure 16.25). Since the Pacific–Farallon ridge was oriented at an angle to the margin of North America, the continent–ridge collision occurred first in northern Canada during the Eocene and later during the Oligocene in southern California. In southern California, two triple junctions formed, one at the intersection of the North American, Juan de Fuca, and Pacific plates and the other at the inter-

section of the North American, Cocos, and Pacific plates. Continued westward movement of North America over the Pacific plate caused the triple junctions to migrate, one to the north and the other to the south, giving rise to the **San Andreas transform fault** (Figure 16.25). A similar occurrence along the coast of Canada resulted in the origin of the Queen Charlotte transform fault.

Where the San Andreas transform fault cuts through coastal California, it forms a complex zone of shearing. Additionally, faults such as this are seldom straight; they

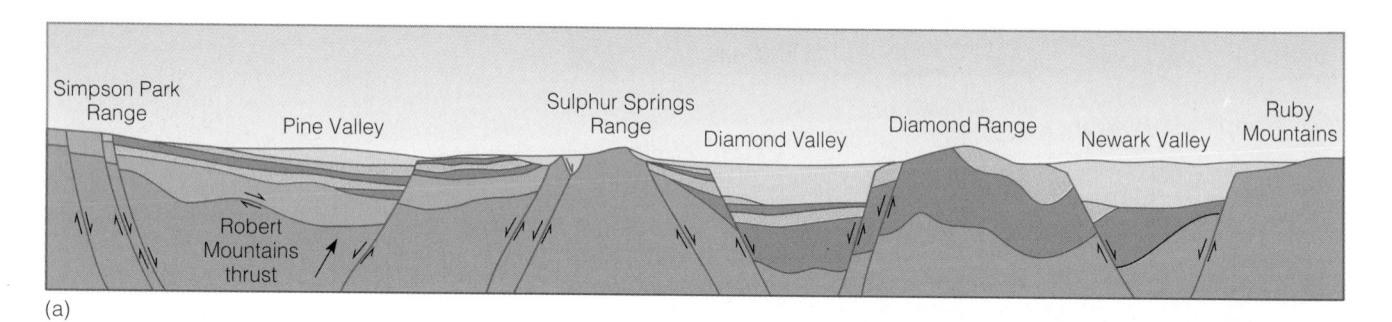

(a)

FIGURE 16.22. (a) Cross section of part of the Basin and Range Province in Nevada. The ranges and valleys are bounded by normal faults. (b) View of the western margin of the Basin and Range Province. The Sierra Nevada are bounded on the east by normal faults and rise 3000 m above the basins to the east.

(b)

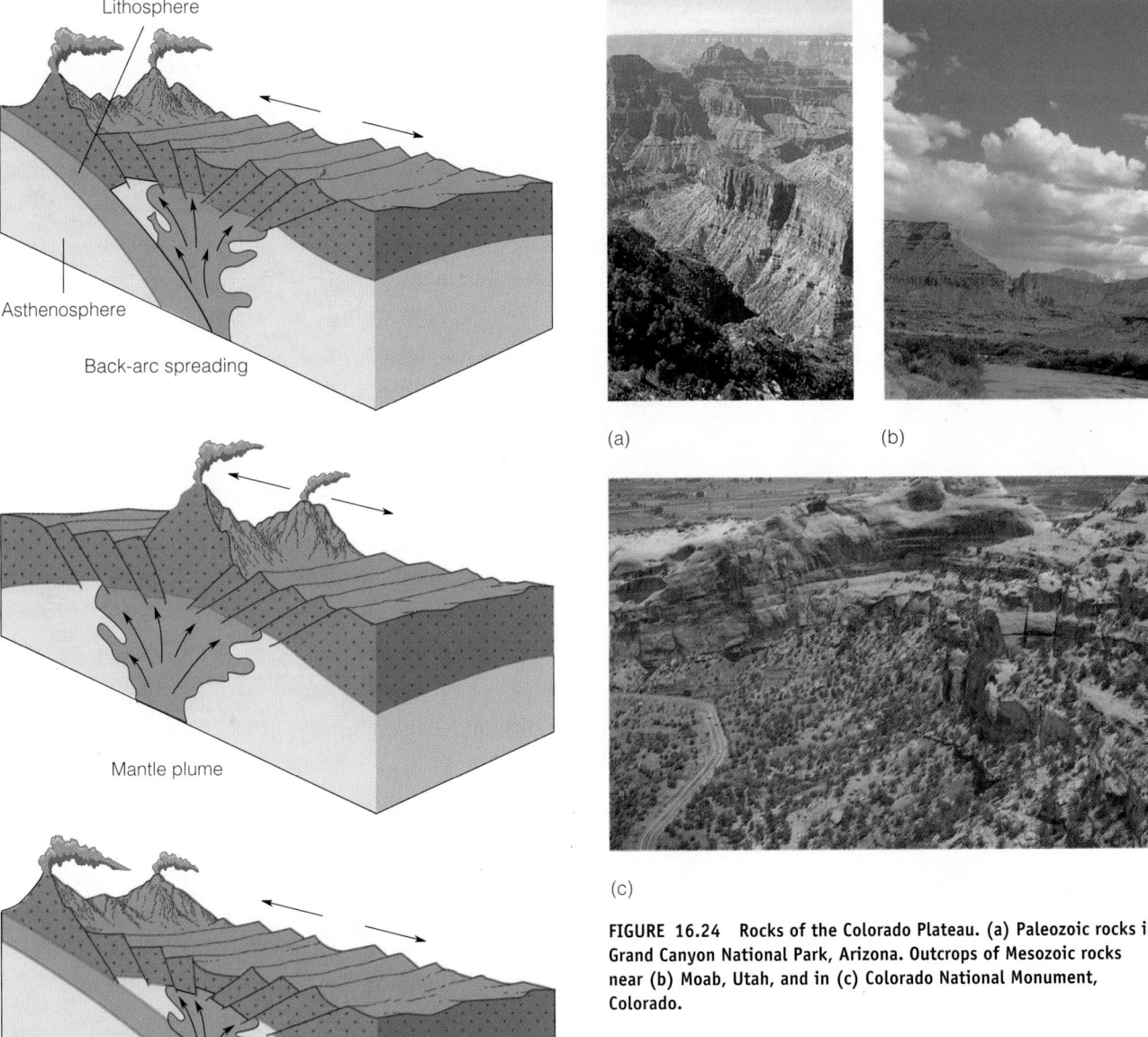

FIGURE 16.23 Geologists agree that tensional forces caused by crustal extension are responsible for basin-and-range structure. There is no agreement, however, on what caused the crustal extension. These illustrations show three models that have been proposed to account for crustal extension and the resulting basin-and-range structure.

FIGURE 16.24 Rocks of the Colorado Plateau. (a) Paleozoic rocks in Grand Canyon National Park, Arizona. Outcrops of Mesozoic rocks near (b) Moab, Utah, and in (c) Colorado National Monument, Colorado.

Labels in Figure 16.23 diagrams: Lithosphere, Asthenosphere, Back-arc spreading; Mantle plume; Overriding the East Pacific rise

Labels in Figure 16.24: (a), (b), (c)

curve and branch into smaller faults. Movements on complex fault systems of this type subject the crustal blocks within and adjacent to the shear zone to extensional and compressional stresses. Those areas subjected to extension form strike-slip basins, while the areas of compression are uplifted and supply sediments to the basins.

Cenozoic movements on the San Andreas transform fault account for the origin of numerous basins and uplifts in California, and in adjacent areas off the southern California coast (Figure 16.26). Many of these fault-bounded basins subsided below sea level and soon filled with turbidites and other deposits. Many of these basins are areas of prolific oil and gas production.

The Interior Lowlands

The Interior Lowlands is a vast region bounded by the Cordillera, the Canadian Shield, the Appalachians, the Ozark and Interior Low Plateaus, the Ouachita Mountains, and the Gulf Coastal Plain (Figure 16.12). During the Cretaceous, most of the western part of this region, an area called the Great Plains, was covered by the Zuni epeiric sea. By Early Tertiary time, the Zuni Sea had withdrawn from most of North America, but a sizable remnant arm of the sea was still present in North Dakota during the Paleocene Epoch. Sediments derived from Laramide high-

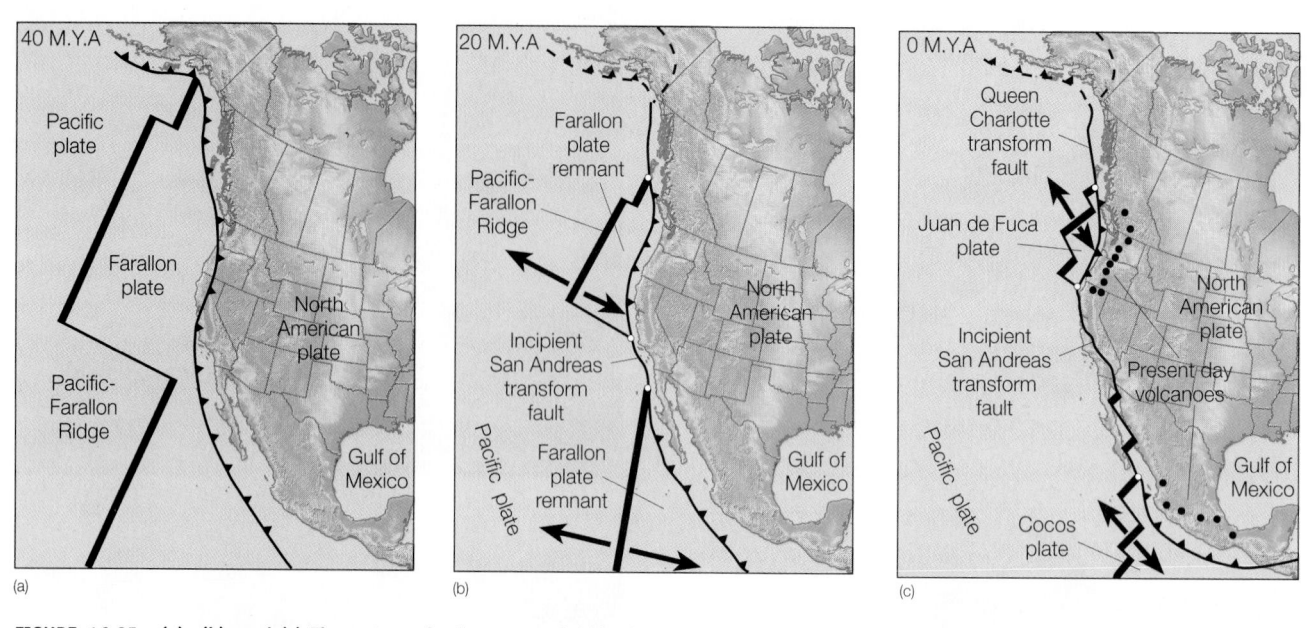

FIGURE 16.25 (a), (b), and (c) Three stages in the westward drift of North America and its collision with the Pacific–Farallon Ridge. As the North American plate overrode the ridge, its margin became bounded by transform faults rather than a subduction zone.

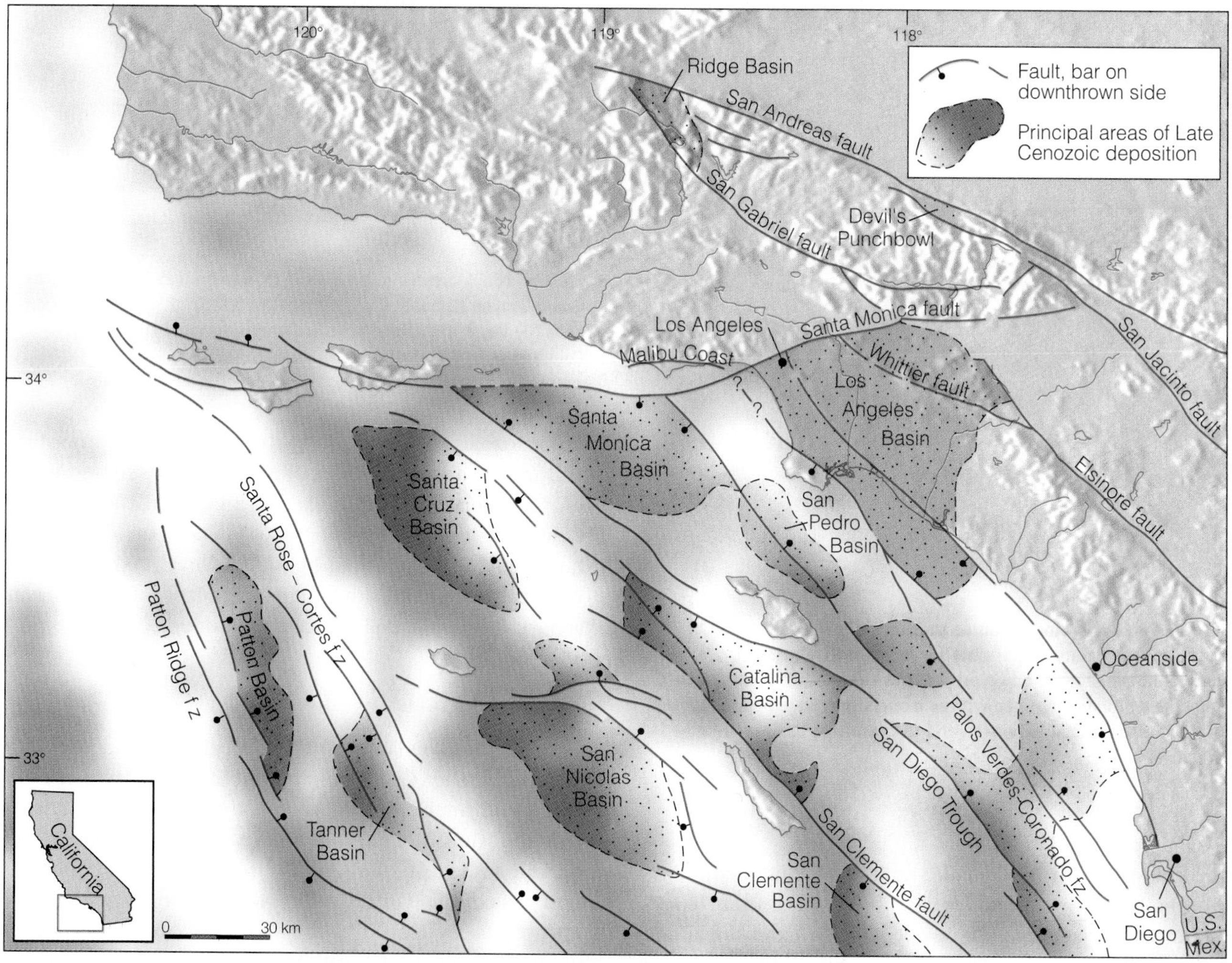

FIGURE 16.26 Major onshore faults in southern California and offshore Cenozoic basins. The principal areas of Late Cenozoic deposition are shown.

lands to the west and southwest were transported to this remnant sea where they were deposited in transitional and marine environments.

Elsewhere within the western Interior Lowlands, sediment was also transported eastward from the Cordillera. These sediments were deposited in terrestrial environments, mostly in fluvial systems, and form large, eastward-thinning wedges that now underlie the Great Plains (Figure 16.27). The Black Hills, the easternmost uplift associated with Laramide deformation, were a sediment source for fluvial and lacustrine (lake) deposits, which were later intricately eroded into the White River Badlands of South Dakota (see Perspective 4.2).

Igneous activity was not widespread in the western Interior Lowlands, but in some local areas it was significant. In northeastern New Mexico, for example, Late

(a)

FIGURE 16.27 Cenozoic sedimentary rocks of the Great Plains exposed at (a) Scott's Bluff, Nebraska, and (b) Theodore Roosevelt National Park, North Dakota.

(b)

Tertiary extrusive volcanism produced volcanoes and numerous lava flows. As a matter of fact, explosive volcanism in this area may have produced the volcanic ash that fell far to the northeast in Nebraska. A number of small intrusive bodies were emplaced in Colorado, Montana, South Dakota, and Wyoming. One of several in the Black Hills of South Dakota–Wyoming is called Devil's Tower (see the Prologue).

Eastward, beyond the Great Plains, Cenozoic deposits, other than those of Pleistocene glaciers, are uncommon in the rest of the Interior Lowlands. Much of the Interior Lowlands was subjected to erosion during the Tertiary. Of course, the eroded material had to be deposited somewhere, and that was on the Gulf Coastal Plain (Figure 16.12).

The Gulf Coastal Plain

Following the final withdrawal of the Cretaceous Zuni Sea, the Cenozoic **Tejas epeiric sea** (see Figure 10.3) made a brief appearance on the continent. But even at its maximum extent, this sea was largely restricted to the Atlantic and Gulf Coastal Plains and parts of coastal California. In fact, its greatest incursion onto North America was in the

area of the Mississippi Valley, where it extended as far north as southern Illinois. Epeiric seas were likewise restricted on the other continents except during the Early Tertiary of Europe.

Sedimentary facies development on the Gulf Coastal Plain was controlled largely by a regression of the Cenozoic Tejas epeiric sea. This sea extended far up the Mississippi River Valley during the Early Tertiary but then began its long withdrawal toward the Gulf of Mexico. Its regression, however, was periodically reversed by minor transgressions; eight transgressive–regressive episodes are recorded in Gulf Coastal Plain sedimentary rocks.

The general Gulf Coast sedimentation pattern was established during the Jurassic and persisted through the Cenozoic. Sediments derived from the eastern Cordillera, western Appalachians, and Interior Lowlands were transported toward the Gulf of Mexico. Deposition occurred in terrestrial, transitional, and marine environments. In general, the sediments form seaward-thickening wedges that grade from terrestrial facies in the north to progressively more offshore marine facies in the south (Figure 16.28). They also clearly record the shoreward and offshore migration of the shoreline.

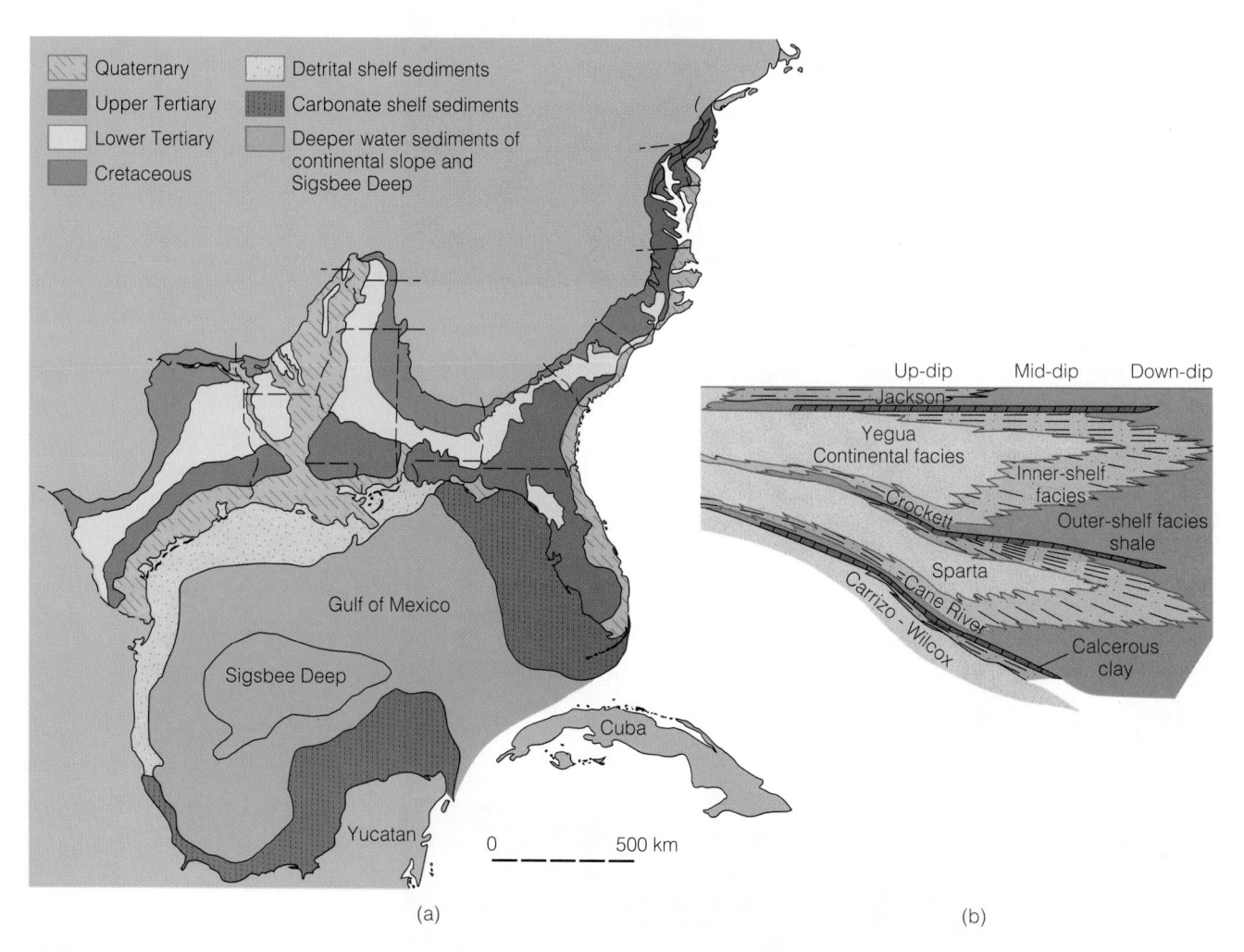

(a) (b)

FIGURE 16.28 Cenozoic deposition on the Gulf Coastal Plain. (a) Depositional provinces in the Gulf of Mexico and surface geology of the northwestern Gulf Coastal Plain. (b) Cross section of the Eocene Claiborne Group showing facies changes and the seaward thickening of the deposits.

Gulf Coastal Plain sediments—mostly sand, silt, shale, and lesser amounts of carbonate rocks—have been studied in considerable detail because many contain large quantities of oil and natural gas. Even though many of these rock units have no surface exposures, thousands of wells have been drilled through them, so their compositions, geometries, and stratigraphic relationships are well known. Many of the hydrocarbon reservoirs are nearshore marine sands and deltaic sands that lie updip from the fine-grained source rocks that supply the hydrocarbons. Such oil and gas traps are called *stratigraphic traps* because they owe their existence to variations in the strata. The Eocene Wilcox Group is an excellent example. Detailed studies of subsurface rocks and outcrops in Texas indicate deposition in fluvial, deltaic,

barrier-island, and beach environments. In fact, a large area of what is now southern Texas was the site of a complex of depositional environments much like those of the present-day Mississippi River delta and adjacent areas (Figure 16.29).

Structural traps—in which hydrocarbons are trapped as a result of folding, faulting, or both—are also common in the Gulf Coastal Plain. In the Gulf Coast, the Jurassic Louann Salt was deposited in the ancestral Gulf of Mexico when North America first separated from North Africa (see Figure 14.2). Salt is a low-density sedimentary rock, and when buried beneath more dense sediment, it tends to rise toward the surface in pillars known as **salt domes** (Figure 16.30). As the salt rises, it penetrates and deforms the overlying strata, creating structures that may trap oil or gas.

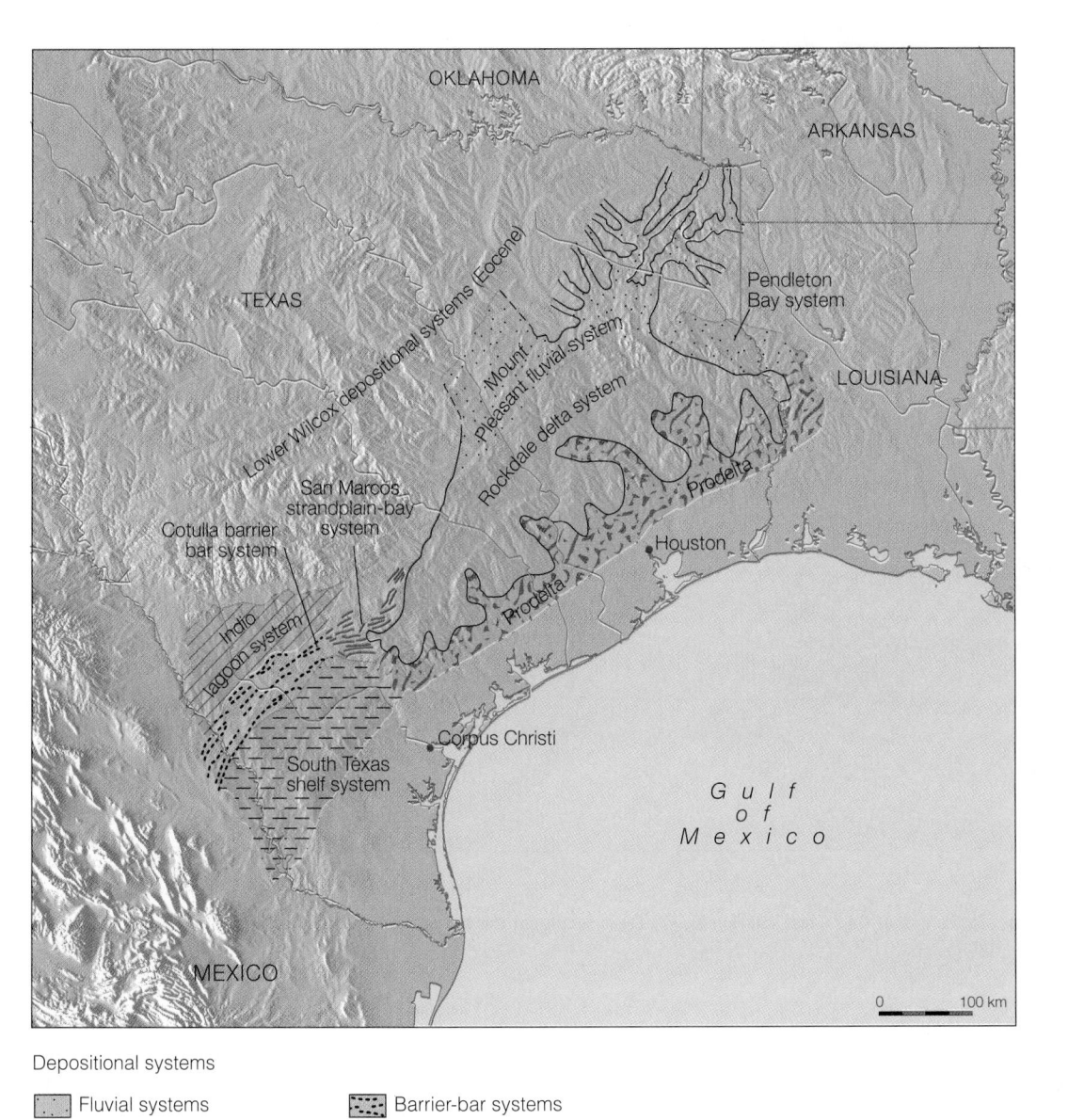

Depositional systems

Fluvial systems		Barrier-bar systems	
Delta systems		Bay-lagoon systems	
Chenier-strandplain systems		Shelf systems	
Marginal embayed systems			

FIGURE 16.29 Facies of the Eocene Wilcox Group in the northwestern Gulf of Mexico.

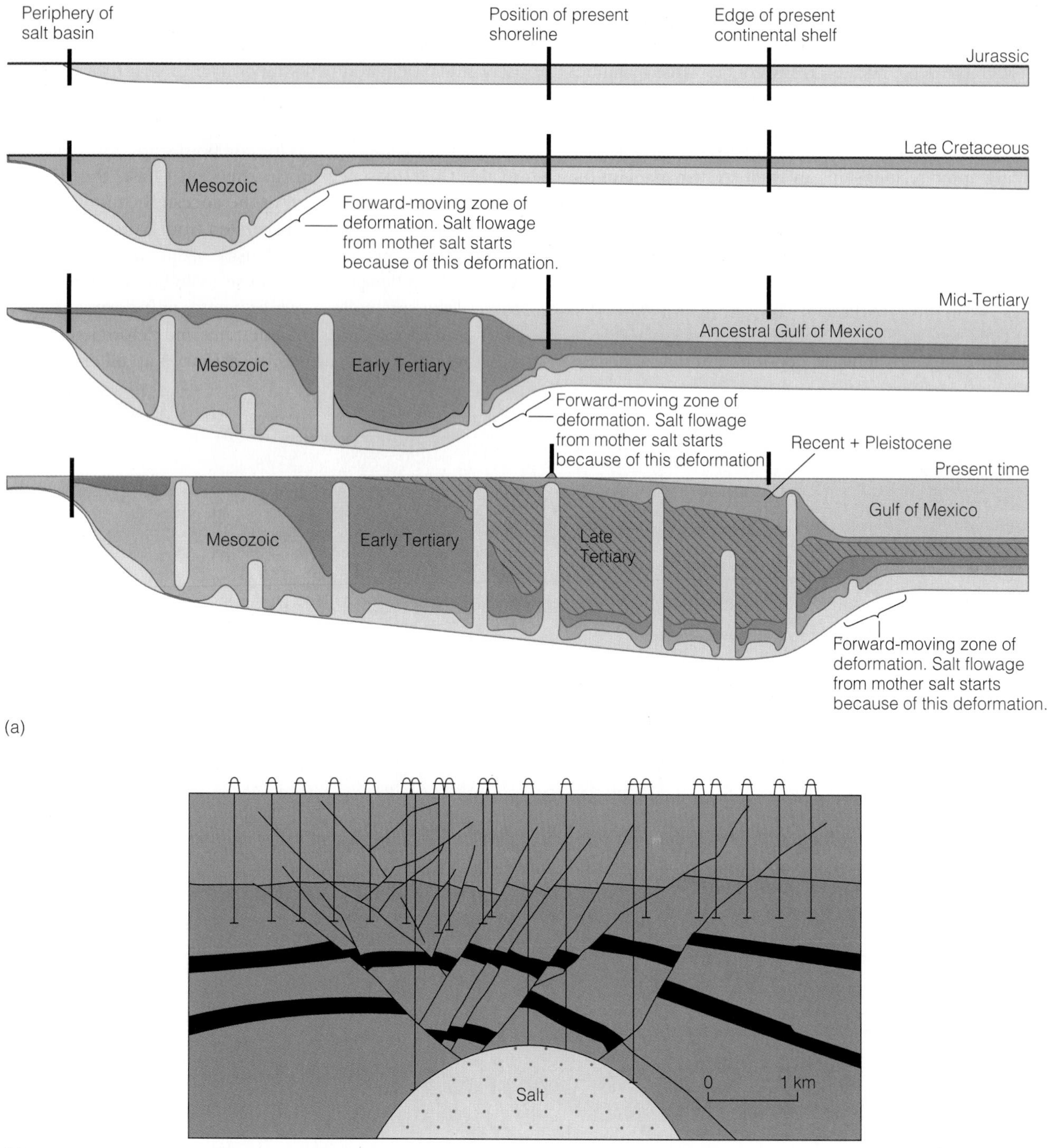

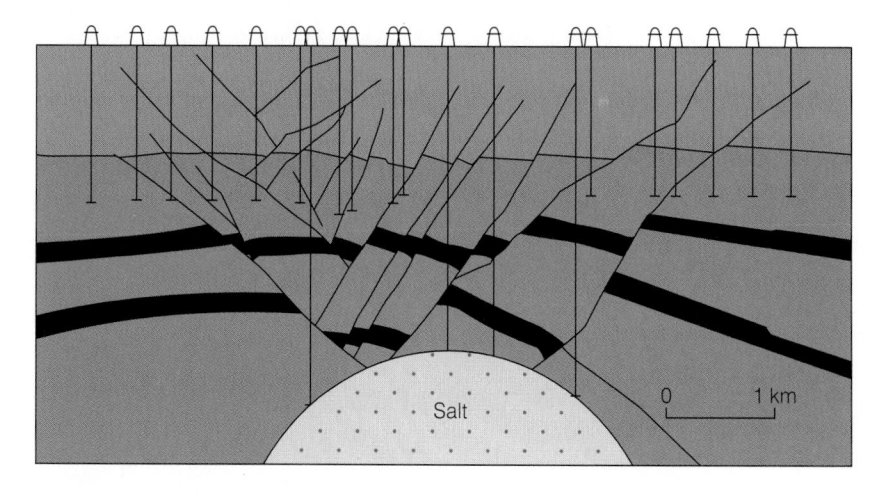

FIGURE 16.30 (a) Four stages in the rise of Gulf Coast salt domes. (b) Cross section of the top of the Heideberg salt dome, Mississippi, showing deformation of overlying strata.

Besides producing structures for the entrapment of oil and gas, the salt domes are natural resources in their own right. For example, the Avery Island salt dome in Louisiana has been mined for salt since the time of the Civil War. A number of salt domes have a cap composed in part of sulfur, which is also mined. Salt domes have also been proposed as sites for the storage of nuclear waste products.

Much of the Gulf Coastal Plain was dominated by detrital sediment deposition during the Cenozoic. In the Florida section of the Coastal Plain and the Gulf Coast of Mexico, however, significant carbonate deposition occurred. A carbonate platform was established in Florida during the Cretaceous, and shallow-water carbonate deposition continued through the Early Tertiary. During the Miocene

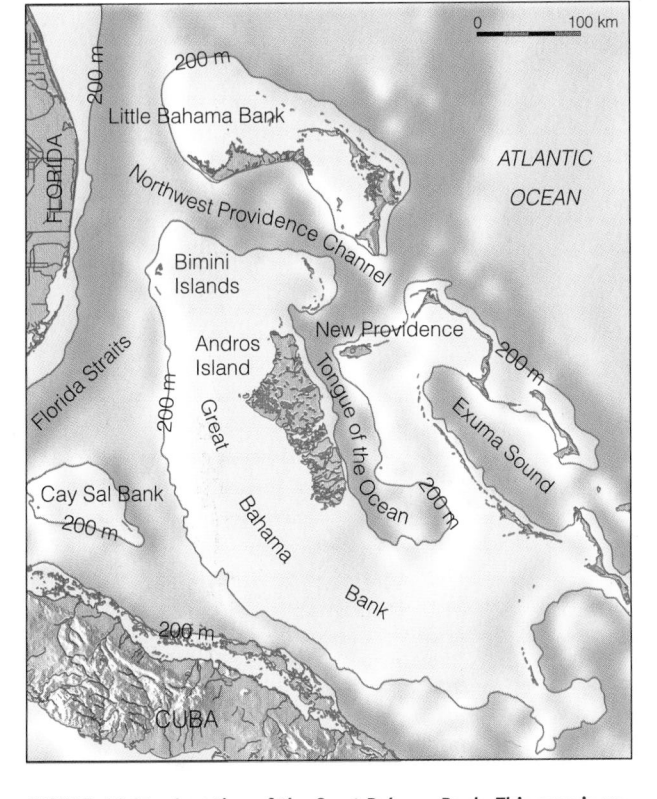

FIGURE 16.31 Location of the Great Bahama Bank. This area is underlain by thick, shallow-water carbonate rocks. Carbonate deposition continues at the present.

Epoch, parts of northern Florida formed islands that were populated by a variety of mammals including horses and camels. Although this is not the only Tertiary-aged mammalian fauna east of the Mississippi, it is the only one of Miocene age. Carbonate deposition continues in Florida at the present, but now it occurs only in Florida Bay and the Florida Keys.

Southeast of Florida, across the 80-km-wide Florida Strait, lies the Great Bahama Bank. This area has been a carbonate bank from the Cretaceous to the present. Thick, shallow-water carbonates accumulated there, and the region has been used for a long time as a modern-day laboratory to help us understand the conditions under which limestone is deposited (Figure 16.31).

Eastern North America

The eastern seaboard has been a passive continental margin since Late Triassic rifting separated North America from North Africa and Europe. Some seismic activity still occurs there; a major earthquake struck Charleston, South Carolina, on August 31, 1886, killing 60 people and causing $23 million in property damage. Overall, however, the region lacks the geologic activity characteristic of active, convergent margins.

CENOZOIC EVOLUTION OF THE APPALACHIANS

The present distinctive topography of the Appalachian Mountains is the product of Cenozoic uplift and erosion. Recall that the Appalachians originally formed during the Late Paleozoic closure of the Iapetus Ocean, an event called the Allegheny orogeny. Block-faulting began during the Late Triassic in response to extensional forces related to the breakup of Pangaea (see Figure 14.1a). In short, in what is now the eastern part of North America, an episode of rifting began that was probably very similar to the rifting now occurring in the Red Sea and Gulf of Aden (Figure 16.5). Since that time, the area has been a passive continental margin.

By the end of the Mesozoic, the Appalachian Mountains had been eroded to a plain (Figure 16.32). Cenozoic uplift rejuvenated the streams, which responded by renewed downcutting. As the streams eroded downward, they were superposed on resistant strata and cut large canyons across these strata. For example, the distinctive topography of the Valley and Ridge Province is the product of Cenozoic erosion and preexisting geologic structures. It consists of northeast–southwest trending ridges of resistant upturned strata and intervening valleys eroded into less resistant strata. Large canyons were cut directly across the ridges, the result of superposition of streams.

The origin of erosion surfaces at different elevations in the Appalachians has been a subject of debate. Some geologists think these erosion surfaces are evidence of uplift followed by extensive erosion and renewed uplift. Others are convinced that they represent differential response to weathering and erosion. According to this view, a low-elevation erosion surface developed on softer strata and was eroded more or less uniformly. Higher surfaces represent the weathering and erosion response of more resistant strata.

THE ATLANTIC CONTINENTAL MARGIN

The Atlantic continental margin of North America includes the Atlantic Coastal Plain (Figure 16.12) and extends seaward across the continental shelf, slope, and rise (Figure 16.33). It is a classic example of a passive continental margin. When Pangaea fragmented during the Early Mesozoic, North America separated from North Africa and Europe, continental crust was rifted, and a new ocean basin formed. The Atlantic continental margin possesses a number of Mesozoic and Cenozoic sedimentary basins that formed as a consequence of this rifting (Figure 16.33). Deposition in these basins began during the Jurassic, and even though sediments of this age are known only from a few deep wells, they are presumed to underlie the entire margin of the continent. The distribution of Cretaceous and Cenozoic sediments is better known because both crop out in the Atlantic Coastal Plain, and both have been penetrated by wells on the continental shelf.

The sediments of the Atlantic Coastal Plain were derived from the Appalachian Mountains, transported east,

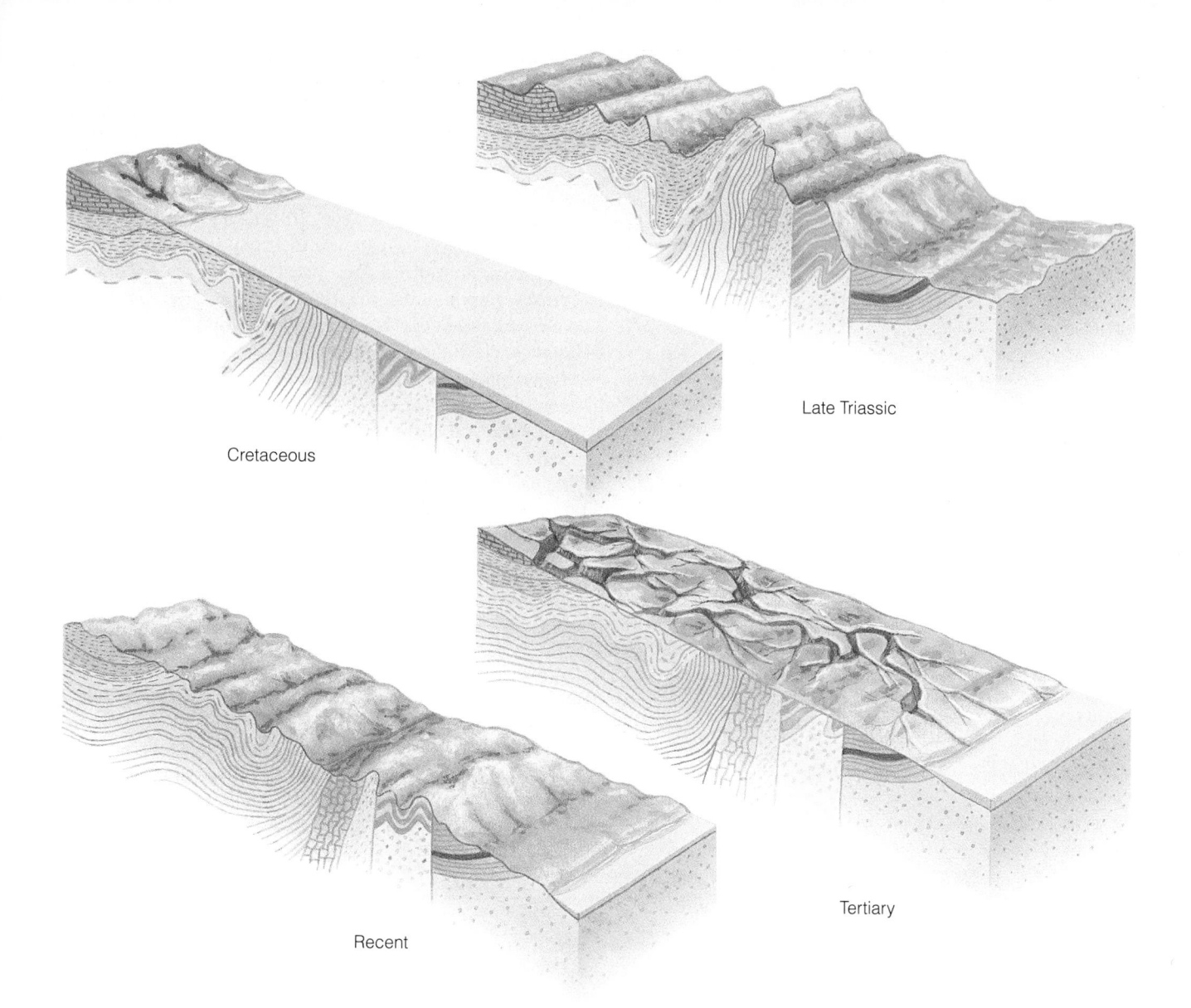

Cretaceous

Late Triassic

Recent

Tertiary

FIGURE 16.32 The origin of the present topography of the Appalachian Mountains. Erosion in response to Cenozoic uplift accounts for this topography.

and deposited in fluvial, nearshore marine and marine shelf environments. In Maryland, for example, the rocks composing the Calvert Cliffs on the western shore of Chesapeake Bay represent marginal marine deposits (see Figure 16.1b). These deposits are interesting because they contain an assemblage of fossil vertebrates that includes sea cows, whales, sharks, bony fish, birds, and reptiles. In general, the sedimentary rocks of the Atlantic Coastal Plain are part of a seaward-thickening wedge that dips gently seaward. In some places, such as off the coast of New Jersey, these deposits are up to 14 km thick. The best-studied seaward-thickening wedge of sedimentary rocks is in the Baltimore Canyon Trough, an area that also exhibits the structures typical of a passive continental margin (Figure 16.34).

Tertiary Mineral Resources

Deposition in two large lakes in what are now parts of Wyoming, Utah, and Colorado resulted in the origin of the Green River Formation. This formation is well known for its fossil fish, as well as for its huge reserves of oil shale and evaporites (see Prologue to Chapter 4). The evaporite mineral trona is mined from the formation for sodium compounds, but no oil is currently being produced. However, about 80 billion barrels of oil could be recovered from Green River oil shale with present recovery processes.

More than half of Florida's total mineral value is obtained from mining phosphate-bearing rock in the Central Florida Phosphate District. Phosphorus from phosphate rock has a variety of uses in metallurgy, preserved foods,

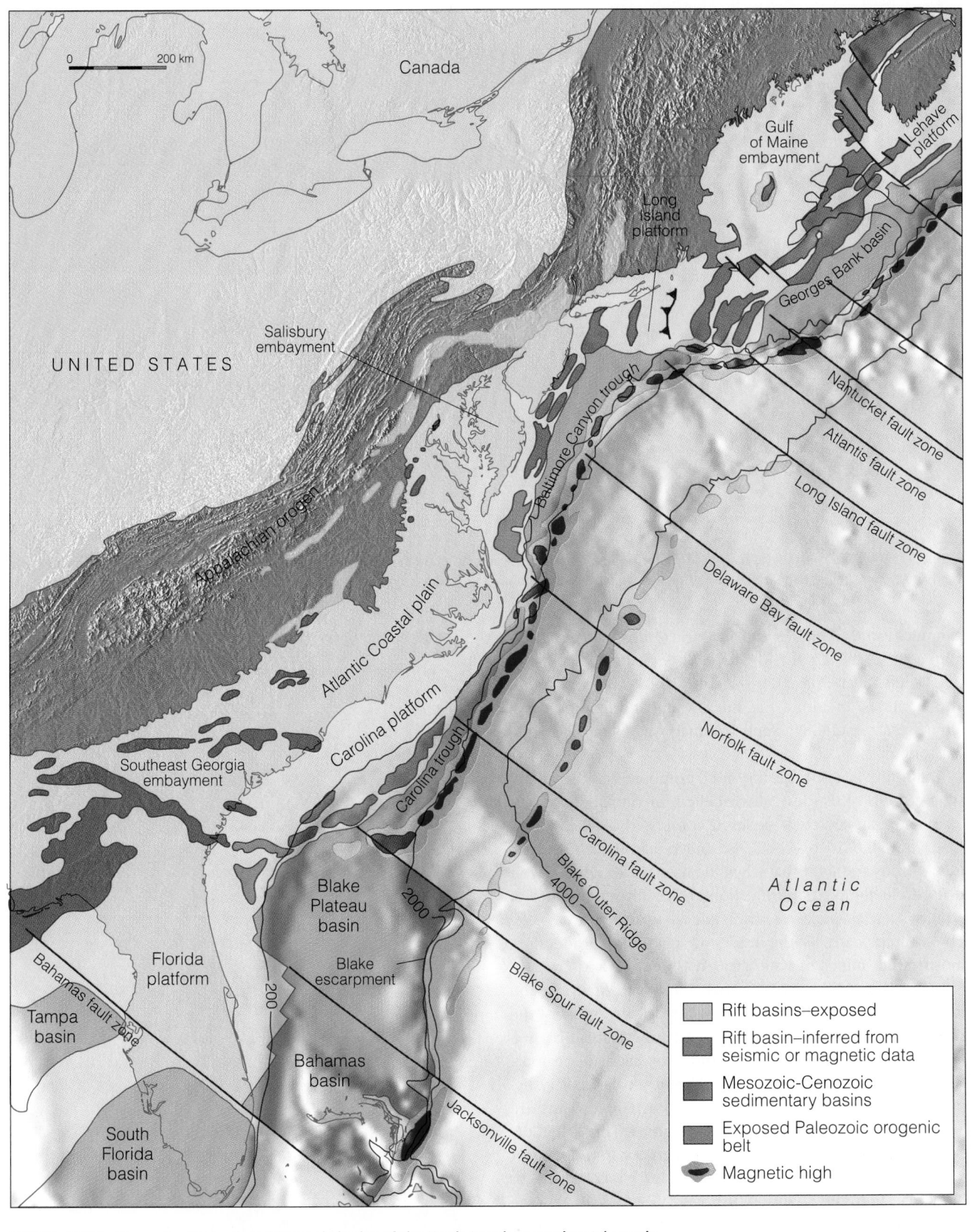

FIGURE 16.33 The major Mesozoic and Cenozoic basins of the North American continental margin.

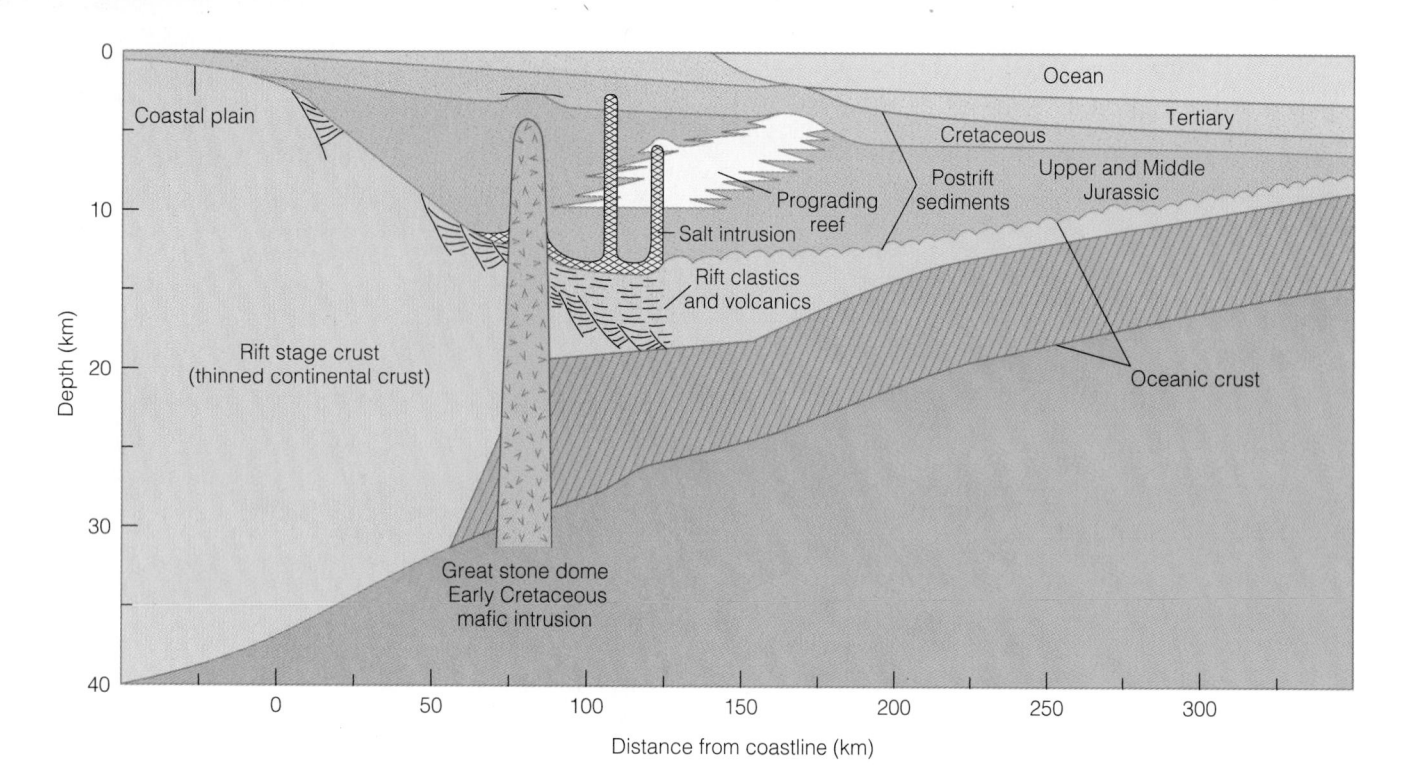

FIGURE 16.34 The continental margin in eastern North America. The coastal plain and the continental shelf are covered mostly by Cenozoic sandstones and shales. Beneath these sediments are Cretaceous and probably Jurassic sedimentary rocks.

ceramics, and matches. The Florida-phosphate rock is mined from the Upper Miocene Bone Valley Member of the Peace River Formation (Figure 16.35); this rock unit also contains an interesting marine and nonmarine fossil fauna that includes such animals as stingrays and mastodons.

Diatomite is a sedimentary rock composed of the microscopic shells of diatoms, single-celled, marine and freshwater plants that secrete a skeleton composed of silica (SiO_2) (see Figure 4.6a). This rock, also called diatomaceous earth, is so porous and light that it will float on water when dry. It is used chiefly in gas purification and to filter a number of liquids such as molasses, fruit juices, water, and sewage. The United States is the leader in diatomite production, mostly from mines in California, Washington, and Oregon.

About 30% of the world's coal is in North America, and of that 97% is in the United States. Historically, most of the coal mined in this country has been Pennsylvanian bituminous coal from the interior and Appalachian regions. However, huge deposits of lower-grade coals, mostly lignite and subbituminous, in the Northern Great Plains are becoming increasingly important resources. These coal deposits are Late Cretaceous to Early Tertiary and are most extensive in the Williston and Powder River basins of Montana, Wyoming, and North and South Dakota (see Figure 11.19). Besides having a low-sulfur content, some of these coal beds are 30–60 m thick!

Gold production from the Pacific Coast, particularly California, comes mostly from Tertiary and Quaternary gravels. The gold occurs as placer deposits, which formed as concentrations of minerals separated from weathered

FIGURE 16.35 Bone Valley Member of the Peace River Formation in the IMC Four Corners Mine, Polk County, Florida.

debris by fluvial processes. These placers were derived by erosion of Mesozoic quartz veins in the Sierra Nevada batholith and adjacent country rocks, which were the host rocks for the gold.

Reservoir rocks of Tertiary age contain a large proportion of the world's petroleum. Some of the petroleum production in the Persian Gulf region, where more than 50% of all petroleum reserves are located, comes from Tertiary reservoir rocks. In fact, oil production from Tertiary reservoir rocks occurs on all continents except Antarctica.

Barely 90 years after the first well was drilled in 1859, the United States became a net petroleum importer and now imports more than half of all the petroleum it consumes, much of it from the Persian Gulf region. Nevertheless, the United States is still the second leading producer, accounting for more than 18% of the total. Much of this production comes from Tertiary reservoirs of the Gulf Coast basin, which includes the Gulf Coastal Plain and the adjacent continental shelf. For example, the Eocene Wilcox Group of Texas (Figure 16.29) contains consider-able petroleum and natural gas. Much of the Gulf Coast petroleum is in structural traps related to salt domes (Figure 16.30) and other structures.

Several Tertiary basins in southern California are also important areas of petroleum production. These fault-bounded basins formed as a consequence of shearing related to movements on the San Andreas fault (Figure 16.26). Six oil accumulations in these basins exceed 1 billion barrels (1 barrel = 42 gallons).

A variety of other Tertiary mineral deposits are important. For example, the United States must import almost all manganese used in the manufacture of steel. The largest manganese deposits are in Lower Tertiary rocks of Russia. A number of small copper- and molybdenum-bearing stocks in Utah, Nevada, Arizona, and New Mexico were emplaced during the Tertiary. One Tertiary molybdenum deposit in Colorado accounts for much of the world production of this element. Tertiary sand and gravel as well as evaporites, building stone, and clay deposits are quarried from many areas around the world.

Summary

1. The Late Triassic rifting of Pangaea continued through the Cenozoic and accounts for the present distribution of continents and oceans.

2. Cenozoic orogenic activity was concentrated mostly in two major belts: the Alpine–Himalayan orogenic belt and the circum-Pacific orogenic belt. Each belt is composed of smaller units called orogens.

3. The Alpine orogeny resulted from convergence of the African and Eurasian plates. Mountain building took place in southern Europe, the Middle East, and North Africa. Plate motions also caused the closure of the Mediterranean basin, which became a site of evaporite deposition.

4. India separated from Gondwana, moved north, and eventually collided with Asia, causing deformation and uplift of the Himalaya.

5. Arc orogens characterize the western and northern Pacific Ocean basin. Back-arc spreading appears to be responsible for back-arc marginal basins such as the Sea of Japan.

6. Subduction of oceanic lithosphere occurred along the western margins of the Americas during much of the Cenozoic. This process continues beneath Central and South America, but the North American plate is now bounded mostly by transform faults.

7. The North American Cordillera is a complex mountainous region extending from Alaska into Mexico. Its Cenozoic evolution included deformation during the Laramide orogeny, extensional tectonics that formed the basin-and-range structures, intrusive and extrusive volcanism, and uplift and erosion.

8. Apparently shallow subduction of the Farallon plate beneath North America resulted in the vertical uplifts of the Laramide orogeny. The Laramide orogen is centered in the middle and southern Rockies, but Laramide deformation occurred from Alaska to Mexico.

9. Cordilleran volcanism was more or less continuous through the Cenozoic, but it varied in eruptive style and location. The Columbia River basalts represent one of the world's greatest eruptive events. Volcanism continues in the Cascade Range of the northwestern United States and British Columbia, Canada.

10. Tensional tectonics in the Basin and Range Province yielded north-south–oriented, normal faults. Differential movement on these faults produced uplifted ranges separated by broad, sediment-filled basins.

11. The Colorado Plateau was deformed less than other areas in the Cordillera. Late Tertiary uplift and erosion were responsible for the present topography of the region.

12. The westward drift of North America resulted in its collision with the Pacific–Farallon ridge. Subduction ceased, and the continental margin became bounded by major transform faults, except where the Juan de Fuca plate continues to collide with North America.

13. Sediments eroded from Laramide uplifts were deposited in intermontane basins, on the Great Plains, and in a remnant of the Cretaceous epeiric sea in North Dakota.

14. Gulf Coastal Plain facies patterns were controlled by transgressions and regressions of the Cenozoic epeiric sea. A seaward-thickening wedge of sediments pierced by salt domes on the Gulf Coastal Plain contains large quantities of oil and natural gas.

15. Cenozoic uplift and erosion were responsible for the present topography of the Appalachian Mountains. Much of the sediment eroded from the Appalachians was deposited on the Atlantic Coastal Plain.

Important Terms

Alpine–Himalayan orogenic belt

Alpine orogeny

arc orogen

back-arc marginal basin

Basin and Range Province

circum-Pacific orogenic belt

collision orogen

Farallon plate

Laramide orogeny

North American Cordillera

orogen

Quaternary

salt dome

San Andreas transform fault

Tejas epeiric sea

Tertiary

Review Questions

1. A collision between the Indian plate and Eurasian plate was responsible for the origin of the:
   a. _____ Andes Mountains.
   b. _____ East-African Rift system.
   c. _____ Himalaya.
   d. _____ Farallon plate.
   e. _____ Tejas epeiric sea.

2. Uplift and erosion were the main processes responsible for the present-day topographic expression of the:
   a. _____ Appalachian Mountains.
   b. _____ Cascade Range.
   c. _____ Gulf Coastal Plain.
   d. _____ Juan de Fuca plate.
   e. _____ San Andreas transform fault.

3. The Basin and Range Province of North America is an area of:
   a. _____ active volcanism.
   b. _____ plate convergence.
   c. _____ deposition in marine environments.
   d. _____ compression and folding.
   e. _____ block-faulting.

4. A vast area of overlapping lava flows in the northwestern United States is known as the:
   a. _____ Columbia River basalts.
   b. _____ Laramide orogen.
   c. _____ Basin and Range Province.
   d. _____ Andes Mountains.
   e. _____ Colorado Plateau.

5. The last of the epeiric seas to invade North America was the _____ Sea.
   a. _____ Sauk
   b. _____ Allegheny
   c. _____ Antler
   d. _____ Tejas
   e. _____ Absaroka

6. Tertiary facies relationships of the Gulf Coastal Plain were controlled largely by:
   a. _____ back-arc spreading.
   b. _____ transgressions and regressions.
   c. _____ carbonate deposition and the origin of salt domes.

d. _____ seafloor spreading.

e. _____ continental divergence.

7. Some geologists think that a hot spot now beneath Yellowstone National Park, Wyoming, was responsible for volcanism in the:

a. _____ Snake River Plain.

b. _____ Colorado Plateau.

c. _____ Laramide intermontane basins.

d. _____ Northwest American orogen.

e. _____ Interior Lowlands.

8. Much of the present-day tectonic activity in the Middle East can be accounted for by the northward movement of the _____ plate.

a. _____ Farallon

b. _____ Kula

c. _____ Nazca

d. _____ Arabian

e. _____ Antler

9. The classic model for orogenies was modified for the Laramide orogeny because it is:

a. _____ the oldest orogeny known in North America.

b. _____ farther inland than are most orogenies.

c. _____ characterized by the emplacement of vast batholiths.

d. _____ an example of mountains made up mostly of extrusive igneous rocks.

e. _____ similar to the arc orogens of the western Pacific Ocean basin.

10. The Cenozoic Era is typically divided into the _____ Periods.

a. _____ Tertiary and Quaternary

b. _____ Pleistocene and Eocene

c. _____ Miocene and Cretaceous

d. _____ Triassic and Recent

e. _____ Paleocene and Mesozoic

11. As a result of subduction of the Juan de Fuca plate beneath North America, volcanism continues in the:

a. _____ Appalachian Mountains.

b. _____ Canadian Shield.

c. _____ Cascade Range.

d. _____ Great Plains.

e. _____ Atlantic Coastal Plain.

12. When the Farallon plate was finally consumed beneath North America, the continental margin, rather than being bounded by a subduction zone, was bounded by:

a. _____ a volcanic arc.

b. _____ badlands topography.

c. _____ transform faults.

d. _____ a basin-and-range structure.

e. _____ back-arc marginal basins.

13. How does the Laramide orogen differ from other typical arc orogens?

14. Why are granitic rocks found in Tibet, and why are shallow marine sedimentary rocks present in the higher parts of the Himalaya?

15. Briefly outline the Cenozoic history of the Appalachian Mountains and the Atlantic Coastal Plain.

16. Describe the structures typical of the Basin and Range Province and explain how they might have formed.

17. How did the Queen Charlotte and San Andreas transform faults originate?

18. Describe the facies of the Gulf Coastal Plain and explain what events or processes controlled facies patterns.

19. What kinds of Tertiary deposits are found in the Interior Lowlands? What was the source of these sediments?

20. Where in North America are salt domes particularly common? Also, how do salt domes form, and why are oil and natural gas commonly found adjacent to them?

21. Why do some geologists think that migration of North America over a mantle plume can account for the volcanic rocks of the Snake River Plain? Where is this proposed mantle plume now?

22. How does an arc orogen differ from a collision orogen? Give an example of each.

23. Briefly summarize the Cenozoic events leading to the Alpine orogeny. What geologic phenomenon indicate that tectonism is still active in this region?

24. What kind of continental margin is the eastern edge of North America? How did it form?

Points to Ponder

1. The West Coast of North America was first bounded by a subduction zone during the Cenozoic and then transform plate boundaries, except were subduction still occurs in the Pacific Northwest. Speculate on a sequence of future events that would bring about a similar situation for North America's East Coast.

2. What kinds of evidence would you look for on another planet to demonstrate that it has areas with histories similar to those of the Snake River Plain, the Basin and Range Province, and Cascade Range?

3. Why is Cenozoic Earth and life history so much better understood than the histories of previous eras? Compare especially the Cenozoic and Precambrian records.

Additional Readings

Baars, D. L. 1983. *The Colorado Plateau: A geologic history.* Albuquerque: University of New Mexico Press.

———. 1995. *Navajo country: A geologic and natural history of the Four Corners region.* Albequerque: University of New Mexico Press.

Bengeyfield, P. 1996. *Mountains and mesas: The northern Rockies and the Colorado Plateau.* Flagstaff, Ariz.: Northland.

Burchfield, B. C., P. W. Lipman, and M. L. Zoback (eds.). 1992. The Cordilleran orogen, conterminous United States. In *Geology of North America.* Vol. G3. Boulder, Colo.: Geological Society of America.

Coleman, R. G. 1993. *Geologic evolution of the Red Sea.* New York: Oxford University Press.

Fiero, B. 1986. *Geology of the Great Basin.* Reno: University of Nevada Press.

Flores, R. M., and S. S. Kaplan (eds.). 1985. Cenozoic paleogeography of the west-central United States. *Rocky Mountain Paleogeography Symposium 3.* Denver: Society of Economic Paleontologists and Mineralogists.

Fritsche, A. E. (ed.). 1995. *Cenozoic paleogeography of the western United States II.* Fullerton, Calif.: Pacific Section, Society of Economic Paleontologists and Mineralogists.

Lipman, P. W., C. E. Chapin, and M. A. Dungan (eds.).

1989. *Cenozoic volcanism in the western United States.* Washington, D.C.: American Geophysical Union.

Molnar, P. 1986. The geologic history and structure of the Himalaya. *American Scientist* 74, no. 2: 144-154.

Monger, J. W. H., and J. Franscheteau (eds.). 1987. *Circum-Pacific orogenic belts and evolution of the Pacific Ocean basin.* Washington, D.C.: American Geophysical Union.

Murphy, J. B., G. L. Oppliger, G. H. Brimhall, Jr., and A. Haynes. 1999. Mantle Plumes and Mountains, *American Scientist* 87, no. 2, pp. 146-153.

Orr, E. L., and W. N. Orr. 1996. *Geology of the Pacific Northwest.* New York: McGraw-Hill.

Osborne, R., and D. Tarling (eds.). 1996. *The historical atlas of the Earth.* New York: Holt.

Parnell, J. (ed.). 1992. *Basins of the Atlantic seaboard, petroleum geology, sedimentology and basin evolution.* London: Geological Society.

Reidel, S. P., and P. R. Hooper (eds.). 1989. *Volcanism and tectonism in the Columbia River flood-basalt province.* Boulder, Colo.: Geological Society of America.

Sinha, A. K. 1992. *Himalayan orogen and global tectonics.* Rotterdam: Balkema.

Smith, A. G. 1994. *Atlas of Mesozoic and Cenozoic coastlines.* New York: Cambridge University Press.

For these web site addresses, along with
current updates and exercises, log on to

http://www.brookscole.com/geo/

▶ **THE SNAKE RIVER PLAIN AND THE YELLOWSTONE HOT SPOT**

This site has explanations for the volcanism in the Snake River Plain and the Yellowstone National Park, Wyoming region. Maps, images, and a glossary of terms are helpful in understanding the geology of this region. The site also has a link to Yellowstone National Park where one can see good images of present-day hydrothermal processes and features.

1. What surface features indicate some source of heat beneath the surface in this region?

2. What types of rocks make up the Snake River Plain? How do the ages of the rocks vary within the area? Give an explanation to account for the age distribution.

3. What types of magma were erupted in the Yellowstone caldera area during the last few millions of years? How does the volume of these eruptions compare with that of Mt. St. Helens, Washington, in 1980?

▶ **COLUMBIA RIVER GORGE**

Another site maintained by the U.S. Geological Survey, Cascades Volcano Observatory, Vancouver, Washington. The site has good descriptions, images, and maps of the volcanic rocks of the Columbia River Plateau as well as links to other related sites. Click on *REPORT: The Geology of the Columbia River Gorge,* then click *Beacon Rock,* and then *Description.* What is Beacon Rock and what is its age? Why are there so few volcanic cones in the Columbia River Plateau? Return to the main page and click *Basalt floods,* which takes you to a site titled *Lava Plateaus and Flood Basalts.*

1. What is the composition of the flows of the Columbia River Plateau?

2. How thick are the flows here and what is their age?

3. Read about the Missoula floods and explain what impact they had on the Columbia River basalts.

▶ **BRYCE CANYON NATIONAL PARK**

The National Park Service maintains this site about Bryce Canyon National Park, Utah. It has images of rocks in the park as well as descriptions of the park's features, natural history, and cultural history. Scroll down and read the section on *geology.* See the images of present-day erosion features. Also, what rock unit is presently being eroded, what is it composed of, and when and how was it deposited?

▶ **CALVERT CLIFFS OF MARYLAND**

This site, maintained by the Maryland Geological Survey, consists of an article by Jeanne D. McLennan in which the rocks, fossils, and geologic history of the Calvert Cliffs area are described. It also has images of the cliffs and fossils.

1. What three formations make up the Chesapeake Group, and what is each composed of, and what are their ages?

2. What evidence indicates these formations were deposited in a marine environment?

▶ **THE CENOZOIC ERA**

This site is part of the "Continents, Oceans, and Life—A New View of . . . The Third Planet," which is maintained by the Milwaukee Public Museum. Take the tour through *Grasslands of Western Nebraska, The Glacier's Edge in Wisconsin 12,000 Years Ago* and others to see what the environment and life was like during the Cenozoic Era.

▶ **UC BERKELEY MUSEUM OF PALEONTOLOGY**

The University of California at Berkeley maintains this huge site with information on a variety of topics. On the main page click *Time Periods,* which brings up a geologic time scale. On the time scale click *Cenozoic Era* and see why the Cenozoic is sometimes called the Age of Mammals. What are the two main subdivisions of the Cenozoic? Click *Stratigraphy* at the bottom of the page. Read about and explain how the Tertiary and Quaternary Periods were named. Also, click *Localities.* What is the age of the fossils in the La Brea Tar Pits? How and when was the Monterey Formation deposited?

CHAPTER

17

Cenozoic Geologic History: Quaternary Period

Several valley glaciers are visible on the south face of Denali (Mount McKinley), Alaska. During the Pleistocene, valley glaciers were more numerous than at present and continental glaciers covered vast areas, especially in the Northern Hemisphere.

Prologue

Since the end of the Great Ice Age about 10,000 years ago, Earth has experienced several climatic fluctuations. During the Holocene Maximum about 6000 years ago, temperatures were slightly warmer on average than at present, and some arid regions such as the Sahara Desert of North Africa supported lush vegetation, swamps, and lakes. Then followed a time of cooler temperatures, but from about A.D. 1000 to 1300, Europe experienced what is known as the Medieval Warm Period during which wine grapes were grown more than 480 km north of their present limits. However, a cooling trend beginning about 1300 led to the **Little Ice Age,** a time of expansion of glaciers in mountain valleys, overall colder winters, cooler, wetter summers, and the persistence of sea ice at high latitudes for long periods.

During the preceding Medieval Warm Period, the climates of Europe and North America were rather mild, and the North Atlantic Ocean was warmer and more storm free than it is now. Taking advantage of the mild conditions, the Norse, commonly called the Vikings, discovered and settled Iceland, and by A.D. 1200 about 80,000 people were living there. They also sailed to Greenland and North America where they established two colonies on the former and one on the latter. Unfortunately for these settlers, the climate deteriorated, the North Atlantic became stormier, and sea ice moved farther south and persisted longer each year. The poor sea conditions coupled with political problems in Norway brought a halt to all shipping across the North Atlantic, and the colonies in Greenland and North American eventually disappeared.

The Little Ice Age was also a time of expansion of small glaciers in mountain valleys. In Europe and Iceland, glaciers moved far down their valleys, reaching their greatest historic extent by the early 1800s. In 1744, one observer in Norway wrote: "When at times [the glacier] pushes forward a great sound is heard, like that of an organ and it pushes in front of it unmeasurable soil, grit and rocks bigger than any house could be, which it then crushes small like sand."* A small ice cap formed in Iceland where none had existed previously, and glaciers in Alaska, Canada, and the mountains of the western United States expanded to their greatest limits during historic time. During the early 1600s, people in the European Alps watched as

FIGURE 17.1   During the Little Ice Age, many of the glaciers in Europe, such as this one in Switzerland, extended much farther down their valleys than they do at present. This view titled *The Unterer Grindelwald* was painted in 1826 by Samuel Birmann (1793–1847).

the Mer de Glace (Sea of Ice Glacier) advanced and overwhelmed farms and villages. Elsewhere glaciers advanced across roadways and pastures, destroyed some villages in Scandinavia, and threatened other villages.

Advancing glaciers caused some problems, but their overall impact on humans was minimal. Far more important from the human perspective was that, during much of the Little Ice Age, summers in the Northern Hemisphere were cooler and wetter. Particularly hard hit were the Scandinavian countries and the Norse colonies in Greenland and Iceland, but at times much of Northern Europe was affected (Figure 17.1). The harsher summer weather resulted in shorter growing seasons, reduced crop yields, food shortages, and a number of famines in Iceland, Scotland, and mainland Europe. Iceland's population was down by half to about 40,000 by 1700. Between 1610 and 1870, sea ice was seen near Iceland for as much as three months a year, and each year the sea ice persisted for long periods, poor growing seasons and food shortages followed.

Although cool, wet summers caused most of the problems, winter temperatures were colder too, and sea ice was a problem in the North Atlantic. Occasionally, Eskimos paddled their kayaks as far south as Scot-

---

*Quoted in C. Officer and J. Page, *Tales of the Earth* (New York: Oxford University Press, 1993), p. 99.

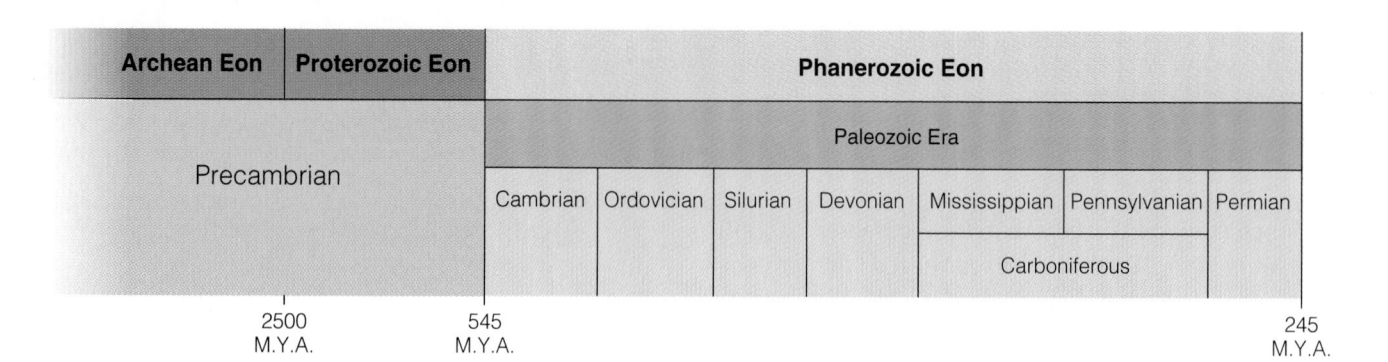

Archean Eon	Proterozoic Eon	Phanerozoic Eon						
		Paleozoic Era						
Precambrian		Cambrian	Ordovician	Silurian	Devonian	Mississippian	Pennsylvanian	Permian
						Carboniferous		

2500 M.Y.A. 545 M.Y.A. 245 M.Y.A.

land, and the canals of Holland froze over during some winters. In 1607 the first "Frost Fair" was held in London on the River Thames, which had begun to freeze over nearly every winter. Even New York harbor in the newly founded United States froze over. And 1816 is known as the "year without a summer," when unusually cold temperatures persisted into June and July in New England and northern Europe; both areas experienced frost and even some snow during these months. Eruptions of Tambora in Indonesia in 1815 and Mayon Volcano in the Philippines in 1814 contributed to the cool spring and summer of 1816.

Exactly when the Little Ice Age ended is debatable. Some authorities put the end at 1880, whereas others think it ended as early as 1850. In either case, during the mid- to late-1800s, sea ice was retreating northward, glaciers began retreating back from their valleys, and summer weather became more moderate.

Introduction

The most recent 1.6 million years of geologic time is designated the **Quaternary Period,** which is divided into two epochs, the **Pleistocene** (1.6 million to 10,000 years ago) and the **Holocene** (or Recent). From the perspective of 4.6 billion years, this period is but a mere moment in geologic time. In previous chapters we used a 24-hour day to represent all geologic time (see Figure 8.2), and noted, for instance, that the entire Cenozoic Era makes up only 1.4% of this 24-hour day, or slightly more than 20 minutes. The Quaternary constitutes only 38 seconds of our 24-hour day! Nevertheless, from our perspective, these last 38 seconds are extremely important. Not only did our own species (*Homo sapiens*) evolve, but the Pleistocene was also one of the few times during all of geologic time that widespread glaciation took place. Extensive ice sheets, particularly on the Northern Hemisphere continents, advanced and retreated over the land surface, directly and indirectly forming much of the present-day topography of the glaciated and nearby regions. Furthermore, the influence of glaciers was felt far beyond the areas covered by ice, because precipitation was greater during times when glaciers were present. Thus, areas such as Sahara Desert of North Africa supported lush vegetation, streams, and lakes.

It now seems hard to believe that so many competent naturalists of the last century were skeptical that widespread glaciation had occurred on the northern continents during the recent past. Many naturalists invoked the biblical flood to explain the large boulders found throughout Europe far from their source. Others believed the boulders were rafted to their present location by icebergs floating in floodwaters. It was not until 1837 that the Swiss naturalist Louis Agassiz argued that the large displaced boulders (called *erratics*), coarse-grained sedimentary deposits, polished and striated bedrock, and U-shaped valleys found throughout parts of Europe were the result of huge ice masses moving over the land. Based on Agassiz's observations and arguments, geologists soon accepted that an ice age had indeed occurred during the recent geologic past.

We know today that the last Ice Age began about 1.6 million years ago and consisted of several intervals of

Phanerozoic Eon										
Mesozoic Era			Cenozoic Era							
Triassic	Jurassic	Cretaceous	Tertiary						Quaternary	
			Paleocene	Eocene	Oligocene	Miocene	Pliocene		Pleistocene	Holocene

245
M.Y.A.

66
M.Y.A.

glacial expansion separated by warmer interglacial periods. Based on the best available evidence, it appears that the present interglacial period began about 10,000 years ago. Geologists do not know, however, whether we are still in an interglacial period or are entering another colder glacial interval.

THE ONSET OF THE ICE AGE

The onset of glacial conditions really began about 40 million years ago when surface ocean waters at high southern latitudes rapidly cooled, and the water in the deep-ocean basins soon cooled to about 10° C colder than it had been previously. The gradual closing of the Tethys Sea during the Oligocene further limited the flow of warm water to higher latitudes. By the Middle Miocene, an Antarctic ice cap had formed, accelerating the formation of very cold waters. Following a brief warming trend during the Pliocene, ice sheets began forming in the Northern Hemisphere about 1.6 million years ago, and the Pleistocene Ice Age was underway (Table. 17.1).

Pleistocene Paleogeography and Climate

PALEOGEOGRAPHY

While the Pleistocene Epoch is well known for glaciation, it was also a time of tectonic unrest. Folding, faulting, and uplifts were common occurrences, resulting from Pacific and North American plate interaction along the San Andreas fault system. Many of California's coastal oil- and

---

TABLE 17.1

Summary of Events Leading to the Pleistocene Ice Age

TIME (MILLIONS OF YEARS AGO)	EVENTS
1.6	Ice Age begins. Ice sheets form in Northern Hemisphere.
5–3	Gulf Stream intensified. Isthmus of Panama closes.
6	Rapid cooling occurs.
15	Present abyssal oceanic circulation established. Iceland–Faroe Ridge sinks; North Atlantic water wells up around Antarctica; more moisture and copious snowpack. Antarctic ice cap develops.
30–25	Antarctica much colder but still no ice cap. Circum-Antarctic current established.
35–30	Partial circulation around Antarctica. Barriers in the western Pacific and closing of Tethys Sea in Middle East and Near East breaks up circum-equatorial circulation pattern.
38	First glaciers on Antarctica. Deep-water circulation speeds up. Rapid cooling of surface water in the south and of deep water everywhere.
48–45	Slight cooling takes place.
>50	No glaciers or ice cap on Antarctica. Deep-water much warmer than at present; circulates slowly. Uniform climate and a warm ocean. Ocean waters flow freely around the world at the equator.

FIGURE 17.2 Marine terraces on the west side of San Clemente Island, California. Each terrace represents a period when that area was at sea level. The highest terrace is now about 400 m above sea level. *(Photo © John Shelton.)*

gas-producing fold structures formed during the Pleistocene. Marine terraces covered with Pleistocene sediments along the Pacific Coast testify to repeated uplifts during this time (Figure 17.2).

Pleistocene tectonic activity took place in the Sierra Nevada in California, the Teton Range of Wyoming, and portions of the central and northern Rocky Mountains as well as in the Alpine–Himalayan orogenic belt (see Chapter 16). Many of these uplifts, particularly the Sierra Nevada, influenced weather conditions by creating rain shadows that led to present-day deserts on the landward sides of these mountains.

Much of the Pleistocene volcanic activity in the western part of the United States and Mexico resulted from continued subduction along the western continental margin of North America. The Cascade Range of California, Oregon, Washington, and British Columbia (see Perspective 16.2), Yellowstone National Park, Wyoming, and Craters of the Moon National Monument, Idaho (Figure 17.3), are three

well-known areas of North American Pleistocene volcanism. Elsewhere, volcanic activity was occurring around the rim of the Pacific Ocean basin—along South America's western margin and in Japan, the Philippines, and the East Indies—and in other locations such as Iceland, Spitzbergen, and the Himalayas as well.

CLIMATE

The climatic effects responsible for Pleistocene glaciation were, as one would expect, worldwide. Nevertheless, the world was not as frigid as it is commonly portrayed in cartoons and movies, nor was the onset of glacial conditions as rapid as many people believe. In fact, evidence from various lines of research indicates that the world's climate gradually cooled from the beginning of the Eocene through the Pleistocene (Figure 17.4). Oxygen isotope data (the ratio of O^{18} to O^{16}) from deep-sea cores reveal that during the past 2 million years there were at least 20 major warm–cold cycles in which the world's mean temperature ranged from 6°C to 10°C. Furthermore, evidence from glacial deposits indicates at least four major episodes of Pleistocene glaciation in North America and six or seven major glacial advances and retreats in Europe.

During times of glacier growth, those areas in the immediate vicinity of the glaciers experienced short summers and long, wet winters. Areas outside the glaciated regions experienced varied climates. During times of glacial growth, lower ocean temperatures reduced evaporation so that most of the world was drier than it is today. Some areas that are arid today, however, were much wetter during times of glacial growth. For example, the expansion of the cold belts at high latitudes caused the temperate, subtropical, and tropical zones to be compressed toward the equator, and the rain that now falls on the Mediterranean shifted so that it fell on the Sahara of North Africa, enabling lush forests to grow in what is now desert. California and the arid Southwest were also wetter because a high-pressure zone over the northern ice sheet deflected Pacific winter storms southward.

Pollen analysis is an especially useful tool in determining past climates (Figure 17.5). Produced by the male reproductive bodies of seed plants, pollen grains have a resistant waxy coating that enables them to be easily preserved as microfossils. Because of its structure, most pollen is transported by wind and eventually deposited in streams, lakes, swamps, bogs, or the nearshore marine realm where it is incorporated into the accumulating sediments. Once the pollen is recovered from sediment samples, the family or type of plants that produced it can usually be identified. By counting and statistically analyzing the types of pollen present, scientists can determine the general floral composition of the area and make some general climatic interpretations as well.

Pollen diagrams (Figure 17.5), tree-ring analysis (see Chapter 2), and measurements of the fluctuations of mountain glaciers have produced a wealth of data for determin-

FIGURE 17.3 Craters of the Moon National Monument in Idaho is one area of Pleistocene and Holocene volcanism. This is a view of a spatter cone on the surface of a lava flow.

ing the climatic history of the Northern Hemisphere since the last major ice sheets retreated 10,000 years ago. Pollen data indicate a continuous trend toward warmer, more temperate climatic conditions until around 6000 years ago. In many areas the interval between 8000 and 6000 years ago was a time of very warm temperatures. After this warm period, conditions gradually became cooler and moister, favoring the growth of valley glaciers on the Northern Hemisphere continents. Careful studies of the deposits along the margins of present-day glaciers reveal that glaciers expanded several times during the last 6000 years (a time called the *Neoglaciation*). The last expansion, which took place between 1500 and the mid- to late-1800s, was the *Little Ice Age* (see the Prologue).

This brief overview of climatic conditions reveals just how complex climatic patterns were during the last 1.6 million years. Intermediate fluctuations were superimposed on large-scale warm–cold cycles, and those fluctua-tions in turn were composed of even shorter variations, such as the Little Ice Age, which had profound effects on the social and economic fabric of human society.

Pleistocene and Holocene Geologic History

Geologists still debate which section of rocks should serve as the Pleistocene stratotype, but the currently accepted date for the beginning of the Pleistocene Epoch is 1.6 million years ago. The Pleistocene-Holocene boundary, placed at 10,000 years ago, is based on a climatic change from cold to warmer conditions corresponding to the melting of the most recent ice sheets. This climatic change is well established by pollen and vegetation changes and variations in the ratio of O^{18} to O^{16} as preserved in the shells of various marine organisms.

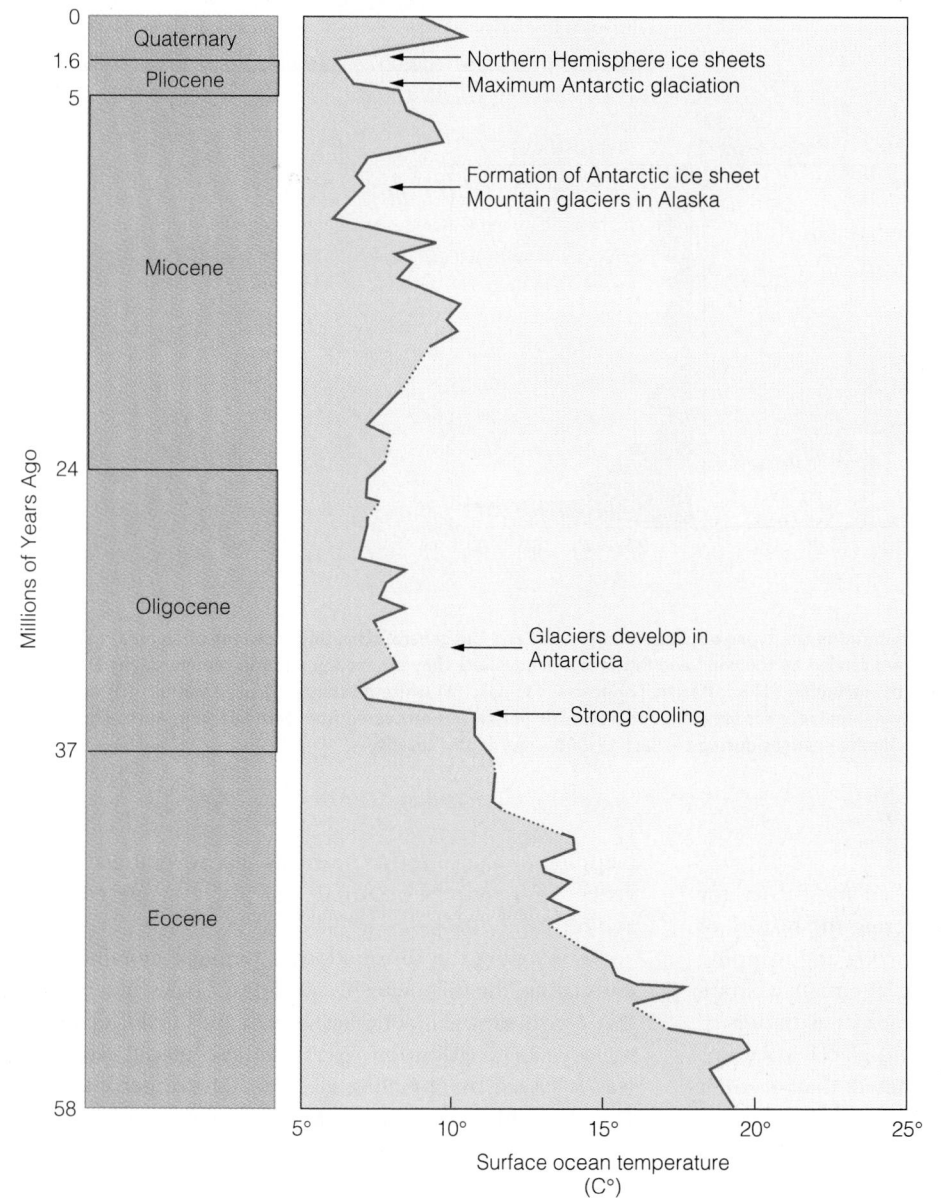

FIGURE 17.4 Fluctuations in the ratio of O^{18} to O^{16} isotopes from a sediment core in the western Pacific Ocean reveal changes in surface ocean temperatures during the past 58 million years. A change from warm surface waters to colder conditions occurred around 35 million years ago.

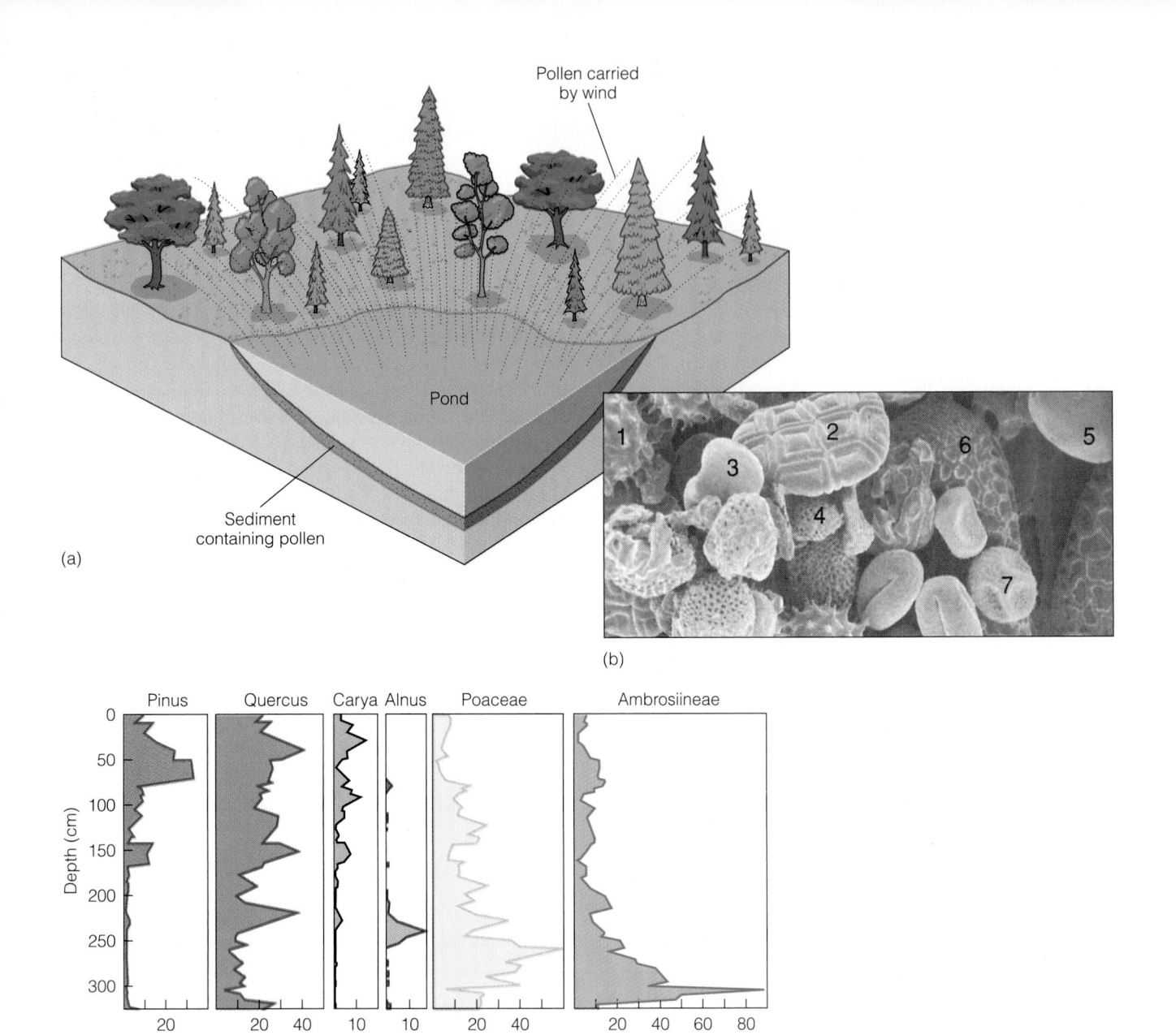

FIGURE 17.5 Geologists use fossil pollen to determine the types of vegetation present and the general climatic conditions of an area. (a) Pollen grains from a variety of nearby sources are carried by the wind and fall into a pond where they are incorporated in sediment. (b) Scanning electron micrograph of present-day pollen grains, including (1) sunflower, (2) acacia, (3) oak, (4) white mustard, (5) little walnut, (6) agave, and (7) ash juniper. (c) Pollen diagrams showing abundance for six different trees. The pollen was recovered from Ferndale Bog, Atoka County, Oklahoma. Changes in pollen abundance show climate changes during the last 12,000 years at this locality.

TERRESTRIAL STRATIGRAPHY

Shortly after Louis Agassiz first proposed his theory for glaciation, research focused on deciphering the history of the Ice Age. This work involved recognizing and mapping terrestrial glacial features and placing them in a stratigraphic sequence. From such glacial features as the distribution of moraines, erratic boulders, and glacial striations (Figure 17.6), geologists have determined that at their greatest extent Pleistocene glaciers covered about three times as much of Earth's surface as they do now and were up to 3 km thick (Figure 17.7). Furthermore, detailed mapping of glacial features reveals that several glacial advances and retreats occurred, not just a single advance and retreat.

By mapping the distribution of terminal moraines and correlating the till deposits, geologists have determined that North America alone has at least four major episodes of Pleistocene glaciation. Each of these glacial advances was followed by retreating glaciers and warmer climates. The four **glacial stages,** the *Wisconsinan, Illinoian, Kansan,* and *Nebraskan,* are named for the states representing the farthest advance where deposits are well ex-

(a)

(b)

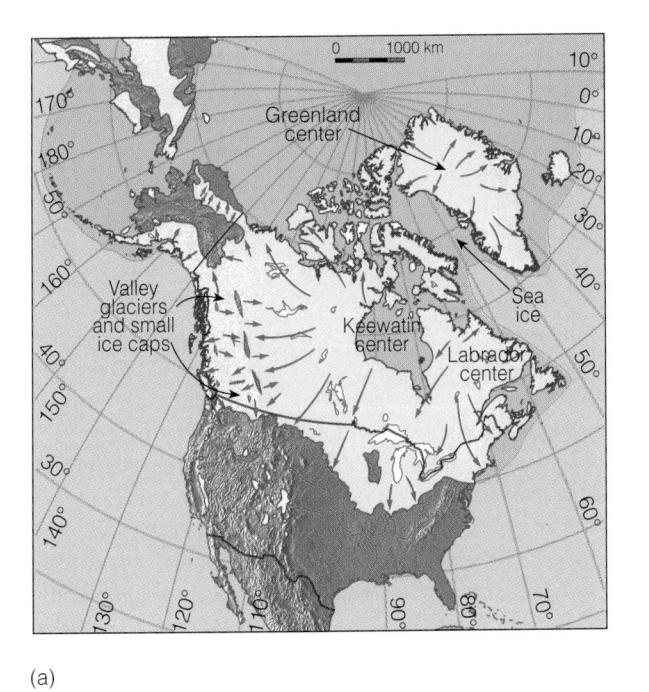

(c)

FIGURE 17.6 Features characteristic of glaciated areas. (a) Glacial till is unsorted sediment deposited by a glacier. Till from the Mount cook area, South Island, New Zealand. (b) Glacial striations and polish on rocks near Marquette, Michigan. (c) This glacial erratic, an ice-transported boulder far from its source, is in Indiana.

(a)

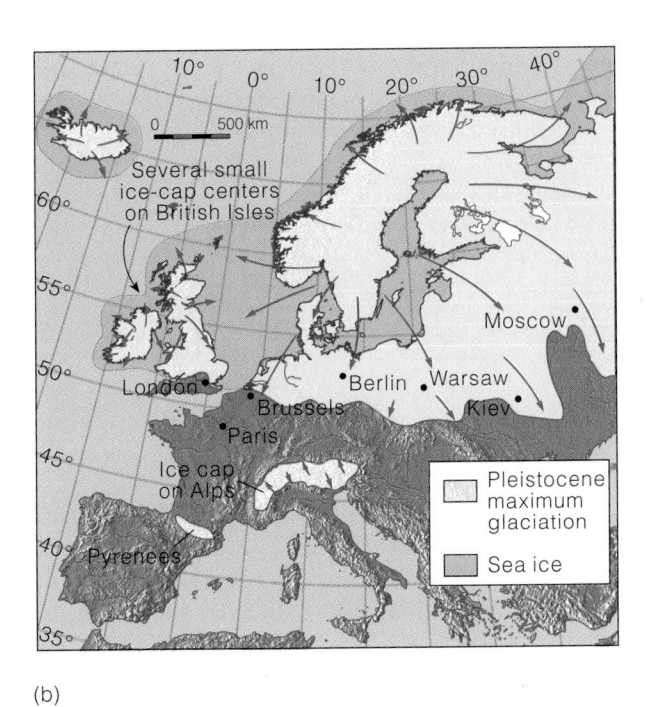

(b)

FIGURE 17.7 (a) Centers of ice accumulation and maximum extent of Pleistocene glaciation in North America. (b) Centers of ice accumulation and directions of ice movement in Europe during the maximum extent of Pleistocene glaciation.

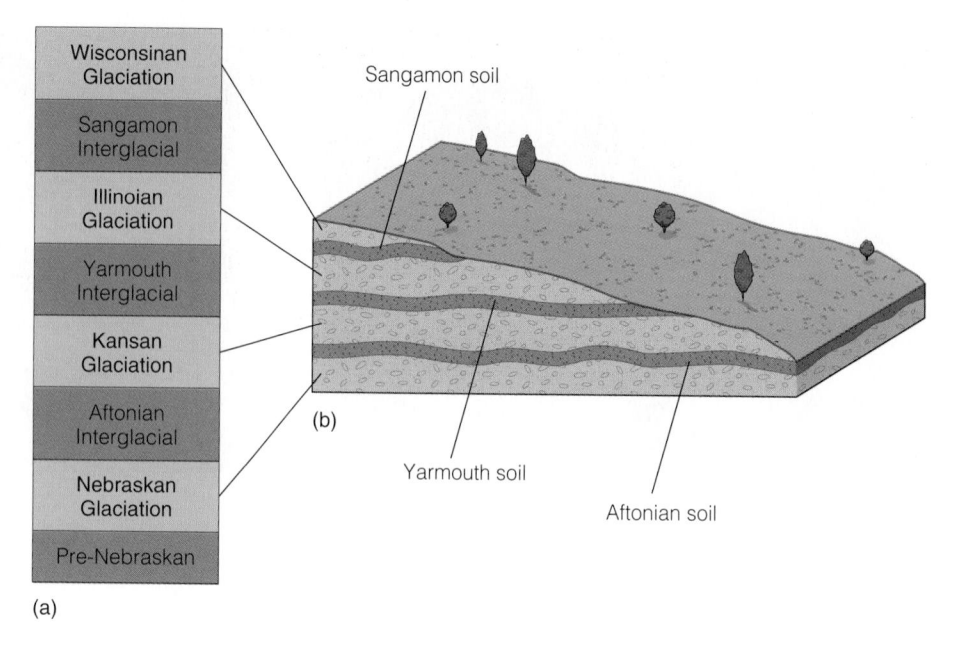

FIGURE 17.8 (a) Standard terminology for Pleistocene glacial and interglacial stages in North America. (b) An idealized succession of deposits and soils developed during the glacial and interglacial stages.

(a)

| Wisconsinan Glaciation |
| Sangamon Interglacial |
| Illinoian Glaciation |
| Yarmouth Interglacial |
| Kansan Glaciation |
| Aftonian Interglacial |
| Nebraskan Glaciation |
| Pre-Nebraskan |

Sangamon soil

(b)

Yarmouth soil

Aftonian soil

posed. The three **interglacial stages,** the *Sangamon, Yarmouth,* and *Aftonian,* are named for localities of well-exposed interglacial soil and other deposits (Figure 17.8). Recent detailed studies of glacial deposits indicate, however, that there were an as yet undetermined number of pre-Illinoian glacial events and that the history of glacial advances and retreats in North America is more complex than previously thought. In view of these data, the traditional four-part subdivision of the Pleistocene of North America must be modified.

While at least four major Pleistocene episodes of glaciation are recognized in North America, six or seven major glacial advances and retreats are recognized in Europe, and at least 20 major warm–cold cycles can be detected in deep-sea cores. Why isn't there better correlation among the different areas if glaciation was a worldwide event? Part of the problem is that glacial deposits are typically chaotic mixtures of coarse materials that are difficult to correlate. Furthermore, advances and retreats of the ice sheets usually destroy the sediment left by the previous advance, thus obscuring earlier chronological evidence. Even within a single major glacial advance, several minor advances and retreats may have occurred. For example, careful study of deposits from the Wisconsinan glacial stage reveals that at least four distinct fluctuations of the ice margin occurred during the last 70,000 years in Wisconsin and Illinois.

Geologists using sophisticated radiometric-dating techniques and detailed pollen studies can now construct an accurate and detailed chronology for terrestrial Pleistocene glacial and interglacial episodes. However, because of the complexity of glacial deposits and the difficulty of separating similar-appearing terrestrial sediments, they cannot always recognize small-scale fluctuations in climate and correlate them with deposits in other areas.

PLEISTOCENE DEEP-SEA STRATIGRAPHY

Until recently, the traditional view of Pleistocene chronology was based on terrestrial sequences of glacial sediments. During the early 1960s, however, new evidence from ocean sediment samples indicated that numerous climatic fluctuations had occurred during the Pleistocene. Evidence for these climatic fluctuations comes from changes in surface ocean temperature recorded in the shells of planktonic foraminifera, which sink to the seafloor after they die and accumulate as sediment. These shells can be recovered for study by coring the ocean sediments. After the sediment is washed through screens, the foraminiferal shells are concentrated, the different species are identified, and the coiling directions and O^{18} to O^{16} ratios of the shells are analyzed.

One way to determine past changes in ocean surface temperatures is simply to determine whether the planktonic foraminifera were warm- or cold-water species. Many planktonic foraminifera are sensitive to variations in temperature and migrate to different latitudes when the surface water temperature changes. For example, the tropical species *Globorotalia menardii* is present or absent within Pleistocene cores, depending on what the surface water temperature was at the time. During periods of cooler climate, it is found only near the equator, while during times of warming its range extends into the higher latitudes.

Some planktonic foraminifera species change the direction they coil during growth in response to temperature fluctuations. For example, the Pleistocene species *Globorotalia truncatulinoides* predominantly coils to the right in water temperatures above 10°C but predominantly coils to the left in water below 8°–10°C (Figure 17.9a). On the basis of changing coiling ratios, detailed climatic curves have been constructed for the Pleistocene and earlier epochs (Figure 17.9b).

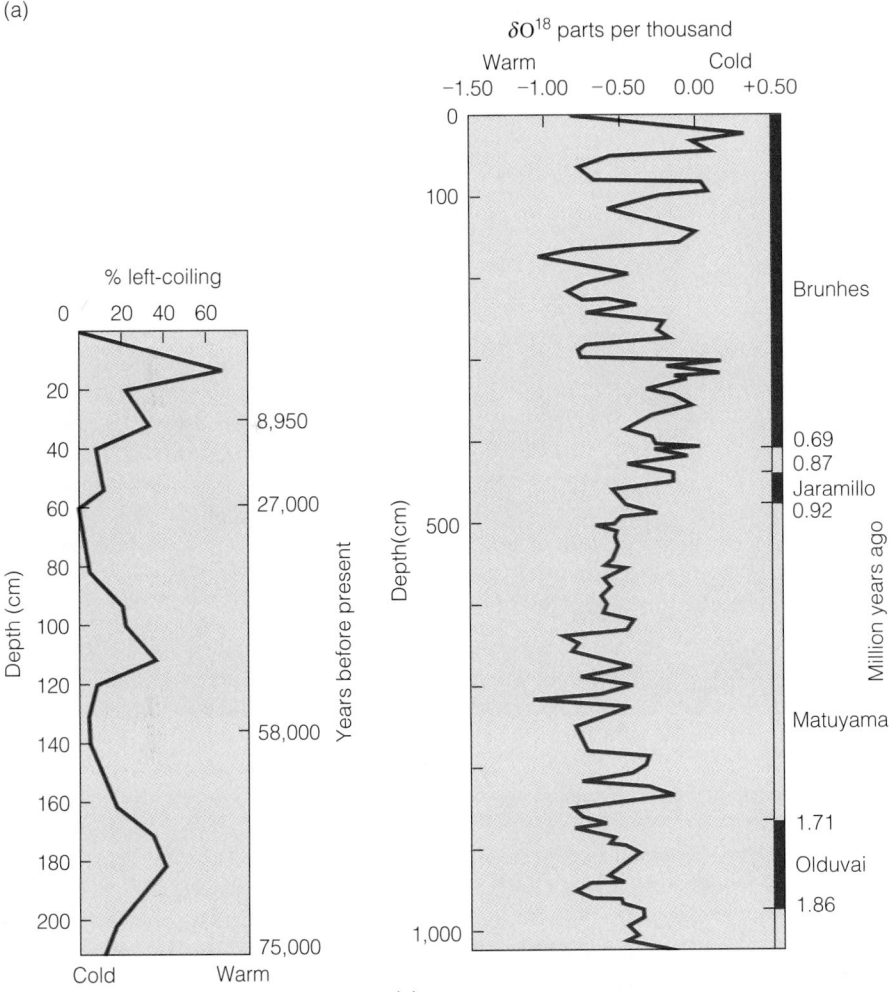

(a)

FIGURE 17.9 Two methods used to determine changes in ocean surface water temperatures. (a) Some planktonic foraminifera species coil in different directions, depending on water temperatures. *Globorotalia truncatulinoides* coils to the left in water temperatures below 8°–10° C. (b) Changes in abundance of coiling direction can be used to determine surface water temperatures and hence oceanic climatic conditions. These data, from a core in the Caribbean, show evidence of three intervals of mild climate during the Wisconsinan glacial stage. (c) Changes in the O^{18} to O^{16} ratio as preserved in planktonic foraminifera shells also reflect surface water temperature fluctuations and thus climatic changes due to glaciation.

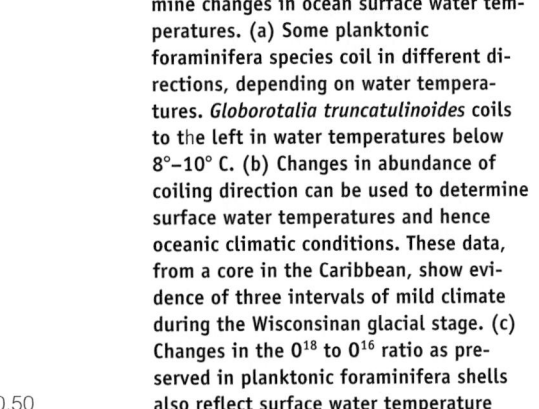

Climatic events can also be determined by analyzing changes in the O^{18} to O^{16} ratio preserved in the shells of planktonic foraminifera. The abundance of these two oxygen isotopes in the calcareous ($CaCO_3$) shells of foraminifera is a function of the oxygen isotope ratio in water molecules and water temperature when the shell is secreted. The ratio of these two isotopes reflects the amount of ocean water stored in glacial ice. Seawater has a higher O^{18} to O^{16} ratio than glacial ice because water containing the lighter O^{16} isotope is more easily evaporated than water containing the O^{18} isotope. Therefore, Pleistocene glaciers are enriched in O^{16} relative to O^{18},

while the heavier O^{18} isotope is concentrated in seawater. The declining percentage of O^{16} and consequent rise of O^{18} in seawater during times of glaciation is preserved in the shells of planktonic foraminifera. Consequently, oxygen isotope fluctuations in planktonic foraminifera accurately indicate surface water temperature changes and thus climatic changes due to glaciation (Figure 17.9c).

Unfortunately, geologists have not yet been able to correlate these detailed climatic changes with corresponding changes recorded in the terrestrial sedimentary record. The time lag between the onset of cooling and the resulting glacial advance produces discrepancies between the marine and terrestrial records. Thus, it is unlikely that all the minor climatic fluctuations recorded in deep-sea sediments will ever be correlated with the continental stratigraphic units.

The Effects of Glaciation

Pleistocene glaciers had many direct and indirect effects. Massive glaciers moving over the surface altered the previously existing topography and yielded a variety of distinctive glacial landforms. As glaciers formed and wasted away, sea level rose and fell; thus, the continental margins were alternately exposed and water covered. Furthermore, cli-

matic changes during times when glaciers were widespread had effects far beyond the glaciated regions. A number of large lakes, for instance, existed in areas—such as Death Valley, California—that are now quite arid. Another legacy of the Pleistocene is that areas once covered by vast glaciers are still responding to isostatic adjustments.

GLACIAL LANDFORMS

As moving solids, glaciers are particularly effective agents of erosion, and they efficiently transport and deposit large amounts of sediment. Accordingly, distinctive landscapes develop in areas over which glaciers have moved. A good example of erosion by continental glaciers is seen in large

FIGURE 17.10 (a) Erosion by continental glaciers formed this ice-scoured plain in the Northwest Territories of Canada. Notice the low relief and extensive bedrock exposures. (b) The large depression known as a cirque on Mt. Wheeler in Great Basin National Park, Nevada, was eroded by a Pleistocene valley glacier.

(a)

(b)

parts of Canada and some northern states where *ice-scoured plains* are present (Figure 17.10a). These regions are characterized by poor surface drainage, numerous lakes and swamps, striated and polished bedrock, low relief, and little or no soil. Erosion by valley glaciers, which were much more common during the Pleistocene, also yielded distinctive landforms such as U-shaped glacial troughs and cirques in the Teton Range of Wyoming, Glacier National Park, Montana, Waterton Provincial Park, Alberta, Canada (see Perspective 9.1), and many other areas (Figure 17.10b).

The most notable deposits of both continental glaciers and valley glaciers are various types of moraines, which are chaotic mixtures of sediment deposited directly by glacial ice. Terminal moraines are present in southern Ohio, Indiana, and Illinois marking the greatest southerly extent of Pleistocene glaciers, whereas recessional moraines mark various positions where the ice front stabilized during a general retreat to the north (Figure 17.11). Moraines deposited by continental glaciers are also present in many other areas, including those responsible for the origin of Cape Cod, Massachusetts (see Perspective 17.1). Valley

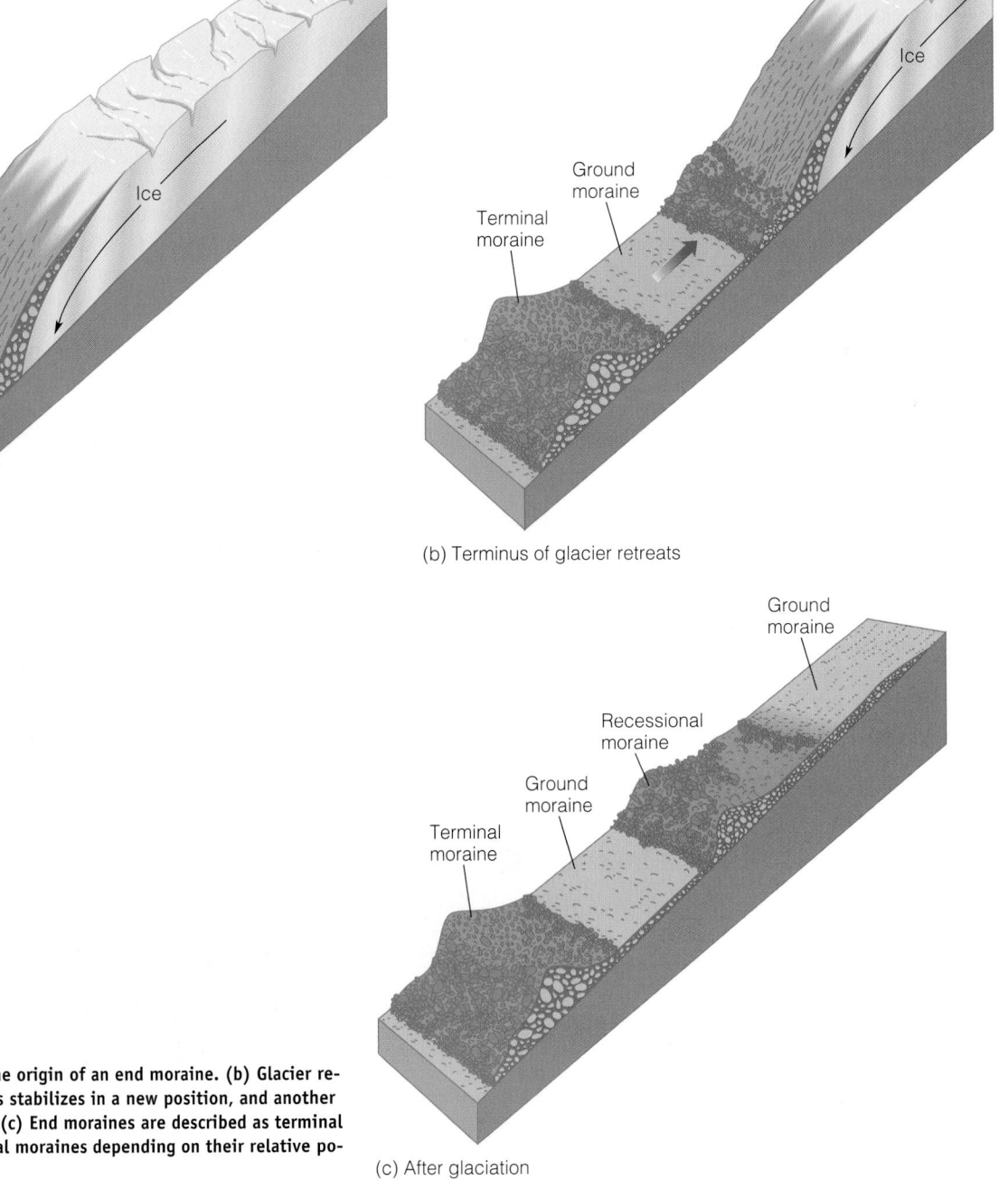

(a) During glaciation

(b) Terminus of glacier retreats

(c) After glaciation

FIGURE 17.11   **(a) The origin of an end moraine. (b) Glacier retreats and its terminus stabilizes in a new position, and another moraine is deposited. (c) End moraines are described as terminal moraines or recessional moraines depending on their relative positions.**

perspective *17.1*

The Geologic History of Cape Cod, Massachusetts

Why do geologists readily embrace the idea that glaciers covered much of the Northern Hemisphere continents in the not-too-distant past? The evidence consists of a combination of features, any one of which taken alone would not be convincing. However, the association of unsorted, unstratified, ridgelike accumulations of debris, coupled with areas of striated and polished bedrock provide compelling evidence. Furthermore, observations of present-day continental glaciers and valley glaciers provide insights on the dynamics of ice movement and erosion, transport, and deposition by moving ice. In short, we are afforded an excellent opportunity to apply the principle of uniformitarianism.

The evidence for Pleistocene and more recent glaciers is found throughout Canada and Alaska and in the entire northern tier of states. For instance, glaciers remain active but diminished in Glacier National Park, Montana, and Waterton Provincial Park, Alberta, Canada (see Perspective 9.1), and the evidence of Pleistocene valley glaciers is obvious in Yosemite National Park, California, and many other areas (see Figure 17.10b). Wisconsin is home to the Ice Age National Scenic Trail where a variety of glacial landforms can be observed. Glacial grooves several meters deep are found in Kelleys Island State Park, Ohio, and striated bedrock is exposed in New York City's Central Park. Also in New York are the scenic Finger Lakes—stream valleys broadened and deepened by glaciers and dammed at their southern ends by moraines.

Another excellent area to observe the effects of glaciation is Cape Cod, Massachusetts, which is designated a National Seashore. Cape Cod resembles a large human arm extending into the Atlantic Ocean from the coast of Massachusetts (Figure 1). It projects 65 km east to the "elbow" and then extends another 65 km north where it resembles a half-curled hand. Cape Cod and the nearby Elizabeth Islands, Martha's Vineyard, and

Nantucket Island all owe their existence to deposition by Late Pleistocene glaciers and the subsequent modification of the deposits by waves. Granite bedrock lies at depths of 90–150 m and forms the foundation on which the cape and nearby islands were built.

Between about 23,000 and 16,000 years ago, during the greatest southward advance of the Cape Cod Bay Lobe of the continental glacier in this region, the ice

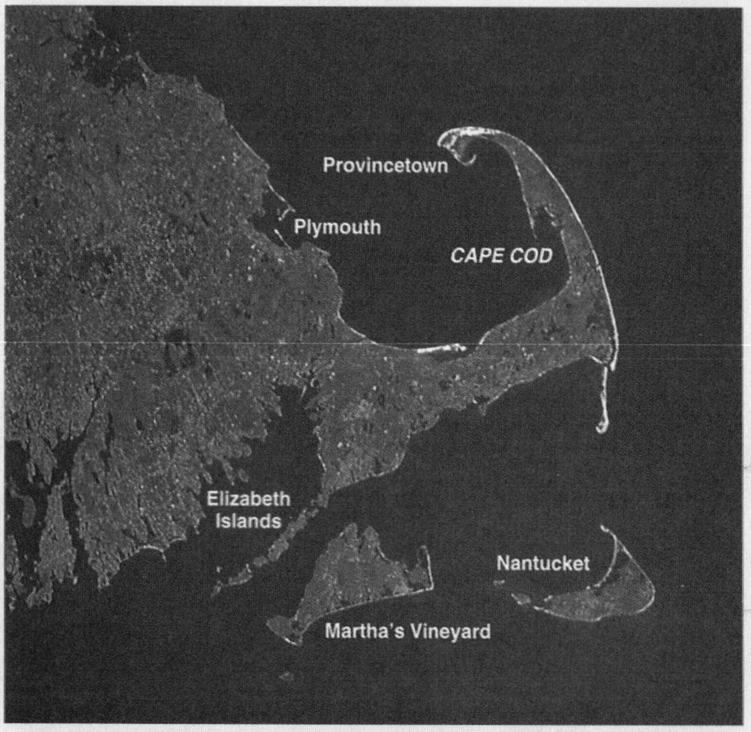

FIGURE 1 Satellite view of Cape Cod and adjacent islands, all of which are composed mostly of sediment deposited by the Cape Cod Bay Lobe of a continental glacier. The dark circular to oval areas on Cape Cod are kettle lakes, or ponds as they are called locally.

front stabilized in the area of present-day Martha's Vineyard and Nantucket Island (Figure 2a). When a glacier's terminus is stabilized in one location, it means the glacier has a balanced budget. That is, additions in its zone of accumulation are balanced by losses in its zone of wastage; thus, its terminus remains stationary but flow continues within the glacier. Accordingly, sediment is transported to the glacier's terminus where it is deposited as an end moraine. This particular end moraine, of which Martha's Vineyard and Nantucket Island are composed, is also called a terminal moraine because it is the most southerly of all the moraines deposited by this glacier.

As the climate became warmer, the Cape Cod Bay Lobe of this vast glacier had a negative budget and began retreating north. About 14,000 years ago, its terminus again became stabilized in the area of present-day Cape Cod and the Elizabeth Islands where it deposited a large recessional moraine (Figure 2b). Thus, it had a balanced budget and as before deposited debris at its terminus. The ice front began retreating again; as it did so, meltwater trapped between the ice front and recessional moraine formed Glacial Lake Cape Cod covering about 1000 km^2 (Figure 2c). Deposits of mud, silt, and sand accumulated in this lake, and these, too, make up part of Cape Cod.

Other distinctive glacial features on Cape Cod are its numerous circular to oval

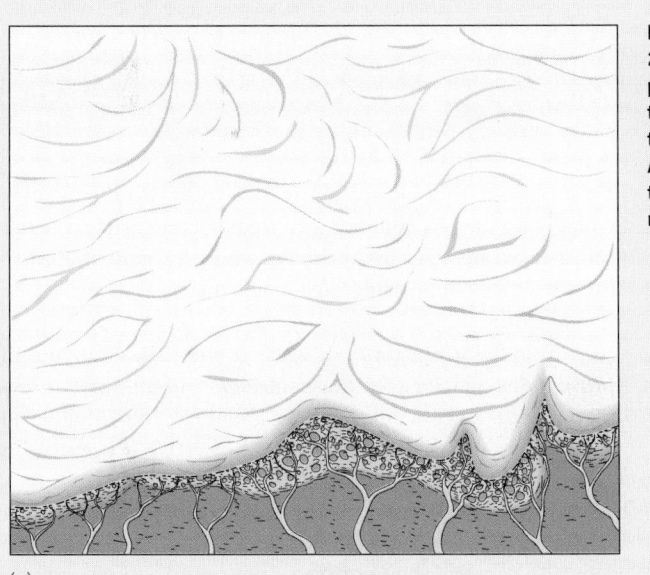

(a)

FIGURE 2 (a) Position of the Cape Cod Bay Lobe of glacial ice 23,000 to 16,000 years ago when its terminal moraine was deposited. Martha's Vineyard and Nantucket Island are composed of this moraine material. (b) Position of the Cape Cod Bay Lobe when the recessional moraine was deposited that now forms Cape Cod. (c) About 5000 years ago, rising sea level covered the lowlands between the moraines and beaches and sand spits formed on the cape and its nearby islands.

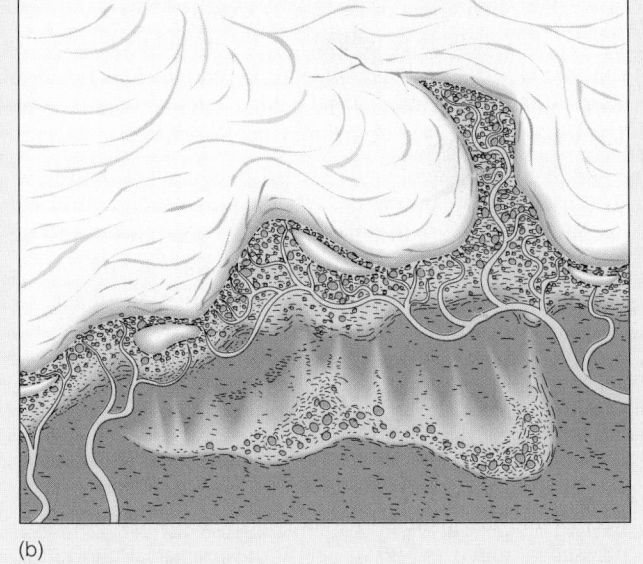

(b)

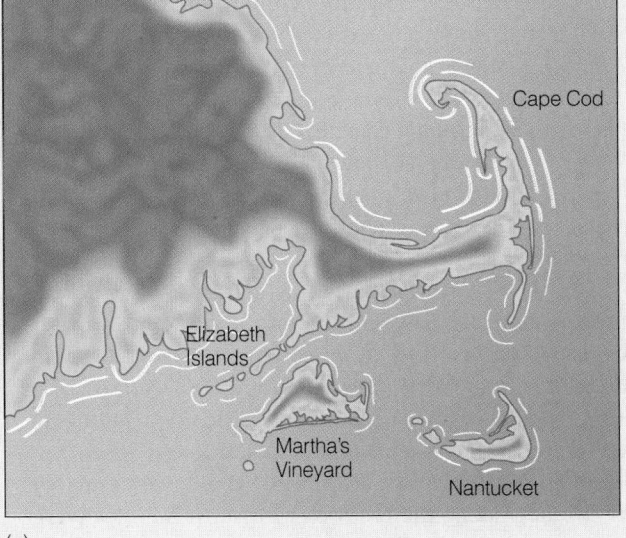

(c)

ponds. These ponds occupy kettles that formed when large blocks of glacial ice were partly or completely buried by outwash deposits. When these ice masses finally melted, they left depressions measuring up to 0.8 km across that filled with water when the water table rose as a result of rising sea level.

When the Cape Cod Bay Lobe of the continental glacier withdrew entirely from this region, Cape Cod looked much different than it does now. On its east side were several headlands and embayments, but by 6000 years ago sea level had risen enough so that waves began smoothing the shoreline by redistributing the sediment. Waves eroded the headlands and deposited sediment in the embayments, and the shoreline has eroded from 1 to 4 km landward from its former location. Even today many steep wave-cut cliffs are present, and those facing east are eroding at nearly 1 m per year, mostly during storms. Redistribution of sand by waves has yielded several baymouth bars and spits such as the ones extending south from the "elbow" and from the north end of the cape (Figure 1).

Native Americans lived on Cape Cod for thousands of years before Europeans arrived. Unfortunately, by 1764 these earliest inhabitants had nearly ceased to exist, mostly because of diseases. The first European settlers in this region were the Pilgrims who first landed on Cape Cod near what is now Provincetown, not at Plymouth Rock despite the persistent myth that this was their first landfall.

glaciers also deposit terminal and recessional moraines, but theirs are not nearly as extensive as those of continental glaciers. Streams discharging from melting glaciers carry considerable sediment that is eventually deposited as *outwash* consisting of layers of sand and gravel.

CHANGES IN SEA LEVEL

More than 70 million km³ of snow and ice covered the continents during the maximum extent of Pleistocene glaciers. This had a major effect not only on the glaciated areas but also on regions far removed from the glaciers. The storage of ocean waters in glaciers lowered sea level 130 m and exposed large areas of the present-day continental shelves, which were soon blanketed by vegetation. Indeed, a land bridge existed across the Bering Strait from Alaska to Siberia. Native Americans crossed the Bering land bridge, and various animals migrated between the continents; the American bison, for example, migrated from Asia. The British Isles were connected to Europe during the glacial intervals because the shallow floor of the North Sea was above sea level. When the glaciers melted, these areas were again flooded, drowning the plants and forcing the animals to migrate inland.

Lowering of sea level caused by glacier growth during the Pleistocene also affected the base level of most major streams. When sea level dropped, streams eroded downward as they sought to adjust to a new lower base level. Stream channels in coastal areas were extended and deepened along the emergent continental shelves. When the glaciers melted and sea level rose, the lower ends of stream valleys along the East Coast of North America were flooded and are now important harbors, while just off the West Coast, they form impressive submarine canyons. Great amounts of sediment eroded by the glaciers were transported by streams to the sea and thus contributed to the growth of submarine fans along the base of the continental slope.

A tremendous quantity of water is still stored on land in present-day glaciers. If these glaciers should melt completely, sea level would rise by about 70 m, flooding all coastal areas where many of the world's large population centers are located (Figure 17.12).

GLACIERS AND ISOSTASY

The fact that Earth's crust floats, in a manner of speaking, in the more dense mantle below is well established in geology. This phenomenon, known as **isostasy,** is easy to understand by analogy to an iceberg. Ice is slightly less dense that water, so an iceberg sinks to its equilibrium position with only about 10% of its volume above water level. Likewise, Earth's crust sinks into the mantle to an equilibrium position and remains at that level unless additional mass is added or subtracted. When enough sediment or ice accumulates in a particular area, the crust responds by gradually subsiding under the weight of the load. Conversely, if erosion of sediment or melting of ice removes the load, the crust responds by slowly rising. No one doubts that this phenomenon, known as **isostatic rebound,** has taken place in areas formerly covered by continental glaciers.

When the Pleistocene ice sheets formed and increased in size, the weight of the ice caused the crust to slowly subside deeper into the mantle. In some places, the surface was depressed as much as 300 m below the preglacial elevations.

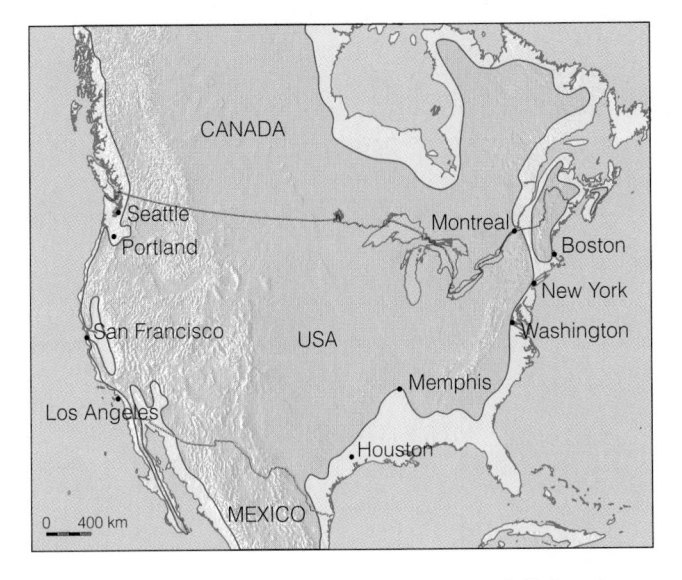

FIGURE 17.12 Large areas of North America—and of all the other continents—would be flooded by a 70-m rise in sea level that would result if all Earth's glacial ice were to melt.

As the ice sheets retreated by melting, the downwarped areas gradually rebounded to their former positions. Evidence of isostatic rebound can be found in such formerly glaciated areas as parts of Scandinavia and the North American Great Lakes region (Figure 17.13). In Canada, for instance, the land has risen as much as 100 m during the last 6000 years

as a result of isostatic rebound, and coastal cities in Scandinavia have been uplifted so that docks built only several centuries ago are now far from the shore.

PLUVIAL AND PROGLACIAL LAKES

The climate of the present-day southwestern United States is hot and dry, and the area has few major streams or lakes. During the Winconsinan glacial stage, however, many large lakes existed in what are now dry basins. These lakes formed as a result of greater precipitation and overall cooler temperatures (especially during the summer), which lowered the evaporation rate. At the same time, increased precipitation and runoff helped maintain high water levels. Lakes that formed during those times are called **pluvial lakes,** and they existed when glaciers were present elsewhere (Figure 17.14a). Ancient shoreline features such as wave-cut cliffs, beaches, and deltas found today on the sides of the enclosing slopes are evidence of these former lakes (Figure 17.14b). The largest of these lakes was Lake Bonneville, which attained a maximum size of 50,000 sq km² and a depth of at least 335 m. The vast salt deposits of the Bonneville Salt Flats west of Salt Lake City, Utah, formed as parts of this ancient lake dried up; Great Salt Lake is simply the remnant of this once vast lake.

Lake Manly, another large pluvial lake, existed in Death Valley, California, which is now the hottest, driest place in North America. During the Pleistocene, however, that area received enough rainfall to maintain a lake 145 km long

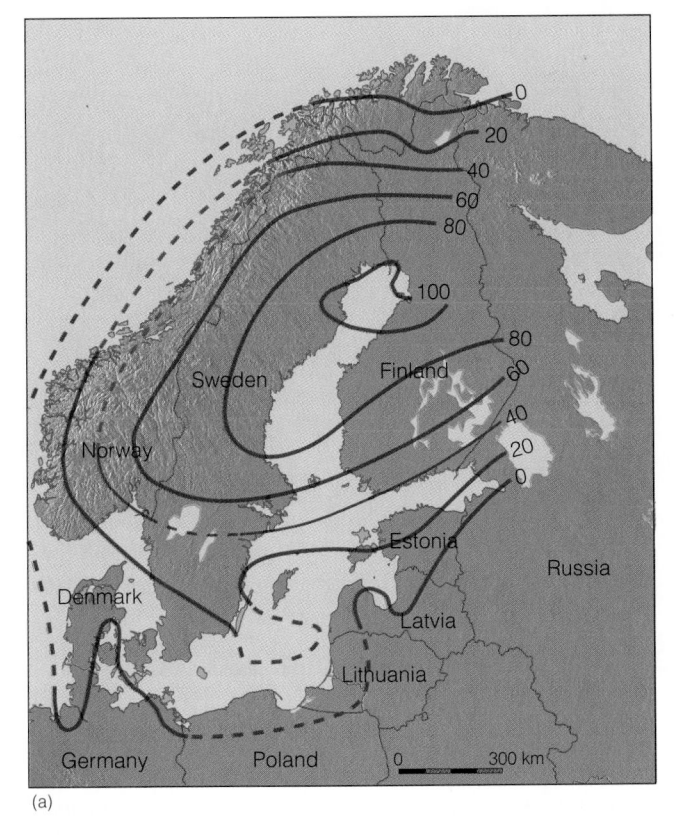

(a)

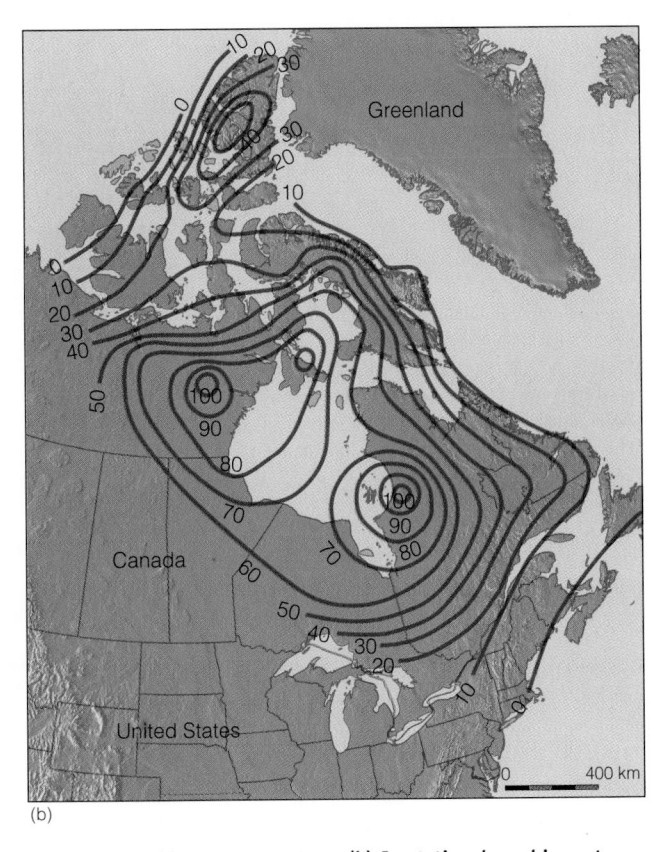

(b)

FIGURE 17.13 (a) Isostatic rebound in Scandinavia. The lines show rates of uplift in centimeters per century. (b) Isostatic rebound in eastern Canada in meters during the last 6000 years.

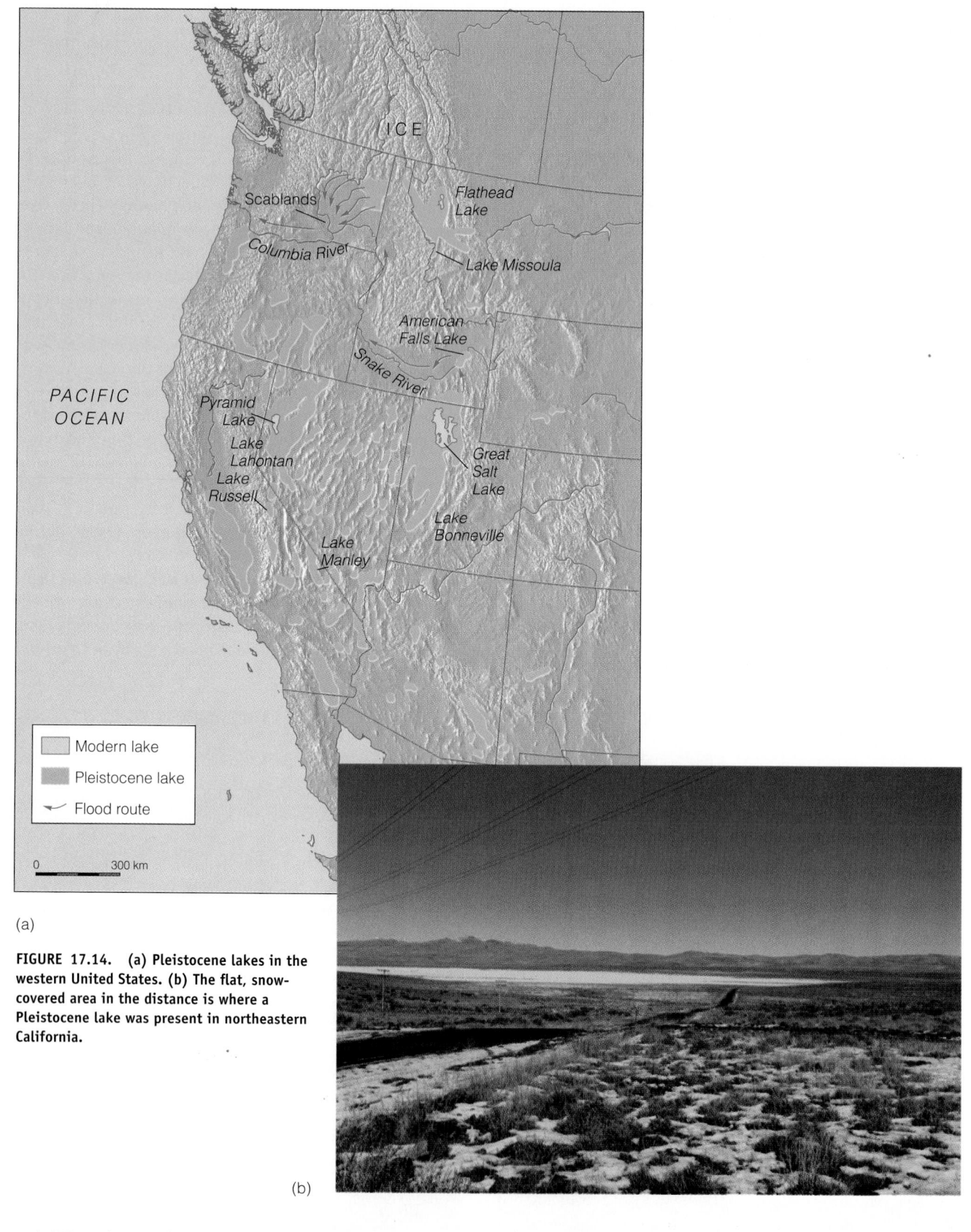

(a)

FIGURE 17.14. (a) Pleistocene lakes in the western United States. (b) The flat, snow-covered area in the distance is where a Pleistocene lake was present in northeastern California.

(b)

and 178 m deep. When the lake evaporated, the dissolved salts were precipitated on the valley floor; some of these evaporite deposits, especially borax, which is used for ceramic glazes, fertilizers, glass, solder, and pharmaceuticals, are important mineral resources (Figure 17.15).

In contrast to pluvial lakes, which form far from glaciers, **proglacial lakes** are formed by the meltwater accumulating along the margins of glaciers. In fact, in many proglacial lakes, one shoreline is the ice front itself, while the other shorelines consist of moraines. Lake Agassiz,

FIGURE 17.15 Twenty-mule teams carried borax out of Death Valley during the late 1800s.

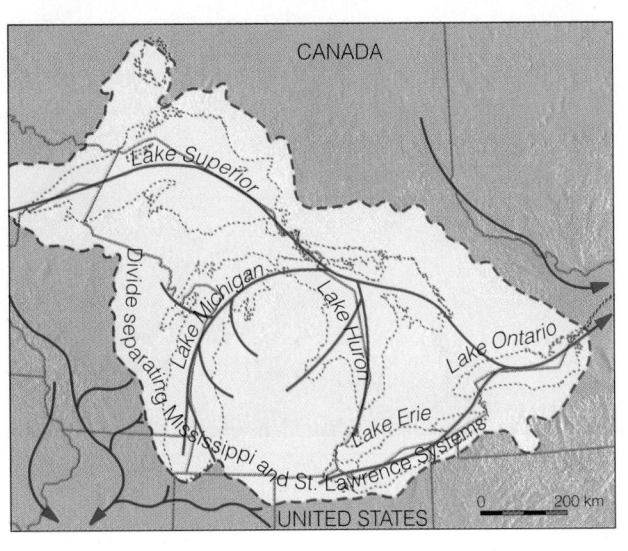

FIGURE 17.16 Theoretical preglacial drainage in the Great Lakes region. The divide separating the preglacial Mississippi and St. Lawrence drainage basins was probably near its present location. The future sites of the Great Lakes are outlined by dashed lines.

named in honor of the French naturalist Louis Agassiz, was a large proglacial lake covering about 250,000 km² of North Dakota and Manitoba, Saskatchewan, and Ontario, Canada. It persisted until the glacial ice along its northern margin melted, at which time the lake was able to drain northward into Hudson Bay. One of the greatest floods in history resulted when the ice dam impounding Glacial Lake Missoula in Montana collapsed (see Perspective 17.2).

Numerous proglacial lakes existed during the Pleistocene, but most of them eventually disappeared. Lake Agassiz and a number of smaller lakes drained when the glaciers disappeared, but others simply filled with sediment. Notable exceptions, however, are the Great Lakes, all of which first formed as proglacial lakes.

A variety of sediments are found in glacial lakes, but of special interest are the finely laminated mud deposits consisting of alternating light and dark layers. Each light–dark couplet is a *varve* (see Figure 4.15d) and represents an annual episode of deposition. The light layer of silt and clay formed during the spring and summer, and the dark layer consisting of smaller particles and organic matter formed during the winter when the lake froze over. Varved deposits commonly contain gravel-sized particles known as *dropstones,* which were probably released from melting icebergs.

HISTORY OF THE GREAT LAKES

The Great Lakes of North America provide us with an excellent example of the direct effects of glaciation. Before the Pleistocene, no large lakes existed in the Great Lakes region, which was then an area of generally flat lowlands with broad stream valleys (Figure 17.16). As the glaciers advanced southward, they eroded the stream valleys more deeply, forming what were to become the basins of the Great Lakes. During these glacial advances, the ice front moved forward as a series of lobes, some of which flowed into the preexisting lowlands where the ice became thicker and moved more rapidly. As a consequence, the lowlands were deeply eroded—four of the five Great Lakes basins were eroded below sea level.

At their greatest extent, the glaciers covered the entire Great Lakes region and extended far to the south (Figure 17.7a). As the ice sheet retreated northward during the late Pleistocene, the ice front periodically stabilized, and numerous recessional moraines were deposited. By about 14,000 years ago, parts of the Lake Michigan and Lake Erie basins were ice free, and glacial meltwater began forming proglacial lakes (Figure 17.17). As the retreat of the ice sheet continued—although periodically interrupted by minor readvances—the Great Lakes basins were uncovered, and the lakes expanded until they eventually reached their present size and configuration (Figure 17.17). Currently, the Great Lakes contain nearly 23,000 km³ of water, about 18% of the water in all freshwater lakes.

Although the history of the Great Lakes just presented is generally correct, it is oversimplified. For instance, the areas and depths of the evolving Great Lakes fluctuated widely in response to minor readvances of the ice front. Furthermore, as the lakes filled, they spilled over the lowest parts of their margins, thus cutting outlets that partly drained them. And finally, as the glaciers retreated northward, isostatic rebound initially raised the southern parts of the Great Lakes region, greatly altering their drainage systems.

Causes of Pleistocene Glaciation

Thus far we have examined some of the effects of glaciation but have not addressed the central questions of what causes large-scale glaciation and why so few episodes of widespread glaciation have occurred. For more than a century, scientists have been attempting to develop a comprehensive theory explaining all aspects of ice ages but have not yet been completely successful. One reason for their lack of success is that the climatic changes responsible for

The Channeled Scablands of Eastern Washington— A Catastrophic Pleistocene Flood

*T*he term *scabland* is used in the Pacific Northwest to describe areas from which the surface deposits have been scoured, thus exposing the underlying rock. Such an area exists in a large part of eastern Washington where numerous deep and generally dry channels are present. Some of these channels, cut into basalt lava flows, are more than 70 m deep, and their floors are covered by gigantic "ripple marks" as much as 10 m high and 70–100 m apart. Additionally, a number of high hills in the area are arranged such that they appear to have been islands in a large braided stream.

In 1923 J Harlan Bretz proposed that the channeled scablands of eastern Washington were formed during a single, gigantic flood. Bretz's unorthodox explanation was rejected by most geologists who preferred a more traditional interpretation based on normal stream erosion over a long period of time. In contrast, Bretz held that the scablands were formed rapidly during a flood of glacial meltwater that lasted only a few days.

The problem with Bretz's hypothesis was that he could not identify an adequate source for his floodwater. He knew that the glaciers had advanced as far south as Spokane, Washington, but he could not explain how so much ice melted so rapidly. The answer to Bretz's dilemma

came from western Montana where an enormous ice-dammed lake (Lake Missoula) had formed. Lake Missoula formed when an advancing glacier plugged the Clark Fork Valley at Ice Cork, Idaho, causing the water to fill the valleys of western Montana (Figure 1). At its highest level, Lake Missoula covered about 7800 km^2 and contained and estimated 2090 km^3 of water (about 42% of the volume of present-day Lake Michigan). The shorelines of Lake Missoula are still clearly visible on the mountainsides around Missoula, Montana (Figure 2a).

When the ice dam impounding Lake Missoula failed, the water rushed out at tremendous velocity and drained south and southwest across Idaho and into Washington. The maximum rate of flow is estimated to have been nearly 11 million m^3/sec, about 55 times greater than the average discharge of the Amazon River. When these raging floodwaters reached eastern Washington, they stripped away the soil and most of the surface sediment, carving out huge valleys in solid bedrock. The currents were so powerful and turbulent they plucked out and moved pieces of basalt measuring 10 m across. Within the channels, sand and gravel were shaped into huge ridges, the so-called giant ripple marks (Figure 2b).

Bretz originally thought that one mas-

sive flood formed the channeled scablands, but geologists now know that Lake Missoula formed, flooded, and re-formed several times. The largest lake formed 18,000 to 20,000 years ago, and its draining produced the last great flood.

How long did the flood last, and did humans witness it? According to one estimate, approximately one month passed from the time the ice dam first broke and water rushed out onto the scablands to the time the scabland streams returned to normal flow. No one knows for sure if anyone witnessed the flood. The oldest known evidence of humans in the region is from the Marmes Man site in southeastern Washington dated at 10,130 years ago, nearly 2000 years after the last flood from Lake Missoula. However, it is now generally accepted that Native Americans were present in North American at least 15,000 years ago.

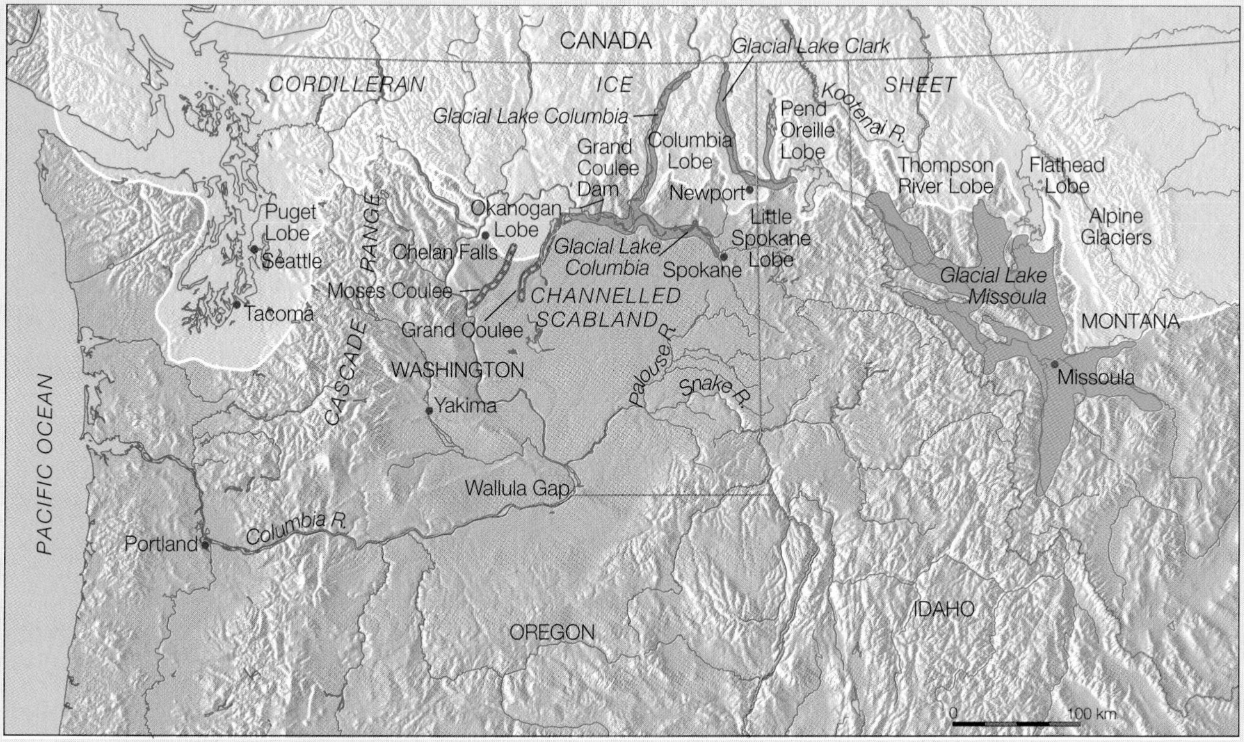

[Figure 1 map]

FIGURE 1 Location of glacial Lake Missoula and the channeled scablands of eastern Washington.

(a)   (b)

FIGURE 2 (a) The horizontal lines on Sentinel Mountain at Missoula, Montana, are wave-cut shorelines of glacial Lake Missoula. (b) These gravel ridges are the so-called giant ripple marks that formed when glacial Lake Missoula drained across this area near Camas Hot Springs, Montana.

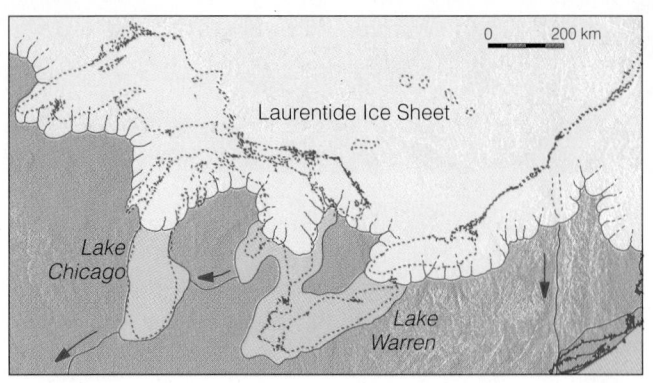

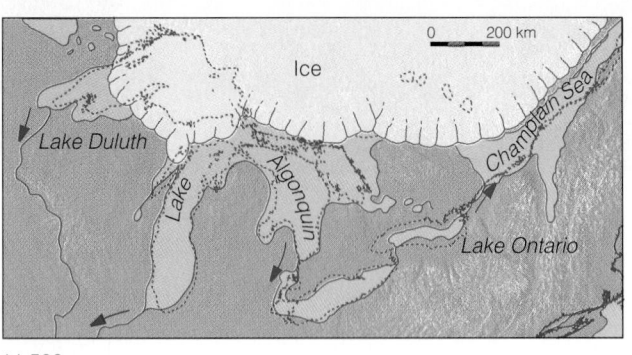

13,000 years ago

11,500 years ago

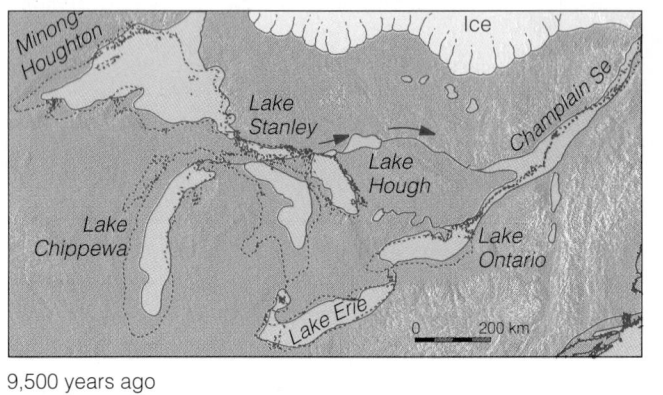

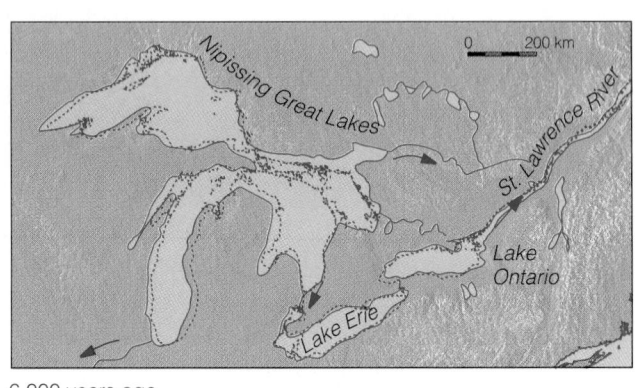

9,500 years ago

6,000 years ago

FIGURE 17.17 Four stages in the evolution of the Great Lakes. As the glacial ice retreated northward, the lake basins began filling with meltwater. The dotted lines indicate the present-day shorelines of the lakes.

glaciation, the cyclic occurrence of glacial–interglacial episodes, and short-term events such as the Little Ice Age operate on vastly different time scales.

Only a few periods of glaciation are recognized in the geologic record, each separated from the others by long intervals of mild climate. Such long-term climatic changes probably result from slow geographic changes related to plate tectonic activity. Moving plates can carry continents to high latitudes where glaciers can exist, provided that they receive enough precipitation as snow. Plate collisions, the subsequent uplift of vast areas far above sea level, and changing atmospheric and oceanic circulation patterns caused by the changing shapes and positions of plates also contribute to long-term climatic change.

Intermediate-term climatic events, such as the glacial–interglacial episodes of the Pleistocene, occur on time scales of tens to hundreds of thousands of years. The cyclic nature of this most recent episode of glaciation has long been a problem in formulating a comprehensive theory of climatic change.

THE MILANKOVITCH THEORY

Changes in Earth's orbit as a cause of intermediate-term climatic events was first proposed during the mid-1800s, but it was made popular during the 1920s by the Serbian astronomer Milutin Milankovitch. He proposed that minor irregularities in Earth's rotation and orbit are sufficient to alter the amount of solar radiation received at any given lat-

itude and hence bring about climate changes. Now called the **Milankovitch theory,** it was initially ignored but has received renewed interest since the 1970s and is widely accepted.

Milankovitch attributed the onset of the Pleistocene Ice Age to variations in three aspects of Earth's orbit. The first is *orbital eccentricity,* which is the degree to which Earth's orbit departs from a circle (Figure 17.18a). When the orbit is nearly circular, both the Northern and Southern Hemispheres have similar contrasts between the seasons. However, if the orbit is more elliptic, hot summers and cold winters will occur in one hemisphere, whereas warm summers and cool winters will take place in the other hemisphere. Calculations indicate a roughly 100,000-year cycle between times of maximum eccentricity, which corresponds closely to the 20 warm–cold climatic cycles that took place during the Pleistocene.

Milankovitch also pointed out that the angle between Earth's axis and a line perpendicular to the plane of the ecliptic shifts about 1.5° from its current value of 23.5° during a 41,000-year cycle (Figure 17.18b). Although changes in *axial tilt* have little effect on equatorial latitudes, they strongly affect the amount of solar radiation received at high latitudes and the duration of the dark period at and near Earth's poles. Coupled with the third aspect of Earth's orbit, precession of the equinoxes, high latitudes might receive as much as 15% less solar radiation, certainly enough to affect glacial growth and melting.

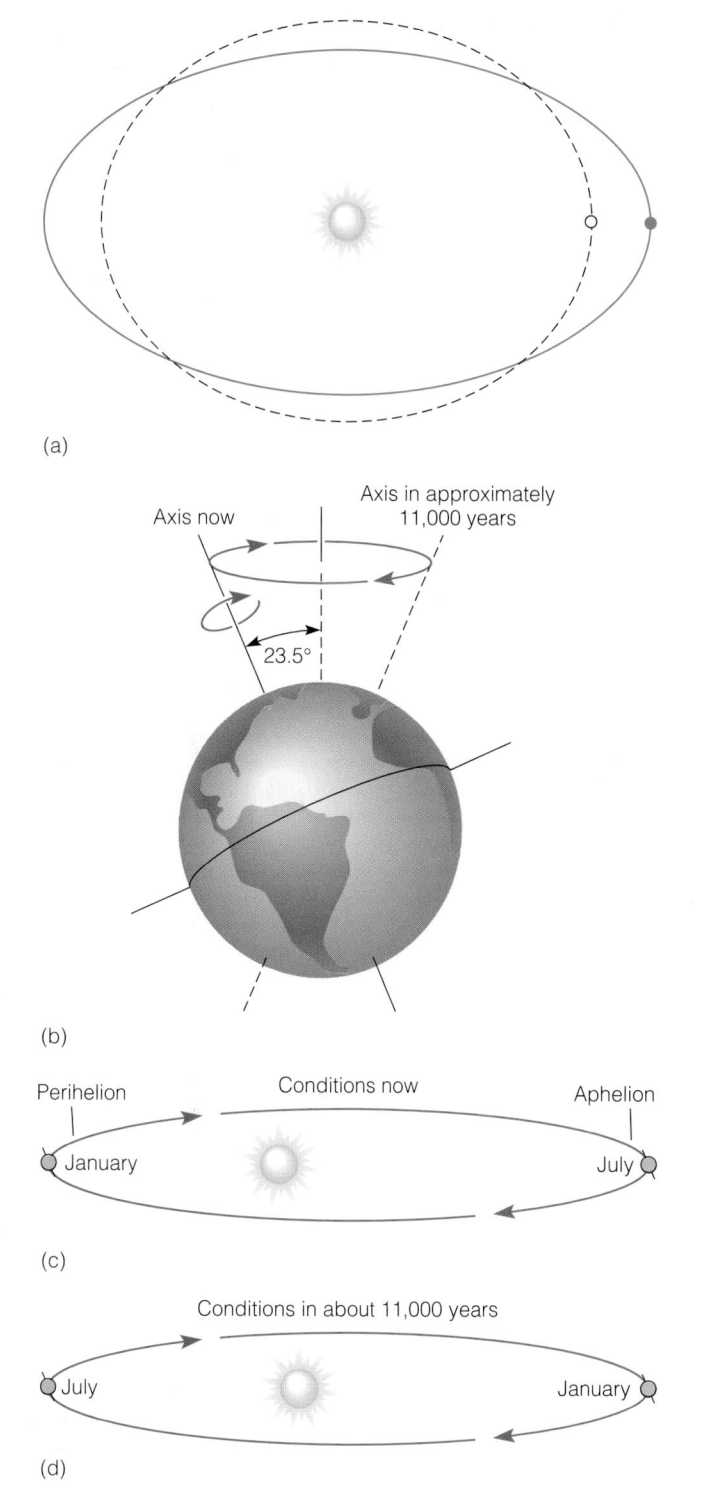

The last aspect of Earth's orbit that Milankovitch cited is *precession of the equinoxes,* which refers to a change in the time of the equinoxes. At present, the equinoxes take place on about March 21 and September 21 when the sun is directly over the equator. But as Earth rotates on its axis, it also wobbles as its axial tilt varies 1.5° from its current value, thus changing the time of the equinoxes. Taken alone, the time of the equinoxes has little climatic effect, but changes in Earth's axial tilt also change the time of *aphelion* and *perihelion,* which are, respectively, when Earth is farthest from and closest to the Sun during its orbit (Figure 17.18c and d). Earth is now at perihelion, closest to the Sun, during Northern Hemisphere winters, but in about 11,000 years perihelion will be in July. Accordingly, Earth will be at aphelion, farthest from the Sun, in January and have colder winters.

Continuous variations in Earth's orbit and axial tilt cause the amount of solar heat received at any latitude to vary slightly through time. The total heat received by the planet changes little, but according to Milankovitch, and now many scientists agree, these changes cause complex climatic variations and provided the triggering mechanism for the glacial–interglacial episodes of the Pleistocene.

SHORT-TERM CLIMATIC EVENTS

Climatic events having durations of several centuries, such as the Little Ice Age, are too short to be accounted for by plate tectonics or Milankovitch cycles. Several hypotheses have been proposed, including variations in solar energy and volcanism.

Variations in solar energy could result from changes within the Sun itself or from anything that would reduce the amount of energy Earth receives from the Sun. The latter could result from the solar system passing through clouds of interstellar dust and gas or from substances in the atmosphere reflecting solar radiation back into space. Records kept over the past 75 years, however, indicate that during this time the amount of solar radiation has varied only slightly. Thus, although variations in solar energy may influence short-term climatic events, such a correlation has not been demonstrated.

During large volcanic eruptions, tremendous amounts of ash and gases are spewed into the atmosphere where they reflect incoming solar radiation and thus reduce atmospheric temperatures. Small droplets of sulfur gases remain in the atmosphere for years and can have a significant effect on the climate. Several such large-scale volcanic events have been recorded, such as the 1815 eruption of Tambora and the 1991 eruption of Mount Pinatubo, and are known to have had climatic effects. However, no relationship between periods of volcanic activity and periods of glaciation has yet been established.

Quaternary Mineral Resources

Most people are probably not aware of the many mineral resources that formed as a result of Quaternary geologic events. Yet both directly and indirectly, glaciation played

FIGURE 17.18 (a) Earth's orbit varies from nearly a circle (dashed line) to an ellipse (solid line) and back again in about 100,000 years. (b) Earth moves around its orbit while spinning about its axis, which is tilted to the plane of the ecliptic at 23.5° and points toward the North Star. Earth's axis of rotation slowly moves and traces out the path of a cone in space. (c) Presently, Earth is closest to the Sun in January when the Northern Hemisphere experiences winter. (d) In about 11,000 years, as a result of precession, Earth will be closer to the Sun in July, when summer occurs in the Northern Hemisphere.

an important role in the formation and distribution of Quaternary mineral resources.

Sand and gravel deposits resulting from glacial activity are a valuable resource in many formerly glaciated areas of the world. In fact, such deposits are one of the most valuable nonmetallic mineral resources in many countries. Most Pleistocene sand and gravel deposits originated as floodplain or terrace gravels, outwash sediment, or esker deposits. The bulk of the sand and gravel in the United States and Canada is used as roadbase and fill for highway and railway construction. Silica sand is also used for glass manufacture, and fine-grained glacial lake sediments are used in the manufacture of bricks and ceramic products.

As mentioned in Chapter 15, placer deposits containing gold are another important mineral resource of the Quaternary. The California Gold Rush of the late 1840s was fueled by the discovery of gold in placer deposits of the American River. The discovery of placer gold deposits in the Yukon Territory of Canada was primarily responsible for the settlement of the area.

The periodic evaporation of pluvial lakes in the Death Valley region of California during the Pleistocene led to the concentration of many evaporite minerals such as borax. During the 1880s, borax was transported from Death Valley to Barstow, California, by the famous 20-mule team wagon trains (Figure 17.15).

Another Quaternary mineral resource is peat, a vast potential energy resource that has been developed in Canada and Ireland. These peatlands formed from plants as the result of particular climate conditions.

Summary

1. The Quaternary Period consists of two unequal epochs: the Pleistocene Epoch, from 1.6 million years ago to 10,000 years ago, and the Holocene Epoch, from 10,000 years ago to the present.

2. In addition to glaciation, the Pleistocene was also a time of volcanism and tectonic unrest during which folding, faulting, and uplifts were common occurrences.

3. During the Pleistocene Epoch, glaciers covered about 30% of the land surface. About 20 warm–cold Pleistocene climatic cycles are recognized from paleontologic and oxygen isotope data derived from deep-sea cores.

4. Several intervals of widespread glaciation, separated by interglacial periods, occurred in North America. The other Northern Hemisphere continents also experienced several episodes of widespread Pleistocene glaciation.

5. Areas far beyond the ice were affected by Pleistocene glaciation: climate belts were compressed toward the equator, large pluvial lakes existed in what are now arid regions, and sea level was as much as 130 m lower than at present.

6. Moraines, striations, outwash, and other features in many areas formed as a result of glacial erosion and deposition.

7. Loading of the crust by Pleistocene glaciers caused isostatic subsidence. When the glaciers disappeared, isostatic rebound began and still continues in some areas.

8. Major glacial intervals separated by tens or hundreds of millions of years probably occur as a consequence of the changing positions of tectonic plates, which in turn cause changes in oceanic and atmospheric circulation patterns.

9. Currently, the Milankovitch theory is widely accepted as the explanation for glacial–interglacial intervals.

10. The reasons for short-term climatic changes, such as the Little Ice Age, are not understood. Two proposed causes for such events are changes in the amount of solar energy received and volcanism.

11. Mineral resources of the Quaternary are mainly sand and gravel, along with some evaporite minerals such as borax.

Important Terms

glacial stage	isostasy	Milankovitch theory	pollen analysis
Holocene Epoch	isostatic rebound	Pleistocene Epoch	proglacial lake
interglacial stage	Little Ice Age	pluvial lake	Quaternary Period

Review Questions

1. Sediment deposited from sediment-laden streams discharging directly from glaciers is:
   a. _____ till.
   b. _____ outwash.
   c. _____ varves.
   d. _____ pluvial.
   e. _____ moraine.

2. Which of the following is one parameter Milankovitch used to explain the glacial-intergracial episodes of the Pleistocene?
   a. _____ The abundance of proglacial lakes
   b. _____ Precession of the equinoxes
   c. _____ Volcanism
   d. _____ Plate collisions
   e. _____ Intensity of sunlight

3. The Pleistocene Epoch began _____ years ago and ended _____ years ago:
   a. _____ 66 million/24 million
   b. _____ 100,000/1350
   c. _____ 1.6 million/10,000
   d. _____ 4.6 billion/225 million
   e. _____ 2 million/4000

4. The phenomenon in which the crust rises following unloading, as when glaciers melt, is known as:
   a. _____ postglacial response.
   b. _____ progradation.
   c. _____ neoglaciation.
   d. _____ glacial maxima.
   e. _____ isostatic rebound.

5. Lower sea level when Pleistocene glaciers were present accounts for:
   a. _____ the origin of placer deposits of gold.
   b. _____ erosion of the Great Lakes basins.
   c. _____ O^{18} to O^{16} ratios in foraminifera shells.
   d. _____ a land bridge between Siberia and Alaska.
   e. _____ advance of glaciers during the Little Ice Age.

6. Which of the following is *not* an important Quaternary resource?
   a. _____ Petroleum
   b. _____ Sand
   c. _____ Gravel
   d. _____ Gold placers
   e. _____ Borax

7. An important area of Quaternary volcanism is:
   a. _____ the Gulf Coastal Plain.
   b. _____ Craters of the Moon National Monument, Idaho.
   c. _____ Cape Cod National Seashore, Massachusetts.
   d. _____ Death Valley, California.
   e. _____ the Thomas Farm Quarry in Florida.

8. From oldest to youngest, the four glacial stages recognized in North America are:
   a. _____ Sangamon-Illinoian-Kentuckian-Algonquin.
   b. _____ Albanian-Mississippian-Pennsylvanian-Canadian.
   c. _____ Nebraskan-Kansan-Illinoian-Wisconsinan.
   d. _____ Montanan-North Dakotan-Yarmouth-Keewatin.
   e. _____ Labradoran-Cordilleran-Greenlandian- Kansan.

9. A large proglacial lake mostly in Canada was:
   a. _____ Lake Manly.
   b. _____ Lake Bonneville.
   c. _____ Lake Missoula.
   d. _____ Lake Agassiz.
   e. _____ Lake Dakota.

10. Although the Pleistocene Epoch began rather recently in geologic terms, the onset of glacial conditions resulted from a cooling trend that began about_____million years ago:
    a. _____ 40
    b. _____ 1.6
    c. _____ 66
    d. _____ 2.5
    e. _____ 225

11. The chaotic deposit of sediment at the end or terminus of a glacier is a(an):
    a. _____ moraine.
    b. _____ cirque.
    c. _____ outwash.
    d. _____ erratic.
    e. _____ esker.

12. When continental glaciers were present during the Pleistocene, many streams eroded deeply as they adjusted to:
    a. _____ less rainfall.
    b. _____ lower base level.
    c. _____ deposition of moraines.
    d. _____ greater precipitation.
    e. _____ colder winters.

13. Explain how planktonic foraminifera and oxygen isotope ratios are used to determine past climatic conditions.

14. What kinds of evidence convinced geologists that widespread glaciation took place during the Pleistocene?

15. Explain the concept of isostatic rebound and cite evidence indicating that rebound has taken place since the end of the Pleistocene.

16. How does the Milankovitch theory explain the onset of glacial episodes?

17. Briefly explain how the Great Lakes formed and evolved.

18. Why is it so difficult to correlate the record of warm–cold intervals recorded in seafloor sediments with the glacial record on land?

19. Why is pollen analysis such a good tool for determining Pleistocene conditions?

20. How did widespread glaciers affect areas far removed from the ice sheets? What were some of these effects?

21. How does a pluvial lake differ from a proglacial lake? Give an example of each.

22. Give examples of erosional and depositional glacial landforms and explain how they form.

23. What was the Little Ice Age, when did it occur, and what impact did it have on humans?

24. What geologic processes, other than glaciation, were going on during the Pleistocene? Give some specific examples.

Points to Ponder

1. We can be sure that the shorelines of the ancient Great Lakes were horizontal when they formed, yet now they are not only elevated above their former level but also dip gently to the south. How can you account for these observations?

2. In a roadside rock exposure, you observe a deposit of alternating light and dark laminated mud containing a few boulders measuring 20–80 cm in diameter. Explain the sequence of events responsible for deposition.

3. How might human activities affect the amount of glacial ice on Earth? What would be the consequence of more and of less glacial ice?

Additional Readings

Benn, D. I., and D. J. A. Evans. 1998. *Glaciers and glaciation.* New York: Arnold.

Bradley, R. S., and P. D. Jones (eds.). 1992. *Climate since A. D. 1500.* New York: Rutledge.

Dawson, A. G. 1992. *Ice Age Earth: Late Quaternary geology and climate.* New York: Rutledge.

Deynoux, M., et al. (eds.). 1994. *Earth's glacial record.* New York: Cambridge University Press.

Elias, S. A. 1995. *The Ice Age history of the Alaskan national parks.* Washington, D.C.: Smithsonian Institution Press.

———. 1996. *The Ice Age history of national parks in the Rocky Mountains.* Washington, D.C.: Smithsonian Institution Press.

Fulton, R. J. (ed.). 1989. *Quaternary geology of Canada and Greenland.* The geology of North America. Vol. K-1. Ottawa: Geological Survey of Canada.

Goudie, A. 1992. *Environmental change.* New York: Oxford University Press.

Grove, J. M. 1988. *The Little Ice Age.* New York: Methuen.

Morrison, R. B. (ed.). 1991. *Quaternary nonglacial geology: Conterminous U.S.* The geology of North America. Vol. K-2. Boulder, Colo: Geological Society of America.

Ray, L. L. 1993. *The Great Ice Age.* U.S. Geological Survey.

Ruddiman, W. F., and H. E. Wright Jr. (eds.). 1987. *North America and adjacent oceans during the last deglaciation.* The geology of North America. Vol. K-3. Boulder, Colo: Geological Society of America.

For these web site addresses, along with current updates and exercises, log on to

http://www.brookscole.com/geo/

▶ **ILLINOIS STATE MUSEUM**

This site, maintained by the Illinois State Museum, contains information about programs, exhibits, collections, and calendar of events at the museum.

1. Click on the *Exhibits* icon, then click on the *Ice Ages.* Here you can learn about the ice ages. What are ice ages? When did they occur? What causes ice ages?

2. Click on the *Exhibits* icon, then click on *The Midwestern U.S. 16,000 Years Ago.* This site shows what the Midwest looked like 16,000 years ago. Check on the *wander through the exhibit* and find out more about the environment, plants, and animals of the midwestern United States during the Ice Age. Here you can click on the *lists of the topics* covered at this site.

▶ **ICE AGES AND GLACIATION**

This site is maintained by the Department of Geological and Environmental Sciences at Hartwick College in New York. Take a virtual field trip through the Ice Age and see maps and images of glacial features.

▶ **WISCONSIN'S GLACIAL LANDSCAPE**

The Ice Age Park and Trail Foundation, Inc., maintains this site, which has information about the Ice Age in Wisconsin, along with maps and illustrations. When did

the most recent continental glacier enter Wisconsin, how long was it present, and how much of the state did it cover? What is the Driftless Area of Wisconsin?

▶ **GLACIAL LAKES**

The United States Geological Survey maintains this site, which is devoted largely to Glacial Lake Missoula. Click *Glacial Lake Missoula,* then click *Description: Glacial Lake Missoula and the Missoula Floods.* How many floods might have taken place? What caused the catastrophic floods of Glacial Lake Missoula?

▶ **GRAND TETON NATIONAL PARK**

The Grand Teton Nation Park, Wyoming, site is maintained by the National Park Service. At the park's home page click *geology,* and then scroll down and click *Journey Through the Past: A Geology Tour.* Next go to the section on *glaciation.* What evidence indicates that glaciers were once much more widespread in the park than at present? What depositional and erosional glacial landforms can be observed in the park? When was the onset of widespread glaciation in this area, and when did the last extensive glaciers begin to melt?

CD-ROM Exploration

Explore the following *In-Terra-Active 2.0* CD-ROM module(s) and increase your understanding of key concepts and processes presented in this chapter.

Chapter Concept: **Glacial Processes and Landforms**

During the Quaternary Period, glaciers covered as much as 30% of the land surface. The depositional and erosional landforms they left behind record several intervals of widespread ice coverage separated by warmer intervals known as interglacials. In this module, you will look for relationships between glacial processes and landforms including those that signal advance and retreat, and then apply this knowledge to the glacial history of a county in Wisconsin.

Section: **Surface**

Module: **Glaciers**

A boulder of granite gneiss from Ontario has been sitting atop Observatory Hill on the University of Wisconsin–Madison campus for thousands of years. Unlike Stonehenge and the Pyramids, there is no evidence that the ancient inhabitants of Wisconsin had the ability to move such stones.

Question: *How could this boulder have gotten on this hill?*

Life of the Cenozoic Era

Well preserved leaves of *Metasequoia* (dawn redwood) from the Oligocene to Miocene John Day Formation of Oregon.

Three widely separated units make up John Day Fossil Beds National Monument, Oregon, where remarkable concentrations of fossil land mammals and plants have been found. Fossils were first discovered by U.S. Cavalry soldiers during the Civil War and have been studied ever since, giving a unique glimpse of what the climate and life was like in central Oregon between 6 and 54 million years ago. The fossil-bearing rock units are well exposed, and along with their interest as repositories of ancient life, their colors and erosion patterns provide some truly inspiring scenery (Figure 18.1).

Many of the rocks in the monument are characterized as *volcaniclastic,* meaning they contain large amounts of volcanic rock debris and/or pyroclastic materials. This, coupled with the fact that they contain few sedimentary structures, resulted in their being largely ignored for many years by geologists interested in sedimentary rocks. Similarly, geologists who study igneous rocks and processes ignored them because

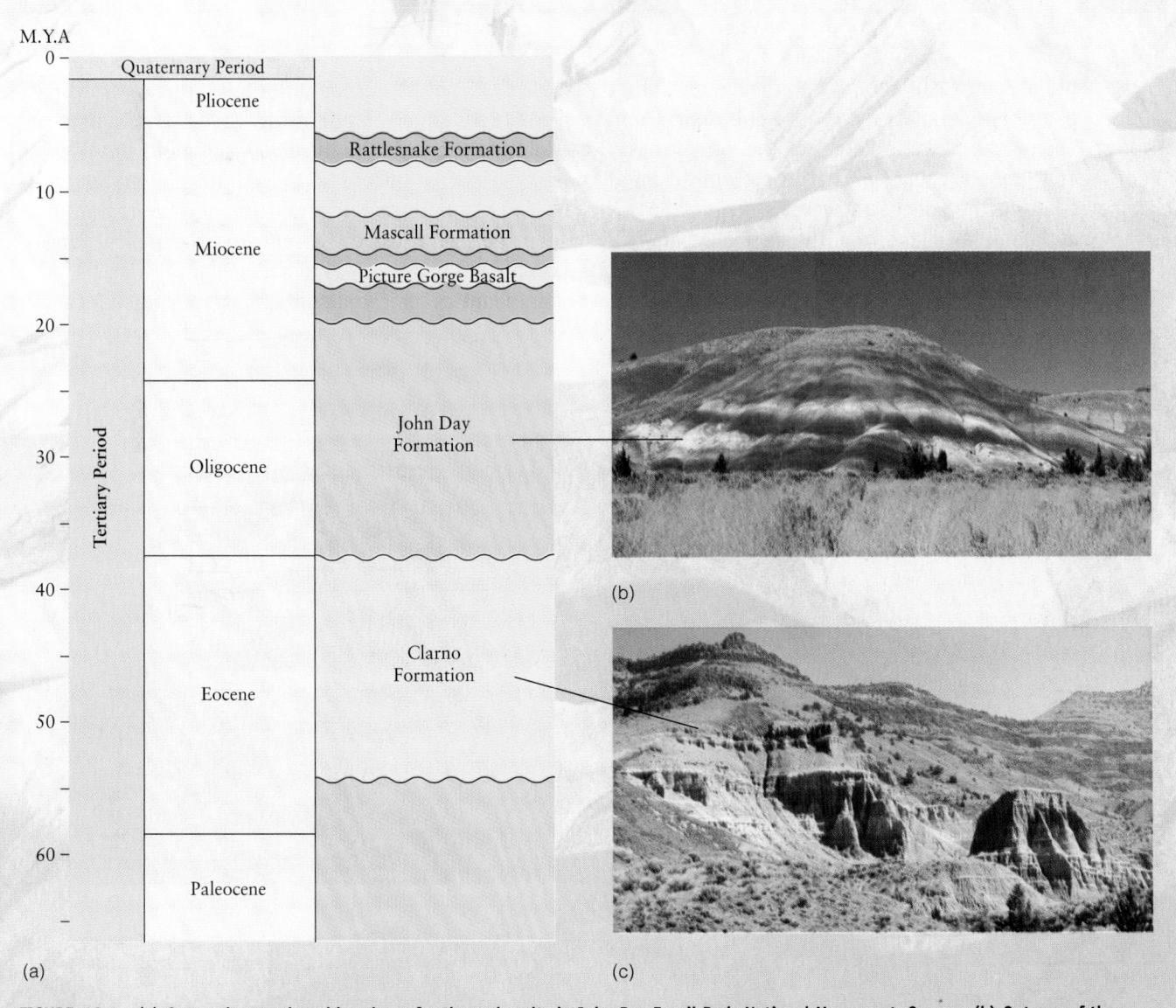

FIGURE 18.1   (a) Composite stratigraphic column for the rock units in John Day Fossil Beds National Monument, Oregon. (b) Outcrop of the Clarno Formation seen from the Painted Cove Trail in the Clarno Unit of the Monument. (c) John Day Formation outcrops at Flood of Fire Trail in the Sheep Rock Unit of the Monument.

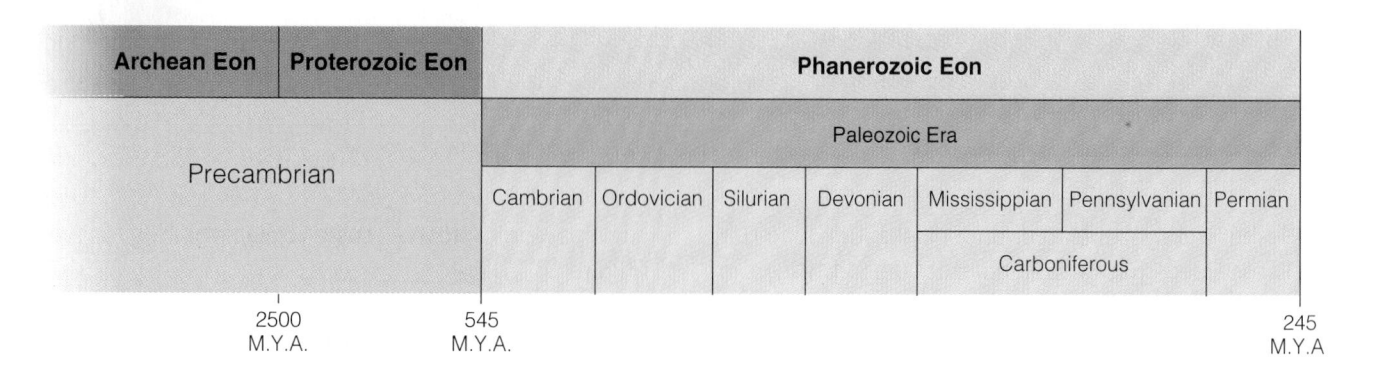

Archean Eon	Proterozoic Eon	Phanerozoic Eon						
		Paleozoic Era						
Precambrian		Cambrian	Ordovician	Silurian	Devonian	Mississippian	Pennsylvanian	Permian
						Carboniferous		

2500 M.Y.A.   545 M.Y.A.   245 M.Y.A

they are sediments. In recent years, though, the rocks have been thoroughly studied, and now their origin is well understood. Because many of the rocks contain volcanic ash-flow deposits, radiometric dating of the fossil-bearing units is rather straightforward.

The Eocene Clarno Formation, the oldest Cenozoic rock unit in the monument (Figure 18.1b), is composed of volcanic ash and volcanic mudflows that buried plants and animals living in a subtropical rain forest. Today, the Cascade Range to the west forms a rainshadow, thus accounting for the present semiarid climate. During the Eocene, however, the climate was warm and humid far inland from the Pacific Ocean. The area was covered by forests of palms, avocados, and ferns that were occupied by early horses, tapirs, rhi-

noceroses, carnivorous mammals, as well as the now extinct oreodonts and titanotheres (Figure 18.2a). Fossil crocodiles and tortoises have also been recovered from the formation.

The John Day Formation consists of a complex of ash-flow tuffs, welded tuffs, basalt and rhyolite lava flows, ash-rich claystones, and coarse-grained volcaniclastic deposits such as tuffaceous conglomerate (Figure 18.1b and c). To the west, the Cascade Range was forming and reducing the amount of moist air reaching the area. Fossil plants in these 20- to 39-million-year-old rocks indicate a drier climate, deciduous trees having replaced those of the previous subtropical forests, and grasslands were more common. Mammals were abundant as indicated by the numerous fossil localities

(a)

FIGURE 18.2 (a) Fossils from the Clarno Formation indicate that the climate was subtropical. Lush forests were occupied by a variety of mammals including (1) titanotheres, (2) a carnivore, (3) ancient horses, (4) tapirs, and (5) early rhinoceroses. (b) This restoration shows the dryer conditions that prevailed when the John Day Formation was deposited. The animals shown are (1) oreodonts, (2) sabertooth catlike animals, (3) three-toed horses, (4) piglike animals known as entelodonts, (5) rhinoceroses, and (6) small deerlike animals.

Phanerozoic Eon									
Mesozoic Era			Cenozoic Era						
Triassic	Jurassic	Cretaceous	Tertiary					Quaternary	
			Paleocene	Eocene	Oligocene	Miocene	Pliocene	Pleistocene	Holocene

245 M.Y.A.   66 M.Y.A.   Page 1   Page 2   3   4

where geologists have found horses, camels, cats, dogs, rhinoceroses, and rodents.

About 16 million years ago, during the Middle Miocene, fluid basalts erupted from fissures flooded a large part of the John Day area, forming the Picture Gorge Basalt (Figure 18.1a). Rather than building volcanic cones, the basalt simply spread out—some flows moving more than 60 km—and formed a basalt plateau. Nearby, though, volcanoes similar to those of the present-day Cascade Range were forming. Erosion of these volcanoes supplied debris that makes up the Mascall Formation of Miocene age (Figure 18.1a). This formation is also well known for its fossil mammals, many of them descendants of the animals found in the older formations, and giants such as early elephants.

Finally, the Rattlesnake Formation formed when debris eroded from nearby highlands covered a large area. The rocks are mostly coarse-grained sandstones and considerable conglomerates that contain far fewer fossils than the older formations. Nevertheless, fossils of grazing mammals are much more common than those of browsers, so widespread grasslands must have been present, and the climate was cooler and dryer than in previous times. Many of the fossil mammals resembled mammals living today, but several varieties still existed that would be unfamiliar to us if we could somehow go back and visit the region during the Late Miocene.

(b)

Introduction

More and more familiar types of plants and animals appeared during the Cenozoic as the world's flora and fauna continued to evolve. Recall from Chapter 15 that mammals evolved from cynodonts during the Late Triassic, but all of them were small and not particularly diverse during the rest of the Mesozoic. As the Cenozoic dawned, though, mammals differentiated rapidly and soon became the most diverse and abundant land-dwelling vertebrate animals. In this chapter we emphasize mammalian evolution, especially of some of the more familiar types such as elephants, hoofed mammals, whales, and carnivores.

Although the emphasis in this chapter is on mammalian evolution, one should be aware of other important events. The angiosperms, or flowering plants, continued to diversify and dominated the land-plant communities; they now account for more than 90% of all land-plant species. The present-day groups of birds appeared during the Early Tertiary, then reached their maximum diversity during the Pleistocene, and declined slightly since then. A land connection formed between North and South America where none existed previously, thus allowing migrations of animals in both directions. In the seas, marine invertebrates continued to diversify, following the extinctions at the end of the Mesozoic. Overall, we can think of the Cenozoic as a time during which the world's flora and fauna became increasingly familiar.

The Cenozoic evolutionary history of mammals is better known than the histories of any other group of vertebrates. One reason for this is that Cenozoic terrestrial deposits are more common and accessible at the surface or in the shallow subsurface. And being younger, these deposits and their contained fossils have had less time to be altered or destroyed. As a result, many mammals such as horses, camels, elephants, rhinoceroses, and whales have good fossil records.

Cenozoic terrestrial deposits are found mostly in western North America; thus, most fossil mammals come from that area. Good examples include the sedimentary rocks of Badlands National Park, South Dakota (see Perspective 4.2), John Day Fossil Beds National Monument, Oregon (see the Prologue), and several other areas. In the east, Pleistocene mammal fossils are fairly common, but those of Tertiary age are rare. Sedimentary rocks exposed in the Calvert Cliffs of Maryland are famous for their invertebrates, shark teeth, and fossil whales and sea cows, but probably the best area in the east to find terrestrial mammal fossils is Florida (see Perspective 18.1).

Marine Invertebrates and Phytoplankton

The Cenozoic marine ecosystem was populated mostly by those plants, animals, and single-celled organisms that survived the terminal Mesozoic extinction. Gone were the ammonites, rudists, and most of the planktonic foraminifera. Especially prolific Cenozoic invertebrate groups were the foraminifera, radiolarians, corals, bryozoans, mollusks, and echinoids. The marine invertebrate community in general became more provincial during the Cenozoic because of changing ocean currents and latitudinal temperature gradients. In addition, the Cenozoic marine invertebrate faunas became more familiar in appearance.

Entire families of phytoplankton became extinct at the end of the Cretaceous. Only a few species in each major group survived into the Tertiary. These species diversified and expanded during the Cenozoic, perhaps because of decreased competitive pressures. Coccolithophores, diatoms, and dinoflagellates all recovered from their Late Cretaceous reduction in numbers to flourish during the Cenozoic. Diatoms were particularly abundant during the Miocene, probably because of increased volcanism during this time. Volcanic ash provided increased dissolved silica in seawater and was used by the diatoms to construct their skeletons. Massive Miocene diatomites have been found in California (Figure 18.3).

The foraminifera comprised a major component of the Cenozoic marine invertebrate community. Though dominated by relatively small forms (Figure 18.4), it included some exceptionally large forms that lived in the warm waters of the Cenozoic Tethys Sea. Shells of these larger forms accumulated to form thick limestones, some of which were used by the ancient Egyptians to construct the Sphinx and the Pyramids of Giza.

Corals were perhaps the main beneficiary of the Mesozoic extinctions. Having relinquished their reef-building role to rudists during the mid-Cretaceous, corals again became the dominant reef builders during the Cenozoic. They formed extensive reefs in the warm waters of the Cenozoic oceans and were particularly prolific in the Caribbean and Indo-Pacific regions (Figure 18.5).

Other suspension feeders such as bryozoans and crinoids were also abundant and successful during the Tertiary as well as the Quaternary. Bryozoans, in particular, were very abundant. Perhaps the least important of the Cenozoic marine invertebrates were brachiopods, with fewer than 60 genera surviving today.

FIGURE 18.3 Outcrop of diatomite from the Miocene Monterey Formation, Newport Lagoon, California.

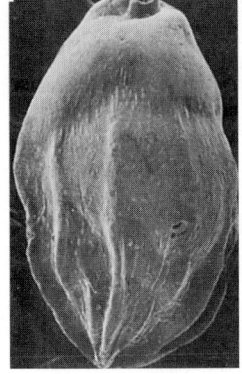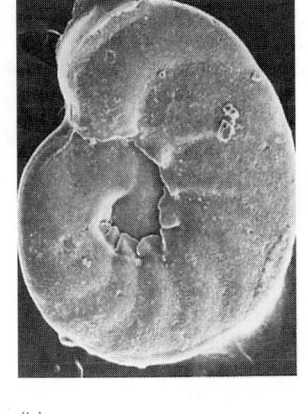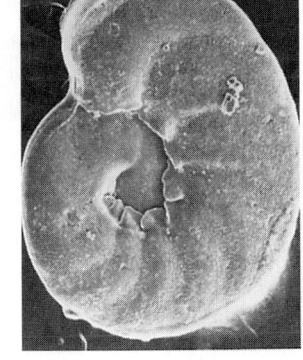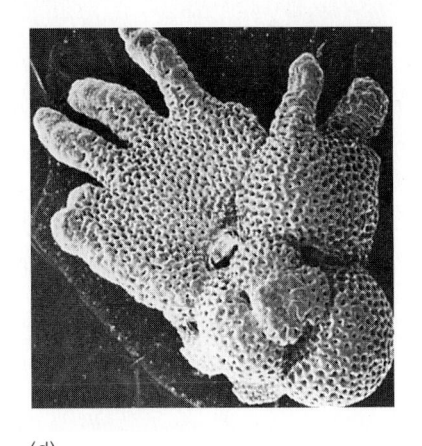

(a) (b) (c) (d)

FIGURE 18.4 Foraminifera of the Cenozoic Era. (a) through (c) are benthonic forms. (a) *Uvigerina cubana*, Late Miocene, California. (b) *Cibicides americanus*, Early Miocene, California. (c) *Lenticulina mexicana*, Eocene, Louisiana. (d) A planktonic form, *Globigerinoides fistulosus*, Pleistocene, South Pacific Ocean.

Just as during the Mesozoic, bivalves and gastropods were two of the major groups of marine invertebrates during the Tertiary, and they had a markedly modern appearance. Following the extinction of ammonites and belemnites at the end of the Cretaceous, the Cenozoic cephalopod fauna consisted of nautiloids and shell-less cephalopods such as squids and octopuses.

Echinoids continued their expansion in the infaunal habitat and were particularly prolific during the Tertiary. New forms such as sand dollars evolved during this time from biscuit-shaped ancestors (Figure 18.6).

Cenozoic Birds

The first members of many living orders and families of birds such as owls, hawks, ducks, penguins, and vultures evolved during the Early Tertiary. A marked increase in variety of songbirds took place beginning during the Miocene, and by 5 to 10 million years ago many present-day genera of birds were present. Today, birds vary considerably in size and adaptations, but their basic skeletal structure has not changed significantly throughout the Cenozoic. This uniformity is not surprising because most birds are fliers and adaptations for flying impose limitations on variations in structure. Birds adapted to numerous habitats and increased in diversity through the Tertiary and into the Pleistocene. Since then, their diversity has decreased slightly.

One of the more remarkable early adaptations was the development of large, flightless predatory birds. One of

FIGURE 18.5 The dominant reef-building animals of the Cenozoic Era were corals such as this present-day colonial scleractinian.

FIGURE 18.6 Echinoids were especially abundant during the Tertiary Period. New infaunal forms such as this sand dollar evolved from biscuit-shaped Mesozoic ancestors.

Florida's Fossil Mammals

Florida is the only area in eastern North America where numerous terrestrial mammals of Cenozoic age have been found. The oldest rocks in the state are marine deposits of Eocene age, which contain fossils of many sea-dwelling creatures including whales and sea cows. Apparently Florida was shallowly submerged during the Eocene, but by Late Oligocene parts of northern Florida rose from the sea and were soon occupied by land-dwelling mammals and other creatures. Unfortunately, Late Oligocene mammal fossils are rare and not well known. However, by Miocene time, considerably more of Florida was above sea level, and Miocene, Pliocene, and Pleistocene mammal fossils are quite common.

Perhaps Florida's most famous mammal fossil site is the Thomas Farm quarry in Gilchrist County. This Early Miocene (approximately 20 million years old) site is owned by the University of Florida, which has carried out systematic excavations of fossils for many years. Apparently, the fossils here were trapped in a sinkhole, a depression in the surface resulting from dissolution of underlying soluble rocks such as limestone or the collapse of a cavern roof. The remains of numerous bats support the sinkhole or cavern interpretation. An interesting side note is that animals are still being trapped in Florida sinkholes, thus providing evidence for how ancient animals were trapped and fossilized.

During the Miocene, Florida was tropical with abundant rainfall and temperatures as much as 4° C warmer than today. It teemed with diverse animals including various reptiles, numerous birds, amphibians, and mammals. In addition to fossil bats, the Thomas Farm quarry fossil site has also yielded birds, lizards, and snakes, and numerous other mammals such as rodents, horses, rhinoceroses, various carnivores, the now extinct oreodonts, and deerlike protoceratids with forked horns on their snouts (Figure 1). The site is not much more than 100 m in diameter, but it has an extraordinary concentration of bones that accumulated in a short time.

Miocene deposits 4 to 15 million years old in central–western Florida include the Bone Valley Member of the Peace River Formation (see Figure 16.35). (Some of the deposits are probably as little as 3 million years old, making them Pliocene.) These deposits are well known and important as sources of phosphate, Florida's most important mineral resource. These phosphates were deposited in marine, estuary (river mouth), freshwater, and terrestrial environments, thus forming complex facies relationships. The marine beds contain the remains of sharks, rays,

FIGURE 1 The Thomas Farm quarry in Gilchrist County, Florida has yielded the remains of numerous mammals including this small fossil horse on display at the Florida Museum of Natural History in Gainesville.

bony fish, and marine mammals, whereas the terrestrial deposits contain, in addition to the same animals found at Thomas Farm, giant sloth and elephant fossils.

Pliocene mammal fossils are also well known, and during the Pleistocene, Florida had an exceptional variety of mammals. Pleistocene glaciers were never closer than 800 km from Florida, which enjoyed a rather mild climate and abounded with mammals. Saber-toothed cats, horses, camels, mastodons, and mammoths were common as were some recent migrants from South America, such as giant sloths more than 6 m long and armored mammals known as glyptodonts that weighed more than 2 metric tons (Figure 2). In fact, the teeth and bones of Pleistocene elephants and various hoofed mammals such as horses and camels are among the most common vertebrate fossils found in the state. Many of these teeth and bones eroded from riverbanks and become concentrated in pools and at river bends. Even casual fossil collectors can find many fossils in this environment.

FIGURE 2 Among the diverse Pliocene and Pleistocene mammals of Florida were 6-m-long giant sloths, and armored mammals known as glyptodonts that weighed more than 2 metric tons.

these flightless predators, *Diatryma,* was more than 2 m tall and had a huge head and beak, toes with large claws, and small vestigial wings (Figure 18.7). Its legs were massive and short, indicating that *Diatryma* did not move very fast, but the early mammals upon which it preyed were slow moving as well. *Diatryma* and related genera were widespread during the Early Tertiary of North America and of Europe, and in South America they were the dominant predator until about 2.5 million years ago. Eventually they became extinct, apparently replaced by the developing mammalian carnivores.

Two remarkable flightless birds known only from Pleistocene deposits are the moas of New Zealand and the elephant birds of Madagascar. The moas were up to 3 m tall, whereas the elephant birds were shorter but more heavily built, weighing as much as 500 kg. Both of these giant flightless birds became extinct soon after humans appeared in New Zealand and Madagascar.

Flightless birds notwithstanding, the true success story among birds belongs to the fliers. They did not undergo much structural change during the Cenozoic, but a bewildering array of adaptive types arose. In fact, birds have been as successful as mammals; they have exploited the aerial habitat as fully as the mammals have adapted to terrestrial habitats.

The Age of Mammals Begins

For more than 100 million years, mammals coexisted with dinosaurs; yet, their fossil record indicates that during this entire time they were neither diverse nor abundant. Even

FIGURE 18.7 Restoration of *Diatryma,* a large flightless, predatory bird that stood more than 2 m tall. It lived during the Paleocene and Eocene in North America and Europe.

during the Late Cretaceous, very near the end of the Age of Reptiles, only a few families of marsupial and placental mammals existed. This situation was soon to change. Mesozoic extinctions eliminated the dinosaurs and many of their relatives, thereby creating numerous adaptive opportunities that were quickly exploited by mammals. The Age of Mammals had begun. *Cenozoic age of mammals*

The Cenozoic evolutionary history of the mammals is better known than the history of any of the other classes of vertebrates. Two factors account for this. First, Cenozoic terrestrial deposits are more common than Mesozoic and Paleozoic deposits. Second, mammal fossils are easier to identify. In Chapter 15 we noted that differentiation of teeth was a trend established in the cynodonts. In mammals the teeth are fully differentiated into distinctive types, and the chewing teeth, called **premolars** and **molars,** differ in each of the mammalian orders. In fact, a single mammal chewing tooth is commonly sufficient to identify the genus from which it came.

Living mammals are assigned to one of two groups, the *prototheria* and the *theria.* Prototherians include the **monotremes,** the only egg-laying mammals, whereas the therians are either **marsupials** (pouched mammals) or **placentals.** The only existing monotremes are the duck-billed platypus and spiny anteater, or echidna, of the Australian region. When their young emerge from eggs, they are rather undeveloped and crawl to a fold in the mother's abdominal skin and develop further. Perhaps

monotremes were more diverse during the past, but their fossil record is poor and their relationships to therians is not clear.

Among the therians, marsupials are now common only in the Australian region where they have been the most diverse mammals throughout the Cenozoic. They continue to dominate the mammalian fauna of that continent. Although marsupials give birth to live young, their young are born in a very immature, almost embryonic condition and then undergo further development in the mother's pouch. Marsupials probably originated in North or South America and migrated to Australia via Antarctica before Pangaea fragmented completely. During much of the Cenozoic, they were also quite successful in South America, being represented by numerous species. However, late in the Cenozoic, when a land connection was finally established between the Americas, all but opossums died out when North American placental mammals migrated south. Recall from Chapter 5 that a number of marsupials in Australia and South America and placental mammals elsewhere developed many similarities, thus once again illustrating the concept of convergent evolution (see Perspective 5.2).

Placental mammals, like marsupials, give birth to live young, but their reproductive method differs in important details. In placentals, the amnion of the amniote egg (see Figure 13.15) has fused with the walls of the uterus, forming a *placenta.* Nutrients and oxygen are carried from mother to embryo through the placenta, permitting the young to develop much more fully before birth. Actually, marsupials also have a placenta, but it is less efficient than that of placental mammals, thus accounting for the fact that their newborn are not as well developed as the young of placentals. A measure of the phenomenal success of placental mammals, which is related in part to this method of reproduction, is that more than 90% of all mammals, fossil and living, are placentals. All of today's placentals and a number of extinct groups evolved from Late Cretaceous and Early Tertiary ancestors (Figure 18.8).

Diversification of Placental Mammals

Mammals began their adaptive radiation during the Mesozoic, and following the demise of dinosaurs and their relatives at the end of the Mesozoic mammals continued to diversify throughout the Cenozoic Era (Figure 18.8). But mammals of the Paleocene are considered *archaic,* meaning that most were not ancestors to any later mammals. In fact, some of these archaic mammals such as marsupials, insectivores, and the rodentlike multituberculates were holdovers from the Mesozoic Era (Figure 18.9). Nevertheless, several new orders of mammals made their first appearance during the Paleocene, including the first rodents, rabbits, primates, and carnivores, but many of these new orders soon became extinct. This was indeed a time of rather rapid diversification.

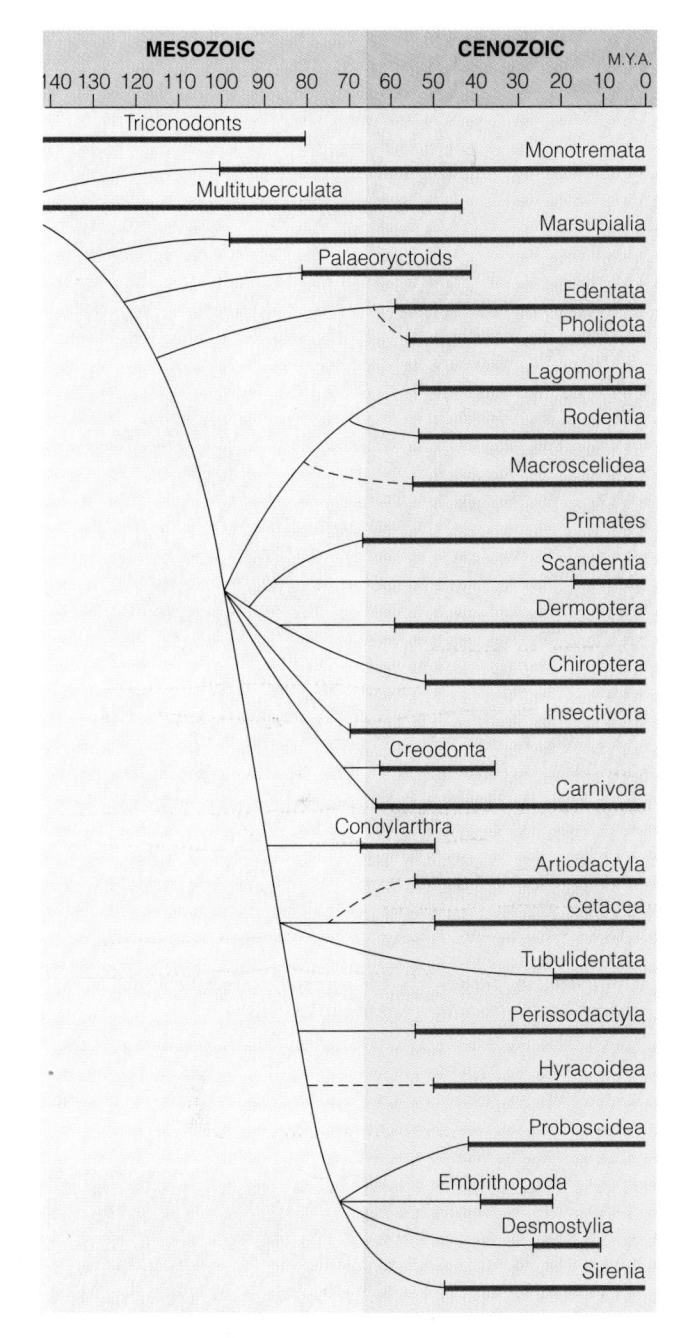

FIGURE 18.8 Evolutionary history of the mammals. Although some orders of placental mammals existed during the Mesozoic, most of them diversified during the Paleocene and Eocene. Several extinct orders of mammals are not shown. Bold lines indicate the actual geologic ranges of the groups, whereas the thinner lines indicate the inferred branching of the groups.

Some large mammals were present by the Late Paleocene, but the first giant terrestrial mammals did not appear until the Eocene (Figure 18.10). Most of the Paleocene mammals, even those belonging to orders that still exist, had not yet become clearly differentiated from their ancestors, and the differences between herbivores and carnivores were slight. Many of these early mammals appear to have dined on insects, although a variety of plant eaters were also present.

Mammal diversification continued during the Eocene, when several more orders evolved, and all but one of these orders still exists. In marked contrast, many of the orders that evolved during the Paleocene are now extinct. Most of today's mammalian orders existed by the Eocene, yet if we could somehow go back for a visit, we would probably not recognize many of these animals. We would certainly know they were mammals, and some would be at least vaguely familiar, but the ancestors of horses, camels, elephants, whales, and rhinoceroses would bear little resemblance to their living descendants.

Climatic changes probably related to the continued fragmentation of Pangaea took place during the Late Eocene. The warm, humid climates of the Paleocene and Eocene changed by Oligocene time and became drier and cooler. During this time, most of the archaic Paleocene mammals became extinct, and a number of groups that made their first appearance during the Eocene also died out. The giant mammals known as titanotheres (Figure 18.2) died out in North America but persisted until the Early Oligocene in Asia, and the Eocene rhinoceros-like uintatheres (Figure 18.10) did not persist into the Oligocene. Early mammals loosely united into a group known as condylarths died out, as did early carnivores called creodonts and a variety of other mammals from primates to most multituberculates. All in all, the Late Eocene to Early Oligocene was a time of considerable biotic change.

By Oligocene time, most existing mammal orders were present, but diversification continued among these orders as more and more familiar genera evolved. If we encountered these animals during our hypothetical trip back in time, we might think some of them looked rather odd, but we would probably have little difficulty recognizing rhinoceroses, elephants, rodents, and a variety of hoofed mammals. Nevertheless, the chalicotheres, large clawed relatives of horses and rhinoceroses, and a few others might be difficult to classify.

By Miocene and certainly Pliocene time, we could easily recognize most mammals (Figure 18.11). If we were to examine some of them closely, we would notice some peculiarities such as three-toed horses, saber-toothed cats, and deerlike animals with forked horns on their snouts, but overall the fauna would be quite familiar. And even though a few rather strange mammals still existed, many of the genera then present were the direct ancestors of Pleistocene and Recent forms. As we complete our hypothetical trip through the Cenozoic, we can conclude that it was a time during which the world's mammalian fauna became increasingly familiar.

INSECTIVORES, RODENTS, RABBITS, AND BATS

Insectivores, rodents, rabbits, and bats are related in that all are placental mammals and thus share a common ancestor. They are discussed together in this section because they constitute the small mammals that have adapted to microhabitats unavailable to larger mammals. When most

FIGURE 18.9. The mammalian fauna of the Paleocene Epoch included the multiturburculate *Ptilodus* (right foreground), shrewlike insectivores (right background), an early carnivore *Protictus* (left background), and the pantodont *Pantolambda* that was about 1 m tall.

FIGURE 18.10 The uintatheres were Eocene rhinoceros-sized mammals with three pairs of bony protuberances on the skull and saberlike upper canine teeth.

FIGURE 18.11 By Pliocene time, most mammals resembled those of today. This mural shows Pliocene mammals of the grasslands of western North America and include *Amebeledon,* a shovel-tusked mastodon *(background); Teleoceras,* a short-legged rhinoceros *(left center); Cranioceros,* a horned, hoofed mammal *(left foreground); Synthetoceros,* a hoofed mammal with a horn on its snout *(right foreground); Merycodus,* an extinct pronghorn *(middle foreground);* and three small mammals in the foreground.

people consider mammals, they think mostly of larger species such as deer, horses, elephants, dogs, and cats, but fail to realize that most mammals are small, weighing less than 1 kg. In fact, the four orders of mammals discussed in this section account for almost 75% of all living species of placental mammals.

The insectivores, represented today by shrews, moles, and hedgehogs, have a history dating back to the Late Cretaceous. As their name implies, their diet is mostly insects. Insectivores have probably not changed much since Mesozoic times. The adaptive success of rodents is manifested by the fact that the order Rodentia alone accounts for more than 40% of all mammal species. They are extremely diverse, having adapted to a wide range of habitats since their first appearance during the Paleocene. The fact that rodents can eat almost anything is also partly responsible for their success. Most rodents, like insectivores, are small, but a few, such as beavers and the capybara of South America, are sizable animals. The latter might be more than 1 m long and weigh in excess of 45 kg.

Rabbits superficially resemble rodents, but they differ in several anatomic details and have had an independent evolutionary history since the Paleocene when they diverged from a common ancestor. Paleocene rabbits are known but did not become abundant until the Oligocene. Rabbits and rodents are both gnawing animals, but details of their continuously growing gnawing teeth, the incisors, and other teeth differ enough to warrant assigning them to separate mammalian orders. Long, powerful hind limbs for hopping and speed are the most obvious evolutionary trends in rabbits.

The oldest known fossil bat is a specimen from the Eocene Green River Formation of Wyoming (Figure 18.12), although a number remarkably well-preserved bats are also known from Middle Eocene deposits in Germany. Apart from having forelimbs modified into wings, bats differ little from their ancestors, which were probably similar to today's shrews. Although bat wings are highly specialized for flying, these wings are simply basic vertebrate forelimbs modified to support a wing membrane. Unlike pterosaurs with their wing-supporting, elongate fourth finger and birds that have wings supported by fused second and third fingers, bats use the entire "hand" for wing support (see Figure 15.17c).

PRIMATES

The order **Primates** includes the tarsiers, lemurs, and lorises (the so-called lower primates) and living monkeys, apes, and humans (the higher primates). Much of the primate story is better told in Chapter 19 where we consider human evolution, so in this chapter we will be brief. Primitive primates may have evolved by the Late

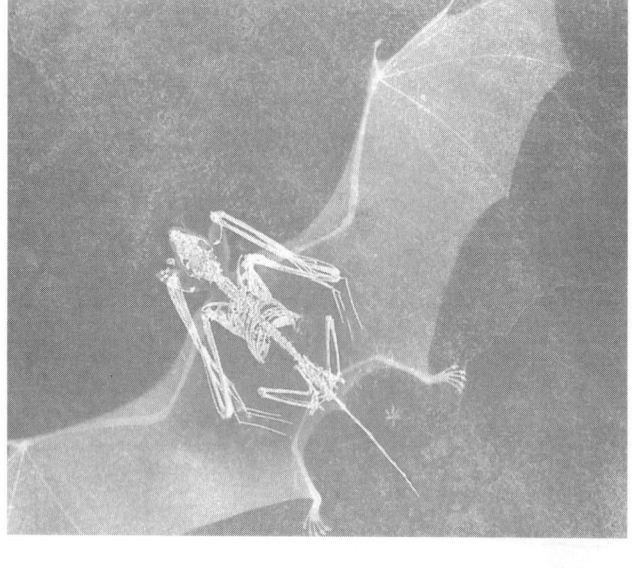

FIGURE 18.12 Fossil bat from the Eocene Green River Formation of Wyoming. The bat's wings are shown as they would have appeared when the animal was alive. Notice that four elongate fingers support each wing.

Cretaceous, but they were undoubtedly present by the Early Paleocene. Most Paleocene primates were mouse-sized, long-snouted animals that looked much like their contemporaries the shrewlike insectivores. By Eocene time, large primates had evolved, and fairly modern-looking lemurs and tarsiers are known from Asia and North America. By Oligocene time, primitive New and Old World monkeys had evolved in South America and Africa, respectively. The hominoids, the group containing apes and humans, evolved during the Miocene (see Chapter 19).

CARNIVORES

Members of the order **Carnivora** are extremely varied in appearance, adaptations, and even diet to some degree. All have well-developed canine teeth used to make the kill and specialized shearing teeth known as **carnassials** for slicing meat (Figure 18.13). A substantial part of their diet is meat, but some carnivores such as cats rarely eat anything but meat, whereas raccoons and bears have varied diets and are thus *omnivorous.* Many land-dwelling carnivores depend on speed, agility, and intelligence to catch their prey, but some employ other tactics. Badgers, for instance, dig prey from burrows, whereas some cats depend on stealth and pouncing rather than speed to catch their meals.

During the Paleocene, carnivorous mammals known as *creodonts* and *miacids* made their appearance. Most were short-limbed, heavy-bodied animals that probably could not run very fast, but then their prey were not particularly fast either. These animals had well-developed carnassials, but they retained such primitive characteristics as short limbs and they were rather flat-footed. Carnivore evolutionary history involves two separate branches, each diverging

from a common ancestor in the Paleocene (Figure 18.14). One branch consisting of creodonts became extinct by Miocene time, but the other branch led to all existing carnivores and includes such diverse animals as cats, raccoons, pandas, seals, weasels, skunks, and hyenas.

One subgroup of living carnivores includes the cats, hyenas, and viverrids (mongooses, civet cats, and genet cats) (Figure 18.14). Despite the obvious differences in appearance of these animals, studies of living species, fossils, and chromosomes clearly indicate that they are more closely related to one another than they are to other carnivores. As a matter of fact, their fossil record is good enough to demonstrate that they differ because they diverged from a common ancestor. Cats, though, have been evolutionarily conservative; they have not changed much since they first evolved, although some remarkable developments in the canine teeth have taken place, giving rise to various saber-toothed cats (see Perspective 5.2).

Dogs, bears, weasels, and pandas (Figure 18.14) have better developed carnassials than their ancestors, and their limbs are longer. Their braincase is larger, too, which probably indicates a higher level of intelligence. Limb elongation was particularly evident in dogs who were no longer flat-footed but rather walked and ran on their toes.

Seals, sea lions, and walruses (Figure 18.14) are adapted to an aquatic life, but their ancestry is not well documented by fossils. The available evidence indicates a relationship to bears. Aquatic adaptations include a streamlined body, a layer of blubber for insulation, and limbs modified into paddles. Most are fish eaters and thus have evolved rather simple, single-cusped teeth, except walruses with their flattened chewing teeth for crushing shells. The largest among these animals are elephant seals, some of which might weight as much as 3.5 metric tons.

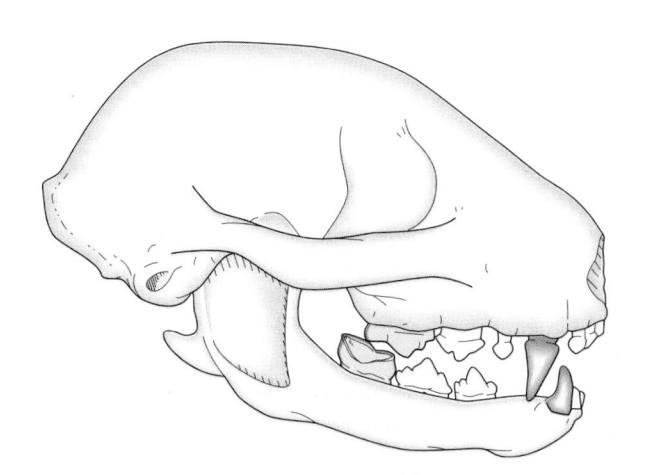

FIGURE 18.13 Jaws and teeth of a cat. Members of the order Carnivora have large canine teeth (brown) and a specialized pair of sharp-crested, bladelike teeth known as carnassials (blue). The upper tooth of a carnassial pair slides past the outer surface of the lower tooth of the pair, giving them a scissors-type action.

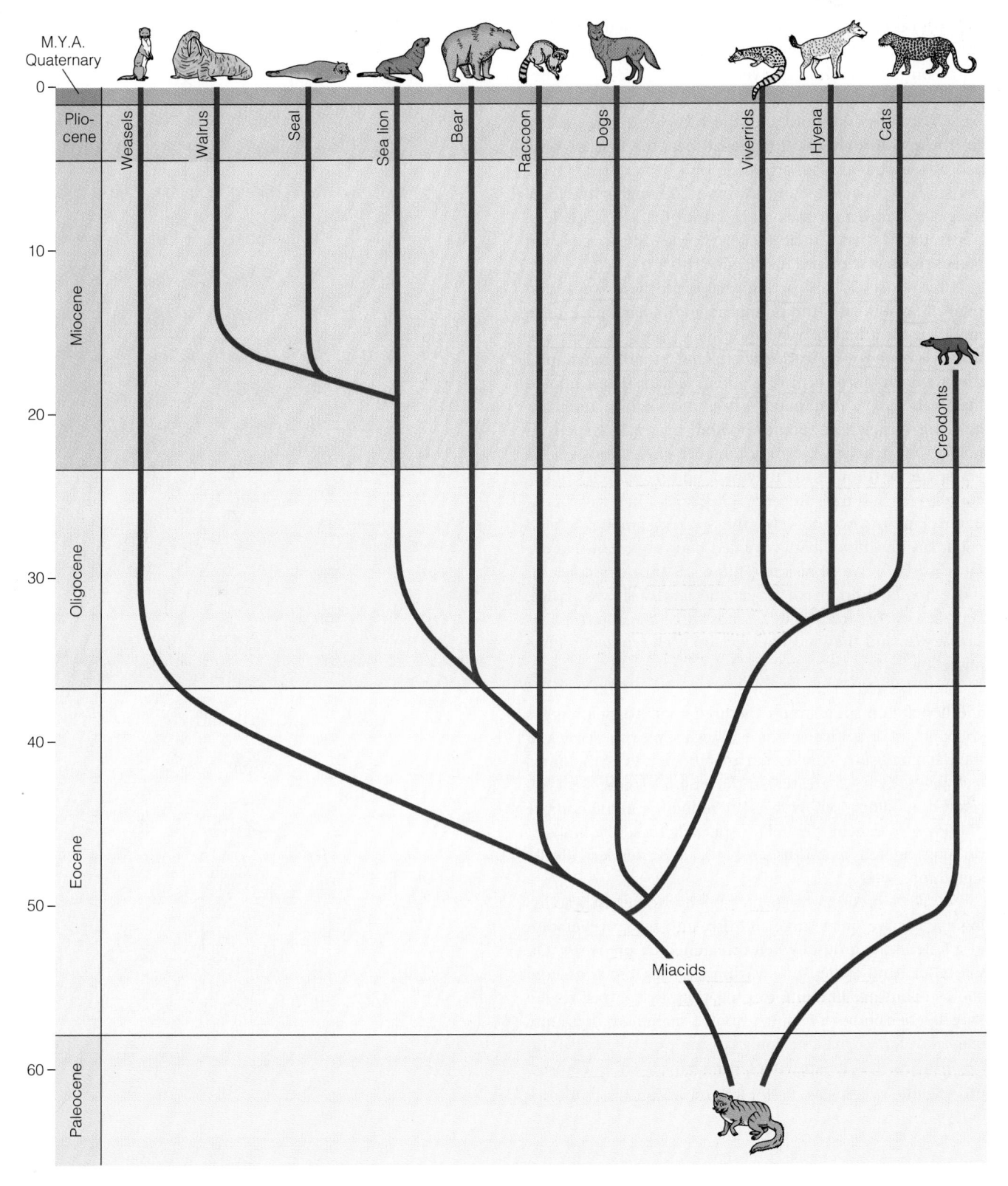

FIGURE 18.14 Evolution of the order Carnivora. Ancestors among the miacids gave rise to all present-day members of this mammalian order. Most of the relationships shown here are well documented by fossils and studies of living animals.

HOOFED MAMMALS

Ungulate is an informal term referring to several types of mammals, the most numerous of which are the orders **Artiodactyla** and **Perissodactyla**. The artiodactyls, or even-toed hoofed mammals, are the most diverse with more than 170 living species of cattle, sheep, goats, swine, antelope, deer, giraffes, hippopotamuses, camels, and several others. Perissodactyls, or odd-toed hoofed mammals, are represented by only 16 living species of horses, rhinoceroses, and tapirs. Many differences exist to differentiate the two orders, but two of the defining characteristics are the number of toes and the way the animal's weight is borne on the toes. Artiodactyls have two or four toes, but more importantly their weight is borne along an axis that passes between the third and fourth digits (Figure 18.15). (In those with only two functional toes, the first, second, and fifth have been lost or remain only as vestiges.) Living perissodactyls have one or three toes, although some fossil species retained four toes on the front feet. Nevertheless, perissodactyls have their weight borne on an axis that passes through the third toe (Figure 18.15b).

All artiodactyls and perissodactyls are herbivores, and their chewing teeth, the premolars and molars, have been modified for a diet of vegetation. One trend in these animals has been **molarization,** involving a transformation of the premolars into teeth resembling molars. Thus, hoofed mammals have a continuous series of molarlike teeth for grinding vegetation. Some hoofed mammals, such as horses, are **grazers,** meaning they eat grass. However, as grasses grow through soil they pick up tiny particles of silicon dioxide and thus are very abrasive, so they wear teeth down rapidly. Accordingly, the grazers among hoofed mammals developed high-crowned chewing teeth that are more resistant to abrasion (Figure 18.15c). In contrast, those hoofed mammals that eat the tender shoots, twigs, and leaves of trees and shrubs are characterized as **browsers,** and they did not develop high-crowned chewing teeth.

Many hoofed mammals live in open-grassland habitats and depend on speed to escape predators. Adaptations for speed in these animals include elongation of the bones of the palm and sole, which increases the length of the limbs and gives them a greater stride length (Figure 18.15a). Another modification involves the reduction in the number of bony elements in the limbs, especially toes. The limbs of running hoofed mammals are long and slender, ideally suited for speed.

Not all hoofed mammals are speedy runners. Some living species are quite small and dart into heavy vegetation or a hole in the ground when threatened by predators. On the other hand, some such as rhinoceroses, are very large and as adults are immune to predation. In contrast to the long slender limbs of running hoofed mammals, these animals developed massive limbs to support their weight. Nevertheless, some large hoofed mammals—rhinoceroses, for example—can run quite fast, at least for short distances.

Artiodactyls—Even-Toed Hoofed Mammals

Rabbit-sized ancestral artiodactyls appeared during the Early Eocene, but at this early stage of development, they differed little from their ancestors. From these tiny ancestors, artiodactyls rapidly diversified into numerous families, many of which are now extinct (Figure 18.16). Among these extinct families were the piglike oreodonts, which were common in North America until their extinction during the Pliocene, and the peculiar protoceratids with forked horns on their snouts.

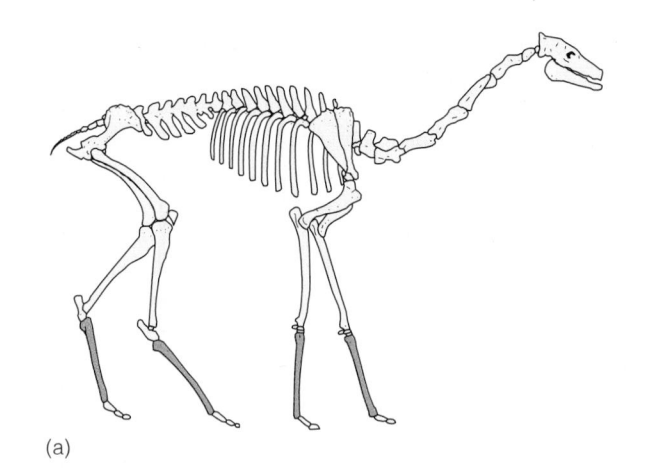
(a)

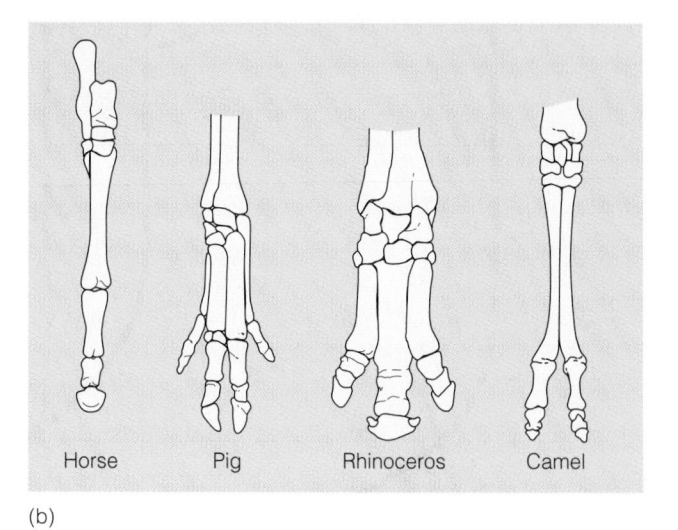

Horse Pig Rhinoceros Camel

(b)

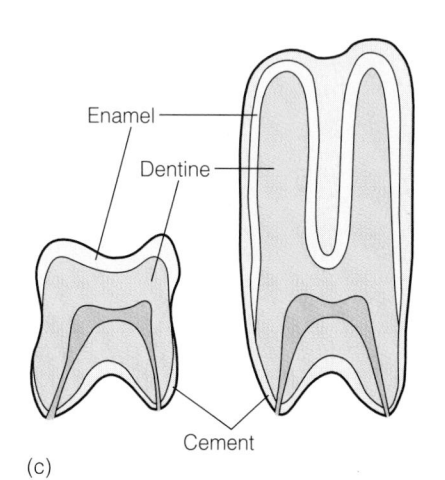

Enamel

Dentine

Cement

(c)

FIGURE 18.15 **(a)** *Oxydactylus,* a Miocene camel, shows the trend of limb elongation seen in many hoofed mammals. The bones between the wrist and toes and the ankle and toes became longer, thereby increasing the length of the limbs. **(b)** The ancestors of hoofed mammals had five-toed feet, but the trend has been a reduction in the number of toes; horses retain only one toe, pigs have four but with reduced side toes, rhinoceroses are three-toed, and camels have two toes. **(c)** Comparison of low-crowned and high-crowned chewing teeth. Grazing hoofed mammals developed high-crowned teeth as an adaptation for eating abrasive grasses. The cusps of high-crowned teeth are elevated into tall, slender pillars, and the teeth are covered with enamel and cement, both of which are hard substances.

Small, four-toed, ancestral camels appeared early in the diversification of artiodactyls (Figure 18.17). By the Oligocene, all camels were two toed, and during the Miocene, they diversified into several distinctive types. Among these were giraffelike camels, gazellelike camels, and giant camels standing 3.5 m at the shoulder. Camels were abundant from Eocene to Pleistocene times in North America, and most of their evolution occurred on this continent. During the Pliocene Epoch, camels migrated to South America and Asia, where their descendants survive. In North America, camels became extinct near the end of the Pleistocene.

The *bovids* (Figure 18.16), the most diverse living artiodactyls, include cattle, bison, sheep, goats, and antelopes. Bovids evolved during the Miocene, but most of their diversification took place during the Pliocene. The most common Tertiary bovid of North America was the pronghorn, which roamed the western interior in vast herds. Bovids evolved on the northern continents but have since migrated to southern Asia and Africa, where they are most common today.

Perissodactyls—Odd-Toed Hoofed Mammals

The living perissodactyls—horses, rhinoceroses, and tapirs—do not look much alike, but they and the extinct ti-

tanotheres and chalicotheres are united by several shared characteristics, and the fossil record indicates they have a common ancestor (Figure 18.18).

Perissodactyls evolved during the Eocene and increased in diversity through the Oligocene, but they have declined markedly since then (Figure 18.19). Today, perissodactyls constitute a minor part of the fauna, and rhinoceroses appear to be on the verge of extinction. It has been suggested that competition with artiodactyls could explain the decline of perissodactyls. Most artiodactyls are **ruminants,** meaning "cud-chewing." They are herbivores with complex three- or four-chambered stomachs. Ruminants use the same resources as perissodactyls, but use them more efficiently.

Evolution of Horses. The earliest member of the horse family was **Hyracotherium** (Figure 18.20) from the Early Eocene of Great Britain and western North America. About the size of a fox, it had four-toed front feet and three-toed hind feet. Each toe, however, was covered by a small hoof. *Hyracotherium* has few of the specializations we associate with horses (Table 18.1).

Since *Hyracotherium* was so unhorselike, how can we be sure it belongs to the horse family at all? Fortunately, the fossil record of horses is exceptionally good, and this record clearly shows that *Hyracotherium* is linked to the

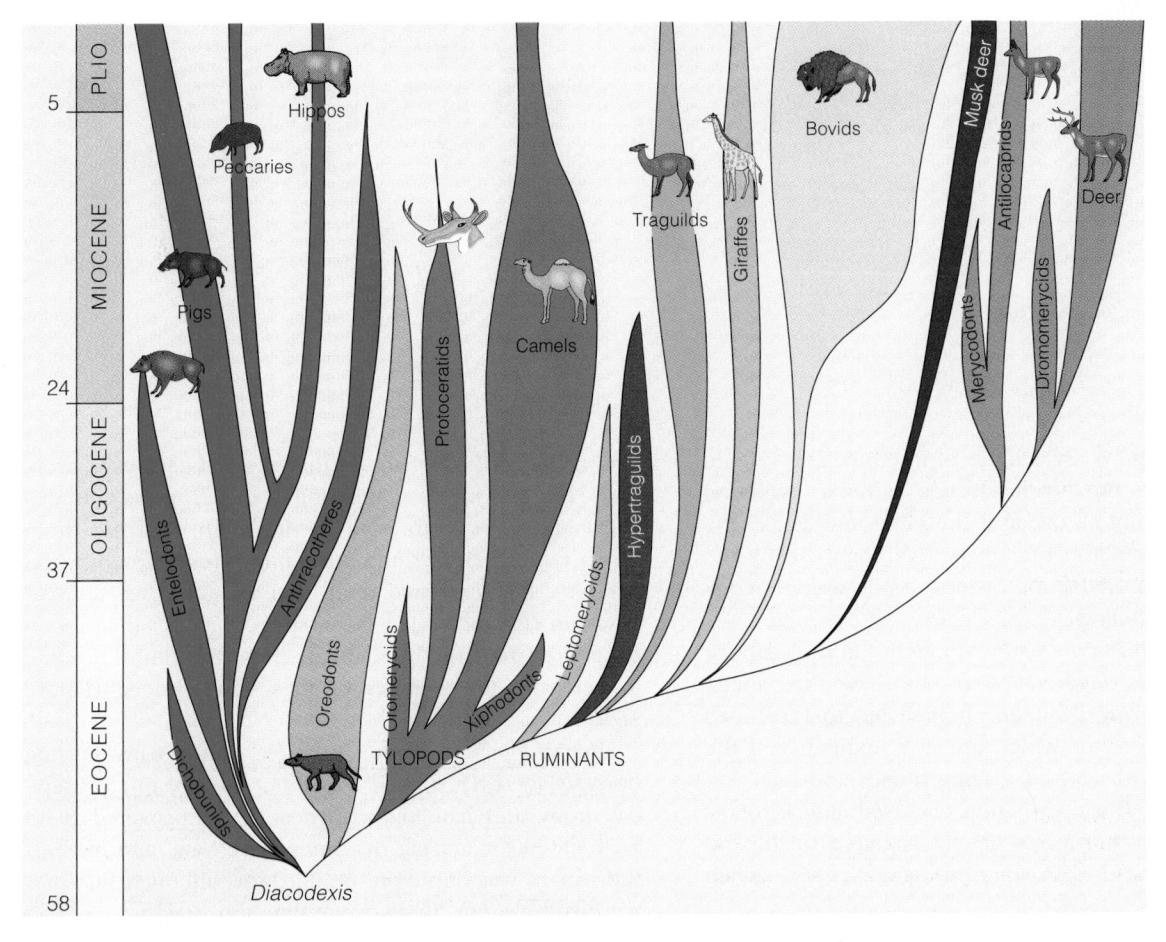

FIGURE 18.16 History of the artiodactyls. Early in their history, artiodactyls split into three major groups: the suids include the pigs, hippopotamuses, and extinct giant hogs; the tylopoda are represented by the camels; and the ruminants consist of the cud-chewing animals.

FIGURE 18.17 (a) Two genera of fossil camels showing the trends of increased size, longer legs, and development of high-crowned chewing teeth. (b) Camels were quite common during the Tertiary in North America where most of their evolution took place. This mural shows giraffe camels *(left background)* and various other Late Miocene hoofed mammals of the western Nebraska plains. Also shown are ancestors of the present-day pronghorn *(left)*, the rhinoceros *Teleoceras,* and ancestors of today's horses *(right foreground)*.

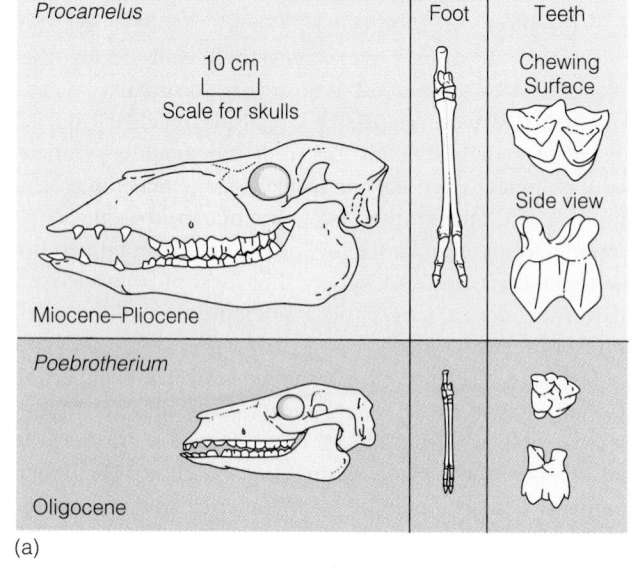

(a)

(b)

modern horse, *Equus,* by a series of intermediates (Figure 18.20). However, the various trends in horse evolution (Table 18.1) did not all occur at the same rate. For example, molarization of the premolars was complete by the Oligocene, but the reduction of toes to one was not complete until the Pliocene.

We can recognize two distinct branches in the adaptive radiation of horses (Figure 18.20a). Both have their ancestry in *Hyracotherium;* but one branch led to three-toed browsing horses, all of which are now extinct, and the other led first to three-toed grazing horses and finally to one-toed grazers. Horse evolution occurred almost totally in North America, but a few genera migrated to the Old World, and during the Pliocene one genus even reached South America.

The appearance of grazing horses during the Miocene coincides with the appearance of the open-prairie, grass-

land habitat. Like all habitats this one had its special problems. For one thing, as mentioned earlier, grass contains silica and is very abrasive to the teeth, wearing them down rapidly. Horses and other ungulates adapted by developing high-crowned teeth that are more resistant to abrasion (Figure 18.15). Also, speed was essential to escape predators in this habitat; thus, the limbs became longer, and the number of toes was reduced (Figure 18.20b).

Merychippus is a good example of the early grazing horses (Figure 18.20b). It was about the size of a present-day pony and had high-crowned, cement-covered teeth well suited for grazing. Its limbs were long, and most of the weight was borne on the third toe, although side toes were still present. By Pliocene time, one-toed grazers had evolved (Figure 18.20b). Compared with *Merychippus,* these horses were larger and had longer limbs and higher-

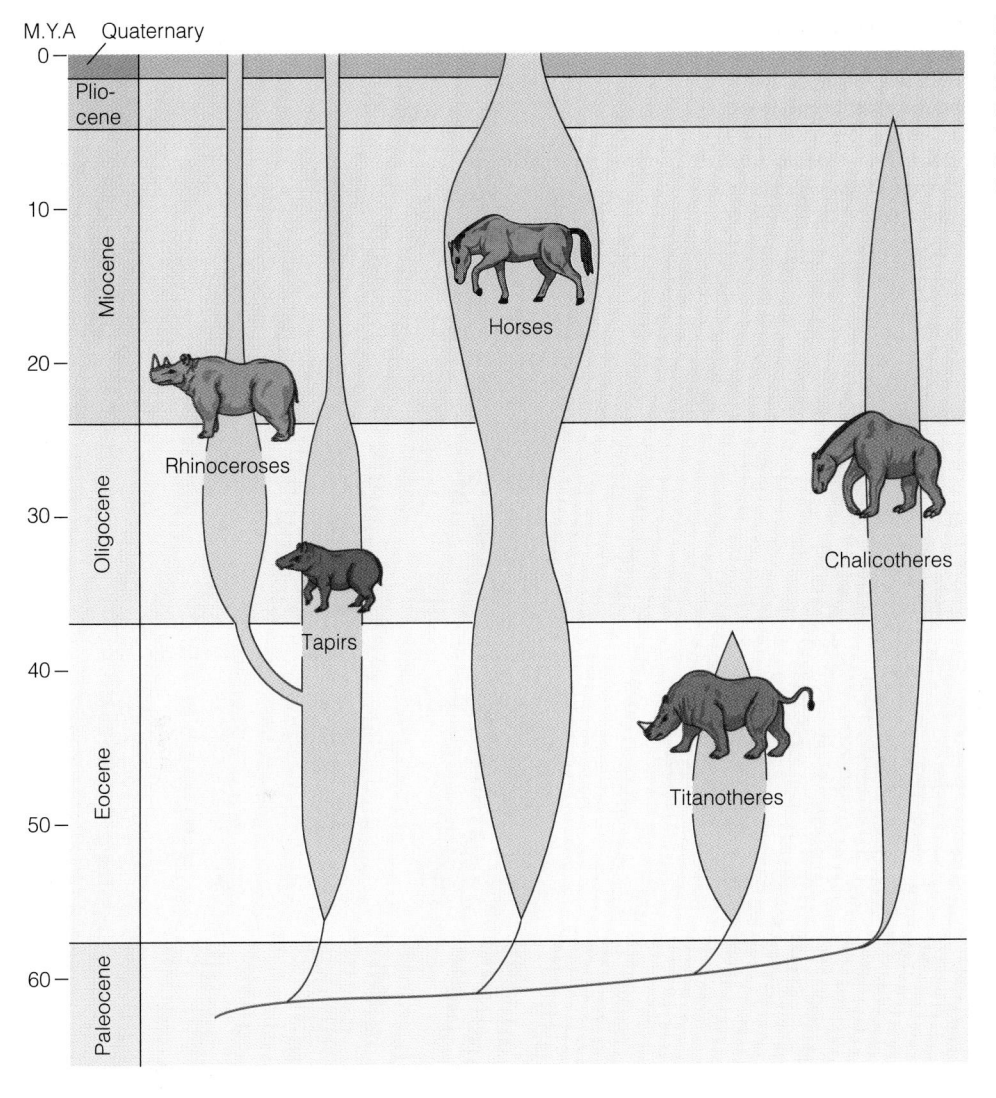

FIGURE 18.18 Evolution of the perissodactyls (odd-toed hoofed mammals). Adaptive radiation from a common ancestor accounts for the facts that perissodactyls share several characteristics yet differ markedly in appearance.

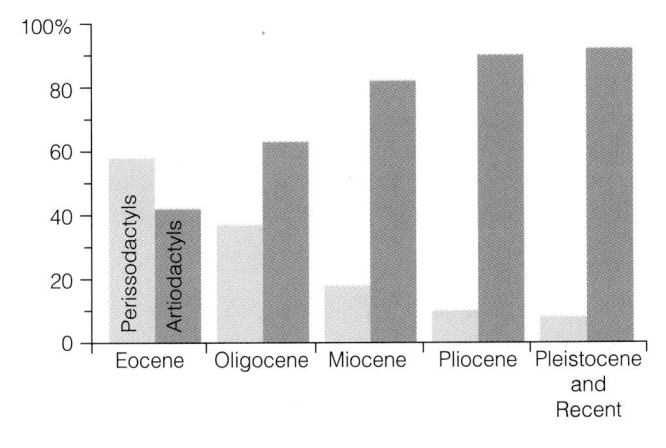

FIGURE 18.19 The relative abundance of perissodactyls and artiodactyls varied throughout the Cenozoic. During the Eocene, perissodactyls accounted for nearly 60% of the hoofed-mammal fauna but have declined and now make up only about 8%. Artiodactyles, on the other hand, have increased from about 40% in the Eocene to more than 90% of the hoofed mammals now.

crowned teeth. Modern *Equus* evolved during the Late Pliocene in North America and migrated to the Old World, where it still lives in the wild.

Other Perissodactyls. The other living perissodactyls—tapirs and rhinoceroses—both had Eocene ancestors that looked much like *Hyracotherium*. In fact, all the earliest perissodactyls, including the extinct titanotheres and chalicotheres, are so similar that it is difficult to differentiate among them.

Both tapirs and rhinoceroses increased in size during the Cenozoic and became more abundant, diverse, and widespread than they are now. An Oligocene–Miocene hornless rhinoceros of Asia was the largest land mammal ever to exist; it was more than 5 m tall at the shoulder and probably weighed 13 or 14 metric tons. Most rhinoceros evolution occurred in the Old World, but North American rhinoceroses were quite common until they became extinct during the Late Pleistocene (Figures 18.2 and 18.11).

FIGURE 18.20 Evolution of horses. (a) Summary chart showing the recognized genera of horses and their evolutionary relationships. Note that during the Oligocene two separate lines emerged, one leading to three-toed browsing horses and the other to one-toed grazers, which includes all present-day horses. **(b)** Simplified diagram showing some of the evolutionary trends from *Hyracotherium* to the present-day horse, *Equus*. Important trends shown here include an increase in size, loss of toes, and development of high-crowned teeth with complex chewing surfaces.

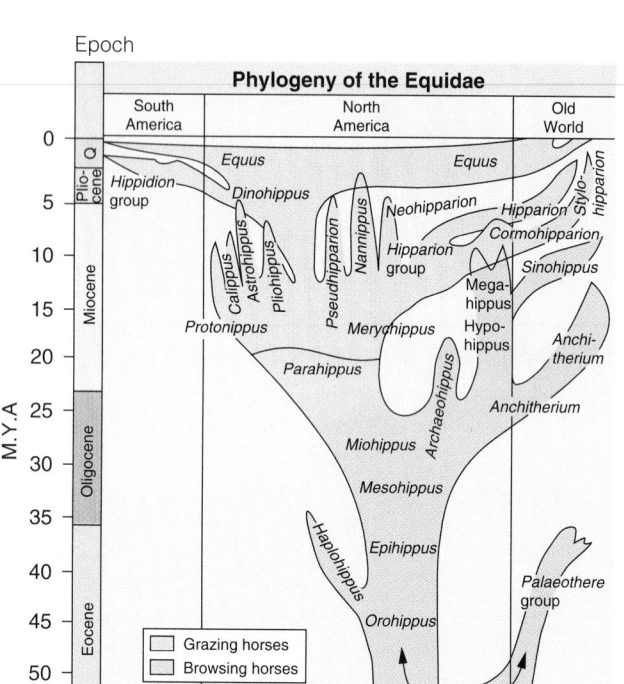

Equus
(Pleistocene)

Pliohippus
(Pliocene)

Merychippus
(Miocene)

Mesohippus
(Oligocene)

Hyracotherium
(Eocene)

(a)

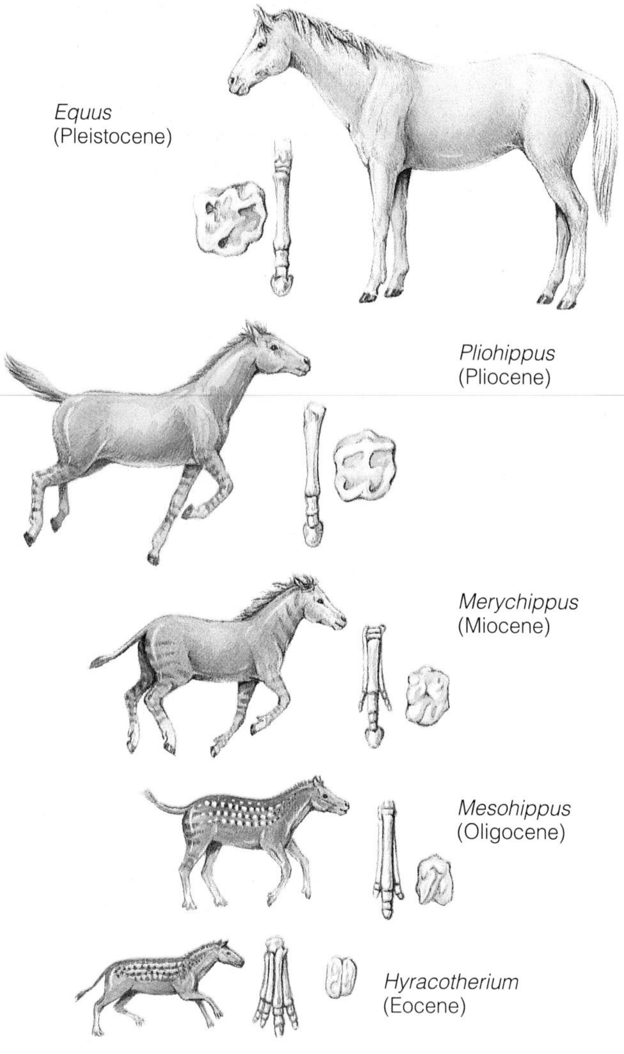

(b)

In addition to horses, rhinoceroses, and tapirs, small *Hyracotherium*-like ancestors also gave rise to titanotheres and chalicotheres (Figure 18.18). Titanotheres existed only from the Early Eocene to Early Oligocene, but they evolved from small ancestors to giants about 2.5 m high at the shoulder (Figure 18.2). The fossil evidence indicates that chalicotheres were never very abundant, although the group existed from the Early Eocene to the Pleistocene. Miocene and later chalicotheres were large animals that superficially resembled horses, but their feet had claws rather than hoofs (Figure 18.21). The prevailing opinion favors the interpretation that these claws were used to hook and pull down branches.

GIANT MAMMALS—ELEPHANTS AND WHALES

We began our discussion of mammals with a consideration of the small mammals and noted that, even though some are not closely related, they constitute most mammals and occupy microhabitats unavailable to larger animals. We end our discussion with a consideration of the largest mammals; these, too, are not particularly closely related, but they are the giants in their respective habitats—elephants on land and whales in the seas.

The African and Indian elephants are the only two surviving species of the once diverse and widespread order **Proboscidea.** During much of the Cenozoic, elephants of one kind or another occupied all the northern continents but now are restricted to only Africa and Southeast Asia. Even though the order has declined, it has a rich fossil record showing that rather small (100- to 200-kg) animals with few characteristics of today's elephants gave rise to the giants now present.

One of the earliest elephant ancestors, *Moeritherium* from the Eocene, had small tusks but resembled a hippopotamus and probably lived a semiaquatic life (Figure 18.22). By Oligocene time, elephants showed the trend to large size and the development of a long snout (proboscis) and large tusks. Incidentally, tusks are modified incisors, not canines as some people think. Among the varied types of fossil elephants were the deinotheres with tusks that curved downward, mastodons that had teeth adapted to browsing, and mammoths with teeth modified for grazing.

TABLE 18.1

Overall Trends in the Cenozoic Evolution of the Present-Day Horse *Equus*. A number of horse genera existed during the Cenozoic that evolved differently. For instance, some horses were browsers rather than grazers and never developed high-crowned chewing teeth, and retained three toes.

TREND
1. Size increase.
2. Legs and feet become longer, an adaptation for running.
3. Lateral toes reduced to vestiges. Only toe three remains functional in *Equus*.
4. Straightening and stiffening of the back.
5. Incisor teeth become wider.
6. Molarization of premolars yielded a continuous row of teeth for grinding vegetation.
7. The chewing teeth, molars and premolars, become high-crowned and cement covered for grinding abrasive grasses.
8. Chewing surfaces of premolars and molars become more complex. Also an adaptation for grinding abrasive grasses.
9. Front part of skull and lower jaw become deeper to accommodate high-crowned premolars and molars.
10. Face in front of eye becomes longer to accommodate high-crowned teeth.
11. Larger, more complex brain.

Mastodons evolved in Africa, but from Miocene to Pleistocene time they spread over the Northern Hemisphere continents and one genus even reached South America. Tusk size and shape varied considerably among mastodons (Figure 18.23).

The mammoths and present-day elephants diverged during the Pliocene and Pleistocene. Most mammoths were no larger than living elephants, and many were in fact smaller, but they had the largest tusks of any elephants. Until their extinction near the end of the Pleistocene, mammoths lived on all the northern continents, as well as in Africa and India.

FIGURE 18.21 Skeletons of perissodactyls known as chalicotheres. These animals were rather horselike but had claws on their feet rather than hoofs.

Living blue whales, at more than 30 m long and weighing an estimated 130 metric tons, are likely the largest animal of any kind to ever exist. Fragmentary remains of some sauropod dinosaurs may indicate animals of comparable size, but this is uncertain. Though one major trend in whale evolution has been large size, not all living or fossil whales are large. Consider, for example, dolphins and porpoises, which are sizable but hardly giants.

River and sea otters, seals, sea lions, and walruses have all returned to an aquatic or semiaquatic existence, but none are so thoroughly adapted to an aquatic life as whales (order **Cetacea**). The fossil records indicates that whales evolved during the Early Eocene from wolf-size, meat-eating, land-dwelling animals known as *mesonychids* (Figure 18.24). Changes that took place during the transition from mesonychids to whales include modification of the front limbs to finlike flippers, loss of the hind limbs, migration of the nostrils to the top of the head to form a blowhole, and development of a large horizontal tail fluke for propulsion.

Fossils discovered during the last 15 years provide a series of intermediates that convince paleontologists that fully aquatic whales evolved from terrestrial ancestors. For instance, the Early Eocene whale *Ambulocetus* still possessed limbs capable of carrying the animal's weight on land, although it was no doubt rather awkward. *Basilosaurus*, a 15-m-long, Late Eocene whale had tiny vestigial hind limbs (Figure 18.24). (See Perspective 5.3 for more information on vestigial structures in whales.) This animal still retained teeth similar to those of mesonychids, and its nostrils (blowhole) were still on the snout. And although fully a whale, it was quite differently proportioned than today's whales. By Oligocene time, both groups of living whales, the toothed whales and baleen whales, had evolved.

Mammals of the Pleistocene Epoch

One of the most remarkable aspects of the Pleistocene mammalian fauna is that so many very large species existed. In North America, for example, there were mastodons and mammoths, giant bison, huge ground sloths, giant camels, and beavers nearly 2 m tall at the shoulder (Figure 18.25). Kangaroos standing 3 m tall, wombats the size of rhinoceroses, leopard-sized marsupial lions, and large platypuses characterize the Pleistocene fauna of Australia. In Europe and parts of Asia lived cave bears, elephants, and the giant deer commonly called the Irish elk with an antler spread of 3.5 m (see Perspective 18.2).

Many smaller mammals were also present, many of which still exist. The major evolutionary trend, however, was toward large body size. Perhaps this was an adaptation to the cooler temperatures of the Pleistocene. Large animals have proportionately less surface area compared to their volume and thus retain heat more effectively than do smaller animals.

The frozen remains of several Pleistocene mammals have been found in Alaska and Siberia. Probably the best-

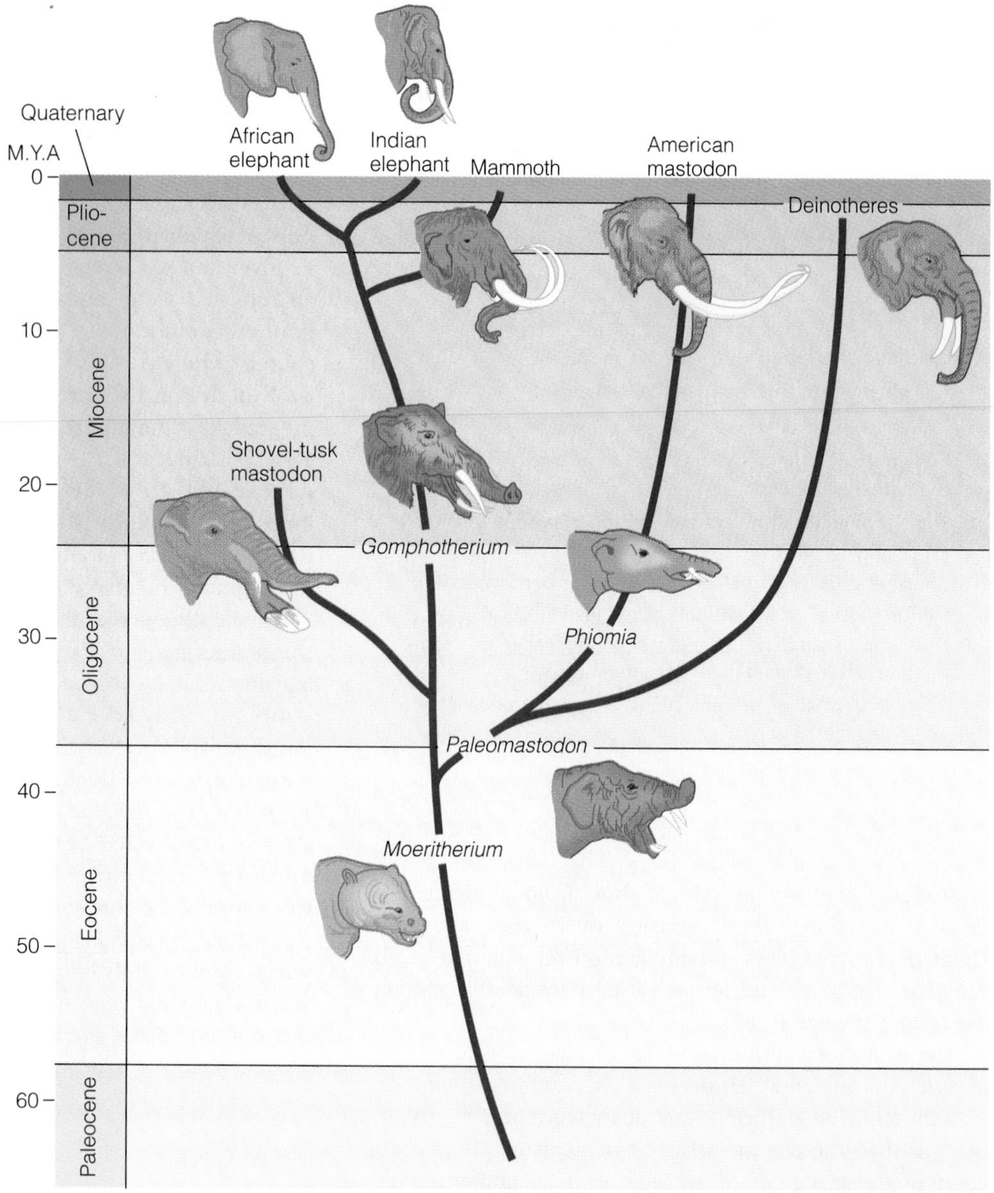

FIGURE 18.22 Phylogeny of present-day elephants and some of their relatives. Increased size and development of large tusks and a long proboscis were some of the evolutionary trends in this group. Several types of fossil elephants are not shown, so they were actually more diverse during much of the Cenozoic than indicated in this illustration.

African elephant Indian elephant Mammoth American mastodon Deinotheres

Quaternary
M.Y.A
0 —
Plio-cene
10 —
Miocene
20 —
Shovel-tusk mastodon
Gomphotherium
30 —
Oligocene
Phiomia
40 —
Paleomastodon
Moeritherium
50 —
Eocene
60 —
Paleocene

FIGURE 18.23 Restorations of three mastodons showing variation in tusk size and shape. (a) *Ambelodon* of the Late Miocene of North America had shovel-like lower tusks and small upper tusks. (b) *Stegodon* had tusks so closely spaced that there was no space between them for the trunk. This mastodon lived during the Pliocene and Pleistocene. (c) *Anancus* from France was a Pliocene–Pleistocene genus with straight tusks 3 m long.

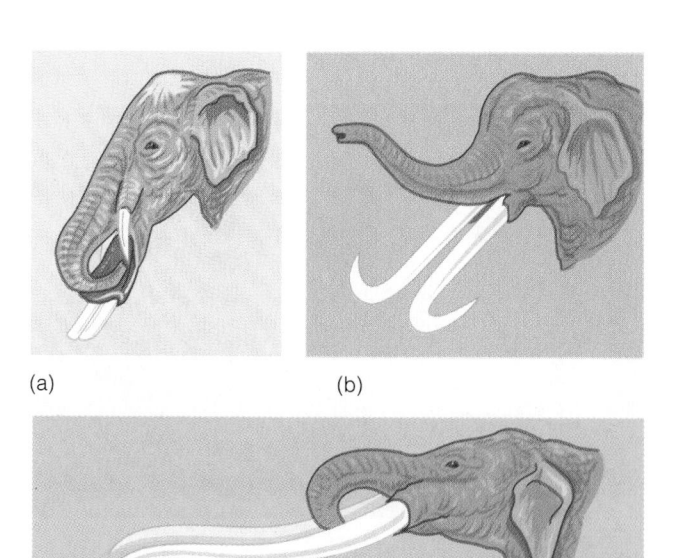

(a)

(b)

(c)

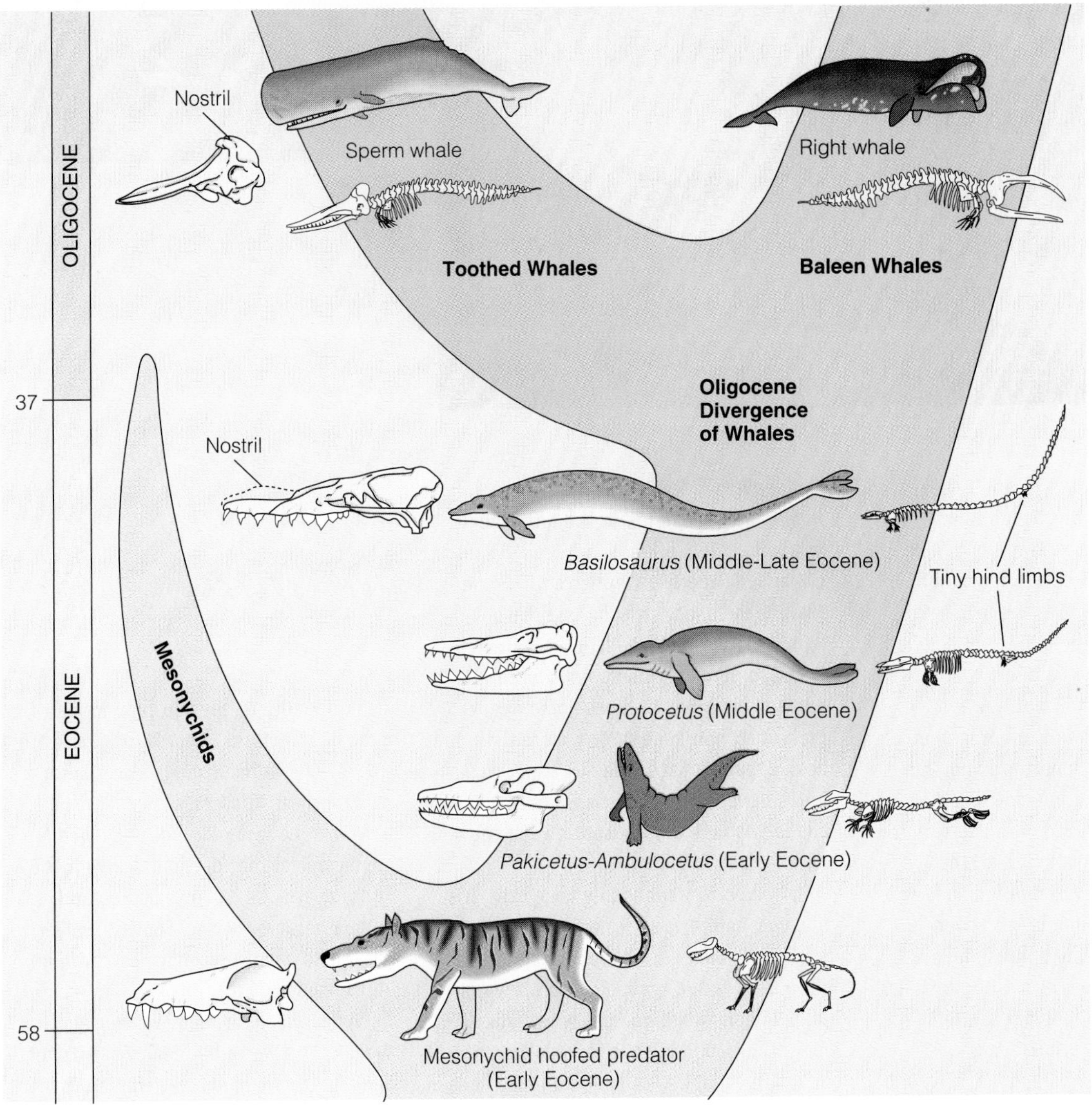

OLIGOCENE

Nostril

Sperm whale

Right whale

Toothed Whales

Baleen Whales

37

**Oligocene
Divergence
of Whales**

Nostril

Basilosaurus (Middle-Late Eocene)

Tiny hind limbs

Mesonychids

Protocetus (Middle Eocene)

EOCENE

Pakicetus-Ambulocetus (Early Eocene)

Mesonychid hoofed predator
(Early Eocene)

58

FIGURE 18.24 Evolution of the whales.

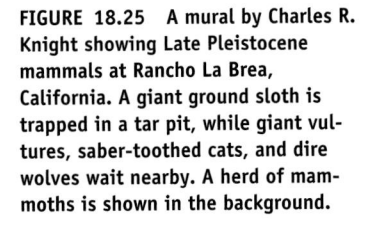

FIGURE 18.25 A mural by Charles R. Knight showing Late Pleistocene mammals at Rancho La Brea, California. A giant ground sloth is trapped in a tar pit, while giant vultures, saber-toothed cats, and dire wolves wait nearby. A herd of mammoths is shown in the background.

The Giant Irish "Elk" and Other Horned or Antlered Mammals

*A*mong the artiodactyls, the ruminants are by far the most diverse and numerous, and most have some kind of horns or antlers. Horns, as in cattle, have a bony core surrounded by a permanent sheath, whereas antlers are bony outgrowths from the skull and are shed annually as in deer, elk, and moose. Some horns and antlers are used for protection from predators, but probably, more important, they serve as displays for attracting mates and for intraspecific tests of strength to establish dominance. Whatever their use, they come in a fantastic variety of shapes and sizes in today's ruminants, and even larger and more unusual horns and antlers are found in a number of extinct animals.

Some of the more unusual antlers are found in such extinct animals as *Synthetoceros*, which belonged to a group of ruminants known as protoceratids, and

in the extinct deer *Eucladoceros* and *Hoplitomeryx* (Figure 1). The latter genus, in addition to having a five-horned skull, possessed saberlike canine teeth. Some small present-day deer in Asia use enlarged canine teeth to strip bark from trees, so perhaps *Hoplitomeryx* used its teeth in a similar fashion. Some extinct cattle had horns measuring more than 4 m across, but probably the most spectacular antlers ever present belonged to the so-called Irish elk that existed during the Late Pleistocene.

Megaloceros, commonly called the Irish elk, was neither an elk nor was it restricted to Ireland, although many fossils have been found there. In fact, it was a fallow deer, with large males weighing more than 700 kg. That is, they were the size of present-day moose, but they had an antler spread of up to 3.5 m (Figure

2)! Their antlers, present only in males, probably weighed 34 to 40 kg. These truly spectacular animals ranged over Europe and much of Asia. *Megaloceros* seems to have been a fast-running, open-woodlands animal that declined markedly beginning about 12,000 years ago and finally became extinct. Some, however, appear to have survived in parts of Europe until about 500 B.C.

As is typical of hoofed mammals, *Megaloceros* was a herbivore, but its teeth indicate that it had a varied diet. A diet of shoots, twigs, and leaves probably sus-

FIGURE 1   Some of the antlered, hoofed mammals of the Cenozoic. (a) *Synthetoceros* from the Late Miocene and Pliocene of North America had two antlers on its skull and a Y-shaped one on its snout. (b) *Eucladoceros* had an antler spread of only 1.7 m, but it had 12 tines on each antler. It lived in Europe during the Pleistocene. (c) *Hoplitomeryx* from the Late Miocene of Italy had five bony outgrowths from its skull and enlarged canine teeth.

(a) *Synthetoceros*   (b) *Eucladoceros*   (c) *Hoplitomeryx*

FIGURE 2 Restoration of the giant deer *Megaloceros giganteus,* commonly called the Irish elk. It lived in Europe and Asia during the Pleistocene. Large males had antler spreads of up to 3.5 m.

tained the animals through part of the year, but incisors similar to those of today's true elk indicate that grazing was also important, perhaps at different times of the year. The overall appearance of *Megaloceros* is well known from numerous fossils and from cave paintings.

As one would expect from the name "Irish elk," fossils of this animal are common in Ireland. Indeed, they are, especially in bog deposits. Probably the largest concentrations of these fossils come from Ballybetagh Bog near Dublin, where the remains of more than 100 males have been found. One explanation for this unusual number of fossils, all of antlered males, is that the animals became mired in mud and drowned because of their heavy antlers.

The fact that all were males might seem to support this hypothesis, but several other observations indicate that it is almost certainly incorrect. For instance, if this hypothesis is correct, one would expect males with relatively large antlers, but they are actually smaller than some 150 other specimens from other areas. In addition, if large animals were trapped and died in mud deposits, one would expect the deposits to be disrupted by thrashing animals in their attempts to escape, and the bones of at least the legs should be found penetrating the mud deposits. Neither of these is found, thus making the miring–drowning hypothesis an unlikely explanation.

Another more likely hypothesis is based on observations of the fossils and of present-day deer that visit the bog. Today, mostly male deer visit the bog, and many of the fossil bones are broken and show signs of scavenging—indicating that the animals probably died near the lakeshore where they were trampled and gnawed. Also, the deaths seem to have been seasonal, probably mostly in winter, and many of the fossils are from animals that were in poor physical condition. One might also expect juveniles to be common in the deposits, but this is not the case. However, this is not too surprising because juveniles with smaller and less durable bones have a lower potential for preservation and are typically underrepresented in most fossil accumulations.

known frozen remains are the woolly mammoths of Siberia (see Figure 3.19b). More than three dozen frozen carcasses have been recovered, but only a few are fairly complete. Nevertheless, they provide a wealth of information unavailable from fossilized bones and teeth. Contrary to the popular myths that these animals were found in blocks of ice or even icebergs, all have been recovered from permafrost (permanently frozen soil).

In addition to mammals, some other Pleistocene vertebrate animals were of impressive proportions. We have already mentioned the giant moas of New Zealand and the elephant birds of Madagascar, but Australia also had giant birds standing 3 m tall and weighing nearly 500 kg and a lizard 6.4 m long and weighing 585 kg. The tar pits of Rancho La Brea in southern California contain the remains of at least 200 kinds of animals. Many of these are fossils of dire wolves, saber-toothed cats, and other mammals, but there are also remains of many birds, especially birds of prey, and a giant vulture with a wing-span of 3.6 m (Figure 18.25).

Pleistocene Extinctions

Many of the large mammals and some large birds of North America, South America, and Australia became extinct about 10,000 years ago, near the end of the Pleistocene. Extinction is a continuous phenomenon in life history, but this was a rather modest mass extinction compared with the ones at the end of the Paleozoic and Mesozoic. It was unusual, though, in that most of its victims were large, land-dwelling mammals weighing more than 40 kg. In North America, 33 (73%) of 45 large genera died out; in South America, 46 (80%) of 58 genera went extinct; and Australia lost 15 (94%) of its 16 large genera. In contrast, Europe lost 7 of 23 large genera, whereas only 2 of 44 died in Africa south of the Sahara. Clearly, extinctions were much more severe in the Americas and Australia.

What caused these extinctions, why they were mostly restricted to large, land-dwelling mammals, and the fact that most extinctions took place in the Americas and Australia has been the subject of debate. According to one hypothesis, these animals died out because they could not adapt to rapid climatic changes at the end of the Pleistocene. A competing hypothesis, known as *prehistoric overkill,* holds that human hunters killed off these mammals and some birds.

Researchers favoring climatic cause for the extinctions point to the rapid changes in climate and vegetation that occurred over much of Earth's surface during the Late Pleistocene. As various glaciers began retreating, the North American and northern Eurasian open-steppe tundras were replaced by conifer and broad-leaf forests as warmer and wetter conditions prevailed. The Arctic region flora changed from a productive herbaceous one that supported a variety of large mammals, including mammoths, to a relatively barren water-logged tundra that supported a far sparser fauna. The southwestern U.S. region also changed from a moist area with numerous lakes, where saber-tooths, giant ground sloths, and mammoths roamed, to a semiarid environment unable to support a diverse large mammalian fauna.

While rapid changes in climate with their accompanying effects on vegetation can certainly result in changes in animal populations, the hypothesis of climate as an agent of extinction presents several problems. First, why didn't the large mammals migrate to more suitable habitats as the climate and vegetation changed? After all, many other animal species did. For example, reindeer and the Arctic fox lived in southern France during the last glaciation and migrated to the Arctic when the climate became warmer. Why didn't the mammoths and other large animals that were adapted to cold climates simply migrate north?

The second argument against the climatic hypothesis is the apparent lack of correlation between extinctions and the earlier glacial advances and retreats throughout the Pleistocene Epoch. Previous changes in climate were not marked by episodes of mass extinctions.

Paul Martin of the University of Arizona at Tucson, the leading proponent of the prehistoric overkill hypothesis, argues that the mass extinctions in North and South America and Australia coincided closely with the arrival of humans in each area. According to Martin, hunters had a tremendous impact on the faunas of North and South America about 11,000 years ago because the animals had no previous experience with humans. The same thing happened much earlier in Australia soon after people arrived about 40,000 years ago. No large-scale extinctions occurred in Africa and most of Europe because animals in those regions had long been familiar with humans.

Several arguments can also be made against the prehistoric overkill hypothesis. One problem is that archaeological evidence indicates that the early human inhabitants of North and South America, as well as Australia, probably lived in small, scattered communities, gathering food and hunting. It is hard to imagine how a few hunters could have decimated so many species of large mammals. On the other hand, humans have caused major extinctions on oceanic islands. For example, in a period of about 600 years after arriving in New Zealand, humans exterminated several species of large, flightless birds called moas.

A second problem is that present-day hunters concentrate on smaller, abundant, and less dangerous animals. The remains of horses, reindeer, and other small animals are found in many prehistoric sites in Europe, while large mammoth and woolly rhinoceros remains are scarce.

Finally, few human artifacts are found among the remains of extinct animals in North and South America, and there is usually little evidence that the animals were hunted. Countering this argument is the assertion that the impact on the previously unhunted fauna was so swift as to leave little evidence.

The reason for the extinctions of large mammals about 10,000 years ago is still unresolved and probably will be for some time. It may turn out that the extinction resulted from a combination of many different circumstances. Populations that were already under stress from climatic changes were perhaps more vulnerable to hunting when humans occupied new areas.

Cenozoic Vegetation and Climate

Angiosperms, or flowering plants, evolved during the Late Jurassic, and their diversification continued into the Tertiary. Some gymnosperms, especially conifers, remained abundant, and seedless vascular plants still occupied many habitats. Overall, the Cenozoic was a time when more and more familiar types of plants evolved.

Many of the Tertiary plants would be quite familiar to us, but their geographic distribution was markedly different than it is now. It has long been known that plant distribution is strongly controlled by climate, and for the Tertiary we see evidence of changing climatic patterns accompanied by shifting distributions of plants. Mean annual temperatures during the Paleocene and Eocene were high, but a marked temperature decrease occurred at the end of the Eocene (Figure 18.26).

Leaf structure is a good climatic indicator. For instance, leaves with entire or complete margins characterize areas of high annual precipitation and high mean annual temperatures, whereas leaves with incised margins typify cooler, less humid regions. If a fossil flora has a high percentage of entire-margined leaves, the climate was probably wet and warm.

The types of plants found in fossil floras are also good climatic indicators. For example, Paleocene strata in the western interior of North America contain fossil ferns and palms, which are common to warmer, more humid climates. All of North America, including most of Alaska, had warm-temperate to subtropical climates throughout the Paleocene.

In fact, sea floor sediments and geochemical data indicate that about 55 million years ago an abrupt warming trend took place. During this event, known as the **Late Paleocene Thermal Maximum,** large-scale oceanic circulation patterns were disrupted so that heat transfer from equatorial regions to the poles diminished or ceased. In any event, deep oceanic waters became considerably warmer, resulting in extinctions of many deep-dwelling foraminifera. Some scientists think that that this deep, warm water also caused the release of methane from sea floor methane hydrates, thus contributing an important greenhouse gas to the atmosphere and either causing or contributing to the temperature increase at this time.

Subtropical conditions continued into the Eocene of North America, probably the warmest of all the Tertiary epochs. The fossil flora of the Eocene beds of the John Day Beds in Oregon includes tropical ferns, figs, and laurels (Figure 18.27; see the Prologue), all of them plants found today only in the humid parts of Mexico and Central America. Yellowstone National Park, Wyoming, does not now have a climate in which avocado, magnolia, laurel, and fig trees could grow. Yet, during the Eocene Epoch, these kinds of trees grew there. Their presence indicates that the climate of Wyoming was considerably warmer during the Eocene than it is now.

A major climatic change occurred at the end of the Eocene Epoch (Figure 18.26). Mean annual temperatures dropped as much as 7° C in about 3 million years. From the Early Oligocene on, mean annual temperatures have varied somewhat worldwide but overall have not changed

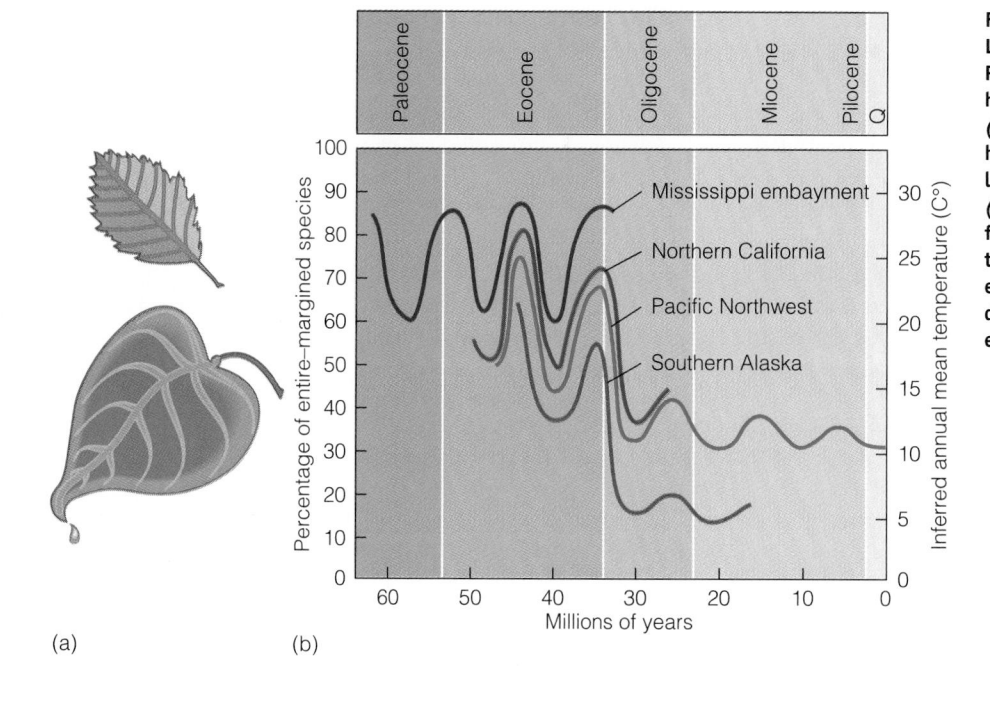

FIGURE 18.26 Cenozoic climate. (a) Leaves are good indicators of climate. Plants adapted to cool climates typically have small leaves with incised margins *(top)*, while plants of humid, warm habitats have larger, entire-margined leaves, and many have pointed drip-tips *(bottom)*. (b) Inferred climatic trends for four areas in North America based on the percentages of plant species with entire-margined leaves. Notice the sharp drop in mean annual temperature at the end of the Eocene.

(a)   (b)

much in the middle latitudes except during the Pleistocene.

A general decrease in precipitation over the last 25 million years took place throughout the midcontinent of North America. As the climate became drier, the vast forests of the Oligocene gave way first to *savannah* conditions (grasslands with scattered trees) and finally to *steppe* environments (short-grass prairie of the desert margin). Herbivorous mammals quickly adapted to the savannah habitat by developing high-crowned chewing teeth suitable for a diet of grass. Among the horses in particular, grazers became more common, and browsers declined to extinction.

Intercontinental Migrations

The mammalian faunas of North America, Europe, and northern Asia exhibited many similarities throughout the Cenozoic. Even today, Asia and North America are only narrowly separated at the Bering Strait, and several times during the Cenozoic the Bering Strait formed a land corridor across which mammals migrated. During the Early Cenozoic, a land connection existed between Europe and North America, allowing mammals to roam across all the northern continents. Many did; camels and horses are only two examples.

On the other hand, the southern continents were largely separate island continents during much of the Cenozoic. Africa remained fairly close to Eurasia, and, at times, faunal interchange between those two continents was possible. For example, elephants first evolved in Africa, but they migrated to all the northern continents.

South America was an island continent from the Late Cretaceous until a land connection with North America was established about 5 million years ago. At that time, the South American mammalian fauna consisted of various marsupials and several orders of placental mammals that lived nowhere else in the world. These animals thrived in isolation and showed remarkable convergence with North American placentals (see Perspective 5.2, Figure 2).

When the Isthmus of Panama formed, however, migrants from North America soon replaced most of the indigenous South American mammals. Among the marsupials only opossums survived, and most of the placentals also died out. The land connection allowed migrations in both directions, and several South American mammals successfully migrated northward (Figure 18.28).

Most of the living species of marsupials are restricted to the Australian region. Recall from Chapter 15 that marsupials occupied Australia before its complete separation from Gondwana, but apparently placentals, other than bats and a few rodents, never did.

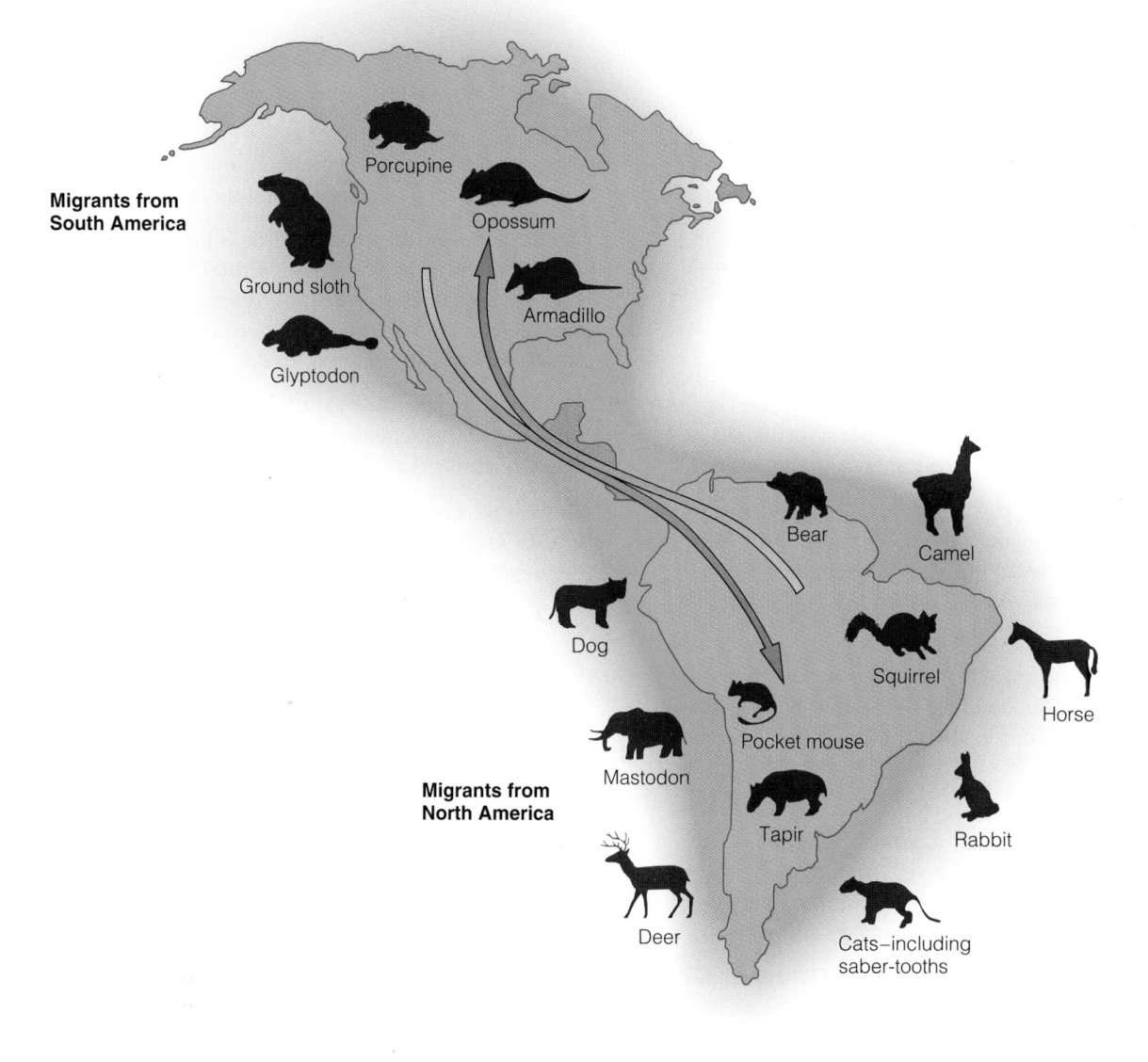

Migrants from South America

Porcupine

Opossum

Ground sloth

Armadillo

Glyptodon

Bear

Camel

Dog

Squirrel

Horse

Pocket mouse

Mastodon

Migrants from North America

Tapir

Rabbit

Deer

Cats–including saber-tooths

FIGURE 18.28 South America existed as an island continent during most of the Cenozoic Era. Its fauna evolved in isolation and consisted of numerous marsupial and placental mammals that were restricted to that continent. When the Isthmus of Panama formed during the Late Pliocene, migrations between North and South America occurred. Many types of placental mammals migrated south, and many South American mammals soon became extinct. Several mammals migrated north and successfully occupied North America.

Summary

1. Marine invertebrate groups that survived the Mesozoic extinctions continued to expand and diversify during the Tertiary.

2. Birds belonging to living orders and families evolved during the Early Tertiary Period.

3. The Paleocene mammalian fauna was composed of Mesozoic holdovers and a number of new orders. This was a time of diversification among mammals, and several orders soon became extinct. Most living mammalian orders were present by the Eocene.

4. Mammals have a good fossil record, and because of their varied teeth, mammal fossils are more easily identified than are fossils of other vertebrates. Consequently, the evolutionary history of mammals is known in considerable detail.

5. Placental mammals owe much of their success to their method of reproduction. Shrewlike placental mammals that appeared during the Cretaceous were the ancestral stock for the placental adaptive radiation of the Cenozoic.

6. Small mammals, especially rodents, occupy microhabitats unavailable to larger animals. Bats, the only flying mammals, have forelimbs modified for wing support, but otherwise they differ little from their ancestors.

7. All carnivorous mammals have well-developed canine teeth and specialized shearing teeth known as carnassials. Speed and agility are needed for some carnivores to catch prey, but others rely on stealth or digging prey from burrows. Aquatic carnivores such as seals appear to be most closely related to bears.

8. The perissodactyls and artiodactyls evolved during the Eocene. Ungulate adaptations include molarization of premolars for grinding vegetation and limb modifications for speed.

9. Artiodactyls diversified throughout the Cenozoic and are now the most abundant ungulates. Perissodactyls were abundant and diverse during the earlier Tertiary but have declined to only 16 living species.

10. The evolutionary history of horses is particularly well documented by fossils. The earliest horse, *Hyracotherium,* and the present-day horse, *Equus,* differ considerably, but a continuous series of intermediate fossils shows that they are related.

11. All perissodactyls—horses, rhinoceroses, tapirs, titanotheres and chalicotheres—diverged from a common ancestor during the Eocene.

12. Studies of fossils and living animals indicate that whales evolved from land-living mammals known as mesonychids.

13. Elephants evolved from rather small ancestors, became diverse and abundant, especially on the northern continents, then dwindled to only two living species.

14. Flowering plants continued their dominance in land-plant communities. Subtropical to tropical climates prevailed through the Paleocene and Eocene in North America. Temperatures declined at the end of the Eocene, and during the Oligocene–Miocene Epochs, the midcontinent region became increasingly arid.

15. Horses, camels, elephants, and other mammals spread across the northern continents because land connections between these continents existed at various times during the Cenozoic. The southern continents were mostly isolated, and their faunas evolving in isolation were unique.

16. The predominant evolutionary trend in Pleistocene mammals was toward giantism. Many of these large mammals became extinct at the end of the Pleistocene.

Important Terms

Artiodactyla	grazer	molar	premolar
browser	*Hyracotherium*	molarization	Primates
carnassials	Late Paleocene Thermal	monotreme	Proboscidia
Carnivora	Maximum	Perissodactyla	ruminant
Cetacea	marsupial	placental mammal	ungulate

Review Questions

1. Animals with teeth adapted to a diet of leaves are known as:
   a. _____ primates.
   b. _____ browsers.
   c. _____ monotremes.
   d. _____ proboscideans.
   e. _____ bryozoans.

2. Which of the following statements is correct?
   a. _____ Primates have well-developed carnassial teeth for defense.
   b. _____ Artiodactyls have one or three functional toes.
   c. _____ High-crowned chewing teeth were present in the oldest known mesonychids.
   d. _____ The large, Early Tertiary predators were flightless birds.
   e. _____ Most Paleocene mammals looked much like those existing now.

3. One feature of Late Eocene whales that indicates they had land-dwelling ancestors is:
   a. _____ incisors enlarged into tusks.
   b. _____ an elongate body and small head.
   c. _____ vestigial rear limbs.
   d. _____ nostrils on top of the head.
   e. _____ enlarged eyes.

4. Prehistoric overkill is a hypothesis to account for:
   a. _____ Pleistocene extinctions.
   b. _____ Eocene climatic change.
   c. _____ diversification of primates.
   d. _____ the fact that artiodactyls are more common than perrisodactyls.
   e. _____ relationships among cats, hyenas, and mongooses.

5. An Oligocene _____ was the largest land mammal to exist:
   a. _____ insectivore
   b. _____ rhinoceros
   c. _____ giraffe
   d. _____ mammoth
   e. _____ wombat

6. All existing orders of mammals had evolved by the:
   a. _____ Oligocene.
   b. _____ Paleocene.
   c. _____ Mesozoic.
   d. _____ Cretaceous.
   e. _____ Eocene.

7. Tertiary bryozoans were particularly abundant and successful _____:
   a. _____ carnivorous mammals.
   b. _____ flightless birds.
   c. _____ marine reptiles.
   d. _____ dinoflagellates.
   e. _____ suspension feeders.

8. The only existing egg-laying mammals are:
   a. _____ therians.
   b. _____ multituberculates.
   c. _____ monotremes.
   d. _____ diatoms.
   e. _____ ungulates.

9. One trend seen in hoofed mammals as they adapted to a diet of vegetation was:
   a. _____ molarization of the premolars.
   b. _____ development of short, stout limbs.
   c. _____ elongation of the snout.
   d. _____ reduction in the number of bones in the limbs.
   e. _____ increased brain size.

10. The ancestors of whales is among a group of Early Cenozoic mammals known as:
    a. _____ deinotheres.
    b. _____ proboscideans.
    c. _____ titanotheres.
    d. _____ mesonychids.
    e. _____ mastodons.

11. A marked temperature decrease took place at the end of the:
    a. _____ Pleistocene.
    b. _____ Mesozoic.
    c. _____ Eocene.
    d. _____ Tertiary.
    e. _____ Miocene.

12. Which continent had the highest extinction rate at the end of the Pleistocene?
    a. _____ Africa
    b. _____ Europe
    c. _____ Australia
    d. _____ South America
    e. _____ North America

13. Summarize how marsupials fared and where they were most abundant throughout the Cenozoic Era.

14. Explain how leaf structure of fossil plants is used to make inferences about ancient climates.

15. What kinds of evidence indicate that whales evolved from land-dwelling ancestors?

16. How does increased volcanism during the Miocene account for the abundance of diatoms?

17. Briefly summarize and cite evidence for and against the hypothesis of prehistoric overkill.

18. Although numerous adaptive types of birds exist, with the exception of flightless birds they vary little in basic structure. Why?

19. How do placental mammals differ from marsupials and monotremes?

20. Describe the modifications for a diet of meat that took place in carnivorous mammals.

21. How does the Paleocene mammalian fauna compare with those of the Cretaceous and Eocene?

22. What changes took place in the limbs and teeth of plains-dwelling, grazing hoofed mammals?

23. How do artiodactyls differ from perissodactyls? Give examples of members of each group. Which is most abundant? Has this abundance always been as it is now?

24. What evolutionary trends characterize camels? Where did they evolve, and where do they live now?

Points to Ponder

1. Scientists recognize three distinct types of mammals—monotremes, marsupials, and placentals. Monotremes lay eggs, so why are they considered mammals at all? Furthermore, marsupials have a placenta, yet they are not classified as placental mammals. Why?

2. What about Cenozoic rocks and mammal fossils makes it possible to decipher a much more complete history of mammals than for any other major group of vertebrates?

3. Describe the changing aspects of North America's mammalian fauna. What effect did climate and vegetation have on the fauna, and what were the effects, if any, of intercontinental migrations?

Additional Readings

Benton, M. J. 1991. *The rise of the mammals*. New York: Crescent Books.

Cox, C. B. 1993. *Biogeography: An ecological and evolutionary approach*. Boston: Blackwell Scientific.

Culotta, E. 1996. It's a long way from *Ambulocetus:* The whale's journey to the sea. *Pacific Discovery* 49, no. 1: 14–18.

Gould, S. J. (ed.). 1993. *The book of life*. New York: Norton.

———. 1994. Hooking leviathan by its past. *Natural History* 103, no. 5: 9–15.

Haynes, G. 1991. *Mammoths, mastodonts, and elephants: Biology, behavior, and the fossil record*. New York: Cambridge University Press.

MacFadden, B. J. 1992. *Fossil horses: Systematics, paleobiology, and evolution of the family Equidae*. New York: Cambridge University Press.

Norman, D. 1994. *Prehistoric life: The rise of the vertebrates*. New York: Macmillan.

Novacek, M. J. 1992. Mammalian phylogeny: Shaking the tree. *Nature* 356: 121–125.

Park, A. 1990. Giants once ruled Australia: Fossil discoveries reveal. *Smithsonian* 10, no. 10: 133–143.

Pielou, E. C. 1991. *After the Ice Age: The return of life to glaciated North America*. Chicago: University of Chicago Press.

Prothero, D. R. 1987. The rise and fall of the American rhino. *Natural History* 96, no. 8: 26–33.

———, and R. M. Schoch (eds.). 1989. *The evolution of perissodactyls*. New York: Oxford University Press.

Savage, R. J. G. 1986. *Mammal evolution: An illustrated guide*. New York: Facts on File.

Stewart, W. N. 1993. *Paleobotany and the evolution of plants*. New York: Cambridge University Press.

Storch, G. 1992. The mammals of island Europe. *Scientific American* 266, no. 2: 64–69.

Trefil, J. 1991. Whale feet. *Discover* 12, no. 5: 44–48.

For these web site addresses, along with current updates and exercises, log on to

http://www.brookscole.com/geo/

World Wide Web Activities

▶ MILWAUKEE PUBLIC MUSEUM

Milwaukee Public Museum's *Continents, Oceans, and Life in Motion—A New View of the Third Planet* is an excellent overview of Earth and life history. Take *A Walk Though Geologic Time* and *The Cenozoic Era* and see maps showing plate position, dioramas of fossil plants and animals, and information about the Ice Age. This site also contains links to other sites of geologic interest.

▶ ROYAL TYRRELL MUSEUM

This site contains considerable information on ancient organisms. It is supported by the Royal Tyrrell Museum Cooperating Society in Drumheller, Alberta, Canada, "a nonprofit society funded to support the scientific, educational, recreational and public operations of the Royal Tyrrell Museum of Paleontology." The home page contains icons for Directions to the Museum, Admissions, Virtual Tours, Programmes, and Educational Programming. Click on the *Age of Mammals* icon. This site has a summary of Cenozoic life, with emphasis on mammals. A variety of restorations of extinct mammals are shown along with explanations of how they lived and what descendants they have, if any. Check out the extinct mammals *Uintatherium* and *Dinictis*. Where did these mammals live and what did they eat?

▶ THE MAMMOTH SAGA

This site—maintained by the Department of Information Technology, Swedish Museum of Natural History, and authored by Dr. Ulf Carlber—is a virtual exhibition of mammoths and other animals and plants of the Ice Age. It is based on an exhibition held at the Swedish Museum of Natural History, Stockholm, Sweden from May 11 to September 18, 1994.

1. Click on *The Ice Ages* site from the Table of Contents. When did the last Ice Age occur? How much of Earth's surface was covered by continental glaciers?

2. Click on the *Sabertooth Cats* site from the Table of Contents. Are sabertooth cats related to tigers? What are the characteristics of sabertooth cats?

3. Click on *The Mammoth* site from the Table of Contents. What are mammoths? How many species of mammoths are recognized? What was their geographic range?

▶ THE AGE OF MAMMALS: LIFE IN THE CENOZOIC ERA

This National Park Service site is an excellent place to find links to any of the Tertiary Period epochs, as well as the Pleistocene. It also has links to other sites at which Cenozoic fossils are found. These include *Agate Fossil Beds, Badlands, Florissant Fossil Beds, Fossil Butte,* and *Hagerman Fossil Beds.*

▶ UC BERKELEY MUSEUM OF PALEONTOLOGY

The University of California at Berkeley maintains this site, which we have visited several times before. On the home page click *Time Periods,* which brings up a geologic time scale. On the time scale click *Pleistocene,* and at the bottom of that page click *The La Brea Tar Pits of Los Angeles, California.* See images of the tar pits, preserved mammals and birds, and excavations of fossils. Why is there such an unusual concentration of fossils in the Tar Pits? From this site you can click the highlighted term *mammals* and go to the UCMP Hall of Mammals and learn about various Cenozoic mammals. Return to the *Pleistocene* and click *The Case of the Irish Elk.* Was this animal actually an Elk? Where and when did it live, and when did it become extinct?

▶ GEOLOGY OF FLORISSANT FOSSIL BEDS NATIONAL MONUMENT

The National Park Service hosts this site regarding the geology of Florissant Fossil Beds National Monument, Colorado. It has excellent images of fossil plants and insects, as well as a brief description of the Monument's geology, and links to U.S. Geological Survey sites of geologic interest.

1. Give a brief explanation of how plants and insects were preserved at this site, the kinds of rocks they are found in, and the age of the rocks.

2. What mammals lived in this area when the plant/insect-bearing layers were deposited?

Evolution of the Primates and Humans

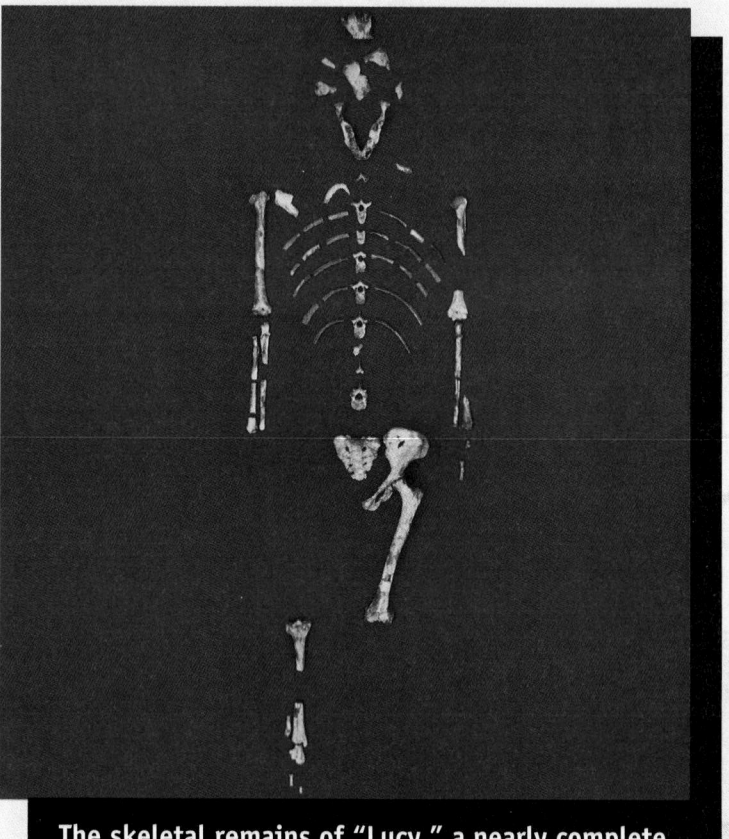

The skeletal remains of "Lucy," a nearly complete specimen of a 3.5-million-year-old *Australopithecus afarensis* female.

Prologue

A hypothesis on the origin and evolution of modern humans has aroused considerable controversy because it challenges the traditional view of human evolution based on fossil evidence. According to the "Eve" or "out of Africa" hypothesis, early modern humans did not evolve slowly in different parts of the world as many traditional paleoanthropologists think. Instead, proponents of this hypothesis claim that early modern humans evolved from a single woman in Africa, whose offspring then migrated from Africa, perhaps as recently as 100,000 years ago and populated Europe and Asia, driving the earlier hominid populations to extinction.

The alternative explanation, the "multiregional" hypothesis, maintains that early modern humans did not have an isolated origin in Africa but rather established separate populations throughout Eurasia. Occasional contact and interbreeding between these populations enabled our species to maintain its overall cohesiveness, while still preserving the regional differences in people we see today.

The current Eve hypothesis arose from an examination of the DNA from the mitochondria of cells. This mitochondrial DNA is useful for tracing family relationships because, unlike nuclear DNA, mitochondrial DNA is inherited only from the mother and so preserves a family phylogeny. Mitochondrial DNA is altered only by mutations that are passed on to the next generation. Each random mutation therefore produces a new type of DNA that is as distinctive as a fingerprint.

Geneticists compared the mitochondrial DNA from many different infants and found surprisingly small but clear differences. There were two general categories of DNA: One type was found in some infants of recent African descent, and the second type was found in everyone else including other Africans. According to the geneticists, these findings mean that modern humans evolved in Africa; then at some point a group of Africans split off, forming a second branch of DNA that was carried and spread throughout the rest of the world.

That the mitochondrial DNA can be traced back to a single woman from whom we all descended is not surprising, considering the statistics of genetic inheritance. What surprised most paleoanthropologists about the Eve hypothesis was how recently modern humans may have evolved. This age is calculated by counting the mutations that have occurred between the present and Eve's DNA. By looking at the DNA types that differ most from one another, and assuming mutations occurred at a regular rate, geneticists have calculated how many times Eve's original DNA would have to have mutated to produce the different types; in other words, they have constructed a molecular clock. These calculations indicate Eve lived about 200,000 years ago.

Where does this leave the traditional paleoanthropologists who base their evolutionary schemes on fossil evidence? As would be expected, many are skeptical of the genetic data, particularly the molecular clock. By changing a few basic assumptions, one might use the geneticists' calculations to show that Eve evolved many hundreds of thousands of years ago, which would be more in line with the fossil data.

Paleoanthropologists point out that if modern humans did originate in Africa and spread throughout the world, then the earliest modern humans in other areas should have African features, a point that they think is not supported by archaeological evidence from Asia or Australasia (Indonesia, New Guinea, and Australia). Furthermore, paleoanthropologists question how modern humans from Africa could have completely replaced all other human groups with whom they came in contact without leaving any evidence in the fossil record of the traits that made them successful or without any detectable genetic mixing between the two groups.

Recent molecular studies of a small segment of DNA called YAP, located on the Y chromosome, has not only provided adherents of the "out of Africa" hypothesis with evidence to support their case, but it has also complicated the evolutionary history of early humans. YAP is the male equivalent of mitochondrial DNA and is passed from father to son. It is altered only by random mutations, and thus populations with similar YAP sequences are more closely related than those that have been separated for extended periods of time.

In examining the YAP DNA in men from Africa, Europe, Asia, and Australia, it was discovered that YAP DNA in present-day Asian populations is much older than YAP DNA in some populations of modern African men. What this implies is that there was a long period of time in which Asian populations evolved their own type of YAP and then spread it to Africa. If this scenario is true, and other genetic studies appear to sup-

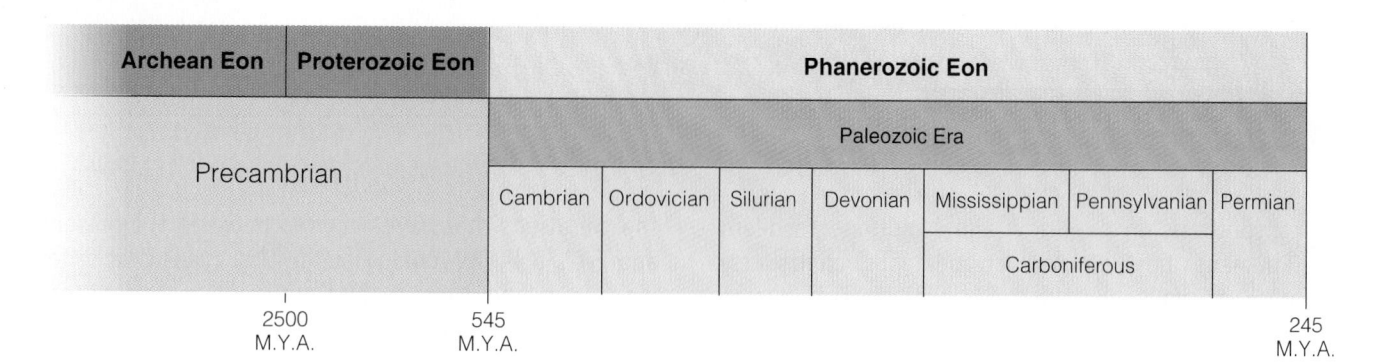

Archean Eon	Proterozoic Eon	Phanerozoic Eon							
		Paleozoic Era							
Precambrian		Cambrian	Ordovician	Silurian	Devonian	Mississippian	Pennsylvanian	Permian	
						Carboniferous			

2500
M.Y.A.

545
M.Y.A.

245
M.Y.A.

port it, then this would mean that early modern humans spent several thousand years in Eurasia after their initial migration from Africa, before some of them returned to Africa between 50,000 and 10,000 years ago. Instead of a simple picture of early modern humans evolving in Africa and then colonizing the rest of the world, a more complex, and probably more realistic, evolutionary history of early modern humans is now emerging.

Introduction

In this chapter we examine the various primate groups, in particular, the origin and evolution of the *hominids*, which are humans and their extinct ancestors. The earliest fossil evidence of hominids is from 4.4-million-year-old rocks in eastern Africa. Fossil hominid remains from rocks slightly younger, about 4.2 million years old, indicate our oldest ancestors walked upright and stood about 1.3 to 1.5 m tall and weighed between 33 and 50 kg. While these hominid fossils are the oldest remains known, many paleoanthropologists think that even older hominid fossils are still to be discovered and will push our ancestry farther into the past.

The study of the origins and evolutionary history of the human species has long fascinated people, but it was not until 1856 that workers discovered the first human-like fossil bones in a quarry cave in the Neander Valley near Düsseldorf, Germany. Even in 1871, when Charles Darwin published *The Descent of Man*, the fossil evidence of humans and their relationship to the apes was still quite meager. Because the fossil record of hominids is sparse compared with that of other vertebrate groups, there has been and continues to be much controversy and debate over our history, even as new fossil discoveries are made and new techniques for scientific analysis are developed (see the Prologue). One of the reasons the study of hominids is so exciting is that new discoveries can completely change our view of hominid evolution.

Primates

Primates are difficult to characterize as an order because they lack the strong specializations found in most other mammalian orders. We can, however, point to several trends in their evolution that help define primates and are related to their *arboreal,* or tree-dwelling, ancestry. These include changes in the skeleton and mode of locomotion, an increase in brain size, a shift toward smaller, fewer, and less specialized teeth, and the evolution of stereoscopic vision and a grasping hand with opposable thumb. Not all of these trends took place in every primate group, nor did they evolve at the same rate in each group. In fact, some primates have retained certain primitive features, whereas others show all or most of these trends.

Phanerozoic Eon										
Mesozoic Era			Cenozoic Era							
Triassic	Jurassic	Cretaceous	Tertiary						Quaternary	
			Paleocene	Eocene	Oligocene	Miocene	Pliocene	Pleistocene	Holocene

245
M.Y.A.

66
M.Y.A.

The primate order is divided into two suborders (Figure 19.1, Table 19.1). The *prosimians,* or lower primates, include the lemurs, lorises, tarsiers, and tree shrews, while the *anthropoids,* or higher primates, include monkeys, apes, and humans.

Prosimians

Prosimians are generally small, ranging from species the size of a mouse up to those as large as a house cat. They are arboreal, have five digits on each hand and foot with either claws or nails, and are typically omnivorous. They have large, forwardly directed eyes specialized for night vision, and hence most are nocturnal (Figure 19.1a and b). As their name implies (*pro* means "before," and *simian* means "ape"), prosimians are the oldest primate lineage, and their fossil record extends back to the Paleocene.

Prosimians were abundant, diversified, and widespread in North America, Europe, and Asia during the Eocene (Figure 19.2). As the continents moved northward during the Cenozoic and the climate changed from warm tropical to cooler midlatitude conditions, the prosimian population decreased in both abundance and diversity. By the Oligocene, there were hardly any prosimians left in the northern continents as the once widespread Eocene populations migrated south to the warmer latitudes of Africa, Asia, and Southeast Asia. Presently, prosimians are found only in the tropical regions of Asia, India, Africa, and Madagascar.

Anthropoids

Anthropoids evolved from a prosimian lineage sometime during the Late Eocene, and by the Oligocene they were well established. Much of our knowledge about the early evolutionary history of anthropoids comes from fossils found in the Fayum district, a small desert area southwest of Cairo, Egypt. During the Late Eocene and Oligocene, this region of Africa was a lush, tropical rain forest that supported a diverse and abundant fauna and flora. Within this forest lived many different arboreal anthropoids as well as various prosimians. In fact, several thousand fossil specimens representing more than 20 species of primates have been recovered from rocks of this region. One of the earliest anthropoids, and a possible ancestor of the Old World monkeys, was *Aegyptopithecus,* a small, fruit-eating, arboreal primate that weighed about 5 kg (Figure 19.3).

Anthropoids are divided into three superfamilies: Old World monkeys, New World monkeys, and hominoids (Table 19.1).

Old World monkeys (Superfamily Cercopithecoidea) are characterized by close-set, downward-directed nostrils (like those of apes and humans), grasping hands, and a nonprehensile tail. They include the macaque, baboon, and proboscis monkey (Figure 19.1d). Present-day Old World monkeys are distributed in the tropical regions of Africa and Asia and are thought to have evolved from a primitive anthropoid ancestor, such as *Aegyptopithecus,* sometime during the Oligocene.

New World monkeys (Superfamily Ceboidea) are found only in Central and South America. They probably evolved from African monkeys that migrated across the widening Atlantic sometime during the Early Oligocene, and they continued evolving in isolation to the present day. No evidence exists of any prosimian or other primitive primates in Central or South America nor of any contact with Old World monkeys after the initial immigration from Africa. New World monkeys are characterized by a prehensile tail, flattish face, and widely separated nostrils and include the howler, spider, and squirrel monkeys (Figure 19.1c).

Hominoids (Superfamily Hominoidea) consist of three families: the *great apes* (Family Pongidae), which includes chimpanzees, orangutans, and gorillas (Figure 19.1e and f); the *lesser apes* (Family Hylobatidae), which are gibbons and siamangs; and the *hominids* (Family Hominidae), which are humans and their extinct ancestors.

The hominoid lineage diverged from Old World Monkeys sometime before the Miocene, but exactly when is still being debated. It is generally accepted, however, that hominoids evolved in Africa, probably from the ancestral group that included *Aegyptopithecus.*

Recall that beginning in the Late Eocene the northward movement of the continents resulted in pronounced climatic shifts. In Africa, Europe, Asia, and elsewhere, a major cooling trend bagan, and the tropical and subtropical rain forests slowly began to change to a variety of mixed forests separated by savannas and open grasslands as tem-

FIGURE 19.1 Primates are divided into two suborders: the prosimians, such as tarsiers (a) and ring-tailed lemurs (b), and (c) through (f) the anthropoids. The anthropoids are further subdivided into three super-families: (c) New World monkeys, (d) Old World monkeys, and the great apes, such as gorillas (e) and chimpanzees (f).

PROSIMIANS

(a)

(b)

ANTHROPOIDS

(c)

(d)

(e)

(f)

TABLE 19.1

Classification of the Primates

Order Primates: Lemurs, lorises, tarsiers, monkeys, apes, humans

  Suborder Prosimii: Lemurs, lorises, tarsiers, tree shrews (lower primates)

  Suborder Anthropoidea: Monkeys, apes, humans (higher primates)

    Superfamily Cercopithecoidea: Macaque, baboon, proboscis monkey

    Superfamily Ceboidea: Howler, spider, and squirrel monkeys

    Superfamily Hominoidea: Apes, humans

      Family Pongidae: Chimpanzees, orangutans, gorillas

      Family Hylobatidae: Gibbons, siamangs

      Family Hominidae: Humans

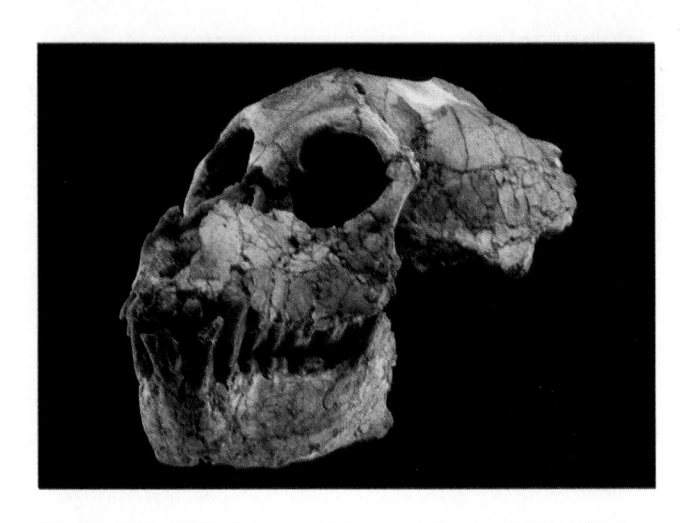

FIGURE 19.3 Skull of *Aegyptopithecus zeuxis,* one of the earliest known anthropoids.

peratures and rainfall decreased. As the climate changed, the primate populations also changed. Prosimians and monkeys became rare, while hominoids diversified in the newly forming environments and became abundant. Ape populations became reproductively isolated from each other within the various forests, leading to adaptive radiation and increased diversity among the hominoids. During the Miocene, Africa collided with Eurasia, producing additional changes in the climate, as well as providing opportunities for migration of animals between the two landmasses.

Two apelike groups evolved during the Miocene that ultimately gave rise to present-day hominoids. While there is still not agreement on the early evolutionary relationships among the hominoids, fossil evidence and molecular DNA similarities between modern hominoid families is providing a clearer picture of the evolutionary pathways and relationships among the hominoids.

The first group, the **dryopithecines,** evolved in Africa during the Miocene and subsequently spread to Eurasia, following the collision between the two continents. The dryopithecines were a varied group of hominoids in size, skeletal features, and lifestyle. The best-known and perhaps ancestor of all later hominoids is *Proconsul,* an apelike fruit-eating animal that led a quadrupedal arboreal existence, with limited activity on the ground (Figure 19.4). The dryopithecines were very abundant and diverse during the Miocene and Pliocene, particularly in Africa.

The second group, the **sivapithecids,** evolved in Africa during the Miocene and then spread throughout Eurasia. The fossil remains of sivapithecids consist mostly of jaws, skulls, and isolated teeth. There are few body or limb bones known, and thus we know little about their body anatomy. All sivapithecids had powerful jaws and teeth with thick enamel and flat chewing surfaces, suggesting a diet of harder foods such as nuts (Figure 19.5). Based on various lines of evidence, the sivapithecids appear to be the ancestral stock from which present-day orangutans evolved.

FIGURE 19.2 *Notharctus,* a primitive Eocene prosimian from North America.

FIGURE 19.4 Probable life appearance of *Proconsul,* a dryopithecine.

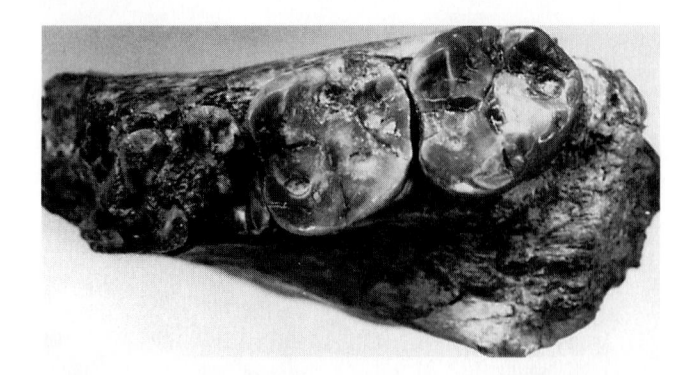

FIGURE 19.5 Molars and upper jawbone of *Ramapithecus,* a sivapithecid, showing the flat chewing surface of the molar teeth.

While there are still many missing pieces, particularly during critical intervals in the African hominoid fossil record, molecular DNA as well as fossil evidence indicates that the dryopithecines, African apes, and hominids form a closely related lineage, while the sivapithecids and orangutans form a different lineage that did not lead to humans.

Hominids

The **hominids** (Family Hominidae), the primate family that includes present-day humans and their extinct ancestors (Table 19.1), have a fossil record extending back 4.4 million years. Several features distinguish them from other hominoids. Hominids are bipedal; that is, they have an upright posture, which is indicated by several modifications in their skeleton (Figure 19.6a and b). In addition, they show a trend toward a large and internally reorganized brain (Figure 19.6c–e). Other features include a reduced face and reduced canine teeth, omnivorous feeding, increased manual dexterity, and the use of sophisticated tools.

Many anthropologists think that these hominid features evolved in response to major climatic changes that began during the Miocene and continued into the Pliocene. During this time, vast savannas replaced the African tropical rain forests where the lower primates and Old World monkeys had been so abundant. As the savannas and grasslands continued to expand, the hominids made the transi-

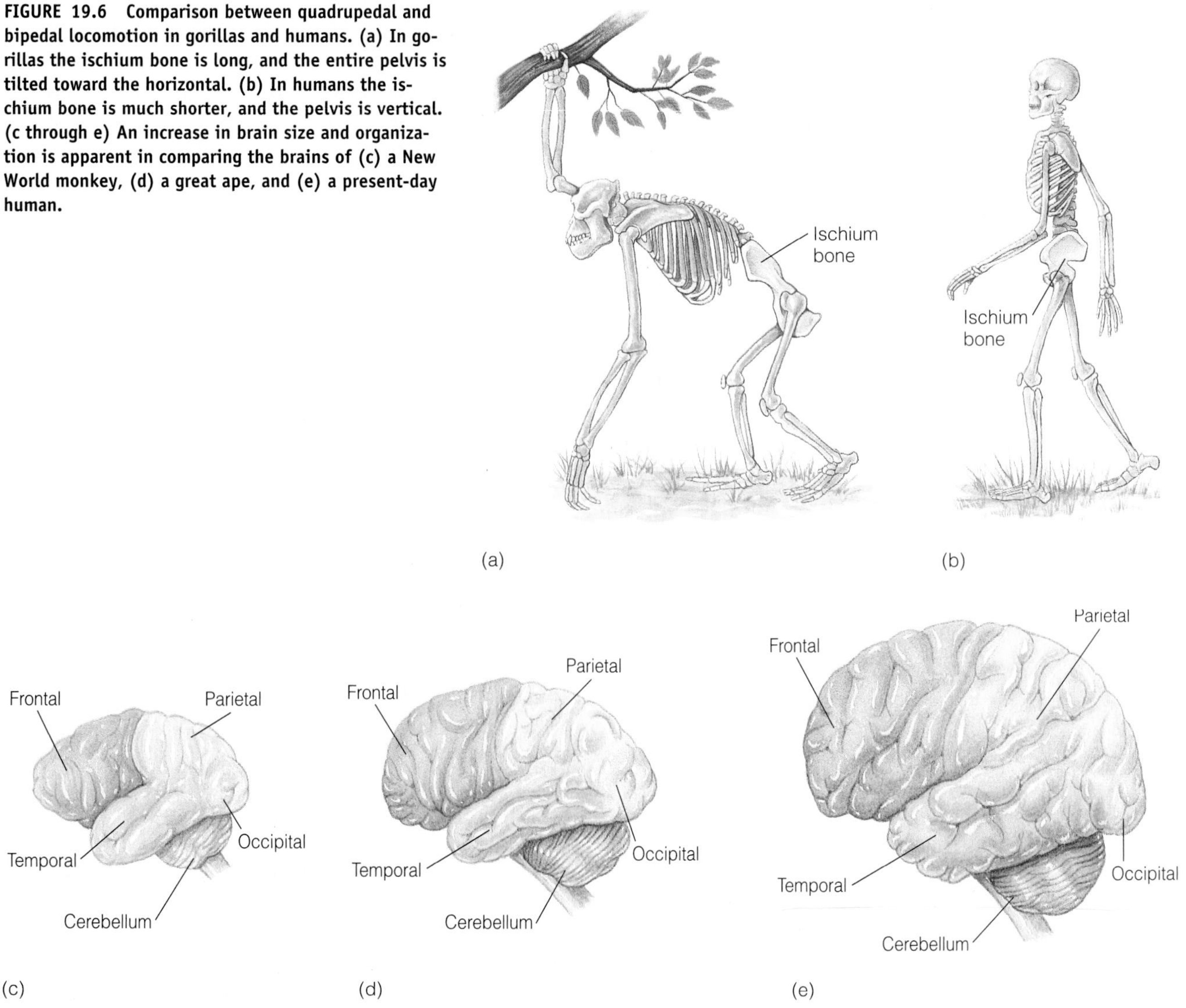

FIGURE 19.6 Comparison between quadrupedal and bipedal locomotion in gorillas and humans. (a) In gorillas the ischium bone is long, and the entire pelvis is tilted toward the horizontal. (b) In humans the ischium bone is much shorter, and the pelvis is vertical. (c through e) An increase in brain size and organization is apparent in comparing the brains of (c) a New World monkey, (d) a great ape, and (e) a present-day human.

Ischium bone

Ischium bone

(a)

(b)

Frontal
Parietal
Temporal
Occipital
Cerebellum

(c)

Frontal
Parietal
Temporal
Occipital
Cerebellum

(d)

Frontal
Parietal
Temporal
Occipital
Cerebellum

(e)

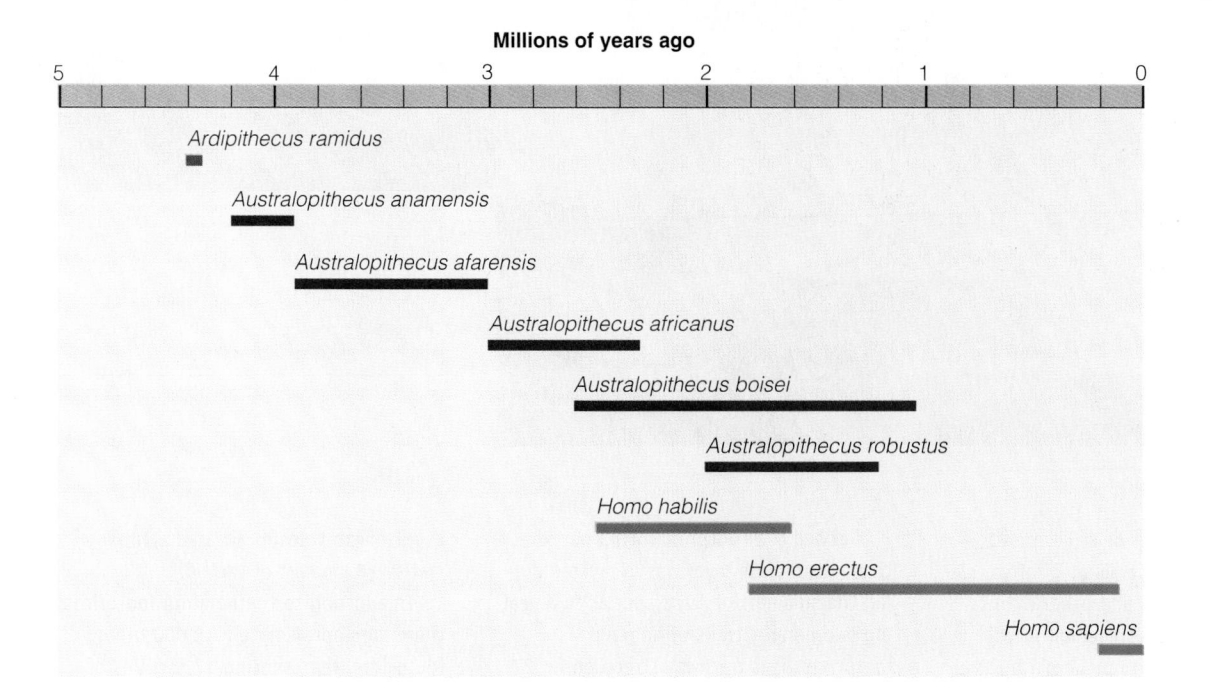

Millions of years ago

Ardipithecus ramidus

Australopithecus anamensis

Australopithecus afarensis

Australopithecus africanus

Australopithecus boisei

Australopithecus robustus

Homo habilis

Homo erectus

Homo sapiens

FIGURE 19.7 The geologic age ranges for the commonly accepted species of hominids.

tion from true forest dwelling to life in an environment of mixed forests and grasslands.

Presently, there is no clear consensus on the evolutionary history of the hominid lineage. This is due, in part, to the incomplete fossil record of hominids and also because some species are known only from partial specimens or fragments of bone. Because of this, there is even disagreement on the total number of hominid species. A complete discussion of all the proposed hominid species and the various competing schemes of hominid evolution is beyond the scope of this chapter. However, we will discuss the generally accepted taxa (Figure 19.7) and present some of the current theories of hominid evolution. It should therefore be remembered that while the fossil record of hominid evolution is not complete, what there is of it is well documented. Furthermore, it is the interpretation of that fossil record that precipitates the often vigorous and sometimes acrimonious debates concerning our evolutionary history.

The oldest known hominid is *Ardipithecus ramidus* (Figure 19.7). Discovered by Tim White and his Ethiopian colleagues at Aramis, Ethiopia, this nearly complete skeleton has been dated at 4.4 million years. As its bones are studied in more detail, its place in the hominid lineage will become clearer.

AUSTRALOPITHECINES

Australopithecine is a collective term for all members of the genus *Australopithecus.* Currently, five species are recognized: *A. anamensis, A. afarensis, A. africanus, A. robustus,* and *A. boisei.* Many paleontologists accept the evolutionary scheme in which *A. anamensis,* the oldest known australopithecine, is ancestral to *A. afarensis,* who

in turn is ancestral to *A. africanus* and the genus *Homo,* as well as the side branch of australopithecines represented by *A. robustus* and *A. boisei.*

The oldest known australopithicine is *Australopithecus anamensis.* Discovered at Kanapoi, a site near Lake Turkana, Kenya, by Meave Leakey of the National Museums of Kenya and her colleagues, this 4.2-million-year-old bipedal species has many features in common with its younger relative, *A. afarensis,* yet is more primitive in other characteristics, such as its teeth and skull. *A. anamensis* is estimated to have been between 1.3 and 1.5 m tall and weighed between 33 and 50 kg.

Australopithecus afarensis (Figure 19.8), which lived 3.9–3.0 million years ago, was fully bipedal (see Perspective 19.1) and exhibited great variability in size and weight. Members of this species ranged from just over 1 m to about 1.5 m tall and weighed between 29 and 45 kg. They had a brain size of 380–450 cubic centimeters (cc), which is greater than the 300–400 cc of a chimpanzee but much smaller than that of present-day humans (1350 cc average).

The skull of *A. afarensis* retained many apelike features, including massive brow ridges and a forward-jutting jaw, but its teeth were intermediate between those of apes and humans. The heavily enameled molars were probably an adaptation to chewing fruits, seeds, and roots (Figure 19.9).

A. afarensis was succeeded by *Australopithecus africanus,* which lived 3.0–2.3 million years ago (Figure 19.10). The differences between the two species is relatively minor. They were both about the same size and weight, but *A. africanus* had a flatter face and somewhat

perspective 19.1

Footprints at Laetoli

During the summer of 1976, fossil footprints of such animals as giraffes, elephants, rhinoceroses, and other extinct mammals were found preserved in volcanic ash at Laetoli in northern Tanzania. Two years later, a member of Mary Leakey's archaeological team, which was searching for early hominid remains, found what appeared to be a human footprint in the same volcanic ash layer.

Dubbed the Footprint Tuff, a portion of this volcanic ash layer was excavated during the summers of 1978 and 1979, revealing two parallel trails of hominid footprints. This trackway stretched for 27 meters and consisted of 54 individual footprints (Figure 1). Radiometric dating of the ash indicates it was deposited between 3.8 and 3.4 million years ago (Pliocene Epoch) during one of several eruptions of ash from the Sadiman volcano, located approximately 20 km east of Laetoli.

In addition to the hominid footprints, there are approximately 18,000 other footprints, representing 17 families of mammals, found in the Laetoli area. Laetoli is part of the eastern branch of the Great Rift Valley of East Africa, a tectonically active area, which is separating from the rest of Africa along a divergent

FIGURE 1   Hominid footprints preserved in volcanic ash at the Laetoli site, Tanzania. Discovered in 1978 by Mary Leakey, these footprints proved that hominids were bipedal walkers at least 3.5 million years ago. The footprints of two adults and possibly those of a child are clearly visible in this photograph.

boundary (see Figure 6.16). During the Pliocene Epoch, Sadiman volcano erupted several times, spewing out tremendous quantities of ash that settled over the surrounding savannah. When light rains in the area moistened the ash, any animals walking over it left their footprints. As the ash dried, it hardened like cement, preserving whatever footprints had been made. Subsequent eruptions buried the footprint-bearing ash layer, thus preserving the footprints.

What makes this find so exciting and scientifically valuable is that the footprints prove early hominids were fully bipedal and had an erect posture long before the advent of stone toolmaking or an increase in the size of the brain. Furthermore, the footprints showed that early hominids walked like modern humans by placing the full weight of the body on the ball of the heel. By examining how deep the impression in the ash is for various parts of the footprint, information about the soft tissue of the feet can be inferred, something that can't be determined from fossil bones alone.

The question of who made the foot-

prints and how many individuals were walking at the time the footprints were made has been debated since they were initially discovered. Most scientists think the footprints were made by *Australopithecus afarensis,* one of the earliest known hominids, whose fossil bones and teeth are found at Laetoli (see Figure 19.9). *A. afarensis* lived from about 3.9 to 3.0 million years ago and exhibited great variability in size and weight. It is estimated that the largest of the hominids making the footprints, a male, was approximately 1.5 m tall, and the smallest hominid, either a female or a child, was approximately 1.2 m tall.

How many individuals made these parallel trails? There seems to be no argument that the individuals making these footprints were walking together, with one walking slightly behind the other. It was originally thought the footprints represented a male and female. However, closer examination of the footprints indicates there were probably three individuals. The larger footprints, which were probably made by a male, have features suggesting they are double prints. In this

scenario, a second individual (possibly another male) followed the first one by deliberately stepping in its tracks, thus producing a double print. The smaller footprints are well defined and were probably made by a female or possibly a child. That the trackway was made by three individuals rather than two is now widely accepted.

To ensure these footprints are not destroyed and will be available for future generations to study, the trackway has been reburied. The Antiquities Department of the Tanzanian government, in cooperation with the Getty Conservation Institute, completely reburied the site under five layers of sand, soil, and erosion-control matting. Some of the layers were treated with root inhibitors to prevent roots from destroying the footprints. The site is currently topped with a bed of lava boulders to provide additional protection against erosion and to mark its location. A sacred ceremony was held in 1996 in which the site was included as a place revered by the Masai people.

larger brain. Furthermore, it appears that the limbs of *A. africanus* might not have been as well adapted for bipedalism as those of *A. afarensis.*

Both *A. afarensis* and *A. africanus* differ markedly from the so-called robust species *A. boisei* (2.6–1.0 million years ago) and *A. robustus* (2.0–1.2 million years ago). *A. boisei* was 1.2–1.4 m tall and weighed between 34 and 49 kg. It had a powerful upper body, a distinctive bony crest on the top of its skull, a flat face, and the largest molars of any hominids. *A. robustus,* on the other hand, was somewhat smaller (1.1–1.3 m tall) and lighter (32–40 kg). It had a flat face, and the crown of its skull had an elevated bony crest that provided additional area for the attachment of strong jaw muscles (Figure 19.11). Its broad flat molars indicated *A. robustus* was a vegetarian.

Most scientists accept the idea that the robust australopithecines form a separate lineage from the other australopithicines that went extinct 1 million years ago.

THE HUMAN LINEAGE

Homo habilis The earliest member of our own genus **Homo** is *Homo habilis,* which lived 2.5–1.6 million years ago (Figure 19.12). Its remains were first found at Olduvai Gorge, but it is also known from Kenya, Ethiopia, and South Africa. *H. habilis* evolved from the *A. afarensis* and *A. africanus* lineage and coexisted with *A. africanus* for about 200,00 years (Figure 19.7). *H. habilis* had a larger brain (700 cc average) than its australopithecine ancestors, but smaller teeth. It was about 1.2–1.3 m tall and only weighed 32–37 kg.

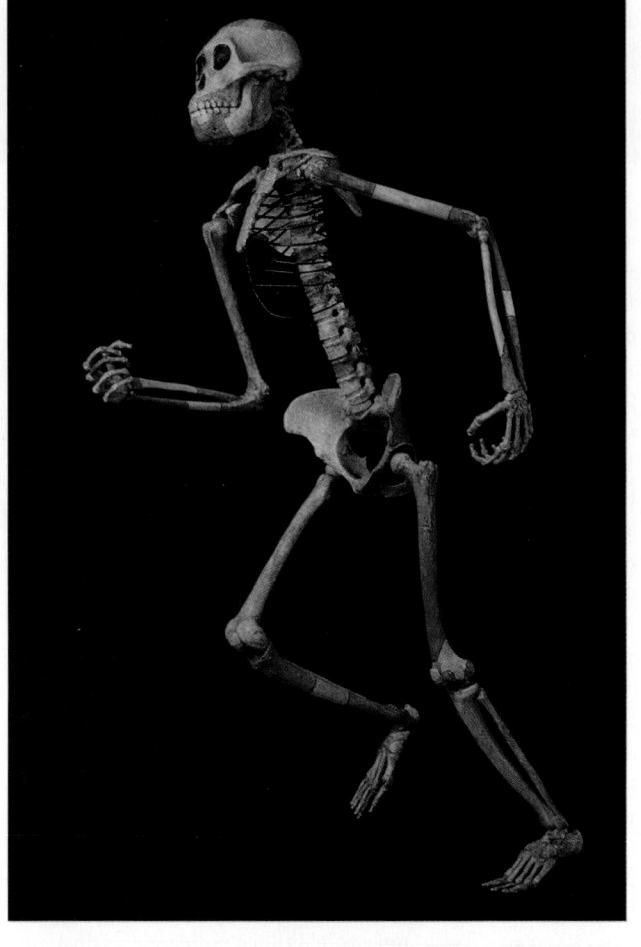

FIGURE 19.8 A reconstruction of Lucy's skeleton by Owen Lovejoy and his students at Kent State University, Ohio. Lucy, whose fossil remains were discovered by Donald Johanson, is an approximately 3.5-million-year-old *Australopithecus afarensis* individual. This reconstruction illustrates how adaptations in Lucy's hip, leg, and foot allowed a fully bipedal means of locomotion.

Homo erectus   In contrast to the australopithecines and *H. habilis,* which are unknown outside Africa, *Homo erectus* was a widely distributed species, having migrated from Africa during the Pleistocene. Specimens have been found not only in Africa but also in Europe, India, China ("Peking Man"), and Indonesia ("Java Man"). *H. erectus* evolved in Africa 1.8 million years ago and by 1 million years ago was present in southeastern and eastern Asia, where it survived until about 100,000 years ago.

Although *H. erectus* developed regional variations in form the species differed from modern humans in several ways. Its brain size of 800–1300 cc, though much larger than that of *H. habilis,* was still less than the average for *Homo sapiens* (1350 cc). *H. erectus*'s skull was thick-walled, its face was massive, it had prominent brow ridges, and its teeth were slightly larger than those of present-day

humans (Figure 19.13). *H. erectus* were comparable in size to modern humans, standing between 1.6 and 1.8 m tall and weighing between 53 and 63 kg.

The archaeological record indicates that *H. erectus* were tool makers. Furthermore, some sites show evidence that they used fire and lived in caves, an advantage for those living in more northerly climates (Figure 19.14).

Currently, a heated debate surrounds the transition from *H. erectus* to our own species, *Homo sapiens.* Paleoanthropologists are split into two camps. On the one side are those who support the "out of Africa" view, and on the other are those supporting the "multiregional" view (see the Prologue). Regardless of which theory turns out to be correct, our species, *H. sapiens* most certainly evolved from *H. erectus.*

NEANDERTHALS

Probably the most famous of all fossil humans are the **Neanderthals,** who inhabited Europe and the Near East between about 200,000 and 30,000 years ago. Some paleoanthropologists regard the Neanderthals as a variety or subspecies of our own species *(Homo sapiens neanderthalensis),* while others regard them as a separate species *(Homo neanderthalensis).* In any case, their name comes from the first specimens found in 1856 in the Neander Valley near Düsseldorf, Germany.

The most notable difference between Neanderthals and present-day humans is in the skull. Neanderthal skulls were long and low with heavy brow ridges, a projecting mouth, and a weak, receding chin (Figure 19.15). Their brain was slightly larger on average than our own, and somewhat differently shaped. The Neanderthal body was somewhat more massive and heavily muscled than ours, with rather short lower limbs, much like those of other cold-adapted people of today.

Based on specimens from more than a hundred sites, we now know that Neanderthals were not much different from us, only more robust. Europe's Neanderthals were the first humans to move into truly cold climates, enduring miserably long winters and short summers as they pushed north into tundra country (Figure 19.16).

The remains of Neanderthals are found chiefly in caves and hutlike rock shelters, which also contain a variety of specialized stone tools and weapons. Furthermore, archaeological evidence indicates that Neanderthals commonly took care of their injured and buried their dead, frequently with such grave items as tools, food, and perhaps even flowers.

CRO-MAGNONS

About 30,000 years ago, humans closely resembling modern Europeans moved into the region inhabited by the Neanderthals and completely replaced them. **Cro-Magnons,** the name given to the successors of the Neanderthals in France, lived from about 35,000 to 10,000

FIGURE 19.9 Re-creation of a Pliocene landscape showing members of *Australopithecus afarensis* gathering and eating various fruits and seeds.

years ago; during this period the development of art and technology far exceeded anything the world had seen before.

Highly skilled nomadic hunters, Cro-Magnons followed the herds in their seasonal migrations. They used a variety of specialized tools in their hunts, including perhaps the bow and arrow (Figure 19.17). They sought refuge in caves and rock shelters and formed living groups of various sizes. Cro-Magnons were also cave painters. Using paints made from manganese and iron oxides, Cro-Magnon people painted hundreds of scenes on the ceilings and walls of caves in France and Spain, where many of them are still preserved today (Figure 19.18).

With the appearance of Cro-Magnons, human evolution has become almost entirely cultural rather than biological. Humans have spread throughout the world by devising means to deal with a broad range of environmental conditions. Since the evolution of the Neanderthals about 200,000 years ago, humans have gone from a stone culture to a technology that has allowed us to visit other planets with space probes and land men on the Moon. It remains to be seen how we will use this technology in the future and whether we will continue as a species, evolve into another species, or become extinct as many groups have before us.

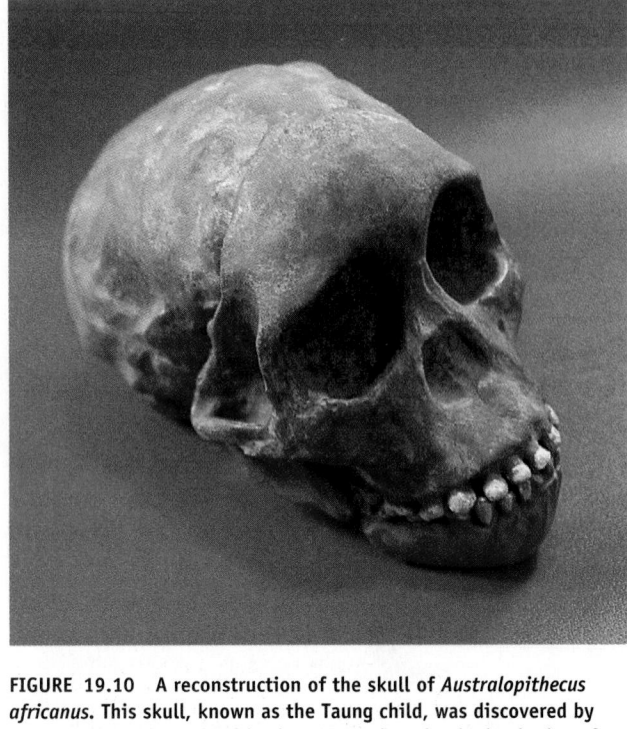

FIGURE 19.10   A reconstruction of the skull of *Australopithecus africanus*. This skull, known as the Taung child, was discovered by Raymond Dart in South Africa in 1924 and marks the beginning of modern paleoanthropology.

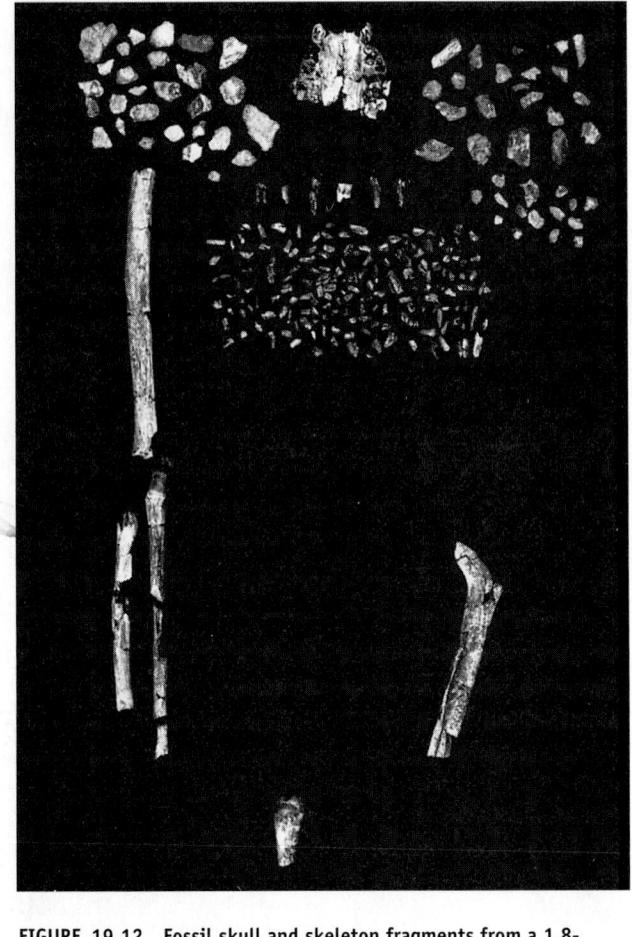

FIGURE 19.12   Fossil skull and skeleton fragments from a 1.8-million-year-old specimen of *Homo habilis*.

FIGURE 19.11   The skull of *Australopithecus robustus*. This species had a massive jaw, powerful chewing muscles, and large, broad, flat chewing teeth apparently used for grinding up coarse plant food.

FIGURE 19.13   A reconstruction of the skull of *Homo erectus,* a widely distributed species whose remains have been found in Africa, Europe, India, China, and Indonesia.

FIGURE 19.14 Re-creation of a Pleistocene setting in Europe in which members of *Homo erectus* are using fire and stone tools.

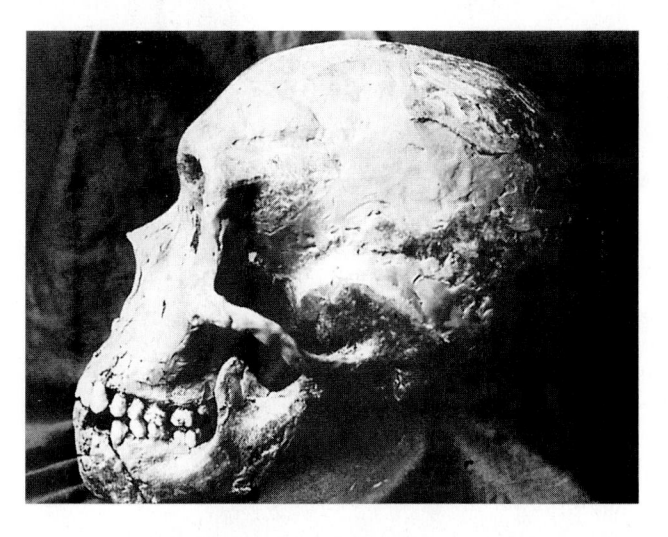

FIGURE 19.15 Reconstructed Neanderthal skull. The Neanderthals were characterized by prominent heavy brow ridges and a weak chin. Their brain was also slightly larger on average than that of modern humans.

FIGURE 19.16 Archeological evidence indicates that Neanderthals lived in caves and participated in ritual burials as depicted in this painting of a burial ceremony that occurred approximately 60,000 years ago at Shanidar Cave, Iraq.

FIGURE 19.17 Re-creation of a Cro-Magnon camp in Europe. Cro-Magnons were highly skilled hunters who formed living groups of various sizes.

FIGURE 19.18 Cro-Magnons were very skilled cave painters. Shown is a painting of a horse from the cave of Niaux, France.

1. The primates evolved during the Paleocene. Several trends help characterize primates and differentiate them from other mammalian orders, including a change in overall skeletal structure and mode of locomotion, an increase in brain size, stereoscopic vision, and evolution of a grasping hand with opposable thumb.

2. The primates are divided into two suborders, the prosimians and anthropoids. The prosimians are the oldest primate lineage and include lemurs, lorises, tarsiers, and tree shrews. The anthropoids include the New and Old World monkeys, apes, and hominids, which are humans and their extinct ancestors.

3. The oldest known hominid is *Ardipithecus ramidus*. It was followed by the australopithecines, a fully bipedal group that evolved in Africa 4.2 million years ago. Currently, five australopithecine species are known: *Australopithecus anamensis, A. afarensis, A. africanus, A. robustus,* and *A. boisei.*

4. The human lineage began about 2.5 million years ago in Africa with the evolution of *Homo habilis,* which survived as a species until about 1.6 million years ago.

5. *Homo erectus* evolved from *H. habilis* about 1.8 million years ago and was the first hominid to migrate out of Africa. Between 1 and 1.8 million years ago, *H. erectus* had spread to Europe, India, China, and Indonesia. *H. erectus* used fire, made tools, and lived in caves.

6. Sometime between 200,000 and 100,000 years ago, *Homo sapiens* evolved from *H. erectus.* These early humans may be ancestors of Neanderthals.

7. Neanderthals were not much different from present-day humans, only more robust and with differently shaped skulls. They made specialized tools and weapons, apparently took care of their injured, and buried their dead.

8. The Cro-Magnons were the successors of the Neanderthals and lived from about 35,000–10,000 years ago. They were highly skilled nomadic hunters, formed living groups of various sizes, and were also skilled cave painters.

9. Modern humans succeeded the Cro-Magnons about 10,000 years ago and have spread throughout the world.

Important Terms

anthropoid

australopithecine

Cro-Magnon

dryopithecine

hominid

hominoid

Homo

Neanderthal

New World monkey

Old World monkey

primate

prosimian

sivapithecid

Review Questions

1. Which DNA is used as the basis for tracing maternal family relationships and is central to the Eve hypothesis?
   a. _____ Nuclear
   b. _____ Mitochondrial
   c. _____ Organelle
   d. _____ Plastid
   e. _____ YAP

2. Which of the following is not a hominoid?
   a. _____ Chimpanzee
   b. _____ Orangutan
   c. _____ Gorilla
   d. _____ Gibbon
   e. _____ Prosimian

3. What is the oldest primate lineage?
   a. _____ Anthropoids
   b. _____ Hominoids
   c. _____ Hominids
   d. _____ Prosimians
   e. _____ None of these

4. Which of the following evolutionary trends characterize primates?
   a. _____ Grasping hand with opposable thumb
   b. _____ Stereoscopic vision
   c. _____ Increase in brain size
   d. _____ Change in overall skeletal structure
   e. _____ All of these

5. The oldest hominids belong to which genus?
   a. _____ *Australopithecus*
   b. _____ *Homo*
   c. _____ *Ramapithecus*
   d. _____ *Ardipithecus*
   e. _____ *Proconsul*

6. The earliest known australopithecine is *Australopithecus* _____.
   a. _____ *afarensis*
   b. _____ *africanus*
   c. _____ *boisei*
   d. _____ *anamensis*
   e. _____ *robustus*

7. Which genus do humans belong to?
   a. _____ *Australopithecus*
   b. _____ *Ardipithecus*
   c. _____ *Homo*
   d. _____ *Proconsul*
   e. _____ *Ramapithecus*

8. According to archaeological evidence, which were the first hominids to use fire?
   a. _____ *Australopithecus boisei*
   b. _____ *Homo habilis*
   c. _____ *Homo erectus*
   d. _____ *Homo sapiens*
   e. _____ *Homo neanderthalensis*

9. Which were the first hominids to migrate out of Africa?
   a. _____ *Australopithecus robustus*

   b. _____ *Australopithecus boisei*
   c. _____ *Homo habilis*
   d. _____ *Homo erectus*
   e. _____ *Homo sapiens*

10. Which extinct lineage of humans were skilled hunters and cave painters?
    a. _____ Neanderthals
    b. _____ Cro-Magnons
    c. _____ Archaic
    d. _____ Acheuleans
    e. _____ None of these

11. To which hominid group do Java Man and Peking Man belong?
    a. _____ *Australopithecus robustus*
    b. _____ *Australopithecus boisei*
    c. _____ *Homo habilis*
    d. _____ *Homo erectus*
    e. _____ *Homo sapiens*

12. Which of the following are not anthropoids?
    a. _____ Old World monkeys
    b. _____ New World monkeys
    c. _____ Apes
    d. _____ Humans
    e. _____ Prosimians

13. Discuss the differences between the prosimians and anthropoids.

14. How do hominoids differ from hominids?

15. Discuss the evolutionary history of the anthropoids.

16. What are the major evolutionary trends that characterize the primates and set them apart from the other orders of mammals?

17. Compare the various features of the five different species of australopithecines.

18. Discuss the evolutionary history of the genus *Homo*.

19. Discuss the Eve and multiregional hypotheses.

20. What are the main differences between the Neanderthals and Cro-Magnons?

Points to Ponder

1. What evidence might help settle the debate over the origin of *Homo sapiens?*

2. Based on all you've learned about evolution and the history of life, what factors do you think will influence the course of human evolution, and can we, as a species, control the direction evolution is taking?

Additional Readings

Agnew, N., and M. Demas. 1998. Preserving the Laetoli footprints. *Scientific American* 279, no. 3: 44-55.

Berger, L. 1998. Redrawing our family tree? *National Geographic* 194, no. 2: 90-99.

Blumenschine, R. J., and J. A. Cavallo, 1992. Scavenging and human evolution. *Scientific American* 267, no. 4: 90-97.

Flanagan, R. 1996. Out of Africa. *Earth* 5, no. 1: 26-35.

Gore, R. 1996. Neanderthals. *National Geographic* 189, no. 1: 2-35.

———. 1997. The first steps. *National Geographic* 191, no. 2: 72-99.

———. 1997. The first Europeans. *National Geographic* 192, no. 1: 96-113.

Johanson, D.C. 1996. Face-to-face with Lucy's family. *National Geographic* 189, no. 3: 96-117.

Johanson, D., and B. Edgar. 1997. From Lucy to language. *Earth* 6, no. 3: 40-47.

Jones, S., R. Martin, and O. Pilbeam. 1992. *The Cambridge encyclopedia of human evolution.* New York: Cambridge University Press.

Larick, R., and R. L. Ciochon. 1996. The African emergence and early Asian dispersals of the genus *Homo. American Scientist* 188, no. 3: 38-51.

Leakey, M. 1995. The dawn of humans. *National Geographic* 188, no. 3: 38-51.

Potts, R. 1997. *Humanity's descent: The consequences of ecological instability.* New York: Avon Books.

Stanley, S. M. 1998. *Children of the Ice Age: How a global catastrophe allowed humans to evolve.* New York: Freeman.

Thomson, K. S. 1992. The challenge of human origins. *American Scientist* 80, no. 6: 519-522.

Thorne, A. G., and M. Wolpoff. 1992. The multiregional evolution of humans. *Scientific American* 266, no. 4: 76-83.

Tudge, C. 1996. The future of *Homo sapiens. Earth* 5, no. 1: 36-40.

Wilson, A. C., and R. L. Cann. 1992. The recent African genesis of humans. *Scientific American* 266, no. 4: 68-73.

World Wide Web Activities

For these web site addresses, along with current updates and exercises, log on to

http://www.brookscole.com/geo/

▶ UNIVERSITY OF CALIFORNIA MUSEUM OF PALEONTOLOGY

This is an excellent site to visit for any aspect of geologic time, paleontology, and evolution. Click on the *Animals, Plants, etc.* icon under the On-Line Exhibits heading that will take you to the *Vertebrata* home page. Click on the *Mammalia* icon that will take you to the home page of all the different mammal groups. Click on the *Primates* icon. This will take you to the *Primates, Apes, monkeys, and you* home page. The museum does not have a proper exhibit on primates but does provide links to various websites concerned with primates and paleoanthropology, which are very helpful and up to date.

▶ ORIGINS OF HUMANKIND

This site is a comprehensive Internet resource concerning human origins. Because this field of research is so broad and changing so rapidly, rather than refer to specific sites to visit here, click on any of the sites listed on the home page to learn more about controversial theories and to find links to other sites concerned with human evolution.

Epilogue

A theme of this book is that Earth is a complex, dynamic planet that has changed continuously since its origin some 4.6 billion years ago. These changes and the present-day features we observe are the result of interactions between the various interrelated internal and external Earth systems and cycles. Furthermore, these interactions have also influenced the evolution of the biosphere.

The rock cycle (see Figure 1.2), with its recycling of Earth materials to form the three major rock groups, illustrates the interrelationships between Earth's internal and external processes. The hydrologic cycle is the continuous recycling of water from the oceans, to the atmosphere, to the land, and eventually back to the oceans again. Changes within this cycle can have profound effects on Earth's topography as well as its biota. For example, a rise in global temperature will cause the ice caps to melt, contributing to rising sea level, which will greatly affect coastal areas where many of the world's large population centers are presently located. We have seen the effect of changing sea level on continents in the past. Such rises and drops in sea level resulted in large-scale transgressions and regressions. Some of these were caused by growing and shrinking continental ice caps when land masses moved over the South Pole as a result of plate movements.

On a larger scale, the movement of plates has had a profound effect on the formation of landscapes, the distribution of mineral resources, and atmospheric and oceanic circulation patterns, as well as the evolution and diversification of life.

The launching in 1957 of Sputnik 1, the world's first artificial satellite, ushered in a new global consciousness in terms of how we view Earth and our place in the global ecosystem. Satellites have provided us with the ability to view not only the beauty of our planet, but also the fragility of Earth's biosphere and the role humans play in shaping and modifying the environment. The pollution of the atmosphere, oceans, and many of our lakes and streams; the denudation of huge areas of tropical forests; the scars from strip mining; the depletion of the ozone layer — all are visible in the satellite images beamed back from space and attest to the impact humans have had on the ecosystem.

Accordingly, we must understand that changes we make in the global ecosystem can have wide-ranging effects that we might not be aware of. For this reason, an understanding of geology, and science in general, is of paramount importance so that disruption to the ecosystem is minimal. On the other hand, we must also remember that humans are part of the ecosystem and like all other life-forms, our presence alone affects the ecosystem. We must therefore act in a responsible manner, based on sound scientific knowledge, so future generations will inherit a habitable environment.

When such environmental issues as acid rain, the greenhouse effect and global warming, and the depletion of the ozone layer are discussed and debated, it is important to remember that they are not isolated topics, but are part of a larger system that involves the entire Earth. Furthermore, it is important to remember that Earth goes through cycles of much longer duration than the human perspective of time. Although they may have disastrous effects on the human species, global warming and cooling are part of a larger cycle that has resulted in numerous glacial advances and retreats during the past 1.6 million years. In fact, geologists can make important contributions to the debate on global warming because of their geologic perspective. Long-term trends can be studied by analyzing deep-sea sediments, ice cores, changes in sea level during the geologic past, and the distribution of plants and animals through time. As we have seen throughout this book, such studies have been done and the results and synthesis of that information can be used to make intelligent decisions about how humans can better manage the environment and the effect we are having in altering the environment.

An example of an environmental imbalance is global warming caused by the greenhouse effect. Carbon dioxide is produced as a by-product of respiration and the burning of organic material. As such, it is a component of the global ecosystem and is constantly being recycled as part of the carbon cycle. The concern in recent years over the increase in atmospheric carbon dioxide has to do with its role in the greenhouse effect. The recycling of carbon dioxide between the crust and atmosphere is an important climatic regulator because carbon dioxide, as well as other gases such as methane, nitrous oxide, chlorofluorocarbons, and water vapor, allow sunlight to pass through them but trap the heat reflected back from Earth's surface. Heat is thus retained, causing the temperature of Earth's surface and, more importantly, the atmosphere to increase, producing the greenhouse effect.

Because of the increase in human-produced greenhouse gases during the last 200 years, many scientists are concerned that a global warming trend has already begun and will result in several global climatic shifts. Most computer models based on the current rate of increase in greenhouse gases show Earth warming as a whole by as much as 5°C during the next 100 years. Such a temperature change will be uneven, however, with the greatest warming occurring in the higher latitudes. As a consequence of this warming, rainfall patterns will shift dramatically. This will have a major effect on the largest grain-producing areas of the world, such as the American Midwest. Drier and hotter conditions will intensify the severity and frequency of droughts, leading to more crop failures and higher food prices. With such shifts in climate, Earth may experience an increase in the expansion of deserts, which will remove valuable crop and grazing lands.

We cannot leave the subject of global warming without pointing out that many scientists are not convinced that the global warming trend is the direct result of increased human activity related to industrialization. They point out that while there has been an increase in greenhouse gases, there is still uncertainty about their rate of generation and rate of removal and about whether the 0.5°C rise in global temperature during the past century is the result of normal climatic variations through time or the result of human activity. Furthermore, they point out that even if there is a general global warming during the next 100 years, it is not certain that the dire predictions made by proponents of global warming will come true. Earth, as we know, is a re-markably complex system, with many feedback mecha-nisms and interconnections throughout its various subsys-tems. It is very difficult to predict all of the consequences that global warming would have for atmospheric and oceanic circulation patterns.

In conclusion, the most important lesson to be learned from the study of historical geology is that Earth is an ex-tremely complex planet in which interactions between its various systems has resulted in changes in the atmosphere, lithosphere, and biosphere through time. By studying how Earth has evolved in the past, we can apply the lessons learned from this study to better understanding how the different Earth systems work and interact with each other, and more importantly, how our actions affect the delicate balance between these systems. Historical geology is not a static science, but one that, like the dynamic Earth which it seeks to understand, is constantly evolving as new infor-mation becomes available.

English-Metric Conversion Chart

ENGLISH UNIT	CONVERSION FACTOR	METRIC UNIT	CONVERSION FACTOR	ENGLISH UNIT
Length				
Inches (in)	2.54	Centimeters (cm)	0.39	Inches (in)
Feet (ft)	0.305	Meters (m)	3.28	Feet (ft)
Miles (mi)	1.61	Kilometers (km)	0.62	Miles (mi)
Area				
Square inches (in²)	6.45	Square centimeters (cm²)	0.16	Square inches (in²)
Square feet (ft²)	0.093	Square meters (m²)	10.8	Square feet (ft²)
Square miles (mi²)	2.59	Square kilometers (km²)	0.39	Square miles (mi²)
Volume				
Cubic inches (in³)	16.4	Cubic centimeters (cm³)	0.061	Cubic inches (in³)
Cubic feet (ft³)	0.028	Cubic meters (m³)	35.3	Cubic feet (ft³)
Cubic miles (mi³)	4.17	Cubic kilometers (km³)	0.24	Cubic miles (mi³)
Weight				
Ounces (oz)	28.3	Grams (g)	0.035	Ounces (oz)
Pounds (lb)	0.45	Kilograms (kg)	2.20	Pounds (lb)
Short tons (st)	0.91	Metric tons (t)	1.10	Short tons (st)
Temperature				
Degrees Fahrenheit (°F)	$-32° \times 0.56$	Degrees Celsius (Centigrade)(°C)	$\times 1.80 + 32°$	Degrees Fahrenheit (°F)

Examples: 10 inches = 25.4 centimeters; 10 centimeters = 3.9 inches

100 square feet = 9.3 square meters; 100 square meters = 1080 square feet

50°F = 10.08°C; 50°C = 122°F

Classification of Organisms

Any classification is an attempt to make order out of disorder and to group similar items into the same categories. All classifications are schemes that attempt to relate items to each other based on current knowledge and therefore are progress reports on the current state of knowledge for the items classified. Because classifications are to some extent subjective, classification of organisms may vary among different texts.

The classification that follows is based on the five-kingdom system of classification of Margulis and Schwartz.* We have not attempted to include all known life forms, but rather major categories of both living and fossil groups.

Kingdom Monera

Prokaryotes
> Phylum Archaebacteria — (Archean – Recent)
> Phylum Cyanobacteria — Blue-green algae or blue-green bacteria (Archean – Recent)

Kingdom Protoctista

Solitary or colonial unicellular eukaryotes
> Phylum Acritarcha — Organic-walled unicellular algae of unknown affinity (Proterozoic – Recent)
> Phylum Bacillariophyta — Diatoms (Jurassic – Recent)
> Phylum Charophyta — Stoneworts (Silurian – Recent)
> Phylum Chlorophyta — Green algae (Proterozoic – Recent)
> Phylum Chrysophyta — Golden-brown algae, silicoflagellates and coccolithophorids (Jurassic – Recent)
> Phylum Euglenophyta — Euglenids (Cretaceous – Recent)
> Phylum Myxomycophyta — Slime molds (Proterozoic – Recent)
> Phylum Phaeophyta — Brown algae, multicellular, kelp, seaweed (Proterozoic – Recent)
> Phylum Protozoa — Unicellular heterotrophs (Cambrian – Recent)
>> Class Sarcodina — Forms with pseudopodia for locomotion (Cambrian – Recent)
>>> Order Foraminifera — Benthonic and planktonic sarcodinids most commonly with calcareous tests (Cambrian – Recent)
>>> Order Radiolaria — Planktonic sarcodinids with siliceous tests (Cambrian – Recent)

> Phylum Pyrrophyta — Dinoflagellates (Silurian?, Permian – Recent)
> Phylum Rhodophyta — Red algae (Proterozoic – Recent)
> Phylum Xanthophyta — Yellow-green algae (Miocene – Recent)

Kingdom Fungi

> Phylum Zygomycota — Fungi that lack cross walls (Proterozoic – Recent)
> Phylum Basidiomycota — Mushrooms (Pennsylvanian – Recent)
> Phylum Ascomycota — Yeasts, bread molds, morels (Mississippian – Recent)

Kingdom Plantae

Photosynthetic eukaryotes
> Division* Bryophyta — Liverworts, mosses, hornworts (Devonian – Recent)
> Division Psilophyta — Small, primitive vascular plants with no true roots or leaves (Silurian – Recent)
> Division Lycopodophyta — Club mosses, simple vascular systems, true roots and small leaves, including scale trees of Paleozoic Era (lycopsids) (Devonian – Recent)
> Division Sphenophyta — Horsetails, scouring rushes, and sphenopsids such as the Carboniferous *Calamites* (Devonian – Recent)
> Division Pteridophyta — Ferns (Devonian – Recent)
> Division Pteridospermophyta — Seed ferns (Devonian – Jurassic)
> Division Coniferophyta — Conifers or cone-bearing gymnosperms (Carbonaceous – Recent)
> Division Cycadophyta — Cycads (Triassic – Recent)
> Division Ginkgophyta — Maidenhair tree (Triassic – Recent)
> Division Angiospermophyta — Flowering plants and trees (Cretaceous – Recent)

Kingdom Animalia

Nonphotosynthetic multicellular eukaryotes (Proterozoic – Recent)
> Phylum Porifera — Sponges (Cambrian – Recent)
>> Order Stromatoporoida — Extinct group of reef-building organisms (Cambrian – Oligocene)

---

*Margulis, L., and K.V.S. Schwartz, 1982. *Five Kingdoms*. New York: W.H. Freeman and Co.

---

*In botany, division is the equivalent to phylum.

Phylum Archaeocyatha — Extinct spongelike organisms (Cambrian)

Phylum Cnidaria — Hydrozoans, jellyfish, sea anemones, corals (Cambrian – Recent)

Class Hydrozoa — Hydrozoans (Cambrian – Recent)

Class Scyphozoa — Jellyfish (Proterozoic – Recent)

Class Anthozoa — Sea anemones and corals (Cambrian – Recent)

Order Tabulata — Exclusively colonial corals with reduced to nonexistent septa (Ordovician – Permian)

Order Rugosa — Solitary and colonial corals with fourfold symmetry (Ordovician – Permian)

Order Scleractinia — Solitary and colonial corals with sixfold symmetry. Most colonial forms have symbionic dinoflagellates in their tissues. Important reef builders today (Triassic – Recent)

Phylum Bryozoa — Exclusively colonial suspension feeding marine animals that are useful for correlation and ecological interpretations (Ordovician – Recent)

Phylum Brachiopoda — Marine suspension feeding animals with two unequal sized valves. Each valve is bilaterally symmetrical (Cambrian – Recent)

Class Inarticulata — Primitive chitino-phosphatic or calcareous brachiopods that lack a hinging structure. They open and close their valves by means of complex muscles (Cambrian – Recent)

Class Articulata — Advanced brachiopods with calcareous valves that are hinged (Cambrian – Recent)

Phylum Mollusca — A highly diverse group of invertebrates (Cambrian – Recent)

Class Monoplacophora — Segmented, bilaterally symmetrical crawling animals with cap-shaped shells (Cambrian – Recent)

Class Amphineura — Chitons. Marine crawling forms, typically with 8 separate calcareous plates (Cambrian – Recent)

Class Scaphopoda — Curved, tusk-shaped shells that are open at both ends (Ordovician – Recent)

Class Gastropoda — Single shelled generally coiled crawling forms. Found in marine, brackish, and fresh water as well as terrestrial environments (Cambrian – Recent)

Class Bivalvia — Mollusks with two valves that are mirror images of each other. Typically known as clams and oysters (Cambrian – Recent)

Class Cephalopoda — Highly evolved swimming animals. Includes shelled sutured forms as well as non-shelled types such as octopus and squid (Cambrian –Recent)

Order Nautiloidea — Forms in which the chamber partitions are connected to the wall along simple, slightly curved lines (Cambrian – Recent)

Order Ammonoidea — Forms in which the chamber partitions are connected to the wall along wavy lines (Devonian – Cretaceous)

Order Coleoidea — Forms in which the shell is reduced or lacking. Includes octopus, squid, and the extinct belemnoids (Mississippian – Recent)

Phylum Annelida — Segmented worms. Responsible for many of the Phanerozoic burrow and trail trace fossils (Proterozoic – Recent)

Phylum Arthropoda — The largest invertebrate group comprising about 80% of all known animals. Characterized by a segmented body and jointed appendages (Cambrian – Recent)

Class Trilobita — Earliest appearing arthropod class. Trilobites had a head, body, and tail and were bilaterally symmetrical (Cambrian – Recent)

Class Crustacea — Diverse class characterized by a fused head and body and an abdomen. Included are barnacles, copepods, crabs, ostracodes, and shrimp (Cambrian – Recent)

Class Insecta — Most diverse and common of all living invertebrates, but rare as fossils (Silurian – Recent)

Class Merostomata — Characterized by four pairs of appendages and a more flexible exoskeleton than crustaceans. Includes the extinct eurypterids, horseshoe crabs, scorpions, and spiders (Cambrian – Recent)

Phylum Echinodermata — Exclusively marine animals with fivefold radial symmetry and a unique water vascular system (Cambrian – Recent)

Subphylum Crinozoa — Forms attached by a calcareous jointed stem (Cambrian – Recent)

Class Crinoidea — Most important class of Paleozoic echinoderms. Suspension feeding forms that are either free-living or attach to sea floor by a stem (Cambrian – Recent)

Class Blastoidea — Small class of Paleozoic suspension feeding sessile forms with short stems (Ordovician – Permian)

Class Cystoidea — Globular to pear-shaped suspension feeding benthonic sessile forms with very short stems (Ordovician – Devonian)

Subphylum Homalozoa — A small group with flattened, asymmetrical bodies with no stems. Also called carpoids (Cambrian – Devonian)

Subphylum Echinozoa — Globose, predominantly benthonic mobile echinoderms (Ordovician – Recent)

Class Helioplacophora — Benthonic, mobile forms, shaped like a top with plates arranged in a helical spiral (Early Cambrian)

Class Edrioasteroidea — Benthonic, sessile or mobile, discoidal, globular- or cylindrical-shaped forms with five straight or curved feeding areas shaped like a starfish (Cambrian – Pennsylvanian)

Class Holothuroidea — Sea cucumbers. Sediment feeders having calcareous spicules embedded in a tough skin (Ordovician – Recent)

Class Echinoidea — Largest group of echinoderms. Globe- or disk-shaped with movable spines. Predominantly grazers or sediment feeders. Epifaunal and infaunal (Ordovician – Recent)

Subphylum Asterozoa — Stemless, benthonic mobile forms (Ordovician – Recent)

Class Asteroidea — Starfish. Arms merge into body (Ordovician – Recent)

Class Ophiuroidea — Brittle star. Distinct central body (Ordovician – Recent)

Phylum Hemichordata — Characterized by a notochord sometime during their life history. Modern acorn worms and extinct graptolites (Cambrian – Recent)

Class Graptolithina — Colonial marine hemichordates having a chitinous exoskeleton. Predominantly planktonic (Cambrian – Mississippian)

Phylum Chordata — Animals with notochord, hollow dorsal nerve cord, and gill slits during at least part of their lifecycle (Cambrian – Recent)

Subphylum Urochordata — Sea squirts, tunicates. Larval forms with notochord in tail region.

Subphylum Cephalochordata — Small marine animals with notochords and small fish-like bodies (Cambrian – Recent)

Subphylum Vertebrata — Animals with a backbone of vertebrae (Cambrian – Recent)

Class Agnatha — Jawless fish. Includes the living lampreys and hagfish as well as the extinct armored ostracoderms (Cambrian – Recent)

Class Acanthodii — Primitive jawed fish with numerous spiny fins (Silurian – Permian)

Class Placodermii — Primitive armored jawed fish (Silurian – Permian)

Class Chondrichthyes — Cartilaginous fish such as sharks and rays (Devonian – Recent)

Class Osteichthyes — Bony fish (Devonian – Recent)

Subclass Actinopterygii — Ray-finned fish (Devonian – Recent)

Subclass Sarcopterygii — Lobe-finned, air-breathing fish (Devonian – Recent)

Order Crossoptergii — Lobe-finned fish that were ancestral to amphibians (Devonian – Recent)

Order Dipnoi — Lungfish (Devonian – Recent)

Class Amphibia — Amphibians. The first terrestrial vertebrates (Devonian – Recent)

Subclass Labyrinthodontia — Earliest amphibians. Solid skulls and complex tooth pattern (Devonian – Triassic)

Subclass Salientia — Frogs, toads, and their relatives (Triassic – Recent)

Subclass Condata — Salamanders and their relatives (Triassic – Recent)

Class Reptilia — Reptiles. A large and varied vertebrate group characterized by having scales and laying an amniote egg (Mississippian – Recent)

Subclass Anapsida — Reptiles whose skull has a solid roof with no openings (Mississippian – Recent)

Order Cotylosauria — One of the earliest reptile groups (Pennsylvanian – Triassic)

Order Chelonia — Turtles (Triassic – Recent)

Subclass Euryapsida — Reptiles with one opening high on the side of the skull behind the eye. Mostly marine (Permian – Cretaceous)

Order Protorosauria — Land living ancestral euryapsids (Permian – Cretaceous)

Order Placodontia — Placodonts. Bulky, paddle-limbed marine reptiles with rounded teeth for crushing mollusks (Triassic)

Order Ichthyosauria — Ichthyosaurs. Dolphin-shaped swimming reptiles (Triassic – Cretaceous)

Subclass Diapsida — Most diverse reptile class. Characterized by two openings in the skull behind the eye. Includes lizards, snakes, crocodiles, thecodonts, dinosaurs, and pterosaurs (Permian – Recent)

Infraclass Lepidosauria — Primitive diapsids including snakes, lizards, and the mosasaurs, a large Cretaceous marine reptile group (Permian – Recent)

Order Mosasauria — Mosasaurs (Cretaceous)

Order Plesiosauria — Plesiosaurs (Jurassic – Cretaceous)

Order Squamata — Lizards and snakes (Triassic – Recent)

Order Rhynchocephalia — The living tuatara *Sphenodon* and its extinct relatives (Jurassic – Recent)

Infraclass Archosauria — Advanced diapsids (Triassic – Recent)

Order Thecodontia — Thecodontians were a diverse group that was ancestral to the crocodilians, pterosaurs, and dinosaurs (Permian – Triassic)

Order Crocodilia — Crocodiles, alligators, and gavials (Triassic – Recent)

Order Pterosauria — Flying and gliding reptiles called pterosaurs (Triassic – Cretaceous)

Infraclass Dinosauria — Dinosaurs (Triassic – Cretaceous)

Order Saurischia — Lizard-hipped dinosaurs (Triassic – Cretaceous)

Suborder Theropoda — Bipedal carnivores (Triassic - Cretaceous)

Suborder Sauropoda — Quadrupedal herbivores, including the largest known land animals (Jurassic - Cretaceous)

Order Ornithischia — Bird-hipped dinosaurs (Triassic - Cretaceous)

Suborder Ornithopoda — Bipedal herbivores, including the duck-billed dinosaurs (Triassic - Cretaceous)

Suborder Stegosauria — Quadrupedal herbivores with bony spikes on their tails and bony plates on their backs (Jurassic - Cretaceous)

Suborder Pachycephalosauria — Bipedal herbivores with thickened bones of the skull roof (Cretaceous)

Suborder Ceratopsia — Quadrupedal herbivores typically with horns or a bony frill over the top of the neck (Cretaceous)

Suborder Ankylosauria — Heavily armored quadrupedal herbivores (Cretaceous)

Subclass Synapsida — Mammal-like reptiles with one opening low on the side of the skull behind the eye (Pennsylvanian - Triassic)

Order Pelycosauria — Early mammal-like reptiles including those forms in which the vertebral spines were extended to support a "sail" (Pennsylvanian - Permian)

Order Therapsida — Advanced mammal-like reptiles with legs positioned beneath the body and the lower jaw formed largely of a single bone. Many therapsids may have been endothermic (Permian - Triassic)

Class Aves — Birds. Endothermic and feathered (Jurassic - Recent)

Class Mammalia — Mammals. Endothermic animals with hair (Triassic - Recent)

Subclass Prototheria — Egg-laying mammals (Triassic - Recent)

Order Docodonta — Small, primitive mammals (Triassic)

Order Triconodonta — Small, primitive mammals with specialized teeth (Triassic - Cretaceous)

Order Monotremata — Duck-billed platypus, spiny anteater (Cretaceous - Recent)

Subclass Allotheria — Small, extinct early mammals with complex teeth (Jurassic - Eocene)

Order Multituberculata — The first mammalian herbivores and the most diverse of Mesozoic mammals (Jurassic - Eocene)

Subclass Theria — Mammals that give birth to live young (Jurassic - Recent)

Order Symmetrodonta — Small, primitive Mesozoic therian mammals (Jurassic - Cretaceous)

Order Upantotheria — Trituberculates (Jurassic - Cretaceous)

Order Creodonta — Extinct ancient carnivores (Cretaceous - Paleocene)

Order Condylartha — Extinct ancestral hoofed placentals (ungulates) (Cretaceous - Oligocene)

Order Marsupialia — Pouched mammals. Opossum, kangaroo, koala (Cretaceous - Recent)

Order Insectivora — Primitive insect-eating mammals. Shrew, mole, hedgehog (Cretaceous - Recent)

Order Xenungulata — Large South American mammals that broadly resemble pantodonts and uintatheres (Paleocene)

Order Taeniodonta — Includes some of the most highly specialized terrestrial placentals of the Late Paleocene and Early Eocene (Paleocene - Eocene)

Order Tillodontia — Large, massive placentals with clawed, five-toed feet (Paleocene - Eocene)

Order Dinocerata — Uintatheres. Large herbivores with bony protuberances on the skull and greatly elongated canine teeth (Paleocene - Eocene)

Order Pantodonta — North American forms are large sheep to rhinoceros-sized. Asian forms are as small as a rat (Paleocene - Eocene)

Order Astropotheria — Large placental mammals with slender rear legs, stout forelimbs, and elongate canine teeth (Paleocene - Miocene)

Order Notoungulata — Largest assemblage of South American ungulates with a wide range of body forms (Paleocene - Pleistocene)

Order Liptoterna — Extinct South American hoofed mammals (Paleocene - Pleistocene)

Order Rodentia — Squirrel, mouse, rat, beaver, porcupine, gopher (Paleocene - Recent)

Order Lagomorpha — Hare, rabbit, pika (Paleocene - Recent)

Order Primates — Lemur, tarsier, loris, monkey, human (Paleocene – Recent)

Order Edentata — Anteater, sloth, armadillo, glyptodont (Paleocene – Recent)

Order Carnivora — Modern carnivorous placentals. Dog, cat, bear, skunk, seal, weasel, hyena, raccoon, panda, sea lion, walrus (Paleocene – Recent)

Order Pyrotheria — Large mammals with long bodies and short columnar limbs (Eocene – Oligocene)

Order Chiroptera — Bats (Eocene – Recent)

Order Dermoptera — Flying lemur (Eocene – Recent)

Order Cetacea — Whale, dolphin, porpoise (Eocene – Recent)

Order Tubulidentata — Aardvark (Eocene – Recent)

Order Perissodactyla — Odd-toed ungulates (hoofed placentals). Horse, rhinoceros, tapir, titanothere, chalicothere (Eocene – Recent)

Order Artiodactyla — Even-toed ungulates. Pig, hippo, camel, deer, elk, bison, cattle, sheep, antelope, entelodont, oreodont (Eocene – Recent)

Order Proboscidea — Elephant, mammoth, mastodon (Eocene – Recent)

Order Sirenia — Sea cow, manatee, dugong (Eocene – Recent)

Order Embrithopodoa — Known primarily from a single locality in Egypt. Large mammals with two gigantic bony processes arising from the nose area (Oligocene)

Order Desmostyla — Amphibious or seal-like in habit. Front and hind limbs well developed, but hands and feet somewhat specialized as paddles (Oligocene – Miocene)

Order Hyracoidea — Hyrax (Oligocene – Recent)

Order Pholidota — Scaly anteater (Oligocene – Recent)

Earth Materials: Minerals and Rocks

Minerals

The term *mineral* brings to mind dietary substances such as calcium, potassium, iron, and magnesium. These are actually chemical elements, not minerals in the geologic sense, but minerals are in fact composed of these and other chemical elements. That is, chemical elements bond together to form compounds such as silicon dioxide (SiO_2), which is the mineral quartz. Most minerals are compounds of two or more elements, but a few, known as *native elements,* are made up of only one element — gold (Au), silver (Ag), and carbon (C) as either diamond or graphite are a few examples.

A mineral is formally defined as a naturally occurring, inorganic, crystalline solid with a narrowly defined chemical composition, and characteristic physical properties. "Naturally occurring" excludes from minerals all manufactured substances such as synthetic diamonds and rubies. The term "inorganic" reminds us that animal and vegetable matter are not minerals. Nevertheless, clams, oysters, corals, and many other animals and some plants have skeletons made of minerals. "Crystalline solids" have their constituent atoms arranged in a regular three-dimensional framework, as opposed to a random distribution of atoms in, for instance, natural and human-made glass. Under ideal conditions, minerals grow and form perfect crystals with planar surfaces (crystal faces), sharp corners, and straight edges (Figure C1).

For many minerals, the chemical composition is fixed: quartz is always composed of silicon and oxygen (SiO_2), and halite contains only sodium and chlorine (NaCl). But some minerals have "narrowly defined chemical compositions" because one element can substitute for another if the atoms of two or more elements are about the same size and have the same electrical charge. In the mineral olivine [$(Mg,Fe)_2SiO_4$)], iron and magnesium substitute for one another, so in addition to silicon and oxygen it might have only iron, only magnesium, or both.

More than 3500 minerals have been identified and described, but only a few — perhaps two dozen — are particularly common. Among these, the most common by far are those classed as *silicates.* Silicate minerals are composed of at least silicon and oxygen atoms, and most contain other elements as well. Silicates are further characterized as *ferromagnesian* if they contain iron (Fe), magnesium (Mg), or both and *nonferromagnesian* if they lack these elements (Figure C2). Geologists also recognize several other mineral groups or classes, but the only other minerals common in rocks are the *carbonates,* all of which possess the carbonate ion (CO_3); the mineral calcite ($CaCO_3$) is a good example.

Chemical composition and internal structure determine the "characteristic physical properties" of minerals. Many of these properties are remarkably constant for a given mineral species, but some, especially color, might vary considerably. Physical properties such as crystal form, hardness, color, luster, specific gravity, cleavage, and several others are used to identify most common minerals (Table C1).

To identify minerals use the Mineral Identification Table (Table C3), which is arranged with minerals having a metallic luster grouped separately from those with a nonmetallic luster. Therefore, luster alone eliminates many minerals from consideration. Next, determine the hardness and note that each part of the table is arranged with minerals in order of increasing hardness. Thus, if you have a non-

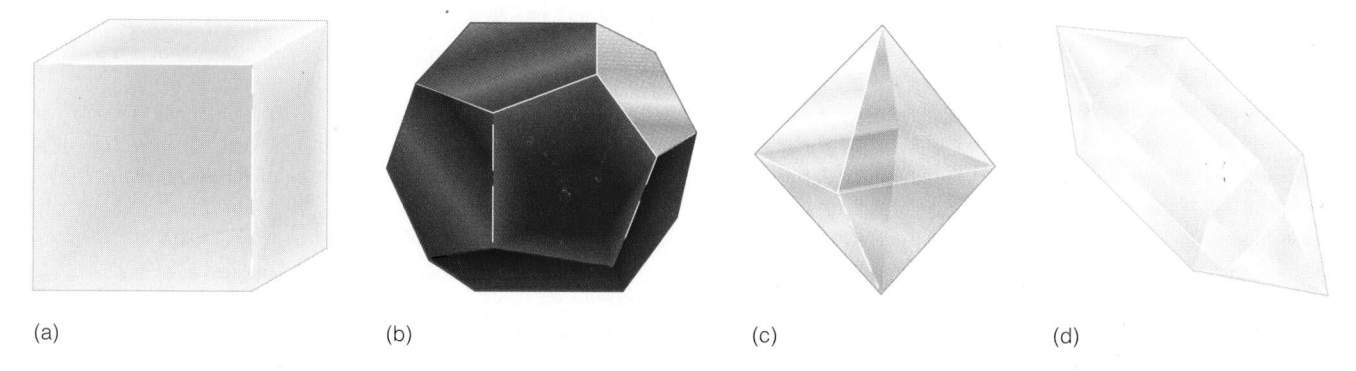

(a)　　　　(b)　　　　(c)　　　　(d)

FIGURE C1 **Examples of several mineral crystals. (a) Cubic crystals typically develop in halite, galena, and pyrite. (b) Dodecahedron crystals such as those of garnet have 12 sides. (c) Diamond has octahedral, or 8-sided, crystals. (d) A prism terminated by pyramids is found in quartz.**

(a)

(b)

FIGURE C2 Silicates, the most common minerals in Earth's crust, are divided into (a) ferromagnesian silicates (clockwise from top left: olivine, augite, biotite, and hornblende) and (b) nonferromagnesian silicates (clockwise from top left: quartz, orthoclase, muscovite, and plagioclase). Most of the so-called rock-forming minerals are included in these images. The only particularly common rock-forming minerals not shown here are various clay minerals and the carbonate minerals calcite and dolomite.

metallic mineral with a hardness of about 6, it must be augite, hornblende, plagioclase, or one of the two potassium feldspars (microcline or orthoclase). If this hypothetical mineral is dark green or black, it must be either augite or hornblende. Reference to other properties listed in the table should allow you to make a final determination.

Rocks

Rocks are solid aggregates of one or more minerals. Only a few of the minerals in Table C3 are *rock-forming minerals* — that is, minerals common enough in rocks to be important in naming and classifying them. Several properties of the three basic families of rocks, igneous, sedimentary,

and metamorphic, were discussed in Chapter 1, so the following discussions of these rock families are brief. Recall that all rock types are related and one type can be derived from any of the others (see the rock cycle in Figure 1.2).

IGNEOUS ROCKS

All igneous rocks are derived from magma that crystallizes within the crust to form *plutonic* or *intrusive igneous rock,* or magma that reaches the surface as lava or pyroclastic materials such as ash, thus forming *extrusive igneous rocks* or *volcanic rocks* (Table C4). Most igneous rocks are classified on the basis of their composition and texture, although for a few, texture only is the primary criteria for their identification.

TABLE C1

Physical Properties Used to Identify Minerals

MINERAL PROPERTY	COMMENTS
Color	Color in minerals having the appearance of metals is rather constant, whereas color in many nonmetallic-appearing minerals varies.
Streak	Rubbing a mineral on an unglazed porcelain plate (streak plate) leaves a fine trail of powder, the color of which is usually more consistent than the color of the unpowdered mineral (Figure C3a).
Luster	The appearance of a mineral in reflected light. Minerals with a metallic luster have the appearance of metals; those with a nonmetallic luster do not look like metals (Figure C3b).
Crystal form	If mineral crystals are visible, they are useful in mineral identification. For example, cubic crystals of pyrite are common, as are 12-sided garnet crystals (Figure C1b).
Cleavage	Not all minerals possess cleavage, but those that do tend to break along a smooth plane or planes of weakness determined by the strengths of bonds in mineral crystals (Figure C3c).
Fracture	The breakage unrelated to internal planes of weakness in mineral crystals. A fracture might be uneven or conchoidal (smoothly curved).
Hardness	A mineral's resistance to abrasion. Hardness values range from 1 (softest) to 10 (hardest) (Table C2).
Specific gravity	The ratio of a mineral's weight to an equal volume of water. A specific gravity of 3.0 indicates a mineral is three times heavier than an equal volume of water.
Reaction with HCl (hydrochloric acid)	Calcite reacts vigorously with HCl, whereas dolomite reacts only when powdered.
Other properties	Talc has a soapy feel; graphite writes on paper; a double image is seen through transparent calcite; magnetite is magnetic; closely spaced, parallel lines are visible on some specimens of plagioclase.

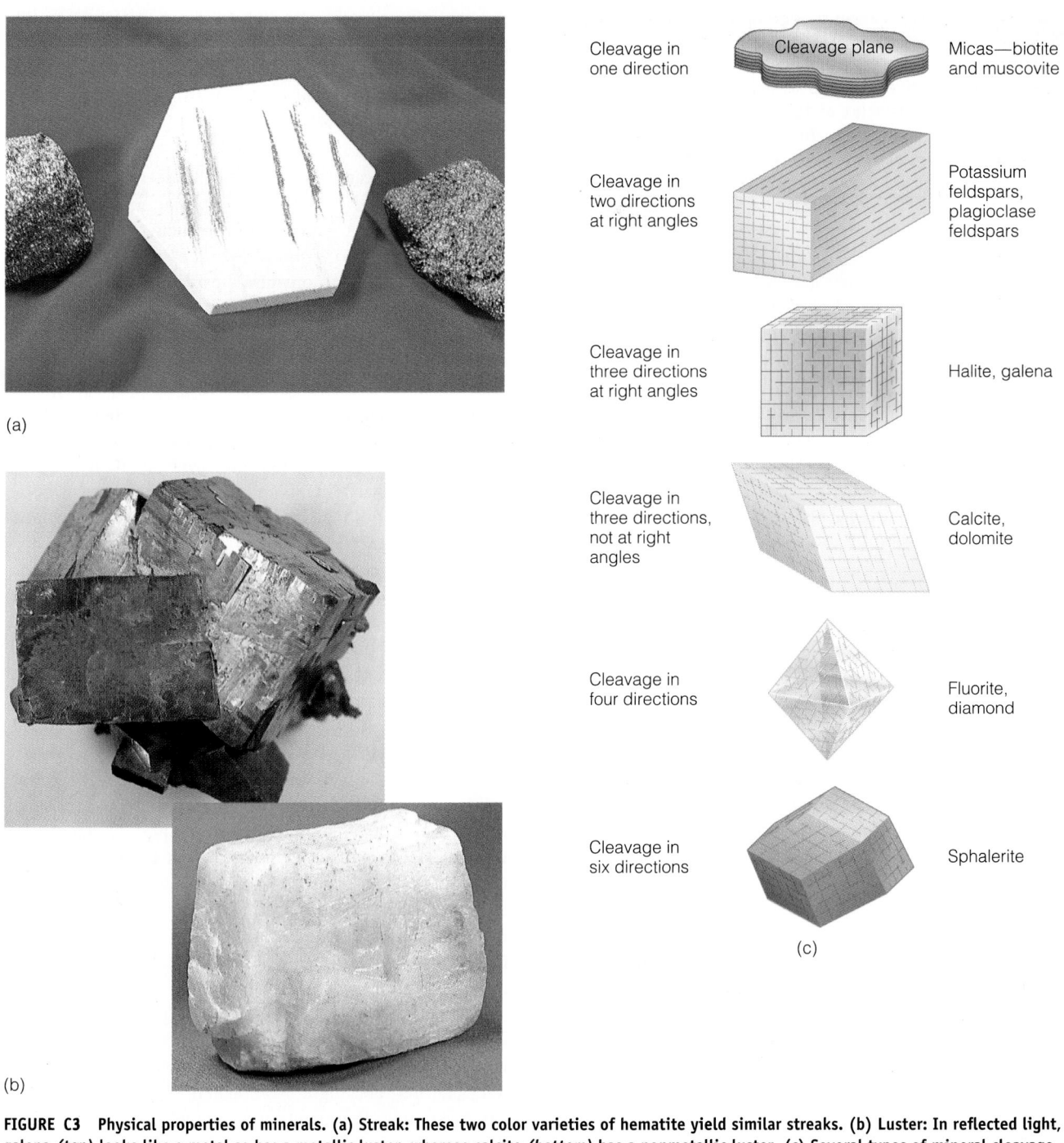

Cleavage in one direction — Micas—biotite and muscovite

Cleavage plane

Cleavage in two directions at right angles — Potassium feldspars, plagioclase feldspars

Cleavage in three directions at right angles — Halite, galena

Cleavage in three directions, not at right angles — Calcite, dolomite

Cleavage in four directions — Fluorite, diamond

Cleavage in six directions — Sphalerite

(a)

(b)

(c)

FIGURE C3 **Physical properties of minerals. (a) Streak: These two color varieties of hematite yield similar streaks. (b) Luster: In reflected light, galena *(top)* looks like a metal so has a metallic luster, whereas calcite *(bottom)* has a nonmetallic luster. (c) Several types of mineral cleavage.**

TABLE C2

Mohs Hardness Scale. Austrian geologist Frederich Mohs devised this relative hardness scale for ten minerals. He assigned a hardness value of 10 to diamond, the hardest mineral known, and lesser values to the other minerals. Relative hardness of a mineral is determined by scratching one mineral with another or by using objects of known hardness.

HARDNESS	MINERAL	HARDNESS OF SOME COMMON OBJECTS
10	Diamond	
9	Corundum	
8	Topaz	
7	Quartz	
		Steel file (6 ½)
6	Orthoclase	
		Glass (5 ½ – 6)
5	Apatite	
4	Fluorite	
3	Calcite	Copper penny (3)
		Fingernail (2 ½)
2	Gypsum	
1	Talc	

In terms of composition, igneous rocks are characterized as *felsic* (>65% silica), *intermediate* (54–65% silica), and *mafic* (45–52% silica). Felsic rocks are dominated by nonferromagnesian silicate minerals and are thus generally light colored, whereas mafic rocks contain a large proportion of ferromagnesian silicates and are mostly dark colored.

Igneous rock textures include *phaneritic* (minerals clearly visible), *aphanitic* (minerals too small to be seen without magnification), and *porphyritic,* meaning that minerals of markedly different sizes are present. Igneous rocks with a *vesicular texture* have numerous small cavities formed by gas trapped in cooling lava, and those with a *glassy texture* have the appearance of glass. Igneous rocks composed of particles ejected during explosive eruptions are said to have a *pyroclastic* or *fragmental texture.*

SEDIMENTARY ROCKS

Mechanical and chemical weathering yield the raw materials for sedimentary rocks (see Figures 1.2 and 4.3).

Detritus—consisting of solid particles such as gravel, sand, silt, and clay-size particles—might be lithified to form *detrital sedimentary rocks,* whereas minerals derived from solution yield *chemical sedimentary rocks.* All detrital sedimentary rocks have a *clastic texture,* meaning they are composed of broken particles, but chemical sedimentary rocks can have either a clastic texture or *crystalline texture* consisting of a mosaic of interlocking mineral crystals. A subcategory of chemical sedimentary rocks known as *biochemical sedimentary rocks* includes those rocks for which organisms play an important role in their origin (Table C5).

Composition is irrelevant for classification of most detrital sedimentary rocks, but those composed of gravel contain mostly rock fragments and quartz. Most sandstones are made up of quartz, but significant amounts of feldspar minerals are present in the sandstone known as arkose (Table C5). Mudstone and claystone contain variable amounts of clay minerals (a particular group of silicate minerals), most of which are clay-sized, meaning they measure less than 1/256 mm.

Among the chemical sedimentary rocks, limestone and dolostone are by far the most abundant. However, in some regions the evaporite rocks known as rock gypsum and rock salt are abundant. Note in Table C5 that limestone is listed among both chemical and biochemical sedimentary rocks, but most no doubt belong in the latter category.

METAMORPHIC ROCKS

Metamorphic rocks result from the transformation in the solid state of other rocks, including previously formed metamorphic rocks, by heat, pressure, and chemically active fluids. Rocks subjected to great heat and high pressure typically develop a layered look known as *foliation,* resulting from the parallel alignment of platy and elongate minerals. *Nonfoliated* metamorphic rocks lack this distinctive feature and consist of a mosaic of roughly equidimensional minerals (Table C6).

Foliated metamorphic rocks typically develop in areas subjected to *regional metamorphism,* which occurs over large regions deep within the crust where pressure and temperatures are high. Nonfoliated metamorphic rocks can form under these conditions as well. One area of obvious regional metamorphism is where plates collide along convergent plate margins. However, many nonfoliated metamorphic rocks also form by *contact metamorphism* adjacent to cooling bodies of magma. In this environment, heat and chemical fluids bring about metamorphism.

Mineral Identification Tables

METALLIC LUSTER					
Mineral	Chemical Composition	Color	Hardness Specific Gravity	Other Features	Comments
Graphite	C	Black	1–2 2.09–2.33	Greasy feel; writes on paper; 1 direction of cleavage	Used for pencil "leads" and as a dry lubricant. Mostly in metamorphic rocks.
Galena	PbS	Lead gray	$2\frac{1}{2}$ 7.6	Cubic crystals; 3 cleavages at right angles	The ore of lead. Mostly in hydrothermal rocks.
Chalcopyrite	$CuFeS_2$	Brassy yellow	$3\frac{1}{2}$–4 4.1–4.3	Usually massive; greenish black streak; iridescent tarnish	The most common copper mineral and an important source of copper. Mostly in hydrothermal rocks.
Magnetite	Fe_3O_4	Black	$5\frac{1}{2}$–$6\frac{1}{2}$ 5.2	Strong magnetism	An important ore of iron. An accessory mineral in many rocks.
Hematite	Fe_2O_3	Red brown	6 4.8–5.3	Usually granular or massive; reddish brown streak	Most important ore of iron. An accessory mineral in many rocks.
Pyrite	FeS_2	Brassy yellow	$6\frac{1}{2}$ 5.0	Cubic and octahedral crystals	The most common sulfide mineral. Found in some igneous and hydrothermal rocks and in sedimentary rocks associated with coal.

NONMETALLIC LUSTER					
Mineral	Chemical Composition	Color	Hardness Specific Gravity	Other Features	Comments
Talc	$Mg_3Si_4O_{10}(OH)_2$	White, green	1 2.82	1 cleavage direction; usually in compact masses	Formed by the alteration of magnesium silicates. Mostly in metamorphic rocks. Used in ceramics, cosmetics, and as a filler in paints.
Smectite	$(Al,Mg)_8(Si_4O_{10})_3$ $(OH)_{10}12H_2O$	Gray, buff, white	1–$1\frac{1}{2}$ 2.5	Earthy masses; particles too small to observe properties	A clay mineral with the unique property of swelling and contracting as it absorbs and releases water.
Illite	$(Ca,Na,K)(Al,Fe^{+3},$ $Fe^{+2},Mg)_2(Si,Al)_4$ $O_{10}(OH)_2$	White, light gray, buff	1–2 2.6–2.9	Earthy masses; particles too small to observe properties	A clay mineral common in soils and clay-rich sedimentary rocks.
Chlorite	$(Mg,Fe)_3(Si,Al)_4O_{10}$ $(Mg,Fe)_3(OH)_6$	Green	2 2.6–3.4	1 cleavage; occurs in scaly masses	Common in low-grade metamorphic rocks such as slate.
Gypsum	$CaSO_4 \cdot 2H_2O$	Colorless, white	2 2.32	Elongate crystals; fibrous and earthy masses	The most common sulfate mineral. Found mostly in evaporite deposits. Used to manufacture plaster of Paris and cements.
Kaolinite	$Al_2Si_4O_{10}(OH)_8$	White	2 2.6	Massive; earthy odor	A common clay mineral formed by chemical weathering of aluminum-rich silicates. The main ingredient of kaolin clay used for the manufacture of ceramics.
Muscovite (Mica)	$KAl_2Si_3O_{10}(OH)_2$	Colorless	2–$2\frac{1}{2}$ 2.7–2.9	1 direction of cleavage; cleaves into thin sheets	Common in felsic igneous rocks, metamorphic rocks, and some sedimentary rocks. Used as an insulator in electrical appliances.

(Continued)

Mineral Identification Tables

			Hardness		
			NONMETALLIC LUSTER		
Mineral	Chemical Composition	Color	Specific Gravity	Other Features	Comments
Biotite (Mica)	$K(Mg,Fe)_3AlSi_3O_{10}(OH)_2$	Black, brown	$2\frac{1}{2}$ 2.9–3.4	1 cleavage direction; cleaves into thin sheets	Occurs in both felsic and mafic igneous rocks, in metamorphic rocks, and in clay-rich sedimentary rocks.
Barite	$BaSO_4$	Colorless, white, gray	3 4.5	Tabular crystals; high specific gravity for a nonmetallic mineral	Commonly found with ores of a variety of metals and in limestones and hot-spring deposits. A source of barium.
Calcite	$CaCO_3$	Colorless, white	3 2.71	3 cleavages at oblique angles; cleaves into rhombs; reacts with dilute hydrochloric acid	The most common carbonate mineral. Main component of limestone and marble. Also common in hydrothermal rocks.
Anhydrite	$CaSO_4$	White, gray	$3\frac{1}{2}$ 2.9–3.0	Crystals with 2 cleavages; usually in granular masses	Found in limestones, evaporite deposits, and the cap rock of salt domes. Used as a soil conditioner.
Halite	$NaCl$	Colorless, white	3–4 2.2	3 cleavages at right angles; cleaves into cubes; cubic crystals; salty taste	Occurs in evaporite deposits. Used as a source of chlorine and in the manufacture of hydrochloric acid, many sodium compounds, and food seasoning.
Dolomite	$CaMg(CO_3)_2$	White, yellow, gray, pink	$3\frac{1}{2}$–4 2.85	Cleavage as in calcite; reacts with dilute hydrochloric acid when powdered	The main constituent of dolostone. Also found associated with calcite in some limestones and marble.
Sphalerite	ZnS	Yellow, brown, black	$3\frac{1}{2}$–4 4.0–4.1	6 cleavages; cleaves into dodecahedra	The most important ore of zinc. Commonly found with galena in hydrothermal rocks.
Fluorite	CaF_2	Colorless, purple, green, brown	4 3.18	4 cleavage directions; cubic and octahedral crystals	Occurs mostly in hydrothermal rocks and in some limestones and dolostones. Used in the manufacture of steel and the preparation of hydrofluoric acid.
Siderite	$FeCO_3$	Yellow, brown	4 3.8–4.0	3 cleavages at oblique angles; cleaves into rhombs	Found mostly in concretions and sedimentary rocks associated with coal.
Apatite	$Ca_5(PO_4)_3F$	Blue, green, brown, yellow, white	5 3.1–3.2	6-sided crystals; in massive or granular masses	An accessory mineral in many rocks. The main constituent of bone and dentine. A source of phosphorous for fertilizer.

(Continued)

Mineral Identification Tables

NONMETALLIC LUSTER					
Mineral	Chemical Composition	Color	Hardness / Specific Gravity	Other Features	Comments
Augite	$Ca(Mg,Fe,Al)(Al,Si)_2O_6$	Black, dark green	6 / 3.25–3.55	Short 8-sided crystals; 2 cleavages; cleavages nearly at right angles	The most common pyroxene mineral. Found mostly in mafic igneous rocks.
Hornblende	$NaCa_2(Mg,Fe,Al)_5$ $(Si,Al)_8O_{22}(OH)_2$	Green, black	6 / 3.0–3.4	Elongate, 6-sided crystals; 2 cleavages intersecting at 56° and 124°	A common rock-forming amphibole mineral in igneous and metamorphic rocks.
Plagioclase feldspars	Varies from $CaAl_2Si_2O_8$ to $NaAlSi_3O_8$	White, gray, brown	6 / 2.56	2 cleavages at right angles	Common in igneous rocks and a variety of metamorphic rocks. Also in some arkoses.
Potassium Feldspars — Microcline	$KAlSi_3O_8$	White, pink green	6 / 2.56	2 cleavages at right angles	Common in felsic igneous rocks, some metamorphic rocks, and arkoses. Used in the manufacture of porcelain.
Potassium Feldspars — Orthoclase	$KAlSi_3O_8$	White, pink	6 / 2.56	2 cleavages at right angles	
Cassiterite	SnO_2	Brown to black	6½ / 7.0	High specific gravity for a nonmetallic mineral	The main ore of tin. Most is concentrated in alluvial deposits because of its high specific gravity.
Olivine	$(Fe,Mg)_2SiO_4$	Olive green	6½ / 3.3–3.6	Small mineral grains in granular masses; conchoidal fracture	Common in mafic igneous rocks.
Quartz	SiO_2	Colorless, white, gray, pink, green	7 / 2.67	6-sided crystals; no cleavage; conchoidal fracture	A common rock-forming mineral in all rock groups and hydrothermal rocks. Also occurs in varieties known as chert, flint, agate, and chalcedony.
Garnet	$Fe_3Al_2(SiO_4)_3$	Dark red, green	7–7½ / 4.32	12-sided crystals common; uneven fracture	Found mostly in gneiss and schist. Used as a semiprecious gemstone and for abrasives.
Zircon	Zr_2SiO_4	Brown, gray	7½ / 3.9–4.7	4-sided, elongate crystals	Most common as an accessory in granitic rocks. An ore of zirconium and used as a gemstone.
Topaz	$Al_2SiO_4(OH,F)$	Colorless, white, yellow, blue	8 / 3.5–3.6	High specific gravity; 1 cleavage direction	Found in pegmatites, granites, and hydrothermal rocks. An important gemstone.
Corundum	Al_2O_3	Gray, blue, pink, brown	9 / 4.0	6-sided crystals and great hardness are distinctive	An accessory mineral in some igneous and metamorphic rocks. Used as a gemstone and for abrasives.

Classification of Igneous Rocks

INTRUSIVE IGNEOUS ROCKS (PLUTONIC ROCKS)			
Composition	Texture	Other Features	Rock Name
K feldspar, quartz, biotite, hornblende	Phaneritic	Light colored, pink and white feldspars visible	Granite
Plagioclase, K feldspar, hornblende, biotite, little quartz	Phaneritic	Light and dark minerals yield salt-and-pepper look	Diorite
Plagioclase, pyroxene, olivine	Phaneritic	Commonly dark, cleavage surfaces of plagioclase visible	Gabbro
Olivine, pyroxene, plagioclase	Phaneritic	Dark, commonly greenish	Peridotite

EXTRUSIVE IGNEOUS ROCKS (VOLCANIC ROCKS)			
Composition	Texture	Other Features	Rock Name
K feldspar, quartz, biotite, hornblende	Aphanitic	White to gray, quartz phenocrysts	Rhyolite
Plagioclase, K feldspar, hornblende, biotite, little quartz	Aphanitic	Commonly gray to dark gray	Andesite
Plagioclase, pyroxene, olivine	Aphanitic	Commonly gray to black	Basalt
———	Vesicular	Gray, low density, numerous vesicles	Pumice
———	Vesicular	Red, black, low density, numerous vesicles	Scoria
———	Glassy	Black, red, brown, dense, hard, glassy	Obsidian
———	Pyroclastic	Generally gray, low density, composed of volcanic ash	Tuff
———	Pyroclastic	Composed of particles >2 mm	Volcanic breccia

Note: If granite, diorite, gabbro, rhyolite, andesite, or basalt have an obvious porphyritic texture, they are referred to as granite porphry, diorite porphry, and so on.

Classification of Sedimentary Rocks

DETRITAL SEDIMENTARY ROCKS			
Sediment Name and Size	Composition	Other Features	Rock Name
Gravel >2 mm	Rock fragments and quartz common	Rounded gravel	Conglomerate
Gravel >2 mm	Rock fragments and quartz common	Angular gravel	Sedimentary breccia
Sand 1/16–2 mm	Mostly quartz	Sand grains visible, feels like sandpaper	Sandstone
Sand 1/16–2 mm	Quartz and >25% feldspar	Pink feldspar visible	Arkose
Mud <1/16 mm (1/256–1/16 mm)	Mostly silt	Gritty feel	Mudrocks Siltstone
	Silt and clay	Slightly gritty feel	Mudstone*
(<1/256 mm)	Mostly clay	Smooth feel	Claystone*

CHEMICAL SEDIMENTARY ROCKS				
Texture	Composition	Other Features	Rock Name	
Varies	Calcite ($CaCO_3$)	Commonly gray, hardness 3, reacts with HCl	Limestone	Carbonate rocks†
Varies	Dolomite ($CaMg(CO_3)_2$)	Resembles limestone, reacts with HCl when powdered	Dolostone	
Crystalline	Halite (NaCl)	Cubic crystals, salty taste	Rock salt	Evaporites‡
Crystalline	Gypsum ($CaSO_4 \cdot 2H_2O$)	Colorless, white, pink, softer than a fingernail	Rock gypsum	

BIOCHEMICAL SEDIMENTARY ROCKS			
Texture	Composition	Other Features	Rock Name
Clastic (broken particles)	Calcite ($CaCO_3$)	Fossils or other particles visible, commonly gray, reacts with HCl	Limestone (various types such as chalk and coquina)
Microcrystalline	Silica (SiO_2)	Various colors, dense, very hard, conchoidal fracture	Chert
Not applicable	Carbonaceous	Dull to shiny black, smudges fingers	Coal

*Mudrocks possessing the property of fissility, meaning they break along closely spaced planes, are commonly called shale.
†Carbonate rocks are composed of carbonate minerals such as calcite and dolomite.
‡Evaporites form when solutions with dissolved materials evaporate.

Classification of Metamorphic Rocks

FOLIATED METAMORPHIC ROCKS

Composition	Other Features	Parent Rock	Rock Name
Clays, muscovite, biotite, chlorite; minerals too small to identify	Splits easily into flat pieces, dull luster on flat surfaces	Mudrocks, volcanic ash	Slate
Clays, muscovite, biotite, chlorite; minerals large enough to identify	Resembles slate but has glossy sheen on flat surfaces	Mudrocks	Phyllite
Micas, quartz, talc, hornblende, garnet	Distinct but wavy foliation	Mudrocks, carbonates, mafic igneous rocks	Schist
Quartz, feldspars, micas, hornblende	Light and dark color bands	Mudrocks, sandstone, felsic igneous rocks	Gneiss
Hornblende, plagioclase	Dark, weak foliation	Mafic igneous rocks	Amphibolite
Quartz, hornblende, micas, feldspars	Streaks or lenses of granite intermixed with gneiss	Felsic igneous rocks mixed with sedimentary rocks	Migmatite

NONFOLIATED METAMORPHIC ROCKS

Composition	Other Features	Parent Rock	Rock Name
Calcite or dolomite	Large interlocking crystals, light color, reacts with HCl same as limestone and dolostone	Limestone or dolostone	Marble
Quartz	Interlocking grains, hard, dense	Quartz sandstone	Quartzite
Chlorite, epidote, hornblende	Green, fine-grained	Mafic igneous rocks	Greenstone
Micas, garnets, quartz	Fine-grained, grains all same size, hard, dense	Mudrocks	Hornfels
Carbon	Black, lustrous	Coal	Anthracite

Glossary

A

Absaroka sequence A widespread sequence of Pennsylvanian and Permian sedimentary rocks bounded above and below by unconformities; deposited during a transgressive–regressive cycle of the Absaroka Sea.

absolute dating The process of assigning an actual age in years before the present to geologic events. Various radioactive decay dating techniques yield absolute ages. (See *relative dating.*)

Acadian orogeny A Devonian orogeny in the northern Appalachian mobile belt resulting from a collision of Baltica with Laurentia.

acanthodian Any of the fish first having a jaw or jawlike mechanism; a class of fishes (class Aconthodii) appearing during the Early Silurian and becoming extinct during the Permian.

acondrite A type of stony meteorite lacking condrules; composition similar to that of terrestrial basalt.

acritarch Organic-walled microfossil that probably represents the cyst of a planktonic algae; appeared in the fossil record about 1.4 billion years ago and became abundant during the Late Proterozoic through Devonian.

Alleghenian orogeny Pennsylvanian to Permian mountain building during which the Appalachian mobile belt from New York to Alabama was deformed; occurred in the area of the present-day Appalachian Mountains.

allele A variant form of a single gene.

allopatric speciation Model for the origin of a new species from a small population that became geographically isolated from its parent population.

alluvial fan A cone-shaped deposit of alluvium; generally deposited where a stream flows from mountains onto an adjacent lowland.

alpha decay A type of radioactive decay involving the emission of a particle consisting of two protons and two neutrons from the nucleus of an atom; emission of an alpha particle decreases the atomic number by 2 and atomic mass number by 4.

Alpine–Himalayan orogenic belt A linear belt of deformation extending from the Atlantic Ocean eastward across southern Europe and northern Africa, through the Middle East, and into Southeast Asia; one of two major Mesozoic–Cenozoic orogenic belts. (See *circum-Pacific orogenic belt.*)

Alpine orogeny A Late Mesozoic–Cenozoic episode of mountain building affecting southern Europe and northern Africa.

amniote egg An egg in which the embryo develops in a liquid-filled cavity called the amnion. The embryo is also supplied with a yolk sac and a waste sac. The amniote egg is shelled in reptiles, birds, and egg-laying mammals and is retained but modified in all other mammals.

anaerobic A term referring to organisms that are not dependent on oxygen for respiration.

analogous organ Body part, such as the wing of insects and birds, that serve the same function but differ in structure and development. (See *homologous organ.*)

Ancestral Rockies Late Paleozoic uplift in the southwestern part of the North American craton.

angiosperm Vascular plants having flowers and seeds; the flowering plants.

angular unconformity An unconformity below which older strata dip at a different angle (usually steeper) than the overlying younger strata. (See *disconformity* and *nonconformity.*)

anthropoid Any member of the primate suborder Anthropoidea; includes New World and Old World monkeys, apes, and humans.

Antler orogeny A Late Devonian to Mississippian orogeny that affected the Cordilleran mobile belt; deformation extended from Nevada to Alberta, Canada.

Appalachian mobile belt A mobile belt along the eastern margin of the North American craton; extends from Newfoundland to Georgia; probably continuous to the Southwest with the Ouachita mobile belt.

arc orogen An area of deformation, such as an island arc, that results from subduction of an oceanic plate; characterized by deformation and igneous activity.

archaeocyathid A benthonic sessile suspension feeder that lived during the Cambrian and constructed reeflike structures.

Archean Eon A part of Precambrian time beginning 4.0 billion years ago, corresponding to the age of the oldest known rocks on Earth, and ending 2.5 billion years ago. (See *Proterozoic Eon.*)

archosaur The archosaurs, or ruling reptiles, include dinosaurs, pterosaurs, crocodiles, and birds.

artificial selection The practice of selective breeding of plants and animals for desirable traits.

Artiodactyla The order composed of even-toed hoofed mammals. Living artiodactyls include swine, sheep, goats, camels, deer, bison, antelope, moose, musk oxen, and hippopotamuses.

asthenosphere Part of the upper mantle lying below the lithosphere; behaves plastically and flows.

australopithecine A term referring to several extinct species of the genus *Australopithecus* that existed in South and East Africa during the Pliocene and Pleistocene epochs.

autotrophic Describes organisms that synthesize their organic nutrients from inorganic raw materials; photosynthesizing bacteria and plants are autotrophs. (See *heterotrophic.*)

B

back-arc marginal basin A basin formed between a continent and a volcanic island arc; thought to form by back-arc spreading; the site of a marginal sea, e.g., the Sea of Japan.

Baltica One of six major Paleozoic continents; composed of Russia west of the Ural Mountains, Scandinavia, Poland, and northern Germany.

ded iron formation (BIF) Sedimentary rocks consisting of alternating thin layers of silica (chert) and iron minerals (mostly the iron oxides hematite and magnetite).

barrier island An elongate sand body oriented parallel to a shoreline but separated from the shoreline by a lagoon.

Basin and Range Province An area centered on Nevada but extending into adjacent states and northern Mexico; characterized by Cenozoic block-faulting.

bedding (stratification) The layering in sedimentary rocks. Layers less than 1 cm thick are laminae, whereas beds are thicker.

benthos Any organism that lives on the bottom of seas or lakes; may live upon the bottom or within bottom sediments.

beta decay A type of radioactive decay during which a fast-moving electron is emitted from a neutron and thus is converted to a proton; results in an increase of 1 atomic number, but no change in atomic mass number.

Big Bang A model for the evolution of the universe from a dense, hot state followed by expansion, cooling, and a less dense state.

biogenic sedimentary structure Any feature in sedimentary rocks produced by the activities of organisms, e.g., tracks, trails, burrows. (See *trace fossil*.)

biostratigraphic unit A unit of sedimentary rock defined by its fossil content.

bioturbation The process of churning or stirring of sediments by organisms.

biozone The fundamental biostratigraphic unit (e.g., range zone and concurrent range zone).

bipedal Walking on two legs as a means of locomotion.

body fossil The actual remains of any prehistoric organism; includes shells, teeth, bones, and, rarely, the soft parts of organisms. (See *trace fossil*.)

bony fish A class of fish (class Osteichthyes) that evolved during the Devonian; the most common fish; characterized by an internal skeleton of bone; divided into two subgroups, the ray-finned fish and lobe-finned fish.

brachiopod Any member of a group of bivalved, suspension-feeding, marine, invertebrate animals.

browser An animal that eats tender shoots, twigs, and leaves. Compare with *grazer*.

C

Caledonian orogeny A Silurian – Devonian orogeny that occurred along the northwestern margin of Baltica, resulting from the collision of Baltica with Laurentia.

Canadian shield The Precambrian shield of North America; exposed mostly in Canada, but crops out in Minnesota, Wisconsin, Michigan, and New York.

carbon 14 dating technique An absolute-dating method that relies on determining the ratio of C^{14} to C^{12} in a sample; useful back to about 70,000 years ago; can be applied only to organic substances.

carbonaceous chondrite A type of stony meteorite; same as ordinary chondrites except they contain about 5% organic compounds including inorganically produced amino acids.

carnassials A pair of specialized upper and lower shearing teeth in members of the mammal order Carnivora.

Carnivora An order of mammals characterized by well developed canine teeth and shearing teeth known as carnassials. The meat-eating mammals such as dogs, cats, bears, weasels, and seals.

carnivore-scavenger Any animal that depends on other animals, living or dead, as a source of nutrients.

cartilaginous fish Fish such as living sharks, rays, and skates and their extinct relatives that have a skeleton composed of cartilage.

cast A replica of an object such as a shell or bone formed when a mold of that object is filled by sediment or mineral matter. (See *mold*.)

catastrophism The concept proposed by Baron Georges Cuvier that explained Earth's physical and biological history by a series of sudden, widespread catastrophes.

Catskill Delta The Devonian clastic wedge deposited adjacent to the highlands that formed during the Acadian orogeny.

Cetacea The mammal order that includes whales, porpoises, and dolphins.

China One of six major Paleozoic continents; composed of all of Southeast Asia, including China, Indochina, part of Thailand, and the Malay Peninsula.

chondrite A stony meteorite that contains chondrules.

chondrule A small, round mineral body formed by rapid cooling; found in chondritic meteorites.

chordate Any member of the phylum Chordata; characterized by a notochord, a dorsal, hollow nerve cord, and gill slits at some time during the animal's life cycle.

chromosome Complex, double-stranded, helical molecule of deoxyribonucleic acid (DNA); specific segments of chromosomes are genes.

circum-Pacific orogenic belt One of two major Mesozoic – Cenozoic orogenic belts; located around the margins of the Pacific Ocean basin; includes the orogens of South and Central America, the Cordillera of western North America, and the Aleutian, Japan, and Philippine arcs. (See *Alpine – Himalayan orogenic belt*.)

clastic texture Sedimentary rocks consisting of the broken particles of preexisting rocks or broken shells have a clastic texture.

clastic wedge An extensive accumulation of mostly clastic sediments eroded from and deposited adjacent to an uplifted area; clastic wedges are coarse grained and thick near the uplift and become finer grained and thinner away from the uplift, e.g., the Queenston Delta.

collision orogen An orogen produced as a result of a collision of two continents, e.g., the collision of India with Asia that resulted in the formation of the Himalaya.

concurrent range zone A type of biozone established by plotting the overlapping ranges of fossils that have different geologic ranges; the first and last occurrences of fossils are used to establish concurrent range zone boundaries.

conformable The relationship between beds in a sedimentary sequence containing no discontinuities. (See *unconformity*.)

conodont A small, toothlike fossil composed of calcium phosphate. Conodont elements were located behind the mouth of an elongate animal and probably functioned in feeding. Geologic range is Cambrian through Triassic.

continental–continental plate boundary A type of convergent plate boundary along which two continental litho-spheric plates collide, e.g., the collision of India with Asia.

continental drift The theory that the continents were once joined into a single landmass that broke apart with the various fragments (continents) moving with respect to one another; the theory of continental movement proposed by Alfred Wegener in 1912.

convergent evolution   The development of similarities in two or more distantly related organisms as a consequence of adapting to a similar lifestyle, e.g., ichthyosaurs and porpoises. (See *parallel evolution.*)

convergent plate boundary   The boundary between two plates that are moving toward one another; three types of convergent plate boundaries are recognized. (See *continental–continental plate boundary, oceanic–continental plate boundary,* and *oceanic–oceanic plate boundary.*)

coprolite   Fossilized feces.

Cordilleran orogeny   A protracted episode of deformation affecting the western part of North America from Jurassic to Early Cenozoic time; typically divided into three separate phases called the Nevadan, Sevier, and Laramide orogenies.

Cordilleran mobile belt   A mobile belt in western North America bounded on the west by the Pacific Ocean and on the east by the Great Plains; extends north-south from Alaska into central Mexico. (See *North American Cordillera.*)

core   The interior part of Earth beginning at a depth of about 2900 km; probably composed mostly of iron and nickel; divided into an outer liquid core and an inner solid core.

correlation   Demonstration of the physical continuity of rock units or biostratigraphic units in different areas, or demonstration of time equivalence as in time-stratigraphic correlation.

craton   A name applied to the relatively stable part of a continent; consists of a Precambrian shield and a platform, a buried extension of a shield; the ancient nucleus of a continent.

cratonic sequence   A widespread sequence of sedimentary rocks bounded above and below by unconformities; deposited during a transgressive–regressive cycle of an epeiric sea, e.g., the Sauk sequence.

Cretaceous Interior Seaway   An interior seaway that existed during the Late Cretaceous; formed as northward-transgressing waters from the Gulf of Mexico joined with southward-transgressing water from the Arctic; effectively divided North America into two large landmasses.

Cro-Magnon   A race of *Homo sapiens* that lived mostly in Europe from 35,000 to 10,000 years ago.

cross-bedding   Strata containing beds that were deposited at an angle to the surface upon which they were accumulating, as in desert dunes, are said to be cross-bedded.

crossopterygian   A specific type of lobe-finned fish; possessed lungs; ancestral to amphibians.

crust   The outer part of Earth; the upper part of the lithosphere; separated from the mantle by the Moho; divided into continental and oceanic crust.

crystalline texture   A texture of rocks consisting of an interlocking mosaic of mineral crystals.

Curie point   The temperature at which iron-bearing minerals in a cooling magma attain their magnetism.

cyclothem   A vertical sequence of cyclically repeated sedimentary rocks resulting from alternating periods of marine and nonmarine deposition; commonly contain a coal bed.

cynodont   A type of therapsid (advanced mammal-like reptile); the ancestors of mammals.

D

daughter element   An element formed by the radioactive decay of another element, e.g., argon 40 is the daughter element of potassium 40. (See *parent element.*)

delta   An alluvial deposit at the mouth of a river.

depositional environment   Any area in which sediment is deposited; a depositional site that differs in physical aspects, chemistry, and biology from adjacent environments.

desiccation crack   A crack formed in clay-rich sediments in response to drying and shrinkage.

disconformity   An unconformity above and below which the strata are parallel. (See *angular unconformity* and *nonconformity.*)

divergent evolution   The diversification of a species into two or more descendant species.

divergent plate boundary   The boundary between two plates that are moving apart; new oceanic lithosphere forms at the boundary; characterized by volcanism and seismicity.

DNA (deoxyribonucleic acid)   The substance of which chromosomes are composed; the genetic material of all organisms except bacteria.

Doppler effect   The apparent change in frequency of a wave resulting from motion of the source of the wave, the receiver, or both.

drift   A collective term for all sediment deposited by glacial activity, including material deposited directly by glacial ice (till) and material deposited in streams derived from melting ice (outwash).

dryopithecine   Any of the members of a Miocene family of apelike primates; possible ancestors of apes and humans.

E

ectotherm   Any of the so-called cold-blooded vertebrates such as amphibians and reptiles; animals dependent on external heat to regulate body temperature. (See *endotherm.*)

Ediacaran faunas   A collective name for all Late Proterozoic faunas containing animal fossils similar to those of the Ediacara fauna of Australia.

electromagnetic force   A combination of electricity and magnetism into one force; binds atoms into molecules.

electron capture decay   A type of radioactive decay involving the capture of an electron by a proton and its conversion to a neutron; results in a loss of 1 atomic number, but no change in atomic mass number.

Ellesmere orogeny   A Mississippian orogeny affecting the northern margin of Laurentia.

encounter theory   A theory for the evolution of the solar system; involves a star passing close to the Sun and thus pulling away gaseous filaments that later accreted into solid bodies.

endotherm   Any of the warm-blooded vertebrates (mammals and birds) whose body temperature is maintained within narrow limits by metabolic processes; animals dependent on internal rather than external heat to regulate body temperature. (See *ectotherm.*)

epeiric sea   A broad shallow sea that covers part of a continent; six epeiric seas covered parts of North America during the Phanerozoic Eon, e.g., the Sauk Sea.

eukaryotic cell   A type of cell with a membrane-bounded nucleus containing chromosomes; also contains such organelles as plastids and mitochondria that are absent in prokaryotic cells. (See *prokaryotic cell.*)

eupantothere   Any member of a group of mammals that included the ancestors of both marsupial and placental mammals.

F

Farallon plate   A Late Mesozoic – Cenozoic oceanic plate that was largely subducted beneath North America; remnants of the Farallon plate are the Juan de Fuca and Cocos plates.

fission track dating   The process of dating samples by counting the number of small linear tracks (fission tracks) that result from damage to a mineral crystal by rapidly moving alpha particles generated by radioactive decay of uranium.

fluvial   A term referring to streams, stream action, and the deposits of streams.

formation   The basic lithostratigraphic unit; a unit of strata that is mappable and that has distinctive upper and lower boundaries.

fossil   The remains or traces of prehistoric organisms preserved in rocks of the crust. (See *body fossil* and *trace fossil.*)

Franklin mobile belt   The most northerly mobile belt in North America; extends from northwestern Greenland westward across the Canadian Arctic islands.

G

gene   The basic unit of inheritance; a specific segment of a chromosome. (See *allele.*)

gene pool   All genes available in an interbreeding population.

geologic record   The record of past events preserved in rocks.

geologic time scale   A geologic chart arranged so that the designation for the earliest part of geologic time appears at the bottom, and progressively younger designations appear in their proper chronologic sequence.

geology   The science concerned with the study of Earth; includes studies of Earth materials (minerals and rocks), surface and internal processes, and Earth history.

glacial stage   A time of extensive glaciation. At least four glacial stages are recognized in North America, and six or seven are recognized in Europe.

Glossopteris flora   A Late Paleozoic flora found only on the Southern Hemisphere continents and India; named after its best known genus, *Glossopteris.*

Gondwana   One of six major Paleozoic continents; composed of the present-day continents of South America, Africa, Antarctica, Australia, and India and parts of other continents such as southern Europe, Arabia, and Florida; began fragmenting during the Triassic Period.

graded bedding   A layer of sediment in which grain size decreases from bottom to top.

grain size   A term relating to the size of the particles making up sediment or sedimentary rock.

granite–gneiss complex   One of the two main types of rock bodies characteristic of areas underlain by Archean rocks.

graptolite   A small, planktonic animal belonging to the phylum Hemichordata. Geologic range is from Cambrian through Mississippian and make good guide fossils.

gravity   The attractive force that acts between all objects, e.g., between Earth and the Moon.

grazer   Any animal that eats low-growing vegetation, especially grasses. Compare with *browser.*

greenstone belt   A linear or podlike association of volcanic and sedimentary rocks particularly common in Archean terranes; typically synclinal and consists of lower and middle volcanic units and an upper sedimentary rock unit.

Grenville orogeny   An area in the eastern United States and Canada that was accreted to Laurentia during the Late Proterozoic.

guide fossil   Any easily identifiable fossil that has a wide geographic distribution and short geologic range; used to determine the geologic ages of strata and to correlate strata of the same age.

gymnosperm   The flowerless, seed-bearing land plants.

H

half-life   The time required for one-half of the original number of atoms of a radioactive element to decay to a stable daughter product, e.g., the half-life of potassium 40 is 1.3 billion years.

herbivore   An animal that is dependent on plants as a source of nutrients.

Hercynian orogeny   Pennsylvanian to Permian orogeny in the Hercynian mobile belt of southern Europe.

heterotrophic   Describes organisms such as animals that depend on pre-formed organic molecules from the environment as a source of nutrients. (See *autotrophic.*)

Holocene Epoch   The last of two epochs comprising the Quaternary Period. Began 10,000 years ago. Also called Recent.

hominid   Abbreviated form of Hominidae; the family of hominoids to which humans belong. Such bipedal primates as *Australopithecus* and *Homo* are hominids. (See *hominoid.*)

hominoid   Abbreviated form of Hominoidea, the superfamily of primates to which apes and humans belong. (See *hominid.*)

Homo   The genus of hominids consisting of present-day humans, *Homo sapiens,* and their ancestors, *Homo erectus* and *Homo habilis.*

homogeneous accretion   A model for the differentiation of Earth into a core, mantle, and crust; holds that Earth was originally compositionally homogeneous but then heated up, allowing heavier elements to sink to the core.

homologous organ   Body part in different organisms that has a similar structure, similar relationships to other organs, and similar development but does not necessarily serve the same function, e.g., the wing of a bird and the forelimbs of whales and dogs. (See *analogous organ.*)

hot spot   Localized zone of melting below the lithosphere; detected by volcanism at the surface.

hypothesis   A provisional explanation for observations. Subject to continual testing and modification. If well supported by evidence, hypotheses are then generally called theories.

Hyracotherium   A small Early Eocene mammal; the oldest genus assigned to the horse family (family Equidae).

I

Iapetus Ocean   A Paleozoic ocean basin that separated North America from Europe; the Iapetus Ocean began closing when North America and Europe started moving toward one another, and it was eliminated when the continents collided during the Late Paleozoic.

ichthyosaur Any of the porpoiselike, Mesozoic marine reptiles.

igneous rock Any rock formed by cooling and crystallization of magma or formed by the consolidation of volcanic ejecta such as ash.

inheritance of acquired characteristics A mechanism proposed by Jean Baptiste de Lamarck to account for evolution; states that characteristics acquired during an organism's lifetime could be passed on to its offspring.

inhomogeneous accretion A model for the differentiation of Earth into a core, mantle, and crust by the sequential condensation of core, mantle, and crust from hot nebular gases.

interglacial stage A time between glacial stages when glaciers cover much less area and global temperatures are warmer than during a glacial stage.

intracontinental rift A linear zone of deformation within a continent produced by tensional forces; characterized by normal faults and volcanism, e.g., the East African Rift System.

iron meteorite A type of meteorite consisting of iron and nickel.

isostasy The concept of Earth's crust "floating" on a more dense underlying layer. As a result of isostasy, thicker, less dense continental crust stands higher than thinner, more dense oceanic crust.

isostatic rebound The phenomenon in which unloading of the crust causes it to rise, as when continental glaciers melt, until it attains equilibrium with the denser underlying mantle.

J

Jovian planet Any of the four planets (Jupiter, Saturn, Uranus, and Neptune) that resemble Jupiter. They are all large and have low mean densities, indicating they are composed mostly of lightweight gases, such as hydrogen and helium, and frozen compounds, such as ammonia and methane.

K

Kaskaskia sequence A widespread sequence of Devonian and Mississippian sedimentary rocks bounded above and below by unconformities; deposited during a transgressive–regressive cycle of the Kaskaskia Sea.

Kazakhstania One of six major Paleozoic continents; a triangular-shaped continent centered on Kazakhstan (part of Asia).

L

labyrinthodont Any of the amphibians, from the Devonian to the Triassic, characterized by the labyrinthine wrinkling and folding of their teeth.

Laramide orogeny The Late Cretaceous to Early Cenozoic phase of the Cordilleran orogeny; responsible for many of the structural features of the present-day Rocky Mountains. In contrast to the preceding Nevadan and Sevier orogenies, the Laramide orogeny occurred much further inland from the continental margin.

Late Paleocene Thermal Maximum A warming trend that began abruptly about 55 million years ago.

Laurasia A Late Paleozoic, Northern Hemisphere continent composed of the present-day continents of North America, Greenland, Europe, and Asia.

Laurentia The name given to a Proterozoic continent that was composed mostly of North America and Greenland, parts of northwestern Scotland, and perhaps parts of the Baltic shield of Scandinavia.

lithology The physical characteristics of rocks.

lithosphere The outer, rigid part of Earth, consisting of the upper mantle, oceanic crust, and continental crust; lies above the asthenosphere.

lithostratigraphic unit A unit of sedimentary rock, such as a formation, defined by its physical characteristics rather than its biologic content or time of origin.

Little Ice Age An interval of nearly four centuries (1300 to the mid- to late-1800s) during which glaciers expanded to their greatest historic extents.

living fossil A living organism that has descended from ancient ancestors with little apparent change.

lobe-finned fish A type of fish with fins containing a fleshy shaft and a series of articulating bones; one of the two main groups of bony fish.

M

magnetic anomaly Any change, such as a change in average strength, of Earth's magnetic field.

magnetic reversal The phenomenon of the complete reversal of the north and south magnetic poles.

mantle The zone surrounding the core and comprising about 83% of Earth's volume; it is less dense than the core and is thought to be composed largely of peridotite.

marine regression The withdrawal of the sea from a continent or coastal area resulting in the emergence of land as sea level falls or the land rises with respect to sea level.

marine transgression The invasion of coastal areas or much of a continent by the sea resulting from a rise in sea level or subsidence of the land.

marsupial mammal Any of the pouched mammals such as opossums, kangaroos, and wombats. At present, marsupials are common only in Australia.

mass extinction A time during which extinction rates are greatly accelerated, thus resulting in a marked decrease in the diversity of organisms — e.g., the terminal Cretaceous extinction event.

meiosis A type of cell division during which the number of chromosomes is reduced by one-half. The cell division process that yields sex cells, sperm and eggs in animals, and pollen and ovules in plants. (See *mitosis*.)

metamorphic rock Any rock altered in the solid state from previously existing rock by heat, pressure, and chemically active fluids.

meteorite A mass of matter of extraterrestrial origin that has fallen to Earth's surface.

micrite Microcrystalline calcium carbonate mud; a limestone composed of micrite.

microplate Small lithospheric plate.

midcontinent rift A Late Proterozoic intracontinental rift within Laurentia; contains thick accumulations of volcanic rocks and detrital sedimentary rocks.

Milankovitch theory A theory that explains cyclic variations in climate and the onset of glaciation as a consequence of irregularities in Earth's rotation and orbit.

mineral A naturally occurring, inorganic, crystalline solid having characteristic physical properties and a narrowly defined chemical composition.

mitosis A type of cell division during which two cells resulting from division receive the same number of chromosomes as the parent cell possessed. (See *meiosis*.)

mobile belt Elongated area of deformation as indicated by folds and faults; generally located adjacent to a craton, e.g., the Appalachian mobile belt.

modern synthesis A synthesis of the ideas of geneticists, paleontologists, population biologists, and others to yield a neo-Darwinian view of evolution; includes chromosome theory of inheritance, mutation as a source of variation, and gradualism.

molar Any of a mammal's teeth used for grinding and chewing.

molarization An evolutionary trend in the hoofed mammals (ungulates) during which the premolars became more like molars.

mold A cavity or impression of an organism or part thereof in sediment or sedimentary rock, e.g., a mold of a clam shell. (See *cast*.)

monomer A comparatively simple organic molecule, such as an amino acid, that is capable of linking with other monomers to form polymers. (See *polymer*.)

monotreme The egg-laying mammals; only two types of monotremes now exist, the platypus and spiny anteater of the Australian region.

morphology The form or structure of an organism or any of its parts.

mosaic evolution The concept that all features of an organism do not evolve at the same rate; all organisms are mosaics of characteristics, some of which are retained from the ancestral condition and some of which are more recently evolved.

mosasaur A group of Mesozoic marine lizards.

multicellular organism Any organism consisting of many cells as opposed to a single cell; possesses cells specialized to perform specific functions such as reproduction and respiration.

mutation Any change in the genes of organisms; some of the inheritable variation in populations upon which natural selection acts arises from mutations in sex cells.

N

natural selection The mechanism proposed by Charles Darwin and Alfred Russell Wallace to account for evolution; natural selection results in the more likely survival to reproductive age of organisms with favorable variations.

Neanderthal A type of human that inhabited Europe and the Near East from 200,000 to 33,000 years ago; considered by some to be a variety or subspecies (*Homo sapiens neanderthalensis*) of *Homo sapiens* and by some as a separate species (*Homo neanderthalensis*).

nekton Actively swimming organisms, e.g., fish, whales, and squids. (See *plankton*.)

Neptunism The long-ago discarded concept held by Abraham Gottlob Werner and others that all rocks were precipitated from the waters of a worldwide ocean. According to this idea, one could determine the age of rocks by their composition.

Nevadan orogeny Late Jurassic to Cretaceous phase of the Cordilleran orogeny; most strongly affected the western part of the Cordilleran mobile belt.

New World monkey Any of the monkeys native to Central and South America. A superclass of the anthropoids.

nonconformity An unconformity in which stratified sedimentary rocks above an erosion surface overlie igneous or metamorphic rocks. (See *angular unconformity* and *disconformity*.)

nonvascular Plants lacking specialized tissues for conducting fluids.

North American Cordillera A complex mountainous region in western North America extending from Alaska into central Mexico. (See *Cordilleran mobile belt*.)

O

oceanic–continental plate boundary A type of convergent plate boundary along which oceanic lithosphere and continental lithosphere collide; characterized by subduction of the oceanic plate beneath the continental plate and by volcanism and seismicity.

oceanic–oceanic plate boundary A type of convergent plate boundary along which two oceanic lithospheric plates collide.

Old World monkey Any of the monkeys native to Africa, Asia, and southern Europe. A superfamily of the anthropoids.

ophiolite A sequence of igneous rocks thought to represent a fragment of oceanic lithosphere; composed of peridotite overlain successively by gabbro, basalt dikes, and pillow basalts.

ordinary chondrite The most abundant type of stony meteorite; composed of high-temperature ferromagnesian silicate minerals such as olivine and some pyroxenes.

organic evolution See *theory of evolution*.

organic reef A wave-resistant limestone structure with a framework of animal skeletons, e.g., stromatoporoid reef or coral reef.

Ornithischia An order of dinosaurs characterized by a bird-like pelvis; includes ornithopods, stegosaurs, ankylosaurs, pachycephalosaurs, and ceratopsians. (See *Saurischia*.)

orogen A linear part of Earth's crust that was deformed during an orogeny. (See *orogeny*.)

orogeny The process of forming mountains, especially by folding and thrust faulting; an episode of mountain building, e.g., the Acadian orogeny.

ostracoderm The "bony-skinned" fish; first appeared during the Late Cambrian and thus are the oldest known vertebrates; characterized by a lack of jaws and teeth and presence of bony armor.

Ouachita mobile belt A mobile belt located along the southern margin of the North American craton; probably continuous with the Appalachian mobile belt.

Ouachita orogeny An orogeny that deformed the Ouachita mobile belt during the Pennsylvanian Period.

outgassing The process whereby gases released from Earth's interior by volcanism formed an atmosphere.

outwash A term for the sediment deposited by streams that issue from glaciers. (See *drift*.)

P

paleoclimate A climate that existed during the past.

paleocurrent A term referring to an ancient current direction; determined by measuring the orientations of various sedimentary structures such as cross-bedding.

paleogeography   The study of Earth's geography throughout geologic time; paleogeography may be determined for the entire planet, such as the position of continents through time, or for a local area.

paleomagnetism   The remanent magnetism in ancient rocks that records the direction and strength of Earth's magnetic field at the time of their formation.

Pangaea   The name proposed by Alfred Wegener for a supercontinent that existed at the end of the Paleozoic Era and consisted of all Earth's landmasses.

Panthalassa Ocean   The Late Paleozoic worldwide ocean that surrounded the supercontinent Pangaea.

parallel evolution   The development of similarities in two or more closely related but separate lines of descent as a consequence of similar adaptations. (See *convergent evolution.*)

parent element   An unstable element that by radioactive decay is changed into a stable daughter element. (See *daughter element.*)

pelycosaur   Pennsylvanian to Permian "finback reptiles"; possessed some mammalian characteristics.

period   The fundamental unit in the hierarchy of time units; a part of geologic time during which a particular sequence of rocks designated as a system was deposited.

Perissodactyla   The order composed of odd-toed hoofed mammals; includes the present-day horses, tapirs, and rhinoceroses.

photochemical dissociation   A process whereby water molecules in the upper atmosphere are disrupted by ultraviolet radiation, yielding oxygen (O_2) and hydrogen (H).

photosynthesis   The metabolic process of synthesizing organic molecules from water and carbon dioxide, using the radiant energy of sunlight captured by chlorophyll-containing cells.

phyletic gradualism   An evolutionary concept holding that a species evolves gradually and continuously through time to give rise to new species. (See *punctuated equilibrium.*)

placental mammal   Any of the mammals that have a placenta to nourish the embryo, as opposed to egg-laying mammals (monotremes) and pouched mammals (marsupials).

placoderm   The "plate-skinned" fish; Late Silurian through Permian; characterized by jaws and bony armor, especially in the head–shoulder region.

plankton   Animals and plants that float passively, e.g., phytoplankton and zooplankton. (See *nekton.*)

plate   A segment of Earth's crust and upper mantle (lithosphere) varying from 50 to 250 km thick.

plate tectonics (plate tectonic theory)   The theory that large segments of the outer part of Earth (lithospheric plates) move relative to one another; lithospheric plates are rigid and move over the asthenosphere, which behaves much like a very viscous fluid.

Pleistocene Epoch   The first of two epochs comprising the Quaternary Period. Commonly called the Ice Age. Occurred between 1.6 million and 10,000 years ago.

plesiosaur   A group of Mesozoic marine reptiles; short-necked and long-necked plesiosaurs existed.

pluvial lake   Any lake that formed beyond the areas directly affected by glaciation during the Pleistocene as a result of increased precipitation and lower evaporation rates, e.g., Lake Bonneville.

point bar   The sediment body deposited on the gently sloping side of a meander loop.

pollen analysis   Identification and statistical analysis of pollen from sedimentary rocks; such analyses provide information about ancient floras and climates.

polymer   A comparatively complex organic molecule, such as nucleic acids and proteins, formed by monomers linking together. (See *monomer.*)

Precambrian   A widely used informal term referring to all rocks stratigraphically beneath strata of the Cambrian System and to all geologic time preceding the Cambrian Period.

Precambrian shield   A vast area of exposed Precambrian rocks on a continent; areas of relative stability for long periods of time, e.g., the Canadian shield.

premolar   Any of a mammal's teeth behind the canines but ahead of the molars; premolars and molars together are called the chewing teeth.

primary producer   Those organisms in a food chain, such as green plants and bacteria, upon which all other members of the food chain depend directly or indirectly; those organisms not dependent on an external source of nutrients. (See *autotrophic.*)

Primates   The order of mammals that includes prosimians (lemurs and tarsiers), monkeys, apes, and humans; characteristics include large brain, stereoscopic vision, and grasping hand.

principle of cross-cutting relationships   A principle used to determine the relative ages of events; holds that an igneous intrusion or fault must be younger than the rocks that it intrudes or cuts.

principle of fossil succession   The principle holding that fossils, and especially assemblages of fossils, succeed one another through time in a regular and determinable order.

principle of inclusions   A principle that holds that inclusions, or fragments, in a rock unit are older than the rock unit itself, e.g., granite fragments in a sandstone are older than the sandstone rock unit.

principle of lateral continuity   A principle that holds that sediment layers extend outward in all directions until they terminate.

principle of original horizontality   According to this principle, sediment is deposited in essentially horizontal layers.

principle of superposition   A principle that holds that younger strata are deposited on top of older layers.

principle of uniformitarianism   A principle that holds that we can interpret past events by understanding present-day processes; based on the assumption that natural laws have not changed through time.

Proboscidea   The mammal order that includes elephants and their extinct relatives.

proglacial lake   A lake formed by meltwater accumulating along the margins of a glacier.

progradation   The seaward (or lakeward) advance of the shoreline as a result of nearshore sedimentation.

prokaryotic cell   A type of cell having no nucleus and lacking such organelles as plastids and mitochondria; cells of bacteria and cyanobacteria (blue-green algae) are prokaryotic. (See *eukaryotic cell.*)

prosimian   Any member of the primate suborder Prosimii; includes tree shrews, lemurs, tarsiers, and lorises; commonly called lower primates.

Proterozoic Eon That part of Precambrian time beginning 2.5 billion years ago and ending 545 million years ago. (See *Archean Eon.*)

protorothyrid A loosely grouped category of all early reptiles. Protorothyrids are all small and about the same size and shape as living lizards.

pterosaur Any of the Mesozoic flying reptiles.

punctuated equilibrium An evolutionary concept that holds that a new species evolves rapidly, perhaps in a few thousands of years, then remains much the same during its several millions of years of existence. (See *phyletic gradualism.*)

Q

quadrupedal Referring to locomotion on all four legs; as opposed to bipedal.

Quaternary Period A term for a geologic period or system comprising all geologic time or rocks from the end of the Tertiary to the present; consists of two epochs or series, the Pleistocene and the Recent (Holocene).

Queenston Delta The clastic wedge resulting from the erosion of the highlands formed during the Taconic orogeny; deposited on the west side of the Taconic Highlands.

R

radioactive decay The spontaneous decay of an atom by emission of a particle from its nucleus (alpha and beta decay) or by electron capture, thus changing the atom to an atom of a different element.

range zone Biostratigraphic unit defined by the occurrence of a single type of organism such as a species or genus.

ray-finned fish A subclass (Actinopterygii) of the bony fish (class Osteichthyes). A term describing the way the fins are supported by thin bones that project from the body.

red beds Sedimentary rocks, mostly sandstone and shale, with red coloration due to the presence of ferric oxides.

reef See *organic reef.*

relative dating The process of determining the age of an event relative to other events; involves placing geologic events in their correct chronologic order but involves no consideration of when the events occurred in terms of numbers of years ago. (See *absolute dating.*)

ripple mark Wavelike (undulating) structure produced in granular sediment such as sand; formed by wind, unidirectional water currents, or wave currents.

rock A consolidated aggregate of minerals or rock fragments. Although volcanic glass and coal are not composed of minerals, they are nevertheless considered to be rocks, too.

rock cycle A sequence of processes through which Earth materials may pass as they are transformed from one rock type to another.

Rodinia The name given to a Late Proterozoic supercontinent.

rounding The process by which the sharp corners and edges of sedimentary particles are abraded during transport.

ruminant Any of the cud-chewing placental mammals such as deer, cattle, camels, bison, sheep, and goats.

S

salt dome A structure resulting from the upward movement of a mass of salt through overlying layers of sedimentary rocks. Oil and gas fields are commonly associated with salt domes.

San Andreas transform fault A major transform fault extending through part of California; connects with spreading ridges in the Gulf of Mexico and in the Pacific Ocean off the northwest coast of the United States.

sandstone–carbonate–shale assemblage A suite of sedimentary rocks characteristic of passive continental margins, but also known from intracratonic basins and back-arc basins.

Sauk sequence A widespread sequence of sedimentary rocks bounded above and below by unconformities; deposited during a latest Proterozoic to Early Ordovician transgressive–regressive cycle of the Sauk Sea.

Saurischia An order of dinosaurs characterized by a lizard-like pelvis; includes prosauropods, sauropods, and theropods. (See *Ornithischia.*)

scientific method A logical, orderly approach that involves data gathering, formulating and testing hypotheses, and proposing theories.

seafloor spreading The theory that the seafloor moves away from spreading ridges and is eventually consumed at subduction zones.

sedimentary facies Any aspect of a sedimentary rock unit that makes it recognizably different from adjacent sedimentary rocks of the same, or approximately same, age, e.g., a sandstone facies.

sedimentary rock Any rock composed of sediment. The sediment may be particles of various sizes such as gravel or sand, the remains of animals or plants as in coal and some limestones, or chemical compounds extracted from solution by organic or inorganic processes.

sedimentary structure Any feature in sedimentary rock such as cross-bedding, desiccation cracks, and animal burrows.

sediment-deposit feeder Any animal that ingests sediment and extracts nutrients from it.

seedless vascular plant A type of land plant with vascular tissues for transporting fluids and nutrients throughout the plant; reproduces by spores rather than seeds, e.g., ferns and horsetail rushes.

sequence stratigraphy The study of rock relationships within a time-stratigraphic framework of related facies bounded by widespread unconformities.

Sevier orogeny The Cretaceous phase of the Cordilleran orogeny that affected the continental shelf and slope areas of the Cordilleran mobile belt.

Siberia One of six major Paleozoic continents; composed of Russia east of the Ural Mountains, and Asia north of Kazakhstan and south of Mongolia.

sivapithecids A group of apelike hominoids that evolved in Africa during the Miocene. Probable ancestors of present-day orangutans.

solar nebular theory A theory for the evolution of the solar system from a rotating cloud of gas.

Sonoma orogeny A Permian–Triassic orogeny caused by the collision of an island arc with the southwestern margin of North America.

sorting The process whereby sedimentary particles are selected by size during transport; if all particles in a deposit or rock are about the same size, it is well sorted as opposed to those having a wide range of sizes and hence poor sorting.

species   A population of similar individuals that in nature can reproduce and produce fertile offspring.

spreading ridge   A long elevated part of the crust, mostly beneath the oceans, where plates separate.

stony meteorite (stone)   Meteorite composed of silicate minerals.

stony-iron meterorite (stony-iron)   Meteorite composed of iron–nickel and silicate minerals.

stratigraphy   The branch of geology concerned with the composition, origin, and areal and age relationships of stratified rocks.

stromatolite   A structure in sedimentary rocks, especially limestones, produced by entrapment of sediment grains on sticky mats of photosynthesizing bacteria; a biogenic sedimentary structure.

strong nuclear force   The force that holds protons together within the nucleus of an atom.

subduction zone   An elongate, narrow zone at a convergent plate boundary where an oceanic plate descends relative to another plate, e.g., the subduction of the Nazca plate beneath the South American plate.

submarine fan   A cone-shaped sedimentary deposit that accumulates on the continental slope and rise.

Sundance Sea   A wide seaway that existed in western North America during the Middle Jurassic Period.

suspension feeder   An animal that consumes microscopic plants, animals, or dissolved nutrients from water.

system   The fundamental unit in the time-stratigraphic hierarchy of units; the Devonian System refers to rocks deposited during a specific interval of geologic time, the Devonian Period.

T

Taconic orogeny   An Ordovician orogeny that resulted in deformation of the Appalachian mobile belt.

Tejas epeiric sea   A Cenozoic epeiric sea that was largely restricted to the Atlantic and Gulf Coastal Plains and parts of coastal California, but did extend into the continental interior in the Mississippi Valley.

terrestrial planet   Any of the four innermost planets (Mercury, Venus, Earth, and Mars). They are all small and have high mean densities, indicating they are composed of rock and metallic elements.

Tertiary Period   A term for a geologic period or system comprising all geologic time or rocks from the end of the Cretaceous to the beginning of the Quaternary.

theory   An explanation for some natural phenomenon that has a large body of supporting evidence; to be considered scientific, a theory must be testable, e.g., plate tectonic theory.

theory of evolution   The theory that all living things are related and that they descended with modification from organisms that lived during the past.

therapsid   Permian to Triassic reptiles that possessed mammalian characteristics and thus are called mammal-like reptiles; one group of therapsids, the cynodonts, gave rise to mammals.

thermal convection cell   A type of circulation of material in the asthenosphere during which hot material rises, moves laterally, cools and sinks, and is reheated and reenters the cycle.

till   All sediment deposited directly by glacial ice.

time-stratigraphic unit   A unit of strata that was deposited during a specific interval of geologic time, e.g., the Devonian

System, a time-stratigraphic unit, was deposited during that part of geologic time designated as the Devonian Period.

time transgressive   Refers to a rock unit that was deposited in an environment that shifted with time, as during a marine transgression. Thus, the age of the rock unit varies over its geographic extent.

time unit   Any of the units such as eon, era, period, epoch, and age used to refer to specific intervals of geologic time.

Tippecanoe sequence   A widespread sequence of sedimentary rocks bounded above and below by unconformities; deposited during an Ordovician to Early Devonian transgressive–regressive cycle of the Tippecanoe Sea.

trace fossil   Any indication of prehistoric organic activity such as tracks, trails, burrows, borings, or nests. (See *body fossil.*)

Transcontinental Arch   An area extending from Minnesota to New Mexico that stood above sea level as several large islands during the Cambrian transgression of the Sauk Sea.

transform fault   A type of fault that changes one type of motion between plates into another type of motion. Recognized on land as a strike-slip fault. (See *San Andreas transform fault.*)

transform plate boundary   Plate boundary along which plates slide past one another and crust is neither produced nor destroyed.

tree-ring dating   The process of determining the age of a tree or wood in structures by counting the number of annual growth rings.

trilobite   A group of benthonic, detritus-feeding, extinct marine invertebrate animals (phylum Arthropoda), having skeletons of an organic compound called chitin.

trophic level   The complex interrelationships among producers, consumers, and decomposers in a community of organisms. There are several trophic levels of production and consumption through which energy flows in a feeding hierarchy.

U

unconformity   An erosion surface that separates younger strata from older rocks. (See *angular unconformity, disconformity,* and *nonconformity.*)

ungulate   An informal term referring to the hoofed mammals, especially the orders Artiodactyla and Perissodactyla.

V

varve   A couplet of sedimentary laminae representing deposition that occurred in one year, e.g., the dark (winter) and light (summer) layers in varved sediments deposited in glacial lakes.

vascular   A term referring to some land plants possessing specialized tissues for transporting fluids.

vertebrate   Any animal having a segmented vertebral column; members of the subphylum Vertebrata; includes fish, amphibians, reptiles, mammals, and birds.

volcanic island arc   A curved chain of volcanic islands parallel to a deep-sea trench where oceanic lithosphere is subducted, causing volcanism and the origin of volcanic islands.

W

Walther's law   The observations by Johannes Walther serve as the basis for a principle that holds that facies deposited in adjacent environments will be superposed one upon the other when the environments migrate laterally; in a conformable vertical sequence of facies, the same facies that occur vertically will replace one another laterally.

weak nuclear force   A force responsible for the breakdown of atomic nuclei, thus producing radioactive decay.

Wilson cycle   The relationship between mountain building (orogeny) and the opening and closing of ocean basins.

Z

Zuni sequence   An Early Jurassic to Early Paleocene sequence of sedimentary rocks bounded above and below by unconformities; deposited during a transgressive – regressive cycle of the Zuni Sea.

Answers to Multiple-Choice Review Questions

Chapter 1
1. c; 2. d; 3. a; 4. d; 5. b; 6. d; 7. c; 8. c; 9. e; 10. e; 11. c; 12. a.

Chapter 2
1. c; 2. a; 3. d; 4. d; 5. b; 6. a; 7. c; 8. b; 9. b; 10. e; 11. d; 12. c.

Chapter 3
1. c; 2. b; 3. b; 4. a; 5. a; 6. e; 7. b; 8. d; 9. e; 10. b; 11. e; 12. d.

Chapter 4
1. a; 2. c; 3. e; 4. b; 5. a; 6. d; 7. b; 8. a; 9. b; 10. a; 11. c; 12. b.

Chapter 5
1. b; 2. c; 3. d; 4. a; 5. c; 6. a; 7. b; 8. a; 9. d; 10. e; 11. a; 12. c.

Chapter 6
1. d; 2. e; 3. b; 4. c; 5. c; 6. a; 7. d; 8. a; 9. d; 10. d; 11. e; 12. e.

Chapter 7
1. d; 2. c; 3. d; 4. e; 5. a; 6. d; 7. c; 8. a; 9. c; 10. d; 11. a; 12. e.

Chapter 8
1. a; 2. c; 3. e; 4. b; 5. a; 6. c; 7. a; 8. d; 9. b; 10. a; 11. b; 12. c.

Chapter 9
1. c; 2. a; 3. e; 4. c; 5. b; 6. a; 7. b; 8. d; 9. a; 10. b; 11. e; 12. c.

Chapter 10
1. e; 2. d; 3. a; 4. b; 5. e; 6. c; 7. b; 8. a; 9. e; 10. a.

Chapter 11
1. c; 2. c; 3. b; 4. b; 5. a; 6. a; 7. d; 8. c; 9. b; 10. e.

Chapter 12
1. a; 2. d; 3. e; 4. c; 5. d; 6. e; 7. a; 8. d; 9. e; 10. e.

Chapter 13
1. c; 2. d; 3. d; 4. e; 5. d; 6. c; 7. d; 8. c; 9. c; 10. c; 11. a; 12. b.

Chapter 14
1. e; 2. c; 3. c; 4. b; 5. d; 6. e; 7. a; 8. e; 9. b; 10. a; 11. b.

Chapter 15
1. b; 2. e; 3. a; 4. b; 5. c; 6. b; 7. b; 8. d; 9. c; 10. e; 11. b; 12. c.

Chapter 16
1. c; 2. a; 3. e; 4. a; 5. d; 6. b; 7. a; 8. d; 9. b; 10. a; 11. c; 12. c.

Chapter 17
1. b; 2. b; 3. c; 4. e; 5. d; 6. a; 7. b; 8. c; 9. d; 10. a; 11. a; 12. b.

Chapter 18
1. b; 2. d; 3. c; 4. a; 5. b; 6. a; 7. e; 8. c; 9. a; 10. d; 11. c; 12. c.

Chapter 19
1. b; 2. e; 3. d; 4. e; 5. d; 6. d; 7. c; 8. c; 9. d; 10. b; 11. d; 12. e.

Credits

This page constitutes an extension of the copyright page. We have made every effort to trace the ownership of all copyrighted material and to secure permission from copyright holders. In the event of any question arising as to the use of any material, we will be pleased to make the necessary corrections in future printings. Thanks are due to the following authors, publishers, and agents for permission to use the material indicated.

Chapter 1
Chapter opening photo: Photo courtesy of NASA. **Figure 1.2:** Modified from Figure 12, Dietrich, R. V., 1979, *Geology and Michigan: Forty-nine Questions and Answers.* **Figure 1.3:** Photo courtesy of Sue Monroe. **Figure 1.4:** Photo Courtesy of Sue Monroe. **Figure 1.5:** Photo courtesy of Sue Monroe. **Figure 1.6:** Photo courtesy of Wayne E. Moore. **Table 1.4:** Adapted from Stephen Dutch, James S. Monroe, and Joseph Moran, *Earth Science.* **Figure 1.13:** Modified from A. R. Palmer, "The Decade of North American Geology, 1983 Geologic Time Scale." *Geology* (Geological Society of America, 1983), p. 504.

Chapter 2
Chapter opening photo: © Mike Surowiak/Tony Stone Images. **Figure 2.1:** Modified from A. R. Palmer, "The Decade of North American Geology, 1983 Geologic Time Scale." *Geology* (Geological Society of America, 1983), p. 504. **Perspective 2.1, Figure 2(a) and (b):** Photos courtesy of Reed Wicander. **Figure 2.2:** Photo courtesy of Reed Wicander. **Figure 2.3(a):** Photo courtesy of Stew Monroe. **Figure 2.3(b):** Photo courtesy of Reed Wicander. **Figure 2.4, bottom right:** Photo courtesy of Dorothy L. Stout. **Figure 2.6:** Data from S. M. Richardson and H. Y. McSween, Jr., *Geochemistry—Pathways and Processes,* Prentice-Hall, 1989. **Figure 2.10:** Photo courtesy of Charles W. Naeser, U.S. Geological Survey. **Figure 2.12:** Reprinted by permission from Stokes and Smiley: *An Introduction to Tree-Ring Dating,* p. 6. Copyright © 1968 University of Chicago Press.

Chapter 3
Chapter opening photo: Photo courtesy of Stew Monroe. **Figure 3.2(a):** Photo courtesy of Stew Monroe. **Figure 3.2(b):** Photo courtesy of Wayne Moore. **Figure 3.2(c):** Photo courtesy of Stew Monroe. **Figure 3.3(a):** Photo by Stew Monroe. **Figure 3.3(b):** Photo by Stew Monroe. **Figure 3.8(d):** Photo courtesy of R. V. Dietrich. **Figure 3.16(a):** Photo by Stew Monroe. **Figure 3.16(b):** Photo courtesy of Sue Monroe. **Figure 3.16(c):** Photo Courtesy of Sue Monroe. **Figure 3.16(d):** Photo courtesy of Sue Monroe. **Figure 3.16(e):** Photo courtesy of Sue Monroe. **Perspective 3.1, Figure 1:** © New York State Historical Association. **Perspective 3.1, Figure 2:** Photo courtesy of the Bobst Library, New York University. **Perspective 3.1, Figure 3:** Reproduced with permission from *Nature,* Vol 35 (1942) p. 214. **Figure 3.17(a):** Photo © J. Koivula/Science Source, Photo Researchers, Inc.

Figure 3.17(b): Photo © J. Koivula/Science Source, Photo Researchers, Inc. **Figure 3.18:** Photo courtesy of Los Angeles County Museum of Natural History. **Figure 3.19(a):** Photo courtesy of AMNH. **Figure 3.19(b):** Photo © Sovfoto/V. Khristoforov. **Figure 3.20(a):** Photo courtesy of Sue Monroe. **Figure 3.20(b):** Photo courtesy of Sue Monroe. **Figure 3.21(a):** Photo courtesy of Stephen R. Manchester. **Figure 3.22(d):** Photo courtesy of Sue Monroe. **Figure 3.25:** Reprinted with permission from Leigh W. Mintz, *Geology: The Science of a Dynamic Earth,* third edition, 1981, Charles E. Merrill Publishing Co., p. 27 (Fig. 2.18). **Figure 3.26:** Photo Courtesy of Molly Fritz Miller, Vanderbilt University. **Figure 3.27:** Photo by Stew Monroe. **Figure 3.28:** Photo courtesy of Stew Monroe.

Chapter 4
Chapter opening photo: Photo courtesy of Stew Monroe. **Figure 4.1(a):** Photo courtesy of Stew Monroe. **Figure 4.1(b):** Photo courtesy of Stew Monroe. **Figure 4.2:** K.L Elnerkin, U.S. Geological Survey. **Figure 4.4(a):** Photo courtesy of Sue Monroe. **Figure 4.4:** Photo courtesy of Sue Monroe. **Figure 4.5(a):** Photo courtesy of Stew Monroe. **Figure 4.5(b):** Photo courtesy of Stew Monroe. **Figure 4.6(a):** Photo courtesy of Sue Monroe. **Figure 4.6(b):** Photo courtesy of Sue Monroe. **Perspective 4.1, Figure 1(a):** Photo courtesy of Mike Johnson. **Perspective 4.1, Figure 1(b):** Photo courtesy of Mike Johnson. **Perspective 4.1, Figure 2(a):** Photo courtesy of Sue Monroe. **Perspective 4.1, Figure 2(b):** Photo courtesy of Sue Monroe **Perspective 4.1, Figure 3(a):** Photo courtesy of Stew Monroe. **Perspective 4.1, Figure 3(b):** Photo courtesy of Stew Monroe. **Perspective 4.1, Figure 4:** Photo courtesy of Stew Monroe. **Figure 4.8(b):** Photo courtesy of Sue Monroe. **Figure 4.8(d):** Photo courtesy of Sue Monroe. **Figure 4.9(c):** Photo by Stew Monroe. **Figure 4.9(e):** Photo by Stew Monroe. **Figure 4.10(a):** Photo courtesy of Sue Monroe. **Figure 4.10(b):** Photo courtesy of Sue Monroe. **Figure 4.13(a):** Photo courtesy of Stew Monroe. **Figure 4.13(c):** Photo courtesy of Sue Monroe. **Figure 4.13(b) & (d):** Reprinted with permission from J. R. L. Allen, "Review of the Origins and Characteristics of Recent Alluvial Sediments," *Sedimentology,* V. 5, 1965 Blackwell Scientific Publications, Ltd. **Figure 4.14(a):** Photo by Reed Wicander. **Figure 4.15:** Photo courtesy of the Geological Survey of Canada, Neg. 179852. **Figure 4.17(c):** From W. L. Fisher et al., 1969, *Delta Systems in the Exploration for Oil and Gas—A Research Colloquium.* Reproduced with permission of Bureau of Economic Geology, The University of Texas at Austin. **Figure 4.20:** From Davies and Gorsline, "Oceanic Sediments and Sedimentary Processes," in Riley and Chester (eds) *Chemical Oceanography,* v. 5, 2nd ed, reprinted with permission of Academic Press, Inc., London. **Figure 4.21:** Photo courtesy of Rex Elliot. **Figure 4.22 (a):** From L. S. Fichter and D. J. Poche, *Ancient Environments and Interpretation of Geologic History.* Copyright © 1979 Macmillan Publishing Company. **Figure 4.22(b):** From *Origin of Sedimentary Rocks,* 2/e by

Blatt/Middleton/Murray, © 1980. Reprinted by permission of Prentice-Hall, Inc., Upper Saddle River, N.J. **Figure 4.24:** Reprinted with permission from Young, Fiddler and Jones, "Carbonate Facies in Ordovician of Northern Arkansas," AAPG *Bulletin*, Vol. 56, No. 1, AAPG © 1972. **Figure 4.25:** Reprinted with permission from *Geologic Atlas of the Rocky Mountain Region*, 1972, copyright © Rocky Mountain Association of Geologists. **Perspective 4.2, Figure 1:** Photo by Stew Monroe.

Chapter 5

Chapter opening photo, top: © Francois Gohier/Photo Researchers Inc.; **bottom:** The Field Museum of Natural History Chicago, and the artist, Charles R. Knight. Neg. #CKL21T. **Figure 5.4:** © Historical Pictures/Stock Montage. **Figure 5.5, top left:** Photo © G.C. Kelly/Photo Researchers Inc; **bottom left:** © Tom McHugh/Photo Researcher; **top and bottom right:** © Kenneth Fink/Photo Researchers 114: **Figure 5.12:** From *Principles of Paleontology* by Raup and Stanley © 1971, 1978 by W. H. Freeman and Company. Used with permission. **Perspective 5.2, Figure1(a):** © Edward Ross. **Perspective 5.2, Figure 1(b):** © Edward Ross. **Figure 5.17:** Reprinted with permission from L. S. Dillon, *Evolution: Concepts and Consequences,* 2nd ed., 1978, p. 250. **Figure 5.18:** Reproduced by permission of the Society for the Study of Evolution. From *Evolution,* v. 28, p. 448. **Figure 5.19(b):** Photo © Peter Scones/Planet Earth Pictures. **Figure 5.20(a):** Reprinted with permission from Starr and Taggart, *Biology: The Unity and Diversity of Life,* 1989, p. 556. Copyright © Wadsworth Publishing Company. **Figure 5.20(b):** Reprinted with permission from Weishampel et al. (eds), *Dinosauria: Paleobiology of the Terrible Lizards.* Copyright © 1990 The Regents of the University of California. **Perspective 5.3, Figure 2:** Reprinted with permission from O. C. Marsh, "Recent Polydactyle Horses," *American Journal of Science,* V. XLIII, no. 256, 1892, p. 343, Fig 6. **Figure 5.26:** Photo © M. W. Tweedie/Photo Researchers Inc. **Figure 5.27:** Reprinted with permission from R. L. Carroll, Patterns and *Processes of Vertebrate Evolution,* p. 156 (Fig 7.4). Copyright © 1988 Cambridge University Press.

Chapter 6

Chapter opening photo: Photo courtesy of D. D. Trent. **Figure 6.2(a):** Photo courtesy of Patricia G. Gensel, University of North Carolina. **Figure 6.2(b):** Photo courtesy of Patricia G. Gensel, University of North Carolina. **Figure 6.3:** Bildarchiv Preussischer Kulturbesitz. **Figure 6.4:** From *General Geology,* 5/e, by Foster, © 1988. Reprinted by permission of Prentice-Hall, Inc., Upper Saddle River, N.J. **Figure 6.6:** Modified from E. H. Colbert, *Wandering Lands and Animals,* 1973, 72, Figure 31. **Figure 6.9(b):** Reproduced with permission from *Science,* Vol 163, January 17, 1969. Copyright © 1969 American Association for the Advancement of Science. **Figure 6.10:** Photo courtesy of ALCOA. **Figure 6.11 (left):** Reprinted with permission ASCE. From G. A. Kiersch, "Vaiont Reservoir Disaster," *Civil Engineering* 34(1964). **Figure 6.11 (right):** Reproduced with permission from *Science,* Vol 163, January 17, 1969. Copyright © 1969 American Association for the Advancement of Science. **Figure 6.12:** Based on *The Bedrock Geology of the World,* by R. L. Larson and W. C. Pitman, III. Copyright © 1985 by W. H. Freeman and Company. **Figure 6.13:** Data from J. B. Minster and T. H. Jordan,

"Present-day Plate Motions," *Journal of Geophysical Research* 83 (1978) 5331–51. **Perspective 6.1, Figure 1:** Photo courtesy of NASA. **Figure 6.14:** Photo courtesy of Woods Hole Oceanographic Institution. **Figure 6.16:** © Robert Caputo/Auora. **Figure 6.17(b):** Photo courtesy of John M. Faivre. **Figure 6.23, inset:** © James Balog/Tony Stone Images. **Figure 6.25:** Data from J. B. Minster and T. H. Jordan, "Present-day Plate Motions," *Journal of Geophysical Research* 83 (1978) 5331–51.

Chapter 7

Chapter opening photo: Photo courtesy of NASA. **Figure 7.1:** Reprinted with permission from Henbest and Couper, *The Restless Universe,* 1982, Published by George Philip and Son, Ltd., © Nigel Henbest and Heather Couper. **Figure 7.7:** Photo courtesy of Dana Berry. **Figure 7.8(b):** Fotosmith. **Figure 7.8(c):** Fotosmith. **Figure 7.8(d):** Fotosmith. **Perspective 7.1, Figure 2:** Photo courtesy of TASS from Sovfoto. **Perspective 7.1, Figure 3:** D.J. Roddy, US Geological Survey. **Figure 7.11:** Photo courtesy of NASA. **Figure 7.14(a):** Photo courtesy of NASA. **Figure 7.14(b):** Photo courtesy of NASA. **Figure 7.15:** Photo courtesy of NASA. **Figure 7.16:** Photo courtesy of NASA/JPL. **Figure 7.17:** Photo courtesy of NASA/JPL. **Figure 7.19(a):** Photo courtesy of NASA. **Figure 7.19(b):** Photo courtesy of NASA. **Figure 7.20:** Photo courtesy of Finley Holiday Film Corp./JPL/NASA. **Figure 7.21:** Photo courtesy NASA. **Figure 7.22:** Photo courtesy of NASA.

Chapter 8

Chapter opening photo: Photo courtesy of Stew Monroe. **Figure 8.1:** Photo courtesy Herb Orth, Life Magazine © Time Warner Inc.; painting by Chesley Bonstell © The Estate of Chesley Bonestell. **Table 8.1:** Reprinted with permission from H. L. James, "Stratigraphic Commission: Note 40—Subdivision of Precambrian: An Interim Scheme to be Used by USGS," *AAPG,* v. 56, no. 6, 1972, p. 1130 (Table 1), and from Harrison and Peterman, "North American Commission on Stratigraphic Nomenclature: Report 9–Adoption of Geochronometric Units for Division of Precambrian Time," *AAPG,* v. 66, no. 6, 1982, p. 802 (Fig 1), American Association of Petroleum Geologists. **Figure 8.4:** Reprinted with permission from A.M. Goodwin, "The Most Ancient Continental Margins," in *The Geology of Continental Margins,* Burk and Drake (eds), 1974, p. 768 (Fig 1), © Springer-Verlag. Figure 8.6: Reprinted with permission from "Review of Archean Clastic Sedimentation, Canadian Shield," by R. W. Ojakangas, Geological Association of Canada Special Paper 28, 1985. **Figure 8.5(a):** Photo courtesy of R.V. Deitrich. **Figure 8.5(b):** Photo courtesy of Stew Monroe. **Figure 8.5(c):** Photo courtesy of Stew Monroe. **Figure 8.5(d):** Photo courtesy of Stew Monroe. **Figure 8.7(a):** Reprinted from K. C. Condie, *Plate Tectonics and Crustal Evolution,* 2nd ed., © 1982, p. 218, (Fig. 10.2). **Figure 8.7(c):** Photo courtesy of Stew Monroe. **Figure 8.9:** Reprinted with permission from Condie and Hunter, "Trace Element Geochemistry of Archean Granitic Rocks from the Barberton Region, South Africa," *Earth and Planetary Science Letters,* v. 29, © 1976, p. 398 (Fig. 7), Elsevier Science Publishers, and with permission from Eriksson, "Tidal Deposits from the Archean Moodies Group, Barberton Mountain Land, South Africa," *Sedimentary Geology,* v. 18, © 1977, p. 278 (Fig. 23), Elsevier Science Publications. **Figure 8.10:** Reprinted from K. C. Condie, *Plate Tectonics and Crustal Evolution,* 2nd ed.,

© 1982, p. 87 (Fig. 5.10). **Figure 8.11:** Reprinted with permission from Dickinson and Luth, "A Model for Plate Tectonic Evolution of Mantle Layers," *Science,* v. 174, no. 4007, 1971, p. 402 (Fig 1), © 1992 by the AAAS. **Figure 8.14:** Reprinted with permission from S. L. Miller, "The Formation of Organic Compounds on the Primitive Earth," in *Modern Ideas of Spontaneous Generation,* Nigrelli (ed.), Annuals of the New York Academy of Sciences, v. 69, Art. 2, Aug, 30, 1957, p. 261 (Fig 1). **Figure 8.15(a):** Photo courtesy of Sidney W. Fox, Distinguished Research Professor, Southern Illinois University at Carbondale. **Figure 8.15(b):** Photo courtesy of Sidney W. Fox, Distinguished Research Professor, Southern Illinois University at Carbondale. **Perspective 8.1, Figure 1(a):** © Peter Ryan/Scrippa/Science Photo Library/Photo Researchers Inc. **Perspective 8.1, Figure 1(b):** Visuals Unlimited/WHOI-D. Foster. **Figure 8.16(a):** Photo courtesy of Phillip E. Playford, Geological Survey of Western Australia. **Figure 8.16 (b):** From R. L. Anstey and T. L. Chase, *Environments Through Time.* Copyright © 1974 Macmillan Publishing Company. **Figure 8.17(a):** Photo courtesy of J. William Schopf, University of California, Los Angeles. **Figure 8.17(b):** Photo courtesy of J. William Schopf, University of California, Los Angeles.

Chapter 9
Chapter opening photo: © Frank Oberle/Tony Stone Images. **Figure 9.1(b):** Photo courtesy of Asko Kontinen, Geological Survey of Finland. **Figure 9.1(c):** Photo courtesy of Asko Kontinen, Geological Survey of Finland. **Table 9.1:** Reprinted with permission from Kontinen, *Precambrian Research,* © 1987, Elsevier Science Publishers. **Figure 9.2:** Reprinted from K. C. Condie, *Plate Tectonics and Crustal Evolution,* 4th edition, p. 65 (Fig. 2.26), copyright © 1997 Butterworth-Heinemann. **Figure 9.4:** Reprinted with permission from J. L. Anderson "Proterozoic Anorogenic Granite Plutonism of North America," Proterozoic Geology: Selected Papers from an International Symposium, p. 135 (Fig 1). **Figure 9.5:** Photo courtesy of R.V. Dietrick. **Figure 9.6 (a):** Reprinted with permission from J. C. Green, *Tectonophysics,* v. 94, p. 414 (Fig 1), © 1983, Elsevier Science Publishers. **Figure 9.6 (b):** Reprinted with permission from Daniels, "Upper Precambrian Sedimentary Rocks: Ononto Group, Michigan-Wisconsin," in *Geology and Tectonics of the Lake Superior Basin,* 1982, Wold and Heinze (eds.), p. 110 (Fig 3), GSA Memoir 156. **Figure 9.6(c):** Photo courtesy of Stew Monroe. **Figure 9.6(d):** Photo courtesy of Sue Monroe. **Figure 9.7:** Reprinted from K. C. Condie, *Plate Tectonics and Crustal Evolution,* 2nd ed., © 1982, p. 228, (Fig. 10.8). **Figure 9.8:** Reprinted with permission from Ian W. D. Dalziel, "Pacific margins of Laurentia and East-Antarctica-Australia as a conjugate rift pair: Evidence and implications for an Eocambrian supercontinent," *Geology,* v. 19, June 1991, p. 600 (Fig. 3), Geological Society of America. **Figure 9.9 (a):** Reprinted with permission from G. M. Young, "Tectono-Sedimentary History of Early Proterozoic Rocks of the Northern Great Lakes Region," *in Early Proterozoic Rocks Geology of the Northern Great Lakes Region,* Medaris (ed.), GSA Memoir 160, 1983, p. 16 (Fig. 1), Geological Society of America. **Figure 9.9(b):** Photo courtesy of Sue Monroe. **Figure 9.9(c):** Photo courtesy of Sue Monroe. **Perspective 9.1, Figure 1:** Photo courtesy of Stew Monroe. **Perspective 9.1, Figure 2(a):** Photo courtesy of Stew Monroe. **Perspective 9.1, Figure 2(b):** Photo courtesy of Jeff Mayo. **Figure 9.10(a):** Photo courtesy of Stew Monroe. **Figure 9.10(b):** Photo courtesy of Stew Monroe.

Figure 9.11(a): Photo courtesy of Anna Siedlecka, the Geological Survey of Norway. **Figure 9.12:** Reprinted from L. A. Frakes, *Climates Throughout Geologic Time,* p. 39. Copyright © 1979 Elsevier Science Publishers. **Figure 9.13:** Reprinted from L. A. Frakes, *Climates Throughout Geologic Time,* p. 88. Copyright © 1979 Elsevier Science Publishers. **Figure 9.14(a):** Photo courtesy of Sue Monroe. **Figure 9.14(b):** Photo courtesy of Stew Monroe. **Figure 9.15:** Reprinted with permission from A. M. Goodwin, "Distribution and Origin of Precambrian Banded Iron Formation," *Revista Brasileira de Geosciencias,* v. 12 (1-3), 1982, p. 462 (Fig. 4). **Figure 9.16:** Photo courtesy of Sue Monroe. **Figure 9.18(a):** Photo courtesy of Preston Cloud. **Table 9.3:** Reprinted with permission from Kontinen, Precambrian Research, © 1987, Elsevier Science Publishers. **Figure 9.19(a):** Photo courtesy of Andrew H. Knoll, Harvard University. **Figure 9.19(b):** Photo courtesy of Andrew H. Knoll, Harvard University. **Figure 9.19(c):** Photo courtesy of Bonnie Bloeser. **Figure 9.20:** Reprinted with permission from Kontinen, Precambrian Research, © 1987, Elsevier Science Publishers. **Figure 9.21:** Photo courtesy of Robert Horodyski. **Figure 9.22(a):** Photo courtesy of Neville Pledge, South Australia Museum. **Figure 9.22(b):** Photo courtesy of Neville Pledge, South Australia Museum. **Figure 9.22(c):** Photo courtesy of Neville Pledge, South Australia Museum. **Figure 9.22(d):** Photo courtesy of Neville Pledge, South Australia Museum. **Figure 9.22(e):** Photo courtesy of Neville Pledge, South Australia Museum. **Figure 9.23(a):** Photo courtesy of Sun Weiguo, Nanjing Institute of Geology and Paleontology, Academia Sinica, Nanjing, People's Republic of China. **Figure 9.23(b):** Photo courtesy of Sun Weiguo, Nanjing Institute of Geology and Paleontology, Academia Sinica, Nanjing, People's Republic of China. **Figure 9.23(c):** Photo courtesy of Sun Weiguo, Nanjing Institute of Geology and Paleontology, Academia Sinica, Nanjing, People's Republic of China. **Figure 9.24(b):** Photo courtesy of Stew Monroe. **Figure 9.24(c):** Photo courtesy of Sue Monroe.

Chapter 10
Chapter opening photo: Courtesy of National Park Service. **Figure 10.2:** Based on data from The Geological Society Publishing House and American Scientist. **Figure 10.6(b):** Photo courtesy of the Wisconsin Dells Visitor and Convention Bureau. **Figure 10.8(b):** Wisconsin Department of Natural Resources. **Figure 10.9(a):** Photo courtesy of L.J. Lipke, Amoco Production Company. **Figure 10.11(a):** From K. J. Mesolella, J. D. Robinson, L. M. McCormick, and A. R. Ormiston, "Cyclic Deposition of Silurian Carbonates and Evaporites in Michigan Basin," *AAPG Bulletin,* Vol. 58, No. 1, AAPG © 1958, Fig 6, p. 40. Reprinted by permission of the American Association of Petroleum Geologists whose permission is required for future use. **Figure 10.11(b):** Photo courtesy of Sue Monroe. **Figure 10.11(c):** Photo courtesy of Sue Monroe. **Figure 10.11(d):** Photo courtesy of Sue Monroe.

Chapter 11
Chapter opening photo: Photo courtesy of Sue Monroe. **Figure 11.1:** Based on data from The Geological Society Publishing House and American Scientist. **Figure 11.2:** Based on data from The Geological Society Publishing House and American Scientist. **Perspective 11.1, Figure 1:** Photo courtesy of Geoffrey Playford, The University of Queensland Brisbane, Australia. **Perspective 11.1, Figure 2:** Photo courtesy of Geoffrey Playford, The University of Queensland Brisbane,

Australia. **Figure 11.10(a):** Photo courtesy of Wayne E. Moore. **Figure 11.10(b):** Photo © Patricia Caulfield/Photo Researchers, Inc. **Figure 11.11(c):** © Tom Bean, 1993/Tom & Susan Bean Inc. **Figure 11.14:** Rubin's Studio of Photography, the Petroleum Museum, Midland, Texas. **Figure 11.15:** Photo Courtesy of Bill Cornell, The University of Texas at El Paso. **Figure 11.19:** From the U.S. Geological Survey.

Chapter 12

Chapter opening photo: Photo courtesy of the Smithsonian Institution. Neg. # 86-13471A. **Figure 12.1(a):** Photo courtesy of Simon Conway Morris and Stefan Bengston, University of Cambridge, England. **Figure 12.1(b):** Photo courtesy of Simon Conway Morris and Stefan Bengston, University of Cambridge, England. **Figure 12.1(c):** Photo courtesy of Simon Conway Morris and Stefan Bengston, University of Cambridge, England. **Figure 12.2(c):** Photo courtesy of the Geological Survey of Canada, Photo 202109-5. **Figure 12.4:** Reproduced with permission from H. Tappan, "Proceedings of the North American Paleontological Convention, Part H," E. Yochelson (ed.), p. 1064 (Fig.2), 1970. **Figure 12.5:** Carnegie Museum of Natural History. **Figure 12.7:** Photo by Porter M. Kier, courtesy of J. Wyatt Durham, University of California, Berkeley. **Figure 12.8(a):** Douglas H. Erwin/National Museum of Natural History. **Figure 12.8(b):** Douglas H. Erwin/National Museum of Natural History. **Figure 12.8(c):** Douglas H. Erwin/National Museum of Natural History. Douglas H. Erwin/National Museum of Natural History. **Figure 12.9:** © The Field Museum, Chicago #Geo80820c. **Figure 12.12(a):** Photo courtesy of Sue Monroe. **Figure 12.12(b):** Photo courtesy of Sue Monroe. **Figure 12.12(c):** Photo courtesy of Stig M. Bergstrom, Ohio State University. **Figure 12.13:** Photo by J.K. Ingham, supplied courtesy of R.J. Aldridge, University of Leicester, Leicester, England. **Figure 12.14:** © The Field Museum, Chicago, #Geo80821c. **Figure 12.15:** © The Field Museum, Chicago, #Geo80819c. **Figure 12.16:** Photo courtesy of the Smithsonian Institution. **Figure 12.17:** Photo courtesy of American Museum of Natural History, Trans. #K10257. **Figure 12.18:** © American Museum of Natural History, Trans.#K10269. **Figure 12.19(a):** Photo courtesy of G.A. Sanderson, Amoco Production Co. **Figure 12.19(b):** Photo courtesy of G.A. Sanderson, Amoco Production Co. **Figure 12.20:** Reproduced with permission from "Mass Extinction in the Marine Fossil Record," by D. M. Raup and J. J. Sepkoski, *Science,* Vol 215, p. 1502 (Fig 2). Copyright © 1982 American Association for the Advancement of Science.

Chapter 13

Chapter opening photo: Photo courtesy of NAGT. **Figure 13.1:** Photo courtesy of Reed Wicander. **Figure 13.2:** Reproduced with permission from "The Evolution of Early Land Plants," by P. Genseld and S. Andrews, *American Scientist,* Vol 75, no. 5, p. 480 (Fig 2). Copyright © 1987 American Scientist. **Perspective 13.1, Figure 1:** Photo courtesy of Jane Gray, University of Oregon. **Figure 13.13:** Photo courtesy of Dr. Lars Ramskold. **Figure 13.15:** Photo courtesy of John E. Repetski, U.S.G.S. **Figure 13.2:** Photo courtesy of Dianne Edwards, University College, England. **Figure 13.23(a):** Photo courtesy of Patricia G. Gensel, University of North Carolina. **Figure 13.23(c):** Photo courtesy of Patricia G. Gensel, University of North Carolina. **Figure 13.23(d):** Photo courtesy of Patricia G. Gensel, University of North Carolina.

Chapter 14

Chapter opening photo: Corbis/Bettmann. **Figure 14.1:** Reproduced with permission from R. S. Dietz and J. C. Holden, *Journal of Geophysical Research,* Vol 75, no. 26, pp. 4939-4956. Copyright © 1970 by the American Geophysial Union. **Figure 14.2:** Reproduced with permission from K. Burke, *Geology,* Vol 3, no. 614, p. 614. Copyright © 1975 by the Geological Society of America. **Figure 14.7:** Photo courtesy of John M. Faivre. **Figure 14.8:** Photo courtesy of Dinosaur State Park. **Figure 14.9:** From D. G. Bebout and R. G. Loucks, "Lower Cretaceous Reefs, South Texas," *AAPG Memoir Series,* No. 33, AAPG © 1983, Figs 1 and 3, pp. 441-442. Reprinted by permission of the American Association of Petroleum Geologists whose permission is required for future use. **Figure 14.11:** Reprinted with permission from R. A. Schweickert and D. S. Cowan, "Tectonic Evolution of the Western Sierra Nevada, California," in *GSA Bulletin* 86, 1975, p. 1334 (Fig. 3), Geological Society of America. **Figure 14.12(a):** Reprinted from B. M. Page, "Effects of Late Jurassic/Early Tertiary Subduction in California," *San Joaquin Geological Society Short Course 1977,* p. 66 (Figs. 5-9). **Perspective 14.1, Figure 1:** Stephen J. Kraseman/Photo Researchers Inc. **Figure 14.16(b):** Photo © John Shelton. **Perspective 14.2, Figure 1:** Photo courtesy of the Carnegie Museum of Natural History. **Perspective 14.2, Figure 2(a):** Photo courtesy of Dinosaur National Monument. **Perspective 14.2, Figure 2(b):** Photo courtesy of Dinosaur National Monument. **Figure 14.21:** Reproduced with permission from "The Movement of Continents," by Zvi Ben-Avraham, *American Scientist,* Vol 69, no. 3, p. 298 (Fig 9). Copyright © 1981 American Scientist.

Chapter 15

Chapter opening photo: James Gurney/NGS Image Collection. **Figure 15.1:** Copyright © The Natural History Museum, London. **Figure 15.2:** Watercolor by Henry De la Beche, © National Museum of Wales, Cardiff. **Figure 15.3:** Photo courtesy of Sue Monroe. **Figure 15.4:** Photo courtesy of Sue Monroe. **Figure 15.6(a):** Photo courtesy of Merton E. Hill. **Figure 15.6(b):** Photo courtesy of John Barron, United States Geological Survey. **Figure 15.6(c):** Photo courtesy of John H. Wrenn, Amoco Production Co. **Figure 15.7:** Photo courtesy of Sue Monroe. **Figure 15.8:** © Tom McHugh/Photo Researchers Inc. **Figure 15.9:** Photo courtesy of B.A. Masters. **Figure 15.10(b):** Photo courtesy of Leo J. Hickey, Yale University. **Figure 15.10(c):** Photo courtesy of Leo J. Hickey, Yale University. **Perspective 15.1, Figure 1:** From Alexander, "Size and Scaling," in *Encyclopedia of Dinosaurs,* p. 666 (Figure 1). Copyright © 1977 Academic Press. **Figure 15.14(b):** Photo courtesy of Stew Monroe. **Figure 15.15(b):** Photo © Douglas Henderson. **Figure 15.16:** Photo courtesy of Stew Monroe. **Perspective 15.2, Figure 1:** Photo courtesy of the American Museum of Natural History, Trans. # 37603. **Perspective 15.2, Figure 2:** © Archive Photos. **Figure 15.20:** © Tom McHugh/Photo Researchers Inc. **Figure 15.21:** Photo courtesy of the Canadian Museum of Nature, Ottawa, Canada; painting by Eleanor M. Kish. **Figure 15.22:** Reproduced with permission from Colbert, *Evolution of the Vertebrates,* 3rd edition published by Wiley-Liss, Inc., a subsidiary of John Wiley & Sons, Inc. **Figure 15.23:** Reproduced with permission from J. S. Hobson, "The Mammal-Like Reptiles: A Study of Transitional Fossils," *The American Biology Teacher,* v. 49, no. 1, 1987, pp. 18-22. National Association of Biology Teachers, Reston, Virginia. **Figure 15.24:** Reproduced with permission from John C. McLoughlin, *Synapsida: A New Look into the Origin of Mammals.* Copyright © 1980 by John C. McLoughlin. Used by permission of Viking Penguin, a division of Penguin Books USA. **Figure 15.25:** Reproduced with permission from J. S. Hobson, "The Mammal-Like Reptiles: A Study of Transitional Fossils," *The American Biology Teacher,* v. 49, no. 1, 1987, pp. 18-22. National Association of Biology Teachers, Reston, Virginia. **Figure 15.27:** Adapted with permission from J. Benes, *Prehistoric Animals and Plants,* 1979, pp. 174-175, illustration by Zdenek Burian, published by Hippocrene Books, Inc. **Figure 15.29(a):** From A. Montanari, R. I. Hay, W. Alvarez, F. Asaro, H. Michel, L. W. Alvarez, and J. Dmiy, "Spheroids at the Cretaceous-Tertiary Boundary are Altered Impact Droplets of Basaltic Composition," *Geology,* Vol 11, p. 669 (Fig 2). Geological Society of America, 1983. Figure 16.4: Reprinted with permission from Gordon and Jurdy, *Cenozoic Global Plate Motions,* v. 91, no. B12, pp. 12, 394-5, 1986, Copyright © American Geophysical Union. **Figure 15.29(b):** Photo courtesy of J. Nichols, U.S. Geological Survey. **Figure 15.29(c):** Photo courtesy of J. Nichols, U.S. Geological Survey. **Figure 15.30:** Photo courtesy of the Canadian Museum of Nature, Ottawa, Canada: painting by Eleanor M. Kish.

Chapter 16

Chapter opening photo: Photo courtesy of Stew Monroe. **Figure 16.1(a):** Photo courtesy of Frank Hanna. **Figure 16.1(b):** Photo courtesy Maryland Geological Survey. **Figure 16.2(a):** Photo courtesy of Stew Monroe. **Figure 16.2(b):** Photo courtesy of Frank Hanna. **Figure 16.5:** Reprinted with permission from Guennoc and Thisse, "Origins of the Red Sea Rift, the Axial Troughs, and their Mineral Deposits," v. 51, 1982, Editions BRGM BP 6009 45060, Orleans Cedex 2, and with permission from Baker, Mohr, and Williams, *Geology of the Eastern Rift System of Africa, Special Paper 136,* 1972, p. 2 (Fig 1) Geological Society of America. **Figure 16.6:** Reprinted with permission from A. M. Spencer (ed), *Mesozoic-Cenozoic Orogenic Belts,* 1974, pp. xii-xiii, The Geological Society Publishing House, Bath, England. **Figure 16.7 (a):** Modified from A. Heim, 1919, *Geologie der Schweiz,* C. H. Tauchnitz: Leipsig. **Figure 16.7(b):** Photo courtesy of Stew Monroe. **Figure 16.8:** From P. Molnar, "The Geologic History and Structure of the Himalaya," *American Scientist,* Vol 74, no. 2, 1986, pp. 148-149. Reprinted with permission of Sigma Xi, The Scientific Research Society. **Figure 16.9:** Reprinted with permission from A. D. Miall, *Principles of Sedimentary Basic Analysis,* p. 432 (Fig 960). Copyright © 1984 Springer-Verlag. **Figure 16.13:** Reprinted with permission from Dickinson and Snyder, "Plate Tectonics of the Laramide Orogeny," in *Laramide Folding Associated With Basement Block Faulting in the Western United States, GSA Memoir 151,* 1978, p. 359, The Geological Society of America. **Figure 16.14:** Reproduced with permission from J. B. Murphy, G. L. Oppliger, G. H. Brimhall, Jr., and A. Hynes, "Mantle Plumes and Mountains," *American Scientist,* Vol 87, no. 2, p. 152 (March-April 1999). **Figure 16.15:** From P. B. King, *The Evolution of North America,* 1977, p. 116 (Fig 74). Copyright © 1977 Princeton University Press. **Figure 16.18(a):** Photo courtesy of Wayne Moore. **Figure 16.17:** From D. L. Blackstone, Jr., *Traveler's Guide to the Geology of Wyoming:* Geological Survey of Wyoming Bulletin 67, 1988, pp. 43-44. Figure 16.18 (a): Reprinted with permission from Christianson and Lipman, "Conozoic Volcanism and Plate-Tectonic Evolution of Western

United States," *Philosophical Transactions,* The Royal Society, London, v. A271, 1972, p. 259. **Figure 16.18(a):** Photo courtesy of Wayne Moore. **Figure 16.19 (a):** Reprinted with permission from P. R. Hooper, "The Columbian River Basalts," *Science,* v. 215, 1982, p. 1465 (Fig 3). Copyright © 1982 by the AAAS. **Figure 16.19(b):** Photo courtesy of Ward's Natural Science Establishment, Inc. **Figure 16.19(c):** Photo courtesy of Stew Monroe. **Figure 16.20(b):** Photo courtesy of Stew Monroe. **Perspective 16.2, Figure 2(a):** Photo courtesy of U.S. Geological Survey. **Perspective 16.2, Figure 2(b):** Photo courtesy of the U.S. Geological Survey. **Perspective 16.2, Figure 3(a):** Photo courtesy of Stew Monroe. **Perspective 16.2, Figure 3(b):** Photo courtesy of Stew Monroe. **Figure 16.21(b):** F. Gohier 1982/Photo Researchers Inc. **Figure 16.22(b):** Photo courtesy of Stew Monroe. **Figure 16.24(a):** Photo courtesy of Stew Monroe. **Figure 16.23:** Reprinted with permission from John H. Stewart, "Basin-Range Structure in Western North America: A Review," *Cenozoic Tectonics and Regional Geophysics of Western Cordillera, GSA Bulletin 152,* 1978, p. 24 (Fig 1-16). **Figure 16.24(b):** Photo courtesy of Stew Monroe. **Figure 16.24(c):** Photo courtesy of Stew Monroe. **Figure 16.25:** Reprinted with permission from W. R. Dickinson, "Cenozoic Plate Tectonic Setting of the Cordilleran Region in the Western United States," *Cenozoic Paleogeography of the Western United States,* Pacific Coast Symposium 3, 1979, p. 2 (Fig 1). **Figure 16.27(a):** Photo courtesy of Stew Monroe. **Figure 16.27(b):** Photo courtesy of Stew Monroe. **Figure 16.28(a):** Reprinted with permission from Walker and Coleman, "Atlantic and Gulf Coastal Province," in DNAG, Centennial Special, Vol. 2, *Geomorphic Systems of North America,* 1987, p. 52 (Fig 1). **Figure 16.28 (b):** Reprinted with permission from S. W. Lowman, "Sedimentary Facies in the Gulf Coast," *AAPG* v. 33, no 12 1949, p. 1972 (Fig 23). Reprinted by permission of the American Association of Petroleum Geologists whose permission is required for future use. **Figure 16.29:** Reprinted with permission from W. L. Fisher and J. H. McGowen, "Depositional Systems in Wilcox Group (Eocene) of Texas and Their Relation to Occurrence of Oil and Gas," *AAPG Bulletin,* v. 53, no 1, 1969, p. 42 (Fig 7). Reprinted by permission of the American Association of Petroleum Geologists whose permission is required for future use. **Figure 16.30(a):** Reprinted with permission from M. A. Hanna "Salt Domes: Favorite Home for Oil," *Oil and Gas Journal,* v. 57, no 6, 1959, p. 140. **Figure 16.30(b):** From D. J. Hughes, "Faulting Associated with Deep-Seated Salt Domes in the Northeast Portion of the Mississippi Salt Basin," *Gulf Coast Association of Geological Societies,* v. 10, 1960, p. 161 (Fig 5). **Figure 16.31:** Reprinted with permission from G. M. Friedman and J. E. Sanders, *Principles of Sedimentology,* 1978, Fig 12-25, p. 371. **Figure 16.32:** From *Stream Sculpture on the Atlantic Slope,* by Douglas Johnson. Copyright © 1931 Columbia University Press. Reprinted with the permission of the publisher. **Figure 16.33:** Reprinted with permission from Grow and Sheridan, "U.S. Continental Margin," *The Geology of North America,* v. 1-2, p. 2-4, USGS. **Figure 16.34:** Reprinted with permission from Grow and Sheridan, "U.S. Continental Margin," *The Geology of North America,* v. 1-2, p. 2-4, USGS.

Chapter 17

Chapter opening photo: V. Clevenger/H. Armstrong Roberts. **Table 17.1:** Reprinted with permission from T. H. Van Andel, *New Views on an Old Planet,* p. 175 (table 11.1).

Index